T0142153

Lecture Notes in Computer Science 13802

More information about this series at https://link.springer.com/bookseries/558

Leonid Karlinsky · Tomer Michaeli ·
Ko Nishino (Eds.)

Computer Vision – ECCV 2022 Workshops

Tel Aviv, Israel, October 23–27, 2022
Proceedings, Part II

Editors
Leonid Karlinsky
IBM Research - MIT-IBM Watson AI Lab
Massachusetts, USA

Tomer Michaeli
Technion – Israel Institute of Technology
Haifa, Israel

Ko Nishino
Kyoto University
Kyoto, Japan

ISSN 0302-9743 ISSN 1611-3349 (electronic)
Lecture Notes in Computer Science
ISBN 978-3-031-25062-0 ISBN 978-3-031-25063-7 (eBook)
https://doi.org/10.1007/978-3-031-25063-7

This Springer imprint is published by the registered company Springer Nature Switzerland AG
The registered company address is: Gewerbestrasse 11, 6330 Cham, Switzerland

Foreword

Organizing the European Conference on Computer Vision (ECCV 2022) in Tel-Aviv during a global pandemic was no easy feat. The uncertainty level was extremely high, and decisions had to be postponed to the last minute. Still, we managed to plan things just in time for ECCV 2022 to be held in person. Participation in physical events is crucial to stimulating collaborations and nurturing the culture of the Computer Vision community.

There were many people who worked hard to ensure attendees enjoyed the best science at the 17th edition of ECCV. We are grateful to the Program Chairs Gabriel Brostow and Tal Hassner, who went above and beyond to ensure the ECCV reviewing process ran smoothly. The scientific program included dozens of workshops and tutorials in addition to the main conference and we would like to thank Leonid Karlinsky and Tomer Michaeli for their hard work. Finally, special thanks to the web chairs Lorenzo Baraldi and Kosta Derpanis, who put in extra hours to transfer information fast and efficiently to the ECCV community.

We would like to express gratitude to our generous sponsors and the Industry Chairs Dimosthenis Karatzas and Chen Sagiv, who oversaw industry relations and proposed new ways for academia-industry collaboration and technology transfer. It's great to see so much industrial interest in what we're doing!

Authors' draft versions of the papers appeared online with open access on both the Computer Vision Foundation (CVF) and the European Computer Vision Association (ECVA) websites as with previous ECCVs. Springer, the publisher of the proceedings, has arranged for archival publication. The final version of the papers is hosted by SpringerLink, with active references and supplementary materials. It benefits all potential readers that we offer both a free and citeable version for all researchers, as well as an authoritative, citeable version for SpringerLink readers. Our thanks go to Ronan Nugent from Springer, who helped us negotiate this agreement. Last but not least, we wish to thank Eric Mortensen, our publication chair, whose expertise made the process smooth.

October 2022

Rita Cucchiara
Jiří Matas
Amnon Shashua
Lihi Zelnik-Manor

Foreword

Preface

Welcome to the workshop proceedings of the 17th European Conference on Computer Vision (ECCV 2022). This year, the main ECCV event was accompanied by 60 workshops, scheduled between October 23–24, 2022. We received 103 workshop proposals on diverse computer vision topics and unfortunately had to decline many valuable proposals because of space limitations. We strove to achieve a balance between topics, as well as between established and new series. Due to the uncertainty associated with the COVID-19 pandemic around the proposal submission deadline, we allowed two workshop formats: hybrid and purely online. Some proposers switched their preferred format as we drew near the conference dates. The final program included 30 hybrid workshops and 30 purely online workshops. Not all workshops published their papers in the ECCV workshop proceedings, or had papers at all. These volumes collect the edited papers from 38 out of the 60 workshops. We sincerely thank the ECCV general chairs for trusting us with the responsibility for the workshops, the workshop organizers for their hard work in putting together exciting programs, and the workshop presenters and authors for contributing to ECCV.

October 2022

Tomer Michaeli
Leonid Karlinsky
Ko Nishino

Preface

Organization

General Chairs

Rita Cucchiara	University of Modena and Reggio Emilia, Italy
Jiří Matas	Czech Technical University in Prague, Czech Republic
Amnon Shashua	Hebrew University of Jerusalem, Israel
Lihi Zelnik-Manor	Technion – Israel Institute of Technology, Israel

Program Chairs

Shai Avidan	Tel-Aviv University, Israel
Gabriel Brostow	University College London, UK
Giovanni Maria Farinella	University of Catania, Italy
Tal Hassner	Facebook AI, USA

Program Technical Chair

Pavel Lifshits	Technion – Israel Institute of Technology, Israel

Workshops Chairs

Leonid Karlinsky	IBM Research - MIT-IBM Watson AI Lab, USA
Tomer Michaeli	Technion – Israel Institute of Technology, Israel
Ko Nishino	Kyoto University, Japan

Tutorial Chairs

Thomas Pock	Graz University of Technology, Austria
Natalia Neverova	Facebook AI Research, UK

Demo Chair

Bohyung Han	Seoul National University, South Korea

Social and Student Activities Chairs

Tatiana Tommasi	Italian Institute of Technology, Italy
Sagie Benaim	University of Copenhagen, Denmark

Diversity and Inclusion Chairs

Xi Yin	Facebook AI Research, USA
Bryan Russell	Adobe, USA

Communications Chairs

Lorenzo Baraldi	University of Modena and Reggio Emilia, Italy
Kosta Derpanis	York University and Samsung AI Centre Toronto, Canada

Industrial Liaison Chairs

Dimosthenis Karatzas	Universitat Autònoma de Barcelona, Spain
Chen Sagiv	SagivTech, Israel

Finance Chair

Gerard Medioni	University of Southern California and Amazon, USA

Publication Chair

Eric Mortensen	MiCROTEC, USA

Workshops Organizers

W01 - AI for Space

Tat-Jun Chin	The University of Adelaide, Australia
Luca Carlone	Massachusetts Institute of Technology, USA
Djamila Aouada	University of Luxembourg, Luxembourg
Binfeng Pan	Northwestern Polytechnical University, China
Viorela Ila	The University of Sydney, Australia
Benjamin Morrell	NASA Jet Propulsion Lab, USA
Grzegorz Kakareko	Spire Global, USA

W02 - Vision for Art

Alessio Del Bue	Istituto Italiano di Tecnologia, Italy
Peter Bell	Philipps-Universität Marburg, Germany
Leonardo L. Impett	École Polytechnique Fédérale de Lausanne (EPFL), Switzerland
Noa Garcia	Osaka University, Japan
Stuart James	Istituto Italiano di Tecnologia, Italy

W03 - Adversarial Robustness in the Real World

Angtian Wang	Johns Hopkins University, USA
Yutong Bai	Johns Hopkins University, USA
Adam Kortylewski	Max Planck Institute for Informatics, Germany
Cihang Xie	University of California, Santa Cruz, USA
Alan Yuille	Johns Hopkins University, USA
Xinyun Chen	University of California, Berkeley, USA
Judy Hoffman	Georgia Institute of Technology, USA
Wieland Brendel	University of Tübingen, Germany
Matthias Hein	University of Tübingen, Germany
Hang Su	Tsinghua University, China
Dawn Song	University of California, Berkeley, USA
Jun Zhu	Tsinghua University, China
Philippe Burlina	Johns Hopkins University, USA
Rama Chellappa	Johns Hopkins University, USA
Yinpeng Dong	Tsinghua University, China
Yingwei Li	Johns Hopkins University, USA
Ju He	Johns Hopkins University, USA
Alexander Robey	University of Pennsylvania, USA

W04 - Autonomous Vehicle Vision

Rui Fan	Tongji University, China
Nemanja Djuric	Aurora Innovation, USA
Wenshuo Wang	McGill University, Canada
Peter Ondruska	Toyota Woven Planet, UK
Jie Li	Toyota Research Institute, USA

W05 - Learning With Limited and Imperfect Data

Noel C. F. Codella	Microsoft, USA
Zsolt Kira	Georgia Institute of Technology, USA
Shuai Zheng	Cruise LLC, USA
Judy Hoffman	Georgia Institute of Technology, USA
Tatiana Tommasi	Politecnico di Torino, Italy
Xiaojuan Qi	The University of Hong Kong, China
Sadeep Jayasumana	University of Oxford, UK
Viraj Prabhu	Georgia Institute of Technology, USA
Yunhui Guo	University of Texas at Dallas, USA
Ming-Ming Cheng	Nankai University, China

W06 - Advances in Image Manipulation

Radu Timofte — University of Würzburg, Germany, and ETH Zurich, Switzerland
Andrey Ignatov — AI Benchmark and ETH Zurich, Switzerland
Ren Yang — ETH Zurich, Switzerland
Marcos V. Conde — University of Würzburg, Germany
Furkan Kınlı — Özyeğin University, Turkey

W07 - Medical Computer Vision

Tal Arbel — McGill University, Canada
Ayelet Akselrod-Ballin — Reichman University, Israel
Vasileios Belagiannis — Otto von Guericke University, Germany
Qi Dou — The Chinese University of Hong Kong, China
Moti Freiman — Technion, Israel
Nicolas Padoy — University of Strasbourg, France
Tammy Riklin Raviv — Ben Gurion University, Israel
Mathias Unberath — Johns Hopkins University, USA
Yuyin Zhou — University of California, Santa Cruz, USA

W08 - Computer Vision for Metaverse

Bichen Wu — Meta Reality Labs, USA
Peizhao Zhang — Facebook, USA
Xiaoliang Dai — Facebook, USA
Tao Xu — Facebook, USA
Hang Zhang — Meta, USA
Péter Vajda — Facebook, USA
Fernando de la Torre — Carnegie Mellon University, USA
Angela Dai — Technical University of Munich, Germany
Bryan Catanzaro — NVIDIA, USA

W09 - Self-Supervised Learning: What Is Next?

Yuki M. Asano — University of Amsterdam, The Netherlands
Christian Rupprecht — University of Oxford, UK
Diane Larlus — Naver Labs Europe, France
Andrew Zisserman — University of Oxford, UK

W10 - Self-Supervised Learning for Next-Generation Industry-Level Autonomous Driving

Xiaodan Liang — Sun Yat-sen University, China
Hang Xu — Huawei Noah's Ark Lab, China

Fisher Yu	ETH Zürich, Switzerland
Wei Zhang	Huawei Noah's Ark Lab, China
Michael C. Kampffmeyer	UiT The Arctic University of Norway, Norway
Ping Luo	The University of Hong Kong, China

W11 - ISIC Skin Image Analysis

M. Emre Celebi	University of Central Arkansas, USA
Catarina Barata	Instituto Superior Técnico, Portugal
Allan Halpern	Memorial Sloan Kettering Cancer Center, USA
Philipp Tschandl	Medical University of Vienna, Austria
Marc Combalia	Hospital Clínic of Barcelona, Spain
Yuan Liu	Google Health, USA

W12 - Cross-Modal Human-Robot Interaction

Fengda Zhu	Monash University, Australia
Yi Zhu	Huawei Noah's Ark Lab, China
Xiaodan Liang	Sun Yat-sen University, China
Liwei Wang	The Chinese University of Hong Kong, China
Xiaojun Chang	University of Technology Sydney, Australia
Nicu Sebe	University of Trento, Italy

W13 - Text in Everything

Ron Litman	Amazon AI Labs, Israel
Aviad Aberdam	Amazon AI Labs, Israel
Shai Mazor	Amazon AI Labs, Israel
Hadar Averbuch-Elor	Cornell University, USA
Dimosthenis Karatzas	Universitat Autònoma de Barcelona, Spain
R. Manmatha	Amazon AI Labs, USA

W14 - BioImage Computing

Jan Funke	HHMI Janelia Research Campus, USA
Alexander Krull	University of Birmingham, UK
Dagmar Kainmueller	Max Delbrück Center, Germany
Florian Jug	Human Technopole, Italy
Anna Kreshuk	EMBL-European Bioinformatics Institute, Germany
Martin Weigert	École Polytechnique Fédérale de Lausanne (EPFL), Switzerland
Virginie Uhlmann	EMBL-European Bioinformatics Institute, UK

Peter Bajcsy	National Institute of Standards and Technology, USA
Erik Meijering	University of New South Wales, Australia

W15 - Visual Object-Oriented Learning Meets Interaction: Discovery, Representations, and Applications

Kaichun Mo	Stanford University, USA
Yanchao Yang	Stanford University, USA
Jiayuan Gu	University of California, San Diego, USA
Shubham Tulsiani	Carnegie Mellon University, USA
Hongjing Lu	University of California, Los Angeles, USA
Leonidas Guibas	Stanford University, USA

W16 - AI for Creative Video Editing and Understanding

Fabian Caba	Adobe Research, USA
Anyi Rao	The Chinese University of Hong Kong, China
Alejandro Pardo	King Abdullah University of Science and Technology, Saudi Arabia
Linning Xu	The Chinese University of Hong Kong, China
Yu Xiong	The Chinese University of Hong Kong, China
Victor A. Escorcia	Samsung AI Center, UK
Ali Thabet	Reality Labs at Meta, USA
Dong Liu	Netflix Research, USA
Dahua Lin	The Chinese University of Hong Kong, China
Bernard Ghanem	King Abdullah University of Science and Technology, Saudi Arabia

W17 - Visual Inductive Priors for Data-Efficient Deep Learning

Jan C. van Gemert	Delft University of Technology, The Netherlands
Nergis Tömen	Delft University of Technology, The Netherlands
Ekin Dogus Cubuk	Google Brain, USA
Robert-Jan Bruintjes	Delft University of Technology, The Netherlands
Attila Lengyel	Delft University of Technology, The Netherlands
Osman Semih Kayhan	Bosch Security Systems, The Netherlands
Marcos Baptista Ríos	Alice Biometrics, Spain
Lorenzo Brigato	Sapienza University of Rome, Italy

W18 - Mobile Intelligent Photography and Imaging

Chongyi Li	Nanyang Technological University, Singapore
Shangchen Zhou	Nanyang Technological University, Singapore

Ruicheng Feng	Nanyang Technological University, Singapore
Jun Jiang	SenseBrain Research, USA
Wenxiu Sun	SenseTime Group Limited, China
Chen Change Loy	Nanyang Technological University, Singapore
Jinwei Gu	SenseBrain Research, USA

W19 - People Analysis: From Face, Body and Fashion to 3D Virtual Avatars

Alberto Del Bimbo	University of Florence, Italy
Mohamed Daoudi	IMT Nord Europe, France
Roberto Vezzani	University of Modena and Reggio Emilia, Italy
Xavier Alameda-Pineda	Inria Grenoble, France
Marcella Cornia	University of Modena and Reggio Emilia, Italy
Guido Borghi	University of Bologna, Italy
Claudio Ferrari	University of Parma, Italy
Federico Becattini	University of Florence, Italy
Andrea Pilzer	NVIDIA AI Technology Center, Italy
Zhiwen Chen	Alibaba Group, China
Xiangyu Zhu	Chinese Academy of Sciences, China
Ye Pan	Shanghai Jiao Tong University, China
Xiaoming Liu	Michigan State University, USA

W20 - Safe Artificial Intelligence for Automated Driving

Timo Saemann	Valeo, Germany
Oliver Wasenmüller	Hochschule Mannheim, Germany
Markus Enzweiler	Esslingen University of Applied Sciences, Germany
Peter Schlicht	CARIAD, Germany
Joachim Sicking	Fraunhofer IAIS, Germany
Stefan Milz	Spleenlab.ai and Technische Universität Ilmenau, Germany
Fabian Hüger	Volkswagen Group Research, Germany
Seyed Ghobadi	University of Applied Sciences Mittelhessen, Germany
Ruby Moritz	Volkswagen Group Research, Germany
Oliver Grau	Intel Labs, Germany
Frédérik Blank	Bosch, Germany
Thomas Stauner	BMW Group, Germany

W21 - Real-World Surveillance: Applications and Challenges

Kamal Nasrollahi	Aalborg University, Denmark
Sergio Escalera	Universitat Autònoma de Barcelona, Spain

Radu Tudor Ionescu	University of Bucharest, Romania
Fahad Shahbaz Khan	Mohamed bin Zayed University of Artificial Intelligence, United Arab Emirates
Thomas B. Moeslund	Aalborg University, Denmark
Anthony Hoogs	Kitware, USA
Shmuel Peleg	The Hebrew University, Israel
Mubarak Shah	University of Central Florida, USA

W22 - Affective Behavior Analysis In-the-Wild

Dimitrios Kollias	Queen Mary University of London, UK
Stefanos Zafeiriou	Imperial College London, UK
Elnar Hajiyev	Realeyes, UK
Viktoriia Sharmanska	University of Sussex, UK

W23 - Visual Perception for Navigation in Human Environments: The JackRabbot Human Body Pose Dataset and Benchmark

Hamid Rezatofighi	Monash University, Australia
Edward Vendrow	Stanford University, USA
Ian Reid	University of Adelaide, Australia
Silvio Savarese	Stanford University, USA

W24 - Distributed Smart Cameras

Niki Martinel	University of Udine, Italy
Ehsan Adeli	Stanford University, USA
Rita Pucci	University of Udine, Italy
Animashree Anandkumar	Caltech and NVIDIA, USA
Caifeng Shan	Shandong University of Science and Technology, China
Yue Gao	Tsinghua University, China
Christian Micheloni	University of Udine, Italy
Hamid Aghajan	Ghent University, Belgium
Li Fei-Fei	Stanford University, USA

W25 - Causality in Vision

Yulei Niu	Columbia University, USA
Hanwang Zhang	Nanyang Technological University, Singapore
Peng Cui	Tsinghua University, China
Song-Chun Zhu	University of California, Los Angeles, USA
Qianru Sun	Singapore Management University, Singapore
Mike Zheng Shou	National University of Singapore, Singapore
Kaihua Tang	Nanyang Technological University, Singapore

W26 - In-Vehicle Sensing and Monitorization

Jaime S. Cardoso	INESC TEC and Universidade do Porto, Portugal
Pedro M. Carvalho	INESC TEC and Polytechnic of Porto, Portugal
João Ribeiro Pinto	Bosch Car Multimedia and Universidade do Porto, Portugal
Paula Viana	INESC TEC and Polytechnic of Porto, Portugal
Christer Ahlström	Swedish National Road and Transport Research Institute, Sweden
Carolina Pinto	Bosch Car Multimedia, Portugal

W27 - Assistive Computer Vision and Robotics

Marco Leo	National Research Council of Italy, Italy
Giovanni Maria Farinella	University of Catania, Italy
Antonino Furnari	University of Catania, Italy
Mohan Trivedi	University of California, San Diego, USA
Gérard Medioni	Amazon, USA

W28 - Computational Aspects of Deep Learning

Iuri Frosio	NVIDIA, Italy
Sophia Shao	University of California, Berkeley, USA
Lorenzo Baraldi	University of Modena and Reggio Emilia, Italy
Claudio Baecchi	University of Florence, Italy
Frederic Pariente	NVIDIA, France
Giuseppe Fiameni	NVIDIA, Italy

W29 - Computer Vision for Civil and Infrastructure Engineering

Joakim Bruslund Haurum	Aalborg University, Denmark
Mingzhu Wang	Loughborough University, UK
Ajmal Mian	University of Western Australia, Australia
Thomas B. Moeslund	Aalborg University, Denmark

W30 - AI-Enabled Medical Image Analysis: Digital Pathology and Radiology/COVID-19

Jaime S. Cardoso	INESC TEC and Universidade do Porto, Portugal
Stefanos Kollias	National Technical University of Athens, Greece
Sara P. Oliveira	INESC TEC, Portugal
Mattias Rantalainen	Karolinska Institutet, Sweden
Jeroen van der Laak	Radboud University Medical Center, The Netherlands
Cameron Po-Hsuan Chen	Google Health, USA

Diana Felizardo	IMP Diagnostics, Portugal
Ana Monteiro	IMP Diagnostics, Portugal
Isabel M. Pinto	IMP Diagnostics, Portugal
Pedro C. Neto	INESC TEC, Portugal
Xujiong Ye	University of Lincoln, UK
Luc Bidaut	University of Lincoln, UK
Francesco Rundo	STMicroelectronics, Italy
Dimitrios Kollias	Queen Mary University of London, UK
Giuseppe Banna	Portsmouth Hospitals University, UK

W31 - Compositional and Multimodal Perception

Kazuki Kozuka	Panasonic Corporation, Japan
Zelun Luo	Stanford University, USA
Ehsan Adeli	Stanford University, USA
Ranjay Krishna	University of Washington, USA
Juan Carlos Niebles	Salesforce and Stanford University, USA
Li Fei-Fei	Stanford University, USA

W32 - Uncertainty Quantification for Computer Vision

Andrea Pilzer	NVIDIA, Italy
Martin Trapp	Aalto University, Finland
Arno Solin	Aalto University, Finland
Yingzhen Li	Imperial College London, UK
Neill D. F. Campbell	University of Bath, UK

W33 - Recovering 6D Object Pose

Martin Sundermeyer	DLR German Aerospace Center, Germany
Tomáš Hodaň	Reality Labs at Meta, USA
Yann Labbé	Inria Paris, France
Gu Wang	Tsinghua University, China
Lingni Ma	Reality Labs at Meta, USA
Eric Brachmann	Niantic, Germany
Bertram Drost	MVTec, Germany
Sindi Shkodrani	Reality Labs at Meta, USA
Rigas Kouskouridas	Scape Technologies, UK
Ales Leonardis	University of Birmingham, UK
Carsten Steger	Technical University of Munich and MVTec, Germany
Vincent Lepetit	École des Ponts ParisTech, France, and TU Graz, Austria
Jiří Matas	Czech Technical University in Prague, Czech Republic

W34 - Drawings and Abstract Imagery: Representation and Analysis

Diane Oyen — Los Alamos National Laboratory, USA
Kushal Kafle — Adobe Research, USA
Michal Kucer — Los Alamos National Laboratory, USA
Pradyumna Reddy — University College London, UK
Cory Scott — University of California, Irvine, USA

W35 - Sign Language Understanding

Liliane Momeni — University of Oxford, UK
Gül Varol — École des Ponts ParisTech, France
Hannah Bull — University of Paris-Saclay, France
Prajwal K. R. — University of Oxford, UK
Neil Fox — University College London, UK
Ben Saunders — University of Surrey, UK
Necati Cihan Camgöz — Meta Reality Labs, Switzerland
Richard Bowden — University of Surrey, UK
Andrew Zisserman — University of Oxford, UK
Bencie Woll — University College London, UK
Sergio Escalera — Universitat Autònoma de Barcelona, Spain
Jose L. Alba-Castro — Universidade de Vigo, Spain
Thomas B. Moeslund — Aalborg University, Denmark
Julio C. S. Jacques Junior — Universitat Autònoma de Barcelona, Spain
Manuel Vázquez Enríquez — Universidade de Vigo, Spain

W36 - A Challenge for Out-of-Distribution Generalization in Computer Vision

Adam Kortylewski — Max Planck Institute for Informatics, Germany
Bingchen Zhao — University of Edinburgh, UK
Jiahao Wang — Max Planck Institute for Informatics, Germany
Shaozuo Yu — The Chinese University of Hong Kong, China
Siwei Yang — Hong Kong University of Science and Technology, China
Dan Hendrycks — University of California, Berkeley, USA
Oliver Zendel — Austrian Institute of Technology, Austria
Dawn Song — University of California, Berkeley, USA
Alan Yuille — Johns Hopkins University, USA

W37 - Vision With Biased or Scarce Data

Kuan-Chuan Peng — Mitsubishi Electric Research Labs, USA
Ziyan Wu — United Imaging Intelligence, USA

W38 - Visual Object Tracking Challenge

Matej Kristan	University of Ljubljana, Slovenia
Aleš Leonardis	University of Birmingham, UK
Jiří Matas	Czech Technical University in Prague, Czech Republic
Hyung Jin Chang	University of Birmingham, UK
Joni-Kristian Kämäräinen	Tampere University, Finland
Roman Pflugfelder	Technical University of Munich, Germany, Technion, Israel, and Austrian Institute of Technology, Austria
Luka Čehovin Zajc	University of Ljubljana, Slovenia
Alan Lukežič	University of Ljubljana, Slovenia
Gustavo Fernández	Austrian Institute of Technology, Austria
Michael Felsberg	Linköping University, Sweden
Martin Danelljan	ETH Zurich, Switzerland

Contents – Part II

W05 - Learning With Limited and Imperfect Data

W06 - Advances in Image Manipulation

W05 - Learning With Limited and Imperfect Data

W05 - Learning With Limited and Imperfect Data

Learning from limited or imperfect data (L^2ID) refers to a variety of studies that attempt to address challenging pattern recognition tasks by learning from limited, weak, or noisy supervision. Supervised learning methods including Deep Convolutional Neural Networks have significantly improved the performance in many problems in the field of computer vision, thanks to the rise of large-scale annotated data sets and the advance in computing hardware. However, these supervised learning approaches are notoriously "data hungry", which makes them sometimes not practical in many real-world industrial applications. This issue of availability of large quantities of labeled data becomes even more severe when considering visual classes that require annotation based on expert knowledge (e.g., medical imaging), classes that rarely occur, or object detection and instance segmentation tasks where the labeling requires more effort. To address this problem, many efforts, e.g., weakly supervised learning, few-shot learning, self/semi-supervised, cross-domain few-shot learning, domain adaptation, etc., have been made to improve robustness to this scenario. The goal of this workshop, which builds on the successful CVPR 2021 L2ID workshop, was to bring together researchers across several computer vision and machine learning communities to navigate the complex landscape of methods that enable moving beyond fully supervised learning towards limited and imperfect label settings.

October 2022

Noel C. Codella
Zsolt Kira
Shuai Zheng
Judy Hoffman
Tatiana Tommasi
Xiaojuan Qi
Sadeep Jayasumana
Viraj Prabhu
Yunhui Guo
Ming-Ming Cheng

SITTA: Single Image Texture Translation for Data Augmentation

Boyi Li[1(✉)], Yin Cui[2], Tsung-Yi Lin[3], and Serge Belongie[4]

[1] Cornell University, Cornell Tech, New York, USA
bl728@cornell.edu
[2] Google Research, Mountain View, USA
[3] NVIDIA, Santa Clara, USA
[4] University of Copenhagen, Copenhagen, Denmark

Abstract. Recent advances in data augmentation enable one to translate images by learning the mapping between a source domain and a target domain. Existing methods tend to learn the distributions by training a model on a variety of datasets, with results evaluated largely in a subjective manner. Relatively few works in this area, however, study the potential use of image synthesis methods for recognition tasks. In this paper, we propose and explore the problem of image translation for data augmentation. We first propose a lightweight yet efficient model for translating texture to augment images based on a single input of source texture, allowing for fast training and testing, referred to as Single Image Texture Translation for data Augmentation (SITTA). Then we explore the use of augmented data in long-tailed and few-shot image classification tasks. We find the proposed augmentation method and workflow is capable of translating the texture of input data into a target domain, leading to consistently improved image recognition performance. Finally, we examine how SITTA and related image translation methods can provide a basis for a data-efficient, "augmentation engineering" approach to model training.

1 Introduction

"The Forms are not limited to geometry. For any conceivable thing or property there is a corresponding Form, a perfect example of that thing or property. The list is almost inexhaustible."

—Plato "Theory of Forms"

Recent years have witnessed a breakthrough in deep learning based image synthesis such as image translation [9,21,27] that manipulates or synthesizes images using neural networks rather than hand-crafted techniques, such as guided filtering [13] or image quilting [7]. However, few of them study the potential use of semantic image synthesis methods as an effective data augmentation tool for recognition tasks. The progress is primarily limited by two bottlenecks: the validity of synthetic data for target labels and the running time. With regard

---

Supplementary Information The online version contains supplementary material available at https://doi.org/10.1007/978-3-031-25063-7_1.

L. Karlinsky et al. (Eds.): ECCV 2022 Workshops, LNCS 13802, pp. 3–20, 2023.
https://doi.org/10.1007/978-3-031-25063-7_1

to data validity, many works such as Stylized-ImageNet [10] propose to apply style transfer to the original dataset for pre-training to improve the model's robustness, but the synthetic images lack the natural appearance of the original images. Also, since it deconstructs texture and content, it will hurt the recognition performance if trained with the augmented data, and needs to be fine-tuned using only the original dataset to obtain a benefit. Though many approaches [17,30,56] have proposed advanced algorithms to translate style and texture, most of them still focus on a subjective evaluation. With regard to running time, current techniques [24,34] usually need to train for at least several hours. This seriously hinders its use for data augmentation in real applications. While in the domain of data augmentation, current methods mainly focus on pixel-level or geometric operations such as blur or crop [35]. The potential of augmenting data into different domains, however, is relatively unexplored.

In light of this, we propose to explore, design and study an efficient Single Image Texture Translation for data Augmentation (SITTA) in image recognition tasks. It enables texture translation or swapping trained with one-shot texture source input. We believe an ideal image synthesis method for data augmentation should yield visually appealing results, improved recognition performance as well as time efficiency. SITTA is the first of few methods that explore this problem and try to balance all these factors. Our model is lightweight and permits fast training (< 5 min) and testing (9 ms) on a 288×288 image using a single Geforce GTX 1080 Ti GPU. In Fig. 2, we illustrate an example of generated 'bubble milk' images by translating the texture of bubble milk to milk images. We visualize the image distribution using t-SNE [26]. We collect 100 natural milk images and 100 natural bubble milk images from the Web, shown on the left. On the right, we can see that the SITTA 'bubble milk' images align well with the original bubble milk images, which hints the efficacy of SITTA for semantic data augmentation. The intuition of replacing texture is conceptually similar to the arithmetic properties of word embeddings [2], e.g., $\overrightarrow{Milk} - \overrightarrow{\text{texture}(Milk)} + \overrightarrow{\text{texture}(BubbleMilk)} = \overrightarrow{BubbleMilk}$.

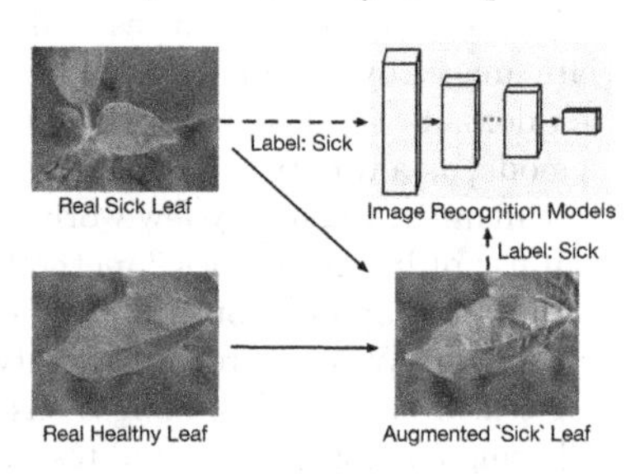

Fig. 1. Single image texture translation for data augmentation (SITTA) workflow. We first obtain textures from sick leaf and content from healthy leaf, then we feed them into SITTA model to generate augmented data. We set the corresponding label as 'sick.'

Our key contributions in this work are as follows.

1. We introduce a new lightweight single image texture translation method for data augmentation (SITTA) that translates textures from a single image to another.
2. We explore the use of SITTA for augmenting data towards the target texture domain. As an example, we translate texture from rare diseased leaves to

abundant healthy leaves to augment training data and improve image classification results in the plant pathology challenge dataset [40].

3. We dig deeper into the use of SITTA for augmenting data and demonstrate its effectiveness in several image recognition tasks (e.g., improving the performance of ResNet-18 by 1.4% for 5-shot classification on CUB-200-2011), which sheds light on the potential direction of image synthesis for data augmentation in the wild.

2 Related Work

Texture Translation . Deep learning based image synthesis manipulates images through the design of various generative neural networks. Fueled by the explorations in generative models such as GAN [11] and VAE [19], researchers have explored ideas such as neural style transfer and image translation. A neural algorithm of artistic style (ArtStyle) [9] can separate and recombine the image content and style of natural images. CycleGAN [56] investigates the use of conditional adversarial networks as a general-purpose solution to image-to-image translation problems. SinGAN [34] demonstrates success in single image retargeting. Few-Shot Unsupervised Image-to-Image Translation (FUNIT) [25] focuses on previously unseen target classes that are specified at test time only by a few example images. TuiGAN [24] aims to learn versatile image-to-image translation with two unpaired images designed in a coarse-to-fine manner. Furthermore, [29,30,49] propose elegant methods for better preserving their structural information and statistical properties. However, these methods focus more on a subjective manner instead of considering all factors including decent outputs, time efficiency, improved recognition results, as well as flexibility with various input resolutions for data augmentation.

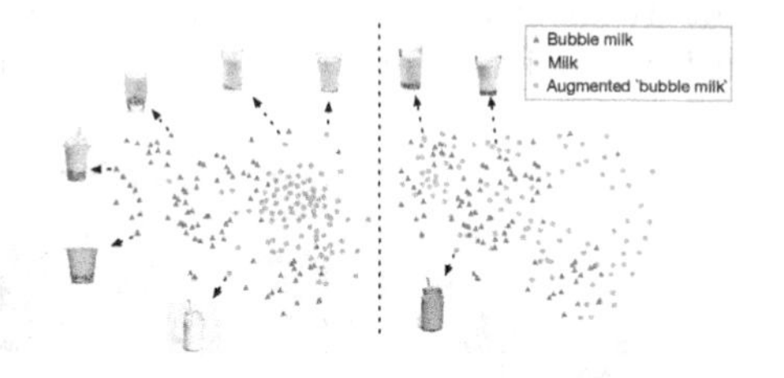

Fig. 2. t-SNE embedding before and after using SITTA. Left: milk (red) and bubble milk (blue) images. Right: corresponding augmented 'bubble milk' (gray) and original bubble milk images. (Color figure online)

Data Augmentation. Data augmentation plays a critical role in various image recognition tasks [4]. On the one hand, basic image manipulation methods have been widely used as effective pre-processing tools [3,10,12,22,35,36,45], with examples including flipping, rotation, jittering, grayscale, and Gaussian blur. Several works [5,42] explore effective combined choices of basic image manipulations. LOOC [48] searches for an optimal augmentation strategy among a large pool of candidates based on specific datasets or tasks. Mixup [51] interpolates two training inputs in feature and label space simultaneously. CutMix [50] uses a copy-paste strategy and mixes the labels in proportion to the number of pixels contributed by each input image to the final composition. MoEx [22] exchange the moments of the learned features of one training image by those of another, and also interpolate the target labels. On the other hand, many researchers have begun to explore the impact of semantic data augmentation for real-world image recognition tasks [1,36,45,52]. Geirhos, et al [10] find that CNNs like ResNet-50 favor texture rather than shape, and appropriate use of augmented ImageNet via style transfer could alleviate this bias and improve the model performance. In addition, [32,38,54,55] aim to learn robust shape-based features for domain generalization. RL-CycleGAN [33] introduces a consistency loss for simulation-to-real-world transfer for reinforcement learning. However, few of them explicitly augment data with natural outputs that could be used to solve real problems. In this paper, we propose SITTA that generates new data semantically via domain-specific texture translation. We provide a simplified illustration of the data augmentation landscape in Fig. 3.

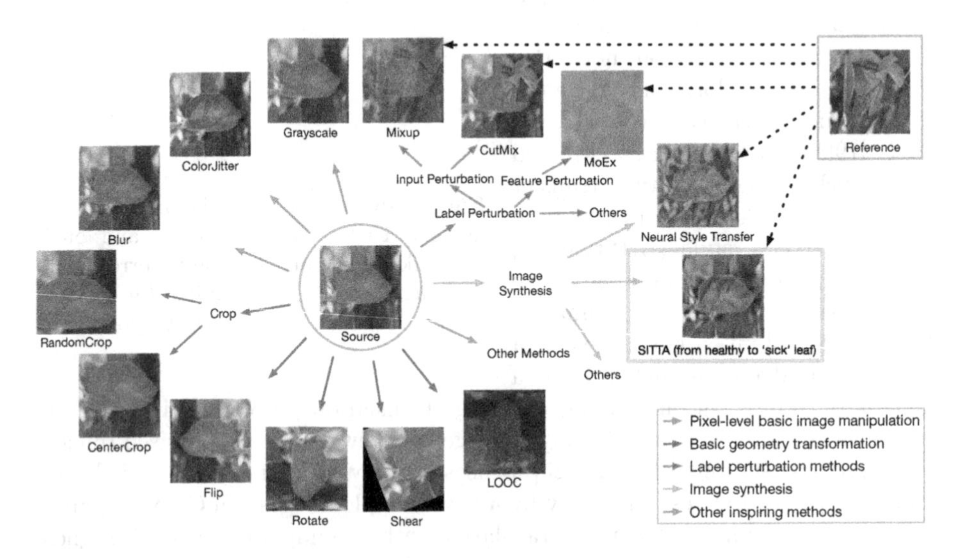

Fig. 3. Overview of the data augmentation landscape.

3 Method

3.1 Workflow

SITTA aims to augment images that align with the target texture domain and could be used as extra training data for various recognition tasks. Assume we have plenty of source domain images SetA but a few target domain images SetB, the only difference between SetA and SetB is that they contain different textures. Our goal is to synthesize extra 'B' data by replacing textures of SetA with the new textures from SetB, we call the augmented dataset as AugSetB. Based on SITTA, we have two ways to achieve this: i) *Single to Single*. Since SITTA enables translating textures between two single images, we train and generate new image I'_B based on every input I_A and I_B from SetA and SetB, respectively. ii) *Single to Multi*. We train and generate new image I'_B with a single source texture image I_B from SetB and all content images from SetA. Please see Section A in the Appendix for workflow details.

3.2 Model Design

Figure 4 shows a brief sketch of our framework. In the illustrated example, SITTA aims to translate texture from a single source texture image (Input B: Parulidae) to a content image (Input A: House Sparrow) and get a synthetic 'Parulidae' image. SITTA consists of three parts: shared Texture Encoder (En_T), shared Content Encoder (En_C) and corresponding Decoder De_A and De_B for input A and B. We first feed input B into En_T to obtain the texture latent vector and feed input A into En_C to get the content (structure) matrix. Then we concatenate the texture and content features and feed them into Decoder to generate the output image.

To efficiently preserve structure and extract textures with an optimal number of layers, we apply two downsampling operations during training. To achieve efficient model design, we emphasize three critical components: input augmentation, structural information re-injection, texture latent regression.

(1) Input augmentation. We propose to directly augment input by Horizontal Flip, CenterCrop, and RandomCrop. We find such input augmentation could effectively guarantee the diversity of image scales for fast and efficient training.
(2) Structural information re-injection. To better preserve the structural information and optimize training, we apply Positional Norm [23] in En_C to extract intermediate normalization constants mean μ and standard deviation σ as structural features β and γ and re-inject them into the later layers of Decoder to transfer structural information. Given the activations $X \in \mathbb{R}^{B\times C\times H\times W}$ (where B denotes the batch size, C the number of channels, H the height, and W the width) in a given layer of a neural net, the extracting operations[1] are listed as:

$$\beta = \mu_{b,h,w} = \frac{1}{C}\sum_{c=1}^{C} X_{b,c,h,w}, \tag{1}$$

$$\gamma = \sigma_{b,h,w} = \sqrt{\frac{1}{C}\sum_{c=1}^{C}\left(X_{b,c,h,w} - \mu_{b,h,w}\right)^2 + \epsilon}. \tag{2}$$

The re-injecting operation after ith intermediate layer is listed as:

$$\mathrm{Out_i}(\mathbf{x}) = \gamma_i F_i(\mathbf{x}) + \beta_i, \tag{3}$$

where the function F is modeled by the intermediate layers. We summarize the SITTA model design in Appendix Section B.
(3) Texture latent regression. To achieve better representation disentanglement [20] for texture and content, we use a latent regression loss L_{idt} to encourage the invertible mapping between the latent texture vectors and the corresponding outputs and enforce the reconstruction based on the latent texture vectors.

[1] The ϵ is a small stability constant (*e.g.*, $\epsilon = 10^{-5}$) to avoid divisions by zero and imaginary values due to numerical inaccuracies.

3.3 Loss Function

SITTA is based on GAN framework [11], the loss function is composed of 5 parts: adversarial loss L_{adv}, latent regression loss L_{idt}, content matrix reconstruction loss L_{rec}, texture vector KL divergence loss L_{kl} and perceptual loss L_f. During training, we have inputs I_A and I_B, extracted texture vectors T_A and T_B, content matrix C_A and C_B, corresponding normalization constants β_A, γ_A and β_B, γ_B, as well as outputs I'_A and I'_B.

Adversarial loss is GAN standard loss function for matching the distribution of translated image to the target domain:

$$\begin{aligned} L_{adv} = & \mathbb{E}[\log D_B(I_B)] + \mathbb{E}[1 - \log D_B(I'_B)))] \\ & + \mathbb{E}[\log D_A(I_A)] + \mathbb{E}[1 - \log D_A(I'_A)))] \end{aligned} \tag{4}$$

we have two discriminators D_A, D_B for distinguishing between real and generated image for A and B domain separately.

L_{idt} makes sure the decoder is able to reconstruct it based on extracted content matrix and texture latent vector:

$$L_{idt} = \mathbb{E}[||I_{BB} - I_B||_1] + \mathbb{E}[||I_{AA} - I_A||_1], \tag{5}$$

where

$$I_{AA} = De_A(T_A, Enc_C(I_A)), I_{BB} = De_B(T_B, Enc_C(I_B)).$$

L_{rec} is based on Cycle-Consistency loss [56] that optimizes the training for this under-constrained problem and regularize the translated image to preserve semantic structure of the input image:

$$L_{rec} = \mathbb{E}[||I'_{BA} - I_A||_1] + \mathbb{E}[||I'_{AB} - I_B||_1], \tag{6}$$

where

$$I'_{BA} = De_A(T_A, Enc_C(I'_B)), I'_{AB} = De_B(T_B, Enc_C(I'_A)).$$

To better preserve the structural information, we use perceptual loss L_f [18] based on VGG19 [37] between the generated outputs and original Inputs. We use KL divergence loss L_{kl} [16,20] to minimize the distribution variance between extracted texture vectors from the target image and generated image.

Our final objective function is:

$$L_{all} = L_{adv} + \lambda_{idt} L_{idt} + \lambda_{rec} L_{rec} + \lambda_{kl} L_{kl} + \lambda_f L_f, \tag{7}$$

where λ_{adv}, λ_{idt}, λ_{rec}, λ_{kl} and λ_f are weights assigned for each loss, respectively.

4 Experiments

4.1 Experimental Setup

During training, we use Adam with 0.0005 learning rate, $\beta_1 = 0.5$ and $\beta_2 = 0.999$. We augment the inputs and resize them to 288×288. Empirically, we

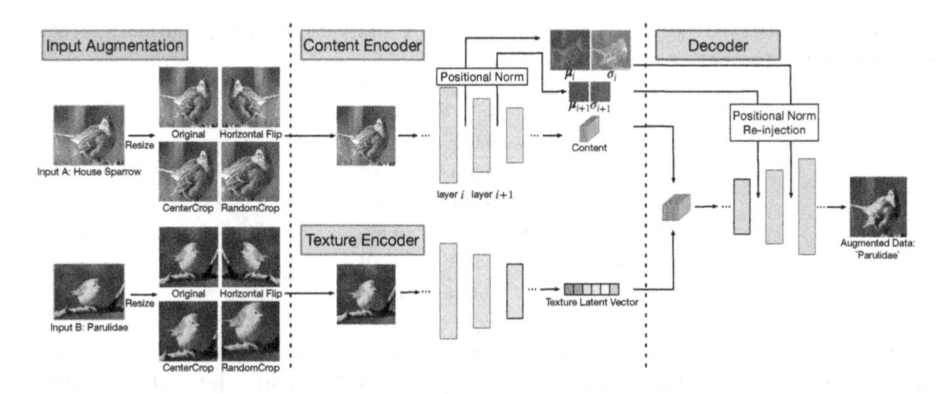

Fig. 4. SITTA Framework. SITTA learns to encode the content and texture from a single content and texture image respectively, then decodes an output image with the texture translated onto the content.

find the model starts to converge after 600 iterations and become stable after 800 iterations for every input pair. We use widely-used dataset [56] as well as images collected from the Internet (including VanillaCake↔ChocolateCake, Milk↔BubbleMilk, etc.). We will release our Pytorch [31] implementation and associated data to facilitate future research.

4.2 Low-level Evaluation on Augmented Data

We refer to the experimental setup of TuiGAN [24] that randomly selects 8 unpaired images, and generate 8 translated images for 4 unpaired image to image translation tasks including Horse↔Zebra, Apple↔ Orange, Milk ↔ BubbleMilk, and VanillaCake↔ChocolateCake. We train and test with a single source image and target texture image. We compare our results with various popular or most recent image synthesis methods with their official code and setting: ArtStyle [8], CycleGAN [56], SinGAN [34][2], FUNIT [25] and TuiGAN [24]. For evaluation, we use three of most popular metrics: Fréchet Inception Distance (FID) [15], Learned Perceptual Image Patch Similarity (LPIPS) [53] and VGG Loss (perceptual loss) [18]. FID aims to capture the similarity of generated images to real ones, LPIPS is used to estimate how likely the outputs are to belong to the target domain, VGG Loss (with VGG19) is used to estimate how much the outputs preserve the structural information in the inputs. We randomly select 32 images and run the experiments for 3 times with a single source and target image, we report the average score in Table 1. We could notice that SITTA could achieve comparatively better scores. In Fig. 5, we show some of the corresponding qualitative comparison results of VanillaCake ↔ ChocolateCake and Milk ↔ BubbleMilk. We could observe that SITTA enables clear and reasonable output. However, ArtStyle fails to learn the bubble or cake patterns, CycleGAN works better while losing detailed structural or textural information, FUNIT generates unnatural images, SinGAN doesn't change the textures appropriately and

[2] We first train a SinGAN model and then apply it to Paint to Image.

Table 1. Quantitative comparison between SITTA and various image synthesis methods. The lower the better.

Method	FID ↓	LPIPS ↓	VGG Loss ↓
ArtStyle	237.8	0.682	0.972
CycleGAN	209.6	0.514	0.805
SinGAN	224.5	0.537	0.819
FUNIT	221.8	0.567	0.698
TuiGAN	223.7	0.513	0.791
SITTA	**197.7**	**0.509**	**0.391**

TuiGAN fails to generate reasonable samples based on multiple scales, leading to unreasonable or distorted outputs. Instead, SITTA doesn't depend on multi-scale training, which is more robust for random inputs of different resolutions. We also display other qualitative comparison in Appendix Section D.1.

(a) VanillaCake ↔ ChocolateCake. (b) Milk ↔ BubbleMilk.

Fig. 5. Qualitative comparison between SITTA and various image synthesis methods.

Ablation Study. Given that SITTA aims to learn a texture mapping between source image and target image but preserve general content, we utilize Positional Norm to extract structural information and re-inject them into intermediate layers of decoder. To evaluate the impact of this operation, we compare our results with or without Positional Norm re-injection. In Fig. 6, we show comparisons of similar structural modes and different structural modes that two inputs share similar or different structures. It could be noticed that Positional Norm re-injection effectively regularizes the model to preserve its original content and avoid obvious unclear regions, color distortion and weird spots for an reasonable output.

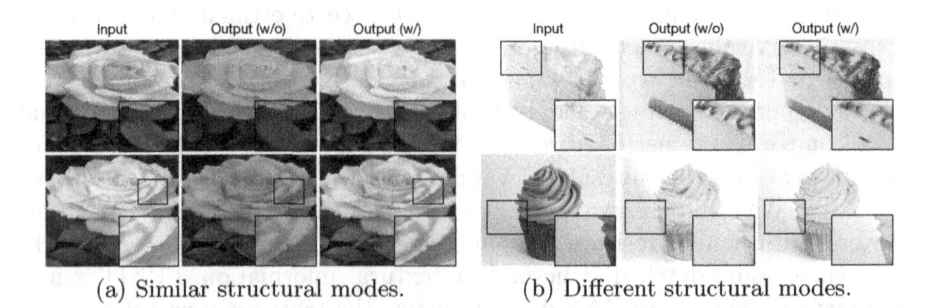

(a) Similar structural modes. (b) Different structural modes.

Fig. 6. Compare w/ or w/o Positional Norm re-injection.

Table 2. Comparisons of testing time and training time (seconds per iteration) and corresponding training details.

Method	Testing Time ↓	Training Time↓	Training Iters↓	Output Type
ArtStyle	0.031	**0.036**	1000	Single
CycleGAN	0.053	0.260	**800**	Pair
SinGAN	0.047	0.099	20000	Single
FUNIT	0.080	–	pre-train	Pair
TuiGAN	0.051	0.372	20000	Pair
SITTA	**0.009**	0.250	**800**	Pair

Running Time. SITTA is a lightweight yet efficient network. Here we show the comparison of average training and testing time per iteration in an epoch using the same single Geforce GTX 1080 Ti GPU without acceleration in Table 2. For testing, we report the time of forward operation. For training, we report the time of forward, backward and optimizer, scheduler update operations. We report the number of training iterations (iters) needed for each method.[3] For output type, we report whether the model generates a pair of outputs ('pair') or a single output ('single'). The implementation of various methods are based on Pytorch [31]. We set image size as 288×288 in all code for fair comparison. We could notice that SITTA achieves fastest testing with an obvious edge. Though ArtStyle could be trained very fast, it cannot achieve texture translation very well. While SITTA shows its superior advantage considering training iterations, training time (seconds/iter) as well as model performance.

4.3 Augmented Data for Image Classification

Table 3. Healthy / sick leaves classification results (Top-1 accuracy %) with different training dataset.

Training data	ResNet-18	VGG16
Baseline	55.2	55.2
+ Repeat	65.7	64.6
+ SITTA	**70.7**	**71.3**

Long-Tailed Image Classification. In the wild, it is very hard and impractical to collect balanced datasets for training recognition models. For example, collecting healthy leaves is easy, while collecting sick leaves could be comparatively very difficult and expensive. In this section, we use SITTA to translate the texture from the few sick leaves to the healthy leaves to obtain more 'sick' data. We use Plant Pathology 2020 dataset [40] that provides data covering a number of category of foliar diseases in apple trees. We select 'healthy' and 'multi-disease' classes as 'healthy' and 'sick' and randomly split the dataset for

[3] Note: We utilize the official code or the widely used github code. For CycleGAN, we train it for the same number of epochs of SITTA to make fair comparisons. For FUNIT, we test the one-shot translation using the offical pre-trained model that has been trained for 100,000 iterations, so the 'training time' is '-', 'training iters' is 'pre-train'. We report the training time of SinGAN and TuiGAN at Scale = 0, which costs less time than other scales.

training and testing. The training set consists of 416 healthy images and 1 sick image, the test set consists of 100 healthy images and 81 sick images. On the whole, we have three types of training data. We refer to the single sick image as *Baseline*. To solve the imbalanced data problem, We repeat the single sick image to match the 'healthy' data count, and refer to this operation as *Repeat*. Also we apply the texture from the single sick leaf to all healthy leaves via SITTA (Please see Fig. 7(a) for examples). To validate the augmented dataset, we use t-SNE [26] to visualize the distribution of Repeat, targeted test set, and SITTA. In Fig. 8, we see that the augmented data shares a similar t-SNE distribution with the targeted test set.

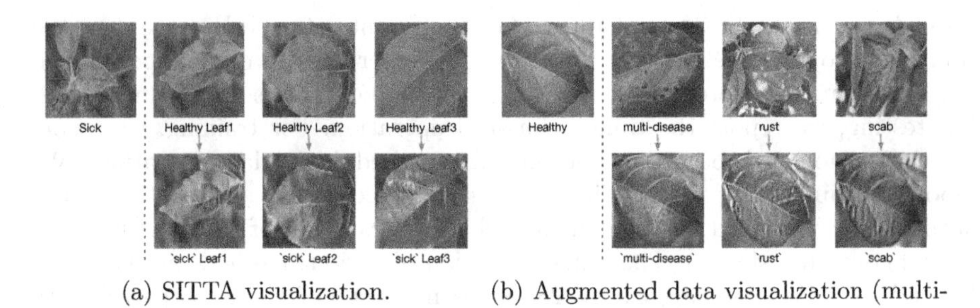

(a) SITTA visualization. (b) Augmented data visualization (multiple classes).

Fig. 7. Compare w/ or w/o Positional Norm re-injection.

We train and test with these datasets on ResNet-18 [14] and VGG16 [37] using cosine learning rate starting from 0.01 for 90 epochs with standard augmentation RandomResizedCrop and RandomHorizontalFlip. We run experiments for three times and report the average score for fair comparison. In Table 3, we could observe consistent superior results of SITTA over the baselines.

Comparison with Various Image Synthesis Methods for Data Augmentation. We emphasize that it is not easy to synthesize useful data for image recognition. To further clarify the concerns, based on Table 3 setting, we conduct an ablation study based on Healthy/Sick leaves classification generated by various image synthesis methods. Please note that most methods cannot be finished within a short time and is impractical for large-scale data augmentation as shown in Table 2. We display the results in Fig. 9. We could observe that SITTA significantly outperforms other methods for data augmentation.

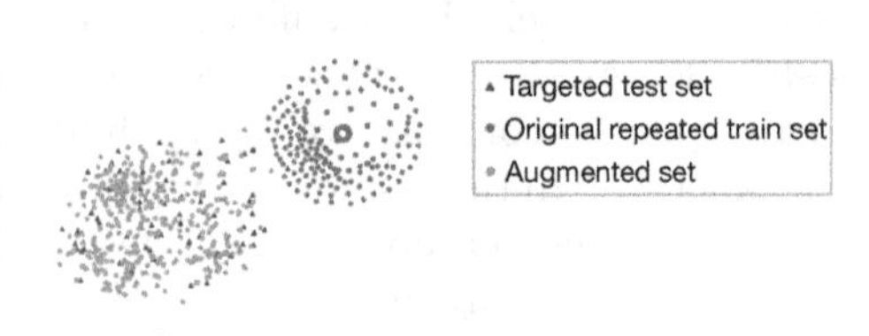

Fig. 8. t-SNE visualization with leaf images.

Complementarity with Other Augmentation Methods. SITTA is a simple yet efficient method to generate new data semantically, and we may regard it as complimentary with other existing augmentation methods. To justify this, we further explore leaf classification in the multiple classes setting. We add the other classes 'rust' and 'scab' from Plant Pathology 2020 [40], aiming to classify which category the leaves belong to. We randomly split the dataset for training and testing. The training set consists of 416 healthy leaf images, 1 multi-disease, 1 rust and 1 scab leaf image, the test set consists of 100 healthy, 81 multi-disease, 612 rust and 582 scab leaf images. We take all four categories for 4-class classification and healthy, rust, scab leaf images for 3-class classification, as well as healthy, multi-disease for 2-class classification. Same as previous setting, we treat the original training data as baseline. We augment the training data via Repeat and SITTA respectively. We display SITTA augmented examples in Fig. 7(b). Since we witnessed no improvement for most data augmentation methods for only a single input, we compare our results with more competitive settings: using different data augmentation methods based on repeated data. We combine Repeat and SITTA with different widely-used and recent augmentation methods [4] including Colorjitter, GaussianBlur, Grayscale, Mixup [51], CutMix [50] and MoEx [22] using the official recommended hyper-parameters for ResNet. In Table 4, we show the comparison of various augmentation strategies based on ResNet-18. Consistent with previous results, SITTA improves the baseline with a large margin and is even highly competitive with state-of-the-art methods Mixup, CutMix and MoEx. Also, since SITTA generates new data that could be easily used with any augmentation methods. It could be observed that SITTA is compatible with other augmentation methods and consistently help boost the model performance with obvious edge.

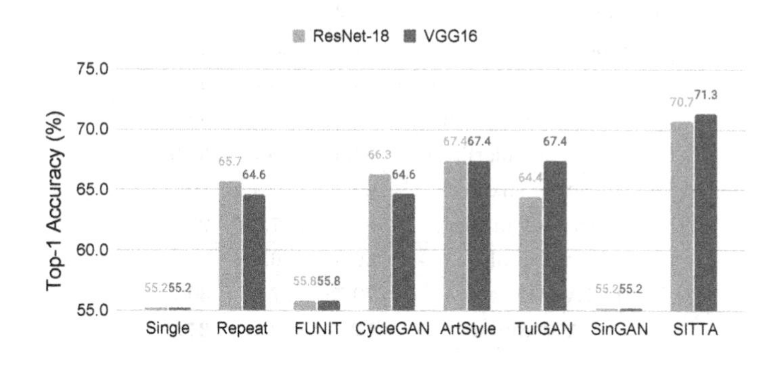

Fig. 9. Healthy / Sick leaves classification results (Top-1 accuracy %) with different image synthesis methods.

Few-shot Image Classification. Modern recognition systems are data-intensive and often need many examples of each class to saturate performance.

However, it is impractical and hampered when the data set is small. Few-shot learning [46,47] is proposed to solve such kind of problem and improve the recognition performance by training with few samples. In this section, SITTA aims to generate additional training images for improving few-shot image classification.

Oxford 102 Flowers. Here we set up all-way few-shot (few images of all classes) classification based on Oxford 102 flowers dataset [28] that consists of 102 flower categories, each class consists of between 40 and 258 images. We strictly follow the official data split rule. For training with SITTA, we randomly select 5 images as 5 shots for each class (Baseline), for testing, we evaluate the model on all test images. We augment the data within each category, each image is augmented to 4 additional images based on the new content of other 4 images (Please see Appendix Section D.3 for a bunch of visualization examples), therefore we have 25 images for each class. To relieve the concern of imbalanced class, we repeat 5 images of each category for 4 times to ensure the same numbers of train set. Since all-way 5-shot 102 classes classification is very difficult to train from scratch [39], we fine-tune the pre-trained model and test on the official testset based on ResNet-18, VGG16 and ResNet-50 (all of which have been pretrained on the 1000-class ImageNet [6]). For fine-tuning, we fine-tune the Pytorch default pre-trained model for 90 epochs and set cosine learning rate starting from 0.01. We run the experiments for 3 times and report the average score in Table 5. It

Table 4. Multi-class leaves classification results (Top-1 accuracy %, average of 3 runs) with various augmentation strategies based on ResNet-18.

Method	2-class	3-class	4-class
Baseline	55.2	8.4	7.8
+ Repeat	61.9	23.4	28.8
+ Repeat + GaussianBlur	63.0	23.7	29.6
+ Repeat + Grayscale	65.2	28.9	32.9
+ Repeat + Colorjitter	64.6	**35.5**	35.5
+ Repeat + Mixup	60.2	17.5	20.1
+ Repeat + CutMix	65.2	27.0	26.7
+ Repeat + MoEx	60.8	26.8	35.6
+ SITTA	70.7	27.4	30.7
+ SITTA + MoEx	**72.9**	32.3	**37.5**

Table 5. Comparison of all-way 5-shot classification (Top-1 accuracy %) results on Oxford 102 flowers.

Method	ResNet-18	VGG16	ResNet-50
Baseline	73.8	75.0	79.6
+ Repeat	76.2	77.7	81.8
+ SITTA	**77.7**	**79.2**	**82.7**

could be obviously noticed that SITTA is able to improve the classification on 102 classes without any additional fine-tuning on real images. The results are very inspiring and show the consistent edge regarding different model architectures.

Caltech-UCSD Birds 200. To further verify the validity of SITTA, we strictly follow the same procedure of Oxford flowers setting and apply SITTA to Caltech-UCSD Birds-200-2011 (CUB-200-2011) dataset [44] for data augmentation (Please see Appendix Section D.4 for a gallery of visualization examples). CUB-200-2011 consists of 11,788 photos of 200 bird species. We run the experiments for 3 times and report the average classification score on the official testset in Table 6. We observe the consistent and competitive improvement of SITTA, which sheds light on the feasibility of SITTA for few-shot learning tasks.

Table 6. Comparison of all-way 5-shot classification (Top-1 accuracy %) results on CUB-200-2011.

Method	ResNet-18	VGG16	ResNet-50
Baseline	30.9	38.9	38.3
+ Repeat	31.1	40.4	38.6
+ SITTA	**32.5**	**40.5**	**38.9**

iNaturalist Birds. iNaturalist (iNat) [43] is a large-scale species classification and detection dataset that features visually similar species, captured in a wide variety of situations from all over the world. We select two genera House Sparrow and Parulidae under the Aves (or bird) supercategory from iNat 2018 (training images) that provides data and labels based on biology taxonomy. We randomly split the dataset into train and test. The training set consists of 518 Parulidae images and 1 House Sparrow image, both test set consists of 105 images separately. Since the two categories share similar structures, we are able to use SITTA to generate synthetic images with translated textures (Please see Appendix Section D.2 for illustration). Following previous long-tailed classification settings, we compare the classification results and display them in Table 7. It could be observed that augmented birds are able to help improve the classification performance, which is consistent with the results of leaf classification.

5 Discussion

Camouflage. Mimicry or Camouflage in natural world provides some real examples for texture swapping that creatures make the textures of their body similar to the environment's to avoid danger or hunt food [41]. For instance, we show the case of mantis and orchid. To prey for the insects, mantis will adaptively change their texture similar to

Table 7. Comparison of iNat bird classification results(Top-1 accuracy %).

Training data	ResNet-18	VGG16
Baseline	50.0	50.0
+Repeat	57.1	63.8
+SITTA	**61.4**	**68.6**

the orchids. In Fig. 10, we give an illustration and find SITTA could translate the orchids' texture to the mantis and obtain reasonable and natural outputs that looks very close to the real example.

Task-specific Augmentation. In this paper, we introduce task- and dataset-specific augmentation. We aim to swap the texture between a target object (for shape) and an exemplar object (for texture). Therefore, we don't want the shape with the new texture confuses with other classes in the dataset. In traditional image recognition pipelines, image augmentation includes rotation, crop, jittering, flip, etc. Recently, [22,51] revised input data or features to improve image classification performance. On the one hand, these methods change the way of representation while keeping the texture the same. On the other hand, image synthesis brings a new direction to change the texture while keeping the structure unchanged. Such kind of augmentation engineering opens a door for understanding objects from textures to structure, as well as augmenting data by 'destroy' the original textures while replacing them with sensible or other resources.

Label Assignments for Augmented Data. Image synthesis brings a new world for data augmentation. Previous methods mainly focus on leading networks to learn more shape information by applying different styles to the original images [10,38,55]. In these cases, the augmented image will be assigned with the original label. However, label assignment should be serious considered given different tasks and datasets. For leaves classification, textures make sick and healthy leaves different. Therefore, we generate a bunch of sick leaves by translating the textures of sick leaves to healthy leaves and assign the new image as 'sick' instead of 'healthy.' While for few-shot learning experiments, we augment the data within each category and assign the same label to them. For future potential directions, more advanced label assignment strategies such as label perturbation [22,50,51] could be considered for synthetic dataset.

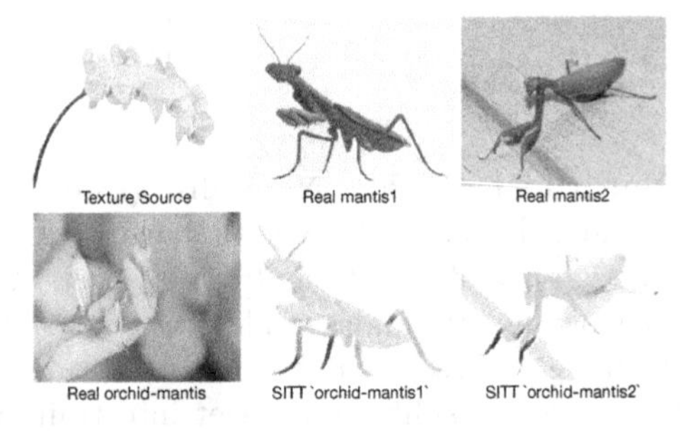

Fig. 10. Augmented Orchid-Mantis for camouflage.

Limitations. Although SITTA brings an obvious improved margin for recognition models, sometimes there are very few provided texture source images. Therefore the extracted texture information via SITTA might not be able to cover all texture patterns in testset, leading to limited accuracy improvement. Besides, in spite of faster running time, there still is a processing time gap between image synthesis and traditional data augmentation such as flip and crop. For the future work, we would explore on both optimally increasing the diversity based on the single or very few texture source and speeding up the image translation model. We believe these signals shed light on the new research direction of semantic image synthesis for data augmentation.

6 Conclusion

In this paper, we explore the problem of image synthesis for recognition tasks. We propose a lightweight, fast and efficient Single Image Texture Translation for data Augmentation (SITTA). Images generated by SITTA not only look appealing but also help visual recognition tasks including long-tailed and few-shot image classification. We hope our work could open the door of using image synthesis for data augmentation in various computer vision tasks and make image synthesis one step closer to solving real problems in the wild.

Acknowledgement. This work was supported in part by the Pioneer Centre for AI, DNRF grant number P1.

References

1. Antoniou, A., Storkey, A., Edwards, H.: Data augmentation generative adversarial networks. arXiv preprint arXiv:1711.04340, 2017
2. Bolukbasi, T., Chang, K.-W., Zou, J.Y., Saligrama, V., Kalai, A.T.: Man is to computer programmer as woman is to homemaker? debiasing word embeddings. Adv. Neural Inf. Process. Syst. **29**, 4349–4357 (2016)
3. Chen, T., Kornblith, S., Norouzi, M., Hinton, G.: A simple framework for contrastive learning of visual representations. arXiv preprint arXiv:2002.05709 (2020)
4. Chen, X., Fan, H., Girshick, R., He, K.: Improved baselines with momentum contrastive learning. arXiv preprint arXiv:2003.04297 (2020)
5. Cubuk, E.D., Zoph, B., Mane, D., Vasudevan, V., Le, Q.V.: AutoAugment: learning augmentation strategies from data. In: Proceedings of the IEEE Conference on Computer Vision and Pattern Recognition, pp. 113–123 (2019)
6. Deng, J., Dong, W., Socher, R., Li, L.-J., Li, K., Fei-Fei, L.: ImageNet: a large-scale hierarchical image database. In: 2009 IEEE Conference on Computer Vision and Pattern Recognition, pp. 248–255. IEEE (2009)
7. Efros, A.A., Freeman, W.T.: Image quilting for texture synthesis and transfer. In: Proceedings of the 28th Annual Conference on Computer Graphics and Interactive Techniques, pp. 341–346 (2001)
8. Gatys, L.A., Ecker, A.S., Bethge, M.: A neural algorithm of artistic style. arXiv preprint arXiv:1508.06576 (2015)
9. Gatys, L.A., Ecker, A.S., Bethge, M.: Image style transfer using convolutional neural networks. In: Proceedings of the IEEE Conference on Computer Vision and Pattern Recognition, pp. 2414–2423 (2016)
10. Geirhos, R., Rubisch, P., Michaelis, C., Bethge, M., Wichmann, F.A., Brendel, W.: Imagenet-trained CNNs are biased towards texture; increasing shape bias improves accuracy and robustness. arXiv preprint arXiv:1811.12231 (2018)

11. Goodfellow, I., et al.: Generative adversarial nets. Adv. Neural. Inf. Process. Syst. **27**, 2672–2680 (2014)
12. He, K., Fan, H., Wu, Y., Xie, S., Girshick, R.: Momentum contrast for unsupervised visual representation learning. In: Proceedings of the IEEE/CVF Conference on Computer Vision and Pattern Recognition, pp. 9729–9738 (2020)
13. He, K., Sun, J., Tang, X.: Guided image filtering. IEEE Trans. Pattern Anal. Mach. Intell. **35**(6), 1397–1409 (2012)
14. He, K., Zhang, X., Ren, S., Sun, J.: Deep residual learning for image recognition. In: Proceedings of the IEEE Conference on Computer Vision and Pattern Recognition, pp. 770–778 (2016)
15. Heusel, M., Ramsauer, H., Unterthiner, T., Nessler, B., Hochreiter, S.: GANs trained by a two time-scale update rule converge to a local Nash equilibrium. arXiv preprint arXiv:1706.08500 (2017)
16. Huang, X., Liu, M.-Y., Belongie, S., Kautz, J.: Multimodal unsupervised image-to-image translation. In: Ferrari, V., Hebert, M., Sminchisescu, C., Weiss, Y. (eds.) ECCV 2018. LNCS, vol. 11207, pp. 179–196. Springer, Cham (2018). https://doi.org/10.1007/978-3-030-01219-9_11
17. Isola, P., Zhu, J.-Y., Zhou, T., Efros, A.A.: Image-to-image translation with conditional adversarial networks. In: Proceedings of the IEEE Conference on Computer Vision and Pattern Recognition, pp. 1125–1134 (2017)
18. Johnson, J., Alahi, A., Fei-Fei, L.: Perceptual losses for real-time style transfer and super-resolution. In: Leibe, B., Matas, J., Sebe, N., Welling, M. (eds.) ECCV 2016. LNCS, vol. 9906, pp. 694–711. Springer, Cham (2016). https://doi.org/10.1007/978-3-319-46475-6_43
19. Kingma, D.P., Welling, M.: Auto-encoding variational Bayes. arXiv preprint arXiv:1312.6114 (2013)
20. Lee, H.-Y., Tseng, H.-Y., Huang, J.-B., Singh, M., Yang, M.-H.: Diverse image-to-image translation via disentangled representations. In: Ferrari, V., Hebert, M., Sminchisescu, C., Weiss, Y. (eds.) ECCV 2018. LNCS, vol. 11205, pp. 36–52. Springer, Cham (2018). https://doi.org/10.1007/978-3-030-01246-5_3
21. LeGendre, C., et al.: DeepLight: learning illumination for unconstrained mobile mixed reality. In: Proceedings of the IEEE Conference on Computer Vision and Pattern Recognition, pp. 5918–5928 (2019)
22. Li, B., Wu, F., Lim, S.-N., Belongie, S., Weinberger, K.Q.: On feature normalization and data augmentation. In: Proceedings of the IEEE/CVF Conference on Computer Vision and Pattern Recognition, pp. 12383–12392 (2021)
23. Li, B., Wu, F., Weinberger, K.Q., Belongie, S.: Positional normalization. In: Advances in Neural Information Processing Systems, pp. 1622–1634 (2019)
24. Lin, J., Pang, Y., Xia, Y., Chen, Z., Luo, J.: TuiGAN: learning versatile image-to-image translation with two unpaired images. In: Vedaldi, A., Bischof, H., Brox, T., Frahm, J.-M. (eds.) ECCV 2020. LNCS, vol. 12349, pp. 18–35. Springer, Cham (2020). https://doi.org/10.1007/978-3-030-58548-8_2
25. Liu, M.-Y., et al.: Few-shot unsupervised image-to-image translation. In: arxiv, Timo Aila (2019)
26. van der Maaten, L., Hinton, G.: Visualizing data using t-SNE. J. Mach. Learn. Res. **9**, 2579–2605 (2008)
27. Mildenhall, B., Srinivasan, P.P., Tancik, M., Barron, J.T., Ramamoorthi, R., Ng, R.: NeRF: representing scenes as neural radiance fields for view synthesis. arXiv preprint arXiv:2003.08934 (2020)

28. Nilsback, M.E., Zisserman, A.: Automated flower classification over a large number of classes. In: 2008 Sixth Indian Conference on Computer Vision, Graphics & Image Processing, pp. 722–729. IEEE (2008)
29. Park, T., Efros, A.A., Zhang, R., Zhu, J.-Y.: Contrastive learning for unpaired image-to-image translation. In: Vedaldi, A., Bischof, H., Brox, T., Frahm, J.-M. (eds.) ECCV 2020. LNCS, vol. 12354, pp. 319–345. Springer, Cham (2020). https://doi.org/10.1007/978-3-030-58545-7_19
30. Park, T., Zhu, J.-Y., Wang, O., Lu, J., Shechtman, E., Efros, A., Zhang, R.: Swapping autoencoder for deep image manipulation. In: Advances in Neural Information Processing Systems 33 (2020)
31. Paszke, A., et al. PyTorch: an imperative style, high-performance deep learning library. arXiv preprint arXiv:1912.01703 (2019)
32. Qin, Z., Liu, Z., Zhu, P., Xue, Y.: A GAN-based image synthesis method for skin lesion classification. Comput. Methods Programs Biomed. **195**, 105568 (2020)
33. Rao, K., Harris, C., Irpan, A., Levine, S., Ibarz, J., Khansari, M.: RL-CycleGAN: reinforcement learning aware simulation-to-real. In: Proceedings of the IEEE/CVF Conference on Computer Vision and Pattern Recognition, pp. 11157–11166 (2020)
34. Shaham, T.R., Dekel, T., Michaeli, T.: SinGAN: learning a generative model from a single natural image. In: Proceedings of the IEEE International Conference on Computer Vision, pp. 4570–4580 (2019)
35. Shorten, C., Khoshgoftaar, T.M.: A survey on image data augmentation for deep learning. J. Big Data **6**(1), 1–48 (2019)
36. Shrivastava, A., Pfister, T., Tuzel, O., Susskind, J., Wang, W., Webb, R.: Learning from simulated and unsupervised images through adversarial training. In: Proceedings of the IEEE conference on computer vision and pattern recognition, pp. 2107–2116 (2017)
37. Simonyan, K., Zisserman, A.: Very deep convolutional networks for large-scale image recognition. arXiv preprint arXiv:1409.1556 (2014)
38. Somavarapu, N., Ma, C.-Y., Kira, Z.: Frustratingly simple domain generalization via image stylization. arXiv preprint arXiv:2006.11207 (2020)
39. Sun, Q., Liu, Y., Chua, T.-S., Schiele, B.: Meta-transfer learning for few-shot learning. In: Proceedings of the IEEE/CVF Conference on Computer Vision and Pattern Recognition (CVPR) (2019)
40. Thapa, R., Snavely, N., Belongie, S., Khan, A.: The plant pathology 2020 challenge dataset to classify foliar disease of apples. arXiv preprint arXiv:2004.11958 (2020)
41. Théry, M., Casas, J.: Predator and prey views of spider camouflage. Nature **415**(6868), 133 (2002)
42. Touvron, H., Vedaldi, A., Douze, M., Jégou, H.: Fixing the train-test resolution discrepancy. In: Advances in Neural Information Processing Systems, pp. 8250–8260 (2019)
43. Van Horn, G., et al.: The iNaturalist species classification and detection dataset. In: Proceedings of the IEEE Conference on Computer Vision and Pattern Recognition, pp. 8769–8778 (2018)
44. Wah, C., Branson, S., Welinder, P., Perona, P., Belongie, S.: The caltech-UCSD birds-200-2011 dataset (2011)
45. Wang, Y., Pan, X., Song, S., Zhang, H., Huang, G., Wu, C.: Implicit semantic data augmentation for deep networks. In: Advances in Neural Information Processing Systems, pp. 12635–12644 (2019)
46. Wang, Y., Yao, Q., Kwok, J.T., Ni, L.M.: Generalizing from a few examples: a survey on few-shot learning. ACM Comput. Surv. (CSUR), **53**(3), 1–34 (2020)

47. Wang, Y.-X., Girshick, R., Hebert, M., Hariharan, B.: Low-shot learning from imaginary data. In: Proceedings of the IEEE Conference on Computer Vision and Pattern Recognition, pp. 7278–7286 (2018)
48. Xiao, T., Wang, X., Efros, A.A., Darrell, T.: What should not be contrastive in contrastive learning. In International Conference on Learning Representations (2021)
49. Yoo, J., Uh, Y., Chun, S., Kang, B., Ha, J.-W.: Photorealistic style transfer via wavelet transforms. In: International Conference on Computer Vision (ICCV) (2019)
50. Yun, S., Han, D., Oh, S.J., Chun, S., Choe, J., Yoo, Y.: CutMix: regularization strategy to train strong classifiers with localizable features. In: Proceedings of the IEEE International Conference on Computer Vision, pp. 6023–6032 (2019)
51. Zhang, H., Cisse, M., Dauphin, Y.N., Lopez-Paz, D.: mixup: beyond empirical risk minimization. In: International Conference on Learning Representations (2018)
52. Zhang, R., Isola, P., Efros, A.A.: Colorful image colorization. In: Leibe, B., Matas, J., Sebe, N., Welling, M. (eds.) ECCV 2016. LNCS, vol. 9907, pp. 649–666. Springer, Cham (2016). https://doi.org/10.1007/978-3-319-46487-9_40
53. Zhang, R., Isola, P., Efros, A.A., Shechtman, E., Wang, O.: The unreasonable effectiveness of deep features as a perceptual metric. In: CVPR (2018)
54. Zhang, Y., Zhang, Y., Qinwei, X., Zhang, R.: Learning robust shape-based features for domain generalization. IEEE Access **8**, 63748–63756 (2020)
55. Zheng, X., Chalasani, T., Ghosal, K., Lutz, S., Smolic, A.: STaDa: style transfer as data augmentation. arXiv preprint arXiv:1909.01056 (2019)
56. Zhu, J.-Y., Park, T., Isola, P., Efros, A.A.: Unpaired image-to-image translation using cycle-consistent adversarial networks. In: 2017 IEEE International Conference on Computer Vision (ICCV) (2017)

Learning from Noisy Labels with Coarse-to-Fine Sample Credibility Modeling

Boshen Zhang[1], Yuxi Li[1], Yuanpeng Tu[2], Jinlong Peng[1], Yabiao Wang[1](✉), Cunlin Wu[3], Yang Xiao[3], and Cairong Zhao[2]

[1] YouTu Lab, Shenzhen, China
{boshenzhang,yukiyxli,jeromepeng,caseywang}@tencent.com
[2] Tongji University, Shanghai, China
{2030809,zhaocairong}@tongji.edu.cn
[3] Key Laboratory of Image Processing and Intelligent Control, Ministry of Education, School of Artificial Intelligence and Automation, Huazhong University of Science and Technology, Wuhan, China
{cunlin_wu,Yang_Xiao}@hust.edu.cn

Abstract. Training deep neural network (DNN) with noisy labels is practically challenging since inaccurate labels severely degrade the generalization ability of DNN. Previous efforts tend to handle part or full data in a unified denoising flow via identifying noisy data with a coarse small-loss criterion to mitigate the interference from noisy labels, ignoring the fact that the difficulties of noisy samples are different, thus a rigid and unified data selection pipeline cannot tackle this problem well . In this paper, we first propose a coarse-to-fine robust learning method called *CREMA*, to handle noisy data in a divide-and-conquer manner. In coarse-level, clean and noisy sets are firstly separated in terms of credibility in a statistical sense. Since it is practically impossible to categorize all noisy samples correctly, we further process them in a fine-grained manner via modeling the credibility of each sample. Specifically, for the clean set, we deliberately design a memory-based modulation scheme to dynamically adjust the contribution of each sample in terms of its historical credibility sequence during training, thus alleviating the effect from noisy samples incorrectly grouped into the clean set. Meanwhile, for samples categorized into the noisy set, a selective label update strategy is proposed to correct noisy labels while mitigating the problem of correction error. Extensive experiments are conducted on benchmarks of different modality, including image classification (CIFAR, Clothing1M etc.) and text recognition (IMDB), with either synthetic or natural semantic noises, demonstrating the superiority and generality of *CREMA*.

Keywords: Robust learning · Label noise · Divide-and-conquer

B. Zhang, Y.Li and Y.Tu—Contributed equally to this work.

L. Karlinsky et al. (Eds.): ECCV 2022 Workshops, LNCS 13802, pp. 21–38, 2023.
https://doi.org/10.1007/978-3-031-25063-7_2

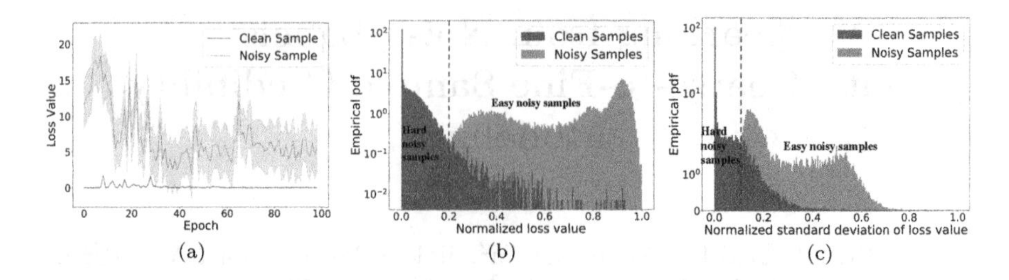

Fig. 1. Training on MNIST with 50% symmetric noise. (a) Compared with noisy samples, clean samples yield relatively smaller loss value and more consistent predictions. (b) Empirical PDF (Probability Density Function) of loss values and (c) their standard deviation justify the above conclusion. That is, clean and noisy samples possess distinctive statistical properties. However, noisy samples can not be completely identified via simple threshold filter strategy (blue dotted line in (b) and (c)) with the statistical metrics. Similar experimental conclusion can also be found on CIFAR-10/100. (Color figure online)

1 Introduction

Deep learning has achieved significant progress in the recognition of multimedia signals (e.g. images, text, speeches). The key to its success is the availability of large-scale datasets with reliable manual annotations. Collecting such datasets, however, is time-consuming and expensive. Some alternative ways to obtain labeled data, such as web crawling [44], inevitably yield samples with noisy labels, which are not appropriate to be directly utilized to train DNN since these complex models can easily over-fitting (i.e., memorizing) noisy labels [2,48].

To handle this problem, classical Learning with Noisy Label (LNL) approaches focus on either identifying and dropping noisy samples (i.e., sample selection) [10,14,43,47] or adjusting the objective term of each sample during training (i.e., loss adjustment) [29,36,46]. The former usually make use of small-loss trick to select clean samples, and then take them to update DNNs. However, the procedure of sample selection cannot guarantee that the selected clean samples are completely clean. In contrast, as indicated in Fig. 1, division relied on statistic metrics can still involve some hard noisy samples in the training set, which will be treated equally as other normal samples in the following training stages. Thus the negative impact brought by wrongly grouped noisy samples may confuse the optimization process and lower the test performance of DNNs [47]. On the other hand, the latter schemes reweight loss values or update labels by estimating the confidence on how clean a sample is. Typical methods include loss correction via an estimated noise transition matrix [8,12,29]. However, estimating an accurate noise transition matrix is practically challenging. Recently, there are approaches directly correcting the labels of all training samples [36,46]. However, we empirically find that unconstrained label correction in full data can do harm to clean samples and reversely hinder the model performance.

Towards the problems above, we propose a simple but effective method called *CREMA* (*Coarse-to-fine sample cREdibility Modeling and Adaptive loss reweighting*), which adaptively reduces the impact of noisy labels via modeling the credibility (i.e., quality) of each sample. In the coarse-level, with the estimated sample credibility by simple statistic metrics, clean and noisy samples can be roughly separated and handled in a divide-and-conquer manner. Since it is practically impossible to separate these samples perfectly, for the selected clean samples, we take their historical credibility sequences to adjust the contribution of each sample to their objective, thus mitigating the negative impact of hard noisy samples (i.e, noisy samples incorrectly grouped into the clean set) in a fine-grained manner. As for the separated noisy samples, some of them are actually clean (i.e., hard clean samples) and can be helpful for model training. Thus instead of discarding them as previous sample selection methods [10,43], we make use of them via a selective label correction scheme.

The insight behind CREMA is from the observation on the loss value during training on noisy data (illustrated in Fig 1), it can be found that **clean and noisy samples manifest distinctive statistical properties during training, where clean samples yield relatively smaller loss value** [32] **and more consistent prediction**. Hence these statistical features can be utilized to coarsely model the sample credibility. However, Fig 1 also shows that the full data can not be perfectly separated by simple statistical metrics. This inspires us to adaptively cope with noises of different difficulty levels with more fine-grained design. For easily recognized noisy samples, we can directly apply certain label correction schemes while avoiding erroneous correction on normal samples. For samples that fall into the confusing area and hybrid with clean ones, since the coarsely estimated credibility in the current epoch is not informative enough to identify noisy samples, *CREMA* applies a fine-grained likelihood estimator of noisy samples by resorting the historical sequence of sample credibility. This is achieved by maintaining a historical memory bank along with the training process and estimating the likelihood function through a consistency metric and assumption of markov property of the sequence.

CREMA is built upon a classic co-training framework [10,43]. The fine-grained sample credibility estimated by one network is used to adjust the loss term of credible samples for the other network. Extensive experiments are conducted on benchmarks of different modality, including image classification (CIFAR, MNIST, Clothing1M etc.) and text recognition (IMDB), with either synthetic or natural semantic noises, demonstrating the superiority and generality of the proposed method. In a nutshell, the key contributions of this paper include:

- *CREMA*: a novel LNL algorithm that combats noisy labels via coarse-to-fine sample credibility modeling. In coarse-level, clean and noisy sets are roughly separated and handled respectively, in the spirit of the idea of divide-and-conquer. Easily recognized noisy samples are handled via a selective label update strategy;
- In *CREMA*, likelihood estimation of historical credibility sequence is proposed to help identify hard noisy samples, which naturally plays as the dynamical weight to modulate loss term of each training sample in a fine-grained manner;

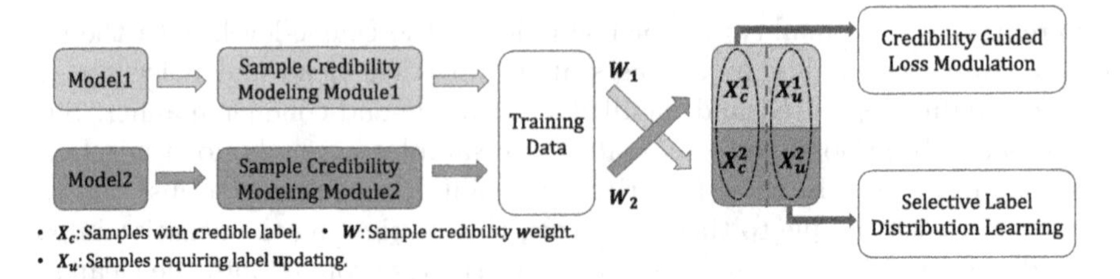

Fig. 2. The pipeline of *CREMA*. *CREMA* trains two parallel networks simultaneously. Clean samples (mostly clean) X_c and noisy samples (mostly noisy) X_u are separated via estimating the credibility of each training data. A selective label distribution learning scheme is applied for easily distinguishable noisy samples in X_u. As for the clean set X_c, likelihood estimation of historical credibility sequence is proposed to handle the hard noisy samples via adaptively modulating their loss term during training.

- *CREMA* is evaluated on six synthetic and real-world noisy datasets with different modality, noise type and strength. Extensive ablation studies and qualitative analysis are provided to verify the effectiveness of each component.

2 Related Works

The existing LNL approaches can be mainly categorized into three groups: loss adjustment, label correction, and noisy sample detection. Next, we will introduce and discuss existing works for training DNN with noisy labels.

Loss Adjustment. Adjusting the loss values of all training samples is able to reduce the negative impact of noisy labels. To do this, many approaches seek to robust loss functions, such as Robust MAE [7], generalized cross entropy [53], symmetric cross entropy [42], Improved MAE [40] and curriculum loss [23]. Rather than treat all samples equally, some methods rectify the loss of each sample through estimating the label transition matrix [8,9,12,19,29,44,50] or imposing different importance on each sample to formulate a weighted training procedure [4,22,38,41]. The noise transition matrix, however, is relatively hard to be estimated and many approaches [5,6,12,14,20,21,33,34,37,54] often make assumptions that a small clean-labeled dataset exists. In real-world scenarios, such condition is not always fulfilled, thus limiting the applications of these approaches.

Label Correction. Label correction methods seek to refurbish the ground-truth of noisy samples, thus preventing DNN overfits to false labels. The most common ways to obtain the updated label include bootstrapping (i.e., a convex combination of the noisy label and the DNN prediction) [1,11,32,39] and label replacing [35,36,46,52]. One critical problem of label Correction methods is to define the confidence of each label being clean, that is, samples with high clean probability should keep their labels almost unchanged, and vice versa. Previous solutions including cross-validation [32], fitting a two-component mixture model [1], local intrinsic dimensionality measurement [13,24] and leveraging the prediction consistency of DNN models [35,39]. However, updating the labels of all training sets is challenging, and well-designed regularization terms are important to prevent DNN from falling into trivial solutions [36,46].

Noisy Sample Detection. One common knowledge used to discover noisy samples is the *memorization effects* (i.e., DNN fits clean samples first and then noisy ones). As a result, after a warm-up training stage with all noisy samples, DNN is able to identify the clean samples by taking the small-loss ones. The *small-loss trick* is exploited by many sample selection methods [10,14,26,28,35,43,45,47]. After separating the noisy samples from the clean ones, Co-teaching [10] and corresponding variants [10,43,45,47] update two parallel network parameters with the clean samples and abandoned the noisy ones. The idea of training two deep networks simultaneously is effective to avoid the confirmation bias problem (i.e., a model would accumulate its error through the self-training process) [10,14,16]. Other noisy measurement metrics such as Area Under the Margin (AUM) [31] is also proposed to better distinguish and remove noisy samples, which hypothesizes that noisy samples will in expectation have smaller AUM than clean samples. However, discarding the noisy samples means that valuable data may be lost, which leads to slow convergence of DNN models [4]. Instead, there are methods that utilize both clean and noisy samples to formulate a semi-supervised learning problem, by discarding only the labels of identified noisy samples. Thus naturally converting LNL problem into a semi-supervised learning one, for which powerful semi-supervised learning methods can be leveraged to boost performance [3,16,49,55].

Hybrid. There are also researches taking two or more techniques above into account to boost performance of robust learning. For example, RoCL [56] and SELFIE [35] propose to dynamically discover informative (refurbished) samples and correct their labels with model predictions. Accordingly, *CREMA* belongs to hybrid group and it differs from existing methods in (1) it adaptively cope with noises of different difficulty levels via estimating the sample credibility in a coarse-to-fine fashion, easy and hard noisy samples are handled in a divide-and-conquer strategy; (2) it estimates likelihood of historical sample credibility sequence to dynamically modulate loss term of hard noisy samples; (3) it explores a selective label correction scheme to deal with hard clean samples while mitigating the correction error; (4) it is end-to-end trainable and does not require extra computation or any modification to the model.

3 Method

In this section, we introduce *CREMA*, an end-to-end approach for LNL problem. The technical pipeline of the approach is shown in Fig. 2. The training process is built upon a classic co-training framework [10,43] to avoid confirmation bias and separate credible samples (mostly clean) and noisy samples (mostly noisy) via per-sample loss value in the coarse stage. The separated samples are handled by the succeeding fine-grained processes in a divide-and-conquer manner.

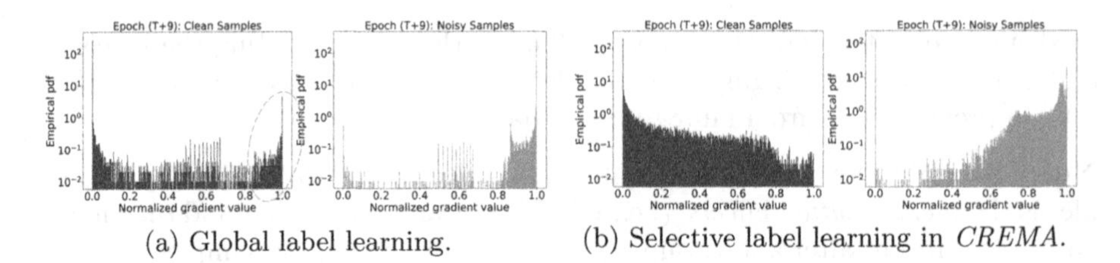

(a) Global label learning. (b) Selective label learning in *CREMA*.

Fig. 3. Training on MNIST with 50% symmetric noise, warm up (i.e., training on all samples with original noisy labels) for T epochs. (a) Global learning procedure requires updating all training samples' label, causes relatively large gradient values even on clean samples (areas within blue dotted lines), making it hard to focus on correcting noisy labels. (b) Training on *CREMA* can effectively identify noisy samples and focus on correcting noisy labels with relatively large gradient values. (Color figure online)

3.1 Coarse Level Separation

Formally, for multi-class classification with noisy label problem, let $\mathcal{D} = \{(x_i, y_i)\}_{i=1}^N$ denote the training data, where x_i is data sample and $y_i \in \{0,1\}^C$ is the one-hot label over C classes. $f(x_i;\theta)$ denotes sample feature extracted by DNN model. With the loss $\mathcal{L}(x,y)$ from the DNN model, clean set $\mathcal{X}_c$, and noisy set $\mathcal{X}_u$ are separated via the widely used low-loss criterion [10,43],

$$
\begin{aligned}
\mathcal{X}_c &= \{(x,y)|\mathcal{L}(x,y) < \tau(x,y) \in \mathcal{D}\}, \\
\mathcal{X}_u &= \{(x,y)|\mathcal{L}(x,y) \geq \tau(x,y) \in \mathcal{D}\},
\end{aligned}
\tag{1}
$$

where τ is the threshold, determined by a dynamic memory rate $R(t) \in [0,1]$. Which is set for DNN to gradually distinguish $(1-R(t))$ data with highest loss value as noisy samples while keeping other samples as the clean set. The loss value simply serves as credibility of each sample in a coarse level. However, as illustrated in Fig. 1, this simple separation criterion can not strictly eliminate noisy samples. Hence we choose to handle them respectively, where $\mathcal{X}_c$ is exploited to update DNN parameters via fine-grained sample credibility guided loss adjustment (Sec. 3.2), and $\mathcal{X}_u$ is leveraged via a label learning scheme (Sect. 3.3).

3.2 Fine-grained Sequential Credibility Modeling

Sequential Credibility Analysis. Previous works prefer to assess the data reliability purely based on its statistical property on a single point of time (e.g. the loss value in current epoch) during training process, i.e. they regard the credibility $w(x,y)$ of the i-th data sample (x,y) proportional to the joint distribution or likelihood of its data-label pair,

$$
w(x,y) \propto P(x,y) \quad \textbf{or} \quad w(x,y) \propto \log P(x,y). \tag{2}
$$

However, as shown in Fig. 1, the training curve of normal and noisy samples usually yield different statistic information, where noisy samples usually have relatively larger loss values and poorer prediction consistency compared with clean ones, therefore the historical record of data training is also informative enough to help distinguish noisy and clean data.

This observation inspires us to estimate the data credibility in a sequential manner. To be specific, we define a sequence with length n as:

$$\mathbf{L}_t^n = [\mathbf{f}_t, \mathbf{f}_{t-1}, \cdots, \mathbf{f}_{t-n+1}], \quad \mathbf{f}_t = f(x; \theta_t). \tag{3}$$

Equation (3) illustrates a sliding window covering the feature snapshot of data from previous n epochs to current time point, where θ_t denotes the model parameters at the t-th epoch, and we model the data credibility with the likelihood and consistency of these historical sequences,

$$w(x, y) \propto C(\mathbf{L}_t^n, y) \log P(\mathbf{L}_t^n | \mathbf{f}_{t-n}, y). \tag{4}$$

The Eq. (4) can be decoupled into two items, where $C(\mathbf{L}_t^n, y)$ measures the stability of training sequence given its label y, while $\log P(\mathbf{L}_t^n | \mathbf{f}_{t-n}, y)$ denotes log-likelihood of sequence generated from the $(t-n)$-th data observation of neural network training process. To estimate the sequential log-likelihood, we further assume that the observation in sequence $\mathbf{L}_t^n$ conforms to a certain markov property as:

$$\mathbf{f}_t \perp \mathbf{f}_i | (\mathbf{f}_{t-1}, y) \quad \forall \quad i < t-1. \tag{5}$$

The assumption in Eq. (5) is reasonable since in most iterative learning algorithm like SGD, the data feature distribution is only decided by the last observation and its label. With this assumption, we can further derive the likelihood as:

$$\begin{aligned} \log P(\mathbf{L}_t^n | \mathbf{f}_{t-n}, y) &= \log P\left(\mathbf{f}_t | \mathbf{L}_{t-1}^n, y\right) + \log P\left(\mathbf{L}_{t-1}^{n-1} | \mathbf{f}_{t-n}, y\right) \\ &= \sum_{i=0}^{n-1} \log P\left(\mathbf{f}_{t-i} | \mathbf{L}_{t-i-1}^{n-i}, y\right) \\ &= \sum_{i=0}^{n-1} \log P(\mathbf{f}_{t-i} | \mathbf{f}_{t-i-1}, y). \end{aligned} \tag{6}$$

With Eq. (6), we can represent the sequential likelihood as the summation of the conditional likelihood of data observation at each adjacent epochs within a sliding window of length n. In the implementation, we can apply a normalized mixture model like GMM [16] or BMM [1] as estimator to estimate the conditional probability $P(\mathbf{f}_{t-i} | \mathbf{f}_{t-i-1}, y)$ in Eq. (6) via modeling the sample-wise loss value distribution. Meanwhile, with the conditional probability estimation, the stability measurement $C(\mathbf{L}_t^n, y)$ is further designed as a modulator to suppress loss on training sequence with intense fluctuation,

$$C(\mathbf{L}_t^n, y) = 1 - \sqrt{\frac{1}{n} \sum_{i=0}^{n-1} \left(P(\mathbf{f}_{t-i} | \mathbf{f}_{t-i-1}, y) - \bar{P}(\mathbf{L}_t^n, y)\right)^2}. \tag{7}$$

$$\bar{P}(\mathbf{L}_t^n, y) = \frac{1}{n}\sum_{i=0}^{n-1} P(\mathbf{f}_{t-i}|\mathbf{f}_{t-i-1}, y). \tag{8}$$

Adaptively Loss Adjustment. The sequential likelihood $\bar{P}(\mathbf{L}_t^n)$ and stability measurement $C(\mathbf{L}_t^n, y)$ reflects how confident of the sample being clean. With the estimated credibility we reweight loss to update DNN as:

$$\theta_{t+1} = \theta_t - \eta\nabla\Big(\frac{1}{|\mathcal{X}_c|}\sum_{(x,y)\in\mathcal{X}_c} w(x,y)\mathcal{L}\big(f(x;\theta_t), y\big)\Big). \tag{9}$$

Where $\mathcal{L}$ is the objective function. $w(x, y)$ is the sample credibility and it modulates the contribution of each sample through gradient descending algorithm. Note that Eq (9) is only applied on clean set $\mathcal{X}_c$, in this way, the negative impact of hard noisy samples within $\mathcal{X}_c$ can be mitigated.

Objective Function. Inspired by the design of symmetric cross entropy (SCE) function [42], a symmetric JS-divergence function with a co-regularization term is leveraged in *CREMA* as:

$$\begin{aligned}\mathcal{L} = & D_{\text{JS}}(y||h(f_1(x;\theta))) + D_{\text{JS}}(y||h(f_2(x;\theta))) \\ & + D_{\text{JS}}\left(h(f_1(x;\theta))||h(f_2(x;\theta))\right),\end{aligned} \tag{10}$$

Where $h(x)$ is the softmax function, $f_1(x)$ and $f_2(x)$ are features extracted by two models. The reason we choose JS-divergence [27] instead of cross entropy (CE) as loss function is that CE tends to over-fit noisy samples as these samples contribute relatively large gradient values during training. While JS-divergence mitigates this problem via using predictions of the current model as supervising signals as well. Since for noisy samples, DNN predictions are usually more reliable than its label. Following previous work [36], a prior label distribution term and a negative entropy term are included to regularize training and further alleviate the over-fitting problem.

3.3 Selective Label Distribution Learning

As discussed above, some hard clean samples are blended with the separated noisy set $\mathcal{X}_u$. Thus instead of discarding these data as in most sample detection methods [10,43], we resort to label correction approaches [36,46] to exploit them with gradually corrected labels and further boost performance.

Specifically, labels of $\mathcal{X}_u$ are treated as extra parameters and updated through back-propagation to optimize a certain objective, this means both the network parameters and labels are updated simultaneously during the training process, where original one-hot labels y will turn into a soft label distribution $\tilde{y} = h(y)$ after updating. Formally, $\tilde{y}$ is updated as $\tilde{y} \leftarrow \tilde{y} - \lambda(\partial\mathcal{L}_l/\partial\tilde{y})$. Where λ is learning rate and $\mathcal{L}_l$ is the objective to supervise the label correction process as $\mathcal{L}_l = D_{\text{JS}}(h(f_1(x;\theta)||\tilde{y}) + D_{\text{JS}}(h(f_2(x;\theta)||\tilde{y})$.

Algorithm 1: *CREMA*. Line 5-9: sequential credibility modeling; Line 10-12: selective label update.

1 **Input:** network parameters $\theta^{(1)}$ and $\theta^{(2)}$, training dataset $\mathcal{D}$, dynamic memory rate $R(t)$, soft label distribution $\tilde{y}$, memory sequence $\mathbf{L}^n_{1,t}$ and $\mathbf{L}^n_{2,t}$.
2 **while** $t <$ MaxEpoch **do**
3 Fetch mini-batch $\mathcal{D}_n$ from $\mathcal{D}$;
4 Divide $\mathcal{D}_n$ into $\mathcal{X}_c$ and $\mathcal{X}_u$ based on $R(t)$; `// divide samples into clean and noisy set based on low-loss criterion`
5 **for** $x_c \in \mathcal{X}_c$ **do**
6 Calculate $w(x_c, y_c)$ based on Eq (6) and Eq (7); `// sample credibility modeling`
7
8 Update $\theta^{(1)}$ and $\theta^{(2)}$ based on Eq. (9); `// adaptive loss adjustment`
9 **end**
10 **for** $x_u \in \mathcal{X}_u$ **do**
11 Update $\tilde{y}$, $\theta^{(1)}$, and $\theta^{(2)}$ through gradient descent; `// update soft label distribution and model parameters`
12 **end**
13 Update $R(t)$;
14 Update $\mathbf{L}^n_{1,t}$ and $\mathbf{L}^n_{2,t}$; `// enqueue feature snapshot of current epoch`
15 **end**
16 **Output:** $\theta^{(1)}$ and $\theta^{(2)}$.

Empirical Insight. In our experiments. we find that global label learning strategy (i.e., correcting labels of all training data) suffers from correction error in clean data. This can be observed from Fig. 3 (a), large gradient value will also be imposed on lots of correctly-labeled samples. Consequently, labels for these clean samples are unnecessarily updated. Compared with global label correction manners, we choose to only update the separated noisy samples $\mathcal{X}_u$ (mostly noisy). As shown in Fig. 3 (b), the proposed selective label correction strategy focuses more on learning noisy labels. The number of correctly-labeled samples with large gradient value is way less than a global correction scheme. Indicating that the selective label correction scheme can mitigate the problem of correction error. Experiments in Sect. 4.3 also quantitatively verify the effectiveness of the selective label learning strategy over the global label learning manner.

Putting this all together, Algorithm 1 delineates the proposed *CREMA* in detail. In a nutshell, *CREMA* is built on a divide-and-conquer framework. Firstly, clean set $\mathcal{X}_c$ and noisy set $\mathcal{X}_u$ are separated based on the low-loss criterion [10]. For $\mathcal{X}_c$, we compute the likelihood of historical credibility sequence, which helps to adaptively modulate the loss term of each training sample. As for $\mathcal{X}_u$, a selective label correction scheme is leveraged to update label distribution and model parameters simultaneously. After each training epoch, memory sequence $\mathbf{L}^n_{1,t}$ and $\mathbf{L}^n_{2,t}$ are updated with the feature snapshot of the most current epoch.

4 Experiments

4.1 Datasets and Implementation Details

Datasets. To validate the effectiveness of the proposed method, we experimentally investigate on four synthetic noisy datasets, i.e., IMDB [25], MNIST, CIFAR-10, CIFAR-100 [15] and two real-world label noise datasets, i.e., Clothing1M [44], and Animal10N [35]. IMDB [25] is a collection of highly polarized movie reviews (positive/negative). It consists of 25,000 training samples and 25,000 samples for testing. The task is formalized as a binary classification task to decide the polarity of sentiment for a review. MNIST consists of 70,000 images for 10 classes, in which 60,000 images for training and the left 10,000 images for testing. Both CIFAR-10 and CIFAR100 contain 50,000 training images and 10,000 testing images. For the Clothing1M, it is a large-scale real-world noisy dataset which is collected from multiple online shopping websites. It contains 1 million training images and clean training subsets (47K for training, 14K for validation and 10K for test) with 14 classes. The noise rate for this dataset is around 38.5%. Animal-10N contains 55,000 human-labeled online images for 10 confusing animals. It includes approximately 8% noisy-labeled samples.

Implementation Details. For the three synthetic image classification noisy datasets, MNIST, CIFAR-10 and CIFAR-100, we follow the setting in previous works [10,43,47], experiments with three kinds of noise types are considered, i.e., symmetric noise (uniformly random), asymmetric noise, and pairflip noise. For the IMDB text classification dataset, we tokenize each sentence and the word embeddings is of dimension 10,000. Varying noise rates τ are conducted to fully evaluate the proposed method, where for symmetric label noise, we set $\tau \in \{20\%, 50\%, 80\%\}$ on image datasets, $\tau \in \{20\%, 40\%\}$ on text dataset, $\tau = 40\%$ for asymmetric noise and $\tau \in \{40\%, 45\%\}$ for pairflip label noise. For real-world noisy Clothing1M dataset, following [46,52], we do not use the 50K clean data, and a randomly sampled pseudo-balanced subset includes about 260K images is leveraged as training data.

For the network structure, a 2-layer bi-directional LSTM network is adopted for IMDB. It is of 128 embedding size and 128 hidden size. A 9-layer CNN with Leaky-ReLU activation function [10] is used for MNIST, CIFAR-10, and CIFAR-100, while ResNet-50 is adopted for Clothing1M and Animal-10N datasets. The batch size is set as 64 for all the datasets. For fair comparisons, we train our model for 200 epochs in total and choose the average test accuracy of last 10 epochs as the final result in three image synthetic noisy datasets. For IMDB dataset, we set total training epochs as 100 and also test the accuracy of last ten epochs. Total training epochs for Clothing1M and Animal-10N are 80 and 150 respectively. Moreover, we use Adam optimizer for all the experiments and set the initial learning rate as 0.001, then it is degraded by a factor of 5 every 30 epochs for Clothing1M and 50 epochs for Animal-10N. The two classifiers in our methods are two networks with the same structure but different initialization parameters. Following [10], $R(t)$ is linearly decreased along with training until

Table 1. Average test accuracy (%) on *MNIST* over the last ten epochs.

Noise rates τ	Standard	PENCIL	Co-teaching	Co-teaching+	JoCoR	CREMA (ours)
Symmetry-20%	79.94 ± 0.10	97.20 ± 0.53	97.40 ± 0.09	97.81 ± 0.03	97.98 ± 0.02	$\mathbf{98.40} \pm 0.14$
Symmetry-50%	52.92 ± 0.21	96.22 ± 0.13	92.47 ± 0.14	95.80 ± 0.09	96.35 ± 0.02	$\mathbf{98.07} \pm 0.24$
Symmetry-80%	23.95 ± 0.18	87.64 ± 0.25	82.04 ± 0.43	58.92 ± 0.37	85.51 ± 0.08	$\mathbf{92.02} \pm 0.54$
Asymmetry-40%	78.80 ± 0.09	94.39 ± 0.37	90.57 ± 0.04	93.28 ± 0.43	94.14 ± 0.12	$\mathbf{97.15} \pm 0.26$
Pairflip-40%	58.51 ± 0.29	94.06 ± 0.09	90.73 ± 0.22	89.91 ± 0.31	93.47 ± 0.10	$\mathbf{95.80} \pm 0.51$
Pairflip-45%	54.54 ± 0.30	90.73 ± 0.29	89.42 ± 0.22	85.81 ± 0.30	91.30 ± 0.25	$\mathbf{94.12} \pm 0.58$

Table 2. Average test accuracy (%) on *CIFAR-10* over the last ten epochs.

Noise rates τ	Standard	PENCIL	Co-teaching	Co-teaching+	JoCoR	CREMA (ours)
Symmetry-20%	68.67 ± 0.11	78.78 ± 0.15	82.56 ± 0.24	82.27 ± 0.21	85.73 ± 0.19	$\mathbf{86.32} \pm 0.16$
Symmetry-50%	42.31 ± 0.18	64.71 ± 0.27	72.97 ± 0.22	63.01 ± 0.33	79.53 ± 0.10	$\mathbf{81.63} \pm 0.13$
Symmetry-80%	15.94 ± 0.07	26.96 ± 0.37	24.03 ± 0.18	17.96 ± 0.06	27.30 ± 0.08	$\mathbf{29.66} \pm 0.16$
Asymmetric-40%	70.04 ± 0.08	70.06 ± 0.28	75.96 ± 0.15	72.21 ± 0.43	76.31 ± 0.21	$\mathbf{82.49} \pm 0.13$
Pairflip-40%	51.66 ± 0.11	75.26 ± 0.18	75.10 ± 0.23	57.59 ± 0.45	68.56 ± 0.16	$\mathbf{85.00} \pm 0.13$
Pairflip-45%	45.78 ± 0.13	71.18 ± 0.28	70.68 ± 0.23	49.60 ± 0.23	57.68 ± 0.21	$\mathbf{82.94} \pm 0.12$

Table 3. Average test accuracy (%) on *CIFAR-100* over the last ten epochs.

Noise rates τ	Standard	PENCIL	Co-teaching	Co-teaching+	JoCoR	CREMA (ours)
Symmetry-20%	34.72 ± 0.07	52.11 ± 0.21	50.48 ± 0.24	49.27 ± 0.03	53.41 ± 0.09	$\mathbf{57.21} \pm 0.25$
Symmetry-50%	16.86 ± 0.09	39.89 ± 0.30	38.24 ± 0.26	40.04 ± 0.70	43.37 ± 0.09	$\mathbf{43.95} \pm 0.42$
Symmetry-80%	4.60 ± 0.12	16.08 ± 0.15	11.78 ± 0.12	13.44 ± 0.37	12.33 ± 0.13	$\mathbf{17.10} \pm 0.19$
Asymmetric-40%	26.93 ± 0.10	32.81 ± 0.23	33.36 ± 0.28	33.62 ± 0.39	32.66 ± 0.13	$\mathbf{38.61} \pm 0.25$
Pairflip-40%	27.48 ± 0.12	33.83 ± 0.52	33.94 ± 0.18	33.80 ± 0.25	33.89 ± 0.12	$\mathbf{38.06} \pm 0.34$
Pairflip-45%	24.21 ± 0.11	29.01 ± 0.28	29.57 ± 0.15	26.93 ± 0.34	28.83 ± 0.10	$\mathbf{32.50} \pm 0.29$

reach a lower bound value σ, for Clothing1M and Animal-10N datasets, we empirically set lower bound σ as 0.8 and 0.92 respectively.

4.2 Comparison with State-of-the-Art Methods

Results on Synthetic Noisy Datasets. Table 1,2,3 and Table 5 show the detailed results of the proposed *CREMA* and other methods in multiple synthetic noisy cases on four widely used datasets, i.e., MNIST, CIFAR-10, CIFAR-100 IMDB. Specifically, four state-of-the-art LNL methods that are highly related to our work are chosen for comparison: PENCIL [46], Co-teaching [10], Co-teaching+ [47], JoCoR [43]. From these tables, we can observe that the proposed *CREMA* can achieve consistent improvements over other methods on four benchmarks across various noise settings. In the Pairflip-40% and Pairflip-45% cases, the proposed method outperforms other baselines by a large margin. Specifically, *CREMA* can achieve 16.44% and 25.26% improvement in accuracy over JoCoR

Table 4. Test accuracy on Animal-10N. "LA", "LC" and "ND" denote "Loss Adjustment", "Label Correction" and "Noisy sample Detection" respectively.

Method	Category			Accuracy
	LA	LC	ND	
Cross-Entropy				79.4
ActiveBias [4]	✓			80.5
PLC [52]		✓		83.4
Co-teaching [10]			✓	80.2
SELFIE [35]		✓	✓	81.8
CREMA (Ours)	✓	✓	✓	**84.2**

Table 5. Average test accuracy (%) on IMDB dataset over the last ten epochs.

Method	Noise rates τ	
	Sym-20%	Sym-40%
Standard	74.08 ± 0.23	58.37 ± 0.26
PENCIL [46]	73.73 ± 0.21	58.07 ± 0.30
Co-teaching [10]	82.07 ± 0.07	73.25 ± 0.19
Co-teaching+ [47]	82.27 ± 0.23	53.56 ± 3.04
JoCoR [43]	84.82 ± 0.07	76.12 ± 0.17
CREMA (ours)	$\mathbf{86.44} \pm 0.04$	$\mathbf{78.39} \pm 0.14$

Table 6. Comparison with state-of-the-art methods in test accuracy on Clothing1M. † means the result without model ensemble [18].

Method	Category			Accuracy
	LA	LC	ND	
Cross-Entropy				69.21
GCE [53]	✓			69.75
IMAE [40]	✓			73.20
cre SCE [42]	✓			71.02
DM [41]	✓			73.30
F-correction [29]	✓			69.84
M-correction [1]	✓			71.00
Masking [9]	✓			71.10
Joint-Optim [36]		✓		72.23
Meta-Cleaner [51]		✓		72.50
Meta-Learning [17]		✓		73.47
PENCIL [46]		✓		73.49
PLC [52]		✓		74.02
Self-Learning [11]		✓		74.45
ProSelfLC [39]		✓		73.40
Co-teaching [10]			✓	70.15
JoCoR [43]			✓	70.30
C2D [55]			✓	74.30
DivideMix† [16]			✓	74.48
CREMA (Ours)	✓	✓	✓	**74.53**

on CIFAR-10. When dealing with extremely noisy scenario, e.g. Symmetry-80%, *CREMA* can also perform generally better than other compared methods. The result demonstrates the superiority and generality of the proposed robust learning method across various types and levels of label noise on multimedia (i.e., image and text) datasets.

Results on Real-World Noisy Datasets. The baseline methods are chosen from recently proposed LNL methods. Including loss adjustment methods [1,4,9,29,40–42,53], label correction methods [11,36,39,46,52], noisy sample detection methods [10,16,43,55] and hybrid methods [35]. Table 6, 4 show results on two real-world noisy datasets respectively. On the large-scale Clothing1M dataset, *CREMA* outperforms all compared methods. Note that *CREMA* follows the standard DNN training procedure, and is similar to other co-training methods [10,43,47] in terms of training time since the time cost for sample credibility modeling is negligible compared with DNN update. It is worth noting that the proposed method outperforms these co-teaching methods by a large margin. Indicating that the discarded samples by co-teaching methods are actually valuable, and *CREMA* well utilized all training samples. The best test accuracy is achieved by *CREMA* among the compared methods in Animal-10N as well. The results indicate that the proposed method can work well on high noise level (i.e., Clothing1M) and fine-grained (i.e., Animal-10N) real-world noisy datasets.

Table 7. Ablation studies of each component within *CREMA* on Clothing1M dataset.

Method	Test Accuracy (%)
Baseline	72.81
+ Selective label update	73.25
+ Sequential likelihood	74.00
+ Stability measurement	**74.53**

Table 8. Investigations on different mixture models on Clothing1M dataset.

Estimator	Test Accuracy (%)
BMM	74.09
GMM	**74.53**

Table 9. Investigations on length of sequence n on Clothing1M dataset.

Length of sequence n	1	2	3	4	5	6
Test Accuracy (%)	73.25	73.99	**74.53**	74.40	74.27	73.96

4.3 Ablation Studies

- **Component Analysis.** *CREMA* contains several important components, including selective label learning strategy, sequential likelihood $\log P(\mathbf{L}_t^n|\mathbf{f}_{t-n}, y)$ and stability measurement $C(\mathbf{L}_t^n, y)$. To verify the effectiveness of each component, we conduct experiments on large-scale noisy dataset Clothing1M. The baseline method is built upon a simple co-teaching framework [43] combined with global label correction schemes (as in [46]), without the credibility guided loss adjustment strategy. The results are shown in Table 7, we can see that, conform to the observation in Fig. 3, the proposed selective label learning strategy achieves better results compared with the global correction counterpart. The sequential likelihood and stability measurement further boost the model performance with 0.75% and 0.53% accuracy gain, this indicates that the proposed sequential sample credibility modeling can effectively combat hard noisy samples mixed with clean ones. With all the three key components above, *CREMA* can achieve 74.53% test accuracy on Clothing1M.
- **Length of sequence** n**.** We also conduct experiments to investigate how the length of sequence n affects the performance. Fig 4(a) shows results on Clothing1M with various values (in $\{1, 2, 3, 4, 5, 6\}$) of n. It can be observed that increasing the length of sequence helps achieve higher accuracy at first but turn poor after hitting the peak value. Intuitively, when no temporal information is provided when $n = 1$, *CREMA* can not utilize consistency metric to identify hard noisy samples that blended with clean ones, thus leading to an inferior result. When n is larger than 4, we also notice that performance degrades, this is probably due to unreliable model inside the very long sequence can harm sample credibility modeling and reversely hinder the final result.

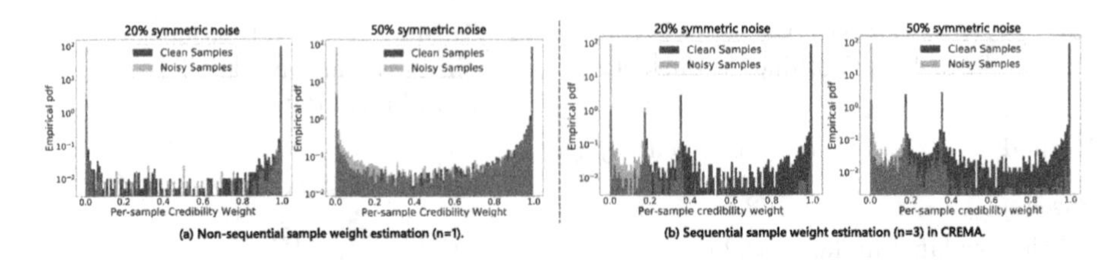

Fig. 4. Empirical PDF of the learned sample credibility weight between typical non-sequential (i.e., sequential length n=1) and sequential manner in CREMA. The model is trained on MNIST with 20% and 50% symmetric label noise.

- **Effect of different estimators.** The probabilistic model plays the role of estimating the conditional probability $P(\mathbf{f}_t|\mathbf{f}_{t-1}, y)$ in Eq. (6). We compare two different estimators, Gaussian Mixture Model (GMM) [30] and Beta Mixture Model (BMM) [1] on Clothing1M. Table 8 shows the results. We can see that GMM obtains a relatively higher test accuracy, but BMM can also achieve good results (74.09%) as well. This indicates that the choice of normalized mixture model is not sensitive to the final result.
- **Reliability of the estimated sample credibility.** Sample credibility plays the role of dynamical weight to modulate the loss term of each training sample, as in Eq. (9). In Fig. 4 we visualize the empirical PDF of the learned credibility weight of all training samples between (b) the sequential estimation manner within *CREMA* and (a) its non-sequential counterpart (i.e., $n = 1$) under two different noisy settings. It can be observed that the overall credibility weight of training samples is distinguishable for clean and noisy data in (b). Specifically, clean samples possess larger weight, thus contributing more gradients during training. Noisy samples are assigned with a relatively small weight to alleviate their negative impact. However, the non-sequential weights yield significantly amount of samples that cannot be correctly-separated via a fixed threshold. Indicating that the proposed fine-grained sequential credibility estimation is more effective for reliable sample weight modeling.

5 Conclusion

In this paper, we propose a novel end-to-end robust learning method, called *CREMA*. Towards the problem that previous works lack the consideration of intrinsic difference among difficulties of noisy samples. We follow the idea of divide-and-conquer that separates clean and noisy samples via estimating the credibility of each training sample. Two branches are designed to handle the imperfectly separated sample sets respectively. For easily recognizable noisy samples, we apply a selective label correction scheme avoiding erroneous label updates on clean samples. For hard noisy samples blended with clean ones, likelihood estimation of historical credibility sequence adaptively modulates the loss term of each sample during training. Extensive experiments conducted on several synthetic and real-world noisy datasets verify the superiority of the proposed method.

References

1. Arazo, E., Ortego, D., Albert, P., O'Connor, N.E., McGuinness, K.: Unsupervised label noise modeling and loss correction. In: Proc. International Conference on Machine Learning (ICML) (2019)
2. Arpit, D., et al.: A closer look at memorization in deep networks. In: Proc. International Conference on Machine Learning (ICML), pp. 233–242 (2017)
3. Berthelot, D., Carlini, N., Goodfellow, I.J., Papernot, N., Oliver, A., Raffel, C.: Mixmatch: a holistic approach to semi-supervised learning. Proc. Advances in Neural Information Processing Systems (NeurIPS) (2019)
4. Chang, H.S., Learned-Miller, E., McCallum, A.: Active Bias: training more accurate neural networks by emphasizing high variance samples. In: Proc. Advances in Neural Information Processing Systems (NeurIPS), pp. 1002–1012 (2017)
5. Dehghani, M., Severyn, A., Rothe, S., Kamps, J.: Avoiding your teacher's mistakes: training neural networks with controlled weak supervision. (2017) arXiv preprint arXiv:1711.00313
6. Dehghani, M., Severyn, A., Rothe, S., Kamps, J.: Learning to learn from weak supervision by full supervision. In: Proc. Advances in Neural Information Processing Systems Workshop (NeurIPSW) (2017)
7. Ghosh, A., Kumar, H., Sastry, P.: Robust loss functions under label noise for deep neural networks. In: Proc. Association for the Advancement of Artificial Intelligence (AAAI) (2017)
8. Goldberger, J., Ben-Reuven, E.: Training deep neural-networks using a noise adaptation layer. In: Proc. International Conference on Learning Representations (ICLR) (2017)
9. Han, B., et al.: Masking: A new perspective of noisy supervision. In: Proc. Advances in Neural Information Processing Systems (NeurIPS), pp. 5836–5846 (2018)
10. Han, B., et al.: Co-teaching: robust training of deep neural networks with extremely noisy labels. In: Proc. Advances in Neural Information Processing Systems (NeurIPS), pp. 8527–8537 (2018)
11. Han, J., Luo, P., Wang, X.: Deep self-learning from noisy labels. In: Proc. IEEE International Conference on Computer Vision (ICCV), pp. 5138–5147 (2019)
12. Hendrycks, D., Mazeika, M., Wilson, D., Gimpel, K.: Using trusted data to train deep networks on labels corrupted by severe noise. In: Proc. Advances in Neural Information Processing Systems (NeurIPS), pp. 10456–10465 (2018)
13. Houle, M.E.: Local intrinsic dimensionality I: an extreme-value-theoretic foundation for similarity applications. In: Proc. International Conference on Similarity Search and Applications (SISAP), pp. 64–79 (2017)
14. Jiang, L., Zhou, Z., Leung, T., Li, L.J., Fei-Fei, L.: MentorNet: learning data-driven curriculum for very deep neural networks on corrupted labels. In: Proc. International Conference on Machine Learning (ICML) (2018)
15. Krizhevsky, A., Hinton, G., et al.: Learning multiple layers of features from tiny images (2009)
16. Li, J., Socher, R., Hoi, S.C.: Dividemix: learning with noisy labels as semi-supervised learning. (2020) arXiv preprint arXiv:2002.07394

17. Li, J., Wong, Y., Zhao, Q., Kankanhalli, M.: learning to learn from noisy labeled data. In: Proc. IEEE Conference on Computer Vision and Pattern Recognition (CVPR), pp. 5046–5054 (2019)
18. Li, J., Xiong, C., Hoi, S.C.: Learning from noisy data with robust representation learning. In: Proc. IEEE International Conference on Computer Vision (ICCV), pp. 9485–9494 (2021)
19. Li, X., Liu, T., Han, B., Niu, G., Sugiyama, M.: Provably end-to-end label-noise learning without anchor points. Proc. International Conference on Machine Learning (ICML) (2021)
20. Li, Y., Yang, J., Song, Y., Cao, L., Luo, J., Li, L.J.: Learning from noisy labels with distillation. In: Proc. IEEE International Conference on Computer Vision (ICCV), pp. 1910–1918 (2017)
21. Litany, O., Freedman, D.: Soseleto: a unified approach to transfer learning and training with noisy labels. (2018) arXiv preprint arXiv:1805.09622
22. Liu, T., Tao, D.: Classification with noisy labels by importance reweighting. IEEE Trans. Pattern Anal. Mach. Intell. (TPAMI) **38**(3), 447–461 (2015)
23. Lyu, Y., Tsang, I.W.: Curriculum loss: robust learning and generalization against label corruption. In: Proc. International Conference on Learning Representations (ICLR) (2020)
24. Ma, X., et al.: Dimensionality-driven learning with noisy labels. In: Proc. International Conference on Machine Learning (ICML) (2018)
25. Maas, A., Daly, R.E., Pham, P.T., Huang, D., Ng, A.Y., Potts, C.: Learning word vectors for sentiment analysis. In: Proceedings of the 49th annual meeting of the association for computational linguistics: human language technologies, pp. 142–150 (2011)
26. Malach, E., Shalev-Shwartz, S.: Decoupling when to update from how to update. In: Proc. Advances in Neural Information Processing Systems (NeurIPS), pp. 960–970 (2017)
27. Manning, C., Schutze, H.: Foundations of statistical natural language processing. MIT press (1999)
28. Nguyen, D.T., Mummadi, C.K., Ngo, T.P.N., Nguyen, T.H.P., Beggel, L., Brox, T.: SELF: learning to filter noisy labels with self-Ensembling. In: Proc. International Conference on Learning Representations (ICLR) (2020)
29. Patrini, G., Rozza, A., Krishna Menon, A., Nock, R., Qu, L.: Making deep neural networks robust to label noise: a loss correction approach. In: Proc. IEEE Conference on Computer Vision and Pattern Recognition (CVPR), pp. 1944–1952 (2017)
30. Permuter, H., Francos, J.M., Jermyn, I.: A study of gaussian mixture models of color and texture features for image classification and segmentation. Pattern Recogn. (PR). **39**, 695–706 (2006)
31. Pleiss, G., Zhang, T., Elenberg, E.R., Weinberger, K.Q.: Identifying mislabeled data using the area under the margin ranking. In: Proc. Advances in Neural Information Processing Systems (NeurIPS) (2020)
32. Reed, S., Lee, H., Anguelov, D., Szegedy, C., Erhan, D., Rabinovich, A.: Training deep neural networks on noisy labels with bootstrapping. In: Proc. International Conference on Learning Representations (ICLR) (2015)
33. Ren, M., Zeng, W., Yang, B., Urtasun, R.: Learning to reweight examples for robust deep learning. In: Proc. International Conference on Machine Learning (ICML) (2018)
34. Shu, J., et al.: Meta-Weight-Net: learning an explicit mapping for sample weighting. In: Proc. Advances in Neural Information Processing Systems (NeurIPS), pp. 1917–1928 (2019)

35. Song, H., Kim, M., Lee, J.G.: SELFIE: refurbishing unclean samples for robust deep learning. In: Proc. International Conference on Machine Learning (ICML), pp. 5907–5915 (2019)
36. Tanaka, D., Ikami, D., Yamasaki, T., Aizawa, K.: Joint optimization framework for learning with noisy labels. In: Proc. IEEE Conference on Computer Vision and Pattern Recognition (CVPR), pp. 5552–5560 (2018)
37. Veit, A., Alldrin, N., Chechik, G., Krasin, I., Gupta, A., Belongie, S.: Learning from noisy large-scale datasets with minimal supervision. In: Proc. IEEE Conference on Computer Vision and Pattern Recognition (CVPR) (2017)
38. Wang, R., Liu, T., Tao, D.: Multiclass learning with partially corrupted labels. IEEE Trans. Neural Networks. Learn. Syst. **29**(6), 2568–2580 (2017)
39. Wang, X., Hua, Y., Kodirov, E., Clifton, D.A., Robertson, N.M.: Proselflc: progressive self label correction for training robust deep neural networks. In: Proc. IEEE Conference on Computer Vision and Pattern Recognition (CVPR), pp. 752–761 (2021)
40. Wang, X., Hua, Y., Kodirov, E., Robertson, N.M.: Imae for noise-robust learning: mean absolute error does not treat examples equally and gradient magnitude's variance matters. (2019) arXiv preprint arXiv:1903.12141
41. Wang, X., Kodirov, E., Hua, Y., Robertson, N.M.: Derivative manipulation for general example weighting. (2019) arXiv preprint arXiv:1905.11233
42. Wang, Y., Ma, X., Chen, Z., Luo, Y., Yi, J., Bailey, J.: Symmetric cross entropy for robust learning with noisy labels. In: Proc. IEEE International Conference on Computer Vision (ICCV), pp. 322–330 (2019)
43. Wei, H., Feng, L., Chen, X., An, B.: Combating noisy labels by agreement: a joint training method with co-regularization. In: Proc. IEEE Conference on Computer Vision and Pattern Recognition (CVPR), pp. 13726–13735 (2020)
44. Xiao, T., Xia, T., Yang, Y., Huang, C., Wang, X.: Learning from massive noisy labeled data for image classification. In: Proc. IEEE Conference on Computer Vision and Pattern Recognition (CVPR), pp. 2691–2699 (2015)
45. Yao, Q., Yang, H., Han, B., Niu, G., Kwok, J.: Searching to exploit memorization effect in learning with noisy labels. In: Proc. International Conference on Machine Learning (ICML) (2020)
46. Yi, K., Wu, J.: Probabilistic end-to-end noise correction for learning with noisy labels. In: Proc. IEEE Conference on Computer Vision and Pattern Recognition (CVPR), pp. 7017–7025 (2019)
47. Yu, X., Han, B., Yao, J., Niu, G., Tsang, I.W., Sugiyama, M.: How does disagreement help generalization against label corruption? In: Proc. International Conference on Machine Learning (ICML) (2019)
48. Zhang, C., Bengio, S., Hardt, M., Recht, B., Vinyals, O.: Understanding deep learning requires rethinking generalization. In: Proc. International Conference on Learning Representations (ICLR) (2017)
49. Zhang, H., Cissé, M., Dauphin, Y.N., Lopez-Paz, D.: mixup: Beyond empirical risk minimization. In: Proc. International Conference on Learning Representations (ICLR) (2018)
50. Zhang, M., Lee, J., Agarwal, S.: Learning from noisy labels with no change to the training process. In: Proc. International Conference on Machine Learning (ICML), pp. 12468–12478 (2021)
51. Zhang, W., Wang, Y., Qiao, Y.: Metacleaner: Learning to hallucinate clean representations for noisy-labeled visual recognition. In: Proc. IEEE Conference on Computer Vision and Pattern Recognition (CVPR), pp. 7365–7374 (2019)

52. Zhang, Y., Zheng, S., Wu, P., Goswami, M., Chen, C.: Learning with feature-dependent label noise: a progressive approach. In: Proc. International Conference on Learning Representations (ICLR) (2021)
53. Zhang, Z., Sabuncu, M.: Generalized cross entropy loss for training deep neural networks with noisy labels. In: Proc. Advances in Neural Information Processing Systems (NeurIPS), pp. 8778–8788 (2018)
54. Zhang, Z., Zhang, H., Arik, S.O., Lee, H., Pfister, T.: Distilling effective supervision from severe label noise. In: Proc. IEEE Conference on Computer Vision and Pattern Recognition (CVPR), pp. 9294–9303 (2020)
55. Zheltonozhskii, E., Baskin, C., Mendelson, A., Bronstein, A.M., Litany, O.: Contrast to divide: self-supervised pre-training for learning with noisy labels. (2021) arXiv preprint arXiv:2103.13646
56. Zhou, T., Wang, S., Bilmes, J.: Robust curriculum learning: From clean label detection to noisy label self-correction. In: Proc. International Conference on Learning Representations (ICLR) (2021)

PLMCL: Partial-Label Momentum Curriculum Learning for Multi-label Image Classification

Rabab Abdelfattah[1(✉)], Xin Zhang[1], Zhenyao Wu[2], Xinyi Wu[2], Xiaofeng Wang[1], and Song Wang[2]

[1] Department of Electrical Engineering, University of South Carolina, Columbia, SC, USA
{rabab,xz8}@email.sc.edu, wangxi@cec.sc.edu

[2] Department of Computer Science, University of South Carolina, Columbia, SC, USA
{zhenyao,xinyiw}@email.sc.edu, songwang@cec.sc.edu

Abstract. Multi-label image classification aims to predict all possible labels in an image. It is usually formulated as a partial-label learning problem, given the fact that it could be expensive in practice to annotate all labels in every training image. Existing works on partial-label learning focus on the case where each training image is annotated with only a subset of its labels. A special case is to annotate only one positive label in each training image. To further relieve the annotation burden and enhance the performance of the classifier, this paper proposes a new partial-label setting in which only a subset of the training images are labeled, each with only one positive label, while the rest of the training images remain unlabeled. To handle this new setting, we propose an end-to-end deep network, PLMCL (Partial-Label Momentum Curriculum Learning), that can learn to produce confident pseudo labels for both partially-labeled and unlabeled training images. The novel momentum-based law updates soft pseudo labels on each training image with the consideration of the updating velocity of pseudo labels, which help avoid trapping to low-confidence local minimum, especially at the early stage of training in lack of both observed labels and confidence on pseudo labels. In addition, we present a confidence-aware scheduler to adaptively perform easy-to-hard learning for different labels. Extensive experiments demonstrate that our proposed PLMCL outperforms many state-of-the-art multi-label classification methods under various partial-label settings on three different datasets.

1 Introduction

With the advances in deep learning, significant progress has been made on single-label image classification problems [11] where each image only has one label. However, in many real applications, one image may contain multiple objects and/or exhibit multiple attributes that cannot be well described by a single

L. Karlinsky et al. (Eds.): ECCV 2022 Workshops, LNCS 13802, pp. 39–55, 2023.
https://doi.org/10.1007/978-3-031-25063-7_3

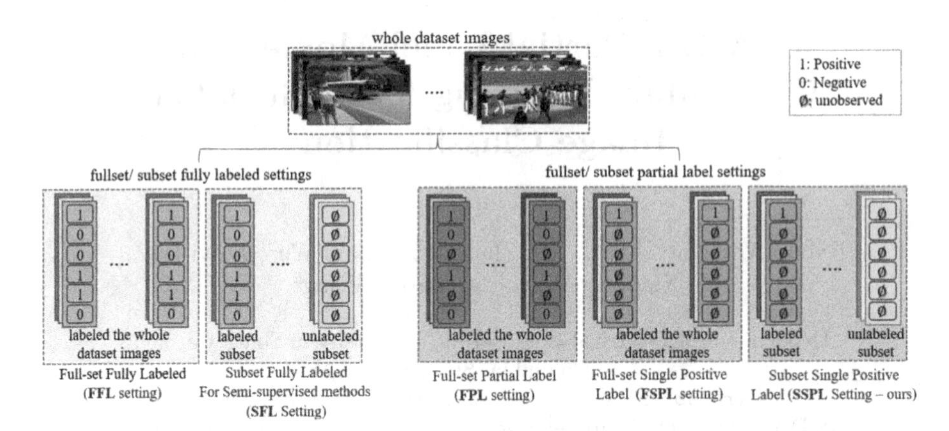

Fig. 1. An illustration of various label settings. **Full-set fully labeled (FFL)**: All positive and negative labels are annotated on each training image. **Full-set partial labels (FPL)** [13]: At least one positive or negative label is annotated on each training image. **Full-set single positive labels (FSPL)** [10]: Exactly one positive label is observed on each training image. **Subset fully labeled (SFL)**: All positive and negative labels are annotated on a subset of training images and the rest of the images are unlabeled. **Subset single positive labels (SSPL-ours)**: Exactly one positive label is annotated on each image from a subset of the training dataset while the rest of the images are unlabeled.

label. For instance, a scene may contain multiple objects and a CT scan may indicate multiple conditions. This leads to an important computer-vision task of multi-label image classification that aims to identify all the labels on an image. A great challenge in multi-label image classification comes from the need of a large number of labeled training images. In particular, many supervised-learning algorithms require all labels on every training image to be accurately annotated, which can be very difficult and laborious [12].

To relieve the annotation burden of fully labeling, recent works on multi-label classification consider training the network with partial labels [10,13,17,22,29, 42]. One typical partial-label setting is **full-set partial labels (FPL)** [13], which annotates only a subset of labels on each training image. Following this setting, a special case is **full-set single positive labels (FSPL)** [10], which annotates only one positive label on each image (Fig. 1). Although these partial-label settings can, to some extent, mitigate the annotation burden, we still need to go through all training images for annotation. To go one step further, we consider a new partial-label setting of **subset single positive labels (SSPL)**, where the training dataset consists of both labeled and unlabeled images, while, for the subset of labeled images, only one positive label is annotated on each image. Clearly, this new setting can further reduce the annotation cost for large-scale multi-label datasets and practically our setting can be utilized by only annotating a limited percentage of total images with one single positive label per image. For instance, COCO dataset has 2.5 M labels which takes 20 K worker hours to identify categories [24], while 60% SSPL only requires 49.2k labels and the annotation time is up to 394 h. Here we did not even account for the fact that detecting absence can be more difficult than detecting presence. However, the inclusion of unlabeled images significantly increases the difficulty of the network design and training. In particular, those existing partial-label learning methods based on label correlation [17] and label matrix completion [5] are not applicable to our new setting since they cannot handle single positive labels and unlabeled data simultaneously. From another point of view, although our proposed setting and semi-supervised learning both use unlabeled subsets, the performance of semi-supervised models may degrade when training on the SSPL setting. This is because semi-supervised methods usually assume access to a subset of fully-labeled images (SFL setting) to initialize the training process, while such a fully-labeled subset is not available on the SSPL setting as shown in Fig. 1. Such degradation can be observed in the comparison experiments presented in Sect. 4.4.

This paper presents a new Partial-Label Momentum Curriculum Learning (PLMCL) method for end-to-end training of the multi-label classifier under the proposed new partial-label setting. In particular, a set of soft pseudo labels for each image are estimated and dynamically updated by momentum. The momentum combines the classifier predictions and the pseudo labels to identify the direction for updating the pseudo labels. Meanwhile, to update the multi-label classifier parameters, we introduce a scheduled loss function based on standard cross-entropy with which the learning gradually moves from observed (easy) labels to unobserved (hard) labels. Our proposed PLMCL can also be easily extended to handle other partial-label settings [10,13,17] for multi-label image classification.

Our contributions are summarized as follows:

- We introduce a new partial-label setting, SSPL, for multi-label image classification by allowing the inclusion of unlabeled images as training images. This setting can be as simple as the case where only a subset of images have labels and only one single positive label is annotated per image in this subset. This new setting can significantly reduce the annotation burden as well as leverage more unlabeled data for multi-label image classification.
- We propose a new PLMCL framework for multi-label image classification under different partial-label settings. The major novelty comes from the introduction of the momentum and the confidence levels into the partial-label setting. Instead of accelerating learning as discussed in [15], the momentum in this paper actually brings a stronger stability requirement into training, which focuses on the convergence of not only the pseudo labels, but also the updating velocity of pseudo labels that indicates high confidence on the pseudo labels. This is important, especially at the early stage of training when the observed labels are lack and the confidence levels on the pseudo labels are low. Otherwise, premature convergence could be achieved with low confidence levels due to lack of observed labels.
- Extensive experimental results show that our method outperforms the state-of-the-arts under three different partial-label settings on three widely-used

datasets. With fewer observed labels in our proposed setting, our method can still get comparable classification results as those methods using full-set single positive labels (FSPL) setting, as shown in Fig. 1.

The rest of the paper is organized as follows. Section 2 briefly reviews the related works. Section 3 presents the details of the proposed method. The experiment results are demonstrated in Sect. 4. Finally, the conclusions are drawn in Sect. 5.

2 Related Work

This section briefly reviews the related work on partial-label learning under different settings (Fig. 1) as well as semi-supervised learning and curriculum learning methods.

2.1 Partial-label Learning Under Different Settings

Full-set Partial Labels (FPL). Early work on multi-label image classification under the FPL setting assumes that the unobserved labels are negatives [4,7,27,33]. This assumption inevitably introduces false negative labels during the training and therefore will lead to a significant performance degradation in classification. Another direction is to estimate the unobserved labels based on the correlation between the labels [27,41], label matrix completion [5,39], low-rank learning [43,44], and probabilistic models [8,20]. Most of these models, however, have to solve an optimization problem during the training, which might be computationally expensive for deep learning given the high complexity of deep neural networks [13]. Some recent work uses end-to-end deep learning models to predict unobserved labels. For instance, the method in [17] takes advantage of image/label similarity graphs and builds the dependencies among the labels based on the label co-occurrence information. Nevertheless, it requires more than one label per image to build these relations and therefore is not applicable for single-label settings. Another work in [13] integrates graph neural network with curriculum learning. The proposed strategy predict the unobserved labels based on network models and add those predicted labels into the training dataset using a threshold-based strategy. Those selected labels are called "weakly labels", which may change over epochs. Therefore, there is a possibility that the predictions of some unobserved labels are never selected into the training, which may result in information loss. In contrast, our model learns the pseudo labels dynamically based on the momentum throughout the training process. Different from [13], every pseudo label will more or less contribute to the training in each epoch, depending on its confidence level. This can guarantee the continuity in pseudo label updates, which is lack from the threshold-based strategy.

Full-set Single Positive Label (FSPL). This setting is introduced in [10], where the authors present an end-to-end model that jointly trains the image classifier and the label estimator for online estimation of unobserved labels. The label estimator is implemented to build a simple look-up table which requires to store a matrix containing the whole dataset images, including all the previously estimated labels per image, in memory and update it in each epoch during the entire training process. Obviously, this method requires ample memory space to store the full label matrix for all training images, which makes it almost infeasible for large datasets or large numbers of labels. Moreover, this method relies on random initialization for the label estimator and assigns equal weights for the observed and unobserved terms in the suggested loss function, which may lead to degradation in the classifier's performance. In contrast, our method initializes the pseudo labels with unbiased probability values guaranteeing stable training using a self-guided momentum factor. Therefore, the self-guided factor works to (i) identify the amount of change should the labels get to move towards 1 or 0 starting from unbiased probability, and (ii) reduce the amount of change for the labels with a high confidence score. In addition, our momentum-based updating method does not require keeping the full labels in the memory since we only keep B vectors in memory per each batch, where B is the batch size. Finally, our method uses a scheduled loss function to give different weights for the unobserved term, which can gradually move the learning from the observed labels to the unobserved labels.

Subset Single Positive Labels (SSPL). This is our proposed setting which assumes that each labeled image is annotated with only a single positive label, while the unlabeled images are totally unlabeled. A practical scenario is iNaturalist dataset, where each image only has one positive label while the other classes are unlabeled [10]. Adding new unlabeled images to enrich this dataset directly leads to the SSPL setting. Actually, this applies to most datasets for multi-class classification. Adding new unlabeled images makes those datasets suitable for the multi-label problem under the SSPL setting without annotation costs. To the best of our knowledge, there are no approaches specifically designed for SSPL. However, some semi-supervised models can be modified to fit SSPL setting. We will go through these methods in the next subsection. Although these methods [10,22,26] can be easily applied to the SSPL setting, these methods may still not handle the unlabeled set of the SSPL properly. Since these models usually assign assumptions for the unobserved labels, such as ignoring the unobserved labels, assuming negative, giving them a down-weight, or starting with a random probability. Therefore, the whole labels per image, in unlabeled set, are addressed based on any of the previously mentioned assumptions during the training. Consequently, the SSPL unlabeled set's images negatively impact the performance of these models.

2.2 Semi-supervised Learning

Most of semi-supervised learning (SSL) methods on multi-class classification, while a few of them study multi-label classification [1,3,6,23,25,28,30,32,36–38]. Among those methods, it is usually assumed that some of the training images are fully labeled while the others are totally unlabeled, i.e., the SFL setting in Fig. 1. Most state-of-the-art methods create pseudo labels for the unlabeled data based on different approaches such as self-training-based and consistency-based approaches. Self-training-based approaches follow these steps to train the semi-supervised model: (i) train the model based on the labeled data, (ii) use the trained model to get the predictions for the unlabeled data, and (iii) apply the threshold-based strategy to the predictions to select the pseudo labels for unobserved data, which are considered only for the predictions greater than the threshold value i.e., [1,30]. Therefore, the selected pseudo labels are used in the loss function for training in the next epoch. On the other hand, the consistency-based approaches follow this strategy: (i) find the supervised loss based on the labeled data, (ii) apply the data perturbations, i.e., data augmentation and stochastic regularization, to produce two versions for the same image, and (iii) use the prediction of one image-version after applying the threshold as the pseudo label for the other image-version [3,32,36]. To be concluded, most of semi-supervised models generate the pseudo labels based on the predictions that greater than the threshold value. To make the semi-supervised models applicable for SSPL setting, the unobserved labels in the labeled set of SSPL are ignored since the labeled data of SSPL only contains one single label per image. Therefore, although the semi-supervised can be directly trained on SSPL setting, the performance of these models significantly decreased since the supervised loss function only works with a single label as discussed in Subsect. 4.4. While in our case, all the unobserved labels, in labeled and unlabeled sets of SSPL, initiate with unbiased probabilities to consider as initial pseudo labels and then improve their probabilities during the training using our momentum-based method. We also adapt some semi-supervised models to be applicable for multi-label classification problems, by replacing Softmax layer with Sigmoid, that are designed for multi-class classification, such as [32].

2.3 Curriculum Learning

Curriculum learning [2] has been applied to many applications related to multi-labels or multi-classes. Given the training data with fully observed labels, different iterative self-paced learning algorithms [19,21] have been developed to enhance traditional image classification, object localization, and multimedia event detection – in each iteration, easy samples are selected and the model parameters are updated accordingly. Recently, CurriculumNet [16] is proposed to learn knowledge from large-scale noisy web images by following the curriculum learning strategy. However, this strategy employs a clustering-based model to assess the complexity of the image and cannot be used for multi-label image classification [13]. Our PLMCL also follows the general idea of curriculum learning [40] by starting the learning on observed labels, and then progressively moving to the unobserved labels.

3 The PLMCL Method

Assume that the classification network is trained on the image dataset $\mathcal{I}$ for multi-label classification. Since we focus on FSPL and SSPL settings, let $\mathcal{D}$

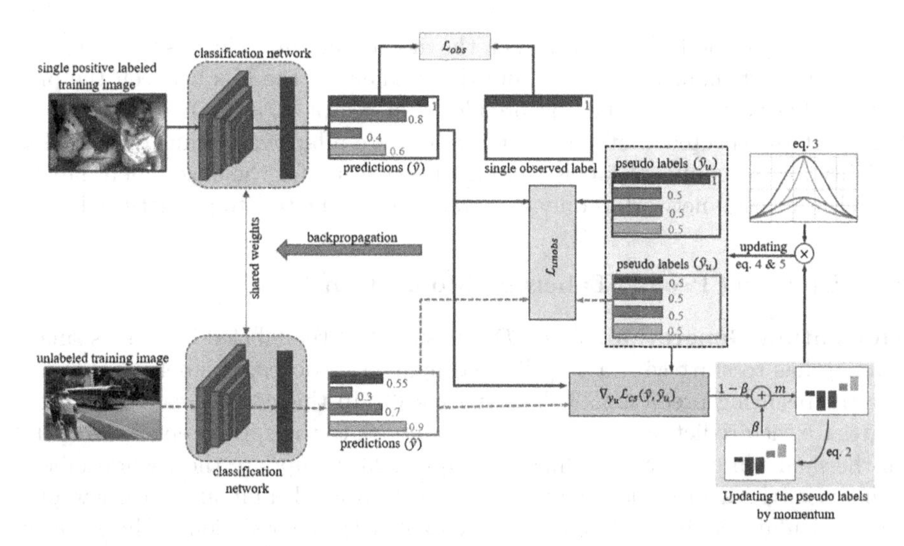

Fig. 2. An overview of the PLMCL method that consists of classification network, backbone and one-layer classifier, to generate a prediction for an input image. The classification network is end-to-end trained using the scheduled cross-entropy loss function. The pseudo labels $\hat{y}_{u,t}$ are updated based on the momentum m_t, which is the weighted sum of the last momentum m_{t-1} and the gradient of the loss between the predicted labels $\hat{y}_{t-1}$ and the pseudo labels $\hat{y}_{u,t-1}$ (Orange region). At the test, only the classification network is applied. (Color figure online)

be the subset of $\mathcal{I}$, in which each image has only one positive label and the images in $\mathcal{I}\backslash\mathcal{D}$ are totally unlabeled. Note that $\mathcal{D} \subset \mathcal{I}$ means SSPL setting and $\mathcal{D} = \mathcal{I}$ means FSPL setting. Let L be the total number of classes. For each image $i \in \mathcal{D}$, we denote $y_o^i \in \{\emptyset, 1\}^L$ as the observed label vector where each entry of y_o^i can be 1 (observed positive) or $\emptyset$ (unlabeled). Meantime, since the network is trained over epochs, we use $\hat{y}_{u,t}^i \in [0,1]^L$ to denote the soft pseudo label vector of of image i at epoch t [35]. Let f denote the classification network and θ_t be the network parameter obtained at epoch t. Given an image $i \in \mathcal{I}$, the predicted labels from the network is defined as $\hat{y}_t^i = f(i, \theta_t) \in [0,1]^L$. The binary cross-entropy loss function between two scalars $p, q \in [0,1]$ is defined as $\mathcal{L}(p,q) = -p\log(q) - (1-p)\log(1-q)$. Given $v \in \mathbb{R}^L$, we use $[v]_j$ to denote the jth entry of v. For notation convenience, we drop the index t if it is clear in context.

3.1 Overview

The PLMCL method is proposed for multi-label classification to deal with the case where each image in $\mathcal{D}$ only contains a single positive label. The procedure of PLMCL is shown in Fig. 2. During training, the pseudo label vector $\hat{y}_{u,0}$ of each image in $\mathcal{I}$ will be initialized with the same values of y_o for the observed labels and the unbiased probability 0.5 for the unobserved labels. Then PLMCL updates the pseudo label $\hat{y}_{u,t}$ based on the momentum m_t, which is the weighted sum of the last momentum m_{t-1} and the gradient of the loss between the predicted label vector $\hat{y}_{t-1}$ and the pseudo label vector $\hat{y}_{u,t-1}$ (Orange region in Fig. 2). It is expected that the pseudo labels are adaptively refined to provide stability for the confident labels and high momentum for the unconfident labels. The classification network is only employed during the test to predict the labels.

3.2 Updating Pseudo Labels by Momentum

Momentum Generation. When $\mathcal{D}$, the subset of the labeled images, is small, training has to start with a very limited number of observed labels, and therefore the resulting predictions will be inaccurate. In this case we cannot directly aggregate the predictions from the previous epoch to the current epoch using the off-the-shelf temporal ensembling techniques [23] to update the pseudo labels. Otherwise, the aggregation may soon arrive at a local minimum with low prediction confidence. Instead, our idea is to exploit the aggregation of the gradient (the momentum [15]) to update the pseudo label vector (Sect. 4.4 includes ablation study to compare these two approaches). By this way, we can help avoid undesired local minima [31].

To do so, we consider the binary cross-entropy loss function

$$\mathcal{L}_{cs}(\hat{y}_{t-1}, \hat{y}_{u,t-1}) = \frac{1}{L}\sum_{j=1}^{L} \mathcal{L}\left([\hat{y}_{t-1}]_j, [\hat{y}_{u,t-1}]_j\right) \tag{1}$$

and introduce the momentum vector m_t, which is updated based on the gradient of the loss function at epoch $t-1$, i.e., $\nabla\mathcal{L}_{cs}(\hat{y}_{t-1}, \hat{y}_{u,t-1})$:

$$m_t = \beta_1 m_{t-1} + (1-\beta_1)\nabla\mathcal{L}_{cs}(\hat{y}_{t-1}, \hat{y}_{u,t-1}), \tag{2}$$

where β_1 is moving-average decay scalar and m_0 is initialized by zero. Notice that even when $\nabla\mathcal{L}_{cs} = 0$ (which means that the loss reaches its local minima), the momentum will still be updated, which can therefore avoid premature convergence. This is especially important at the early stage of training. Another observation is that, when one entry of the momentum vector consistently moves towards the same direction, the estimation of the associated pseudo label will be more confident and the variations can be well smoothed over consequence epochs which leads to a more stable training behavior.

Self-Guided Momentum Factor. Intuitively, when the confidence of the pseudo label for a class is low, we want the corresponding momentum to contribute more to the pseudo label's iteration, in order to drive the pseudo label away from its current value to avoid undesired local minima. Otherwise, we want to keep the pseudo label, in which case the momentum should contribute less. With this idea, we introduce the self-guided momentum factor $\psi(\hat{y}_{u,t}) \in \mathbb{R}^L$ at epoch t:

$$\psi(\hat{y}_{u,t}) = \alpha\exp(-\lambda|2\hat{y}_{u,t} - \mathbf{1}|^n), \tag{3}$$

where $\alpha > 0$ is the curriculum learning rate, $\lambda > 0$ and n are tuned parameters, and $\mathbf{1} \in \mathbb{R}^L$ is the vector with 1 at each entry. The exponential function exp and the absolute value $|\cdot|$ are operated element-wisely. In all our experiments, we assign n with order 2. Notice that the confidence of a pseudo label is reflected by the term $|2\hat{y}_{u,t} - \mathbf{1}|$ in the definition of ψ. For instance, if $[\hat{y}_{u,t}]_j$, the jth entry of $\hat{y}_{u,t}$, is 0.5, the related momentum factor will achieve its maximal value α, which means that our module is not confident about the pseudo label and the related momentum should contribute more in the next iteration of the pseudo label. Otherwise, if $[\hat{y}_{u,t}]_j = 0$ or $[\hat{y}_{u,t}]_j = 1$, the related momentum factor reaches its minimum $\alpha \exp(-\lambda)$ (because $[\hat{y}_{u,t}]_j \in [0, 1]$), which indicates high confidence on the current pseudo label and the momentum should contribute less so that the value of the pseudo label can be maximally kept.

With the self-guided momentum factor, we can update the pseudo label latent parameters $y_{u,t} \in \mathbb{R}^L$ by:

$$y_{u,t} = y_{u,t-1} - \psi(\hat{y}_{u,t-1}) \circ m_t \tag{4}$$

where $\circ$ means the element-wise multiplication. From this equation, we can see that the momentum m_t actually indicates a weighted difference between $y_{u,t}$ and $y_{u,t-1}$. Keeping this in mind, Eq. (2) yields an interesting observation: Our approach expects not only the convergence of the pseudo labels (as it is in [23]), but also the convergence of the updating velocity of the pseudo labels. By doing so, unexpected local minima can be effectively avoided [15].

Notice that the entry of $y_{u,t}$ might be outside the interval $[0, 1]$. So we need to regulate $y_{u,t}$ to obtain $\hat{y}_{u,t}$ by

$$[\hat{y}_{u,t}]_j = \sigma([y_{u,t}]_j) \tag{5}$$

for $j = 1, 2, \cdots, L$, where $\sigma : \mathbb{R} \to [0, 1]$ is the sigmoid function that maps each entry of $y_{u,t}$ into $[0, 1]$. In practice, the gradient in Eq. (2) is actually with respect to $y_{u,t-1}$, i.e., $\nabla \mathcal{L}_{cs} = \nabla_{y_{u,t-1}} \mathcal{L}_{cs}(\hat{y}_{t-1}, \hat{y}_{u,t-1})$.

3.3 Scheduled Loss Function

To learn the parameters of the primary multi-label CNN, the following loss function must be minimized

$$\mathcal{L}_{PLMCL} = \mathcal{L}_{obs} + \mathcal{L}_{unobs}, \tag{6}$$

where $\mathcal{L}_{obs}$ is the cross-entropy loss between the predicted labels and the ground truth from the observed labels plus the regularizer [10] and $\mathcal{L}_{unobs}$ is the weighted cross-entropy loss between the predicted labels and the pseudo labels for the unobserved labels

$$\mathcal{L}_{unobs} = \sum_{i \in \mathcal{I}} \sum_{j \in \mathcal{U}_i} \xi([\hat{y}_{u,t}]_j, \varphi_t) \mathcal{L}\left([\hat{y}_t]_j, [\hat{y}_{u,t}]_j\right). \tag{7}$$

Algorithm 1. Momentum Curriculum Algorithm

Require: input image $i \in \mathcal{I}$;
Require: observed label y_o;
Require: neural network $f(i,\theta)$ with trainable parameters θ and input i.
Require: $[\hat{y}_u]_j = 1$ or 0, if labeled
$[\hat{y}_u]_j = 0.5$, if unlabeled
1: **Repeat**
2: $\hat{y} \longleftarrow f(i,\theta)$;
3: $m \longleftarrow \beta_1 m + (1-\beta_1)\nabla_{y_u}\mathcal{L}_{cs}(\hat{y},\hat{y}_u)$;
4: $y_u \longleftarrow y_u - \psi(\hat{y}_u)m$;
5: $\mathcal{L}_{PLMCL} \longleftarrow \mathcal{L}_{obs} + \mathcal{L}_{unobs}$;
6: update θ via backpropagation to improve the classification model based on curriculum framework;
7: $\hat{y}_u \longleftarrow \sigma(y_u)$;
8: **until** the max iteration or convergence;
9: **Output** optimal CNN parameters, $\hat{y}_u$ able to guide the prediction $\hat{y}$ to be very close from true label.

Here $\mathcal{U}_i$ is the set of the unobserved labels on image i and $\xi([\hat{y}_{u,t}]_j, \varphi_t)$ is a new scheduler that weights the unobserved loss, defined by

$$\xi([\hat{y}_{u,t}]_j, \varphi_t) = \beta_2 \frac{1-\gamma\exp(-10|2[\hat{y}_{u,t}]_j - 1|)}{1+\gamma\exp(-10|2[\hat{y}_{u,t}]_j - 1|)}, \tag{8}$$

where β_2 is the positive hyper-parameter, $\gamma = 1 - \varphi_t$, and φ_t is the current epoch t divided by the total number of epochs. This scheduler plays a critical role in balancing the contributions of unobserved labels to the total loss and the potential negative impacts of false pseudo labels. When a pseudo label has low confidence level, the related loss is in appropriate, the scheduler will assign a small weight to this loss such that the total loss $\mathcal{L}_{unobs}$ will not be significantly affected. This is especially important at the beginning of the training process when the pseudo labels are still not indicative. Different from most previous schedulers relying on the training steps or amount of available training data [18], our scheduler is adaptive, counting for not only the training steps, but the validation performance during training as well. To be specific, our scheduler ξ uses the confidence $|2[\hat{y}_{u,t}]_j - 1|$ to determine the scheduler value. When $[\hat{y}_{u,t}]_j$ is close to 0.5, $|2[\hat{y}_{u,t}]_j - 1|$ will be close to 0 and therefore the value of ξ will be small. Over epochs, the pseudo labels can gradually build up their confidence and start to converge. In that case, the value of ξ will increase.

We use soft pseudo labels, instead of sharp 1 or 0, in $\mathcal{L}_{unobs}$, because the latter may make the predictions of the output layer over-confident [34]. In contrast, using the pseudo labels as the distribution scores helps smooth the predictions, which is consistent with the idea of label-smoothing regularization [34].

4 Experiments

4.1 Datasets Preparation

Datasets. The **PASCAL VOC** dataset [14] contains natural images from 20 different classes, including 5,717 training images and 5,823 images in the official validation set for testing. The **MS-COCO** dataset [24] consists of 80 classes, including 82,081 training images and 40,137 testing images. The **NUS-WIDE** dateset [9] consists of nearly 150K color images with various resolutions for training and 60.2K for testing. This dataset is associated with 81 classes.

Data Preparation. The SSPL setting annotates certain percent of the training images (20% to 80%) with only a single positive label and the rest of the images are totally unlabeled. The FSPL setting has 100% of the training images annotated with only one single positive label. To do so, we follow [10] by assigning randomly only one positive label for each image in the training set.

Implementation Details. For a fair comparison, we use ResNet-50 to initialize the weights of the backbone architecture for all models. The learning rate is 10^{-5} to train the VOC and COCO datasets and 10^{-4} to train the NUS dataset. The batch size is chosen as 8 for the VOC dataset, and 16 for the COCO and NUS datasets. Training is performed in two ways: end-to-end and linear-init. End-to-end training simultaneously updates the parameters of the backbone and the classifier for 10 epochs, while linear-init training first fixes the parameters of the backbone and trains the classifier for 25 epochs, and then fine-tunes the parameters of the classifier and the backbone for 5 epochs [10]. The best mAP on the validation set is reported for all experiments. It is worth mentioning that all experiments on our model share the same set of hyper-parameters. The mean average precision (mAP) is applied to evaluate the performance of different approaches for multi-label classification in our paper, similar to [10,17].

Table 1. Quantitative results (mAP) of multi-label image classification on different subsets of the single observed label (SSPL) setting on COCO, VOC, and NUS datasets. Bold represents the highest mAP and underline represents the second-best.

	COCO dataset							
Losses	end-to-end				linear-init			
	20%	40%	60%	80%	20%	40%	60%	80%
$\mathcal{L}_{AN}$	46.3	53.8	59.5	62.4	53.4	61.1	63.8	65.6
$\mathcal{L}_{AN-LS}$	48.9	57.9	62.3	65.5	58.2	62.8	67.6	68.5
$\mathcal{L}_{WAN}$	57.0	60.9	63.5	64.5	60.2	62.4	64.1	66.1
$\mathcal{L}_{EPR}$	52.7	58.4	61.5	62.6	61.0	63.6	65.3	66.6
$\mathcal{L}_{ROLE}$	47.3	57.6	62.7	65.2	61.9	65.8	67.4	68.6
$\mathcal{L}_{PLMCL}$	**61.0**	**65.1**	**68.4**	**70.4**	**64.8**	**67.8**	**69.2**	**71.5**
	VOC dataset							
Losses	end-to-end				linear-init			
	20%	40%	60%	80%	20%	40%	60%	80%
$\mathcal{L}_{AN}$	51.3	71.7	80.2	82.8	78.6	81.4	83.7	84.9
$\mathcal{L}_{AN-LS}$	70.0	79.0	85.0	86.1	71.5	81.4	84.9	86.4
$\mathcal{L}_{WAN}$	76.4	82.5	85.1	85.6	79.4	83.6	85.9	86.9
$\mathcal{L}_{EPR}$	75.5	81.0	83.9	84.5	82.6	84.9	85.9	86.9
$\mathcal{L}_{ROLE}$	66.9	80.9	85.9	86.8	73.9	82.3	85.7	87.5
$\mathcal{L}_{PLMCL}$	**79.9**	**84.6**	**87.6**	**88.4**	**83.1**	**85.4**	**87.8**	**87.7**
	NUS dataset							
Losses	end-to-end				linear-init			
	20%	40%	60%	80%	20%	40%	60%	80%
$\mathcal{L}_{AN}$	28.5	35.2	38.9	40.8	35.2	42.8	44.8	46.5
$\mathcal{L}_{AN-LS}$	27.8	36.0	39.1	41.4	39.1	44.4	47.2	48.0
$\mathcal{L}_{WAN}$	37.6	41.3	43.7	44.8	41.4	44.5	45.7	46.2
$\mathcal{L}_{EPR}$	34.9	39.5	42.3	44.2	43.4	46.0	47.2	48.2
$\mathcal{L}_{ROLE}$	27.0	32.2	37.1	39.7	36.3	42.8	46.4	47.5
$\mathcal{L}_{PLMCL}$	**37.9**	**42.5**	**44.8**	**46.6**	**46.2**	**49.4**	**50.3**	**50.7**

Table 2. Quantitative results (mAP) of multi-label image classification for FSPL setting on three different datasets with end-to-end and linear-init training. Bold represents the highest mAP and underline represents the second-best.

Supervised by	Losses	COCO dataset		VOC dataset		NUS dataset	
		end-to-end	linear-init	end-to-end	linear-init	end-to-end	linear-init
Fully Labeled	$\mathcal{L}_{BCE}$	75.5	76.8	89.1	90.1	52.6	54.7
	$\mathcal{L}_{BCE-LS}$	76.8	78.8	90	90.9	53.5	54.2
1 Positive (FSPL Setting)	$\mathcal{L}_{AN}$	64.1	67.3	85.1	87.7	42	46.8
	$\mathcal{L}_{AN-LS}$	<u>66.9</u>	<u>69.2</u>	86.7	86.5	44.9	50.5
	$\mathcal{L}_{WAN}$	64.8	67.8	86.5	87.1	<u>46.3</u>	47.5
	$\mathcal{L}_{EPR}$	63.3	67.5	85.5	85.6	46.0	48.9
	$\mathcal{L}_{ROLE}$	66.3	69.0	<u>87.9</u>	<u>88.2</u>	43.1	<u>51.0</u>
	$\mathcal{L}_{PLMCL}$	**71.1**	**72.0**	**89.0**	**88.7**	**48.6**	**51.5**

4.2 Subset Single Positive Label (SSPL)

This subsection compares the performance of our model $\mathcal{L}_{PLMCL}$ with other state-of-the-art approaches that address the unobserved labels in multi-label classification under SSPL settings. The comparison results are listed in Table 1. We observe that $\mathcal{L}_{AN}$ [22] has the worst performance since it assumes that all the unobserved labels are negatives. This assumption adds more noisy labels (false negatives) into the training process. To overcome this negative influence, $\mathcal{L}_{AN-LS}$ is introduced using label smoothing on $\mathcal{L}_{AN}$. Following [10], we use $\mathcal{L}_{AN-LS}$ as a baseline. To go one step further in mitigating the deteriorated effect of noisy labels, some models tend to down-weight the term related to negative labels in the loss function $\mathcal{L}_{WAN}$ [26]. Consequently, $\mathcal{L}_{WAN}$ improves the mAPs on different datasets as compared to $\mathcal{L}_{AN}$ and $\mathcal{L}_{AN-LS}$. Finally, expected positive regularization $\mathcal{L}_{EPR}$ [10] and online estimation of unobserved labels $\mathcal{L}_{ROLE}$ are introduced in [10] to deal with the single positive label, which is considered an extreme case of unobserved labels, in multi-label classification. We notice that $\mathcal{L}_{EPR}$ [10] provides stable performance in different settings since $\mathcal{L}_{EPR}$ ignores the unobserved terms and focuses on finding the positive labels according to its regularizer. Also notice that the performance of $\mathcal{L}_{ROLE}$ degrades a lot under 20% SSPL in end-to-end setting since its classifier and label estimator are randomly initialized [10]. Compared with these methods, our proposed $\mathcal{L}_{PLMCL}$, based on the momentum, is superior in the three different datasets.

4.3 Full-set Single Positive Label (FSPL)

This subsection compares different models under FSPL setting. Table 2 reports the mAPs over three different datasets under two training settings: end-to-end and linear-init. Besides FSPL, we also include in Table 2 the mAPs of two models ($\mathcal{L}_{BCE}$ and $\mathcal{L}_{BCE-LS}$) based on full-set fully labeled (FFL) setting [10], as references to examine the performance of our model under FSPL setting. Compared with the FFL setting using 100% of the labels, the FSPL setting uses only

Table 3. Quantitative results (mAP) of multi-label image classification on different subsets of the single observed label (SSPL) setting and FSPL (100%) setting on two different datasets with end-to-end learning using the Resnet-101 backbone. Bold represents the highest mAP and underline represents the second-best.

Models	**COCO Dataset**					**VOC Dataset**				
	SSPL				FSPL	SSPL				FSPL
	20%	40%	60%	80%	100%	20%	40%	60%	80%	100%
$\mathcal{L}_{AN}$	42.0	52.4	57.9	62.3	63.9	52.9	73.6	82.1	85.1	86.1
$\mathcal{L}_{AN-LS}$	49.2	57.5	61.8	65.6	66.9	67.1	80.6	85.8	86.8	88.3
$\mathcal{L}_{WAN}$	<u>57.3</u>	<u>61.8</u>	<u>64.9</u>	<u>66.0</u>	66.7	<u>76.1</u>	<u>83.1</u>	85.9	86.5	87.2
$\mathcal{L}_{EPR}$	51.8	57.6	60.9	62.8	63.8	75.1	80.3	85.0	84.6	86.7
$\mathcal{L}_{ROLE}$	46.2	58.3	62.8	65.5	<u>67.3</u>	66.7	82.3	<u>86.2</u>	<u>87.4</u>	<u>88.8</u>
$\mathcal{L}_{PLMCL}$ (ours)	**61.4**	**67.3**	**69.5**	**71.5**	**72.2**	**82.0**	**86.2**	**88.4**	**88.8**	**89.7**

around 1.2% of the total number of labels. It is worth mentioning that the mAPs of our model $\mathcal{L}_{PLMCL}$ under FSPL remain at a comparable level. For instance, the mAP of our model on the NUS dataset is 48.6%, which is only 4.0% lower than the FFL setting $\mathcal{L}_{BCE}$. Similar conclusions can be drawn in COCO and VOC datasets. When comparing the approaches under FSPL setting (the bottom part of Table 2), we observe that our model $\mathcal{L}_{PLMCL}$ always achieves the best performance on different datasets.

Backbone ResNet-101. Besides ResNet-50, we also conducted experiments using a different backbone, ResNet-101, in the end-to-end setting on two datasets under different SSPL settings. The results are reported in Table 3. Our model $\mathcal{L}_{PLMCL}$ achieves the hightest mAP scores under all settings. It is worth mentioning that our model $\mathcal{L}_{PLMCL}$ at 60% of SSPL setting achieves higher mAP compared to $\mathcal{L}_{ROLE}$ under FSPL setting (100%) in COCO dataset, which demonstrates the efficiency of our model in saving labeling costs.

4.4 Ablation Study

PLMCL Modules. Our ablation analysis highlights the importance of each component in our method: the momentum m_t, the scheduler ξ, and the self-guided factor ψ. The results are reported in Table 4. We start with the basic variation of the model which uses the cross-entropy loss on the observed labels.

In this case, we assume that all the unobserved labels are negative, since ignoring the unobserved labels may lead to overfitting. When adding components into the model, this assumption can be removed since we use the pseudo labels to deal with the unobserved labels. We first add the momentum generation module to update the soft pseudo labels which improve the mAP for different settings and datasets (Table 4, second row). Adding the confidence-aware scheduler and the self-guided factor further enhances in the overall performance, since they work together to make the training adaptive with respect to the confidence of the pseudo labels (Table 4, last two rows).

Table 4. Ablation study for all proposed model components on SSPL (20%) and FSPL with linear-init training setting on COCO dataset.

Hyper Parameters			COCO Dataset (mAP %)	
m_t	ξ	ψ	20%	100%
			59.0	67.3
✓			62.0	69.9
✓	✓		63.1	70.4
✓	✓	✓	64.8	72.0

Hyper Parameters. The influence of the hyper-parameters is studied on the VOC validation set under the 40% SSPL setting. We run experiments with different hyper-parameters. In all of these experiments, our model remains stable and reliable. Table 5 shows the variations in mAP according to the changes of hyper-parameters β_1, β_2, and λ. Recall that the hyper-parameter β_1 is used to balance the previous network knowledge and the current decision of the network.

We observed that the mAP score improves when the balance is achieved (β_1 is close to 0.5) and decreases in the extreme cases ($\beta_1 = 0.1$ emphasizing the current decision of the network and $\beta_1 = 0.9$ emphasizing the previous network knowledge). λ demonstrates the effectiveness of the self-guided factor ψ that controls the amount of momentum contributing to the update of pseudo levels. The results in Table 5 indicates that the mAP score increases as the λ increases. When increasing the scheduler parameter β_2, the mAP score almost remains the same level. Overall, our model is not very sensitive to the variations in these hyper-parameters.

Table 5. mAP score (%) of our method as a function of β_1, β_2, λ in subset single positive setting SSPL at 40% using end-to-end training setting on the VOC validation set.

β_1	0.1	0.5	0.7	0.9
mAP (%) ($\beta_2 = 0.6$)	84.4	84.5	**84.6**	84.0
β_2	0.1	0.4	0.6	0.9
mAP (%) ($\beta_1 = 0.7$)	83.5	84.5	**84.6**	84.5
λ	1	2	3	4
mAP (%) ($\beta_1 = 0.7$, $\beta_2 = 0.6$)	83.0	83.2	83.9	**84.6**

Comparison with Semi-supervised Models. We compare our PLMCL with the existing semi-supervised models such as FixMatch [32] and UPS [30] on COCO dataset. Figure 3 shows the comparison of the mAP results under different SSPL settings. Notice that PLMCL always outperforms the other two methods.

This is because FixMatch and UPS are trained based on the supervised loss over the set of observed labels while ignoring the unobserved labels of the labeled set as explained in Subsect. 2.2. Therefore, there is a lack of continuity in pseudo label updating. On the contrary, PLMCL can guarantee such continuity and capture the temporal information during training. So it is not surprising that PLMCL achieves the best mAP score.

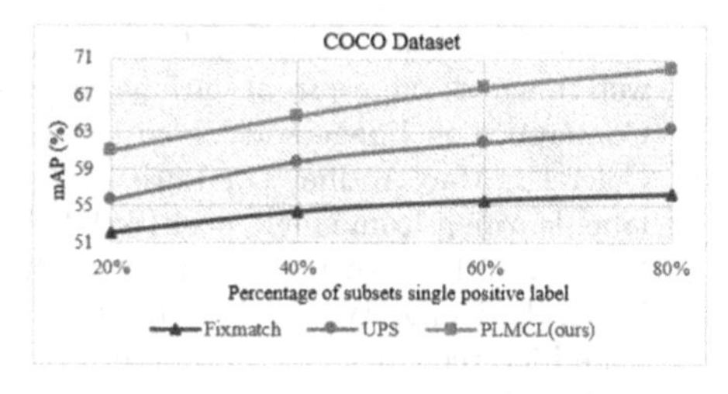

Fig. 3. Ablation study on the difference between using our method PLMCL and semi-supervised models.

5 Conclusions

This paper presents a momentum-based curriculum learning method for multi-label image classification under partial-label settings. In PLMCL, soft pseudo labels are generated per image and then updated based on the momentum and the confidence levels. Through extensive experiments on three different partial-label settings, including our proposed new SSPL setting, we demonstrated that our PLMCL outperforms the state-of-the-art methods, especially when training on fewer observed labels.

Acknowledgements. The authors gratefully acknowledge the partial financial support of the National Science Foundation (1830512).

References

1. Arazo, E., Ortego, D., Albert, P., O'Connor, N.E., McGuinness, K.: Pseudo-labeling and confirmation bias in deep semi-supervised learning. In: 2020 International Joint Conference on Neural Networks (IJCNN), pp. 1–8. IEEE (2020)
2. Bengio, Y., Louradour, J., Collobert, R., Weston, J.: Curriculum learning. In: 26th annual International Conference on Machine Learning (ICML), pp. 41–48 (2009)
3. Berthelot, D., et al.: ReMixMatch: semi-supervised learning with distribution alignment and augmentation anchoring. arXiv preprint arXiv:1911.09785 (2019)
4. Bucak, S.S., Jin, R., Jain, A.K.: Multi-label learning with incomplete class assignments. In: IEEE/CVF Conference on Computer Vision and Pattern Recognition (CVPR), pp. 2801–2808. IEEE (2011)
5. Cabral, R.S., Torre, F., Costeira, J.P., Bernardino, A.: Matrix completion for multi-label image classification. In: Advances in Neural Information Processing Systems, pp. 190–198 (2011)
6. Chapelle, O., Scholkopf, B., Zien, A.: Semi-supervised learning. IEEE Trans. Neural Netw. **20**(3), 542 (2009)
7. Chen, M., Zheng, A., Weinberger, K.: Fast image tagging. In: International Conference on Machine Learning (ICML), pp. 1274–1282. PMLR (2013)
8. Chu, H.-M., Yeh, C.-K., Wang, Y.-C.F.: Deep generative models for weakly-supervised multi-label classification. In: Ferrari, V., Hebert, M., Sminchisescu, C., Weiss, Y. (eds.) ECCV 2018. LNCS, vol. 11206, pp. 409–425. Springer, Cham (2018). https://doi.org/10.1007/978-3-030-01216-8_25

9. Chua, T.S., Tang, J., Hong, R., Li, H., Luo, Z., Zheng, Y.: Nus-wide: a real-world web image database from national university of Singapore. In: ACM International Conference on Image and Video Retrieval, pp. 1–9 (2009)
10. Cole, E., Mac Aodha, O., Lorieul, T., Perona, P., Morris, D., Jojic, N.: Multi-label learning from single positive labels. In: IEEE/CVF Conference on Computer Vision and Pattern Recognition (CVPR), pp. 933–942 (2021)
11. Deng, J., Dong, W., Socher, R., Li, L.J., Li, K., Fei-Fei, L.: ImageNet: a large-scale hierarchical image database. In: IEEE Conference on Computer Vision and Pattern Recognition (CVPR), pp. 248–255. IEEE (2009)
12. Deng, J., Russakovsky, O., Krause, J., Bernstein, M.S., Berg, A., Fei-Fei, L.: Scalable multi-label annotation. In: SIGCHI Conference on Human Factors in Computing Systems, pp. 3099–3102 (2014)
13. Durand, T., Mehrasa, N., Mori, G.: Learning a deep convnet for multi-label classification with partial labels. In: IEEE/CVF Conference on Computer Vision and Pattern Recognition (CVPR), pp. 647–657 (2019)
14. Everingham, M., Winn, J.: The pascal visual object classes challenge 2012 (voc2012) development kit. Pattern Anal. Statist. Model. Comput. Learn. Tech. Rep **8**, 5 (2011)
15. Goodfellow, I., Bengio, Y., Courville, A.: Deep learning. MIT press (2016)
16. Guo, S., Guo, S., et al.: CurriculumNet: weakly supervised learning from large-scale web images. In: Ferrari, V., Hebert, M., Sminchisescu, C., Weiss, Y. (eds.) ECCV 2018. LNCS, vol. 11214, pp. 139–154. Springer, Cham (2018). https://doi.org/10.1007/978-3-030-01249-6_9
17. Huynh, D., Elhamifar, E.: Interactive multi-label CNN learning with partial labels. In: IEEE/CVF Conference on Computer Vision and Pattern Recognition (CVPR), pp. 9423–9432 (2020)
18. Jean, S., Firat, O., Johnson, M.: Adaptive scheduling for multi-task learning. arXiv preprint arXiv:1909.06434 (2019)
19. Jiang, L., Meng, D., Zhao, Q., Shan, S., Hauptmann, A.G.: Self-paced curriculum learning. In: Twenty-Ninth AAAI Conference on Artificial Intelligence (2015)
20. Kapoor, A., Viswanathan, R., Jain, P.: Multilabel classification using Bayesian compressed sensing. Adv. Neural. Inf. Process. Syst. **25**, 2645–2653 (2012)
21. Kumar, M., Packer, B., Koller, D.: Self-paced learning for latent variable models. Adv. Neural. Inf. Process. Syst. **23**, 1189–1197 (2010)
22. Kundu, K., Tighe, J.: Exploiting weakly supervised visual patterns to learn from partial annotations. Adv. Neural. Inf. Process. Syst. **33**, 561–572 (2020)
23. Laine, S., Aila, T.: Temporal ensembling for semi-supervised learning. arXiv preprint arXiv:1610.02242 (2016)
24. Lin, T.-Y., et al.: Microsoft COCO: common objects in context. In: Fleet, D., Pajdla, T., Schiele, B., Tuytelaars, T. (eds.) ECCV 2014. LNCS, vol. 8693, pp. 740–755. Springer, Cham (2014). https://doi.org/10.1007/978-3-319-10602-1_48
25. Liu, Y., Jin, R., Yang, L.: Semi-supervised multi-label learning by constrained non-negative matrix factorization. In: AAAI, vol. 6, pp. 421–426 (2006)
26. Mac Aodha, O., Cole, E., Perona, P.: Presence-only geographical priors for fine-grained image classification. In: IEEE/CVF International Conference on Computer Vision (ICCV), pp. 9596–9606 (2019)
27. Mahajan, D., et al.: Exploring the limits of weakly supervised pretraining. In: Ferrari, V., Hebert, M., Sminchisescu, C., Weiss, Y. (eds.) ECCV 2018. LNCS, vol. 11206, pp. 185–201. Springer, Cham (2018). https://doi.org/10.1007/978-3-030-01216-8_12

28. Niu, X., Han, H., Shan, S., Chen, X.: Multi-label co-regularization for semi-supervised facial action unit recognition. arXiv preprint arXiv:1910.11012 (2019)
29. Pineda, L., Salvador, A., Drozdzal, M., Romero, A.: Elucidating image-to-set prediction: an analysis of models, losses and datasets. CoRR (2019)
30. Rizve, M.N., Duarte, K., Rawat, Y.S., Shah, M.: In defense of pseudo-labeling: an uncertainty-aware pseudo-label selection framework for semi-supervised learning. In: International Conference on Learning Representations (2021)
31. Sariyildiz, M.B., Cinbis, R.G.: Gradient matching generative networks for zero-shot learning. In: IEEE/CVF Conference on Computer Vision and Pattern Recognition (CVPR), pp. 2168–2178 (2019)
32. Sohn, K., et al.: FixMatch: simplifying semi-supervised learning with consistency and confidence. Adv. Neural. Inf. Process. Syst. **33**, 596–608 (2020)
33. Sun, C., Shrivastava, A., Singh, S., Gupta, A.: Revisiting unreasonable effectiveness of data in deep learning era. In: IEEE International Conference on Computer Vision (ICCV), pp. 843–852 (2017)
34. Szegedy, C., Vanhoucke, V., Ioffe, S., Shlens, J., Wojna, Z.: Rethinking the inception architecture for computer vision. In: IEEE Conference on Computer Vision and Pattern Recognition (CVPR), pp. 2818–2826 (2016)
35. Tanaka, D., Ikami, D., Yamasaki, T., Aizawa, K.: Joint optimization framework for learning with noisy labels. In: IEEE Conference on Computer Vision and Pattern Recognition (CVPR), pp. 5552–5560 (2018)
36. Tarvainen, A., Valpola, H.: Mean teachers are better role models: weight-averaged consistency targets improve semi-supervised deep learning results. In: Advances in Neural Information Processing Systems 30 (2017)
37. Wang, B., Tu, Z., Tsotsos, J.K.: Dynamic label propagation for semi-supervised multi-class multi-label classification. In: IEEE International Conference on Computer Vision (ICCV), pp. 425–432 (2013)
38. Wang, L., Ding, Z., Fu, Y.: Adaptive graph guided embedding for multi-label annotation. In: IJCAI (2018)
39. Wang, Q., Shen, B., Wang, S., Li, L., Si, L.: Binary codes embedding for fast image tagging with incomplete labels. In: Fleet, D., Pajdla, T., Schiele, B., Tuytelaars, T. (eds.) ECCV 2014. LNCS, vol. 8690, pp. 425–439. Springer, Cham (2014). https://doi.org/10.1007/978-3-319-10605-2_28
40. Wang, X., Chen, Y., Zhu, W.: A survey on curriculum learning. In: IEEE Transactions on Pattern Analysis and Machine Intelligence (2021)
41. Wu, B., Lyu, S., Ghanem, B.: ML-MG: multi-label learning with missing labels using a mixed graph. In: IEEE International Conference on Computer Vision (ICCV), pp. 4157–4165 (2015)
42. Xu, M., Jin, R., Zhou, Z.H.: Speedup matrix completion with side information: application to multi-label learning. In: Advances in Neural Information Processing Systems, pp. 2301–2309 (2013)
43. Yang, H., Zhou, J.T., Cai, J.: Improving multi-label learning with missing labels by structured semantic correlations. In: Leibe, B., Matas, J., Sebe, N., Welling, M. (eds.) ECCV 2016. LNCS, vol. 9905, pp. 835–851. Springer, Cham (2016). https://doi.org/10.1007/978-3-319-46448-0_50
44. Yu, H.F., Jain, P., Kar, P., Dhillon, I.: Large-scale multi-label learning with missing labels. In: International Conference on Machine Learning (ICML), pp. 593–601. PMLR (2014)

Open-Vocabulary Semantic Segmentation Using Test-Time Distillation

Nir Zabari(✉) and Yedid Hoshen

School of Computer Science and Engineering, The Hebrew University of Jerusalem, Jerusalem, Israel
nir.zabari@mail.huji.ac.il

Abstract. Semantic segmentation is a key computer vision task that has been actively researched for decades. In recent years, supervised methods have reached unprecedented accuracy; however, obtaining pixel-level annotation is very time-consuming and expensive. In this paper, we propose a novel open-vocabulary approach to creating semantic segmentation masks, without the need for training segmentation networks or seeing any segmentation masks. At test time, our method takes as input the image-level labels of the categories present in the image. We utilize a vision-language embedding model to create a rough segmentation map for each class via model interpretability methods and refine the maps using a test-time augmentation technique. The output of this stage provides pixel-level pseudo-labels, which are utilized by single-image segmentation techniques to obtain high-quality output segmentations. Our method is shown quantitatively and qualitatively to outperform methods that use a similar amount of supervision, and to be competitive with weakly-supervised semantic-segmentation techniques.

Keywords: Semantic segmentation · Language-based segmentation · Open-vocabulary segmentation

1 Introduction

The task of semantic segmentation, which involves assigning a class category to each pixel of an image, has evolved recently using deep neural networks trained on large-scale annotated datasets [10,21,48]. Although the progress has been very impressive, such systems are mainly useful for common semantic categories where large annotated datasets are available. However, in many real-world scenarios, segmentation is required for rare class categories for which training data are unavailable. Annotation of new data for every novel category is expensive, time-consuming, and impractical in most cases. Moreover, current supervised methods do not scale well to work on imbalanced datasets of large vocabularies, where common categories have many examples, while the rare categories follow a long-tailed distribution [19,49].

Supplementary Information The online version contains supplementary material available at https://doi.org/10.1007/978-3-031-25063-7_4.

L. Karlinsky et al. (Eds.): ECCV 2022 Workshops, LNCS 13802, pp. 56–72, 2023.
https://doi.org/10.1007/978-3-031-25063-7_4

Fig. 1. Results of our method on real-world images from rare class categories. As input, our algorithm receives an image along with text-prompts describing the classes that we want to segment. A language-vision model is distilled by generating a relevance map in relation to each prompt category. Further refinement is performed through test time augmentations. Next, the relevance maps are fed into a single image segmentation algorithm, which transforms the relevance maps into a high-quality segmentation.

Where a few annotated examples of a rare class are available, different paradigms have been proposed, including Few-Shot Semantic Segmentation (FSSS) and Weakly Supervised Semantic Segmentation (WSSS). The two settings differ in the expected supervision. FSSS requires a few pixel-annotated images containing the rare category, while WSSS requires very coarse supervision e.g. image level labels, bounding boxes, or scribbles but on many images. These methods cannot be applied in real-world settings where such supervision is unavailable. In order to operate in settings with no labels, Zero Shot Semantic Segmentation (ZSSS) methods were proposed, but current methods are not yet accurate enough for many use cases when applied to rare classes.

We propose a simple and computationally lean method for segmenting objects that does not require new data annotation, is fast (30 s on average, depending mostly on the image resolution), and achieves high accuracy results. Our method utilizes CLIP [37], a recently developed vision-language model that embeds visual and linguistic concepts into a shared space through large-scale contrastive learning on web data. Our method uses the documented zero-shot abilities of CLIP. Given a CLIP model and text prompts describing the semantic classes present in the image, our method first creates a set of per-category relevance maps of the image. The mapping between the text prompts and the per-category relevance maps is performed by utilizing a recently developed transformer interpretation method [9]. Since the relevance maps generated by the explainability methods are noisy and inaccurate, we use custom test-time augmentations to refine the relevance maps. We then combine the individual per-category maps into a multi-class relevance map of the image. The relevance map, which is obtained in a zero-shot manner, is already quite accurate and is used as input to single-image segmentation methods. We present the results using an unsupervised single image clustering technique that we augment with our refined relevance maps as weak pseudo-supervision. We also use our method as input to a supervised single-image supervision method, where we replace the expected supervision with our hard segmentation map.

Our method has been extensively evaluated and ablated. We quantitatively evaluated our method on standard segmentation benchmarks and demonstrate that it performs better than current methods that use a similar amount of supervision. The ability of our method to operate on rare and unique images (e.g., landmarks and abstract concepts) is clearly demonstrated (see Fig. 1 and Fig. 6 and the SM). Finally, we provide an ablation study demonstrating the importance of the different components of the method. Our main contributions are:

1. Introducing a simple, lean, novel method for segmenting any object without pixel-level annotations or multiple sample images
2. Suggesting the idea of distilling pre-trained language-vision networks for semantic segmentation of any object category.
3. An extensive evaluation showing better performance than other methods using the same level of supervision, with impressive results on rare objects.

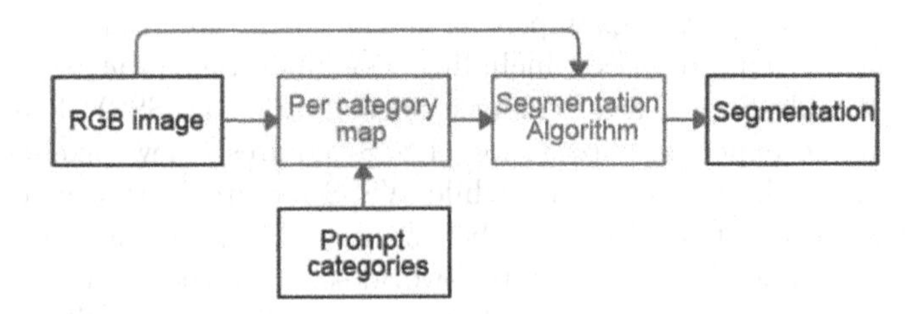

Fig. 2. Segmentation pipeline. A high-level description of our method. We receive an RGB image to process and user-suggested categories for segmentation. This assumes that the prompt categories actually appear in the image. By using an interpretability method, we produce a relevance map for each prompt category. Afterwards, we apply a segmentation algorithm (e.g., a clustering or interactive segmentation algorithm) that is able to segment the image and leverage the pseudo labels induced by the relevance maps. Our final output is a segmentation map.

2 Related Works

Most current semantic segmentation models [10,21,48,50,53] classify pixels into a fixed set of closed categories for which ample supervision is provided. Weakly-supervised semantic segmentation, few-shot semantic segmentation, and zero-shot semantic segmentation extend these methods by generating segmentation masks for semantic classes with simpler-to-obtain annotations, a limited number of annotated images or no supervision, in order to reduce the need for pixel annotations. These methods have evolved rapidly in recent years.

Weakly-Supervised Semantic-Segmentation (WSSS). These approaches use weak forms of supervision, such as image labels [6,20,36], bounding boxes [11,25], and scribbles [31,51]. Image-level labels are probably the most popular form of weak supervision due to their simplicity and ease of obtaining them from public datasets or web-based data. A typical WSSS pipeline begins with generating a pseudo mask, followed by training a new semantic segmentation network [28]. Interpretability techniques, such as CAM [43], are often used to infer incomplete pixel-level annotations automatically. These masks are often inaccurate as interpretability methods usually generate pseudo masks that highlight only the most discriminative parts (e.g., highlight the railroad when classifying a train). They therefore cannot generate a complete semantic map, but can only be used as a seed for segmentation. There have been several proposed methods for improving the segmentation masks generated from CAM, such as PSA [2] and IRN [1], which use boundary information by calculating affinities between pixels. Other methods [22,27] start with seeds and build up to the object boundary iteratively. While our method also uses image-level annotations, it differs from the above techniques as no training images are required.

Few-Shot Semantic-Segmentation (FSSS). Following the success of few-shot and zero-shot in classification [34,37] and object detection [7,38,39], few-shot and zero-shot semantic segmentation methods have emerged. FSSS typically require a small number of segmented training images, usually fewer than ten. The annotated examples serve as the support set, from which later segmentation masks are generated for the input image, called a query set. [52] uses masked average pooling to extract foreground and background information. [13] leverages metric-learning for few-shot segmentation. [40] employs an encoder-decoder architecture with auxiliary conditional encoder branches that concatenate features from both query and support images to feed the decoder.

Zero-Shot Semantic-Segmentation (ZSSS). While FSSS methods use zero-shot learning for recognition at the image level, ZSSS methods use zero-shot learning at the level of individual pixels. These approaches learn, at training time, to classify pixels into a closed set of class categories, and at test-time, apply these models on never-seen class categories. [8] uses a state-of-the-art segmentation network (DeepLab), and propagates information of unseen classes to pixel embeddings using Word2Vec, together with self-training. [47] leverages image captions in order to segment unknown classes. [30] train a generator to

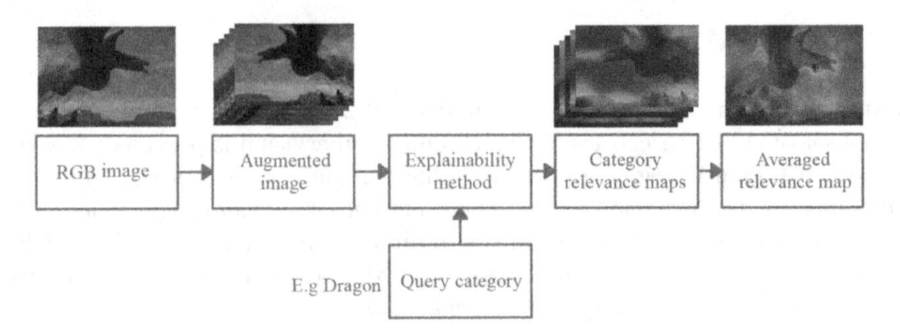

Fig. 3. Relevance maps generation. A sketch of the relevance map generation module, that can generate a relevance map based on a textual object query. We utilize an interpretability method, a language-vision pre-trained model using TTA techniques, to obtain a refined relevance map.

produce visual features from semantic word embeddings, similar to [8], but it alternates between generating "good features", and maintaining the structural-relations between categories in the text latent space, as before.

Shared Language-Vision Latent space. Recently multiple works have proposed learning shared text and image embeddings in the same latent space [12,24,29,37,42,46]. Contrastive Language Image Pre-Training (CLIP) uses 400M image-caption pairs to assess the similarity between a text and an image, as a pre-training task. A variety of computer vision tasks have been shown to benefit from CLIP model embeddings, which have demonstrated robustness and generalization on a wide range of visual concepts. Some notable applications that draw on CLIP's shared embedding space are [35] which use StyleGan latent space to generate images based on a text description. [16] adapts an image generator to other domains with zero examples. Using CLIP's latent space, [17] generates an image from its caption, and vice versa. [32] uses CLIP for video retrieval.

3 Method

The input to our method is an RGB image I of shape (h, w) and a list of K categories, each given by a text prompt $(T_1, T_2, ..., T_K)$. The output of our method is a segmentation mask $S(I) := S \in \{0, .., K\}^{(h,w)}$. A pixel is denoted by $p = (x, y)$. Background pixels p are denoted by $S_p = 0$. Other categories i are denoted by $S_p = i$. Our method consists of two stages as shown in Fig. 2. In the first stage (see Fig. 3), we compute a relevance map for each prompt category $\mathcal{C}$, where a larger value for a pixel indicates a higher likelihood of belonging to category $\mathcal{C}$. A relevance map is created for each category given its text prompt, by utilizing a pre-trained vision-language model (here we use CLIP, due to its availability and its widespread adoption as a baseline for language-vision models). The map is created by finding the image regions that explain the decision made by CLIP. To increase the map accuracy to a high-enough accuracy

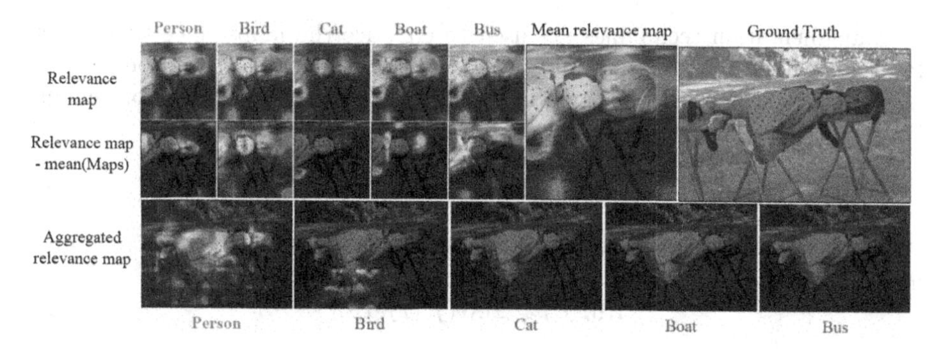

Fig. 4. Crop Augmentation. Crops are selected from a grid overlaid on the image and relevance maps are calculated for each crop in relation to the prompt category. As crops overlap, each pixel's relevance value is averaged across different crops. In order to remove the noise and artifacts produced by the interpretability method, we center each relevance map according to the mean of the relevance maps of both the query category (e.g., Person, in green) and the distractor categories (Bird, Cat, Boat, and Bus, in red). Then, the probability of each class category is obtained using CLIP, which allows us to analyze only relevant labels, even if we do not know what the labels are. When our prompt category class probability is greater than a specified threshold (we use $\mathbb{P}(class) > 0.3$), we add the crop's relevance maps to an aggregated relevance map of the entire image. Finally, the aggregated relevance map is normalized to the range $[0, 1]$, resulting in the final crop view (bottom row). (Color figure online)

so they can serve as a source of synthetic supervision, we employ commonly used test-time-augmentation (TTA) techniques [4,5] to obtain the final relevance map. In the second stage, we convert the image from the averaged relevance map image $SS \in [0, 1]^k$, where $SS_p(\mathcal{C})$ for $1 \leq \mathcal{C} \leq k$ is an indication that pixel p belongs to category $\mathcal{C}$, to the final segmentation image $S \in \{0, .., k\}^{(h,w)}$. The transition from the relevance maps to the segmentation image can be achieved in multiple ways: weakly-supervised clustering by scribble, interactive segmentation techniques, active contours models, etc. The synthetic supervision is generated from the map using stochastic pixel sampling, and is then fed to downstream methods. We experimented with weakly supervised clustering and interactive segmentation models.

3.1 Relevance Map Generation

Relevance Map Mining. Our observation is that we can leverage large pre-trained language-vision models to segment any object. These models were previously trained on very large datasets of image-caption pairs and have demonstrated great performance for zero-shot classification. We suggest that interpretability techniques can be used to infer dense class labels, in an open-vocabulary manner. For each prompt category, we created a relevance map using an interpretability technique. We denote this by the function M, that takes, as input an image I and a prompt $T_{\mathcal{C}}$ and returns a relevance map $M(I, T_{\mathcal{C}})$. Test-time augmentation techniques are used to enhance the maps since they suffer from noise, tend to extend beyond object boundaries, and can be incomplete or imprecise. Hence, for each prompt category $\mathcal{C}$ and input image I, a set of random image augmentation functions $f_1, f_2..f_V$, V views are generated:

$$views = \{f_v(I)\}_{v=1}^{V} \quad (1)$$

We then average our maps to reduce noise and obtain the averaged relevance maps for category $\mathcal{C}$:

$$RM(\mathcal{C}) = \mathbb{E}[M(f_v(I), T_{\mathcal{C}})] \quad (2)$$

Image Views. We denote different image transformations as views. The 4 views that we use are: the original image, horizontal flipping, randomly changing the image contrast, and a random crop of the image. Due to the effectiveness of the cropping operation, we compute results for multiple image crops and then aggregate them back into the complete relevance map (see Fig. 4).

Relevance Map Refinement. Despite averaging over multiple augmentations, the resulting maps are often not specific enough. Consider an image with a person and a doll. A query for the label "person" will result in high values in the generated maps for both the person and the doll pixels, and vice versa. To reduce this inaccuracy, we calibrate the maps by computing relevance maps for several "distractor" prompts. We compute the average relevance map for the distractor and query classes and center each per-category map with respect to the mean relevance map. This significantly improves the specificity of the maps, as demonstrated in Fig. 4.

3.2 From View-Averaged Relevance Maps to Segmentation

The obtained view-averaged per-category relevance maps have a much higher quality than the naive outputs of the interpretability technique. However, they are still not precise enough, e.g., segmentations typically spill over beyond the object boundaries. We would like to transform the multiple maps into a single segmentation, where each pixel is classified with a semantic label, i.e., an integer number $p \in \{0, .., k\}$.

We propose to combine our maps with existing single-image segmentation methods. We investigated both combining them with unsupervised methods (thereby increasing their performance) as well as adapting supervised methods by replacing their pixel-supervision with our maps as 'pseudo-supervision":

Unsupervised Clustering [26]**.** As a baseline, we used the method proposed by [26], which reframes image segmentation as a pixel clustering problem. An image is optimized with a small fully convolutional network to optimize both the feature representation and the pixel labeling together. Using this technique, pixels of similar features will are assigned with the same label, while still being segmented continuously due to the continuity prior. We inject supervision into this technique, utilizing *Stochastic Pixel Sampling* (see below) and adding these "pseudo labels" as an additional classification loss. The network operates directly on the deep features of a pretrained network. We stop the clustering process after a given number of iterations or when the minimum number of classes is reached (usually the number of prompt categories + 1). More details can be found in the SM.

Interactive Segmentation. This segmentation paradigm guides the results by interaction, where users provide boundary seeds, regions of interest (ROI), bounding boxes, regions-seeds, i.e., foreground points and background points. We employed interactive segmentation [45] that takes the region-seeds as an input. The foreground and background points were generated on the fly using our *Stochastic Pixel Sampling.*

Stochastic Pixel Sampling. The simplest option of transforming mean per-category relevance maps into coarse segmentation is by taking the maximum class of every pixel, thresholded with the background probability. We found that better downstream segmentation can be obtained by stochastic pixel sampling. We propose an auxiliary operation to stochastically sample pixels of the prompt categories or of the background. We sample pixels using the following probability (where τ is the temperature):

$$\mathbb{P}(p) \sim Softmax\left(\frac{RM(\mathcal{C})}{\tau}\right)(p) \quad (3)$$

4 Experiments

We conducted a careful evaluation of our method against representative baselines on ImageNet-Segmentation and PASCAL VOC 2012 datasets. Both the qualitative and the quantitative experiments are presented in this paper. Additionally, we present a detailed ablation study of the design choices we made for our method.

4.1 Datasets and Evaluation Metrics

Our evaluation was conducted on two benchmark datasets: PASCAL VOC 2012 [14] validation set and ImageNet-Segmentation [18]. Further qualitative results can be found in the SM.

PASCAL-VOC. A natural scene dataset where the validation set contains 1449 images for 20 class categories. We choose our hyperparameters using the train-set. The method is then evaluated on the validation set.

ImageNet-Segmentation. This dataset consists of a subset of the ImageNet dataset. The dataset contains 4276 images from 445 categories.

The effectiveness of our results is demonstrated in both datasets. We also present the open-vocabulary results of our method on images containing objects from rare class categories. Examples can be found in Fig. 1 and Fig. 6 and in the SM.

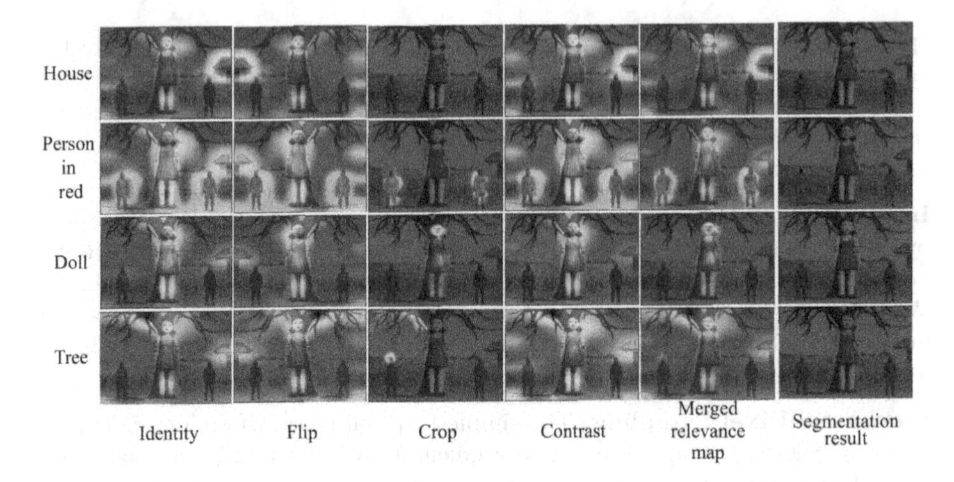

Fig. 5. Obtaining clean relevance maps via Test Time Augmentation. Over a representative set of augmentations, a relevance map is created for each prompt category. By averaging all the relevance maps for each augmentation view, a more subtle map can be obtained for each category. While crop augmentation generally provides a finer map, other augmentations that work over the whole image usually produce a coarser map.

Evaluation Metric: We evaluated our performance using mean intersection over union (mean IOU). Following the [26], mean IOU was calculated as the mean IOU of each segment in the ground truth (GT) and the estimated segment that had the largest IOU with the GT segment. Specifically, each class category was treated as an individual segment.

4.2 Baselines

Transformer Interpretability Based Segmentation (TIBS) [9]. Our method uses this baseline as the atomic building block for generating relevance maps. It proposes multiple improvements over this baseline, the benefit of which is evaluated. Instead of using our method, a segmentation map is generated for this baseline by thresholding its relevance map with its mean.

Pixel-Level K-Means Clustering [33]. K-means clustering on RGB values of 5×5 patches. We use the numbers reported in [26].

Graph-Based Segmentation (GS) [15]. A long-standing graph-based method. We use the reported numbers from [26].

Invariant Information Clustering (IIC) [23]. An unsupervised deep classification and segmentation method that maximizes the mutual information between different views of an image. Image segmentation is accomplished by using the IIC objective on image patches together with local spatial invariance with regards to the pixel coordinates. We used the reported numbers from [26].

Fig. 6. Representative examples demonstrating our segmentation performance on well-known landmarks. The first and third rows show the input images, while the second and fourth rows show their segmentation masks respectively.

Unsupervised Segmentation (US) [26]**.** This approach was described in Sect. 3.1. This version of the method uses spatial continuity loss together with a self-distillation term for feature similarity to train a network per image. It optimizes dense-per-pixel labels along with network parameters, and works within a few minutes for a reasonable size image. The baseline does not use our relevance maps as scribble pseudo-supervision.

4.3 Implementation Details

We provide the main implementation details in this section. More details can be found in the SM.

Relevance Map Generation. We used the ViT-32/B configuration of CLIP for multi-modal embedding. The relevance maps were generated using the transformer interpretability method of Chefer et al. [9]. For test-time augmentations, we used identity, flip, contrast change, and crops. The crop parameters had a crop size of $(224, 224)$, sampled on a regular grid with stride 50. The final map was obtained by averaging all crops, using a custom probability threshold (see the SM for details).

Segmentation Generation. We adapted two methods to use our relevance maps as pseudo-supervision: Unsupervised Clustering (UC), based on [26] and Interactive Segmentation (IS) [45]. In the configuration used, the loss weighting is as follows: continuity loss of 5, feature similarity of 1, and optional scribble loss of 0.5. The configuration used for IS is HRNet-32. More details can be found in the SM.

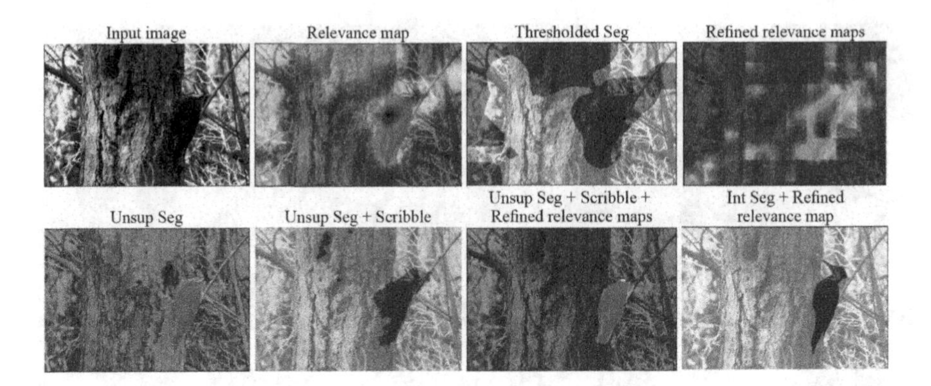

Fig. 7. Visual comparison on an image from PASCAL VOC. Left to right, top row: the input image, the generated relevance map with respect to the text query "a photo of a bird", the segmentation output obtained by thresholding the relevance map, and a refined map generated using test-time augmentation of all mentioned views. Bottom row: We can see that although US [26] does not perform well on its own, combining it with synthetic scribble supervision provided by our relevance maps achieved much better results. On the right, we show that using the refined relevance maps to supervise the scribble achieves even better results. Finally, we present the results of IS [45] with our refined relevance maps as pseudo-supervision, which achieves the best results.

5 Results

5.1 Comparison to Other Methods

PASCAL VOC 2012. A quantitative evaluation is presented in Table 1, with a comparison to other methods that use a similar amount of supervision (Table 1a), the ablation study (Table 1b), and a comparison to other image-label weakly supervised methods (Table 1c). K-means performs poorly as it uses naive pixel color features. Graph-based segmentation can produce too coarse or too fine images, due to the requirement for tuning the granularity τ parameter.

Like K-means, single-image IIC did not achieve strong results, and was slow. US did not achieve strong results on its own but did achieve a 12% IOU gain with our relevance maps and stochastic sampling as synthetic supervision. TIBS is essentially a thresholded interpretability map, which despite its knowledge distillation from CLIP, generated relatively noisy segmentations. Finally, we can see that the Interactive segmentation model, supervised by our method achieved the best segmentation results.

ImageNet-Segmentation. The results are presented on Table 2. The trends are similar to those of PASCAL VOC; our method is an improvement over TIBS. The combination of our method and IS performs much better than the combination of our method and US.

Table 1. Quantitative results on PASCAL VOC 2012 validation set. k denotes the number of categories. HP: hyperparameters. mIOU: Mean IOU.

Method	HP	mIOU
K-means	k = 2	31.6
K-means	k = 17	23.8
Graph-based[15]	$\tau = 100$	26.8
Graph-based [15]	$\tau = 500$	36.4
IIC [23]	k = 2	27.2
IIC[23]	k = 20	20.0
Unsup Seg [26]	super-pixels	30.8
Unsup Seg [26]	Continuity loss $\mu = 5$	35.2
Unsup Seg [26]	Stochastic sampling	47.0
TIBS [9]	Mean thresholding	38.8
TIBS [9]	Clustering [26]	39.7
Ours + US [26]	Identity, crop	**50.2**
Ours + IS [45]	k = 1, 3 clicks, crop	**62.3**

(a)

Method	HP	mIOU
US [26]	Identity	48.6
US [26]	Crop	49.3
US [26]	Identity, crop	50.2
IS[45]	k = 1, 3 clicks, crop	62.3
IS[45]	GT k, 3 clicks, crop	**63.9**

(b)

Method	mIOU	Training time
CRF-RNN [41]	52.8	Days
AffinityNet [3]	61.7	Days
Single Stage [6]	62.7	Days
IRN [3]	63.5	Days
Ours + IS	63.9	**Minutes**
SSDD [44]	**64.9**	Days

(c)

5.2 Analysis

Qualitative Results. We present a qualitative comparison on a single image from PASCAL VOC in Fig. 7. We see that the refined relevance map is a significant improvement over the naive relevance map. We also see that although US on its own does not generate strong results, when combined with our method, it can achieve accurate segmentations. Furthermore, we see that the combination with our refined map as pseudo-scribble supervision improves the results. Finally, we see that the combination of IS and our method obtains the most accurate segmentations.

Test-time augmentations. We experimented with 4 image transformations, where each one generated an independent relevance map. Overall, crops generate the most detailed relevance map. The most common approach in TTA is to aggregate predictions using averaging to obtain a more accurate and detailed relevance map, as can be seen in Table 1 and Fig. 5, 7. We conclude that the view with the highest contribution is the crop view, but using all TTAs improves segmentation performance.

Binarization Threshold and Stochastic Sampling. As the relevance maps already provide some segmentation signal, we provide an evaluation of direct binarization, and then pass the results as pseudo-labels to Unsupervised Clustering. We see in Sect. 5.2 that the threshold needs to be very high due to the inaccuracy of the relevance map. Instead, we suggest using stochastic sampling, which replaces the hard threshold with a softer probabilistic sampling. We stochastically sample labels from the relevance maps in every iteration for unsupervised clustering, or by generating clicks for interactive segmentation. We found that stochastic sampling significantly improves the final quality of the segmentation (Fig. 8).

Unsupervised Clustering Design Choices. We experimented with different relative weightings between the losses but found that the default parameters suggested by the authors showed the best results.

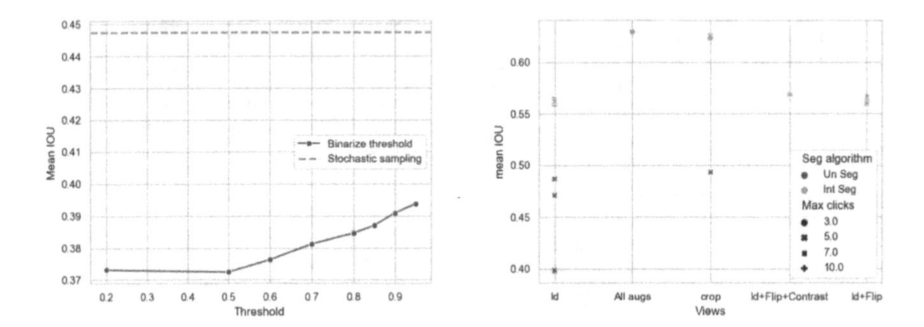

Fig. 8. Binarization threshold analysis (left). The effect of the binarization threshold of the relevance map for a single object on segmentation of the PASCAL VOC 2012 training set. The thresholded map was used without stochastic sampling as scribble supervision to the Unsupervised Clustering method. **Segmentation analysis (right).** The results from the PASCAL VOC validation set for $k = 1$. Over all sets of augmentation views, interactive segmentation is superior to unsupervised segmentation. Additionally, we see that the crop view is better than using identity, flip, and contrast views together, even though using all the views together is better than using the crop view alone.

Interactive Segmentation Design Choices. The interactive segmentation method requires iterations of user clicks. Instead of using user feedback, we generated "pseudo clicks" on the fly using stochastic pixel sampling. We experimented here as well with the number of clicks generated, and found that the best results were obtained with 3 clicks.

6 Discussion

Other Segmentation Methods. We presented a method for obtaining good initial segmentation of an image in a zero-shot manner. It is then used to initialize a downstream segmentation algorithm. We showed how to combine our method with segmentation methods designed for human annotated inputs, replacing manual annotations with our automatic method. Additionally, we showed that our method can be incorporated into an unsupervised segmentation method as extra "pseudo supervision", significantly improving performance. These two methods are just an example of the potential of the method. Our method is general in that it can be incorporated with any single-image segmentation method, supervised or unsupervised. Investigating combining it with other segmentation methods e.g. GrabCut is left for future work.

Augmentations. Within the range of test-time augmentations that we investigated, we found that crop augmentations were the most significant. We believe

Table 2. Quantitative comparison on ImageNet-Segmentation. We can see that all variants of our method outperform TIBS. We can also see that combining our method as pseudo-supervision for Interactive Segmentation is superior to combining it with Unsupervised Segmentation.

Method	HP	mIOU
TIBS [9]	Mean thresholding	51.2
TIBS [9]	Clustering	48.8
Ours + US [26]	Identity	60.6
Ours + IS [45]	k = 1, 3 clicks, Identity	68.9
Ours + IS [45]	k unk., 3 clicks, Identity	69.0
Ours + IS [45]	k = 1, 3 clicks, Crop	**70.3**

that this augmentation can be further adapted to our task by taking crops with a small step size or with different scales in order to create relevance maps with better efficiency for objects with different scales. We did not do this for this paper due to run-time considerations.

7 Conclusions

We have presented a new semantic segmentation method that can segment any object for which a textual description can be provided without requiring pixel supervision or a training dataset with multiple images. We have demonstrated that vision-language models can be used to transform image-level guidance into high accuracy relevance maps. These relevance maps can provide pseudo-supervision for existing single-image segmentation methods that require human annotated pixel-level supervision. We justified our design choices, and demonstrated that our qualitative and quantitative results surpass other methods requiring similar amounts of supervision, while showing promising segmentations on rare-class categories.

References

1. Ahn, J., Cho, S., Kwak, S.: Weakly supervised learning of instance segmentation with inter-pixel relations. In: The IEEE Conference on Computer Vision and Pattern Recognition (CVPR) (2019)
2. Ahn, J., Kwak, S.: Learning pixel-level semantic affinity with image-level supervision for weakly supervised semantic segmentation. In: The IEEE Conference on Computer Vision and Pattern Recognition (CVPR) (2018)
3. Ahn, J., Kwak, S.: Learning pixel-level semantic affinity with image-level supervision for weakly supervised semantic segmentation. In: 2018 IEEE/CVF Conference on Computer Vision and Pattern Recognition, pp. 4981–4990 (2018)
4. Moshkov, N., et al.: Test-time augmentation for deep learning-based cell segmentation on microscopy images. Sci. Rep. **10**, 1–7 (2020)

5. Shanmugam, D., et al.: Better aggregation in test-time augmentation. In: ICCV, pp. 1214–1223 (2021)
6. Araslanov, N., Roth, S.: Single-stage semantic segmentation from image labels. In: 2020 IEEE/CVF Conference on Computer Vision and Pattern Recognition (CVPR), pp. 4252–4261 (2020)
7. Bansal, A., Sikka, K., Sharma, G., Chellappa, R., Divakaran, A.: Zero-shot object detection. In: ECCV (2018)
8. Bucher, M., Vu, T.H., Cord, M., Pérez, P.: Zero-shot semantic segmentation. In: NeurIPS (2019)
9. Chefer, H., Gur, S., Wolf, L.: Transformer interpretability beyond attention visualization. In: Proceedings of the IEEE/CVF Conference on Computer Vision and Pattern Recognition (CVPR), pp. 782–791 (2021)
10. Chen, L.C., Papandreou, G., Kokkinos, I., Murphy, K.P., Yuille, A.L.: DeepLab: Semantic image segmentation with deep convolutional nets, atrous convolution, and fully connected CRFs. IEEE Trans. Pattern Anal. Mach. Intell. **40**, 834–848 (2018)
11. Dai, J., He, K., Sun, J.: BoxSup: exploiting bounding boxes to supervise convolutional networks for semantic segmentation. In: 2015 IEEE International Conference on Computer Vision (ICCV), pp. 1635–1643 (2015)
12. Desai, K., Johnson, J.: VirTex: learning visual representations from textual annotations. In: 2021 IEEE/CVF Conference on Computer Vision and Pattern Recognition (CVPR), pp. 11157–11168 (2021)
13. Dong, N., Xing, E.P.: Few-shot semantic segmentation with prototype learning. In: BMVC (2018)
14. Everingham, M., Gool, L.V., Williams, C.K.I., Winn, J.M., Zisserman, A.: The PASCAL visual object classes (VOC) challenge. Int. J. Comput. Vision **88**, 303–338 (2009)
15. Felzenszwalb, P.F., Huttenlocher, D.P.: Efficient graph-based image segmentation. Int. J. Comput. Vision **59**, 167–181 (2004)
16. Gal, R., Patashnik, O., Maron, H., Chechik, G., Cohen-Or, D.: StyleGAN-NADA: CLIP-guided domain adaptation of image generators. ArXiv abs/2108.00946 (2021)
17. Galatolo, F.A., Cimino, M.G.C.A., Vaglini, G.: Generating images from caption and vice versa via clip-guided generative latent space search. In: IMPROVE (2021)
18. Guillaumin, M., Küttel, D., Ferrari, V.: ImageNet auto-annotation with segmentation propagation. Int. J. Comput. Vision **110**, 328–348 (2014)
19. Gupta, A., Dollár, P., Girshick, R.B.: LVIS: a dataset for large vocabulary instance segmentation. In: 2019 IEEE/CVF Conference on Computer Vision and Pattern Recognition (CVPR), pp. 5351–5359 (2019)
20. Hong, S., Yeo, D., Kwak, S., Lee, H., Han, B.: Weakly supervised semantic segmentation using web-crawled videos. In: 2017 IEEE Conference on Computer Vision and Pattern Recognition (CVPR), pp. 2224–2232 (2017)
21. Huang, Z., et al.: CCNet: criss-cross attention for semantic segmentation. In: 2019 IEEE/CVF International Conference on Computer Vision (ICCV), pp. 603–612 (2019)
22. Huang, Z., Wang, X., Wang, J., Liu, W., Wang, J.: Weakly-supervised semantic segmentation network with deep seeded region growing. In: 2018 IEEE/CVF Conference on Computer Vision and Pattern Recognition, pp. 7014–7023 (2018)
23. Ji, X., Vedaldi, A., Henriques, J.F.: Invariant information clustering for unsupervised image classification and segmentation. In: 2019 IEEE/CVF International Conference on Computer Vision (ICCV), pp. 9864–9873 (2019)

24. Jia, C., et al.: Scaling up visual and vision-language representation learning with noisy text supervision. In: ICML (2021)
25. Kervadec, H., Dolz, J., Wang, S., Granger, É., Ayed, I.B.: Bounding boxes for weakly supervised segmentation: global constraints get close to full supervision. In: MIDL (2020)
26. Kim, W., Kanezaki, A., Tanaka, M.: Unsupervised learning of image segmentation based on differentiable feature clustering. IEEE Trans. Image Process. **29**, 8055–8068 (2020)
27. Kolesnikov, A., Lampert, C.H.: Seed, expand and constrain: three principles for weakly-supervised image segmentation. ArXiv abs/1603.06098 (2016)
28. Lee, S., Lee, M., Lee, J., Shim, H.: Railroad is not a train: saliency as pseudo-pixel supervision for weakly supervised semantic segmentation. In: 2021 IEEE/CVF Conference on Computer Vision and Pattern Recognition (CVPR), pp. 5491–5501 (2021)
29. Li, G., Duan, N., Fang, Y., Jiang, D., Zhou, M.: Unicoder-VL: a universal encoder for vision and language by cross-modal pre-training. In: AAAI (2020)
30. Li, P., Wei, Y., Yang, Y.: Consistent structural relation learning for zero-shot segmentation. In: NeurIPS (2020)
31. Lin, D., Dai, J., Jia, J., He, K., Sun, J.: ScribbleSup: scribble-supervised convolutional networks for semantic segmentation. In: 2016 IEEE Conference on Computer Vision and Pattern Recognition (CVPR), pp. 3159–3167 (2016)
32. Luo, H., et al.: CLIP4Clip: an empirical study of clip for end to end video clip retrieval. ArXiv abs/2104.08860 (2021)
33. MacQueen, J.: Some methods for classification and analysis of multivariate observations (1967)
34. Narayan, S., Gupta, A., Khan, F.S., Snoek, C.G.M., Shao, L.: Latent embedding feedback and discriminative features for zero-shot classification. ArXiv abs/2003.07833 (2020)
35. Patashnik, O., Wu, Z., Shechtman, E., Cohen-Or, D., Lischinski, D.: StyleCLIP: text-driven manipulation of StyleGAN imagery. ArXiv abs/2103.17249 (2021)
36. Pathak, D., Krähenbühl, P., Darrell, T.: Constrained convolutional neural networks for weakly supervised segmentation. In: 2015 IEEE International Conference on Computer Vision (ICCV), pp. 1796–1804 (2015)
37. Radford, A., et al.: Learning transferable visual models from natural language supervision. In: ICML (2021)
38. Rahman, S., Khan, S.H., Barnes, N.: Improved visual-semantic alignment for zero-shot object detection. In: AAAI (2020)
39. Rahman, S., Khan, S.H., Porikli, F.M.: Zero-shot object detection: learning to simultaneously recognize and localize novel concepts. ArXiv abs/1803.06049 (2018)
40. Rakelly, K., Shelhamer, E., Darrell, T., Efros, A.A., Levine, S.: Conditional networks for few-shot semantic segmentation. In: ICLR (2018)
41. Roy, A., Todorovic, S.: Combining bottom-up, top-down, and smoothness cues for weakly supervised image segmentation. In: 2017 IEEE Conference on Computer Vision and Pattern Recognition (CVPR), pp. 7282–7291 (2017)
42. Sariyildiz, M.B., Perez, J., Larlus, D.: Learning visual representations with caption annotations. In: Vedaldi, A., Bischof, H., Brox, T., Frahm, J.-M. (eds.) ECCV 2020. LNCS, vol. 12353, pp. 153–170. Springer, Cham (2020). https://doi.org/10.1007/978-3-030-58598-3_10
43. Selvaraju, R.R., Das, A., Vedantam, R., Cogswell, M., Parikh, D., Batra, D.: Grad-CAM: visual explanations from deep networks via gradient-based localization. Int. J. Comput. Vision **128**, 336–359 (2019)

44. Shimoda, W., Yanai, K.: Self-supervised difference detection for weakly-supervised semantic segmentation. In: 2019 IEEE/CVF International Conference on Computer Vision (ICCV), pp. 5207–5216 (2019)
45. Sofiiuk, K., Petrov, I.A., Konushin, A.: Reviving iterative training with mask guidance for interactive segmentation. ArXiv abs/2102.06583 (2021)
46. Tan, H.H., Bansal, M.: LXMERT: learning cross-modality encoder representations from transformers. In: EMNLP (2019)
47. Tian, G., Wang, S., Feng, J., Zhou, L., Mu, Y.: Cap2Seg: inferring semantic and spatial context from captions for zero-shot image segmentation. In: Proceedings of the 28th ACM International Conference on Multimedia (2020)
48. Wang, J., et al.: Deep high-resolution representation learning for visual recognition. IEEE Trans. Pattern Anal. Mach. Intell. **43**, 3349–3364 (2021)
49. Wang, T., et al.: The devil is in classification: a simple framework for long-tail instance segmentation. In: Vedaldi, A., Bischof, H., Brox, T., Frahm, J.-M. (eds.) ECCV 2020. LNCS, vol. 12359, pp. 728–744. Springer, Cham (2020). https://doi.org/10.1007/978-3-030-58568-6_43
50. Yuan, Y., Chen, X., Wang, J.: Object-contextual representations for semantic segmentation. In: Vedaldi, A., Bischof, H., Brox, T., Frahm, J.-M. (eds.) ECCV 2020. LNCS, vol. 12351, pp. 173–190. Springer, Cham (2020). https://doi.org/10.1007/978-3-030-58539-6_11
51. Zhang, J., Yu, X., Li, A., Song, P., Liu, B., Dai, Y.: Weakly-supervised salient object detection via scribble annotations. In: 2020 IEEE/CVF Conference on Computer Vision and Pattern Recognition (CVPR), pp. 12543–12552 (2020)
52. Zhang, X., Wei, Y., Yang, Y., Huang, T.: SG-one: similarity guidance network for one-shot semantic segmentation. IEEE Trans. Cybern. **50**, 3855–3865 (2020)
53. Zhang, X., Xu, H., Mo, H., Tan, J., Yang, C., Ren, W.: DCNAS: densely connected neural architecture search for semantic image segmentation. In: 2021 IEEE/CVF Conference on Computer Vision and Pattern Recognition (CVPR), pp. 13951–13962 (2021)

SW-VAE: Weakly Supervised Learn Disentangled Representation via Latent Factor Swapping

Jiageng Zhu[1,2,3](✉), Hanchen Xie[2,3], and Wael Abd-Almageed[1,2,3]

[1] USC Ming Hsieh Department of Electrical and Computer Engineering, Los Angeles, USA
[2] USC Information Sciences Institute, Marina del Rey, USA
{jiagengz,hanchenx,wamageed}@isi.edu
[3] Visual Intelligence and Multimedia Analytics Laboratory, Los Angeles, USA

Abstract. Representation disentanglement is an important goal of the representation learning that benefits various of downstream tasks. To achieve this goal, many unsupervised learning representation disentanglement approaches have been developed. However, the training process without utilizing any supervision signal have been proved to be inadequate for disentanglement representation learning. Therefore, we propose a novel weakly-supervised training approach, named as **SW-VAE**, which incorporates pairs of input observations as supervision signal by using the generative factors of datasets. Furthermore, we introduce strategies to gradually increase the learning difficulty during training to smooth the training process. As shown on several datasets, our model shows significant improvement over state-of-the-art (SOTA) methods on representation disentanglement tasks.

1 Introduction

Deep neural network (DNN) has achieved great success in many computer vision tasks, such as image classification [8], face recognition [33] and image generation [13]. Learning a latent representation $\boldsymbol{z}$ from input data $\boldsymbol{x}$ is a critical first step in training DNNs, in which the objective is to learn lower dimensional representations that facilitate downstream tasks, including classification [14], few-shot learning [34] or semantic segmentation [26].

Two of the fundamental challenges for robust latent representations are *over-fitting* and *interpretability*. Since DNNs are massively paramterized models, they require large amounts of training data that sufficiently span all *factors of variations*, such as pose, expression and gender in face recognition models [22]. In the absence of large scale dataset that spans all factors of variations, DNNs tend to overfit to the underlying factors of variations. Moreover, since the training objective is minimizing the empirical risk, standard methods for training

Supplementary Information The online version contains supplementary material available at https://doi.org/10.1007/978-3-031-25063-7_5.

L. Karlinsky et al. (Eds.): ECCV 2022 Workshops, LNCS 13802, pp. 73–87, 2023.
https://doi.org/10.1007/978-3-031-25063-7_5

DNNs (*e.g.*,[18,30]) produce latent representations that lack semantic meaning and hard to interpret without further processing.

Bengio *et al.* [1] define a disentangled representation $\boldsymbol{z}$ such that a change in a given representation dimension z_i corresponds to a change in *one and only one* underlying factor of variation i of the data (*i.e.*, invariant to all other factors). The definition of [1] addresses the semantic interpretability of the representation in standard training methods, and is used as the basis for methods such as Factor Variational Autoencoder (FactorVAE) [17] and β-VAE [15]. Recent work revealed the benefits of disentangled representations of factors of variations in various downstream tasks such as, visual reasoning [31], interpretability [1,15], filtering out nuisance [20], answering counterfactual questions [29], and fairness [7].

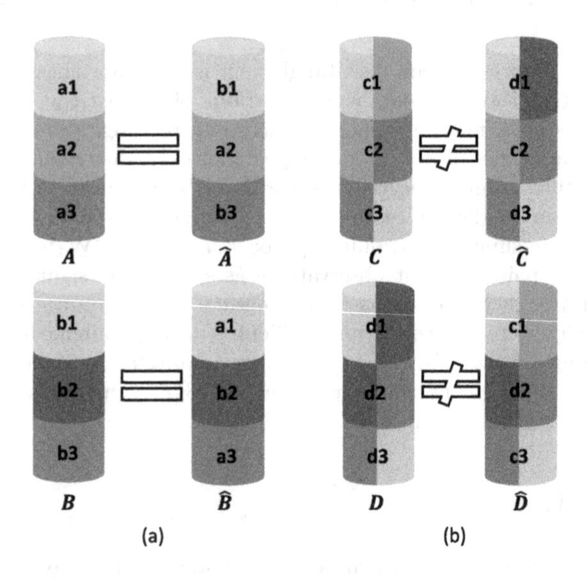

Fig. 1. The basic concept of swapping latent factors of variation, represented by different colors. Left (a): If two representations are completely disentangled (*e.g.*, A and B), then swapping equivalent latent factors (a1 with b1, and a3 with b3) will lead to similar representations. Right (b): If C and D are entangled, then swapping latent factors (c1 with d1 and c3 with d3) may lead to dissimilar representations.

Unsupervised disentanglement methods [5,15–17] relying on variational autoencoder (VAE) [18] which assume that the latent representation follows a normal distribution, where the encoder is used to estimate the posterior $p_\theta(\mathbf{z}|\mathbf{x})$ and the decoder is used to estimate $p_\phi(\mathbf{x}|\mathbf{z})$. The loss function is constructed for comparing the difference between prior and posterior. However, Locatello *et al.* [23] proved that there are an infinite number of entangled models whose latent representation z has the same marginal distribution with the ideal disentangled model, and since unsupervised learning methods only use information of the observations $\mathbf{x}$, they can not discriminate between the disentangled model and other entangled models. Locatello *et al.* [23] further empirically analyzed state-of-the-art (SOTA) unsupervised models on different datasets, which demonstrates the necessity of supervision signal. To address the challenge, many semi-supervised and weakly supervised learning methods have been proposed [4,24,25]. One of these is methods is Ada-VAE [24] which requires the knowledge of exact number of different generative factors between a pair of images in order to achieve disentangled representation with guarantees. Rather than relying on such precise prior knowledge, our method only requires the information of the maximum number of different generative factors, and empirical results demonstrate that the disentangled representation can still be achieved with weaker supervision signal.

In this paper, we introduces a new method for learning latent representations with disentangled factors of variations via introducing weak supervision signals during training to encourage the model to learn disentangled representations. Inspired by the self-supervised learning methods which utilize the supervision provided by pairs of inputs [6,11], we propose using pairs of inputs to introduce supervision signals. After encoding input pairs into latent representations, similar to the swapping method used in image manipulation [28], we encourage the disentanglement by swapping the latent factors and comparing their corresponding reconstructions with the original inputs. As the simple example illustrated in Fig. 1, when the representation is fully disentangled, swapping the same elements does not change their distribution, whereas the distribution of entangled representation changes after swapping. In disentanglement learning, DSD [10], which also adopts swapping concept, is trained under two steps. Firstly, a pair of labeled inputs are encoded, swapped, and decoded. In the second step, a random k-th part of latent representations encoded from unlabelled inputs is swapped and decoded. Therefore, DSD is trained under semi-supervised condition, and as discussed in the DSD [10], more than 20% of labeled data are needed to train the model. However, in many cases, the number of labeled data is limited due to the annotation cost is expensive. Compared to DSD, our method does not require actual label information while merely needing access to the total number of generative factors, and we empirically prove that our method achieves comparable performance. On the occasion of the exact number of different generative factors in the pairs are available, where the supervision strength is still weaker than DSD, our method can outperform DSD. Furthermore, we propose training strategies which progressively adjust the difficulty of task during learning process.

Our contributions in this paper are: (1) **SW-VAE**, a new weakly supervised representation disentanglement framework and (2) extensive quantitative and qualitative experimental evaluation demonstrating that **SW-VAE** outperforms SOTA representation disentanglement methods on various datasets including dSprites [27], 3dshapes [2], MPI3D-toy [12], MPI3D-realistic [12], and MPI3D-real [12].

2 Related Work

Learning Disentanglement Representation: Variational autoencoders (VAE) [18] is the basic framework of most SOTA disentanglement methods. VAE uses a DNN to map the inputs into latent representation modeled as a distribution denoted by $q_\phi(z|x)$. Recent methods modifies VAE by adding implicit or explicit regularization to disentanglement representation. **β-VAE** [15] adds and tunes a hyper-parameter β before KL divergence (D_{KL}) in order to achieve balanced performance on both reconstruction and latent representation disentanglement. In β-VAE, when $\beta > 1$, the distribution $q_\phi(z|x)$ is used to calculate disentanglement regularization by comparing with the assuming prior $p(z)$. Kullback-Leibler divergence (D_{KL}) is used to calculate the distance between $q_\phi(z|x)$ and $p(z)$ for disentanglement regularization. The disentanglement regularization is then combined with the quality of image reconstruction to be the total objective function of the model training.

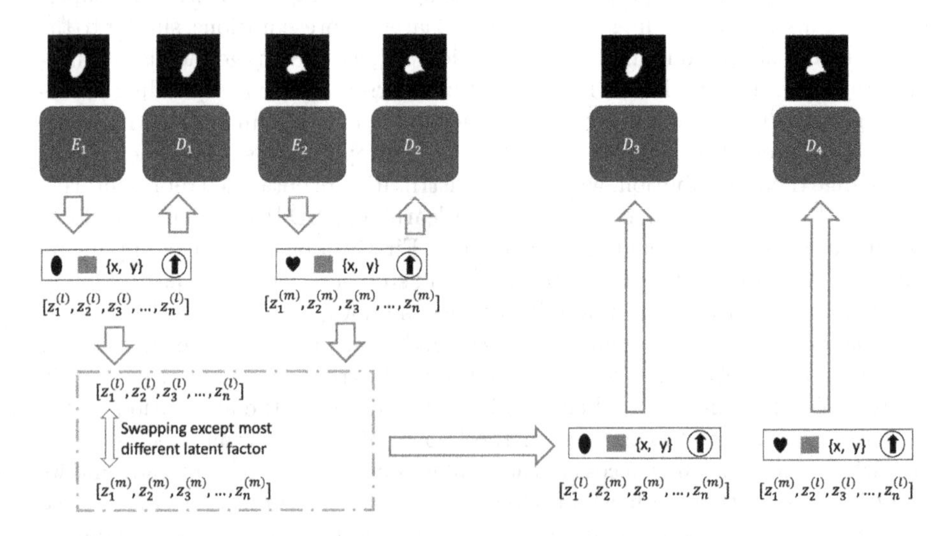

Fig. 2. Framework of **SW-VAE**, E_1 and E_2 share same parameters; D_1,D_2,D_3,D_4 share same parameters

Burgess *et al.* [3] proposed **AnnealedVAE**, a modification of **β-VAE**. Intuitively, **AnnealedVAE** can be viewed as gradually increasing the capacity of latent encoding. When the encoding capacity is low, the model is forced only to encode input data which brings the most significant improvement in reconstructions. As capacity increases during training, the model learns to encode other semantic factors into the latent representation progressively while continues to disentangle the previous learned factors. **FactorVAE** [17] proposes a better trade-off between reconstruction quality and disentanglement by incorporating the discriminator to calculate the Total Correlation (TC) between $q(z)$

and $\prod q(z_i)$. The discriminator in **FactorVAE** has the same function as the discriminator in **GAN** [13]. **DIP-VAE** [20] adds $D(q_\phi(z)||p(z))$ as additional regularization to encourage disentanglement, where $q_\phi(z)$ is the marginal distribution of the latent representation z learned by model and $q_\phi(z) = \int q_\phi(z|x)p(x)dx$. D stands for any suitable distance function. **β-TCVAE** [5] decomposes the D_{KL} regularization used in **β-VAE** into three parts: index-code mutual information, total correlation and dimension-wise KL divergence. Index-code mutual information controls the mutual information between input data and factors in latent representation. Total correlation encourages the model to find the statistically independent factors in latent space and dimension-wise individual latent components are kept from overly diverging from their priors via KL divergence.

Weakly Supervised Learning: Zhou *et al.* [35] conclude that there are three forms of weak supervision in general. The first type is incomplete supervision, *i.e.*, only part of training samples have labels. This condition can be addressed by semi-supervised learning or active learning. The second type is inexact supervision where the labels are less precise than labels used in supervised learning, or only coarse-grained labels are provided. The third type is inaccurate supervision. The labels of some data are wrong in this case, *e.g.*, some images or languages are labelled into the wrong class.

3 SW-VAE Model

As discussed in Sect. 1, Bengio *et al.* [1] define representation disentanglement as learning a latent representation $z \in \mathbb{R}^{d_z}$ of input observation $x \in \mathbb{R}^{d_x}$, in which each dimension z_i changes *if and only if* an underlying factor of variation i of the data changes, and therefore the joint probability can be modeled as $p(z) = \prod_{i=1} p(\mathbf{z_i})$. Meanwhile, each latent element z_i is expected to contain one and only one semantic meaning, such that traversing alone one latent element and fixing other elements will only change one factor of variation within the reconstructed images obtained from the decoder [18]. Thus, for each factor of variation, there is only one highly associated latent factor. When measuring the performance of different disentanglement models, this principle is used to evaluate both the degree of disentanglement and the level of similarity between a latent factor and one semantic generative factor.

The number of generative factors of variations is less than the dimension of latent representation ($v \leq d_z$), such that a subset of z encodes information that is irrelevant to the generative factors of variations, and does not necessarily have semantic meaning; yet still satisfying the independency assumptions. Disentanglement therefore leads to a latent factor z_i that is *invariant* to all other factors of variation of x.

In variational autoencoder (VAE) [19], a variational model $q_\phi(z|x)$ is used to produce a probability distribution q_ϕ given an input sample x. This essentially simulates sampling a latent representation z from a prior distribution $p_\theta(z)$, where θ and ϕ are the generative and variational parameter spaces respectively.

The overall loss function of the VAE is shown in Eq. (1).

$$L_{VAE}(x,z) = -\mathbb{E}_{q_\phi(z|x)}[logp_\theta(x|z)] + \beta D_{KL}(q_\phi(z|x)||p(z)) \tag{1}$$

where $\beta = 1$. **β-VAE** [15] forces the VAE to learn the disentanglement by setting $\beta > 1$. Other unsupervised learning methods [17,20] improve the performance of representation disentanglement by modifying D_{KL}, which serves as disentanglement regularization. Similarly, we add another disentanglement regularization in the loss function by introducing supervision signals using selected pairs of inputs.

Weakly Supervised Swap Variational Autoencoder ***(SW-VAE):*** As illustrated in Fig. 2, the proposed network framework, **SW-VAE**, consists of two encoders and four decoders, where the parameters within all encoders and decoders are shared respectively. During training, the network is fed a pair of samples $x^{(l)}$ and $x^{(m)}$ and generates representations $z^{(l)}$ and $z^{(m)}$ via encoders E_1 and E_2 respectively. Decoders D_1 and D_2 are used to reconstruct $x^{(l)}$ and $x^{(m)}$ as $x^{(l)}_{rec}$ and $x^{(m)}_{rec}$ respectively; essentially simulating generating input samples x from the distribution $p_\theta(x|z)$. Random factors of $z^{(l)}$ and $z^{(m)}$ are swapped to generate two new corresponding latent representations $\hat{z}^{(l)}$ and $\hat{z}^{(m)}$. Decoders D_3 and D_4 are used to decode the new latent representations $\hat{z}^{(l)}$ and $\hat{z}^{(m)}$ to $\hat{x}^{(l)}_{rec}$ and $\hat{x}^{(m)}_{rec}$.

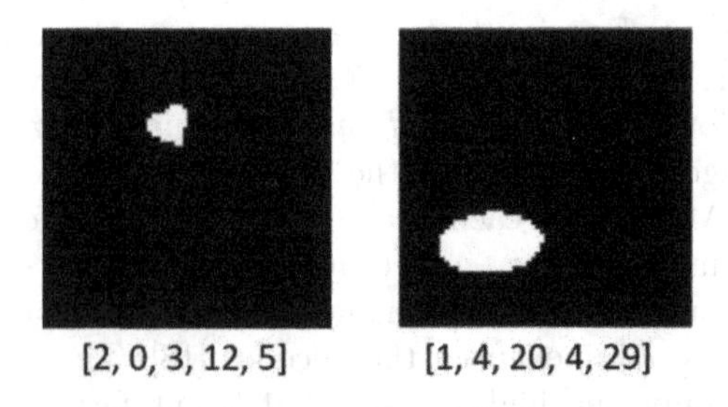

Fig. 3. Illustration of generating a pair of images of dSprites with different generative factors. The true generative factors values are unavailable during training.

Detecting Distinct Generative Factors: Some latent factors in the representations $z^{(l)}$ and $z^{(m)}$ are fed into a *Detect and Swap* module, which attempts to detect the generative factors of variations that are different between $x^{(l)}$ and $x^{(m)}$. As described in Eq. (2), if the latent representations are fully disentangled, $z^{(l)}$ and $z^{(m)}$ will be same with respect to the dimensions where the same generative factors of $x^{(l)}$ and $x^{(m)}$ are encoded in. $z^{(l)}$ and $z^{(m)}$ will be different in the dimensions where different generative factors of $x^{(l)}$ and $x^{(m)}$ are encoded in. We call the set containing all different underlying factors to be DF_z where $DF_z \subseteq \mathbb{R}^{d_z}$.

$$p(z_j^{(l)}|\mathbf{x^{(l)}}) = p(z_j^{(m)}|\mathbf{x^{(m)}});\ j \notin DF_z$$
$$p(z_i^{(l)}|\mathbf{x^{(l)}}) \neq p(z_i^{(m)}|\mathbf{x^{(m)}});\ i \in DF_z \tag{2}$$

According to [15], the posterior distribution of latent representations is assumed to be Multivariate Gaussian, $p(z|x) = q_\theta(z|x) = \mathcal{N}(\mu, \sigma^2\mathbf{I})$; and the stochastic model becomes differentiable by incorporating reparameterization trick. By using the Multivariate Gaussian assumption and reparamterization trick, the mutual information between the corresponding dimensions of two latent representation $z^{(l)}$ and $z^{(m)}$ can be directly measured by computing the KL divergence between the distributions of these two latent elements. In practice, by using the posterior distribution of latent representation, we can calculate the KL divergence as shown in Eq. (3):

$$D_{KL}(q_\phi(z_i^{(l)}|x^{(l)})||q_\phi(z_i^{(m)}|x^{(m)}))$$
$$= \frac{(\sigma_i^{(l)})^2 + (\mu_i^{(l)} - \mu_i^{(m)})^2}{2(\sigma_i^{(l)})^2} + log(\frac{\sigma_i^{(l)}}{\sigma_i^{(m)}}) - \frac{1}{2} \tag{3}$$

Instead of using full labels for training, we generate image pairs with at most k distinct factors of variations, and use k as the weak supervision signals for **SW-VAE**. Since there are at most k distinct generative factors when generating the pair of inputs $x^{(l)}$ and $x^{(m)}$, we expect that there are also at most k latent factors with distinct different values. We assume the top k most distinct underlying factors of variation are the latent factors that produce the highest k KL divergence values for $i = [1, 2, \ldots, d_z]$. After detecting the indices of distinct factors in the latent space, we then swap the factors between two latent representations, except the top k most distinct ones. We create two new *swapped* latent representations $\hat{z}^{(l)}$ and $\hat{z}^{(m)}$ using Eq. (4):

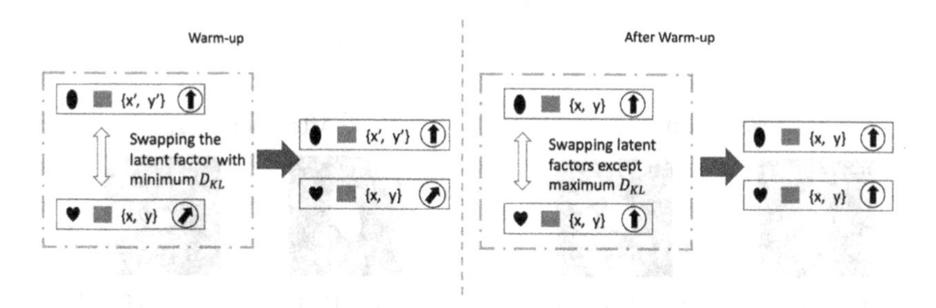

Fig. 4. Increasing number of latent factors to swap as training progresses. Only one latent factor is swapped during warm up. The number of latent factors to be swapped increases gradually

$$\hat{z}_i^{(l)} = z_i^{(m)};\quad \hat{z}_i^{(m)} = z_i^{(l)};\ \forall i \notin DF_z$$
$$\hat{z}_j^{(l)} = z_j^{(l)};\quad \hat{z}_j^{(m)} = z_j^{(m)};\ \forall j \in DF_z \tag{4}$$

The two swapped representations are then fed into decoders D_3 and D_4 to produce two reconstructions of input observations $x^{(l)}$ and $x^{(m)}$ as $\hat{x}^{(l)}_{rec}$ and $\hat{x}^{(m)}_{rec}$ respectively. Intuitively, if the network successfully detects all distinct underlying factors of variations, the swapped representations $\hat{z}^{(l)}$ and $\hat{z}^{(m)}$ will be identical to the original representations $z^{(l)}$ and $z^{(m)}$, and therefore the new reconstructed images will be same with the original reconstructions and similar to input observations. This process simultaneously enforces disentangling latent factors and encoding semantic meaning within into these latent factors. Violating either of these two requirements will lead to differences between the new reconstructions and the original reconstructions. The overall generic loss of the entire network can then be formulated as shown in Eq. (5).

$$\begin{aligned} L =& L_{VAE}(x^{(l)}_{rec}, z^{(l)}) + L_{VAE}(x^{(m)}_{rec}, z^{(m)}) \\ &+ L_g(\hat{x}^{(l)}_{rec}, x^{(l)}_{rec}) + L_g(\hat{x}^{(m)}_{rec}, x^{(m)}_{rec}) \end{aligned} \tag{5}$$

where L_g is the distance function to calculate the difference between x_{rec} and $\hat{x}_{rec}$. For a detailed study of the behavior of various distance functions, we follow both VAE [18] and VAE-GAN [21], and introduce two different instantiations of **SW-VAE** as follows.

***SW-VAE**$_{SIM}$*: In **SW-VAE**$_{SIM}$ we directly compare the similarity of reconstructions x_{rec} and $\hat{x}_{rec}$ by calculating the mean square error (**MSE**) loss or binary cross-entropy (**BCE**) loss, as shown in Eqs. (6) and (7) respectively.

$$\begin{aligned} L_{compare\_MSE} =& L_{VAE}(x^{(l)}_{rec}, z^{(l)}) + L_{VAE}(x^{(m)}_{rec}, z^{(m)}) \\ &+ \gamma||\hat{x}^{(l)}_{rec} - x^{(l)}_{rec}||^2_2 + \gamma||\hat{x}^{(m)}_{rec} - x^{(m)}_{rec}||^2_2 \end{aligned} \tag{6}$$

$$\begin{aligned} L_{compare\_BCE} =& L_{VAE}(x^{(l)}_{rec}, z^{(l)}) + L_{VAE}(x^{(m)}_{rec}, z^{(m)}) \\ &+ \gamma\mathbf{BCE}(\hat{x}^{(l)}_{rec}, x^{(l)}_{rec}) + \gamma\mathbf{BCE}(\hat{x}^{(m)}_{rec}, x^{(m)}_{rec}) \end{aligned} \tag{7}$$

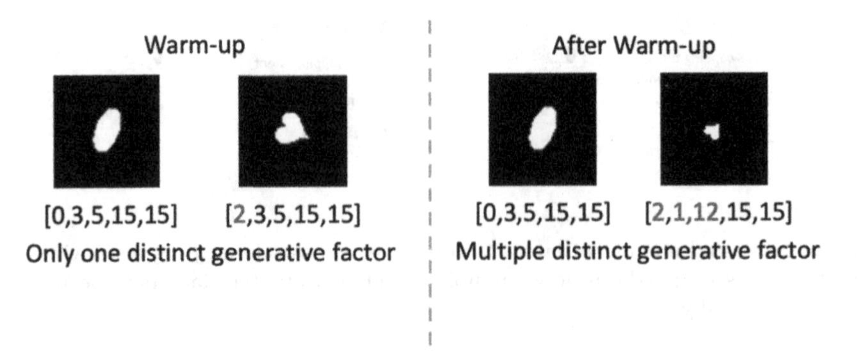

Fig. 5. Increasing number of distinct generative factors used to produce image pairs as training progresses.

***SW-VAE**$_{GAN}$*: In **SW-VAE**$_{GAN}$ we use a discriminator to measure the similarity between reconstruction pairs. During training, the VAE tries to minimize the distance between x_{rec} and $\hat{x}_{rec}$. The discriminator tries to measure the differences between two input images. We use binary cross-entropy loss to train the VAE and discriminator. Image labels are set to *one* when training VAE and labels are set to *zero* when training discriminator. In **SW-VAE**$_{GAN}$, the loss function to train VAE is expressed in Eq. (8) and the loss function to train the discriminator is expressed in Eq. (9).

$$\begin{aligned} L_{GAN} = \; & L_{VAE}(x_{rec}^{(l)}, z^{(l)}) + L_{VAE}(x_{rec}^{(m)}, z^{(m)}) \\ & + \gamma \mathbf{BCE}(C_w\left[\hat{x}_{rec}^{(l)}, x_{rec}^{(l)}\right], \mathbf{1}) + \gamma \mathbf{BCE}(C_w\left[\hat{x}_{rec}^{(m)}, x_{rec}^{(m)}\right], \mathbf{1}) \end{aligned} \tag{8}$$

$$L_{disc} = \mathbf{BCE}(C_w\left[\hat{x}_{rec}^{(l)}, x_{rec}^{(l)}\right], \mathbf{0}) + \mathbf{BCE}(C_w\left[\hat{x}_{rec}^{(m)}, x_{rec}^{(m)}\right], \mathbf{0}) \tag{9}$$

The overall architecture of **SW-VAE** is shown in Fig. 2. The difference between these two algorithms is the measurement criterion of comparing the similarity between the reconstruction output x_{recon} from original the latent representations and the new reconstruction output $\hat{x}_{recon}$ from the new latent representations.

Training Using Maximum Number of Different Generative Factors: As previously discussed, Locatello *et al.* [23] argue that disentanglement of the underlying generative factors of variations is infeasible without any supervision signal. In this work, we use pairs of input $x^{(l)}$ and $x^{(m)}$ as weak supervision signals to learn disentangled representations.

As illustrated in Fig. 3, the pair of training samples $(x^{(l)}, x^{(m)})$ can be generated as follow: we first randomly select a vector $V^{(l)} = [v_1, v_2, ..., v_n]$ to generate observation $x^{(l)} = g(V^{(l)})$, where $g(V)$ is the observation generating function. Then, we randomly change the value of at most k elements in $V^{(l)}$ to form a new vector $V^{(m)}$. Finally, we generate the $x^{(m)} = g(V^{(m)})$. During training, true indices of different generative factors and true value of generative factors are not provided, where the only information provided to the model is the maximum number of changed factors k, as mentioned earlier.

Table 1. Disentanglement metrics on dSprites and 3dShapes; **Bold, Red**: best result, **Bold, Black**: second best result; **SW-VAE**(m) is trained only using the maximum number of generative factors, and **SW-VAE**(e) is trained with knowledge of the exact number of different generative factors; Baseline results are generated by us with official public implementation.

Models	dSprites					3dshapes				
	MIG	SAP	IRS	FVAE	DCI	MIG	SAP	IRS	FVAE	DCI
Unsupervised Disentanglement Learning										
betaVAE	0.1115	0.0373	0.5520	0.7886	0.3263	0.1868	0.0643	0.4731	0.8488	0.2460
AnnealedVAE	0.1253	0.0422	0.5592	0.8228	0.3715	0.2350	0.0867	0.5450	0.8651	0.3428
FactorVAE	0.1308	0.0238	0.5634	0.7503	0.1914	0.2265	0.0437	0.6298	0.7876	0.3031
DIP-VAE-I	0.0667	0.0247	0.4497	0.6897	0.1661	0.1438	0.0273	0.5010	0.7671	0.1372
DIP-VAE-II	0.0212	0.0587	0.5434	0.6354	0.0997	0.1372	0.0204	0.4237	0.7416	
BetaTCVAE	0.2125	0.0582	0.5437	0.8414	0.3295	0.3644	0.0955	0.5942	0.9627	0.6004
Weakly/Semi-Supervised Disentanglement Learning										
DSD	0.3937	0.0771	0.6327	**0.9121**	**0.6015**	0.6343	**0.1518**	**0.7623**	**0.9968**	0.9019
Ada-ML-VAE	0.1150	0.0366	0.5712	0.7010	0.2940	0.5092	0.1273	0.6203	0.9956	**0.9400**
Ada-GVAE	0.2664	0.0735	0.5927	0.8472	0.4790	0.5607	0.1502	0.7076	0.9965	**0.9459**
SW-VAE(m)	**0.4228**	**0.0780**	**0.6572**	0.8571	0.5994	**0.6353**	0.1506	0.7316	0.9958	0.9030
SW-VAE(e)	**0.4637**	**0.1077**	**0.6774**	**0.8913**	**0.6852**	**0.7121**	**0.1564**	**0.7837**	**0.9978**	0.9198

During training, there are more than one factors of variants that need to be swapped. However, in the early training stage, since the model is not well trained yet, exchanging a large number of factors in the latent representations tends to harm model performance, which is discussed in Sect. 4.3. Thus, during *warm-up*, we only swap the factors of latent representation where the model is highly confident. As training progresses, we increase the difficulty by increasing the number of latent representation factors to swap. We show this procedure in Fig. 4 and call this strategy as ISF. We denote **SW-VAE** trained under the supervision of maximum number of changed factors as **SW-VAE**(m).

Training Using Exact Number of Different Generative Factors: As previously discussed, **SW-VAE**(m) is limited to merely know the maximum number k of different generative factors. By slightly increasing the level of supervision strength, where the exact number of different generative factors of the pair of inputs are known, the performance of the model is further boosted though the true indices and values of different generative factors are still inaccessible for the model. We denote **SW-VAE** trained under supervision of exact different generative factors as **SW-VAE**(e). During training, we set the number of different generative factors to one and progressively increasing until it reaches the maximum number. This strategy is illustrated in Fig. 5 and we call this strategy as IGF. The importance of this training strategies is studied in section Sect. 4.3.

Table 2. Disentanglement metrics on MPI3D-toy, MPI3D-realistic, and MPI3D-real; Bold, Red: best result, **Bold, Black**: second best result; **SW-VAE**(m) is trained only using the maximum number of generative factors, and **SW-VAE**(e) is trained with knowledge of the exact number of different generative factors; Baseline results are generated by us with official public implementation.

Models	MPI3d-toy					MPI3d-realistic					MPI3d-real				
	MIG	SAP	IRS	FVAE	DCI	MIG	SAP	IRS	FVAE	DCI	MIG	SAP	IRS	FVAE	DCI
Unsupervised Disentanglement Learning															
betaVAE	0.132	0.061	0.584	0.353	0.274	0.150	0.138	0.574	0.350	0.361	0.137	0.071	0.579	0.368	0.367
AnnealedVAE	0.155	0.107	0.553	0.411	0.360	0.109	0.114	0.531	0.466	0.363	0.099	0.038	0.490	0.396	0.229
FactorVAE	0.183	0.070	0.512	0.398	0.273	0.136	0.069	0.560	0.386	0.215	0.093	0.031	0.529	0.391	0.192
DIP-VAE-I	0.156	0.085	0.484	0.517	0.284	0.167	0.134	0.525	0.566	0.337	0.130	0.074	0.509	0.533	0.264
DIP-VAE-II	0.061	0.027	0.412	0.487	0.163	0.032	0.017	0.432	0.393	0.149	0.131	0.062	0.509	0.544	0.244
BetaTCVAE	0.173	0.084	0.638	0.355	0.342	0.165	0.046	0.571	0.366	0.244	0.181	0.146	0.636	0.431	0.344
Weakly/Semi-Supervised Disentanglement Learning															
DSD	**0.399**	0.193	0.612	**0.510**	0.445	0.402	0.199	0.602	0.621	0.512	0.353	**0.222**	0.601	0.603	0.522
Ada-ML-VAE	0.293	0.093	0.520	0.439	0.392	0.283	0.131	0.582	0.483	0.270	0.240	0.074	0.576	0.476	0.285
Ada-GVAE	0.347	0.238	0.613	0.501	0.427	0.339	0.227	0.609	0.592	0.479	0.264	0.215	0.602	**0.601**	0.401
SW-VAE(m)	0.394	0.210	0.592	0.506	**0.452**	0.401	0.201	0.602	0.591	0.532	**0.360**	0.216	0.593	0.562	**0.542**
SW-VAE(e)	0.467	**0.213**	**0.614**	0.520	0.554	0.452	**0.202**	0.614	**0.596**	0.574	0.484	0.225	**0.617**	0.565	0.566

4 Experimental Evaluation

4.1 Benchmarks, Baseline Methods and Evaluation Metrics

We use following five different datasets where the images are annotated with different underlying factor of variations.

dSprites [15] contains 73,728 binary 64×64 images generated by 6 generative factors.

3dShapes [2] generated by 6 generative factors: floor hue, wall hue, object hue, scale, shape and orientation. The total dataset contains 480,000 RGB $64 \times 64 \times 3$ images.

MPI3D [12] contains 3 different datsets: MPI3D-real, MPI3D-toy and MPI3D-realistic. MPI3D-real is a real world dataset controlled by 7 generative factors: object color, object shape, object size, camera height, background color, horizontal axis and vertical axis. MPI3D-toy, MPI3D-realistic are synthetic versions of MPI3D-real with two levels of realism. Each of the three datasets contains 1,036,800 RGB $512 \times 512 \times 3$ images.

SOTA model used for comparisons are: (1) β-VAE [15], (2) AnnealedVAE [3], (3) FactorVAE [17], (4) DIP-VAE-I [20], (5) DIP-VAE-II [20], (6) β-TCVAE [5], (7) DSD [10], (8) Ada-ML-VAE [25] and (9) Ada-VAE [25].

Following metrics are used to evaluate the performance of **SW-VAE**. All metrics are range from 0 to 1, where the score of 1 indicates the latent factors are fully disentangled.

- **Mutual Information Gap (MIG)** [5] calculates the difference between the top two highest mutual information between latent and generative components.

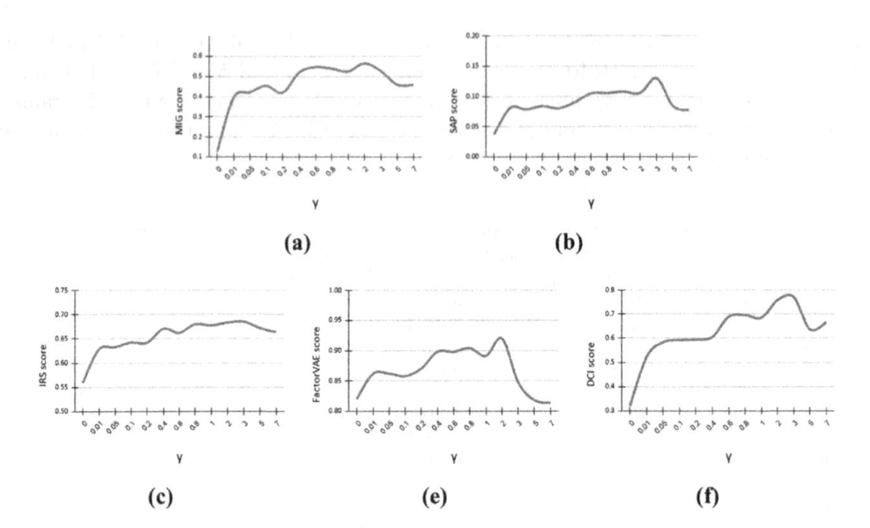

Fig. 6. Disentanglement metrics with varying γ on dSprites. **(a)** Mutual Information Gap (MIG) **(b)** Separated Attribute Predictability (SAP) **(c)** Interventional Robustness Score (IRS) **(d)** FactorVAE score (FVAE) **(e)** DCI-Disentanglement (DCI)

- **Separated Attribute Predictability (SAP)** [20] calculates the average perdition error difference between the top two most predictive latent components.
- **Interventional Robustness Score (IRS)** [32] assesses the degree of dependency between a latent factor and a generative factor, regardless of additional generative factors.
- **FactorVAE (FVAE) score** [17] predicts the index of a fixed generating factor using a majority vote classifier, and the accuracy is the final score value.
- **DCI-Disentanglement (DCI)** [9] computes the entropy of the distribution by normalizing across each dimension of the learned representation in order to predict the value of a generative component.

4.2 Quantitative Results

Tables 1 and 2 show disentanglement metrics results tested on dSprites, 3dshapes, MPI3D-toy, MPI3D-realistic and MPI3D-real respectively. By observing the results, **SW-VAE** outperforms baselines methods in most cases, where **SW-VAE**(e) consistently outperforms **SW-VAE**(m). Mean, median, and variance of all disentanglement metrics scores of all models tested on all datasets are shown Appendix.

4.3 Ablation Study

Effectiveness of New Regularization: To prove the effectiveness of using the proposed regularization as discussed in Eqs. (5) to (8), we evaluated **SW-VAE**(m) on dSprites using different values of γ. We start experimenting by setting $\gamma = 0$, then slowly increase the value of γ. We show the change of disentanglement metrics MIG and DCI scores in Fig. 6. We can observe a great improvement by utilizing swapping concept even if γ is very small. Other metrics results are included in supplementary materials.

Table 3. Disentanglement metrics of **SW-VAE**(m) with different training strategies applied to dSpirits dataset

ISF	IGF	MIG	SAP	IRS	FVAE	DCI
		0.232	0.048	0.617	0.790	0.347
	✓	0.439	0.081	0.652	0.872	0.610
✓	✓	**0.525**	**0.108**	**0.677**	**0.891**	**0.685**

Effectiveness of Different Training Strategies: As we have discussed in Sect. 3, we propose two training strategies (gradually increasing the number of swapped latent factors (ISF) shown in Fig. 4 and gradually increasing the number of different generative factors (IGF) shown in Figure Creffig:finalspstrainingspsmethod to help learning representation disentanglement. We study the importance of these strategies by comparing the results of three different situations: (1) No strategy is used; (2) Only IGF strategy is used ; (3) Both IGF and ISF strategies are used. The result of disentanglement performance evaluated on dSprites with different strategies are shown in Table 3. Compared to none of the proposed training strategies is used, we can observe significant performance improvement by adopting IGF strategy. Furthermore, there is additional performance improvement by utilizing both the ISF and IGF strategies.

5 Conclusion

We introduce **SW-VAE**: a novel weakly-supervised representation disentanglement method. The supervision signals are introduced by utilizing pairs of training observations where the number of different generative factors are controlled. **SW-VAE** uses the maximum or exact number of different generative factors as an instruction to swap the latent factors estimated by the encoder. Further, the comparison between the reconstruction from the original latent representation and the reconstruction from the new latent representation serves as new disentanglement regularization. Experimental evaluation demonstrates that our approach significantly outperforms SOTAs both qualitatively and quantitatively.

Acknowledgement. This material is based on research sponsored by Air Force Research Laboratory under agreement number FA8750-19-1-1000. The views and conclusions contained herein are those of the authors and should not be interpreted as necessarily representing the official policies or endorsements, either expressed or implied, of Air Force Research Laboratory or the US, Government.

References

1. Bengio, Y., Courville, A., Vincent, P.: Representation learning: a review and new perspectives. IEEE Trans. Pattern Anal. Mach. Intell. **35**(8), 1798–1828 (2013). https://doi.org/10.1109/TPAMI.2013.50
2. Burgess, C., Kim, H.: 3d shapes dataset (2018). https://github.com/deepmind/3dshapes-dataset/
3. Burgess, C.P., et al.: Understanding disentangling in β-VAE (2018)
4. Chen, J., Batmanghelich, K.: Weakly supervised disentanglement by pairwise similarities. In: The Thirty-Fourth AAAI Conference on Artificial Intelligence, AAAI 2020, The Thirty-Second Innovative Applications of Artificial Intelligence Conference, IAAI 2020, The Tenth AAAI Symposium on Educational Advances in Artificial Intelligence, EAAI 2020, New York, NY, USA, 7–12 February 2020, pp. 3495–3502. AAAI Press (2020). https://aaai.org/ojs/index.php/AAAI/article/view/5754
5. Chen, R.T.Q., Li, X., Grosse, R., Duvenaud, D.: Isolating sources of disentanglement in variational autoencoders (2019)
6. Chen, T., Kornblith, S., Norouzi, M., Hinton, G.: A simple framework for contrastive learning of visual representations (2020)
7. Creager, E., et al.: Flexibly fair representation learning by disentanglement. CoRR abs/1906.02589 (2019). arxiv.org:1906.02589
8. Deng, J., Dong, W., Socher, R., Li, L., Li, K., Li, F.-F.: Imagenet: a large-scale hierarchical image database. In: 2009 IEEE Conference on Computer Vision and Pattern Recognition, pp. 248–255 (2009). https://doi.org/10.1109/CVPR.2009.5206848
9. Eastwood, C., Williams, C.K.I.: A framework for the quantitative evaluation of disentangled representations. In: International Conference on Learning Representations (2018). https://openreview.net/forum?id=By-7dz-AZ
10. Feng, Z., Wang, X., Ke, C., Zeng, A.X., Tao, D., Song, M.: Dual swap disentangling. In: Advances in Neural Information Processing Systems. Curran Associates, Inc. (2018)
11. Gidaris, S., Singh, P., Komodakis, N.: Unsupervised representation learning by predicting image rotations (2018)
12. Gondal, M.W., et al.: On the transfer of inductive bias from simulation to the real world: a new disentanglement dataset (2019)
13. Goodfellow, I.J., et al.: Generative adversarial networks (2014)
14. He, K., Zhang, X., Ren, S., Sun, J.: Deep residual learning for image recognition (2015)
15. Higgins, I., et al.: Beta-VAE: learning basic visual concepts with a constrained variational framework. In: 5th International Conference on Learning Representations, ICLR 2017, Toulon, France, 24–26 April 2017, Conference Track Proceedings (2017). https://openreview.net/forum?id=Sy2fzU9gl
16. Hinton, G.E., Salakhutdinov, R.R.: Reducing the dimensionality of data with neural networks. Science **313**(5786), 504–507 (2006)
17. Kim, H., Mnih, A.: Disentangling by factorising. In: Dy, J., Krause, A. (eds.) Proceedings of the 35th International Conference on Machine Learning. Proceedings of Machine Learning Research, vol. 80, pp. 2649–2658. PMLR, Stockholmsmässan, Stockholm Sweden, 10–15 July 2018. https://proceedings.mlr.press/v80/kim18b.html
18. Kingma, D.P., Welling, M.: Auto-encoding variational bayes (2014)

19. Kingma, D.P., Welling, M.: Auto-encoding variational bayes. In: Bengio, Y., LeCun, Y. (eds.) 2nd International Conference on Learning Representations, ICLR 2014, Banff, AB, Canada, 14–16 April 2014, Conference Track Proceedings (2014). arxiv.org:1312.6114
20. Kumar, A., Sattigeri, P., Balakrishnan, A.: Variational inference of disentangled latent concepts from unlabeled observations. In: ICLR arxiv:1711.00848 (2018)
21. Larsen, A.B.L., Sønderby, S.K., Larochelle, H., Winther, O.: Autoencoding beyond pixels using a learned similarity metric. In: Balcan, M.F., Weinberger, K.Q. (eds.) Proceedings of The 33rd International Conference on Machine Learning. Proceedings of Machine Learning Research, vol. 48, pp. 1558–1566. PMLR, New York, New York, USA, 20–22 June 2016. https://proceedings.mlr.press/v48/larsen16.html
22. Liu, Z., Luo, P., Wang, X., Tang, X.: Deep learning face attributes in the wild. In: Proceedings of International Conference on Computer Vision (ICCV), December 2015
23. Locatello, F., et al.: Challenging common assumptions in the unsupervised learning of disentangled representations (2019)
24. Locatello, F., Poole, B., Raetsch, G., Schölkopf, B., Bachem, O., Tschannen, M.: Weakly-supervised disentanglement without compromises. In: III, H.D., Singh, A. (eds.) Proceedings of the 37th International Conference on Machine Learning. Proceedings of Machine Learning Research, vol. 119, pp. 6348–6359. PMLR, 13–18 July 2020. https://proceedings.mlr.press/v119/locatello20a.html
25. Locatello, F., Tschannen, M., Bauer, S., Rätsch, G., Schölkopf, B., Bachem, O.: Disentangling factors of variations using few labels. In: International Conference on Learning Representations (2020). https://openreview.net/forum?id=SygagpEKwB
26. Long, J., Shelhamer, E., Darrell, T.: Fully convolutional networks for semantic segmentation. In: 2015 IEEE Conference on Computer Vision and Pattern Recognition (CVPR), pp. 3431–3440 (2015). https://doi.org/10.1109/CVPR.2015.7298965
27. Matthey, L., Higgins, I., Hassabis, D., Lerchner, A.: dSprites: Disentanglement testing sprites dataset. https://github.com/deepmind/dsprites-dataset/ (2017)
28. Park, T., et al.: Swapping autoencoder for deep image manipulation. In: Advances in Neural Information Processing Systems (2020)
29. Siddharth, N., et al.: Learning disentangled representations with semi-supervised deep generative models (2017)
30. Simonyan, K., Zisserman, A.: Very deep convolutional networks for large-scale image recognition (2015)
31. van Steenkiste, S., Locatello, F., Schmidhuber, J., Bachem, O.: Are disentangled representations helpful for abstract visual reasoning? (2019)
32. Suter, R., ore Miladinović, Schölkopf, B., Bauer, S.: Robustly disentangled causal mechanisms: Validating deep representations for interventional robustness (2019)
33. Taigman, Y., Yang, M., Ranzato, M., Wolf, L.: Deepface: closing the gap to human-level performance in face verification. In: 2014 IEEE Conference on Computer Vision and Pattern Recognition, pp. 1701–1708 (2014). https://doi.org/10.1109/CVPR.2014.220
34. Vinyals, O., Blundell, C., Lillicrap, T., Kavukcuoglu, K., Wierstra, D.: Matching networks for one shot learning. In: Lee, D., Sugiyama, M., Luxburg, U., Guyon, I., Garnett, R. (eds.) Advances in Neural Information Processing Systems, vol. 29. Curran Associates, Inc. (2016). https://proceedings.neurips.cc/paper/2016/file/90e1357833654983612fb05e3ec9148c-Paper.pdf
35. Zhou, Z.H.: A brief introduction to weakly supervised learning. Natl. Sci. Rev. **5**(1), 44–53 (2017). https://doi.org/10.1093/nsr/nwx106

Learning Multiple Probabilistic Degradation Generators for Unsupervised Real World Image Super Resolution

Sangyun Lee[1], Sewoong Ahn[2], and Kwangjin Yoon[2](✉)

[1] Soongsil University, Seoul, South Korea
[2] SI Analytics, Daejeon, South Korea
{anse3832,yoon28}@si-analytics.ai

Abstract. Unsupervised real world super resolution (USR) aims to restore high-resolution (HR) images given low-resolution (LR) inputs, and its difficulty stems from the absence of paired dataset. One of the most common approaches is synthesizing noisy LR images using GANs (i.e., degradation generators) and utilizing a synthetic dataset to train the model in a supervised manner. Although the goal of training the degradation generator is to approximate the distribution of LR images given a HR image, previous works have heavily relied on the unrealistic assumption that the conditional distribution is a delta function and learned the deterministic mapping from the HR image to a LR image. In this paper, we show that we can improve the performance of USR models by relaxing the assumption and propose to train the probabilistic degradation generator. Our probabilistic degradation generator can be viewed as a deep hierarchical latent variable model and is more suitable for modeling the complex conditional distribution. We also reveal the notable connection with the noise injection of StyleGAN. Furthermore, we train multiple degradation generators to improve the mode coverage and apply *collaborative learning* for ease of training. We outperform several baselines on benchmark datasets in terms of PSNR and SSIM and demonstrate the robustness of our method on unseen data distribution. Code is available at https://github.com/sangyun884/MSSR.

1 Introduction

Unsupervised real world super resolution (USR) aims to restore high-resolution (HR) images given low-resolution (LR) observations, and its difficulty stems from the absence of paired dataset. While recent progress of deep convolutional neural network-based approaches has shown remarkable results on the bicubic downsampled dataset [3,12,14,20,30], they generalize poorly in the real world as they do not consider the complex degradation process.

In real world, the degradation process is generally unknown and only implicitly represented through the corrupted observations. Since it is difficult to collect

S. Lee—Did this work during the intership at SI Analytics.

L. Karlinsky et al. (Eds.): ECCV 2022 Workshops, LNCS 13802, pp. 88–100, 2023.
https://doi.org/10.1007/978-3-031-25063-7_6

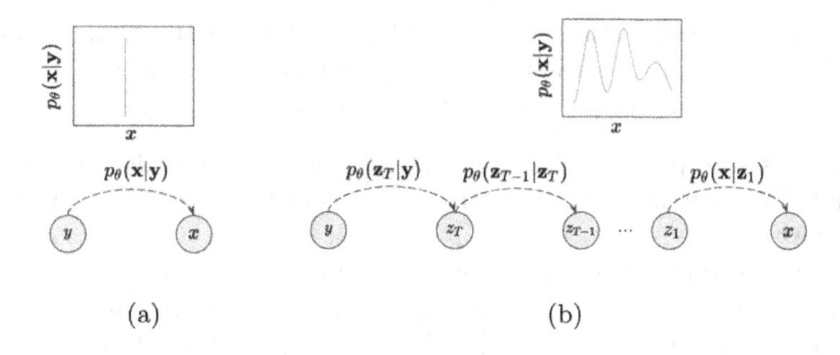

Fig. 1. The most crucial step of USR is to learn the degradation process in which HR image $\mathbf{y}$ transforms into LR image $\mathbf{x}$. (a): Previous GAN-based studies model the $p(\mathbf{x}|\mathbf{y})$ as a delta function. (b): We design the probabilistic generator with multiple latent variables to model the flexible distribution.

paired data consisting of LR images and their intact HR representations, several studies [2,5,15] have focused on generating realistic synthetic paired data using conditional generative models, primarily generative adversarial networks (GANs) [6]. In contrast to the approaches dubbed blind SR [23,24], GAN-based approaches are effective in the real world as they do not assume any functional form of the degradation process, which can actually be a complex and nonlinear.

However, previous GAN-based approaches have heavily relied on the unrealistic assumption that the degradation process is deterministic and learned the deterministic mapping from a HR image to a LR image. Since they fail to capture the complex distribution of LR images due to their deterministic nature (see Fig. 1 (a)), the generalization capability of a SR model trained on the pseudo-paired dataset is sub-optimal.

In this paper, we show that we can improve the performance of USR models by relaxing the assumption and propose to train the probabilistic degradation generator to better approximate the stochastic degradation process. As shown in Fig. 1, our probabilistic degradation generator can be viewed as a deep hierarchical latent variable model and is more suitable for modeling the complex conditional distribution. Our method is conceptually simple and extremely easy to implement: add a Gaussian noise multiplied by the learned standard deviation to intermediate feature maps of the degradation generator. We also reveal that the noise injection of StyleGAN [10] can be viewed as a special case of our method.

Furthermore, we train multiple degradation generators to improve the mode coverage and apply *collaborative learning* for ease of training. In collaborative learning, multiple SR models teach with one another, mutually improving their generalization capabilities. We show that collaborative learning facilitates the training procedure.

In addition, we outperform several baselines on benchmark datasets in PSNR and SSIM and demonstrate the robustness of our method on unseen data distribution. Our contributions are summarized as follows:

- We propose the probabilistic degradation generator to model the stochastic degradation process and reveal the connection with the noise injection of StyleGAN.
- We train multiple degradation generators to improve the mode coverage and apply a collaborative learning strategy to facilitate the training procedure.
- Our model outperforms several baselines on benchmark datasets and shows a strong out-of-domain generalization ability.

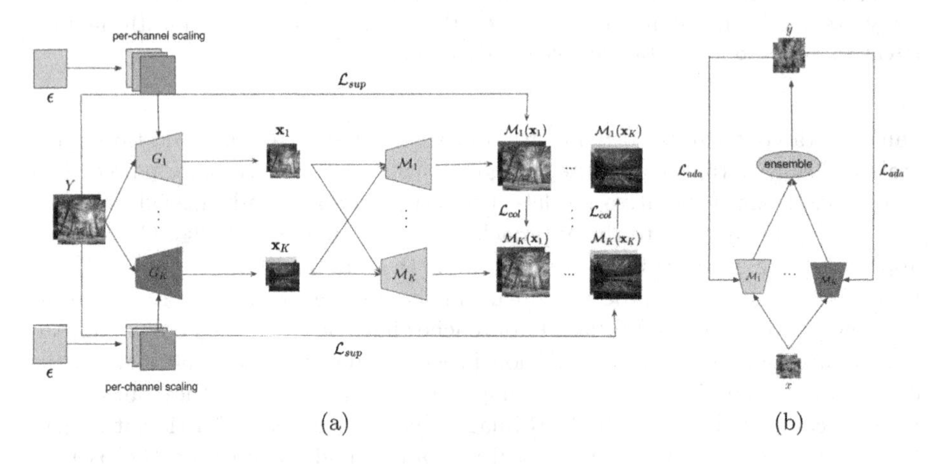

Fig. 2. A schematic of collaborative learning. (a): Collaborative learning on the synthetic LR domain using multiple probabilistic degradation generators. (b): Collaborative learning on the real LR dataset using pseudo-labels. We alternatively train (a) and (b).

2 Related Works

2.1 GAN-Based Unsupervised Real-World Super Resolution

Although previous SR models [3,12,14,20,30] achieved remarkable results on bicubic downsampled data, they perform poorly in the real world, where the degradation process is complex and unknown. Several studies [2,5,15] have attempted to model the degradation process using GANs. They use conditional GANs to generate LR images given HR images and train the SR model on the generated pseudo-paired data. However, they use a deterministic degradation model and fail to approximate the stochastic degradation process, to which most of the real world scenarios belong.

2.2 Stochastic Degradation Operations

There have been studies called blind SR that take account of the stochastic degradation process [23,24]. However, they assume the limited type of degradation operation that consists of blur kernel and noise from the pre-defined distribution. Recently, DeFlow [27] tackled this problem by learning unpaired learning with normalizing flow. In this paper, we propose a method that is more straightforward and easy to plug into any existing degradation network to enhance the expressiveness.

2.3 Noise Injection to Feature Space

Our work is closely related to the approaches that inject the noise into the feature space of the generator. For example, Karras et al. proposed a noise injection module that helps a model generate the pseudo-random details of images [11]. Figuring out the effectiveness of the noise injection is still an open problem [4], and they may be related to modern hierarchical generative models [8,13,21,22] as the reparameterization trick can be viewed as injecting noises to intermediate states. We consider our probabilistic degradation generator as a hierarchical generative model with multiple latent variables and will subsequently reveal the connection with the noise injection of StyleGAN.

3 Proposed Methods

3.1 Probabilistic Degradation Generator

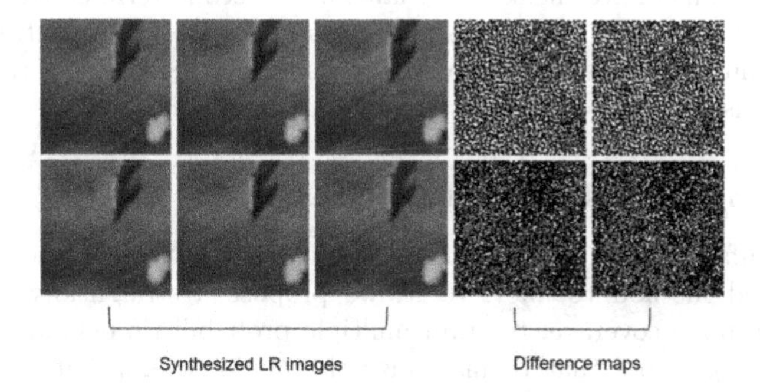

Fig. 3. LR images (first three columns) generated by probabilistic degradation generators and difference maps (last two columns). The images of the first row are generated by probabilistic DeResnet [26], and the images of the second row are generated by probabilistic HAN [19]. The difference maps are calculated as the pixel-wise L1 distance between the first and second images and the first and third images, respectively. The proposed probabilistic degradation generator can synthesize multiple LR images given a single HR observation.

From a probabilistic perspective, training the degradation generator is to estimate the parameter θ that satisfies $p_\theta(\mathbf{x}|\mathbf{y}) \approx p(\mathbf{x}|\mathbf{y})$, where $\mathbf{x} \sim p(\mathbf{x})$ and $\mathbf{y} \sim p(\mathbf{y})$ are a LR and HR image. Our aim is to generalize the deterministic degradation generators in the previous studies to better approximate the stochastic degradation process in the real world. With T latent variables $\mathbf{z}_1, \mathbf{z}_2, ..., \mathbf{z}_T$, the probabilistic degradation generator is defined as:

$$p_\theta(\mathbf{x}|\mathbf{y}) = \int p_\theta(\mathbf{x}|\mathbf{z}_1) p_\theta(\mathbf{z}_T|\mathbf{y}) \prod_{i=2}^{T} p_\theta(\mathbf{z}_{i-1}|\mathbf{z}_i) d\mathbf{z}_{1:T}, \quad (1)$$

which is the Markov chain that consists of following Gaussian transition:

$$p_\theta(\mathbf{z}_{i-1}|\mathbf{z}_i) = \mathcal{N}(\mathbf{z}_{i-1}; \mu_\theta(\mathbf{z}_i, i), \mathbf{\Sigma}_\theta(\mathbf{z}_i, i)). \quad (2)$$

Note that $p_\theta(\mathbf{x}|\mathbf{z}_1)$ and $p_\theta(\mathbf{z}_T|\mathbf{y})$ are delta functions. For simplicity and ease of computation, we set $\mathbf{\Sigma}_\theta(\mathbf{z}_i, i) = diag(\sigma_\theta(\mathbf{z}_i, i))^2$ with $\sigma_\theta(\cdot)$ being a vector-valued function. Therefore, we can sample each latent variable in a differentiable manner as

$$\mathbf{z}_{i-1} = \mu_\theta(\mathbf{z}_i) + diag(\sigma_\theta(\mathbf{z}_i, i))\epsilon, \quad \epsilon \sim \mathcal{N}(\mathbf{0}, \mathbf{I}), \quad (3)$$

which can be seen as injecting the noise into the feature vector $\mu_\theta(\mathbf{z}_i)$. Note that our model is reduced to the deterministic generator in previous studies when $\sigma_\theta(\mathbf{z}_i) = \mathbf{0}$. Since maximum likelihood training is difficult as there is no paired data, we train the probabilistic generator via adversarial learning with cycle-consistency constraint [32].

Connection with StyleGAN. When the variances $\mathbf{\Sigma}_\theta$ of latent variables do not depend on $\mathbf{z}_i$ and are shared within the spatial dimension (i.e., the elements in the same channel have the same variance), the second term of the right-hand side of Eq. 3 is the same as the noise injection in StyleGAN. In fact, this choice reduces the number of parameters without compromising the performance, and thus we extensively use this setting throughout the manuscript.

3.2 Training Multiple Degradation Generators

GANs are infamous for their mode-dropping trait. Although one can utilize a complicated method to alleviate it, we propose a straightforward way to enhance the mode coverage: to train multiple probabilistic degradation generators $G_1, ..., G_K$. Note that to maximize the mode coverage, it is crucial for each generator to exclusively cover a certain area of the $p(\mathbf{x}|\mathbf{y})$ without redundancy. Depending on the fact that the convergence properties of GANs can vary depending on the training algorithm [18], we initially attempted to train each degradation generator with different training schemes and regularizations. However, we empirically found that the diversity of the LR domain can be efficiently achieved by varying model architectures of the degradation generator. Figure 3 shows that each generator synthesizes different noise patterns (zoom in for the best view). We describe detailed information on architectures in Sect. 4.1.

DeResnet [26] Probabilistic DeResnet GT

Method	LPIPS [30]
Gaussian noise	0.215
DeResnet [26]	0.049
Probabilistic DeResnet	**0.026**

Fig. 4. Comparison between generated LR images. Since the validation set of NTIRE2020 dataset [16] consists of paired data, we can measure LPIPS between LR images generated by degradation generators and the ground truth LR image. We also report the result of additive Gaussian noise with standard deviation of 20 followed by bicubic downsampling as a baseline.

Table 1. Comparison between deterministic and probabilistic degradation generators.

Degradation methods	PSNR	SSIM
(a) DeResnet [26]	26.65	0.722
(b) Probabilistic DeResnet	**26.96**	**0.744**

3.3 Collaborative Learning

Although we can improve the mode coverage by training multiple degradation generators, simply aggregating the generated pseudo-paired data may not be the best way [31], implying that there is room for improvement by devising a more clever way to utilize them. As each data generated by different probabilistic generators shares the content of an image except for the degradation function used, knowledge gained in one can be readily applied to others. Inspired by He et al. [7], we propose to utilize a collaborative learning strategy, where multiple SR models teach with one another, mutually improving their generalization capabilities. Figure 2 shows an overview of the collaborative learning strategy. With $p_{\theta_i}(\mathbf{x}|\mathbf{y})$ being the implicit distribution of G_i, we train SR networks $\{\mathcal{M}_i\}_{i=1}^{K}$ by minimizing the objective function

$$\begin{aligned}\mathcal{L}_i = &\lambda_{sup}\mathcal{L}_{sup}(\mathcal{M}_i(\mathbf{x}_i), \mathbf{y}) \\ &+\lambda_{col}\mathcal{L}_{col}(\{(\mathcal{M}_i(\mathbf{x}_j), \mathcal{M}_j(\mathbf{x}_j))\}_{j\neq i}) \\ &+\lambda_{ada} r(p, P)\mathcal{L}_{sup}(\mathcal{M}_i(\mathbf{x}), \hat{\mathbf{y}}),\end{aligned} \tag{4}$$

where $\mathbf{x}_i \sim p_{\theta_i}(\mathbf{x}|\mathbf{y})$, $\mathbf{x} \sim p(\mathbf{x})$, and $\mathbf{y} \sim p(\mathbf{y})$. $\lambda_{sup}, \lambda_{col}$, and λ_{ada} denote the weights that determine the relative strength of each term. P and p represent maximum and current training iteration, and thus $r(p, P) = p/P$ gradually increases during training. For $\mathcal{L}_{sup}$, we used L1-loss, although more complicated loss functions such as adversarial loss or perceptual loss can be plugged in to improve the perceptual quality. $\mathcal{L}_{col}$ is collaborative learning loss, i.e.,

$$\mathcal{L}_{col}(\{(\mathcal{M}_i(\mathbf{x}_j), \mathcal{M}_j(\mathbf{x}_j))\}_{j \neq i}) = \sum_{j \neq i} \mathcal{L}_{sup}(\mathcal{M}_i(\mathbf{x}_j), \mathcal{M}_j(\mathbf{x}_j)). \quad (5)$$

Note that knowledge distillation is applied only in image space. We initially tried to apply knowledge distillation in feature space and found that both yield similar results. $\hat{\mathbf{y}}$ is a ensembled pseudo-label for a real LR image $\mathbf{x}$, i.e.,

$$\hat{\mathbf{y}} = \frac{1}{K} \sum_{l=1}^{K} \mathcal{M}_l(\mathbf{x}), \quad (6)$$

which helps a model to adapt on the real world dataset.

4 Experiments

4.1 Implementation Details

In our experiments, we train our MSSR (Multiple Synthetic data Super Resolution) through two stages. In the first stage, we train two probabilistic degradation generators G_1 and G_2 (i.e., $K = 2$) based on the code of [26], where we use DeResnet [26] architecture for G_1 and Holistic Attention Network (HAN) [19] for G_2. For the architecture of G_2, we remove the up-sampling block from HAN [19], and strides of the first two Conv-layer are set to 2 for 4x down-sampling. In the second stage, we train two SR models with RRDBnet [25] as a SR architecture. While training SR models, learning rate and batch size are set to 1e−5 and 16, and we set P to 1000000. Also, we set $\lambda_{sup} = 1$, $\lambda_{col} = 0.01$, and $\lambda_{ada} = 10$. We use NTIRE2020 track 1, AIM2019 datasets [16,17] in the remaining parts of this section.

Table 2. Effects of collaborative learning. CL: Collaborative learning. For a single degradation generator setting, we report the results of probabilistic DeResnet in Table 1. In a naive combination setting, we simply combine the data generated by two probabilistic degradation generators and train a single SR model. For collaborative learning, we report the result of the best performing one out of two models.

Methods	PSNR	SSIM
Single degradation generator	26.96	0.744
Naive combination	27.11	**0.758**
CL ($\lambda_{ada} = 0$)	27.22	0.757
CL	**27.25**	**0.758**

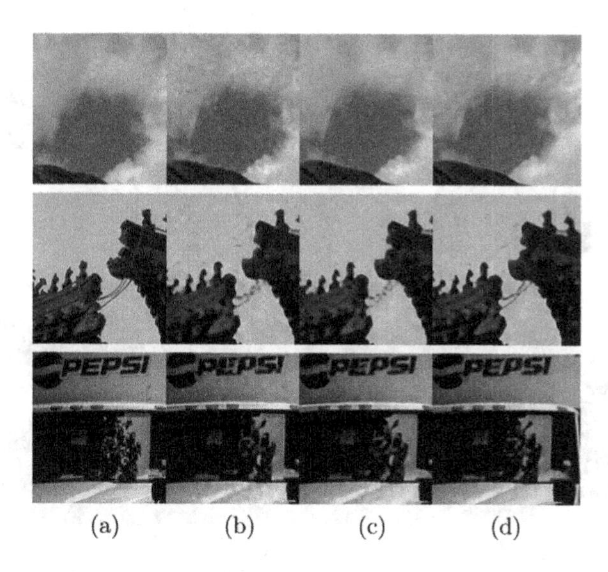

Fig. 5. Qualitative comparison between deterministic and probabilistic degradation generators on NTIRE2020 dataset [16]. Note that the same SR architecture [25] is used, and the only difference is degradation models. (a): GT, (b): DeResnet [26], (c): Probabilistic DeResnet, (d): Probabilistic HAN [19].

4.2 Ablation Studies

Probabilistic Degradation Generator. Table 1 shows the PSNR and SSIM evaluated on NTIRE2020 validation set [16]. As the results showed, the probabilistic generator significantly improves PSNR from 26.65 to 26.96. Figure 4 demonstrates that the deterministic generator fails to synthesize the realistic LR image, which is evidenced by high LPIPS. Figure 5 shows the qualitative comparison between different degradation methods. We can see that (b) fails to remove the noise on the LR image as its degradation model is too simple. On the other hand, (c) and (d) achieves more clear results than (b), demonstrating the superiority of probabilistic degradation generator.

Collaborative Learning. Table 2 presents the effect of collaborative learning on NTIRE2020 dataset. Compared to a single degradation generator setting, training multiple degradation generators largely improves the PSNR from 26.96 to 27.11. In addition, the model further improves the PSNR from 27.11 to 27.22 through the collaborative learning. Utilizing pseudo-labels as shown in (b) of Fig. 2 also improves the PSNR from 27.22 to 27.25.

4.3 Comparison with State-of-the-Arts

Table 3 shows that our model achieves the state-of-the-art PSNR and SSIM on the benchmark datasets. Note that for SimUSR [1], we detach the

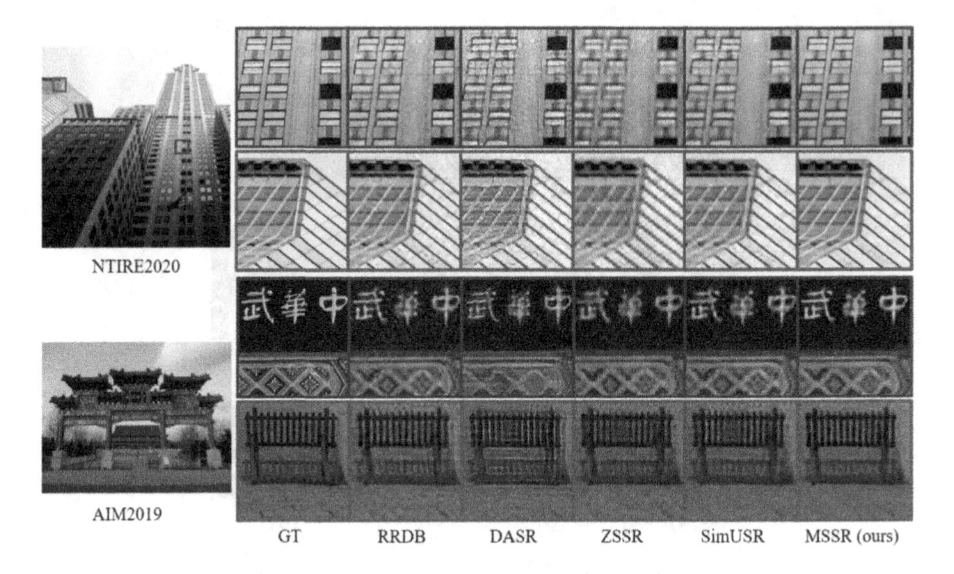

Fig. 6. Qualitative comparison with State-of-the-Arts.

Table 3. Quantitative results of different networks. Results of CinCGAN [29], Impressionism [9] and FSSR [5] are also reported. * denotes that the method do not use the adversarial loss or perceptual loss to improve the perceptual quality (i.e., PSNR-oriented).

Dataset	Metric	CinCGAN	ZSSR*	Impressionism	FSSR	DASR	SimUSR*	MSSR* (ours)
NTIRE 2020	PSNR	24.19	25.33	24.96	22.41	22.95	26.48	**27.25**
	SSIM	0.683	0.662	0.664	0.519	0.491	0.725	**0.758**
AIM 2019	PSNR	21.74	22.42	21.85	20.82	21.60	**22.88**	**22.88**
	SSIM	0.619	0.617	0.598	0.527	0.564	0.646	**0.652**

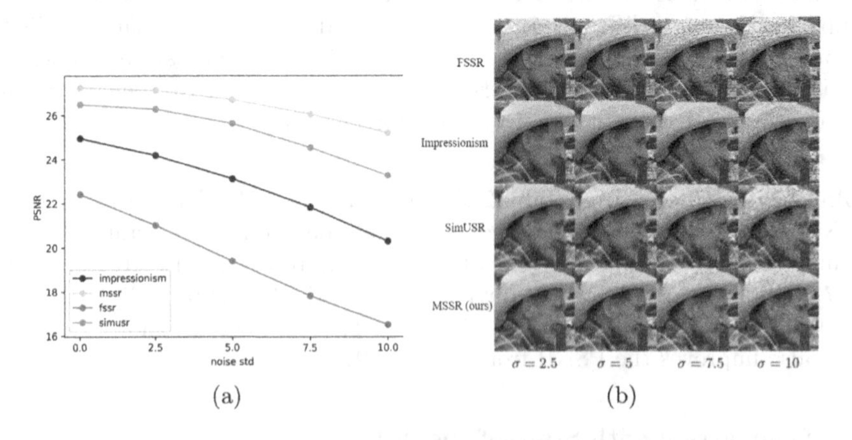

(a) (b)

Fig. 7. Performance degradation on NTIRE2020 dataset due to the increasing magnitude of input perturbation. We inject additive gaussian noise on validation LR images only at the test time to measure the robustness to unseen data.

augmentation [28] used in the original paper for a fair comparison. As shown in Fig. 6, our model synthesizes more convincing results compared to other methods.

4.4 Robustness to Perturbation

In this section, we construct the experiment to test the generalization capabilities of USR models. We train SR models on the NTIRE2020 train set and evaluate them on the validation set with additional Gaussian noise perturbation. As Fig. 7 indicates, performances of existing methods are drastically degraded by input perturbation, limiting the practical use of SR in the real world. In contrast, our model preserves the performance even when the perturbation gets stronger as the probabilistic degradation generators as well as collaborative learning strategy improve the generalization capability of our model to unseen data.

4.5 Discussion

At first glance, collaborative learning appears to be a standard ensemble technique, but it is different from the ensemble as each model exchanges the information via nested knowledge distillation. We consider that the nested knowledge distillation eases the training procedure by providing rich statistics of $p(\mathbf{y}|\mathbf{x})$. Since SR is an ill-posed problem, there are multiple underlying HR representations for a LR observation. For example, if we use mean-squared loss, it means that a model should learn to predict the expectation of the possible HR images $\mathbb{E}[p(\mathbf{y}|\mathbf{x})]$, requiring costly optimization. Given the limited capacity of a model, it can be problematic as we train a model to invert the multiple degradation generators. Contrarily, the proposed collaborative learning enables direct optimization as it can train models with soft labels which contain rich statistics peer models have learned by observing various samples, facilitating the training procedure (see Fig. 8).

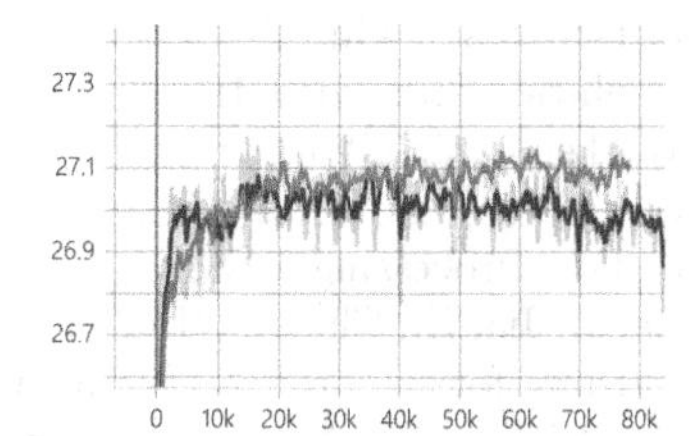

Fig. 8. PSNR of naive combination (blue) and collaborative learning (red) during training. (Color figure online)

5 Conclusion

In this paper, we improve the previous GAN-based USR approaches by training multiple probabilistic degradation generators. Unlike a deterministic generator, the probabilistic generator can model the stochastic degradation process effectively, improving the generalization capability of SR models. In addition, we utilize the collaborative learning strategy to adapt to multiple generated subsets effectively. As a result, MSSR achieved superior results to other methods and showed robustness in unseen data distribution. These results indicate the effectiveness of our approach for real world SR.

Acknowledgement. This work was supported by Institute of Information & communications Technology Planning & Evaluation (IITP) grantfunded by the Korea government (MSIT) (No. 2021-0-02068, Artificial Intelligence Innovation Hub).

References

1. Ahn, N., Yoo, J., Sohn, K.A.: SimUSR: a simple but strong baseline for unsupervised image super-resolution. In: Proceedings of the IEEE/CVF Conference on Computer Vision and Pattern Recognition Workshops, pp. 474–475 (2020)
2. Bulat, A., Yang, J., Tzimiropoulos, G.: To learn image super-resolution, use a GAN to learn how to do image degradation first. In: Ferrari, V., Hebert, M., Sminchisescu, C., Weiss, Y. (eds.) ECCV 2018. LNCS, vol. 11210, pp. 187–202. Springer, Cham (2018). https://doi.org/10.1007/978-3-030-01231-1_12
3. Dong, C., Loy, C.C., He, K., Tang, X.: Image super-resolution using deep convolutional networks. IEEE Trans. Pattern Anal. Mach. Intell. **38**(2), 295–307 (2015)
4. Feng, R., Zhao, D., Zha, Z.J.: Understanding noise injection in GANs. In: International Conference on Machine Learning, pp. 3284–3293. PMLR (2021)
5. Fritsche, M., Gu, S., Timofte, R.: Frequency separation for real-world super-resolution. In: 2019 IEEE/CVF International Conference on Computer Vision Workshop (ICCVW), pp. 3599–3608. IEEE (2019)
6. Goodfellow, I., et al.: Generative adversarial nets. In: Advances in Neural Information Processing Systems, vol. 27 (2014)
7. He, J., Jia, X., Chen, S., Liu, J.: Multi-source domain adaptation with collaborative learning for semantic segmentation. arXiv preprint arXiv:2103.04717 (2021)
8. Ho, J., Jain, A., Abbeel, P.: Denoising diffusion probabilistic models. Adv. Neural. Inf. Process. Syst. **33**, 6840–6851 (2020)
9. Ji, X., Cao, Y., Tai, Y., Wang, C., Li, J., Huang, F.: Real-world super-resolution via Kernel estimation and noise injection. In: Proceedings of the IEEE/CVF Conference on Computer Vision and Pattern Recognition Workshops, pp. 466–467 (2020)
10. Karras, T., Laine, S., Aila, T.: A style-based generator architecture for generative adversarial networks. In: Proceedings of the IEEE/CVF Conference on Computer Vision and Pattern Recognition, pp. 4401–4410 (2019)
11. Karras, T., Laine, S., Aila, T.: A style-based generator architecture for generative adversarial networks. In: Proceedings of the IEEE/CVF Conference on Computer Vision and Pattern Recognition (CVPR) (2019)
12. Kim, J., Lee, J.K., Lee, K.M.: Accurate image super-resolution using very deep convolutional networks. In: Proceedings of the IEEE Conference on Computer Vision and Pattern Recognition, pp. 1646–1654 (2016)
13. Kingma, D.P., Salimans, T., Jozefowicz, R., Chen, X., Sutskever, I., Welling, M.: Improved variational inference with inverse autoregressive flow. Adv. Neural. Inf. Process. Syst. **29**, 4743–4751 (2016)
14. Lim, B., Son, S., Kim, H., Nah, S., Mu Lee, K.: Enhanced deep residual networks for single image super-resolution. In: Proceedings of the IEEE Conference on Computer Vision and Pattern Recognition Workshops, pp. 136–144 (2017)
15. Lugmayr, A., Danelljan, M., Timofte, R.: Unsupervised learning for real-world super-resolution. In: 2019 IEEE/CVF International Conference on Computer Vision Workshop (ICCVW), pp. 3408–3416. IEEE (2019)

16. Lugmayr, A., Danelljan, M., Timofte, R.: NTIRE 2020 challenge on real-world image super-resolution: methods and results. In: Proceedings of the IEEE/CVF Conference on Computer Vision and Pattern Recognition Workshops, pp. 494–495 (2020)
17. Lugmayr, A., et al.: Aim 2019 challenge on real-world image super-resolution: methods and results. In: 2019 IEEE/CVF International Conference on Computer Vision Workshop (ICCVW), pp. 3575–3583. IEEE (2019)
18. Mescheder, L., Geiger, A., Nowozin, S.: Which training methods for GANs do actually converge? In: International Conference on Machine Learning, pp. 3481–3490. PMLR (2018)
19. Niu, B., et al.: Single image super-resolution via a holistic attention network. In: Vedaldi, A., Bischof, H., Brox, T., Frahm, J.-M. (eds.) ECCV 2020. LNCS, vol. 12357, pp. 191–207. Springer, Cham (2020). https://doi.org/10.1007/978-3-030-58610-2_12
20. Shi, W., et al.: Real-time single image and video super-resolution using an efficient sub-pixel convolutional neural network. In: Proceedings of the IEEE Conference on Computer Vision and Pattern Recognition, pp. 1874–1883 (2016)
21. Song, Y., Ermon, S.: Generative modeling by estimating gradients of the data distribution. In: Advances in Neural Information Processing Systems, vol. 32 (2019)
22. Vahdat, A., Kautz, J.: NVAE: a deep hierarchical variational autoencoder. arXiv preprint arXiv:2007.03898 (2020)
23. Wang, L., et al.: Unsupervised degradation representation learning for blind super-resolution. In: Proceedings of the IEEE/CVF Conference on Computer Vision and Pattern Recognition, pp. 10581–10590 (2021)
24. Wang, X., Xie, L., Dong, C., Shan, Y.: Real-ESRGAN: training real-world blind super-resolution with pure synthetic data. In: Proceedings of the IEEE/CVF International Conference on Computer Vision, pp. 1905–1914 (2021)
25. Wang, X., et al.: ESRGAN: enhanced super-resolution generative adversarial networks. In: Leal-Taixé, L., Roth, S. (eds.) ECCV 2018. LNCS, vol. 11133, pp. 63–79. Springer, Cham (2019). https://doi.org/10.1007/978-3-030-11021-5_5
26. Wei, Y., Gu, S., Li, Y., Jin, L.: Unsupervised real-world image super resolution via domain-distance aware training. arXiv preprint arXiv:2004.01178 (2020)
27. Wolf, V., Lugmayr, A., Danelljan, M., Van Gool, L., Timofte, R.: DeFlow: learning complex image degradations from unpaired data with conditional flows. In: Proceedings of the IEEE/CVF Conference on Computer Vision and Pattern Recognition, pp. 94–103 (2021)
28. Yoo, J., Ahn, N., Sohn, K.A.: Rethinking data augmentation for image super-resolution: a comprehensive analysis and a new strategy. In: Proceedings of the IEEE/CVF Conference on Computer Vision and Pattern Recognition, pp. 8375–8384 (2020)
29. Yuan, Y., Liu, S., Zhang, J., Zhang, Y., Dong, C., Lin, L.: Unsupervised image super-resolution using cycle-in-cycle generative adversarial networks. In: Proceedings of the IEEE Conference on Computer Vision and Pattern Recognition Workshops, pp. 701–710 (2018)
30. Zhang, Y., Li, K., Li, K., Wang, L., Zhong, B., Fu, Y.: Image super-resolution using very deep residual channel attention networks. In: Ferrari, V., Hebert, M., Sminchisescu, C., Weiss, Y. (eds.) ECCV 2018. LNCS, vol. 11211, pp. 294–310. Springer, Cham (2018). https://doi.org/10.1007/978-3-030-01234-2_18

31. Zhao, S., Li, B., Xu, P., Keutzer, K.: Multi-source domain adaptation in the deep learning era: a systematic survey. arXiv preprint arXiv:2002.12169 (2020)
32. Zhu, J.Y., Park, T., Isola, P., Efros, A.A.: Unpaired image-to-image translation using cycle-consistent adversarial networks. In: Proceedings of the IEEE International Conference on Computer Vision, pp. 2223–2232 (2017)

Out-of-Distribution Detection Without Class Labels

Niv Cohen(✉), Ron Abutbul, and Yedid Hoshen

School of Computer Science and Engineering, The Hebrew University of Jerusalem, Jerusalem, Israel
nivc@cse.huji.ac.il

Abstract. Out-of-distribution detection seeks to identify novelties, samples that deviate from the norm. The task has been found to be quite challenging, particularly in the case where the normal data distribution consist of multiple semantic classes (e.g. multiple object categories). To overcome this challenge, current approaches require manual labeling of the normal images provided during training. In this work, we tackle multi-class novelty detection *without* class labels. Our simple but effective solution consists of two stages: we first discover "pseudo-class" labels using unsupervised clustering. Then using these pseudo-class labels, we are able to use standard *supervised* out-of-distribution detection methods. We verify the performance of our method by favorable comparison to the state-of-the-art, and provide extensive analysis and ablations.

1 Introduction

Detecting novelties, images that are semantically different from normal ones, is a key ability required by many intelligent systems. Some applications include: detecting unknown, interesting scientific phenomena (e.g. new star categories such as supernovae), or detecting safety-critical events (e.g. alerting an autonomous car when an unexpected object is encountered). In this paper, we assume that a training set entirely consisting of normal images is provided. A trained model is used at test time to classify new samples as normal or anomalous, i.e. similar or different from previously seen training samples. The normal data may consist of one or more semantic classes (e.g. "dog", "cat"), where the case of a single semantic class is of particular interest (Table 1).

Table 1. Comparison of different novelty detection settings

Normal Data	One-class (OCC)	Multi-class (OOD)	Our Setting
Multi-class	✗[a]	✓	✓
Without Labels	✓	✗	✓

[a] A few novelty detection methods do evaluate on multi-class data without labels, and we compare to them in this work

L. Karlinsky et al. (Eds.): ECCV 2022 Workshops, LNCS 13802, pp. 101–117, 2023.
https://doi.org/10.1007/978-3-031-25063-7_7

Most recent works that address out-of-distribution detection (OOD) have relied on *supervised* class labels for each normal sample. On the other hand, methods that do not assume such labels have mostly focused on single-class data novelty detection. It is common to assume a normal-only dataset during training, so methods that do not assume class labels for this set are referred to as unlabelled methods. This setting of out-of-distribution detection with a normal only train set, but without class labels for the normal data, or *multi-class anomaly detection* (AD), received little attention. .

Here, we propose a simple approach for anomaly detection on unlabeled, multi-class normal only data. It is based on the following principle: unlabeled multi-class AD can be approximated using unsupervised clustering followed by supervised multi-class AD. In practice, we perform unsupervised image clustering using SCAN [38], a recent state-of-the-art method. After obtaining the approximate labels for every image in the dataset, we can use them for adapting ImageNet-pre-trained features for the AD task. The adapted features are then used as input to generic AD methods e.g. k nearest neighbor (kNN) or the Mahalanobis distance.

Extensive experiments and ablations are performed to evaluate our method. We find that our method outperforms state-of-the-art self-supervised [37] and pre-trained [1,31] anomaly detection methods. As an additional result, we show that the features learned by the unsupervised clustering method are already competitive with the best self-supervised anomaly detection methods.

Our main contributions are:

1. Presenting a framework that uses unsupervised clustering to obtain pseudo-labels which are then used to adapt feature representations for multi-class anomaly detection.
2. State-of-the-art results for unlabeled, multi-class anomaly detection on popular datasets.
3. Demonstrating that deep self-supervised image clustering methods are excellent feature learners for self-supervised multi-class anomaly detection.

1.1 Related Works

Out-of-Distribution Detection (OOD). In some settings, the normal data are provided together with their semantic class ground truth labels. Using these labels for supervised training highly increases the model's ability to detect OOD samples [13]. A model trained to classify labelled semantic classes, may have less confidence in predicting the class label of a sample coming from a new distribution. Moreover, training a model to distinguish between semantic classes of the normal data enhances the sensitivity of the learnt features for the desired attribute. For example, a model trained to distinguish between animal species may learn a representation sensitive to their attributes (such as skin color and texture, head shape, etc.) and less sensitive to nuisance attributes (angle of view, lighting conditions, etc.). Using such representations, one can detect anomalous samples based on a large Mahalanobis distance in representation space, improving out-of-distribution detection capabilities [11,19].

One Class Classification. In the absence of positive anomaly examples, anomaly detection methods design inductive biases allowing models to detect anomalies without seeing any during training. The common thread behind these methods is learning a strong representation of the data with which separating between normal and anomalous data is easily done. Learning representations cannot be performed by using actual labels of the normal training data. Therefore, anomaly detection algorithms, as other self-supervised method (e.g. image clustering [38] or disentanglement [16]), use a variety of techniques to design the inductive bias of the model toward having desired properties. Such techniques include self-supervised training to increase the model sensitivity to the properties essential for the task at hand [12,14], simulating samples which may resemble expected anomalies [20], and data augmentation used to guide the model to ignore nuisance variation modes [37].

Adaptation of Pre-trained Representations. When pre-trained representations are available, they can be used to significantly improve anomaly detection accuracy. Early attempts suggested a compactness loss to map the normal training data closer together, relying on auto-encoder pretraining [34]. Similar techniques were adopted to ImageNet pre-trained features [28], and later improved with self-supervised feature adaptation [30,31,33]. Pre-trained representations were lately adapted for the OE setting as well [7,30].

Clustering-based Anomaly Detection. Most anomaly detection methods perform some form of density estimation of the normal data. When the normal training data is multi-class i.e. consists of multiple clusters, it is natural that the density model would include this inductive bias. Indeed, multiple clustering-based approaches have been used for anomaly detection including: K-means [24] and GMMs [21]. As clustering the raw data directly is unlikely to achieve strong results for images, deep learning methods have been applied to learn better representations, notably DAGMM [43]. In the past few years, clustering methods have not dominated anomaly detection benchmarks, even for multi-class data. In this paper, we revisit clustering-based anomaly detection and show that it achieves very strong performance for multi-class anomaly detection with and without pre-trained features. We note that SSD, a recently published work, uses K-means clustering for anomaly scoring in the setting of OOD without labels [35]. We show that K-means clustering is less successful than our method on this task.

Self-supervised Image Clustering. The task of clustering images without labels had been extensively studied. Most recent methods learn deep features simultaneously with optimizing the images cluster assignments [3]. Methods free to optimize their own features are prone to cluster according to nuisance properties, which the self-learnt feature might pick up. A large variety of techniques have been used to mitigate this problem, including augmentations [15] and contrastive feature learning [38].

2 Preliminaries

In this section we present the necessary background for this paper.

2.1 Deep Image Clustering

Recent years have seen significant progress in deep image clustering. This is mostly thanks to improvements in self-supervised representation learning. In this paper we take advantage of SCAN [38], a state-of-the-art unsupervised image clustering method. SCAN operates in three steps:

(i) Representation learning. SimCLR [4], a contrastive self-supervised representation learning method, is used to learn a strong feature extractor ϕ_{SimCLR}.
(ii) Classifier training. The feature extractor ϕ_{SimCLR} is used to compute the nearest neighbors for every training image. A classifier C is trained (initialized with the features of ϕ_{SimCLR}) to classify each training image x into one of K clusters. The classifier C is trained under two constraints: (i) an equal number of images are assigned to each cluster. Formally, the entropy of $E_{x \in X_{train}} C(x)$ is maximized (ii) each image should be confidently assigned to a similar cluster as its nearest neighbours. Formally, we minimize $\sum_{\tilde{x} \in kNN(x)} log(C(x) \cdot C(\tilde{x}))$. By the end of training the classifier produces a clustering of the training set.
(iii) Self-training. In the final stage, the clustering is improved using a self-training approach [23].

2.2 Feature-Adaptation for Labeled Multi-class Anomaly Detection

In the labeled multi-class anomaly detection task (also known as OOD), each normal training image x is labeled by its class label y. Standard OOD methods train a classifier C to predict the class probability vector $C(x)$ given x. The standard confidence-based score, measures for a test image x_{test} the maximum prediction probability (probability of the most confident class):

$$Confidence = \max\left(C(x_{test})\right) \tag{1}$$

Images with low maximal confidence, are denoted as anomalous. Many techniques have been proposed to improve the obtained OOD detection [39]. Here, we extend the work of Hendrycks et al. [14] and Fort et al. [11] which suggested initializing the classifier C with weights that are pre-trained on a large-scale datasets. This significantly increases the OOD detection accuracy. We propose a clustering-based technique for adapting these methods to unlabeled datasets.

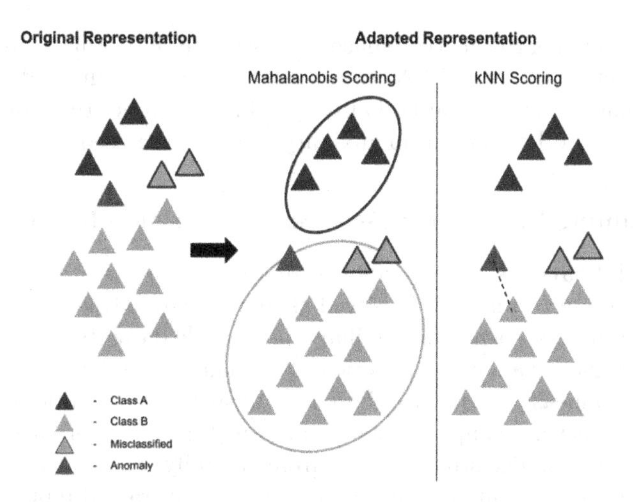

Fig. 1. An illustration of our feature adaptation and scoring procedure. Left - the data original unadpted pre-trained features. Feature adaptation increases the separation between the normal data (blue, green) and the anomaly (red). We illustrate that when labels are noisy w.r.t. the ground truth clusters, Mahalanobis distance scoring may misclassify anomalies (center, anomaly within the ellipses fitted to the normal data), while kNN scoring criterion (right, dashed line shows the anomaly's nearest in-distribution neighbour) may detect anomalies better. (Color figure online)

3 Deep Clustering for Multi-class Anomaly Detection

In this work, we deal with anomaly detection when the normal training data is composed of many different semantic classes but no labels. We suggest taking advantage of the multi-class nature of the normal data as an inductive bias. The knowledge that the normal data comes from different, relatively distinct classes, rather than a single uni-modal class, or other possible distributions, serves us to learn better representations. As illustrated in Fig. 1, adapting features for better inter-cluster separation can provide better discrimination of anomalies, even when the cluster labels are imperfect. We first propose a naive but effective approach for using unsupervised clustering method for anomaly detection. We then propose a more accurate, two-stage approach which combines unsupervised clustering with pre-trained features and supervised multi-class anomaly detection for this task.

3.1 Self-Supervised Clustering for Multi-class Anomaly Detection

We first address the fully self-supervised case where no pre-trained features are available. We suggest a surprisingly simple but effective approach for this setting. First, we run the SCAN clustering algorithm (see Sect.2) on the normal training images. We than use the features from the penultimate layer of the SCAN-trained classifier to represent each image. Anomalous samples are expected to be found in low-density area of this feature space. We estimate the normal samples density around a target image by its kNN distance to the normal training data. We show in Sect.4.2 that this simple method already achieves competitive results with the state-of-the-art methods when no auxiliary datasets are used.

3.2 Finetuning Pre-trained Features with Pseudo Labels

Pre-trained Features. Using effective representations is at the core of deep anomaly detection techniques. It is well established by [11,30] that transferring deep representations learned on auxiliary, large-scale datasets (e.g. ImageNet or CLIP) is effective for anomaly detection. The main reason for their effectiveness is the ability to measure semantic similarity. In such representations, the normal data lies in relatively compact regions and are typically more separated from semantic anomalies. The strong results from the fully self-supervise case above, motivate us to apply similar multi-class priors to pre-trained representation.

Feature Adaptation. Pre-trained representations are typically trained on large datasets, which are not necessarily representative of our normal in-distribution samples. Better AD results are obtained when the pre-trained features are adapted using the normal samples provided for training [30]. A naive approach to adapt pre-trained features to our multi-class normal data is to finetune them using the loss of a clustering method, such as SCAN. However, we found that this form of feature adaptation results in a representation worse than the initial one. We believe this is caused by catastrophic forgetting, where a network extensively trained for a new task loses capabilities and knowledge that it had during pretraining.

To overcome this limitation, we suggest a *two stage approach*. In the *first stage*, we simply train a self-supervised clustering method on the training data X_{train} achieving approximate "pseudo-labels" describing clusters of our normal training data. In the *second stage*, we transfer the knowledge obtained from the self-supervised clustering into the pre-trained network. To do this, we follow [11] and finetune our pre-trained network to classify training images into their pseudo-labels $\widetilde{y}_{train}$. We train the classifier C_{OOD}, using standard cross entropy loss $\mathcal{L}_{CE}$:

$$\mathcal{L}_{CE} = -\sum_{i} (\widetilde{y}_{train})_i \cdot \log \Big(C_{OOD}(X_{train})_i \Big) \tag{2}$$

Model Averaging. Although our two-stage approach is more stable than adaptation using the clustering objective directly, it can still lead to catastrophic forgetting. During finetuning, the network gains knowledge from the clusters we found, but forgets its pre-trained knowledge, which is crucial for the density estimation-based anomaly scoring we use. Although the model achieves its best performance on one particular epoch, it is non-trivial to select that epoch without relying on having anomalous samples during training. Therefore, we choose to score our anomalies with an average model, taken as the moving average of the weights of the model during different training epochs.

Anomaly Scoring. To detect anomalies, we use density estimation with the adapted features. Recent methods suggested a per-class Mahalanobis approach [11] deeming a sample normal if it lies within a small Mahalanobis distance to any of the cluster centers. However, when the provided labels are not accurate this approach might fail. False cluster assignment may distort the Mahalanobis distance which relies on the empirical covariance matrix of each cluster. Instead, we use kNN scoring which makes few assumptions on the distribution of the data. Precisely, we encode all train images and the target images using the feature encoder. We then compute the distance between the features of each target image and all the normal train images. The anomaly score for each target image is set as the minimum of the distances to all nomral train samples. This score is relatively robust and less sensitive to the accuracy of the clustering algorithm (Fig. 1).

4 Results

We evaluate our method on two commonly used OOD detection datasets. We use a variety of other datasets to simulate anomalies:

CIFAR-10 [18]: Contains images from 10 different classes, with a 32×32 resolution. We evaluate it against a variety of dataset supplied in a similar resolution, namely: CIFAR-100 [18], SVHN [25], LSUN [41]. The hardest benchmark here is the CIFAR-100 which contain the most similar classes to CIFAR-10. For LSUN we report both a version with some artifacts used by previous works, and a version suggested by CSI [37][1] where the downscaling was done more carefully to avoid these artifacts. We do not compare on ImageNet [8] data here, as this dataset was used for pretraining.

ImageNet-30: Contain images from 30 classes of high resolution images chosen from the ImageNet [8] dataset. Accordingly, we evaluate it against of a variety of datasets that have similar resolution, namely: CUB-200 [40], Dogs [8], Pets [27], Flowers [26], Food [2], Places [42], Caltech [10].

4.1 Multi-class Anomaly Detection with Pre-Trained Features

In this section we provide results for the case where pre-trained features are available.

Methods: We evaluate anomaly detection methods that utilize pre-trained features, and do not require class labels for the train data. We present a comparison to a naive initialization of unsupervised clustering with pre-trained features in Sect.5.

Deep Nearest Neighbors (DN2) [30]: A kNN density estimation method based on pre-trained features. Test samples are scored according to the distance to their nearest normal training images. A larger distance indicates a low density

[1] https://github.com/alinlab/CSI.

of normal samples, and therefore a high probability of abnormality. We follow *MeanShifted* [31] by using the cosine similarity distance instead of l_2 distance.

Mean Shifted Contrastive Loss (MSCL) [31]: A-state-of-the-art method for adapting pre-trained features for anomaly detection using contrastive learning. It suggests to compute the contrastive loss around the mean of the normal training data features to mitigate catastrophic forgetting.

Ours: The two stage clustering with pseudo-label adaptation method described in Sect.3.1. The hyperparameters and implementation details are described in Sect.4.3.

Comparison: As can be seen in Table 2 and Table 3, feature adaptation has significantly improved our multi-class anomaly detection performance. A substantial part of our improvement comes from using large network architectures. We emphasize that the ability to easily use readily-available pre-trained networks is inherent to our approach. Results were also improved in MSCL, but as MSCL is not explicitly designed to deal with multi-class data, our method significantly outperforms it. Using more powerful pretraining networks (as in [11]) can also improve the results of our method. We find that ViT [9] pre-trained on ImageNet-21 [17] achieves better results than ResNet152 pre-trained on ImageNet. As ImageNet pretraining include the exact ImageNet class labels of the ImageNet-30 dataset we wish to avoid in our setting we used a CLIP [29] pre-trained ViT architecture for all pre-trained methods (including ours) in the ImageNet-30 evaluations.

When we compare between the kNN image retrievals of the original pre-trained features and our adapted features, we observe some differences. As can be seen in Fig. 2 the raw pre-trained features retrieve images from different classes. These images are similar to our target out-of-distribution target sample in various semantic attribute. Adapted features, however, tend to associate our target image with a single normal class, resulting in a lower similarity between the target image and the retrieved normal ones. This is desirable, as we want to distinguish out-of-distribution images from our normal data.

4.2 Multi-class Anomaly Detection Without Pretraining

Although strong pre-trained features are often available, we cannot always assume their availability. In this case, we find that self-supervised clustering on its own can often learn a good enough representation to outperform previous state-of-the-art on unlabelled multi-class anomaly detection without pretraining.

Methods: We compare the following methods:

Density Based Methods: Classical methods use direct density estimation techniques to estimate the likelihood of the data. Different methods suggested modification to this score to account for the dataset statistics (*Likelihood Ratio* [32]) or its complexity (*Input Complexity* [36]).

Fig. 2. ***Raw pre-trained features vs. our finetuned features:*** For each out-of-distribution image from the Caltech-256 dataset, the top 5 nearest neighbors from the in-distribution data (ImageNet-30) are shown according to their order. Note how raw pre-trained features (Raw) neighbors are chosen from all classes. In contrast, finetuned pre-trained features with our method (FT) neighbors belong to a single semantic class.

Rot [14]: A classification based method utilizing an auxiliary task of rotation prediction for self-supervised detection of anomalies.

GOAD [1]: Another rotation-prediction method, that proposed to learn a feature space where inter-class separation of the normal data is relatively small.

CSI [37]: A contrastive learning method which contrasts against distribution-shifted augmentations of the data samples along with other samples.

SSD [35]*:* A self-supervised method with similar features to CSI [37]. It scores anomalies using the Mahalanobis distance with respect to K-means clusters. Although this method is somewhat similar to ours, it relies on Mahalanobis distance and K-means clustering, which are not optimal for feature adaptation without labels (see Sect.5). We note that *SSD* [35] also reports higher results for the ResNet-34 architecture, which is a non-standard evaluation for AD without pretraining methods.

SCAN Features [38]: Features taken from the last stage of of the SCAN clustering methods, used to score anomalies as in DN2 (see Sect.3.2).

Table 2. OOD detection without class labels on CIFAR-10 ROCAUC(%)

		Network	CIFAR-100	SVHN	LSUN	LSUN (FIX)
	Likelihood	Glow	58.2	8.3	-	-
	Likelihood Ratio [32]	PixelCNN++	-	91.2	-	-
	Input Complexity [36]	Glow	73.6	95.0	-	-
	Rot [14]	ResNet-18	82.3	97.8	92.8	81.6
	GOAD [1]	ResNet-18	77.2	96.3	89.3	78.8
	CSI [37]	ResNet-18	89.2	99.8	97.5	90.3
	SSD [35]	ResNet-18	89.6	-	-	-
	SCAN Features	ResNet-18	90.2	94.3	92.4	92.1
Pretrained	DN2 [30]	ResNet-18	83.3	95.0	91.0	88.9
	DN2 [30]	ResNet-152	86.5	96.2	88.7	86.7
	MSCL [31]	ResNet-152	90.0	98.6	90.6	92.6
	Ours	ResNet-18	90.8	98.6	98.6	94.3
	Ours	ResNet-152	93.3	98.8	95.4	95.7
	Ours	ViT	**96.7**	**99.9**	**99.3**	**99.1**

Table 3. OOD detection without class labels on ImageNet-30 ROCAUC(%).

		Network	CUB-200	Dogs	Pets	Flowers-102	Food-101	Places-365	Caltech-256
	Rot	ResNet-18	74.5	77.8	70.0	86.3	71.6	53.1	70.0
	GOAD	ResNet-18	71.5	74.3	65.5	82.8	68.7	51.0	67.4
	CSI	ResNet-18	90.5	**97.1**	85.2	94.7	89.2	78.3	87.1
Pretr.	DN2	CLIP ViT	93.8	94.2	89.7	93.2	95.7	**96.7**	90.3
	Ours	CLIP ViT	**99.4**	95.9	**94.9**	**98.3**	**96.4**	96.1	**94.4**

Comparison: As can be seen in Table 2, simple utilization of SCAN features often performs better or on par with the top competing self-supervised method. We note that the last stage of the SCAN method, namely, self-labelling, is somewhat similar to adapting on pseudo-label performed by our method. Therefore, we do not expect our method to provide further gains over SCAN's final representation.

Although clustering methods such as K-means and GMM have classically been very popular, most recent deep learning methods do not use clustering. The results reported here provide strong motivation for revisiting self-supervised clustering methods for anomaly detection. We conclude that relying on the multi-class distribution prior often allows us to outperform other methods, even in the setting where no pretraining is allowed.

4.3 Implementation Details

Clustering: We use SCAN's official implementation[2] for all clustering tasks unless mentioned otherwise. We ran all of our clustering algorithms with the same number of clusters $K = 10$. We use the SCAN algorithm's default parameters for each dataset. For ImageNet-30 dataset we use the configuration originally provided by the authors for the ImageNet-50 dataset. We note that SCAN unsupervised image clustering use a MoCo [5] pretraining on the entire ImageNet dataset (pre-trained without labels). We therefore do not compare it to self-supervised methods that do not use pretraining [1,14,35,37].

Pretraining: For all models using ResNet152, we use ImageNet [8] pretraining. For ViT, we used ImageNet-21 [17] pretraining. To evaluate our method with ImageNet-30, we used the CLIP [29] ViT visual head rather than pretraining on labelled ImageNet data (Table 3).

Optimization: We ran the adaptation for 5 epochs using an Adam optimizer. We used Cosine Annealing learning rate scheduler with an initial learning rate of $1e-5$ and final learning rate of $1e-6$.

Model Averaging: We average on the model weights at the end of each of the 5 training epochs.

Scoring: We used $k = 1$ for all our kNN evaluations.

Comparison to MSCL: For the *MSCL* comparison we experimented with 5, 10 and 100 training epochs, and chose the best performing number of epochs. The rest of the parameters were left unchanged.

5 Discussion

Do we actually need our two-stage approach? We compare between pre-trained feature adaptation using two approaches: (i) Our two stage clustering with pseudo-class labels supervised adaptation (ii) A single-stage initialization of the unsupervised clustering method with pre-trained features. In the latter, we initialize the SimCLR features used in SCAN with ResNet152 ImageNet pre-trained features. The results are presented in Table 4. The full faeture adaptation

Table 4. Comparison between OOD detection w.o. labels using the representations obtained by different stages of the SCAN clustering method (CIFAR-10 vs. CIFAR-100) ROCAUC (%). The results show that the naive approach of directly using SCAN with pre-trained features falls behind our two-stage approach.

Ours (ResNet-152)	SIMCLR	SCAN	Self-labelling
93.3	87.3	83.4	89.3

[2] https://github.com/wvangansbeke/Unsupervised-Classification.

on the CIFAR-10 vs. CIFAR-100 evaluation results in 89.3% ROCAUC which is far worse than the 93.3% results achieved by our two-stage approach. During the clustering stage, the pre-trained features are used find the data clusters, performance but at the same time, the features deteriorate further away from their pre-trained initialization. This justifies our two stage approach, which better preserves the expressivity of the pre-trained features.

Is kNN density estimation preferable to other anomaly scoring methods? Sometimes. Previous works used a variety of scoring criterion for multi-class anomaly detection [33,34]. Although the Mahalanobis distance was shown to give stronger results in a previous work, it is sensitive to the approximate labels that are typically generated by self-supervised clustering (see Fig. 1 for an illustration). A comparison between the different scoring methods can be seen in Table 5. We find that the optimal scoring method may differ between datasets. Following previous works [6], all our main results are reported using kNN with $k = 1$.

Table 5. Comparison between scoring methods ROCAUC (%)

Scoring method	1NN	2NN	5NN	10NN	Mahalanobis	Confidence
CIFAR10-CIFAR100	96.7	96.8	**96.9**	96.8	96.6	92.9
CIFAR10-SVHN	**99.9**	**99.9**	**99.9**	**99.9**	92.4	97.1
ImageNet30-CUB200	99.3	99.4	99.4	99.4	99.6	**99.7**

Can simple K-means clustering be used to finetune our features? To a small extent. We evaluated the labels obtained by K means clustering as an alternative for the more complicated SCAN clustering method. Although the adaptation yielded small improvements, the results underperformed SCAN significantly. For example the multi-class anomaly detection results only improved from 86.5% ROCAUC to 87.3% (CIFAR-10 vs. CIFAR-100).

Table 6. Comparison of different numbers of clusters K (ImageNet-30 vs. CUB200) ROCAUC (%)

K	10	20	30	No Adaptation
	99.1	98.3	98.9	93.8

Can using a larger model with ImageNet pretraining assist the SCAN clustering performance? Not significantly. We tried to initialize the SCAN algorithm with a ResNet152 pre-trained on ImageNet, but achieved only a minor improvement in the clustering accuracy. Although adaptation of pre-trained features using SCAN results in catastrophic forgetting, it is very likely that image clustering accuracy can be substantially improved with a method designed to take advantage of pre-trained representations. Such future improvements in image clustering are likely to directly enhance the results of our approach too.

Does model averaging result in a good representation with comparison to individual checkpoints? Yes. Model weight averaging can yield a similar (or better) multi-class anomaly detection accuracy compared to that of the optimal epoch along the training process. For example, on CIFAR-10 vs. CIFAR-100 with pre-trained ResNet152, the initial pre-trained features scored anomalies with 86.5% ROCAUC. The 5 individual training epochs scored 92.9%, 88.9%, 92.9%, 92.6%, 92.3%. The averaged model scored 93.3% ROCAUC. This phenomenon has also been noticed by previous works [22].

What is the sensitivity of our method to random repetitions? We do not provide error bars for each of our runs since our results are fairly consistent. As a typical case, we ran 3 repetitions of our approach for the CIFAR10-CIFAR100 and CIFAR10-LSUN (fix) experiments. The standard deviations were 0.4% and 0.5% respectively.

What is the sensitivity of our method to the number of clusters K? While changing K may somewhat vary the results, our method can often perform useful adaptation without knowing the exact ground truth number of classes (Table 6). When available, we advise using the ground-truth number of clusters.

How does adaptation with the pseudo-labels obtained by clustering compare to using the ground-truth labels? There is still a significant gap between the pseudo-labels provided by our clustering algorithm and the ground truth class labels (on CIFAR-10, SCAN clustering accuracy is 88.3%). Yet, the gap in the multi-class anomaly detection results seems to be smaller. For example, in the CIFAR-10 vs. CIFAR-100 multi-class anomaly detection task, our method achieves 96.7% ROCAUC, while a similar method by Fort et al. [11] achieves 98.4% ROCAUC using the ground truth labels. Even though this a significant difference, we find it is reasonable given the significant inaccuracy of self-supervised clustering we used.

Relation to auxiliary-task based anomaly detection. Our work can be seen as an extension of a line of methods utilizing auxiliary tasks to address a one-class-classification setting. Such methods were previously suggested for image anomaly detection [12,14], and also for other data modalities [1]. These methods rely on predefined augmentations to create an auxiliary task in order to guide the model learning toward meaningful properties of our data. The prediction of pseudo-labels from clustering algorithms can be viewed as another example of such auxiliary work. Future works may suggest new kinds of data-adaptive auxiliary tasks for similar settings.

New Data Modalities. Multi-class anomaly detection may be encountered in other modalities beyond images. Data modalities where transfer learning and self-supervised learning show promising results include natural language, video and audio. Therefore, we believe similar methods may provide comparable improvements in these modalities.

Guided Anomaly Detection. Another interesting application of clustering based out-of-distribution detection is using the pseudo labels to guide our anomaly detection algorithm toward the type of samples that we wish to consider as anomalies. For example, if one wishes to define anomalies according to colors, they may use augmentations to guide the clustering procedure toward finding color-based clusters. Similarly, one may wish to ignore an attribute, and guide the clustering algorithm accordingly.

6 Limitations

Highly Unbalanced Multi-class Datasets. Most state-of-the-art clustering algorithms rely on the assumption of an approximately equal split of the data among the classes. As our method relies on such algorithms, performance may decrease on imbalanced datasets. We expect that future clustering methods will overcome this limitations, consequently also freeing our method from this issue.

Fine-grained and Non-standard Classes. Many self-supervised learning algorithms, including self-supervised clustering, heavily rely on augmentations and other sources of inductive bias. The inductive bias in these methods guides the model to be sensitive to a single salient object in the center of the image. In cases when the anomalies or the semantic classes in the normal data are not object-centric, current self-supervised methods may not perform as well. While this is a limitation of our method, it is similarly a limitation of many other anomaly detection techniques reliant on self-supervised learning, including nearly all competing baseline methods. We therefore consider this as a limitation of the field in general, rather than a limitation specific to our method.

Pre-trained Features. The main results of our paper are reliant on pre-trained features. As discussed, pre-trained features achieve strong results on most datasets, but may not be a good choice in some settings. One possible solution to this limitation, is combining pre-trained and fully self-supervised methods in scenarios where the preferred method cannot be determined in advance.

7 Conclusion

We address the problem of multi-class anomaly detection without labels. We propose a conceptually simple but effective method to combine recent improvements in unsupervised clustering with pre-trained features for anomaly detection. Our approach outperforms state-of-the-art feature adaptation methods for multi-class anomaly detection. Future work may utilize future improvements in self-supervised clustering of image datasets to further improve results. We also expect similar improvements in other data modalities such as natural language, video, audio and time-series.

Acknowledgements. This work was partly supported by the Malvina and Solomon Pollack scholarship and, the Federmann Cyber Security Research Center in conjunction with the Israel National Cyber Directorate.

References

1. Bergman, L., Hoshen, Y.: Classification-based anomaly detection for general data. In: ICLR (2020)
2. Bossard, L., Guillaumin, M., Van Gool, L.: Food-101 – Mining Discriminative Components with Random Forests. In: Fleet, D., Pajdla, T., Schiele, B., Tuytelaars, T. (eds.) ECCV 2014. LNCS, vol. 8694, pp. 446–461. Springer, Cham (2014). https://doi.org/10.1007/978-3-319-10599-4_29
3. Caron, M., Bojanowski, P., Joulin, A., Douze, M.: Deep Clustering for Unsupervised Learning of Visual Features. In: Ferrari, V., Hebert, M., Sminchisescu, C., Weiss, Y. (eds.) Computer Vision – ECCV 2018. LNCS, vol. 11218, pp. 139–156. Springer, Cham (2018). https://doi.org/10.1007/978-3-030-01264-9_9
4. Chen, T., Kornblith, S., Norouzi, M., Hinton, G.: A simple framework for contrastive learning of visual representations. In: International Conference on Machine Learning, pp. 1597–1607. PMLR (2020)
5. Chen, X., Fan, H., Girshick, R., He, K.: Improved baselines with momentum contrastive learning. (2020) arXiv preprint arXiv:2003.04297
6. Cohen, N., Hoshen, Y.: Sub-Image anomaly detection with deep pyramid correspondences. (2020) arXiv preprint arXiv:2005.02357
7. Deecke, L., Ruff, L., Vandermeulen, R.A., Bilen, H.: Transfer-based semantic anomaly detection. In: International Conference on Machine Learning, pp. 2546–2558. PMLR (2021)
8. Deng, J., Dong, W., Socher, R., Li, L.J., Li, K., Fei-Fei, L.: ImageNet: A large-scale hierarchical image database. In: 2009 IEEE conference on Computer Vision and Pattern Recognition, pp. 248–255. IEEE (2009)
9. Dosovitskiy, A., et al.: An image is worth 16x16 words: Transformers for image recognition at scale. (2020) arXiv preprint arXiv:2010.11929
10. Fei-Fei, L., Fergus, R., Perona, P.: Learning generative visual models from few training examples: an incremental bayesian approach tested on 101 object categories. In: 2004 Conference on Computer Vision and Pattern Recognition Workshop, pp. 178–178. IEEE (2004)
11. Fort, S., Ren, J., Lakshminarayanan, B.: Exploring the limits of out-of-distribution detection. (2021) arXiv preprint arXiv:2106.03004
12. Golan, I., El-Yaniv, R.: Deep anomaly detection using geometric transformations. (2018) arXiv preprint arXiv:1805.10917
13. Hendrycks, D., Gimpel, K.: A baseline for detecting misclassified and out-of-distribution examples in neural networks. (2016) arXiv preprint arXiv:1610.02136
14. Hendrycks, D., Mazeika, M., Kadavath, S., Song, D.: Using self-supervised learning can improve model robustness and uncertainty. (2019) arXiv preprint arXiv:1906.12340
15. Ji, X., Henriques, J.F., Vedaldi, A.: Invariant information clustering for unsupervised image classification and segmentation. In: Proceedings of the IEEE/CVF International Conference on Computer Vision, pp. 9865–9874 (2019)
16. Kingma, D.P., Welling, M.: Auto-encoding variational bayes. (2013) arXiv preprint arXiv:1312.6114
17. Kolesnikov, A., et al.: Big Transfer (BiT): General Visual Representation Learning. In: Vedaldi, A., Bischof, H., Brox, T., Frahm, J.-M. (eds.) ECCV 2020. LNCS, vol. 12350, pp. 491–507. Springer, Cham (2020). https://doi.org/10.1007/978-3-030-58558-7_29

18. Krizhevsky, A., Hinton, G., et al.: Learning multiple layers of features from tiny images (2009)
19. Lee, K., Lee, K., Lee, H., Shin, J.: A simple unified framework for detecting out-of-distribution samples and adversarial attacks. Advances in Neural Information Processing Systems. **31** (2018)
20. Li, C.L., Sohn, K., Yoon, J., Pfister, T.: Cutpaste: self-supervised learning for anomaly detection and localization. In: Proceedings of the IEEE/CVF Conference on Computer Vision and Pattern Recognition, pp. 9664–9674 (2021)
21. Li, L., Hansman, R.J., Palacios, R., Welsch, R.: Anomaly detection via a gaussian mixture model for flight operation and safety monitoring. Trans. Res. Part C Emerg. Technol. **64**, 45–57 (2016)
22. Matena, M., Raffel, C.: Merging models with fisher-weighted averaging. (2021) arXiv preprint arXiv:2111.09832
23. McLachlan, G.J.: Iterative reclassification procedure for constructing an asymptotically optimal rule of allocation in discriminant analysis. J. Am. Stat. Assoc. **70**(350), 365–369 (1975)
24. Münz, G., Li, S., Carle, G.: Traffic anomaly detection using k-means clustering. In: GI/ITG Workshop MMBnet, pp. 13–14 (2007)
25. Netzer, Y., Wang, T., Coates, A., Bissacco, A., Wu, B., Ng, A.Y.: Reading digits in natural images with unsupervised feature learning (2011)
26. Nilsback, M.E., Zisserman, A.: Automated flower classification over a large number of classes. In: Indian Conference on Computer Vision, Graphics and Image Processing, pp. 722-729 (2008)
27. Parkhi, O.M., Vedaldi, A., Zisserman, A., Jawahar, C.V.: Cats and dogs. In: IEEE Conference on Computer Vision and Pattern Recognition (2012)
28. Perera, P., Patel, V.M.: Learning deep features for one-class classification. IEEE Trans. Image Process. **28**(11), 5450–5463 (2019)
29. Radford, A., et al.: Learning transferable visual models from natural language supervision. In: International Conference on Machine Learning, pp. 8748–8763. PMLR (2021)
30. Reiss, T., Cohen, N., Bergman, L., Hoshen, Y.: Panda: adapting pretrained features for anomaly detection and segmentation. In: Proceedings of the IEEE/CVF Conference on Computer Vision and Pattern Recognition, pp. 2806–2814 (2021)
31. Reiss, T., Hoshen, Y.: Mean-shifted contrastive loss for anomaly detection. (2021) arXiv preprint arXiv:2106.03844
32. Ren, J., et al.: Likelihood ratios for out-of-distribution detection. (2019) arXiv preprint arXiv:1906.02845
33. Rippel, O., Chavan, A., Lei, C., Merhof, D.: Transfer learning gaussian anomaly detection by fine-tuning representations. (2021) arXiv preprint arXiv:2108.04116
34. Ruff, L., et al.: Deep one-class classification. In: International conference on machine learning, pp. 4393–4402. PMLR (2018)
35. Sehwag, V., Chiang, M., Mittal, P.: SSD: a unified framework for self-supervised outlier detection. (2021) arXiv preprint arXiv:2103.12051
36. Serrà, J., Álvarez, D., Gómez, V., Slizovskaia, O., Núñez, J.F., Luque, J.: Input complexity and out-of-distribution detection with likelihood-based generative models. (2019) arXiv preprint arXiv:1909.11480
37. Tack, J., Mo, S., Jeong, J., Shin, J.: Csi: novelty detection via contrastive learning on distributionally shifted instances. In: Adv. Neural Inf. Process. Syst. (NeurIPS) **33**, 11839–11852 (2020)

38. Van Gansbeke, W., Vandenhende, S., Georgoulis, S., Proesmans, M., Van Gool, L.: SCAN: Learning to Classify Images Without Labels. In: Vedaldi, A., Bischof, H., Brox, T., Frahm, J.-M. (eds.) ECCV 2020. LNCS, vol. 12355, pp. 268–285. Springer, Cham (2020). https://doi.org/10.1007/978-3-030-58607-2_16
39. Wei, H., Xie, R., Cheng, H., Feng, L., An, B., Li, Y.: Mitigating neural network overconfidence with logit normalization. (2022) arXiv preprint arXiv:2205.09310
40. Welinder, P., et al.: Caltech-UCSD Birds 200. Tech. Rep. CNS-TR-2010-001, California Institute of Technology (2010)
41. Yu, F., Seff, A., Zhang, Y., Song, S., Funkhouser, T., Xiao, J.: LSUN: Construction of a large-scale image dataset using deep learning with humans in the loop. (2015) arXiv preprint arXiv:1506.03365
42. Zhou, B., Lapedriza, A., Xiao, J., Torralba, A., Oliva, A.: Learning deep features for scene recognition using places database **27** (2014)
43. Zong, B., et al.: Deep autoencoding gaussian mixture model for unsupervised anomaly detection. In: International Conference on Learning Representations (2018)

Unsupervised Domain Adaptive Object Detection with Class Label Shift Weighted Local Features

Andong Tan[1,2(✉)], Niklas Hanselmann[1,3], Shuxiao Ding[1,4], Federico Tombari[2,5], and Marius Cordts[1]

[1] Mercedes-Benz AG R&D, Stuttgart, Germany
andong.tan@tum.de,
{niklas.hanselmann,shuxiao.ding,marius.cordts}@mercedes-benz.com
[2] Technical University of Munich, Munich, Germany
tombari@in.tum.de
[3] University of Tuebingen, Tuebingen, Germany
[4] University of Bonn, Bonn, Germany
[5] Google, Zurich, Switzerland

Abstract. Due to the high transferability of features extracted from early layers (called local features), aligning marginal distributions of local features has achieved compelling results in unsupervised domain adaptive object detection. However, such marginal feature alignment suffers from the class label shift between source and target domains. Existing class label shift correction methods focus on image classification, and cannot be directly applied to object detection due to objects' co-occurrence. Meanwhile, one property of local features is that they have small receptive fields and can be easily mapped back to specific areas of input images. Therefore, to handle object co-occurrence scenarios, we propose to leverage this property to decompose the source feature maps and compute the source domain class distribution at the pixel level. The decomposition is based on each feature pixel's receptive field overlap with ground- truth bounding boxes. In the target domain, where no labels are available, we estimate this distribution using predicted bounding boxes and thus get the estimated class label shift between domains. This estimated shift is further used to re-weight source local features during the feature alignment. To the best of our knowledge, this is the first work trying to explicitly correct class label shift in unsupervised domain adaptive object detection. Experimental results demonstrate that this approach can systematically improve several recent domain adaptive object detectors, such as SW and HTCN on benchmark datasets with different degrees of class label shift.

Keywords: Unsupervised domain adaptation · Object detection · Class label shift

---

Supplementary Information The online version contains supplementary material available at https://doi.org/10.1007/978-3-031-25063-7_8.

L. Karlinsky et al. (Eds.): ECCV 2022 Workshops, LNCS 13802, pp. 118–133, 2023.
https://doi.org/10.1007/978-3-031-25063-7_8

1 Introduction

Deep learning has shown great success in object detection [19,21,22]. However, it relies heavily on large-scale annotated datasets, which are typically very expensive and sometimes difficult to obtain. In a dataset of interest without annotations (called target domain), directly applying models trained in a different annotated dataset (called source domain) typically witnesses performance drop due to domain shift [3]. Therefore, how to combat the domain shift and adapt models trained in the label-rich source domain to unlabeled target domain becomes an important problem. This setting is commonly referred to as unsupervised domain adaptation (UDA). In this work, we focus on the object detection task.

Recent work has shown that features extracted by early layers (local features) are more transferable than features extracted by later layers (global features) [28]. Based on this observation, many domain adaptive object detectors try to transfer local features between domains [4,23]. A popular approach for feature transfer in unsupervised domain adaptation [9] is the adversarial alignment of marginal feature distribution, aiming at encouraging domain-invariant representations. However, such alignment focuses on alleviating the covariate shift (different feature representation of same class) and ignores the class label shift (different class label distribution) between domains [2]. Recent work proves that ignoring the class label shift may even directly cause the failure of the algorithm [29]. In image classification, some methods are already proposed to correct such class label shift based on number of images belonging to each class [1,2,6,18]. These image-number-based mechanisms, however, assume that each image includes exactly one object belonging to one class. Under this assumption, the contribution of each class during the alignment of marginal feature distribution is directly coupled with the number of images from each class. In contrast, object detectors have to handle a much more

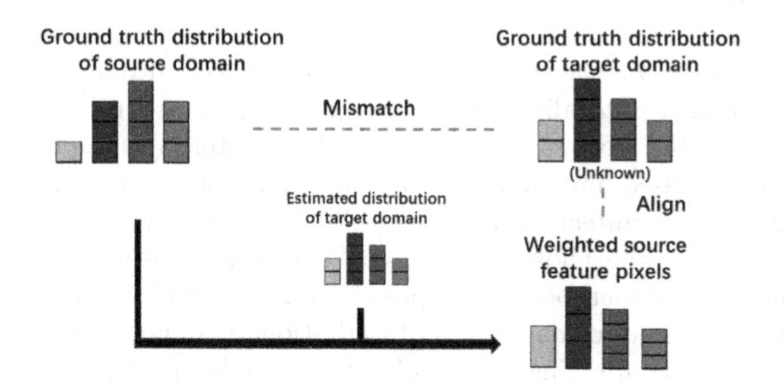

Fig. 1. The class label shift correction mechanism: source feature pixels are weighted according to the estimated class distribution of target domain in pixel level, such that the weighted marginal feature distribution of the source domain could be more similar to the target domain. In this scenario the estimation in the target domain is perfect. Every rectangle block indicates one feature pixel. Larger blocks have larger weights, and each column/color indicates a different class.

complicated scenario: one image might include multiple objects, each object could have a different size, and they may overlap with each other. We name this scenario object co-occurrence for convenience in later discussion. In this scenario, image-number-based methods can not be readily applied anymore.

To approach the class label shift problem in object detection we propose an iterative pixel-level class label shift estimation and correction mechanism based on the source feature maps' decomposition and re-weighting. The decomposition leverages the property that local features are typically characterized by small receptive fields and can be easily mapped back to specific areas of input images (Fig. 2). The re-weighting mechanism is inspired by domain adaptive classifiers trying to correct label shifts by weighting source image samples [1,2,6,18]. Instead of weighting image samples, we weight different areas of feature maps according to what class each feature pixel mainly belongs to. The primary motivation is trying to align the pixel-level class distributions in the source domain with those of the target domain by re-weighting the importance of source features from different classes, as described by Fig. 1.

Our contributions are summarized as follows: (1) To handle object co-occurrence scenarios in unsupervised domain adaptive object detection, we propose an intuitive mechanism to decompose source feature maps based on the assumption that the size of the overlap area between a feature pixel's receptive field and an object's bounding box approximately indicates the portion of information belonging to that object encoded in the current feature pixel. (2) We design an iterative approach to estimate, update, and utilize the pixel level class label shift for re-weighting different areas of the source feature maps. (3) Experimental results show that our method consistently improves the performance of multiple recent domain adaptive object detectors.

2 Related Work

Domain Adaptation in Object Detection. In unsupervised domain adaptation, a popular approach is to align marginal feature distributions from source and target domains using adversarial training [9]. DA-Faster [5] is the first work proposed for unsupervised domain adaptive object detection. It designs a two-stage feature alignment architecture based on global features where feature pixels' receptive fields cover the whole input image. Following this work, some global feature alignment based methods are proposed [17,24,31]. SW improved DA-Faster and achieved a large performance boost by additionally imposing strong local feature alignment while applying weak global feature alignment [23]. The idea was later extended to align features from multiple levels [11,32], or augmented with an auxiliary multi-label classification task [27]. More recently, HTCN retains the design of SW and further explores the transferability and discriminability of different areas of local feature maps [4].

Apart from marginal feature alignment, an orthogonal type of method tries to generate intermediate domains between the source and target domains to decrease the domain discrepancy [12,16]. Nevertheless, all the above methods focus on learning domain-invariant representations to alleviate the covariate shift without considering the class label shift across domains.

Cross-Domain Class Label Shift Correction. Considering neural networks as a black box predictor, the most important works addressing the cross domain class label shift problem in unsupervised visual domain adaptation are proposed in the context of image classification. Recently, Black Box Shift Learning (BBSL) [18] and Regularized Learning under Label Shifts (RLLS) [2] emerge to be the most popular methods in this area, and are further studied or improved by many following works [1,6,25]. However, these methods strongly rely on the assumption that the studied black box predictor outputs a probability distribution across classes given an image including exactly one object belonging to one class. Therefore, they can not be directly applied in object detection, where object co-occurrence scenarios have to be handled. Different from all above, we aim at alleviating the class label shift problem in unsupervised domain adaptive object detection. To the best of our knowledge, we are the first to propose a mechanism to explicitly correct the class label shift for object detection task in unsupervised domain adaptation.

3 Preliminary: Marginal Feature Alignment

The marginal feature alignment is typically achieved by using the adversarial discriminative model, as it fits well to the domain adaptation theory [30]. In this theory, domain adaptation can be formulated as a binary classification problem [3]. Concretely, the discriminator of the model tries to classify whether the current extracted feature belongs to the source or target domain, while the feature extractor tries to fool the discriminator in this binary classification problem. Thus, the optimization process is a 2-player game.

Denote $\mathcal{S}, \mathcal{T}$ as source and target domains, this 2-player game can be described formally as a *minmax* optimization problem [3,5]:

$$\min_f d_H(\mathcal{S}, \mathcal{T}) \iff \max_f \min_{h \in \mathcal{H}} \{\mathrm{err}_S(h(\mathbf{x})) + \mathrm{err}_T(h(\mathbf{x}))\} \tag{1}$$

with

$$\mathrm{err}_S(h(\mathbf{x})) = E_{\mathbf{x} \sim D_S}[||g(\mathbf{x}) - h(\mathbf{x})||], \tag{2}$$

$$\mathrm{err}_T(h(\mathbf{x})) = E_{\mathbf{x} \sim D_T}[||g(\mathbf{x}) - h(\mathbf{x})||]. \tag{3}$$

Here, $\mathbf{x}$ refers to the feature extracted from the original image input, f is a feature extractor that produces $\mathbf{x}$, $d_H(S, T)$ is the H-divergence [15] used to describe domain discrepancy between domain $\mathcal{S}$ and domain $\mathcal{T}$. The feature distributions in the source and target domains are denoted as D_S and D_T. E is the expectation function. Furthermore, g is a ground-truth labeling function $g : \mathbf{x} \mapsto \{0, 1\}$ indicating whether the current $\mathbf{x}$ comes from the source or the target domain. For convenience in later discussion, we define the ground-truth labels for the source- and target domain to be 0 and 1, respectively. The $\mathcal{H}$ is a

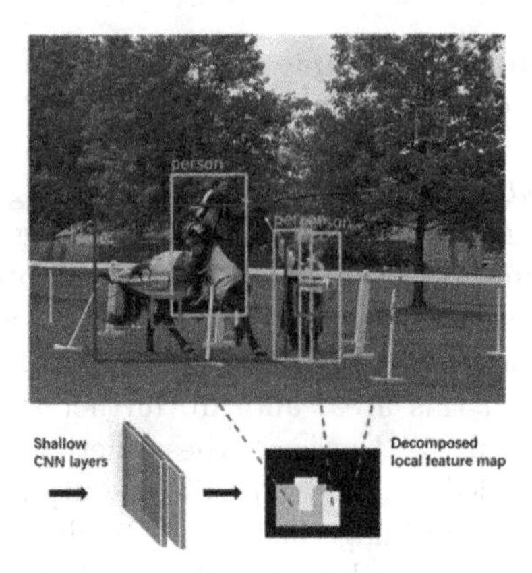

Fig. 2. Feature map decomposition in object co-occurrence scenarios: since the receptive field of every pixel after feature extraction may include information from multiple objects in the original image, we propose to assign each pixel the class label whose ground-truth bounding box has the largest overlap with the feature pixel's receptive field and thus decompose the feature map. Blue and green indicate ground-truth bounding boxes of horse and person, while red indicates the receptive field of a specific feature pixel. In the feature map, dark pink, light pink and black indicate feature pixels assigned to the class horse, person, and background, respectively. (Color figure online)

set of hypothesis h with $h : \mathbf{x} \mapsto [0, 1]$. It can be represented as a discriminator that predicts which domain a given feature $\mathbf{x}$ belongs to. The err_S and err_T are prediction errors in the source and target domains, respectively.

In this *minmax* optimization, the feature extractor f and domain discriminator h are optimized against each other. In other words, the feature extractor f is optimized to maximize the domain discriminator's classification error, while the domain discriminator h is optimized to minimize that error. Upon convergence of this process, the features extracted by f from different domains are expected to have a similar distribution such that the discriminator h fails to classify the domain of any given feature $\mathbf{x}$. In that case, the feature alignment between the source and target domains is achieved.

To further address the class label shift problem in the context of marginal feature alignment, one has to first identify which class any given feature $\mathbf{x}$ belongs to. This step is easy in image classification but difficult in object detection due to object co-occurrence.

4 Method

In this section, we first describe our novel mechanism to decompose the feature pixels (activations) in the feature map into exclusive feature sets according to classes to handle object co-occurrence scenarios. Then we theoretically show how normal adversarial training in pixel-level ignores the cross domain class label shift based on our decomposition. Finally, we propose to alleviate this problem via an iterative estimate, update, and re-weighting mechanism and offer some concrete examples to apply this mechanism in recent methods such as SW [23] and HTCN [4].

4.1 Feature Map Decomposition

In the context of image classification, since every image is assigned only to one label, the receptive field of any extracted feature corresponds only to one class. However, in the context of object detection, the receptive field could overlap with bounding boxes of multiple objects of potentially different semantic classes, as shown in the leftmost red box in Fig. 2. This means that a pixel in a feature map simultaneously represents information from different classes. Therefore, a mechanism is needed to handle this problem, e.g., by assigning a unique class label to each feature pixel. We name our novel mechanism to handle the object co-occurrence as "Feature Map Decomposition".

Class Label Assignment. To decompose the feature maps into different classes exclusively, we propose to assign every feature pixel to the class label of the object which has the largest bounding box overlap with the pixel's receptive field. If there is no overlap with any bounding box, the feature pixel is assigned the "background" label. If multiple objects from different classes have the same size of overlap area with the receptive field, the label is randomly chosen from one of the classes. Here, the underlying assumption is that the size of the overlap area between a pixel's receptive field and the object's bounding box approximately indicates the portion of information belonging to that object encoded in the current feature pixel.

Receptive Field Calculation. The size of the receptive field is calculated using the largest possible receptive field. This means that for design choices such as the skip connections in ResNet [10], where two different network branches are summed, the larger receptive field size of the two is taken. As a concrete example, the local features in the SW [23] and HTCN [4] baselines have a receptive field of 35×35 pixels when using a ResNet101 [10] as the backbone. Since object detectors typically work on high-resolution inputs, the receptive fields of these local features will not cover a large portion of the input, allowing for a reasonable feature map decomposition. This is visualized in Fig. 2. We note that there are some other concepts such as the effective receptive fields [20]. However, for simplicity, we use the size of the conventional theoretical receptive fields.

Using the proposed novel and intuitive feature map decomposition mechanism, addressing the class label shift problem in object detection becomes possible at pixel level because every feature pixel gets assigned a unique class. For example, it can be easily counted how many feature pixels in local feature maps extracted from a domain belong to a specific class. Thus a class distribution in any domain can be calculated.

4.2 Class-aware Re-Weighting from a Mathematical View

Consider all images of a dataset with C classes. We denote the set of all feature pixels extracted from this dataset which mainly include information of class c as F_c. Then all such sets from C classes together describe the feature pixels' class distribution of a dataset, denoted as $\mathcal{F}$:

$$\mathcal{F} = \{F_1, F_2, ..., F_c, ..., F_C\} \tag{4}$$

Given a feature map $\mathbf{x}$ (extracted from one image) with height H and width W, it can be decomposed as:

$$\mathbf{x} = \{\mathbf{x}_1, \mathbf{x}_2, \mathbf{x}_3, ..., \mathbf{x}_i, ..., \mathbf{x}_{HW}\}, \mathbf{x}_i \in F_c, F_c \subset \mathcal{F} \tag{5}$$

where $\mathbf{x_i}$ is a feature pixel. Figure 2 visualizes the feature map decomposition based on one image. We note that the above two definitions are permutation invariant w.r.t. pixels, and therefore ignore the spatial information encoded in the structure of an image. We focus on handling the class label shift in this work and leave exploring spatial information related domain shifts as future work.

Under the assumptions from Eq. 4 and 5, we can make use of the linearity of the expectation function and decompose the error terms in Eqs. 2 and 3 in a class-wise fashion:

$$E_{\mathbf{x_i} \sim D_S}[|h(\mathbf{x_i}) - g(\mathbf{x_i})|] = \sum_{c=1}^{C} \frac{N_c}{N} E_{\mathbf{x_i} \sim F_c^S}[|h(\mathbf{x_i}) - g(\mathbf{x_i})|], \tag{6}$$

$$E_{\mathbf{x_i} \sim D_T}[|h(\mathbf{x_i}) - g(\mathbf{x_i})|] = \sum_{c=1}^{C} \frac{M_c}{M} E_{\mathbf{x_i} \sim F_c^T}[|h(\mathbf{x_i}) - g(\mathbf{x_i})|], \tag{7}$$

where N and M are the total numbers of feature pixels extracted from the source and target domains respectively. N_c and M_c are the total numbers of feature pixels belonging to class c in source and target domains respectively. Therefore $N = \sum_{c=1}^{C} N_c$ and $M = \sum_{c=1}^{C} M_c$. F_c^S and F_c^T are sets of features belonging to class c in the source and target domains, respectively. Overall, after applying our feature map decomposition mechanism, the adversarial training process focuses on optimizing the following term:

$$\max_{f} \min_{h \in \mathcal{H}} \{\sum_{c=1}^{C} \{\frac{N_c}{N} E_{\mathbf{x_i} \sim F_c^S}[|h(\mathbf{x_i}) - g(\mathbf{x_i})|] + \frac{M_c}{M} E_{\mathbf{x_i} \sim F_c^T}[|h(\mathbf{x_i}) - g(\mathbf{x_i})|]\}\}. \tag{8}$$

From the above formulation, one can see that the contribution of features from a specific class is different between the source and target domains, due to the in general unequal terms $\frac{N_c}{N}$ and $\frac{M_c}{M}$. This difference corresponds to the class distribution difference between the source and the target domain, as visualized in Fig. 1, top.

To alleviate this problem, we propose to re-weight the importance of features from different classes such that each class' contribution is as similar as possible in the source and target domains. To this end, we multiply two factors α_c and β_c with $\frac{N_c}{N}$ and $\frac{M_c}{M}$ according to the semantic class. After modification, the optimization objective becomes:

$$\max_{f}\min_{h\in\mathcal{H}}\{\sum_{c=1}^{C}\{\frac{N_c}{N}\alpha_c E_{\mathbf{x_i}\sim F_c^S}[||h(\mathbf{x_i})-g(\mathbf{x_i})||]+\frac{M_c}{M}\beta_c E_{\mathbf{x_i}\sim F_c^T}[||h(\mathbf{x_i})-g(\mathbf{x_i})||]\}\}. \tag{9}$$

If we fix these factors as $\alpha_c = \frac{N}{N_c}\frac{\tilde{M}_c}{M}$, $\beta_c = 1$, where $\tilde{M}_c$ is the estimated number of feature pixels belonging to class c in the target domain, the optimization objective would further become:

$$\max_{f}\min_{h\in\mathcal{H}}\{\sum_{c=1}^{C}\{\frac{\tilde{M}_c}{M}E_{\mathbf{x_i}\sim F_c^S}[||h(\mathbf{x_i})-g(\mathbf{x_i})||]+\frac{M_c}{M}E_{\mathbf{x_i}\sim F_c^T}[||h(\mathbf{x_i})-g(\mathbf{x_i})||]\}\}. \tag{10}$$

Due to the similarity between $\frac{\tilde{M}_c}{M}$ and $\frac{M_c}{M}$, we are able to encourage the contribution of features from the source and target domains of the same class to be as similar as possible by weighting the optimization objective of corresponding feature pixels.

4.3 Target Domain Class Label Distribution Estimation

The term $\frac{N}{N_c}$ in α_c can be directly computed from the source domain ground-truth bounding boxes using the proposed feature map decomposition. In the remaining term $\frac{\tilde{M}_c}{M}$, M can be obtained as the number of images in the target domain multiplied by the height and width of the feature maps. Therefore, the quantity that remains to be estimated is the number of feature pixels belonging to class c in the target domain, $\tilde{M}_c$. This is done by using predicted confident bounding boxes of the model. Concretely, only predictions with a confidence higher than a threshold t (e.g., 0.7) are used to calculate the estimation. The process is going through all predictions of all images in the target domain, and summing the areas with high confidence belonging to a specific class together to estimate the number of pixels belonging to class c. This value is then scaled down according to the relative size of the feature map w.r.t. the original image to estimate $\tilde{M}_c$.

4.4 Overview on the Proposed Iterative Approach

The first step to apply the class label correction approach is calculating the distribution statistics of the source domain dataset. This is done by applying the feature map decomposition mechanism based on the ground-truth annotations of the dataset. This gives us the pixel level source domain distribution $\frac{N_c}{N}$ for every class c.

As the proposed method also depends on the estimated target domain distribution and there is no inference result in the target domain at the start of the training, α_c is initialized to $\frac{N}{CN_c}$ until the first update for the target domain statistics. This is because a uniform distribution is used as a random guess when there is no knowledge of the class distribution in the target domain yet. The next step is the estimation of target domain distribution. This follows exactly what is described in the previous section. In the early training phase, it can be observed that the model might not detect some specific classes of objects at all in the target domain, and thus there is no estimated $\frac{\tilde{M_c}}{M}$ for some classes. In this case, $\alpha_c = \frac{N}{CN_c}$ is used.

In the end, the estimated $\frac{\tilde{M_c}}{M}$ as well as the calculated $\frac{N}{N_c}$ are used to update the weighting factor $\alpha_c = \frac{N}{N_c}\frac{\tilde{M_c}}{M}$ every d iterations using the current on-training model.

4.5 Proposed Modifications of Recent Methods

A core aspect of the proposed class label correction approach is changing the importance of feature pixels belonging to different classes in the domain-adversarial objective. This can be readily integrated into the local feature alignment modules of current methods developed for unsupervised domain adaptive object detection. Following the above paradigm, we offer some concrete examples to modify loss functions of several recent methods.

SW. In local feature alignment, the concrete loss function used in [23] to express the error term in Eq. 2 is

$$\mathcal{L}_s^{SW}(u,v,k) = -\frac{1}{M_s HW}(1-D_k)(1-p_k^{(u,v)})^2. \tag{11}$$

And we propose to modify this to

$$\mathcal{L}_s^{SW}(u,v,k) = -\frac{1}{M_s HW}\alpha_c^{(u,v,k)}(1-D_k)(1-p_k^{(u,v)})^2. \tag{12}$$

In the above formulation, $\alpha_c^{(u,v,k)}$ is the corresponding α_c of a class that the feature pixel of kth image in position (u, v) belongs to.

HTCN. In a similar way we propose to modify the loss function for local feature alignment defined in HTCN [4] by additionally multiplying a correcting factor $\alpha_c^{(u,v,k)}$:

$$\mathcal{L}_s^{HTCN} = -\frac{1}{M_s HW}\alpha_c^{(u,v,k)}(1-D_k)\log(1-p_k{}^{(u,v)})^2. \tag{13}$$

5 Experiments

In this section, we show the effect of our approach in several adaptation scenarios. First, we consider a transfer from a real-world dataset (Pascal VOC [8]) to a synthetic dataset (Clipart [13]) by integrating our method into two representative detectors which use local feature alignment, SW [23] and HTCN [4]. Furthermroe, the approach is evaluated on the adaptation from a synthetic dataset (Sim10k [14]) to a real-world dataset (Cityscapes [7]). Here we create several versions of Cityscapes [7] by sampling subsets to evaluate the method under different degrees of the class label shift w.r.t. Sim10k [14]. For more complex scenarios, we also evaluate the adaptation from Synscapes [26] to Cityscapes [7]. In the end, a detailed ablation study is presented.

For a fair comparison, we use Faster-RCNN [22] as the base detector with ResNet101 [10] as the backbone feature extractor in all experiments. In all HTCN-based experiments, we trained all models with a learning rate of 0.001 for 50000 iterations and decay it to 0.0001 for another 50000 iterations. One iteration means feeding one image from the source domain and one image from the target domain into the model. In all SW-based experiments, the models are trained with a learning rate of 0.001 for 3 epochs (each epoch has 16551 iterations, matching the dataset size of the Pascal VOC [8]), followed by another 7 epochs with a decayed learning rate of 0.0001. Additionally, unless otherwise indicated, the interval d for updating the estimated class distribution in the target domain of all our methods is set to 50 images/update, since smaller update interval such as 10 images/update is computationally more expensive and we didn't observe an obvious performance gain (shown in ablation study). All experiments of our methods except those in the ablation study use 0.7 as a confidence threshold t when choosing confident predictions for estimating the class distribution in the target domain. Other hyperparameters are set exactly the same as SW [23] or HTCN [4] baselines respectively. We report an average of three runs in all experiments. Note that all values are rounded from the average of three runs, so the mAP may not equal the average of mAP from every class.

5.1 Pascal VOC - Clipart

Pascal VOC [8] includes real-world images from 20 classes. Following [4,23], we use both the training and validation datasets as the training set (16551 images). The Clipart [13] is a collection of clip art images consisting of 1000 images. This dataset has the same 20 semantic classes as Pascal VOC [8]. To alleviate the class label shift between domains, we integrate our approach into two representative works that utilize the local feature alignment: SW [23] and HTCN [4]. The accuracy is shown in Table 1. The table includes the mAP reported in the original papers, as well as reproduced results using the respective official codebases for a fair comparison with the results obtained by integrating our proposed method into the baselines. The table shows that our proposed method leads to improved results in both baseline methods. However, the improvement of our method applied in SW [23] is larger than the improvement in HTCN [4]. This might be due to the "transferability map" and feature concatenation techniques introduced in HTCN [4]. For example, the "transferability map" weights the local features in the forward propagation of the detection heads, while our method weights the local features in the feature alignment module. This might

Table 1. Class label shift adapted baselines on adaptation from Pascal VOC [8] to Clipart [13] dataset. We report the original paper's results indicated by $*$ as well as our reproduced values (repr.)

Methods	mAP	aero	bcycle	bird	boat	bottle	bus	car	cat	chair	cow	table	dog	hrs	bike	prsn	plnt	sheep	sofa	train	tv
SW* [23]	38.1	26.2	48.5	32.6	33.7	38.5	54.3	37.1	18.6	34.8	58.3	17.0	12.5	33.8	65.5	61.6	52.0	9.3	24.9	54.1	49.1
SW(repr.)	38.3	32.8	55.0	32.6	26.7	40.8	60.2	37.1	14.3	39.4	37.6	21.7	18.1	35.1	67.4	58.7	45.0	17.9	31.2	48.7	45.8
A-SW(ours)	**40.8**	**35.5**	54.8	**32.7**	26.7	**42.3**	**64.0**	**38.1**	10.9	**40.9**	**49.3**	**22.6**	**20.3**	**36.5**	**82.4**	**60.7**	**46.4**	**23.9**	**31.7**	**50.7**	45.2
HTCN* [4]	40.3	33.6	58.9	34.0	23.4	45.6	57.0	39.8	12.0	39.7	51.3	21.1	20.1	39.1	72.8	63.0	43.1	19.3	30.1	50.2	51.8
HTCN(repr.)	41.0	36.9	53.9	34.0	32.2	38.7	58.0	36.6	16.3	36.9	61.4	22.0	21.6	35.9	78.1	59.6	50.0	21.5	25.3	56.3	45.7
A-HTCN(ours)	**41.4**	**40.2**	**56.1**	**34.9**	31.9	**39.5**	**58.8**	**39.6**	**17.2**	**38.2**	58.7	18.9	19.5	34.6	**80.1**	**61.5**	**51.1**	17.4	**26.6**	54.1	**48.3**

Table 2. Number of images sampled from Cityscapes [7] with car ratio smaller than 10% and larger than 10%, as well as the car ratio of the resulting constructed subsets

< 10% car pixels	> 10% car pixels	overall ratio (%)
100	900	16.64
200	800	14.23
400	600	12.27
1000	0	8.42

reduce the gain obtained by our method. We investigate the interaction between the two mechanisms further in Sect. 5.4.

5.2 Sim10k - Cityscapes

This section studies an adaption scenario from a synthetic dataset to a real-world dataset. Sim10k [14] is a dataset generated by the computer engine as a simulation of real-world traffic scenes. It consists of 10000 training images. Cityscapes [7] is a collection of images from real-world German cities and consists of 2975 training images. Since the only common class between the two datasets is the car, the detection performance is only evaluated on the car class. This also offers the possibility to manually control the class label shift between domains. To show the effect of our approach in different degrees of class label shift between the source and target domains, we use the Sim10k [14] as the source domain while sampling 4 different subsets of Cityscapes [7] as the target domains. Concretely, we first separate the Cityscapes [7] dataset into two partitions: the first partition includes all images where less than 10% of the feature pixels are occupied by cars and the second partition includes the remaining images. By randomly sampling a different number of images from both partitions, different subsets of Cityscapes [7] can be constructed. Each of the subsets has in total of 1000 images. In these sets, the percentage of car feature pixels among all pixels ranges from 8.42% to 16.64%. A larger percentage is difficult to achieve due to the property of the dataset. Table 2 gives an overview of these subsets. In contrast, this percentage in the source domain Sim10k [14] is 5.47%. More statistical details of these subsets are shown in the supplementary.

As indicated by Sect. 5.1 as well as the later analysis in Sect. 5.4, the design of the "transferability" map or feature concatenation in HTCN [4] might interfere

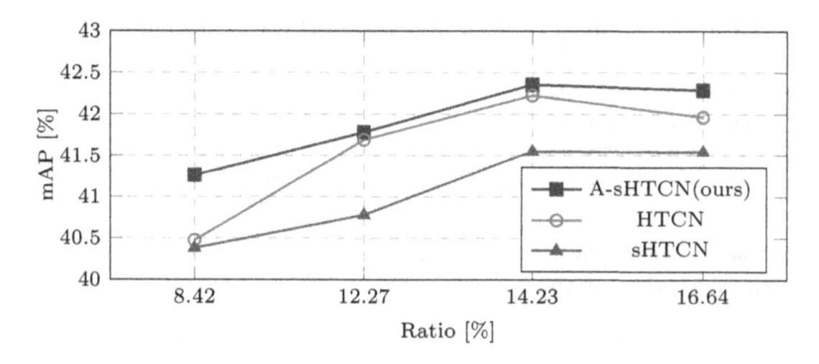

Fig. 3. Performance obtained integrating our method into the baseline sHTCN (HTCN without the transferability map and feature concatenation) under different degrees of class label shift between Sim10k [14] and Cityscapes [7] listed in Table 2. Results using the full HTCN [4] are also shown for comparison

with our method. Therefore, we use a simplified version of HTCN [4] without the "transferability map" and feature concatenation in the following experiments, which we refer to as "sHTCN" as the baseline. We integrate our iterative approach into sHTCN (named as A-sHTCN) and compare the performances. Note that all other design choices in HTCN [4] are kept. The update interval of our A-sHTCN is set to 1000 images/update instead of 50 images/update due to the limit on computation resources. The results are shown in Fig. 3. For comparison, the performance of the complete version of HTCN [4] is also included in the brown curve. Our method (A-sHTCN) consistently improves the baseline sHTCN under all four degrees of class label shift. Surprisingly, it even outperforms the full version of the HTCN method in all degrees of the class label shift although our result is reported using relatively low update frequency. This result indicates that our design might be superior to the removed HTCN [4] design such as the "transferability map".

5.3 Synscapes - Cityscapes

This adaptation scenario also studies the transfer from synthetic to a real-world driving dataset. Compared to the adaptation from Sim10k [14] to Cityscapes [7], this scenario is more complex, since the performance is evaluated in 8 semantic classes rather than just the car class.

Since to the best of our knowledge no prior work reports the performance of Faster-RCNN [22] trained only in the source domain (Synscapes [14]) and directly evaluated in the target domain (Cityscapes [7]), we included this experiment and reported the value in Table 3. Surprisingly, the baseline method HTCN [4] (23.9 mAP) even underperforms the model trained only in the source domain (25.2 mAP). A possible reason is that the disadvantage of marginal feature alignment, such as the class misalignment, is larger than the advantage it brings when two domains already look very similar. Class misalignment problem means features from different semantic classes can be encouraged to have similar

Table 3. Adaptation from Synscapes [14] to Cityscapes [7]

Methods	mAP	bus	bicycle	car	motorcycle	person	rider	train	truck
Source Only	25.2	22.9	18.4	47.2	20.0	31.3	37.8	13.3	10.9
HTCN	23.9	22.5	17.3	47.4	18.2	28.4	37.0	12.2	8.5
A-HTCN(ours)	25.1	24.7	16.9	48.2	19.7	29.9	36.5	12.2	13.0

Table 4. Interaction between "transferability maps" & feature concatenation and our class label shift correction method

Trans. maps + feat. concat	Label shift corr. (ours)	mAP
		39.6
✓		41.0
	✓	41.6
✓	✓	41.4

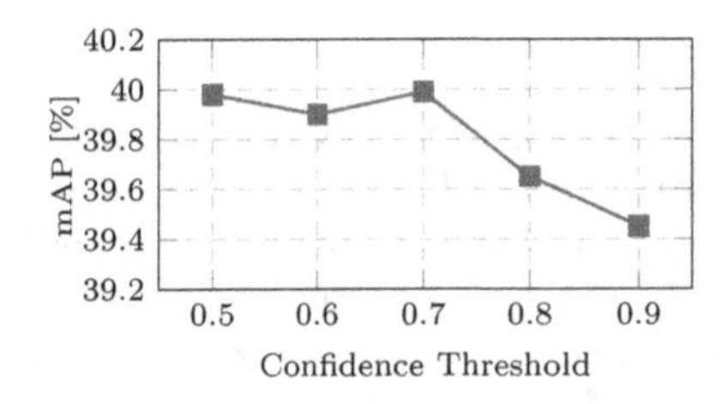

Fig. 4. Performance of our method (A-HTCN) integrated into HTCN [4] using different confidence thresholds t for predictions to estimate the class distribution in the target domain

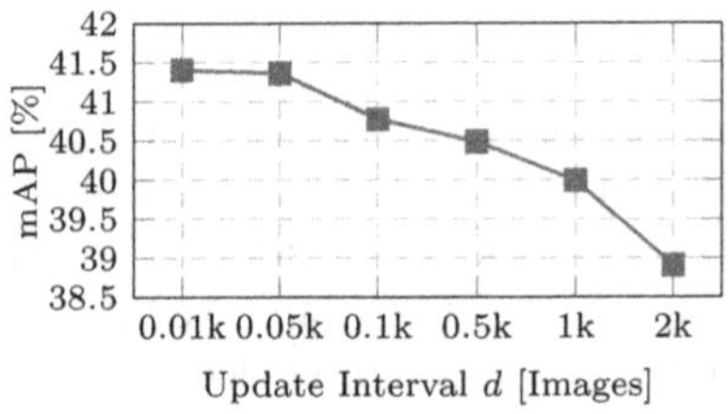

Fig. 5. Performance of our method integrated into HTCN [4] using different distribution estimation update interval d of estimated class distribution

representations due to the unavailability of labels in the target domain, negatively affecting the performance. As the marginal feature alignment is mainly handling the appearance difference between domains, it might bring minor benefits when domains already look similar. Nevertheless, our method (A-HTCN) can still help to increase the performance of the baseline method for 1.2 mAP. This suggests that our method is beneficial independently of whether the appearance difference between domains is large or small.

5.4 Ablation Study

To further study various design choices made in our proposed method, we perform a detailed ablation study to investigate their impacts. Due to the computational expense, all experiments are performed with an update interval d of 1000 images unless otherwise indicated. The confidence threshold t for the target domain class label distribution estimation is fixed to 0.7 unless explicitly indicated. All experiments use the adaptation from Pascal VOC [8] to Clipart [13] dataset. All experiments use HTCN [4] as the baseline method.

Interaction Between Transferability Maps and Class Label Shift Correction. To further study the interaction between the "transferability maps" as well as feature concatenation of the HTCN [4] and our proposed method, additional experiments are conducted. We use $d = 50$ for our method. The results in Table 4 show that both "transferability maps" and feature concatenation from HTCN [4] and our label correction methods could help to increase the performance of the model, while our method offers a larger improvement. This also matches the experimental results presented in Sect. 5.2. Interestingly, additionally applying our method could improve the performance compared to HTCN baseline [4]. However, the overall performance (41.4 mAP) is slightly lower than only using our method (41.6 mAP). This might indicate that our method alone is superior to the mechanism such as "transferability maps" or feature concatenation designed in the HTCH baseline [4].

Influence of Different Confidence Thresholds. Since the pixel level class distribution in the target domain is calculated using predictions with confidence higher than a threshold t in the target domain, we study the effect of this hyperparameter and show the experimental results in Fig. 4. It can be observed that setting a too high confidence threshold decreases the performance. This might be because too many bounding boxes predicted with lower confidence are ignored, negatively impacting the estimation quality for the target domain class label distribution.

Influence of Different Update Intervals. Since our method periodically updates the estimated class distribution in the target domain, it is interesting to study how often such updates should be conducted. One update means using the current on-training model to detect objects over the entire target domain dataset and using the predictions to calculate and update the class distribution. The results shown in Fig. 5 indicate that the more often the estimated distribution is updated, the better the performance. One possible explanation is that it is beneficial to always use the most current estimation available during training.

6 Conclusion

This paper addresses the question: how to alleviate the class label shift problem in unsupervised domain adaptive object detection. To this end, we propose an intuitive and novel feature map decomposition mechanism to handle object co-occurrence scenarios. Further, based on our theoretical analysis, we propose an iterative approach to estimate, update, and utilize the pixel level class distribution in the target domain to re-weight the local features of the source domain. Although our experiments focus on 2D object detection, we expect our method to inspire future works in other vision tasks, as object co-occurrence exists in most in-the-wild scenarios and our assumption that the overlap area between a feature pixel's receptive field and location annotations approximately indicates the portion of information from specific classes is generally true in neural networks.

References

1. Alexandari, A., Kundaje, A., Shrikumar, A.: Maximum likelihood with bias-corrected calibration is hard-to-beat at label shift adaptation. In: ICML (2020)
2. Azizzadenesheli, K., Liu, A., Yang, F., Anandkumar, A.: Regularized learning for domain adaptation under label shifts. ArXiv: abs/1903.09734 (2019)
3. Ben-David, Shai, Blitzer, John, Crammer, Koby, Kulesza, Alex, Pereira, Fernando, Vaughan, Jennifer Wortman: A theory of learning from different domains. Mach. Learn. **79**(1), 151–175 (2009). https://doi.org/10.1007/s10994-009-5152-4
4. Chen, C., Zheng, Z., Ding, X., Huang, Y., Dou, Q.: Harmonizing transferability and discriminability for adapting object detectors. In: 2020 IEEE/CVF Conference on Computer Vision and Pattern Recognition (CVPR), pp. 8866–8875 (2020)
5. Chen, Y., Li, W., Sakaridis, C., Dai, D., Gool, L.: Domain adaptive faster r-cnn for object detection in the wild. In: 2018 IEEE/CVF Conference on Computer Vision and Pattern Recognition, pp. 3339–3348 (2018)
6. des Combes, R.T., Zhao, H., Wang, Y.X., Gordon, G.: Domain adaptation with conditional distribution matching and generalized label shift. ArXiv: abs/2003.04475 (2020)
7. Cordts, M., et al.: The cityscapes dataset for semantic urban scene understanding. In: Proceedings of the IEEE Conference on Computer Vision and Pattern Recognition (CVPR) (2016)
8. Everingham, M., Van Gool, L., Williams, C.K.I., Winn, J., Zisserman, A.: The PASCAL Visual Object Classes Challenge 2012 (VOC2012) Results. http://www.pascal-network.org/challenges/VOC/voc2012/workshop/index.html
9. Ganin, Y., Lempitsky, V.: Unsupervised domain adaptation by backpropagation. ArXiv: abs/1409.7495 (2015)
10. He, K., Zhang, X., Ren, S., Sun, J.: Deep residual learning for image recognition. In: 2016 IEEE Conference on Computer Vision and Pattern Recognition (CVPR), pp. 770–778 (2016)
11. He, Z., Zhang, L.: Multi-adversarial faster-rcnn for unrestricted object detection. In: 2019 IEEE/CVF International Conference on Computer Vision (ICCV), pp. 6667–6676 (2019)
12. Hsu, H.K., et al.: Progressive domain adaptation for object detection. In: Proceedings of the IEEE/CVF Winter Conference on Applications of Computer Vision (WACV) (2020)
13. Inoue, N., Furuta, R., Yamasaki, T., Aizawa, K.: Cross-domain weakly-supervised object detection through progressive domain adaptation. In: Proceedings of the IEEE Conference on Computer Vision and Pattern Recognition (CVPR) (2018)
14. Johnson-Roberson, M., Barto, C., Mehta, R., Sridhar, S.N., Rosaen, K., Vasudevan, R.: Driving in the matrix: can virtual worlds replace human-generated annotations for real world tasks? In: 2017 IEEE International Conference on Robotics and Automation (ICRA), pp. 746–753. IEEE (2017)
15. Kifer, D., Ben-David, S., Gehrke, J.: Detecting change in data streams. In: Proceedings of the Thirtieth International Conference on Very Large Data Bases - VLDB 2004, VLDB Endowment, vol. 30. pp. 180–191 (2004)
16. Kim, T., Jeong, M., Kim, S., Choi, S., Kim, C.: Diversify and match: a domain adaptive representation learning paradigm for object detection. In: 2019 IEEE/CVF Conference on Computer Vision and Pattern Recognition (CVPR), pp. 12448–12457 (2019)

17. Li, C., et al.: Spatial attention pyramid network for unsupervised domain adaptation. ArXiv: abs/2003.12979 (2020)
18. Lipton, Z.C., Wang, Y.X., Smola, A.: Detecting and correcting for label shift with black box predictors. ArXiv: abs/1802.03916 (2018)
19. Liu, W., et al.: SSD: single shot multibox detector. In: Leibe, B., Matas, J., Sebe, N., Welling, M. (eds.) ECCV 2016. LNCS, vol. 9905, pp. 21–37. Springer, Cham (2016). https://doi.org/10.1007/978-3-319-46448-0_2
20. Luo, W., Li, Y., Urtasun, R., Zemel, R.: Understanding the effective receptive field in deep convolutional neural networks. In: NIPS (2016)
21. Redmon, J., Divvala, S., Girshick, R.B., Farhadi, A.: You only look once: unified, real-time object detection. In: 2016 IEEE Conference on Computer Vision and Pattern Recognition (CVPR), pp. 779–788 (2016)
22. Ren, S., He, K., Girshick, R.B., Sun, J.: Faster r-cnn: towards real-time object detection with region proposal networks. IEEE Trans. Pattern Anal. Mach. Intell. **39**, 1137–1149 (2015)
23. Saito, K., Ushiku, Y., Harada, T., Saenko, K.: Strong-weak distribution alignment for adaptive object detection. In: 2019 IEEE/CVF Conference on Computer Vision and Pattern Recognition (CVPR), pp. 6949–6958 (2019)
24. Shan, Y., Lu, W., Chew, C.: Pixel and feature level based domain adaption for object detection in autonomous driving. ArXiv: abs/1810.00345 (2019)
25. Wang, H., Liu, A., Yu, Z., Yue, Y., Anandkumar, A.: Distributionally robust learning for unsupervised domain adaptation. ArXiv: abs/2010.05784 (2020)
26. Wrenninge, M., Unger, J.: Synscapes: A photorealistic synthetic dataset for street scene parsing. ArXiv: abs/1810.08705 (2018)
27. Xu, C., Zhao, X., Jin, X., Wei, X.S.: Exploring categorical regularization for domain adaptive object detection. In: 2020 IEEE/CVF Conference on Computer Vision and Pattern Recognition (CVPR), pp. 11721–11730 (2020)
28. Yosinski, J., Clune, J., Bengio, Y., Lipson, H.: How transferable are features in deep neural networks? In: NIPS (2014)
29. Zhao, H., des Combes, R.T., Zhang, K., Gordon, G.: On learning invariant representation for domain adaptation. ArXiv: abs/1901.09453 (2019)
30. Zhao, S., et al.: A review of single-source deep unsupervised visual domain adaptation. In: IEEE Transactions on Neural Networks and Learning Systems, pp. 1–21 (2020). https://doi.org/10.1109/TNNLS.2020.3028503
31. Zhu, X., Pang, J., Yang, C., Shi, J., Lin, D.: Adapting object detectors via selective cross-domain alignment. In: Proceedings of the IEEE/CVF Conference on Computer Vision and Pattern Recognition (CVPR) (2019)
32. Zhuang, C., Han, X., Huang, W., Scott, M.: ifan: image-instance full alignment networks for adaptive object detection. In: AAAI (2020)

OpenCoS: Contrastive Semi-supervised Learning for Handling Open-Set Unlabeled Data

Jongjin Park(✉), Sukmin Yun, Jongheon Jeong, and Jinwoo Shin

Korea Advanced Institute of Science and Technology (KAIST), Daejeon, South Korea
{jongjin.park,sukmin.yun,jongheonj,jinwoos}@kaist.ac.kr

Abstract. Semi-supervised learning (SSL) has been a powerful strategy to incorporate few labels in learning better representations. In this paper, we focus on a practical scenario that one aims to apply SSL when unlabeled data may contain *out-of-class* samples - those that cannot have one-hot encoded labels from a closed-set of classes in label data, *i.e.*, the unlabeled data is an *open-set*. Specifically, we introduce *OpenCoS*, a simple framework for handling this realistic semi-supervised learning scenario based upon a recent framework of self-supervised visual representation learning. We first observe that the out-of-class samples in the open-set unlabeled dataset can be identified effectively via self-supervised contrastive learning. Then, OpenCoS utilizes this information to overcome the failure modes in the existing state-of-the-art semi-supervised methods, by utilizing one-hot pseudo-labels and soft-labels for the identified in- and out-of-class unlabeled data, respectively. Our extensive experimental results show the effectiveness of OpenCoS under the presence of out-of-class samples, fixing up the state-of-the-art semi-supervised methods to be suitable for diverse scenarios involving open-set unlabeled data.

Keywords: Contrastive learning · Realistic semi-supervised learning · Open-set semi-supervised learning · Class-distribution mismatch

1 Introduction

Despite the recent success of deep neural networks with large-scale labeled data, many real-world scenarios suffer from expensive data acquisition and labeling costs. This has motivated the community to develop *semi-supervised learning* (SSL; [6,14]), *i.e.*, by further incorporating unlabeled data for training. Indeed, recent SSL works [3,4,8,40] demonstrate promising results on several benchmark datasets, as they could even approach the performance of fully supervised learning using only a small number of labels, *e.g.*, 93.73% accuracy on CIFAR-10 with 250 labeled data [3].

J. Park and S. Yun—Equal contribution.

Supplementary Information The online version contains supplementary material available at https://doi.org/10.1007/978-3-031-25063-7_9.

L. Karlinsky et al. (Eds.): ECCV 2022 Workshops, LNCS 13802, pp. 134–149, 2023.
https://doi.org/10.1007/978-3-031-25063-7_9

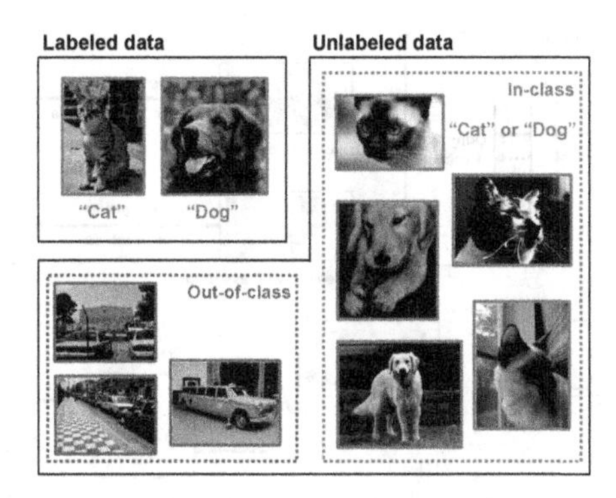

Fig. 1. Illustration of an open-set unlabeled data under class-distribution mismatch in semi-supervised learning, *i.e.*, unlabeled data may contain unknown out-of-class samples.

However, SSL methods often fail to generalize when there is a mismatch between the class-distributions of labeled and unlabeled data [10,15,34,37], *i.e.*, when the unlabeled data contains *out-of-class* samples, whose ground-truth labels are not contained in the labeled dataset (as illustrated in Fig. 1). In this scenario, various label-guessing techniques used in the existing SSL methods may label those out-of-class samples incorrectly, which in turn significantly harms the overall training through their inner-process of entropy minimization [14,24] or consistency regularization [40,46]. This problem may largely hinder the existing SSL methods from being used in practice, considering the *open-set* nature of unlabeled data collected in the wild [1].

Contribution. In this paper, we focus on a realistic SSL scenario, where unlabeled data may contain some unknown *out-of-class samples*, *i.e.*, there is a class distribution mismatch between labeled and unlabeled data [34]. Compared to prior approaches that have bypassed this problem by simply filtering out them [10,32,37], the unique characteristic in our approach is to further leverage the information in out-of-class samples by assigning *soft-labels* to them: some of them may still contain useful features for the in-classes.

Our first finding is that a recent technique of *self-supervised representation learning* [5,7,16,44] can play a key role for our goal. More specifically, we show that a pre-trained representation via self-supervised contrastive learning, namely SimCLR [7], on both labeled and unlabeled data enables us to design (a) an effective score for detecting out-of-class samples in unlabeled data, and (b) a systematic way to assign labels to the detected in- and out-of-class samples separately by modeling *posterior predictive distributions* using the labeled samples. Finally, we found (c) auxiliary batch normalization layers [45] could further help to mitigate the class-distribution mismatch via decoupling batch normalization layers. We propose a generic SSL framework, coined *OpenCoS*, based on the aforementioned techniques for handling open-set unlabeled data, which can be integrated with any existing SSL methods, *e.g.*, ReMixMatch [3] and FixMatch [40].

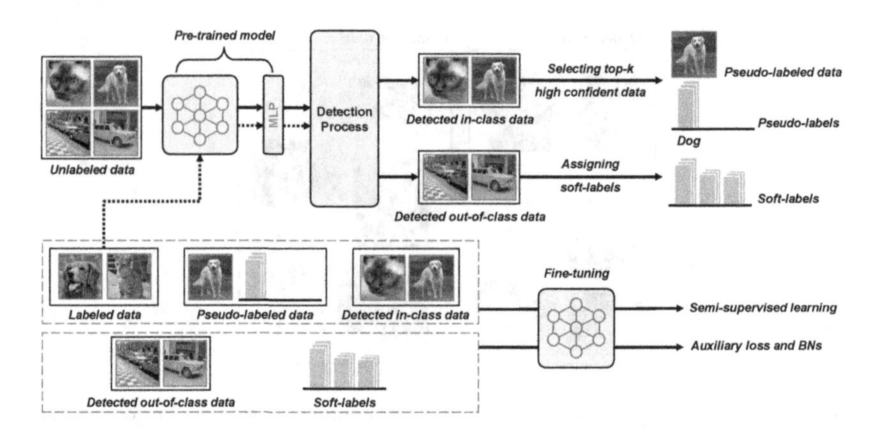

Fig. 2. Overview of our proposed framework, *OpenCoS*. First, our method detects in- and out-of-class samples based on contrastive representation. Then, OpenCoS chooses *top-k* confident in-class samples to assign one-hot encoded pseudo-labels, and integrate them into the original labeled dataset for semi-supervised learning. On the other hand, the out-of-class samples detected by OpenCoS are further utilized via an auxiliary loss and batch normalization layers with soft-labels generated from the representation.

We verify the effectiveness of the proposed method on a wide range of SSL benchmarks based on CIFAR-10, CIFAR-100 [23], and ImageNet [12] datasets, assuming the presence of various out-of-class data, *e.g.*, SVHN [33] and Tiny-ImageNet datasets. Our experimental results demonstrate that OpenCoS greatly improves existing state-of-the-art SSL methods [3,40]. We also compare our method to other recent works [10,15,32,37] addressing the same class distribution mismatch problem in SSL, and again confirms the effectiveness of our framework, *e.g.*, we achieve an accuracy of 69.38% with 40 labels (just 4 labels per class) on CIFAR-10 with TinyImageNet as out-of-class, while the recent baseline, OpenMatch [37] does 62.71%.

Overall, our work highlights the benefit of unsupervised representations in semi-supervised learning, which was also explored by [8] under a different perspective. We newly found that such a *label-free* representation turns out to enhance model generalization due to its (i) robustness on the novel, out-of-class samples, (ii) successful performance on identifying out-of-class samples, and (iii) label-efficient transferability (of high confident samples) to the downstream task.

2 Preliminaries

2.1 Semi-supervised Learning

The goal of *semi-supervised learning* for classification is to train a classifier $f : \mathcal{X} \to \mathcal{Y}$ from a *labeled dataset* $\mathcal{D}_l = \{x_l^{(i)}, y_l^{(i)}\}_{i=1}^{N_l}$ where each label y_l is from a set of classes $\mathcal{Y} := \{1, \cdots, C\}$, and an *unlabeled dataset* $\mathcal{D}_u = \{x_u^{(i)}\}_{i=1}^{N_u}$ where each y_u exists but is assumed to be unknown. In an attempt to leverage the extra information in $\mathcal{D}_u$, a number of techniques have been proposed, *e.g.*, entropy minimization [14,24] and consistency regularization [31,38,40,46]. In general, recent approaches in semi-supervised learning can be distinguished by the prior they adopt for the representation of unlabeled data: for example, the consistency regularization technique attempt to minimize the cross-entropy loss between any two predictions of different augmentations $t_1(x_u)$ and $t_2(x_u)$ from a given unlabeled sample x_u, jointly with the standard training for a labeled sample (x_l, y_l):

$$\mathcal{L}_{\text{SSL}}(x_l, x_u) := \mathbb{H}(y_l, f(x_l)) + \beta \cdot \mathbb{H}(f(t_1(x_u)), f(t_2(x_u))), \tag{1}$$

where $\mathbb{H}$ is a standard cross-entropy loss for labeled data, and β is a hyperparameter. Recently, several "holistic" approaches of various techniques [11,49] have shown remarkable performance in practice, *e.g.*, MixMatch [4], ReMixMatch [3], and FixMatch [40], which we mainly consider in this paper. We note that our scheme can be integrated with any recent semi-supervised learning methods.

2.2 Contrastive Representation Learning

Contrastive learning [7,16,18,35] defines an unsupervised task for an encoder $f_e : \mathcal{X} \to \mathbb{R}^{d_e}$ from a set of samples $\{x_i\}$: assume that a "query" sample x_q is given and there is a positive "key" $x_+ \in \{x_i\}$ that x_q matches. Then the *contrastive loss* is defined to let f_e extract the necessary information to identify x_+ from x_q as follows:

$$\mathcal{L}^{\text{con}}(f_e, x_q, x_+; \{x_i\}) := -\log \frac{\exp(h(f_e(x_q), f_e(x_+))/\tau)}{\sum_i \exp(h(f_e(x_q), f_e(x_i))/\tau)}, \tag{2}$$

where $h(\cdot,\cdot)$ is a pre-defined similarity score, and τ is a temperature hyperparameter. In this paper, we primarily focus on *SimCLR* [7], a particular form of contrastive learning: for a given $\{x_i\}_{i=1}^{N}$, SimCLR first samples two separate data augmentation operations from a pre-defined family $\mathcal{T}$, namely $t_1, t_2 \sim \mathcal{T}$, and matches $(\tilde{x}_i, \tilde{x}_{i+N}) := (t_1(x_i), t_2(x_i))$ as a query-key pair interchangeably. The actual loss is then defined as follows:

$$\mathcal{L}^{\text{SimCLR}}(f_e; \{x_i\}_{i=1}^{N}) := \frac{1}{2N} \sum_{q=1}^{2N} \mathcal{L}^{\text{con}}(f_e, \tilde{x}_q, \tilde{x}_{(q+N) \bmod 2N}; \{\tilde{x}_i\}_{i=1}^{2N} \setminus \{\tilde{x}_q\}), \tag{3}$$

$$h^{\text{SimCLR}}(v_1, v_2) := \text{CosineSimilarity}(g(v_1), g(v_2)) = \frac{g(v_1) \cdot g(v_2)}{||g(v_1)||_2 ||g(v_2)||_2}, \tag{4}$$

where $g : \mathbb{R}^{d_e} \to \mathbb{R}^{d_p}$ is a 2-layer neural network called *projection header*. In other words, the SimCLR loss defines a task to identify a "semantically equivalent" sample to x_q up to the set of data augmentations $\mathcal{T}$.

3 OpenCoS: A Framework for Open-Set SSL

We consider semi-supervised classification problems involving C classes. In addition to the standard assumption of semi-supervised learning (SSL), we assume that the unlabeled dataset $\mathcal{D}_u$ is *open-set, i.e.*, the hidden labels y_u of x_u may not be in $\mathcal{Y} := \{1, \cdots, C\}$. In this scenario, existing semi-supervised learning techniques may degrade the classification performance, possibly due to incorrect label-guessing procedure for those out-of-class samples. In this respect, we introduce *OpenCoS*, a generic method for detecting and labeling in- and out-of-class unlabeled samples in semi-supervised learning. Overall, our key intuition is to utilize the unsupervised representation from *contrastive learning* [5,7,16,44] to leverage open-set unlabeled data in an appropriate manner. We present a brief overview of our method in Sect. 3.1, and describe how our approach can handle a realistic SSL scenario in Sect. 3.2 and Sect. 3.3.

3.1 Overview of OpenCoS

Recall that our goal is to train a classifier $f : \mathcal{X} \to \mathcal{Y}$ from a labeled dataset $\mathcal{D}_l$ and an *open-set* unlabeled dataset $\mathcal{D}_u$. Overall, OpenCoS aims to overcome the presence of out-of-class samples in $\mathcal{D}_u$ through the following procedure:

1. **Pre-training via contrastive learning.** OpenCoS first learns an unsupervised representation of f via SimCLR[1] [7], using both $\mathcal{D}_l$ and $\mathcal{D}_u$ without labels. More specifically, we learn the penultimate features of f, denoted by f_e, by minimizing the contrastive loss defined in (3). We also introduce a projection header g (4), which is a 2-layer MLP as per [7].
2. **Detecting in- and out-of-class samples.** From a learned representation of f_e and g, OpenCoS identifies *out-of-class* dataset $\mathcal{D}_u^{\text{out}}$ from the given unlabeled dataset $\mathcal{D}_u$. Detected out-of-class dataset by our method is denoted as $\widetilde{\mathcal{D}}_u^{\text{out}}$. This detection process is based on the similarity score between $\mathcal{D}_l$ and $\mathcal{D}_u$ in the representation space of f_e and g. Furthermore, OpenCoS constructs a pseudo-labeled dataset $\mathcal{D}_l^{\text{pseudo}}$ to enlarge $\mathcal{D}_l$, by assigning one-hot labels to *top-k* confident samples in detected in-class dataset $\widetilde{\mathcal{D}}_u^{\text{in}} := \mathcal{D}_u \backslash \widetilde{\mathcal{D}}_u^{\text{out}}$ (see Sect. 3.2 for more details).
3. **Semi-supervised learning with auxiliary loss and batch normalization.** Now, one can use any existing semi-supervised learning scheme, *e.g.*, ReMixMatch [3], to train f using labeled dataset $\mathcal{D}_l \cup \mathcal{D}_l^{\text{pseudo}}$ and unlabeled dataset $\widetilde{\mathcal{D}}_u^{\text{in}}$. In addition, OpenCoS minimizes an *auxiliary loss* that assigns a soft-label to each sample in $\widetilde{\mathcal{D}}_u^{\text{out}}$, which is also based on the representation of f_e and g (see Sect. 3.3). Furthermore, we found maintaining auxiliary batch normalization layers [45] for $\widetilde{\mathcal{D}}_u^{\text{out}}$ is beneficial to our loss as they mitigate the distribution mismatch arisen from $\widetilde{\mathcal{D}}_u^{\text{out}}$.

Putting it all together, OpenCoS provides an effective and systematic way to utilize open-set unlabeled data for semi-supervised learning. Figure 2 and Algorithm 1 describe the overall training scheme of OpenCoS.

[1] Nevertheless, our framework is not restricted to a single method of SimCLR: *e.g.*, we also show that OpenCoS can be applicable with DINO [5], another recent self-supervised learning scheme (see Sect. 5).

3.2 Detection Criterion

For a given labeled dataset $\mathcal{D}_l$ and an open-set unlabeled dataset $\mathcal{D}_u$, we aim to detect a subset of the unlabeled training data $\mathcal{D}_u^{\text{out}} \subseteq \mathcal{D}_u$ whose elements are out-of-class, *i.e.*, $y_u \notin \mathcal{Y}$. A popular way to handle this task is to use a confidence-calibrated classifier trained by $\mathcal{D}_l$ to detect $\mathcal{D}_u^{\text{out}}$ [2,19–21,25,26,29,41]. However, training such a good classifier is infeasible in our setup due to the small size of $\mathcal{D}_l$, *i.e.*, lack of labeled data. Instead, we first perform unsupervised contrastive learning (*i.e.*, SimCLR) using both $\mathcal{D}_l$ and the open-set unlabeled dataset $\mathcal{D}_u$. Then, OpenCoS further utilizes the labeled dataset $\mathcal{D}_l$ to estimate the class-wise distributions of (pre-trained) embeddings, and uses them to define a detection score for $\mathcal{D}_u$.

Formally, we assume that an encoder $f_e : \mathcal{X} \to \mathbb{R}^{d_e}$ and a projection header $g : \mathbb{R}^{d_e} \to \mathbb{R}^{d_p}$ pre-trained via SimCLR on $\mathcal{D}_l \cup \mathcal{D}_u$. Motivated by the similarity metric used in the pre-training objective of SimCLR (4), we propose a simple yet effective detection score $s(x_u)$ for unlabeled input x_u based on the cosine similarity between x_u and *class-wise prototypical representations* $\{v_c\}_{c=1}^C$ obtained from $\mathcal{D}_l$. Namely, we first define a *class-wise similarity score* $\text{sim}_c(x_u)$ for each class c as follows:

$$v_c(\mathcal{D}_l; f_e, g) := \frac{1}{N_l^c} \sum_i \mathbf{1}_{y_l^{(i)}=c} \cdot g(f_e(x_l^{(i)})), \quad \text{and} \tag{5}$$

$$\text{sim}_c(x_u; \mathcal{D}_l, f_e, g) := \text{CosineSimilarity}(g(f_e(x_u)), v_c), \tag{6}$$

where $N_l^c := |\{(x_l^{(i)}, y_l^{(i)}) | y_l^{(i)} = c\}|$ is the sample size of class c in $\mathcal{D}_l$. Then, our detection score $s(x_u)$ is defined by the maximal similarity score between x_u and the prototypes $\{v_c\}_{c=1}^C$:

$$s(x_u) := \max_{c=1,\cdots,C} \text{sim}_c(x_u). \tag{7}$$

In practice, we use a pre-defined threshold t for detecting out-of-class samples in $\mathcal{D}_u$, *i.e.*, we detect a given sample x_u as out-of-class if $s(x_u) < t$. In our experiments, we found an empirical value of $t := \mu_l - 2\sigma_l$ performs well across all the datasets tested, where μ_l and σ_l are mean and standard deviation computed over $\{s(x_l^{(i)})\}_{i=1}^{N_l}$, respectively, although more tuning of t could further improve the performance.[2] Then, we perform such detection process to split the open-set unlabeled dataset $\mathcal{D}_u$ into two datasets $\widetilde{\mathcal{D}}_u^{\text{in}}$ and $\widetilde{\mathcal{D}}_u^{\text{out}}$, where $\widetilde{\mathcal{D}}_u^{\text{in}}$ and $\widetilde{\mathcal{D}}_u^{\text{out}} = \mathcal{D}_u \backslash \widetilde{\mathcal{D}}_u^{\text{in}}$ are unlabeled datasets detected to be *in-class* and *out-of-class*, respectively.

Furthermore, to overcome the regime of few labels in semi-supervised learning, we enlarge the labeled dataset $\mathcal{D}_l$ using a subset of $\widetilde{\mathcal{D}}_u^{\text{in}}$ with their one-hot encoded pseudo-labels, *i.e.*, construct an additional (pseudo-) labeled dataset $\mathcal{D}_l^{\text{pseudo}}$ to utilize it for semi-supervised learning. To be specific, we choose *top-k* high confident samples in $\widetilde{\mathcal{D}}_u^{\text{in}}$, under a classifier built upon the pre-trained encoder f_e with the linear evaluation protocol [7,50].

[2] The detection performances with various choices of t are provided in Section D.2 of the supplementary material.

Algorithm 1. OpenCoS: A general framework for open-set semi-supervised learning (SSL).

Input: Classifier f, encoder f_e, projection header g, labeled dataset $\mathcal{D}_l$, open-set unlabeled dataset $\mathcal{D}_u$.

Pre-train f_e and g via contrastive learning using $\mathcal{D}_l \cup \mathcal{D}_u$
Detect out-of-class samples using threshold t as
 // $s(\cdot)$ defined in (7)
 $\widetilde{\mathcal{D}}_u^{\text{in}} \leftarrow \{x_u \in \mathcal{D}_u | s(x_u; f_e, g) > t\}$
 $\widetilde{\mathcal{D}}_u^{\text{out}} \leftarrow \mathcal{D}_u \backslash \widetilde{\mathcal{D}}_u^{\text{in}}$
Construct a pseudo-labeled dataset $\mathcal{D}_l^{\text{pseudo}}$
Update $\mathcal{D}_l \leftarrow \mathcal{D}_l \cup \mathcal{D}_l^{\text{pseudo}}$
for each sample $x_l \in \mathcal{D}_l$, $x_u^{\text{in}} \in \widetilde{\mathcal{D}}_u^{\text{in}}$, and $x_u^{\text{out}} \in \widetilde{\mathcal{D}}_u^{\text{out}}$ **do**
 // $q(\cdot)$ defined in (9)
 $\mathcal{L}_{\text{OpenCoS}} \leftarrow \mathcal{L}_{\text{SSL}}(x_l, x_u^{\text{in}}; f) + \lambda \cdot \mathbb{H}(q(x_u^{\text{out}}), f(x_u^{\text{out}}))$
 Update parameters of f by computing the gradients of the proposed loss $\mathcal{L}_{\text{OpenCoS}}$
end for

3.3 Auxiliary Loss and Batch Normalization

Now, one can train a classifier f on the labeled dataset $\mathcal{D}_l \cup \mathcal{D}_l^{\text{pseudo}}$ and unlabeled dataset $\widetilde{\mathcal{D}}_u^{\text{in}}$ under any existing semi-supervised learning method [3,4,40]. In addition, we propose to further utilize $\widetilde{\mathcal{D}}_u^{\text{out}}$ via an auxiliary loss that assigns a soft-label to each $x_u^{\text{out}} \in \mathcal{D}_u^{\text{out}}$. More specifically, for any semi-supervised learning objective $\mathcal{L}_{\text{SSL}}(x_l, x_u^{\text{in}}; f)$, we consider the following modified loss to optimize:

$$\mathcal{L}_{\text{OpenCoS}} = \mathcal{L}_{\text{SSL}}(x_l, x_u^{\text{in}}; f) + \lambda \cdot \mathbb{H}(q(x_u^{\text{out}}), f(x_u^{\text{out}})), \tag{8}$$

where $x_l \in \mathcal{D}_l \cup \mathcal{D}_l^{\text{pseudo}}$,[3] $\mathbb{H}$ denotes the cross-entropy loss, λ is a hyperparameter, and $q(x_u^{\text{out}})$ defines a specific assignment of distribution over $\mathcal{Y}$ for x_u^{out}. Here, we assign $q(x_u^{\text{out}})$ based on the class-wise similarity scores $\text{sim}_c(x_u^{\text{out}})$ defined in (6), again utilizing the contrastive representation f_e and g:

$$q_c(x_u) := \frac{\exp(\text{sim}_c(x_u; f_e, g)/\tau)}{\sum_i \exp(\text{sim}_i(x_u; f_e, g)/\tau)}, \tag{9}$$

where τ is a temperature hyperparameter.

At first glance, assigning a label of $\mathcal{Y}$ to x_u^{out} may seem counter-intuitive, as the true label of x_u^{out} is not in $\mathcal{Y}$ by definition. However, even when the out-of-class samples cannot be represented as one-hot labels, one can still model their *posterior predictive distributions* over in-classes as a linear combination (*i.e.*, soft-label) of $\mathcal{Y}$: for instance, although "cat" images are out-of-class for CIFAR-100, still there are some classes in CIFAR-100 that is *semantically similar* to

[3] If $\mathcal{D}_l \cup \mathcal{D}_l^{\text{pseudo}}$ is class-imbalanced, we apply oversampling method [22] for balancing class distributions.

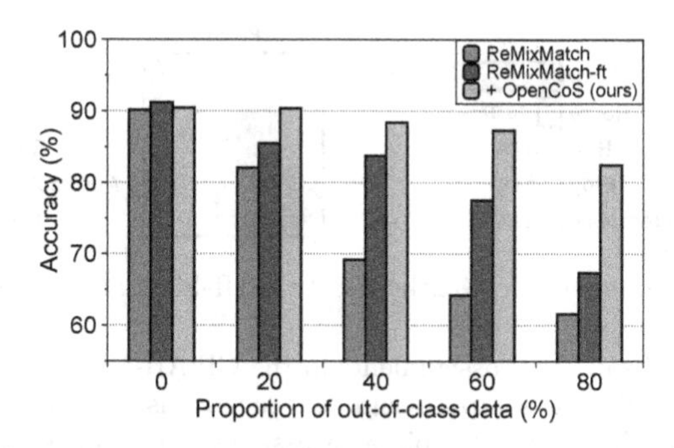

Fig. 3. Comparison of median test accuracy under varying proportions of out-of-class samples on the CIFAR-10 + TinyImageNet benchmark with 25 labels per class.

"cat", *e.g.*, "leopard", "lion", or "tiger". Here, we hypothesize that assigning a soft-label of in-classes is beneficial for such semantically similar classes. Even if out-of-classes are totally different from in-classes, one can assign the uniform labels to ignore them. We empirically found that such soft-labels based on the contrastive representation offer an effective way to utilize out-of-class samples, while they are known to significantly harm in the vanilla semi-supervised learning schemes [10,15,34].

Auxiliary Batch Normalization. Finally, we suggest to handle a *data-distribution shift* originated from the class-distribution mismatch [34], *i.e.*, $\mathcal{D}_l$ and $\mathcal{D}_u^{\text{out}}$ are drawn from the different underlying distribution. This may degrade the in-class classification performance as the auxiliary loss utilizes out-of-class samples. To handle the issue, we use additional batch normalization layers (BN) [39] for training samples in $\widetilde{\mathcal{D}}_u^{\text{out}}$ to disentangle those two distributions. In our experiments, we empirically observe such *auxiliary BNs* are beneficial when using out-of-class samples via the auxiliary loss. Auxiliary BNs also have been studied in adversarial learning literature [45]: decoupling BNs improves the performance of adversarial training by handling a distribution mismatch between clean and adversarial samples. In this paper, we found that a similar strategy can improve model performance in realistic semi-supervised learning.

4 Experiments

In this section, we verify the effectiveness of our method over a wide range of semi-supervised learning (SSL) benchmarks in the presence of various out-of-class data. The full details on experimental setups can be found in the supplementary material.

Datasets. We perform experiments on image classification tasks for several benchmarks in the literature of SSL [3,40]: CIFAR-10, CIFAR-100 [23], and ImageNet [12] datasets. Specifically, we focus on settings where each dataset is extremely label-scarce: only 4 or 25 labels per class are given during training,

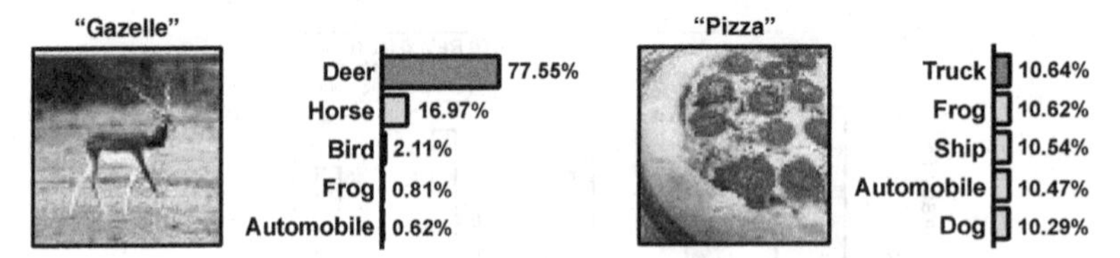

(a) Soft-label assignments of "Gazelle". (b) Soft-label assignments of "Pizza".

Fig. 4. Illustration of soft-label assignments in the CIFAR-10 + TinyImageNet benchmark. Unlabeled out-of-class samples from (a) "gazelle" is assigned with soft-labels of ≈78% confidence for "deer", and (b) "pizza" is assigned with almost uniform soft-labels (≈10% of confidence).

while the rest of the training data are assumed to be unlabeled. To configure realistic SSL scenarios, we additionally assume that unlabeled data contain samples from an external dataset: for example, in the case of CIFAR-10, we use unlabeled samples from SVHN [33] or TinyImageNet[4] datasets.

Baselines. We evaluate MixMatch [4], ReMixMatch [3], and FixMatch [40] as baselines in our experimental setup, which are considered to be state-of-the-art methods in conventional SSL. We also compare our method with four prior works applicable to our setting: namely, we consider Uncertainty-Aware Self-Distillation (UASD; [10]), RealMix [32], DS^{3}L [15], and OpenMatch [37], which propose schemes to detect and filter out out-of-class samples in the unlabeled dataset: *e.g.*, DS^{3}L re-weights unlabeled samples, and OpenMatch learns an outlier detector to reduce the effect of such out-of-class samples. Recall that our method uses SimCLR [7] for pre-training. Unless otherwise noted, we also pre-train the baselines via SimCLR for a fair comparison, denoting those fine-tuned models by "-ft," *e.g.*, MixMatch-ft and UASD-ft. We confirm that fine-tuned models show comparable or better performance compared to those trained from scratch, as presented in Fig. 3. Also, we report the performance purely obtainable from (unsupervised) SimCLR: namely, we additionally consider (a) *SimCLR-le*: a SimCLR model with linear evaluation protocol [7,50], *i.e.*, it additionally learns a linear layer with the labeled dataset, (b) *SimCLR-ft*: the whole SimCLR model is fine-tuned with the labeled dataset, and (c) *SimCLR-sd*: the self-distilled SimCLR model with a distillation loss of [8]. Somewhat interestingly, these models turn out to be strong baselines in our setups; they often outperform the state-of-the-art semi-supervised baselines under large proportions of out-of-class samples (see Table 1). Finally, we remark that our framework can incorporate any conventional semi-supervised methods for training. We denote our method built upon an existing method by "+ OpenCoS", *e.g.*, ReMixMatch-ft + OpenCoS.

Training Details. As suggested by [34], we have re-implemented all baseline methods considered, including SimCLR, under the same codebase and performed

[4] https://tiny-imagenet.herokuapp.com/.

Table 1. Comparison of median test accuracy on various benchmark datasets. We report the mean and standard deviation over three runs with different random seeds and splits, and also report the mean of the best accuracy in parentheses. The best scores are indicated in bold. We denote methods handling unlabeled out-of-class samples (*i.e.*, open-set) as "Open-SSL".

In-class		CIFAR-Animals	CIFAR-10		CIFAR-100	
Out-of-class	Open-SSL	CIFAR-Others	SVHN	TinyImageNet	SVHN	TinyImageNet
# labels per class = 4						
SimCLR-le [7]	-	65.58±3.51	56.89±3.19	58.20±0.88	22.86±0.17	27.93±0.67
SimCLR-ft [7]	-	67.29±2.76 (68.25)	42.16±2.50 (42.67)	54.26±1.26 (55.01)	18.99±0.04 (19.12)	29.57±0.33 (29.57)
SimCLR-sd [8]	-	64.70±3.57 (67.07)	56.18±1.63 (57.22)	49.88±2.46 (52.78)	22.88±0.71 (23.60)	26.23±1.32 (28.26)
UASD-ft [10]	✓	43.92±1.94 (52.87)	42.99±3.05 (44.70)	50.38±2.78 (51.66)	19.66±0.44 (19.92)	25.72±0.69 (26.33)
RealMix-ft [32]	✓	64.42±7.26 (67.99)	38.22±3.41 (41.55)	48.28±5.73 (49.78)	18.48±0.42 (20.04)	22.14±0.71 (26.51)
DS^{3}L-ft [15]	✓	63.98±6.96 (72.20)	36.81±7.67 (47.32)	56.32±1.31 (57.58)	16.35±0.20 (16.97)	23.95±1.43 (25.06)
OpenMatch-ft [37]	✓	52.09±5.98 (69.96)	55.93±2.09 (57.36)	62.71±1.33 (64.27)	14.85±1.17 (22.30)	17.82±2.75 (27.57)
MixMatch-ft [4]	-	44.34±5.13 (65.55)	23.71±8.65 (38.69)	38.90±4.24 (46.59)	13.45±1.23 (16.76)	23.16±1.85 (26.54)
FixMatch-ft [40]	-	34.94±6.18 (75.83)	32.70±6.28 (55.58)	35.99±2.63 (63.35)	23.56±0.68 (24.24)	30.70±3.67 (32.52)
ReMixMatch-ft [3]	-	47.61±6.51 (64.06)	24.56±3.99 (47.65)	28.51±5.87 (55.68)	9.36±1.97 (21.30)	22.33±1.10 (29.77)
+ **OpenCoS (ours)**	✓	**81.29±0.93 (81.77)**	**66.42±7.26 (66.78)**	**69.38±3.64 (70.02)**	**30.29±2.33 (30.62)**	**36.79±0.97 (37.12)**
# labels per class = 25						
SimCLR-le [7]	-	80.03±0.73	70.31±0.14	71.84±0.10	37.74±0.42	43.68±0.26
SimCLR-ft [7]	-	81.44±0.49 (81.61)	64.41±1.37 (64.65)	73.05±0.11 (73.30)	39.61±0.28 (39.87)	49.69±0.30 (49.96)
SimCLR-sd [8]	-	80.94±0.93 (81.82)	70.49±0.60 (71.04)	66.59±0.47 (69.84)	38.47±1.08 (38.98)	44.09±1.31 (45.69)
UASD-ft [10]	✓	82.17±0.85 (82.50)	66.70±1.00 (67.43)	73.97±0.37 (74.54)	39.51±0.76 (39.65)	44.58±0.77 (44.90)
RealMix-ft [32]	✓	80.27±2.64 (81.04)	58.15±5.27 (67.27)	69.19±2.31 (72.29)	44.14±1.01 (44.89)	47.57±1.39 (49.47)
DS^{3}L-ft [15]	✓	81.31±0.50 (83.27)	50.00±8.34 (63.11)	69.13±2.30 (72.23)	29.00±0.97 (30.17)	40.16±0.90 (41.82)
OpenMatch-ft [37]	✓	86.14±0.30 (86.51)	74.74±0.98 (75.02)	77.94±0.76 (78.14)	29.03±1.26 (39.08)	32.15±2.62 (47.38)
MixMatch-ft [4]	-	83.88±1.60 (84.21)	17.98±2.60 (54.19)	69.27±6.59 (75.11)	38.60±1.86 (43.02)	50.23±0.89 (51.38)
FixMatch-ft [40]	-	69.86±1.92 (84.24)	68.02±0.68 (71.91)	70.49±1.15 (77.27)	41.73±1.29 (42.28)	45.94±1.03 (49.96)
ReMixMatch-ft [3]	-	81.62±1.47 (83.90)	37.98±3.43 (65.33)	67.38±7.25 (73.34)	32.75±0.77 (44.62)	49.63±1.10 (53.20)
+ **OpenCoS (ours)**	✓	**87.15±0.70 (87.72)**	**78.97±1.13 (79.24)**	**82.46±1.33 (82.74)**	**49.19±1.16 (49.56)**	**54.01±1.75 (54.32)**

experiments with the same model architecture of ResNet-50 [17].[5] We checkpoint per 2^{16} training samples and report (a) the median test accuracy of the last 5 checkpoints out of 50 checkpoints in total and (b) the best accuracy among all the checkpoints. We simply fix $\tau = 0.1$, the temperature hyperparameter in (9), and $\lambda = 0.5$ in (8) in all our experiments. Finally, we take the top 10% and 1% of confident samples for a pseudo-labeled dataset when 4 and 25 labels per class are given, respectively. The details on model architecture and hyperparameters can be found in the supplementary material.

4.1 Effects of Out-of-Class Unlabeled Samples

We first evaluate the effect of out-of-class unlabeled samples in semi-supervised learning, on varying proportions to the training dataset. We consider CIFAR-10

[5] Note that this architecture is larger than Wide-ResNet-28-2 [47] used in the SSL literature [34]. We use ResNet-50 following the standard of SimCLR.

Table 2. Comparison of median test accuracy on 9 super-classes of ImageNet, which are obtained by grouping semantically similar classes in ImageNet; *Dog*, *Reptile*, *Produce*, *Bird*, *Insect*, *Food*, *Primate*, *Aquatic animal*, and *Scenery*. The open-set unlabeled samples are from the entire ImageNet dataset of 1,000 classes. All the benchmarks have 25 labels per class, and we report the mean and standard deviation over three runs with different random seeds and splits. The best scores are indicated in bold. We denote methods handling unlabeled out-of-class samples (*i.e.*, open-set) as "Open-SSL".

In-class		Dog	Reptile	Produce	Bird	Insect	Food	Primate	Aquatic	Scenery
Number of class	Open-SSL	118	36	22	21	20	19	18	13	11
SimCLR-le [7]	-	$43.02_{\pm 0.56}$	$51.76_{\pm 0.92}$	$64.76_{\pm 0.58}$	$75.81_{\pm 1.01}$	$59.90_{\pm 0.92}$	$56.53_{\pm 0.73}$	$53.67_{\pm 0.69}$	$68.41_{\pm 1.31}$	$64.73_{\pm 0.73}$
SimCLR-ft [7]	-	$46.72_{\pm 0.63}$	$51.76_{\pm 1.42}$	$65.21_{\pm 1.05}$	$77.37_{\pm 0.95}$	$58.93_{\pm 1.50}$	$54.63_{\pm 0.79}$	$55.29_{\pm 1.79}$	$68.82_{\pm 2.15}$	$62.79_{\pm 1.15}$
SimCLR-sd [8]	-	$36.12_{\pm 5.76}$	$49.35_{\pm 1.22}$	$61.88_{\pm 1.38}$	$75.71_{\pm 1.72}$	$57.13_{\pm 2.21}$	$57.49_{\pm 9.99}$	$47.70_{\pm 4.71}$	$68.72_{\pm 0.70}$	$66.31_{\pm 1.03}$
UASD-ft [10]	✓	$45.64_{\pm 0.89}$	$53.07_{\pm 0.73}$	$67.09_{\pm 0.65}$	$78.92_{\pm 0.40}$	$61.53_{\pm 1.56}$	$55.90_{\pm 1.04}$	$56.70_{\pm 0.85}$	$70.31_{\pm 0.86}$	$64.36_{\pm 1.13}$
RealMix-ft [32]	✓	$43.55_{\pm 2.36}$	$45.70_{\pm 0.16}$	$56.06_{\pm 0.65}$	$71.94_{\pm 1.21}$	$53.33_{\pm 1.78}$	$48.25_{\pm 1.30}$	$45.89_{\pm 0.84}$	$58.10_{\pm 1.34}$	$60.79_{\pm 1.36}$
DS^{3}L-ft [15]	✓	$44.12_{\pm 1.37}$	$52.09_{\pm 1.47}$	$65.39_{\pm 1.22}$	$78.00_{\pm 0.25}$	$59.30_{\pm 2.33}$	$54.32_{\pm 1.37}$	$53.44_{\pm 0.80}$	$70.67_{\pm 1.23}$	$62.67_{\pm 2.37}$
OpenMatch-ft [37]	✓	$48.85_{\pm 2.85}$	$54.55_{\pm 1.98}$	$67.21_{\pm 1.37}$	$78.60_{\pm 1.80}$	$63.20_{\pm 1.76}$	$55.86_{\pm 1.85}$	$57.19_{\pm 1.49}$	$70.97_{\pm 0.89}$	$64.65_{\pm 6.73}$
MixMatch-ft [4]	-	$43.24_{\pm 0.65}$	$43.68_{\pm 3.01}$	$56.79_{\pm 1.89}$	$71.04_{\pm 2.28}$	$57.70_{\pm 0.56}$	$52.53_{\pm 0.56}$	$52.78_{\pm 0.84}$	$62.36_{\pm 2.10}$	$60.06_{\pm 0.76}$
ReMixMatch-ft [3]	-	$47.47_{\pm 1.47}$	$54.39_{\pm 0.78}$	$66.88_{\pm 0.38}$	$78.95_{\pm 0.67}$	$62.30_{\pm 1.32}$	$55.48_{\pm 0.76}$	$56.63_{\pm 1.81}$	$68.67_{\pm 1.24}$	$65.58_{\pm 0.86}$
FixMatch-ft [40]	-	$49.69_{\pm 0.86}$	$54.35_{\pm 0.64}$	$67.43_{\pm 1.37}$	$78.73_{\pm 1.21}$	$62.53_{\pm 2.02}$	$54.84_{\pm 0.52}$	$57.70_{\pm 1.62}$	$69.79_{\pm 0.93}$	$64.12_{\pm 1.48}$
+ **OpenCoS (ours)**	✓	$\mathbf{50.54}_{\pm 1.19}$	$\mathbf{56.89}_{\pm 0.73}$	$\mathbf{71.70}_{\pm 0.43}$	$\mathbf{81.68}_{\pm 0.39}$	$\mathbf{65.73}_{\pm 2.18}$	$\mathbf{60.46}_{\pm 1.70}$	$\mathbf{60.89}_{\pm 2.84}$	$\mathbf{73.23}_{\pm 2.12}$	$\mathbf{67.33}_{\pm 1.41}$

and TinyImageNet datasets, and synthetically control the proportion between the two in 50K training samples. For example, 80% of proportion means the training dataset consists of 40K samples from TinyImageNet, and 10K samples from CIFAR-10. In this experiment, we assume that 25 labels per class are always given in the CIFAR-10 side. We compare three models on varying proportions of out-of-class: (a) a ReMixMatch model trained from scratch (ReMixMatch), (b) a SimCLR model fine-tuned by ReMixMatch (ReMixMatch-ft), and (c) our OpenCoS model applied to ReMixMatch-ft (+ OpenCoS).

Figure 3 demonstrates the results. Overall, we observe that the performance of ReMixMatch rapidly degrades as the proportion of out-of-class samples increases in unlabeled data. While ReMixMatch-ft significantly mitigates this problem, however, it still fails at a larger proportion: *e.g.*, at 80% of out-of-class, the performance of ReMixMatch-ft falls into that of ReMixMatch. OpenCoS, in contrast, successfully prevents the performance degradation of ReMixMatch-ft, especially at the regime that out-of-class samples dominate in-class samples.

4.2 Experiments on CIFAR Datasets

In this section, we evaluate our method on several benchmarks where CIFAR datasets are assumed to be in-class: more specifically, we consider scenarios that either CIFAR-10 or CIFAR-100 is an in-class dataset, with an out-of-class dataset of either SVHN or TinyImageNet. Additionally, we also consider a separate benchmark called *CIFAR-Animals + CIFAR-Others* following the setup in the related work [34]: the in-class dataset consists of 6 animal classes from CIFAR-10, while the remaining samples are considered as out-of-class. We fix every benchmark to have 50K training samples. We assume an 80% proportion of out-of-class, *i.e.*, 10K for in-class and 40K for out-of-class samples, except for CIFAR-Animals + CIFAR-Others, which consists of 30K and 20K samples for in- and out-of-class, respectively. We report ReMixMatch-ft + OpenCoS as it tends to outperform FixMatch-ft + OpenCoS in such CIFAR-scale experiments, while FixMatch-ft + OpenCoS does[6] in the large-scale ImageNet experiments in Sect. 4.3. Table 1 shows the results: OpenCoS consistently improves ReMixMatch-ft, outperforming the other baselines simultaneously. For example, OpenCoS improves the test accuracy of ReMixMatch-ft 28.51% → 69.38%, also outperforming the strongest open-set SSL baseline, OpenMatch-ft of 62.71%, on 4 labels per class of CIFAR-10 + TinyImageNet.

Also, we observe large discrepancies between the median and best accuracy of semi-supervised learning baselines, MixMatch-ft, ReMixMatch-ft, and FixMatch-ft, especially in the extreme label-scarce scenario of 4 labels per class, *i.e.*, these methods suffer from over-fitting on out-of-class samples. One can also confirm this significant over-fitting in state-of-the-art SSL methods by comparing other baselines with detection schemes, *e.g.*, USAD-ft, RealMix-ft, DS^{3}L-ft, and OpenMatch-ft, which show less over-fitting but with lower best accuracy.

4.3 Experiments on ImageNet Datasets

We also evaluate OpenCoS on ImageNet to verify its scalability to a larger and more complex dataset. We design 9 benchmarks from ImageNet dataset, similarly to Restricted ImageNet [42]: more specifically, we define 9 super-classes of ImageNet, each of which consists of 11∼118 sub-classes. We perform our experiments on each super-class as an individual dataset. Each of the benchmarks (a super-class) contains 25 labels per sub-class, and we use the full ImageNet as an unlabeled dataset (excluding the labeled ones). In this experiment, we checkpoint per 2^{15} training samples and report the median test accuracy of the last 3 out of 10. We present additional experimental details, *e.g.*, configuration of the dataset, in the supplementary material. Table 2 shows the results: OpenCoS still effectively improves the baselines, largely surpassing SimCLR-le and SimCLR-ft as well. For example, OpenCoS improves the test accuracy on Bird to 81.68% from FixMatch-ft of 78.73%, also improving SimCLR-ft of 77.37% significantly. OpenCoS also outperforms other open-set SSL baselines, such as DS^{3}L-ft of 78.00% and OpenMatch-ft of 78.60% on Bird. This shows the efficacy of OpenCoS in exploiting open-set unlabeled data from unknown (but related) classes or even unseen distribution of another dataset in the real-world.

5 Ablation Study

We perform an ablation study to understand further how OpenCoS works. Specifically, we assess the individual effects of the components in OpenCoS and show

[6] Nevertheless, we observe that OpenCoS also improves the opposite choices, *i.e.*, FixMatch for CIFAR and ReMixMatch for ImageNet as presented in the supplementary materials..

Table 3. Ablation study on four components of our method: the detection criterion ("Detect"), auxiliary loss ("Aux. loss"), auxiliary BNs ("Aux. BNs"), and *top-k* pseudo-labeling ("Top-k PL"). We report the mean and standard deviation over three runs with different random seeds and a fixed split of labeled data.

OpenCoS components				In- + Out-of-class
Detect	Aux. loss	Aux. BNs	Top-k PL	CIFAR-10 + SVHN
–	–	–	–	22.53 ± 2.53
✓	–	–	–	50.87 ± 1.27
✓	✓	–	–	55.25 ± 1.04
✓	✓	✓	-	56.70 ± 1.35
✓	✓	✓	✓	**58.02±0.83**

that each of them has an orthogonal contribution to the overall improvements. We consider CIFAR-10 + SVHN benchmark with 4 labels per class, and Table 3 summarizes the results.

Detecting and Soft-labeling Out-of-Class Samples. We first observe that our detection method ("Detect"), which simply applies SSL using detected in-class unlabeled samples, fixes the failure modes of the baseline, *i.e.*, ReMixMatch-ft, from 22.53% to 50.87%. Interestingly, leveraging out-of-class samples achieve significant improvements; *e.g.*, auxiliary loss ("Aux. loss") and auxiliary BNs ("Aux. BNs") improves the performance from 50.87% (of "Detect") to 56.70%. We emphasize that our soft-labeling scheme can be rather viewed as a more reasonable way to label such out-of-class samples compared to existing state-of-the-art SSL methods, *e.g.*, MixMatch simply assigns its sharpened predictions. **Pseudo-labeling confident samples.** We empirically observed that pseudo-labels of confident in-class samples are more accurate than randomly chosen samples. From this observation, we expect leveraging pseudo-labels of such confident samples overcomes the limitation of few labeled data. Finally, we remark that our pseudo-labeling scheme for confident in-class samples ("Top-k PL") improves the performance from 56.70% to 58.02%. **Actual soft-label assignments.** We also present some concrete examples of our soft-labeling scheme for a better understanding, which are obtained from unlabeled samples in the CIFAR-10 + TinyImageNet benchmark. Overall, we qualitatively observe that out-of-class samples that share some semantic features to the in-classes (*e.g.*, Fig. 4a) have relatively high confidence capturing such similarity, while returning very close to uniform otherwise (*e.g.*, Fig. 4b).

Other Pre-training Scheme. In order to investigate the compatibility of OpenCoS with other self-supervised training schemes for pre-training, we also evaluate OpenCoS with DINO [5] instead of SimCLR, which is built upon a transformer-based architecture, namely Vision Transformer (ViT) [13]. We consider 9 ImageNet benchmarks (see Sect. 4.3) to validate the effectiveness of OpenCoS with FixMatch-ft, and Table 4 summarizes the results. We observe that OpenCoS still

Table 4. Comparison of median test accuracy on 9 super-classes of ImageNet with DINO [5] (instead of SimCLR), which is based on ViT [13] architecture. We report the mean and standard deviation over three runs with different random seeds and splits.

Super-classes	# In-classes	FixMatch-ft	+ OpenCoS
Dog	118	63.29 ± 0.97	$\mathbf{65.92 \pm 0.21}$
Reptile	36	64.35 ± 1.50	$\mathbf{66.76 \pm 0.71}$
Produce	22	79.18 ± 0.81	$\mathbf{80.94 \pm 0.62}$
Bird	21	88.57 ± 0.17	$\mathbf{90.89 \pm 0.38}$
Insect	20	75.13 ± 0.40	$\mathbf{76.90 \pm 1.04}$
Food	19	66.95 ± 0.59	$\mathbf{72.00 \pm 0.94}$
Primate	18	73.07 ± 0.42	$\mathbf{76.37 \pm 0.82}$
Aquatic	13	76.72 ± 1.48	$\mathbf{80.21 \pm 0.62}$
Scenery	11	68.36 ± 1.79	$\mathbf{69.82 \pm 0.31}$

consistently improves FixMatch-ft on the ViT encoder pre-trained via DINO. For example, OpenCoS improves the test accuracy on Food to 72.00% from FixMatch-ft of 66.95%. Remarkably, the performances of ViT outperform that of ResNet (in Table 2), which implies that OpenCos can be further improved under better unsupervised representations for handling open-set unlabeled data.

6 Discussion

Conclusion. In this paper, we propose a simple and general framework for handling novel unlabeled data, aiming toward a more realistic assumption for semi-supervised learning. Our key idea is (intentionally) not to use label information, *i.e.*, by relying on *unsupervised* representation, when handling novel data, which can be naturally incorporated into semi-supervised learning with our framework: OpenCoS. In contrast to previous approaches, OpenCoS opens a way to further utilize those open-set data by guessing their labels appropriately, which are again obtained from unsupervised learning. We hope our work would motivate researchers to extend this framework with a more realistic assumption, *e.g.*, noisy labels [27,43], imbalanced learning [30].

Limitations. As our method benefits from larger models due to unsupervised pre-training [7], the performance would be limited on small or tiny architectures.

Potential Negative Societal Impacts. The open-set unlabeled dataset may contain sensitive data, *e.g.*, facial images, as it is hard to check all collected data by humans. If someone tries to classify such data from web-crawled datasets, our method could make this process more easily. For this reason, it is also important to consider this privacy issue.

References

1. Bendale, A., Boult, T.E.: Towards open set deep networks. In: CVPR (2016)
2. Bergman, L., Hoshen, Y.: Classification-based anomaly detection for general data. In: ICLR (2020)
3. Berthelot, D., et al.: Remixmatch: semi-supervised learning with distribution matching and augmentation anchoring. In: ICLR (2020)
4. Berthelot, D., Carlini, N., Goodfellow, I., Papernot, N., Oliver, A., Raffel, C.A.: Mixmatch: a holistic approach to semi-supervised learning. In: NeurIPS (2019)
5. Caron, M., Touvron, H., Misra, I., Jégou, H., Mairal, J., Bojanowski, P., Joulin, A.: Emerging properties in self-supervised vision transformers. In: ICCV (2021)
6. Chapelle, O., Scholkopf, B., Zien, A.: Semi-supervised learning. IEEE Trans. Neural Networks **20**(3), 542–542 (2009)
7. Chen, T., Kornblith, S., Norouzi, M., Hinton, G.: A simple framework for contrastive learning of visual representations. In: ICML (2020)
8. Chen, T., Kornblith, S., Swersky, K., Norouzi, M., Hinton, G.: Big self-supervised models are strong semi-supervised learners. In: NeurIPS (2020)
9. Chen, X., Fan, H., Girshick, R., He, K.: Improved baselines with momentum contrastive learning. arXiv preprint arXiv:2003.04297 (2020)
10. Chen, Y., Zhu, X., Li, W., Gong, S.: Semi-supervised learning under class distribution mismatch. In: AAAI (2020)
11. Cubuk, E.D., Zoph, B., Shlens, J., Le, Q.V.: Randaugment: practical data augmentation with no separate search. arXiv preprint arXiv:1909.13719 (2019)
12. Deng, J., Dong, W., Socher, R., Li, L.J., Li, K., Fei-Fei, L.: Imagenet: a large-scale hierarchical image database. In: CVPR (2009)
13. Dosovitskiy, A., et al.: An image is worth 16×16 words: transformers for image recognition at scale. arXiv preprint arXiv:2010.11929 (2020)
14. Grandvalet, Y., Bengio, Y.: Semi-supervised learning by entropy minimization. In: NeurIPS (2004)
15. Guo, L.Z., Zhang, Z.Y., Jiang, Y., Li, Y.F., Zhou, Z.H.: Safe deep semi-supervised learning for unseen-class unlabeled data. In: ICML (2020)
16. He, K., Fan, H., Wu, Y., Xie, S., Girshick, R.: Momentum contrast for unsupervised visual representation learning. In: CVPR (2020)
17. He, K., Zhang, X., Ren, S., Sun, J.: Deep residual learning for image recognition. In: CVPR (2016)
18. Hénaff, O.J., et al.: Data-efficient image recognition with contrastive predictive coding. arXiv preprint arXiv:1905.09272 (2019)
19. Hendrycks, D., Gimpel, K.: A baseline for detecting misclassified and out-of-distribution examples in neural networks. In: ICLR (2017)
20. Hendrycks, D., Mazeika, M., Dietterich, T.: Deep anomaly detection with outlier exposure. In: ICLR (2019)
21. Hendrycks, D., Mazeika, M., Kadavath, S., Song, D.: Using self-supervised learning can improve model robustness and uncertainty. In: NeurIPS (2019)
22. Japkowicz, N.: The class imbalance problem: significance and strategies. In: Proceedings of the International Conference on Artificial Intelligence (2000)
23. Krizhevsky, A., Hinton, G., et al.: Learning multiple layers of features from tiny images. Tech. rep, Citeseer (2009)
24. Lee, D.H.: Pseudo-label: The simple and efficient semi-supervised learning method for deep neural networks. In: ICML Workshop (2013)

25. Lee, K., Lee, H., Lee, K., Shin, J.: Training confidence-calibrated classifiers for detecting out-of-distribution samples. In: ICLR (2018)
26. Lee, K., Lee, K., Lee, H., Shin, J.: A simple unified framework for detecting out-of-distribution samples and adversarial attacks. In: NeurIPS (2018)
27. Lee, K., Yun, S., Lee, K., Lee, H., Li, B., Shin, J.: Robust inference via generative classifiers for handling noisy labels. In: ICML (2019)
28. Li, Z., Hoiem, D.: Learning without forgetting. In: ECCV (2016)
29. Liang, S., Li, Y., Srikant, R.: Enhancing the reliability of out-of-distribution image detection in neural networks. In: ICLR (2018)
30. Liu, Y., Gao, T., Yang, H.: Selectnet: learning to sample from the wild for imbalanced data training. In: MSML (2020)
31. Miyato, T., Maeda, S.i., Koyama, M., Ishii, S.: Virtual adversarial training: a regularization method for supervised and semi-supervised learning. IEEE Trans. Pattern Anal. Mach. Intell. **41**(8), 1979–1993 (2018)
32. Nair, V., Alonso, J.F., Beltramelli, T.: Realmix: towards realistic semi-supervised deep learning algorithms. arXiv preprint arXiv:1912.08766 (2019)
33. Netzer, Y., Wang, T., Coates, A., Bissacco, A., Wu, B., Ng, A.Y.: Reading digits in natural images with unsupervised feature learning. In: NeurIPS Workshop (2011)
34. Oliver, A., Odena, A., Raffel, C.A., Cubuk, E.D., Goodfellow, I.: Realistic evaluation of deep semi-supervised learning algorithms. In: NeurIPS (2018)
35. Oord, A.v.d., Li, Y., Vinyals, O.: Representation learning with contrastive predictive coding. arXiv preprint arXiv:1807.03748 (2018)
36. Raina, R., Battle, A., Lee, H., Packer, B., Ng, A.Y.: Self-taught learning: transfer learning from unlabeled data. In: ICML (2007)
37. Saito, K., Kim, D., Saenko, K.: Openmatch: open-set consistency regularization for semi-supervised learning with outliers. In: NeurIPS (2021)
38. Sajjadi, M., Javanmardi, M., Tasdizen, T.: Regularization with stochastic transformations and perturbations for deep semi-supervised learning. In: NeurIPS (2016)
39. Sergey Ioffe, C.S.: Batch normalization: accelerating deep network training by reducing internal covariate shift. In: ICML (2015)
40. Sohn, K., et al.: Fixmatch: simplifying semi-supervised learning with consistency and confidence. In: NeurIPS (2020)
41. Tack, J., Mo, S., Jeong, J., Shin, J.: Csi: novelty detection via contrastive learning on distributionally shifted instances. In: NeurIPS (2020)
42. Tsipras, D., Santurkar, S., Engstrom, L., Turner, A., Madry, A.: Robustness may be at odds with accuracy. In: ICLR (2019)
43. Wang, Y., et al.: Iterative learning with open-set noisy labels. In: CVPR (2018)
44. Wu, Z., Xiong, Y., Yu, S., Lin, D.: Unsupervised feature learning via non-parametric instance-level discrimination. In: CVPR (2018)
45. Xie, C., Tan, M., Gong, B., Wang, J., Yuille, A., Le, Q.V.: Adversarial examples improve image recognition. In: CVPR (2020)
46. Xie, Q., Dai, Z., Hovy, E., Luong, M.T., Le, Q.V.: Unsupervised data augmentation for consistency training. arXiv preprint arXiv:1904.12848 (2019)
47. Zagoruyko, S., Komodakis, N.: Wide residual networks. In: BMVC (2016)
48. Zhai, X., Oliver, A., Kolesnikov, A., Beyer, L.: S4l: Self-supervised semi-supervised learning. In: ICCV (2019)
49. Zhang, H., Cisse, M., Dauphin, Y.N., Lopez-Paz, D.: mixup: Beyond empirical risk minimization. In: ICLR (2018)
50. Zhang, R., Isola, P., Efros, A.A.: Colorful image colorization. In: Leibe, B., Matas, J., Sebe, N., Welling, M. (eds.) ECCV 2016. LNCS, vol. 9907, pp. 649–666. Springer, Cham (2016). https://doi.org/10.1007/978-3-319-46487-9_40

Semi-supervised Domain Adaptation by Similarity Based Pseudo-Label Injection

Abhay Rawat[1,2(✉)], Isha Dua[2], Saurav Gupta[2], and Rahul Tallamraju[2]

[1] International Institute of Information Technology, Hyderabad, India
[2] Mercedes-Benz Research and Development India, Bengaluru, India
{abhay.rawat,isha.dua,saurav.gupta,rahul.tallamraju}@mercedes-benz.com

Abstract. One of the primary challenges in Semi-supervised Domain Adaptation (SSDA) is the skewed ratio between the number of labeled source and target samples, causing the model to be biased towards the source domain. Recent works in SSDA show that aligning only the labeled target samples with the source samples potentially leads to incomplete domain alignment of the target domain to the source domain. In our approach, to align the two domains, we leverage contrastive losses to learn a semantically meaningful and a domain agnostic feature space using the supervised samples from both domains. To mitigate challenges caused by the skewed label ratio, we pseudo-label the unlabeled target samples by comparing their feature representation to those of the labeled samples from both the source and target domains. Furthermore, to increase the support of the target domain, these potentially noisy pseudo-labels are gradually injected into the labeled target dataset over the course of training. Specifically, we use a temperature scaled cosine similarity measure to assign a soft pseudo-label to the unlabeled target samples. Additionally, we compute an exponential moving average of the soft pseudo-labels for each unlabeled sample. These pseudo-labels are progressively injected (or removed) into the (from) the labeled target dataset based on a confidence threshold to supplement the alignment of the source and target distributions. Finally, we use a supervised contrastive loss on the labeled and pseudo-labeled datasets to align the source and target distributions. Using our proposed approach, we showcase state-of-the-art performance on SSDA benchmarks - Office-Home, DomainNet and Office-31. The inference code is available at https://github.com/abhayraw1/SPI

Keywords: Semi-supervised domain adaptation · Contrastive learning · Pseudo-labelling

1 Introduction

Domain Adaptation approaches aim at solving the data distribution shift between training (source) and test (target) data. Unsupervised Domain Adap-

Supplementary Information The online version contains supplementary material available at https://doi.org/10.1007/978-3-031-25063-7_10.

L. Karlinsky et al. (Eds.): ECCV 2022 Workshops, LNCS 13802, pp. 150–166, 2023.
https://doi.org/10.1007/978-3-031-25063-7_10

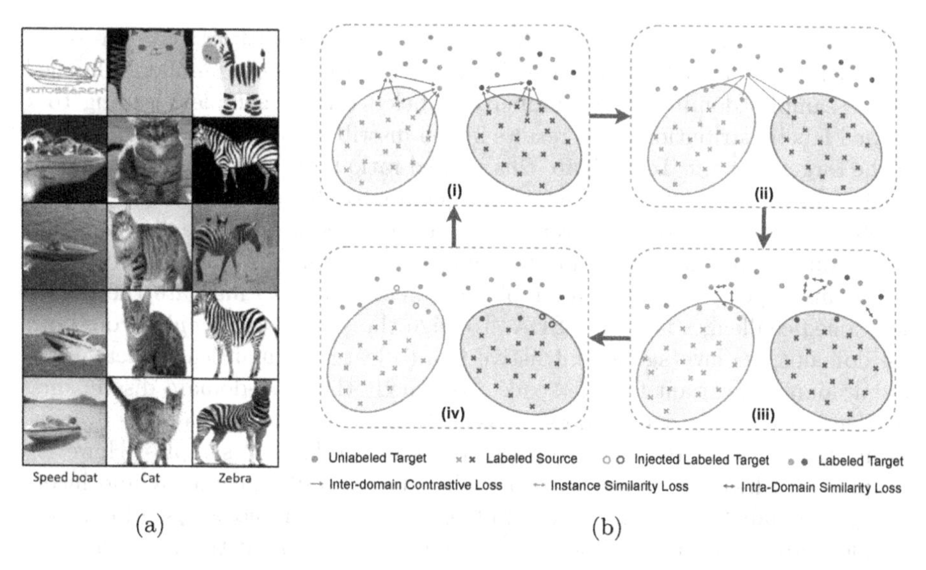

Fig. 1. (a) Nearest neighbours of the target domain (Real) from the source domain (Clipart). (b) High level overview of our proposed method SPI. (i) Inter-domain contrastive loss works on the labeled source and target samples, pulling the samples from the same class closer, (ii) unlabeled samples are drawn towards similar samples from the support set using the instance level similarity loss, (iii) Intra-domain alignment loss focuses on bringing similar samples in the unlabeled target domain closer together, (iv) Confident samples are injected into the labeled target dataset.

tation (UDA) considers the problem of adapting the model trained on a labeled source distribution to an unlabeled target distribution. Most recent approaches in UDA [11,17,23,32,40] aim to learn a domain agnostic feature representation such that the features belonging to similar categories are closer together in the latent space. The underlying assumption here is that learning a mapping to a domain agnostic feature space, along with a classifier that performs sufficiently well on the source domain, could generalize to the target domain. However, recent studies [7,20,42,46] show that these conditions are not sufficient for successful domain adaptation and might even hurt generalization due to the discrepancy between the marginal label distributions of the two domains.

Semi-Supervised learning (SSL) [1,3,36,45] has already proven to be highly efficient in terms of performance per annotation and thus provides a more economical way to train deep learning models. UDA approaches in general, however, do not perform well in a semi-supervised setting where we have access to some labeled samples from the target domain [31]. Semi-Supervised Domain Adaptation (SSDA) [19,21,35], leverages a small amount of labeled samples in the target domain to aid in learning models with low error rate on the target domain. However, as shown by [19], simply aligning the labeled target samples with the labeled source samples can cause intra-domain discrepancy in the target domain. During training, the labeled target samples get pulled towards the corresponding source sample clusters. However, the unlabeled samples with less correlation to the labeled target samples are left behind. This is because the number of labeled source samples dominate the number of labeled tareget samples, leading to a skewed label distribution. This results in sub-distributions within the same class of the target domain. To mitigate this skewed ratio between the labeled samples from the source and target domain, recent approaches [17,40] assign pseudo-labels to unlabeled data. However, these pseudo-labels are potentially noisy and may result in poor generalization to the target domain.

In this paper, we present a simple yet effective way to mitigate the above-mentioned challenges faced in SSDA. To align the *supervised samples* from both the domains, we leverage a contrastive loss to learn a semantically meaningful and a domain invariant feature space. To remedy the intra-domain discrepancy problem, we compute soft pseudo-labels for the unlabeled target samples by comparing their feature representations to those of the labeled samples. However, samples with less correlation with the labeled ones could have noisy and incorrect pseudo-labels. Therefore, we gradually inject (or remove) pseudo-labeled samples into (or from) the labeled target dataset throughout training, based on the model's confidence on the respective pseudo-labels.

Our proposed method, SPI (Similarity based Pseudo-label Injection) for SSDA has the following four components.

1. **Domain alignment - between labeled source and labeled target samples:** To align the feature representations of the two domains, we use a supervised contrastive loss on the labeled samples from both domains. This enforces similar representations for samples belonging to the same class across both domains, as shown in Fig. 1b (i).
2. **Soft Pseudo-Labeling - between different views of the unlabeled samples:** We use a non-parametric, similarity based pseudo labeling technique to generate soft pseudo-labels for the unlabeled samples using the labeled samples from both the domains as shown in Fig 1b (ii). To further enforce consistency of unlabeled target sample representations, we employ a similarity-based consistency between different augmented views of the same instance (Instance Similarity Loss).
3. **Intra-domain similarity - between the unlabeled samples of the target domain:** We compare the positions of the most highly activated feature dimensions of the unlabeled samples to ascertain if the samples are semantically similar. We then minimize the distance between feature representations of similar samples as shown in Fig. 1b (iii).
4. **Pseudo-label Injection:** Finally, an exponential moving average of the sample's soft pseudo-label is updated over training. Samples with pseudo-label confidence above a given threshold are injected into the labeled target dataset as shown in Fig. 1b (iv).

In summary, we present a novel method that leverages feature level similarity between the unlabeled and labeled samples across domains to pseudo-label the unlabeled data. We use a novel loss function that includes supervised contrastive loss, an instance-level similarity loss and an intra-domain consistency loss which together help to bring the features of similar classes closer for both domains. Finally, we propose to gradually inject and remove pseudo-labels into the labeled target dataset to increase the support of the same. We showcase superior classification accuracy across popular image classification benchmarks - Office-Home, Office-31, and DomainNet against state-of-the-art SSDA approaches.

2 Related Works

Unsupervised Domain Adaptation (UDA) methods [6,11,17,21,22,32,40] aims to reduce the cross-domain divergence between the source and target domains by either minimizing the discrepancy between the two domains or, by training the domain invariant feature extractor via adversarial training regimes [6,10,11,22,26,32]. The distribution level distance measures have been adopted to reduce the discrepancy like Correlation Distances [37,44], Maximum Mean Discrepancy (MMD) [17,22], JS Divergence [34] and Wasserstein Distance [33]. SSDA approaches, on the other hand, utilize a small number of labeled samples from the target domain which is a more realistic setting in most vision problems.

Semi-Supervised Domain Adaptation. deals with data sampled from two domains - one with labeled samples and the other with a mix of labeled and unlabeled samples - and hence aims to tackle the issues caused due to domain discrepancy. Recent works [19,31] show that UDA methods do not perform well when provided with some labeled samples due to the bias towards labeled samples from the source domain. Other approaches for SSDA rely on learning domain invariant feature representations [19,21,31,35]. However, studies [7,20,42,46] have shown that learning such a domain agnostic feature space and a predictor with low error rate is not a sufficient condition to assure generalization to the target domain.

Pseudo Labeling techniques have been used in UDA and SSDA approaches to reduce the imbalance in the labeled source and target data. CAN [17] and CDCL [40] uses k-means approach to assign pseudo labels to the unlabeled target samples. DECOTA [43] and CLDA [35] use the predictions of their classifiers as the pseudo-label. CDAC [40] uses confident predictions of the weakly augmented unlabeled samples as pseudo-labels for the strongly augmented views of the same sample, similar to FixMatch [36]. In contrast, our approach uses a similarity based pseudo labeling approach inspired by PAWS [1]. We extend this approach to the SSDA setting by computing the pseudo-labels for the unlabeled samples using labeled samples from both the domains.

Contrastive Learning aims to learn feature representations by pulling the positive samples closer and pushing the negative samples apart [2,14,22,25,38]. In a supervised setting, [18] leverages the labels of samples in a batch to construct positive and negative pairs. But, due to the absence of labeled samples in UDA, methods like CDCL [40], assign pseudo labels to the unlabeled target samples using clustering to help construct the positive and negative pairs. CLDA [35], an SSDA method, uses contrastive loss to align the strongly and weakly augmented views of the same unlabeled image to ensure feature level consistency.

3 Method

In semi-supervised domain adaptation (SSDA), apart from labeled samples from the source distribution $\mathcal{D}_s$ and the unlabeled samples from the target distribution $\mathcal{D}_t$, we also have access to a small amount of labeled samples from $\mathcal{D}_t$. Let $S = \{(x_i^s, y_i^s)\}_{i=1}^{N_s}$ and $T = \{(x_i^t)\}_{i=1}^{N_u}$ denote the set of data sampled i.i.d. from $\mathcal{D}_s$ and $\mathcal{D}_t$ respectively. Additionaly, we also have $\hat{T} = \{(x_i^t, y_i^t)\}_{i=1}^{N_l}$, a small set of labeled samples from $\mathcal{D}_t$. The number of samples in the S, T and $\hat{T}$ are denoted by N_s, N_u, N_l respectively, where $N_l \ll N_u$ Both domains are considered to have the same set of labels $Y = \{1, 2, \cdots C\}$ where C is the number of categories. Our aim is to learn a classifier using S, T and $\hat{T}$ that can accurately classify novel samples from the target domain $\mathcal{D}_t$ during inference/testing.

Preliminaries: Our method consists of a feature extractor $\mathcal{F}$: $\mathcal{X} \to \mathbb{R}^d$ and a classifier $\mathcal{H}$: $\mathbb{R}^d \to \mathbb{R}^C$ parameterized by $\Theta_\mathcal{F}$ and $\Theta_\mathcal{H}$ respectively, where $\mathcal{X}$ denotes the image space, d denotes the dimensionality of the feature space and C is the number of classes. Classifier $\mathcal{H}$ is a single linear layer that maps the features from the feature extractor $\mathcal{F}$ to unnormalized class scores. These scores are converted into respective class probabilities using the softmax function $\sigma(\cdot)$.

$$h(x) = \sigma(\mathcal{H}(z)), \tag{1}$$

where $z = \mathcal{F}(x)$ denotes the feature representation of the input image x. For brevity, we denote the class probability distribution for a sample x_i computed using the classifier as h_i.

Figure 2 depicts a concise overview of our proposed method, SPI. We sample η_{sup} images per class (with replacement) from the labeled source S and the target dataset $\hat{T}$ to construct the mini-batch of labeled samples for each domain. We refer to the concatenated mini-batch of labeled samples from both domains as the *support set* throughout the paper. The support set contains $\eta_{sup}\, C$ samples from both the domains, $2\,\eta_{sup}\, C$ samples in total. For the unlabeled batch B_u, we sample images uniformly from the unlabeled target dataset T. Following the multi-view augmentation technique [1,4,5], each image in the unlabeled batch B_u is further augmented to produce η_g global views and η_l local views. Throughout the paper, we use the term *unlabeled images* to refer to the global views of these unlabeled images, unless specified otherwise. For the sake of simplicity, we use a common symbol τ to denote the temperature in different equations, Eqs. 2, 3 and 4.

An Outline of SPI. In Sect. 3.1, we introduce the supervised contrastive loss used for domain alignment. We explain the process of injecting unlabeled samples into the labeled dataset in detail in Sect. 3.2. Section 3.3 presents our modification of the multi-view similarity loss [1], used in SSL, to the SSDA setting. This loss helps in two ways: a) by ensuring the feature representation for

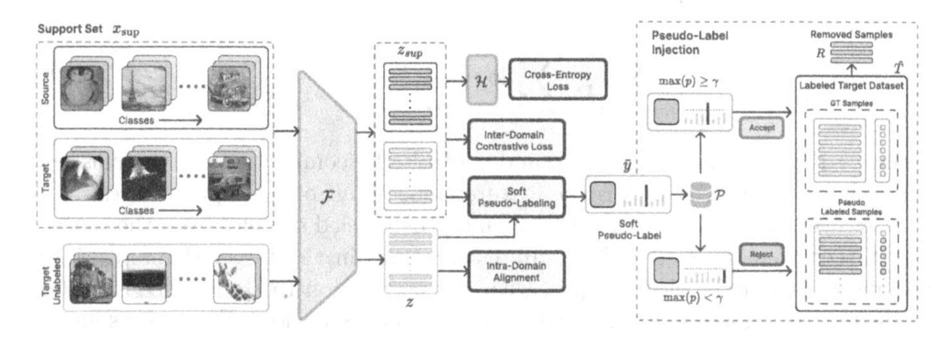

Fig. 2. An overview of our proposed method, SPI. We sample labeled images from both the source (orange) and target (blue) domain to construct the support set $x_{sup}(Color figure online)$. Features from the unlabeled samples z are compared with those from the support set z_{sup} to compute the soft pseudo-labels $\tilde{y}$ for the target unlabeled images. $\mathcal{P}$ maintains a mapping from the image ids to the exponential moving averages of these soft pseudo-labels. After every epoch, we select samples from $\mathcal{P}$ for injection (green) into the target labeled dataset $\hat{T}$ based on the confidence threshold γ (if $p = \mathcal{P}(\text{ID}(.)) \geq \gamma$). Additionally, injected samples are also selected for removal (red) from $\hat{T}$ if their confidence falls below γ. Classifier is denoted as $\mathcal{H}$ (Color figure online)

different views of the same image is similar, and b) by bringing these representations closer to similar samples from the support set. In Sect. 3.4, we present a technique to determine if two samples from the unlabeled target domain are similar at a feature level [13]. Subsequently, we present a feature-level similarity loss between similar samples to bring them closer to each other in the latent space. Finally, we combine these components, presenting SPI as an end-to-end framework in Sect. 3.5.

3.1 Inter-domain Feature Alignment

The premise for most domain adaptation approaches is to learn a domain invariant feature representation for both the source and target domains. CDCL [40] uses contrastive learning in the context of UDA to align the source and target domains. Due to the absence of class labels in UDA, they use pseudo-labels generated using k-means clustering for the contrastive loss. CLDA [35], a method for SSDA, applies contrastive loss on the exponential moving average of the source and target feature centroids to align the two domains. In contrast, we leverage contrastive loss more directly by explicitly treating the samples of the same classes as positives, irrespective of the domain. The feature extractor is then trained to minimize the $\mathcal{L}_{con}$ by maxmizing the similarity between the features of the same class.

Let $A = \{1, 2, \cdots, 2\,\eta_{sup}\,C\}$ denote the set of indices in the support set. Note that the support set contains samples from both the domains and η_{sup} samples from each class. Let P_i denote the set indices of all the images with the same label as i. Then, the supervised contrastive loss $\mathcal{L}_{con}$ can be written as:

$$\mathcal{L}_{con} = \sum_{i \in A} \frac{-1}{|P_i|} \sum_{p \in P_i} \log \frac{\exp(z_i \cdot z_p / \tau)}{\sum_{a \in A \setminus i} \exp(z_a \cdot z_p / \tau)}, \tag{2}$$

where, τ denotes the temperature and controls the compactness of the clusters for each class, and $z = \mathcal{F}(\cdot)$ is the feature embeddings of the input images.

However, as pointed out in [19], aligning the labeled samples from the source and target domains can lead to sub-distributions in the target domain. More specifically, the unlabeled samples with less correlation to the labeled samples in the target domain do not get pulled by the contrastive loss. This causes intra-domain discrepancy and thereby leads to poor performance. To mitigate this issue, we propose to inject unlabeled samples into the labeled target dataset $\hat{T}$ thus, effectively increasing the support of the labeled samples in the target domain. We discuss this approach in more detail Sect. 3.2.

3.2 Pseudo-Label Injection

To reduce the intra-domain discrepancy, we propose injecting samples from the unlabeled target dataset T into the labeled target dataset $\hat{T}$. Using the support set, we first compute the soft pseudo-labels of the unlabeled samples. Throughout the training, we keep an exponential moving average of the sharpened soft pseudo-labels for every sample in the unlabeled target dataset T. This moving average estimates the confidence of our model's prediction of each unlabeled sample. Using this estimate, we inject the highly confident samples into the labeled target dataset $\hat{T}$ with their respective label set to the dominant class after each epoch.

To compute the soft pseudo-labels for the unlabeled samples from the target domain, we take inspiration from PAWS [1], a recent work in semi-supervised learning and extend it to the SSDA setting. We denote the support set x_{sup} and their respective labels as y_{sup}. Let $\hat{z}_{sup}$ be the normalized features representation of the samples in the support set x_{sup} and $\hat{z}_i (= z_i/\|z_i\|)$ denote the normalized feature representation for the unlabeled sample x_i. Then, the soft pseudo-label for ith unlabeled sample can be computed using:

$$\tilde{y}_i = \sigma_\tau(\hat{z}_i \cdot \hat{z}_{sup}^\top)\, y_{sup} \tag{3}$$

where $\sigma_\tau(\cdot)$ denotes the softmax operator with temperature τ. These soft pseudo-labels are then sharpened using the sharpening function π with temperature $\tau > 0$, described as follows:

$$\pi(\tilde{y}) = \frac{\tilde{y}^{1/\tau}}{\sum_{j=1}^{C} \tilde{y}_j^{1/\tau}} \tag{4}$$

Sharpening helps to produce confident predictions from the similarity measure between the unlabeled and labeled samples.

Throughout the training, we keep an exponential moving average (EMA) of the sharpened soft pseudo-labels of each image in the unlabeled target dataset T.

More specifically, we maintain a mapping $\mathcal{P}: \mathbb{I} \rightarrow \mathbb{R}^C$ from image ids of the unlabeled samples to the running EMA of their respective sharpened soft pseudo-labels (class probability distribution). Let $\mathrm{ID}(\cdot)$ denote an operator that returns the image-id corresponding to the input sample in the unlabeled target dataset T and $\mathcal{P}(\mathrm{ID}(x_i))$ be the EMA of the sharpened pseudo-label of x_i. Then, this exponential moving average for a sample x_i in the unlabeled dataset T is updated as follows:

$$\mathcal{P}(\mathrm{ID}(x_i)) \leftarrow \rho\, \pi(\tilde{y}_i) + (1-\rho)\, \mathcal{P}(\mathrm{ID}(x_i)) \tag{5}$$

where ρ denotes the momentum parameter. When a sample is encountered for the first time in the course of training, $\mathcal{P}(\mathrm{ID}(x_i))$ is set to $\pi(\tilde{y}_i)$ and Eq. 5 is used thereafter.

After each epoch, we examine the EMA (class probability distribution) for each sample in $\mathcal{P}$. If the confidence of a particular sample for a class crosses a certain threshold γ, we inject that sample and its corresponding predicted class into the labeled target dataset $\hat{T}$. We define the set of samples considered for injection I as:

$$I_t \triangleq \{(x_i, \arg\max \mathcal{P}(\mathrm{ID}(x_i)) \mid x_i \in T \wedge \max \mathcal{P}(\mathrm{ID}(x_i)) \geq \gamma\}\ , \tag{6}$$

where t denotes the current epoch.

However, these samples could potentially be noisy and might hinder the training process; therefore we also remove samples from the labeled dataset $\hat{T}$ if their confidence falls below the threshold γ. The set of samples to be removed from the labeled target dataset R is defined as:

$$R_t \triangleq \{(x_i, y_i) \mid x_i \in (\hat{T}_t \setminus \hat{T}_0) \wedge \max \mathcal{P}(\mathrm{ID}(x_i)) < \gamma\}\ , \tag{7}$$

where y_i, denotes the corresponding pseudo-label that had been previously assigned to the sample x_i from Eq. 6. Note that the original samples from the labeled target dataset $\hat{T}_0$ are never removed from the dataset as both I and R contain samples only from the unlabeled target dataset T.

The labeled target dataset $\hat{T}$ after each epoch t, is therefore updated as:

$$\hat{T}_{t+1} = \begin{cases} (\hat{T}_t \setminus R_t) \cup I_t & \text{if } t \geq W \\ \hat{T}_t & \text{otherwise} \end{cases} \tag{8}$$

where W represents the number of warmup epochs up to which the labeled target dataset $\hat{T}$ remains unaltered. These warmup epochs allow the feature representations of the source and target domains to be aligned to some extent before the samples are injected into the label target dataset. This prevents false-positive samples from getting into $\hat{T}$ which would otherwise hinder the learning process.

3.3 Instance Level Similarity

We now introduce the Instance Level Similarity loss. Inspired by [1,5], we follow a multi-view augmentation to generate $\eta_g = 2$ global crops and η_l local crops of the unlabeled images. The key insight behind such an augmentation scheme is to enforce the model to focus on the object of interest by explicitly bringing the feature representations of these different views closer. The global crops contain more semantic information about the object of interest, whereas the local crops only contain a limited view of the image (or object). By computing the feature level similarity between the global crops and the support set samples, we compute the pseudo-label for the unlabeled samples using Eq. 3.

The feature extractor is then trained to minimize the cross-entropy between the pseudo-label generated using one global view and the sharpened pseudo-label generated using the other global view. Additionally, the cross-entropy between the pseudo-label generated using the local views and the mean of the sharpened pseudo-label from the global views is added to the loss.

With a slight abuse of notation, given a sample x_i, we define $\tilde{y}_i^{g_1}$ and $\tilde{y}_i^{g_2}$ as the pseudo-label for the two global crops and $\tilde{y}_i^{l_j}$ denote the pseudo-label for the jth local crop. Similarly, we follow the same notation to define the sharpened pseudo-label for these crops denoted by π. The feature extractor is thus trained to minimize the following loss:

$$\mathcal{L}_{ils} = -\sum_{i=1}^{|B_u|} \left(\mathrm{H}(\tilde{y}_i^{g_1}, \pi_i^{g_2}) + \mathrm{H}(\tilde{y}_i^{g_2}, \pi_i^{g_1}) + \sum_{j=1}^{\eta_l} \mathrm{H}(\tilde{y}_i^{l_j}, \pi_i^{g}) \right) , \tag{9}$$

where, $\mathrm{H}(\cdot, \cdot)$ denotes the cross-entropy, $\pi_i^g = (\pi_i^{g_1} + \pi_i^{g_1})/2$ and $|B_u|$ denotes the number of unlabeled samples.

3.4 Intra-domain Alignment

To ensure that the unlabeled samples from the same class in the target domain are closer together in the latent space, we use consistency loss between the unlabeled samples. Due to the absence of labels for these samples, we compute the pairwise feature similarity between the unlabeled samples to estimate whether they could potentially belong to the same class. As proposed by [13], two samples x_i and x_j can be considered similar if the indices of their top-k highly activated feature dimensions are the same. Let top-k (z) denote the set of indices of the top k highly activated feature dimensions of z then, we consider two unlabeled samples i and j similar if:

$$\text{top-k } (z_i) \ominus \text{top-k } (z_j) = \Phi \tag{10}$$

where, z_i and z_j are their respective feature representations, and $\ominus$ is the symmetric set difference operator.

We construct a binary matrix $M \in \{0,1\}^{|B_u| \times |B_u|}$ whose individual entries M_{ij} denote whether the ith sample is similar to jth sample in the unlabeled batch B_u using Eq. 10. Using the similarity matrix M, we compute the intra-domain consistency loss $\mathcal{L}_{ida}$ for the target unlabeled samples as follows:

$$\mathcal{L}_{ida} = \frac{1}{|B_u|^2} \sum_{i=1}^{|B_u|} \sum_{j=1}^{|B_u|} M_{ij} \left\| z_i - z_j \right\|_2 \tag{11}$$

3.5 Classification Loss and Overall Framework

We use the label-smoothing cross-entropy [24] loss to train the classifier layer. For the classifier training, we only use the samples from the labeled source dataset S and the labeled target dataset $\hat{T}$, which is constantly being updated with new samples.

$$\mathcal{L}_{cls} = -\sum_{i=1}^{2\,\eta_{sup}\,C} \mathrm{H}(h_i, \hat{y}_i) \tag{12}$$

where, h_i is the predicted class probabilities (Eq. 1), H denotes the cross-entropy loss and $\hat{y}_i = (1-\alpha)y_i + \alpha/C$ is the smoothened label corresponding to x_i. Here, α is the smoothing parameter and y_i is the one-hot encoded label vector.

Combining the different losses used in our proposed method SPI, $\mathcal{L}_{con}$, $\mathcal{L}_{ils}$ and $\mathcal{L}_{ida}$, yields a single training objective:

$$\mathcal{L}_{SPI} = \lambda\,\mathcal{L}_{con} + \mathcal{L}_{ils} + \mathcal{L}_{ida} + \mathcal{L}_{cls} \tag{13}$$

4 Experiments

4.1 Datasets

We compare the performance of SPI against the existing approaches on the popular image classification benchmarks: Office-Home [39], Office-31 [30] and DomainNet [27]. Office-Home comprises of 4 domains: **A**rt, **C**lipart, **P**roduct, and **R**ealWorld, and a total of 65 different categories. The Office-31 benchmark contains 3 domains - **A**mazon, **W**ebcam, and **D**SLR, containing objects from 31 different categories. Following [19,35], we use a subset of the DomainNet benchmark adapted by [31], which contains a total of 126 classes and 4 domains: **C**lipart, **S**ketch, **P**ainting, and **R**eal.

Following the authors of [19,21,35], we use the publically available train, validation, and test splits provided by [31] to ensure a consistent and fair comparison against the existing approaches.

4.2 Baselines

To quantify the efficacy of our approach, we compare its performance to the previous state-of-the-art methods in SSDA task: **CLDA** [35], **CDAC** [21], **MME** [31], **APE** [19], **DECOTA** [43], **BiAT** [16], **UODA** [28], **ENT** [12] and **Meta-MME** [29]. We also include the results of popular UDA approaches **DANN** [11], **ADR** [32] and **CDAN** [23]. These UDA methods have been modified to utilize the labeled target samples to ensure a fair comparison, as reported in [31]. In addition to this, **S+T** method provides a rudimentary baseline wherein the model is trained only using the labeled samples from both the source and target domains.

Table 1. Accuracy on Office-Home in the 3-shot setting (ResNet-34)

Method	R → C	R → P	R → A	P → R	P → C	P → A	A → P	A → C	A → R	C → R	C → A	C → P	Mean
S + T	55.7	80.8	67.8	73.1	53.8	63.5	73.1	54.0	74.2	68.3	57.6	72.3	66.2
DANN [11]	57.3	75.5	65.2	51.8	51.8	56.6	68.3	54.7	73.8	67.1	55.1	67.5	63.5
ENT [12]	62.6	85.7	70.2	79.9	60.5	63.9	79.5	61.3	79.1	76.4	64.7	79.1	71.9
MME [31]	64.6	85.5	71.3	80.1	64.6	65.5	79.0	63.6	79.7	76.6	67.2	79.3	73.1
Meta-MME [29]	65.2	–	–	–	64.5	66.7	–	63.3	–	–	67.5	–	–
APE [19]	66.4	86.2	73.4	82.0	65.2	66.1	81.1	63.9	80.2	76.8	66.6	79.9	74.0
DECOTA [43]	**70.4**	**87.7**	74.0	82.1	**68.0**	69.9	81.8	64.0	80.5	79.0	68.0	83.2	75.7
CDAC [21]	67.8	85.6	72.2	81.9	**67.0**	67.5	80.3	65.9	80.6	80.2	67.4	81.4	74.2
CLDA [35]	66.0	87.6	**76.7**	**82.2**	63.9	**72.4**	81.4	63.4	81.3	80.3	70.5	80.9	75.5
SPI (Ours)	69.3	86.9	74.3	81.9	66.6	68.6	**82.5**	**66.4**	**81.6**	**80.9**	**71.1**	**83.8**	**76.15**

4.3 Implementation Details

We use ResNet-34 [15] as the feature extractor in our experiments on Office-Home and DomainNet datasets and VGG-16 for experiments on the Office-31 dataset. The feature extractors use pre-trained ImageNet [9] weights as provided by the PyTorch Image Models library [41]. The last layer of the feature extractor is replaced with a linear classification layer according to the dataset of interest.

We set $\lambda = 4.0$ in Eq. 13 to prioritize the effect of cross-domain contrastive loss. The momentum parameter ρ for the EMA update in Eq. 5 is set to 0.7 for all the experiments. The value of the injection threshold γ in Eq. 6 and Eq. 7 is set to 0.8 for the experiments on the Office-Home dataset. For experiments on DomainNet and Office-31, we use a value of 0.9 for the threshold. The label smoothing parameter α is set to 0.1 for all the experiments.

In our experiments, we set the number of warmup epochs $W = 5$ in Eq. 8 for the experiments. For Office-Home and Office-31, we set samples-per-class $(\eta_{sup}) = 4$ and for DomainNet $\eta_{sup} = 2$. We use 2 global η_g and 4 local η_l crops for all the datasets. We set the batch size of unlabeled samples B_u as 128 for Office-Home and DomainNet, and 32 for Office-31.

Similar to [31], we use a Stochastic Gradient Descent optimizer with a learning rate of 0.0002, a weight decay of 0.0005 and a momentum of 0.9. The learning rate is increased linearly from ≈ 0 to its maximum value of 0.0002 during the W warmup epochs. Subsequently, it is decayed to a minimum value of 10^{-5} using a cosine scheduler during training. We utilize RandAugment [8] as the augmentation module in our setup. Specifically, we use the implementation of RandAugment provided by [21] for our experiments and comparisons.

All experiments for Office-Home and Office-31 were done on a single NVIDIA Tesla V100 GPU, whereas we used a distributed setup with 2 GPUs for DomainNet experiments. For more experiments and more detailed report on the implementation details, we refer the reader to the supplementary report.

4.4 Results

Office-Home: Table 1 compares the results of the baseline methods with our approach on the Office-Home benchmark. We use a 3-shot setting, as commonly

Table 2. Accuracy on DomainNet in 1-shot and 3-shot settings (ResNet-34)

Method	R→C		R→P		P→C		C→S		S→P		R→S		P→R		Mean	
	1-shot	3-shot	1-shot	3-shot	1-shot	3-shot	1-shot	3-shot	1-shot	3-shot	1-shot	3-shot	1-shot	3-shot	1-shot	3-shot
S+T	55.6	60	60.6	62.2	56.8	59.4	50.8	55	56	59.5	46.3	50.1	71.8	73.9	56.9	60
DANN [11]	58.2	59.8	61.4	62.8	56.3	59.6	52.8	55.4	57.4	59.9	52.2	54.9	70.3	72.2	58.4	60.7
ADR [32]	57.1	60.7	61.3	61.9	57	60.7	51	54.4	56	59.9	49	51.1	72	74.2	57.6	60.4
CDAN [23]	65	69	64.9	67.3	63.7	68.4	53.1	57.8	63.4	65.3	54.5	59	73.2	78.5	62.5	66.5
ENT [12]	65.2	71	65.9	69.2	65.4	71.1	54.6	60	59.7	62.1	52.1	61.1	75	78.6	62.6	67.6
MME [31]	70	72.2	67.7	69.7	69	71.7	56.3	61.8	64.8	66.8	61	61.9	76.1	78.5	66.4	68.9
UODA [28]	72.7	75.4	70.3	71.5	69.8	73.2	60.5	64.1	66.4	69.4	62.7	64.2	77.3	80.8	68.5	71.2
Meta-MME [29]	–	73.5	–	70.3	–	72.8	–	62.8	–	68	-	63.8	–	79.2	–	70.1
BiAT [16]	73	74.9	68	68.8	71.6	74.6	57.9	61.5	63.9	67.5	58.5	62.1	77	78.6	67.1	69.7
APE [19]	70.4	76.6	70.8	72.1	72.9	76.7	56.7	63.1	64.5	66.1	63	67.8	76.6	79.4	67.6	71.7
DECOTA [43]	**79.1**	**80.4**	74.9	75.2	76.9	78.7	65.1	68.6	72.0	72.7	69.7	71.9	79.6	81.5	73.9	75.6
CDAC [21]	77.4	79.6	74.2	75.1	75.5	**79.3**	67.6	69.9	71	73.4	69.2	72.5	80.4	81.9	73.6	76
CLDA [35]	76.1	77.7	75.1	75.7	71	76.4	63.7	69.7	70.2	73.7	67.1	71.1	80.1	82.9	71.9	75.3
SPI (Ours)	76.56	79.2	**75.6**	**76.16**	**77.13**	79.2	**72.25**	**72.81**	**72.94**	**74.5**	**73.0**	**73.5**	**81.8**	**83.2**	**75.61**	**76.94**

used in most baseline approaches. We observe state-of-the-art performance across the majority of the domain adaptation scenarios. On average, we outperform the existing approaches in terms of classification accuracy. Despite using a 3-shot setting (195 images), we observe an improved performance of 5.87% on A → R, 3.45% on R → P, and 3.76% on P → C over LIRR [20], which uses 5% (> 210 images) of labeled target data.

Office-31: We compare SPI's performance against other baseline approachs on the Office-31 benchmark. We use a VGG-16 backbone for this set of experiments. From the results in Table 3, we observe a superior mean classification accuracy in the 3-shot setting.

Table 3. Accuracy on Office-31 under the 3-shot setting using VGG-16 backbone.

Method	W→A	D→A	Mean
S+T	73.2	73.3	73.25
DANN [11]	75.4	74.6	75
ADR [32]	73.3	74.1	73.7
CDAN [23]	74.4	71.4	72.9
ENT [12]	75.4	75.1	75.25
MME [31]	76.3	77.6	76.95
CLDA [35]	**78.6**	76.7	77.6
SPI (Ours)	78.0	**79.0**	**78.5**

Domain-Net: Table 2 compares the performance of SPI with baseline approaches, showing that SPI surpasses the existing baselines in both 1-shot and 3-shot settings. In some scenarios like R → S and C → S, SPI shows a significant boost in the 1-shot setting, outperforming even the 3-shot accuracies of the state-of-the-art baseline approaches. SPI's mean accuracy for the 1-shot setting (75.6%) is on par with that of the 3-shot setting for CDAC [21] (76.0%). This shows the superiority of our approach. Moreover, we observe that SPI 1-shot is on par with 5-shot setting of CLDA [35] (76.7%). In a 3-shot setting, SPI (76.94%) outperforms the 5-shot setting of CLDA and is on par with CDAC (76.9%).

5 Ablation Study

Momentum Parameter ρ: The momentum parameter ρ used in Eq. 5 controls how the exponential moving average of the class probabilities of an unlabeled sample is updated in the mapping $\mathcal{P}$ during training. Setting the momentum ρ to 1.0 disables the averaging function, and thus $\mathcal{P}(\text{ID}(x_i))$ contains the latest sharpened value of sharpened soft pseudo-label $\pi(\tilde{y}_i)$ (computed using Eq. 3 and Eq. 4). On the other hand, lower values of ρ signify a more conservative approach, wherein weightage to the current sharpened soft pseudo-label is less.

Through our experiments, and as shown in Table 5a, we found that setting ρ= 0.7 for both Office-Home and DomainNet benchmarks worked well across different domain adaptation scenarios

Injection Threshold γ: In Table 4, we compare the effect in the performance of our model on different domain adaptation scenarios for different values of threshold γ. A higher value of threshold indicates that we will be more conservative in injecting samples into the labeled target dataset, and vice versa. In our experiments, we observe better setting $\gamma = 0.8$ for Office-Home and $\gamma = 0.9$ for DomainNet gives us the best performance.

Table 4. Impact of threshold γ

Threshold	Office-Home		DomainNet	
γ	C → P	R → C	R → C	S → P
0.9	82.26	67.36	**79.23**	**74.48**
0.8	**83.82**	**69.33**	72.30	70.36
0.7	82.70	65.73	68.66	66.90

Effect of Different Components of SPI: Table 5b Row 1 presents the top-1 accuracies on different tasks using only $\mathcal{L}_{con}$ to train the model. We observe an increase in the top-1 accuracies when using $\mathcal{L}_{ils}$ (Row 2) and $\mathcal{L}_{ida}$ (Row 3) with the supervised contrastive loss. The performance gain when using $\mathcal{L}_{ils}$ and $\mathcal{L}_{ida}$ separately with $\mathcal{L}_{con}$ is not significant. However, when used together (Row 6), we achieve the best performance with up to 9.6% and 6.7% gain in the top-1 accuracy for R → C and S → P, respectively, in the DomainNet benchmark.

Table 5b (Rows 5 and 6) shows the effect of using EMA for pseudo-label injection. It is evident that using EMA helps improve the overall performance of the model. Lastly, we report the results without using $\mathcal{L}_{con}$ for domain alignment. The stark difference in performance hints at the importance of $\mathcal{L}_{con}$ in inter-domain alignment. We also point out that even without the explicit domain alignment loss, SPI's performance is on par with the 1-shot performance of DECOTA in both DomainNet scenarios. Therefore, it is evident that even without an explicit domain-alignment loss, SPI can leverage the similarity between the feature representations to achieve domain alignment to some extent.

Table 5. (a) Impact of different values of momentum parameter ρ used to compute the exponential moving average in Eq. 5. Results are from the Office-Home benchmark. (b) Different components of SPI. The last row uses EMA to add and remove samples from $\hat{T}$

(a)

Momentum	Office-Home	
ρ	C → P	P → R
1.0	82.2	81.0
0.9	82.1	81.1
0.7	**83.8**	**81.9**
0.5	82.3	81.3
0.3	81.6	80.6
0.1	79.7	79.4

(b)

Losses	Office-Home		DomainNet	
	C → P	R → C	R → C	S → P
$\mathcal{L}_{con}$	82.42	68.35	68.15	67.75
$\mathcal{L}_{con} + \mathcal{L}_{ils}$	83.44	69.04	70.43	68.72
$\mathcal{L}_{con} + \mathcal{L}_{ida}$	83.06	69.02	69.60	67.79
$\mathcal{L}_{ida} + \mathcal{L}_{ils}$	74.84	65.64	59.90	60.37
$\mathcal{L}_{con} + \mathcal{L}_{ida} + \mathcal{L}_{ils}$	82.23	66.35	71.88	70.66
$\mathcal{L}_{con} + \mathcal{L}_{ida} + \mathcal{L}_{ils}$ EMA	**83.82**	**69.33**	**79.23**	**74.48**

Table 6. Performance comparison of SPI to study the effect of (a) removing injected samples from labeled target dataset $\hat{T}$ and, (b) when samples are injected per epoch vs. per iteration

(a)

Removal	Office-Home		DomainNet	
	C → P	R → C	R → C	S → P
No	82.99	68.26	70.28	68.82
Yes	**83.82**	**69.33**	**79.23**	**74.48**

(b)

Updation Method	Office-Home		DomainNet	
	C → P	R → C	R → C	S → P
Iter	83.44	69.04	75.23	72.07
Epoch	**83.82**	**69.33**	**79.23**	**74.48**

Removal of Injected Samples: The key idea of our proposed method is the injection and removal of samples from the labeled target dataset $\hat{T}$. In this section, we study the effect of removing injected samples from $\hat{T}$. We posit that, over the course of training, as the model assigns the pseudo-label using Eq. 3, injected samples in $\hat{T}$ could be incorrectly labeled and need to be removed. From the results presented in Table 6a, it is evident that removing injected samples from $\hat{T}$ provides a boost in performance, which is more profound in DomainNet scenarios.

Pseudo-Label Injection Interval: In SPI, we choose to inject the samples into $\hat{T}$ at the end of every epoch. We empirically show in Table 6b that this setting performs better than injecting/removing samples after every iteration.

6 Conclusions

In this work, we presented an end-to-end framework, SPI which leverages feature level similarity between the unlabeled and labeled samples across domains to pseudo-label the unlabeled data. We introduced a novel loss function that includes 1) a supervised contrastive loss for inter-domain alignment, 2) an instance-level similarity loss to pull the unlabeled samples closer to the similar labeled samples, and, 3) an intra-domain consistency loss that clusters similar unlabeled target samples. We introduced a pseudo-labeling technique to inject (remove) confident (unconfident) samples into (from) the labeled target dataset. We performed extensive experiments and ablations to verify the efficacy of our method. Our framework SPI achieved state-of-the-art accuracies on the popular domain adaptation benchmarks.

References

1. Assran, M., et al.: Semi-supervised learning of visual features by non-parametrically predicting view assignments with support samples. In: Proceedings of the IEEE/CVF International Conference on Computer Vision, pp. 8443–8452 (2021)
2. Bachman, P., Hjelm, R.D., Buchwalter, W.: Learning representations by maximizing mutual information across views. Advances in neural information processing systems 32 (2019)
3. Berthelot, D., Carlini, N., Goodfellow, I., Papernot, N., Oliver, A., Raffel, C.A.: Mixmatch: a holistic approach to semi-supervised learning. Advances in Neural Information Processing Systems 32 (2019)
4. Caron, M., Misra, I., Mairal, J., Goyal, P., Bojanowski, P., Joulin, A.: Unsupervised learning of visual features by contrasting cluster assignments. Adv. Neural. Inf. Process. Syst. **33**, 9912–9924 (2020)
5. Caron, M., Touvron, H., Misra, I., Jégou, H., Mairal, J., Bojanowski, P., Joulin, A.: Emerging properties in self-supervised vision transformers. In: Proceedings of the IEEE/CVF International Conference on Computer Vision, pp. 9650–9660 (2021)
6. Chen, C., Chen, Z., Jiang, B., Jin, X.: Joint domain alignment and discriminative feature learning for unsupervised deep domain adaptation. In: Proceedings of the AAAI Conference on Artificial Intelligence, vol. 33, pp. 3296–3303 (2019)
7. Tachet des Combes, R., Zhao, H., Wang, Y.X., Gordon, G.J.: Domain adaptation with conditional distribution matching and generalized label shift. Advances in Neural Information Processing Systems 33, 19276–19289 (2020)
8. Cubuk, E.D., Zoph, B., Shlens, J., Le, Q.V.: Randaugment: Practical automated data augmentation with a reduced search space. In: Proceedings of the IEEE/CVF Conference on Computer Vision and Pattern Recognition Workshops, pp. 702–703 (2020)
9. Deng, J., Dong, W., Socher, R., Li, L.J., Li, K., Fei-Fei, L.: Imagenet: a large-scale hierarchical image database. In: 2009 IEEE Conference on Computer Vision and Pattern Recognition, pp. 248–255. IEEE (2009)
10. Ganin, Y., Lempitsky, V.: Unsupervised domain adaptation by backpropagation. In: International Conference on Machine Learning, pp. 1180–1189. PMLR (2015)
11. Ganin, Y., et al.: Domain-adversarial training of neural networks. J. Mach. Learn. Res. **17**(1), 2096–2030 (2016)
12. Grandvalet, Y., Bengio, Y.: Semi-supervised learning by entropy minimization. Advances in neural information processing systems 17 (2004)
13. Han, K., Rebuffi, S.A., Ehrhardt, S., Vedaldi, A., Zisserman, A.: Automatically discovering and learning new visual categories with ranking statistics. arXiv preprint arXiv:2002.05714 (2020)

14. He, K., Fan, H., Wu, Y., Xie, S., Girshick, R.: Momentum contrast for unsupervised visual representation learning. In: Proceedings of the IEEE/CVF Conference on Computer Vision and Pattern Recognition, pp. 9729–9738 (2020)
15. He, K., Zhang, X., Ren, S., Sun, J.: Deep residual learning for image recognition. In: Proceedings of the IEEE Conference on Computer Vision and Pattern Recognition, pp. 770–778 (2016)
16. Jiang, P., Wu, A., Han, Y., Shao, Y., Qi, M., Li, B.: Bidirectional adversarial training for semi-supervised domain adaptation. In: IJCAI, pp. 934–940 (2020)
17. Kang, G., Jiang, L., Yang, Y., Hauptmann, A.G.: Contrastive adaptation network for unsupervised domain adaptation. In: Proceedings of the IEEE/CVF Conference on Computer Vision and Pattern Recognition, pp. 4893–4902 (2019)
18. Khosla, P., Teterwak, P., Wang, C., Sarna, A., Tian, Y., Isola, P., Maschinot, A., Liu, C., Krishnan, D.: Supervised contrastive learning. Adv. Neural. Inf. Process. Syst. **33**, 18661–18673 (2020)
19. Kim, T., Kim, C.: Attract, perturb, and explore: learning a feature alignment network for semi-supervised domain adaptation. In: Vedaldi, A., Bischof, H., Brox, T., Frahm, J.-M. (eds.) ECCV 2020. LNCS, vol. 12359, pp. 591–607. Springer, Cham (2020). https://doi.org/10.1007/978-3-030-58568-6_35
20. Li, B., et al.: Learning invariant representations and risks for semi-supervised domain adaptation. In: Proceedings of the IEEE/CVF Conference on Computer Vision and Pattern Recognition, pp. 1104–1113 (2021)
21. Li, J., Li, G., Shi, Y., Yu, Y.: Cross-domain adaptive clustering for semi-supervised domain adaptation. In: Proceedings of the IEEE/CVF Conference on Computer Vision and Pattern Recognition, pp. 2505–2514 (2021)
22. Long, M., Cao, Y., Wang, J., Jordan, M.: Learning transferable features with deep adaptation networks. In: International Conference on Machine Learning, pp. 97–105. PMLR (2015)
23. Long, M., Cao, Z., Wang, J., Jordan, M.I.: Conditional adversarial domain adaptation. Advances in neural information processing systems 31 (2018)
24. Müller, R., Kornblith, S., Hinton, G.E.: When does label smoothing help? Advances in neural information processing systems 32 (2019)
25. Van den Oord, A., Li, Y., Vinyals, O.: Representation learning with contrastive predictive coding. arXiv e-prints pp. arXiv-1807 (2018)
26. Paul, S., Tsai, Y.-H., Schulter, S., Roy-Chowdhury, A.K., Chandraker, M.: Domain adaptive semantic segmentation using weak labels. In: Vedaldi, A., Bischof, H., Brox, T., Frahm, J.-M. (eds.) ECCV 2020. LNCS, vol. 12354, pp. 571–587. Springer, Cham (2020). https://doi.org/10.1007/978-3-030-58545-7_33
27. Peng, X., Bai, Q., Xia, X., Huang, Z., Saenko, K., Wang, B.: Moment matching for multi-source domain adaptation. In: Proceedings of the IEEE/CVF International Conference on Computer Vision, pp. 1406–1415 (2019)
28. Qin, C., Wang, L., Ma, Q., Yin, Y., Wang, H., Fu, Y.: Contradictory structure learning for semi-supervised domain adaptation. In: Proceedings of the 2021 SIAM International Conference on Data Mining (SDM), pp. 576–584. SIAM (2021)
29. Qiu, S., Zhu, C., Zhou, W.: Meta self-learning for multi-source domain adaptation: a benchmark. In: Proceedings of the IEEE/CVF International Conference on Computer Vision, pp. 1592–1601 (2021)
30. Saenko, K., Kulis, B., Fritz, M., Darrell, T.: Adapting visual category models to new domains. In: Daniilidis, K., Maragos, P., Paragios, N. (eds.) ECCV 2010. LNCS, vol. 6314, pp. 213–226. Springer, Heidelberg (2010). https://doi.org/10.1007/978-3-642-15561-1_16

31. Saito, K., Kim, D., Sclaroff, S., Darrell, T., Saenko, K.: Semi-supervised domain adaptation via minimax entropy. In: Proceedings of the IEEE/CVF International Conference on Computer Vision, pp. 8050–8058 (2019)
32. Saito, K., Ushiku, Y., Harada, T., Saenko, K.: Adversarial dropout regularization. arXiv preprint arXiv:1711.01575 (2017)
33. Shen, J., Qu, Y., Zhang, W., Yu, Y.: Wasserstein distance guided representation learning for domain adaptation. In: Thirty-Second AAAI Conference on Artificial Intelligence (2018)
34. Shui, C., Chen, Q., Wen, J., Zhou, F., Gagné, C., Wang, B.: Beyond h-divergence: Domain adaptation theory with jensen-shannon divergence (2020)
35. Singh, A.: Clda: Contrastive learning for semi-supervised domain adaptation. Advances in Neural Information Processing Systems 34 (2021)
36. Sohn, K., Berthelot, D., Carlini, N., Zhang, Z., Zhang, H., Raffel, C.A., Cubuk, E.D., Kurakin, A., Li, C.L.: Fixmatch: simplifying semi-supervised learning with consistency and confidence. Adv. Neural. Inf. Process. Syst. **33**, 596–608 (2020)
37. Sun, B., Feng, J., Saenko, K.: Return of frustratingly easy domain adaptation. In: Proceedings of the AAAI Conference on Artificial Intelligence, vol. 30 (2016)
38. Tian, Y., Krishnan, D., Isola, P.: Contrastive multiview coding. In: Vedaldi, A., Bischof, H., Brox, T., Frahm, J.-M. (eds.) ECCV 2020. LNCS, vol. 12356, pp. 776–794. Springer, Cham (2020). https://doi.org/10.1007/978-3-030-58621-8_45
39. Venkateswara, H., Eusebio, J., Chakraborty, S., Panchanathan, S.: Deep hashing network for unsupervised domain adaptation. In: Proceedings of the IEEE Conference on Computer Vision and Pattern Recognition, pp. 5018–5027 (2017)
40. Wang, R., Wu, Z., Weng, Z., Chen, J., Qi, G.J., Jiang, Y.G.: Cross-domain contrastive learning for unsupervised domain adaptation. IEEE Trans. Multimed. (2022)
41. Wightman, R.: Pytorch image models. https://github.com/rwightman/pytorch-image-models (2019). https://doi.org/10.5281/zenodo.4414861
42. Wu, Y., Winston, E., Kaushik, D., Lipton, Z.: Domain adaptation with asymmetrically-relaxed distribution alignment. In: International Conference on Machine Learning, pp. 6872–6881. PMLR (2019)
43. Yang, L., et al.: Deep co-training with task decomposition for semi-supervised domain adaptation. In: Proceedings of the IEEE/CVF International Conference on Computer Vision, pp. 8906–8916 (2021)
44. Yao, T., Pan, Y., Ngo, C.W., Li, H., Mei, T.: Semi-supervised domain adaptation with subspace learning for visual recognition. In: Proceedings of the IEEE conference on Computer Vision and Pattern Recognition, pp. 2142–2150 (2015)
45. Zhang, B., et al.: Flexmatch: boosting semi-supervised learning with curriculum pseudo labeling. Advances in Neural Information Processing Systems 34 (2021)
46. Zhao, H., Des Combes, R.T., Zhang, K., Gordon, G.: On learning invariant representations for domain adaptation. In: International Conference on Machine Learning, pp. 7523–7532. PMLR (2019)

W06 - Advances in Image Manipulation

W06 - Advances in Image Manipulation

Image manipulation is a key computer vision tasks, aiming at the restoration of degraded image content, the filling in of missing information, or the needed transformation and/or manipulation to achieve a desired target (with respect to perceptual quality, contents, or performance of apps working on such images). Recent years have witnessed an increased interest from the vision and graphics communities in these fundamental topics of research. Not only has there been a constantly growing flow of related papers, but also substantial progress has been achieved.

Each step forward eases the use of images by people or computers for the fulfillment of further tasks, as image manipulation serves as an important frontend. Not surprisingly then, there is an ever growing range of applications in fields such as surveillance, the automotive industry, electronics, remote sensing, medical image analysis, etc. The emergence and ubiquitous use of mobile and wearable devices offer another fertile ground for additional applications and faster methods.

This workshop aimed to provide an overview of the new trends and advances in those areas. Moreover, it offered an opportunity for academic and industrial attendees to interact and explore collaborations.

October 2022

Radu Timofte
Andrey Ignatov
Ren Yang
Marcos V. Conde
Furkan Kınlı

Evaluating Image Super-Resolution Performance on Mobile Devices: An Online Benchmark

Xindong Zhang[1,2], Hui Zeng[2], and Lei Zhang[1,2](✉)

[1] Department of Computing, The Hong Kong Polytechnic University, Hung Hom, Hong Kong
{csxdzhang,cslzhang}@comp.polyu.edu.hk
[2] OPPO Research, Beijing, China

Abstract. Deep learning-based image super-resolution (SR) has shown its strong capability in recovering high-resolution image details from low-resolution inputs. With the ubiquitous use of AI-accelerators on mobile devices (*e.g.*, smartphones), increasing attention has been received to develop mobile-friendly SR models. Because of the complicated and tedious routines to deploy SR models on mobile devices, researchers have to use indirect indices, such as FLOPs, number of parameters, and activations, to evaluate and compare the efficiency of SR models. However, these indices cannot faithfully reflect the real performance of SR models on mobile devices. To mitigate this gap, we develop an online benchmark to automatically evaluate the performance of SR models on mobile devices. With a simple model definition file as input, *e.g.*, PyTorch or ONNX file, our benchmark can generate the on-device evaluation indices and relevant statistics, including latency, memory, and energy consumption within 15 min, freeing the researchers from labor-intensive SR model deployment works. We further comprehensively study current SR models on mobile devices equipped with typical AI accelerators, such as Qualcomm, MediaTek, Hisilicon, and Samsung. Our benchmark provides a common platform for researchers to easily evaluate and compare the practical performance of their SR models on mobile devices. More details can be found at https://github.com/xindongzhang/MobileSR-Benchmark.

1 Introduction

Image super-resolution (SR) aims at reproducing high-resolution (HR) images from their degraded low-resolution (LR) counterparts. Deep convolutional networks (DCNNs) have become prevalent in SR research [13,16,17,39,43,48,58] for their strong capability in recovering or generating high-frequency image details, showing promising values in image and video restoration, enhancement and display. Very recently, transformer-based SR methods [38,56] have also been developed and demonstrated better SR results than DCNNs, yet in the price of higher computational cost.

X. Zhang and H. Zeng—Equal contribution

L. Zhang—This work is supported by the Hong Kong RGC RIF grant (R5001-18) and the PolyU-OPPO Joint Innovation Lab.

L. Karlinsky et al. (Eds.): ECCV 2022 Workshops, LNCS 13802, pp. 169–186, 2023.
https://doi.org/10.1007/978-3-031-25063-7_11

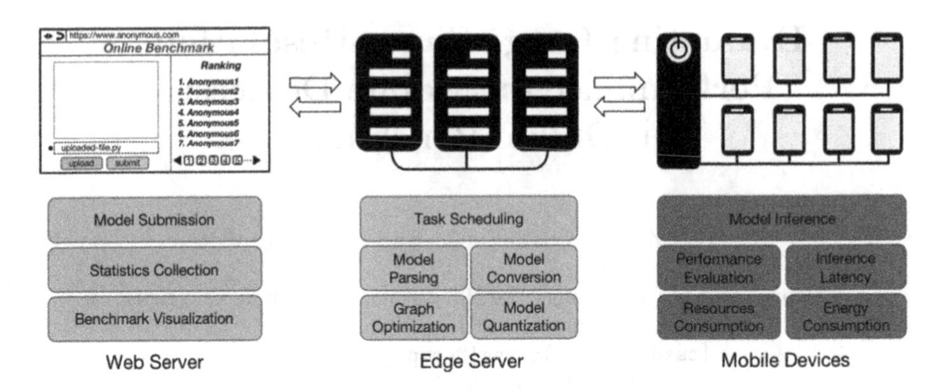

Fig. 1. The overall system structure of our online benchmark.

With the advent of AI-accelerators (*e.g.*, DSP, NPU, APU, etc.) on mobile devices, some of which can approach the computational power of mid-end graphics cards [24], much attention has been attracted on performing real-time SR on mobile devices [23,25]. To achieve this purpose, many techniques have been developed for efficient SR, including lightweight and efficient SR model design and search [11,12], network pruning [34], training strategies [57] and model quantization [33]. Confronted by the complicated and error-prone routines to deploy SR models on mobile devices, most of the existing methods can only evaluate the efficiency of SR models using indirect indices [54,55], such as FLOPs, number of parameters and activations, which however cannot faithfully reflect the real efficiency of SR models on mobile devices.

A few attempts have been made to benchmark AI models on mobile devices. Ignatov *et al.* developed an off-line android application, named AI-benchmark [24], to collect on-device statistics, including latency and memory consumption, for evaluating the performance of AI models on different mobile devices. MLPerf [46] is a large mobile inference benchmark focusing on high-level vision tasks (including image classification, objection detection and semantic segmentation) and natural-language processing. However, both the above two benchmarks require users to follow complicated routines to convert their models into a deployable format and cannot return the efficiency indices on time. For example, MLPerf updates its benchmark once a month. In addition, these benchmarks do not pay enough attention to the computationally more intensive low-level version tasks such as SR. Considering the increasing research interest in real-time SR models, it is highly desirable to develop a benchmark to conveniently evaluate and compare the on-device performance of SR models on mobile devices.

In this work, we develop an online benchmark to fill the gap between model design and performance evaluation, especially the efficiency, on mobile devices. The overall system is illustrated in Fig. 1. It consists of three components, a web server, an edge server, and multiple mobile devices. Users only need to upload their model definition files via the website. Our edge server will automatically download the model files from the web server, conduct a series of model conversion and optimization processes, and pass the deployable TFLITE models to a set of wireless-connected mobile devices for on-device evaluation. The key performance indices include hardware latency, memory, and energy consumption. Finally, the edge server will collect the evaluation results and send them back to the web server, visualizing them to the users. All the above operations are conducted automatically and the whole process can be finished within 15 min.

The proposed benchmark is expected to bridge the gap between model design and performance evaluation on mobile devices and facilitate the development of mobile-friendly AI models for low-level vision tasks. The main contributions of this work can be summarized as follows:

1) We develop an online benchmark to automatically parse, optimize and execute SR models on mobile devices, which significantly reduces the workload of researchers to deploy and evaluate their SR models on mobile devices.
2) With the online benchmark, we perform comprehensive comparison and analyses of the performance (such as hardware latency, memory, and power consumption) of existing SR models on multiple mobile devices.

2 Related Work

2.1 Super-Resolution Methods

We first briefly review the deep learning-based SR methods, including CNN-based, transformer-based, and light-weight and efficient SR networks.

CNN-Based Methods. CNN-based methods have demonstrated impressive performance in the SR tasks [13,16,17,20,29,31,39,40,43,44,44,48,49,59,59]. To build more effective SR models, many methods tend to employ deeper and more complicated architectures as well as the attention techniques. Zhang et al. proposed a residual-in-residual structure coupled with channel attention to train a very deep network over 400 layers. MemNet [49] and RDN [59] are designed by employing the dense blocks [20] to utilize the intermediate features from all layers. In addition to increasing the depth of network, some works, such as SAN [13], NLRN [40], HAN [44] and NLSA [43], excavate the feature correlations along the spatial or channel dimension to boost the SR performance.

Transformer-Based Methods. The breakthrough of transformer networks in natural language processing (NLP) inspired the use of self-attention (SA) in computer vision tasks. The SA mechanism can effectively model the dependency across data, and it has achieved impressive results on some low-level vision problems [9,38,50,52,56]. IPT [9] is a large pre-trained model for various low-level vision tasks based on the standard vision transformer. SwinIR [38] adapts the Swin Transformer [42] to image restoration, taking the advantages of both CNNs and transformers. ELAN [56] employs an elaborately designed efficient long-range attention block to capture long-range and larger window-size of SA, yielding a good balance between image quality and inference efficiency.

Light-Weight and Efficient Methods. While many SR methods [13,39,41] employ deeper and more complicated network architectures for better image quality, the introduced heavy computational costs limit their usage in real-world applications. It is demanded to design lightweight and efficient super-resolution (LESR) models. Hui et al. [22] proposed an information distillation network (IDN) to compress the number of filters per layer. They then extended IDN to information multi-distillation network (IMDN) [21] and won the AIM 2019 constrained image SR challenge [55]. Liu et al. [41] further improved IMDN to residual feature distillation block (RFDB) and won the AIM 2020 [54] SR challenge. Though being validated effective on GPU servers, the above-mentioned methods could be much slower on mobile devices due to the different hardware hierarchies. Zhang et al. [57] proposed an edge-oriented re-parameterizable block for performance boosting at the training phase without introducing extra computation cost in inference, achieving real-time performance on mobile devices with comparable PSNR/SSIM indexes. Zhan et al. [53] proposed a two-stage framework for hardware-aware SR network search and pruning for mobile devices.

2.2 Mobile Super-Resolution Challenges and Benchmarks

Challenges. PIRM 2018 challenge on perceptual image enhancement of mobile devices [25] is the first image enhancement challenge evaluating latency on mobile devices. MAI2020 [23] introduces the first Mobile AI challenge, whose target is to develop end-to-end deep learning-based image SR solutions that can demonstrate real-time performance on mobile or edge NPUs. How to evaluate the performance, especially efficiency, of deep SR models on mobile devices is also a challenging problem. Most of the existing works validate the efficiency of SR models via indirect indices (*e.g.*, FLOPs, number of parameters, activations) or inference latency on GPU server, which may however lead to very different conclusions on mobile devices [23,25,54,57]. What's more, previous challenges require researchers to provide frozen graphs for evaluation, which however may cause biases since different ways of conversion and quantization could introduce non-negligible variations [19].

Benchmarks. AI Benchmark is a comprehensive benchmark suite developed by Andrey et al. [24] for mobile devices. It evaluates both latency and accuracy among various tasks. MLPerf inference benchmark [46] is one of the largest benchmarks contributed by both academic and industry communities. However, the above two benchmarks only focus on part of indices for evaluating AI models, and they do not provide online platforms for researchers to collect the indices on time. In this work, we propose an online benchmark, which can automatically parse, quantize and convert the SR models, and then evaluate them by using indices of on-device latency, memory and energy consumption, etc.

3 Evaluating Image Super-Resolution on Mobile Devices

In this section, we first introduce the overall system of our benchmark, then illustrate each of its components and evaluation settings.

3.1 Overall Pipeline of Benchmark

To ease the burden of deploying and evaluating SR models on mobile devices, we propose an autonomous and hierarchical benchmark system. As shown in Fig. 1, our benchmark system is in a three-level architecture, which includes a web server, an edge server, and multiple mobile devices. The web server is designed to interact with users, including model submission, statistics collection, and visualization. As for the edge server, we design several autonomous tasks interacting with the web server and targeted mobile devices, for model file downloading, targeted model conversion, and model evaluation. The targeted mobile devices are connected with an edge server via wireless ADB, and most modules of mobile devices are turned off for the purpose of energy evaluation.

3.2 Components of Benchmark System

In this part, we reveal and discuss the importance of each component relevant to deploying and evaluating SR models in mobile scenarios, which is illustrated in Fig. 2.

3.2.1 Web Server

Previous research of SR models on mobile devices, evaluate and post their efficiency results by their own implementation [19,53,57]. The evaluation settings and protocols may be not public to other researchers who want to perform fair comparisons. Thus, we develop an edge server providing a public stage for researchers to evaluate and compare their models among others. Specifically, we provide a window on the web server for researchers to upload their model definition files, *i.e.*, PyTorch or ONNX, and the web server will insert the uploaded model files into the database for further processing. The evaluation results will also be returned and visualized to the users via the website window. Our website will also regularly update the ranking pages of evaluated SR models as a reference for interested researchers.

3.2.2 Edge Server

As illustrated in Fig. 2, we design and develop several components in the edge server, including network parsing (graph trace and code generation, as mentioned in Fig. 2), network quantization, graph optimization, and network conversion, each of which will be discussed below.

Network Parsing. Various training frameworks, including Caffe [27], Tensorflow [7], PyTorch [45], and MXNet [10] etc., are vibrant in research and community, and each of which design their own protocol of model definition. Thus it is laborious and challenging to parse the trained CNN models with a case-by-case handling strategy. The parsing step is crucial acting as translating the model to our predefined intermediate representation (IR) for later processing, including graph optimization, network pruning, and quantization. Though ONNX [3] could be used as a general model front-end for deployment, it is unfriendly to be deployed on mobile devices due to the fact that it is partially supported by most inference engines on mobile devices.

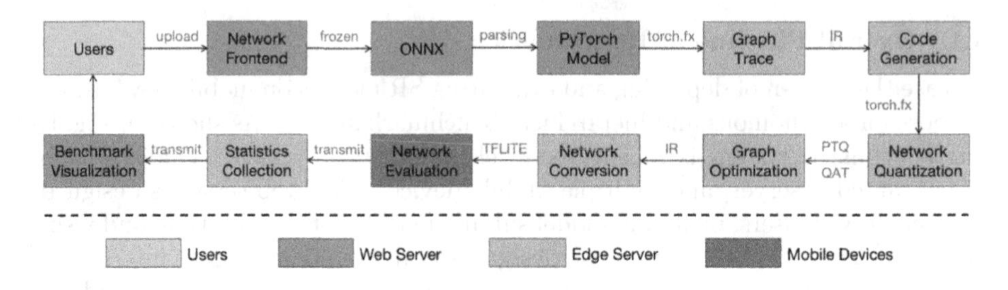

Fig. 2. The sequential routines and components of our benchmark system.

To this end, we design and provide a convenient network parser base on [15] to autonomously parse network definition for further processing, which can be summarized in the following two usage cases. For researchers who use PyTorch [45] for SR model training, they only need to provide their implementation of the network with or without pre-trained weights. Then we use Torch.fx [47] to trace the network, then we translate and generate python-like codes with Torchscript [14] IR illustrating an equivalent graph as the origin. For researchers who use other training frameworks, the first step is to translate the network into the general model front-end ONNX. Finally, following the same routines as PyTorch to parse and generate network codes for further processing.

Network Quantization. AI-specific hardware of commodity mobile devices, like TPU, NPU, and DSP, can be only executed in low-bit arithmetic (such as 8-bit or 16-bit), while most researchers and engineers develop and train their models on GPU-server with FP16 or FP32 arithmetic. Therefore, a further quantization step [26,51] is needed to transfer the FP32 models to INT8/INT16 models with acceptable losses of accuracy. Based on the code and model generated from the network parser, we first freeze and fold the parameter of BatchNorm2d as discussed on [36], and re-implement the post-training quantization method [51] for fair comparison based on the hardware quantization specification introduced in TFLITE [4]. The representative quantization methods [26,51] are often developed and validated on high-level vision tasks. Their effectiveness on low-level vision tasks remains uncertain. It is worth mentioning that, in this work, we focus on evaluating the efficiency of SR models with different arithmetic rather than the performance of different quantization methods. Further research for the improvement of quantization accuracy on low-level vision tasks is also imperative but out of the scope of this paper.

Graph Optimization and Network Conversion. After quantization and tuning, graph optimization could be used to trim the network graph for further acceleration [28]. Since executing SR on mobile is highly IO- and computation-intensive, paying less attention to this step could cause a significant efficiency drop in running SR models on mobile devices. To mitigate the abovementioned issues, we dedicatedly design graph optimization and by-pass strategies[1] for

[1] Including the Conv2d-BN folding, Deconv2d-Gather fusing, Conv2d-Gather fusing, Group-wise Conv2d rewrite, *etc.*

deploying SR models on mobile, and provide a more reasonable and fair benchmark tool for further reducing network redundancy. More details of fusion and by-pass strategies will be discussed on the website. Finally, we select TFLITE as the frontend format for communicating with the common inference engines, including TFLITE [4], MNN [28], and NCNN [2] etc., and it is also easy to extend it to other model front-end like ONNX and Caffe2 [1] providing great flexibility for deploying LESR model on mobile devices.

3.2.3 Mobile Devices

To excavate the behavior of SR models on mobile devices, we select the mainstream mid-end and high-end mobile devices for benchmark evaluation, their specifications are listed in Table 1. Based on mobile devices, there are several dedicated inference engines [24,28]. Some of them are designed for the general deployment of mobile devices, including TFLITE, NCNN and MNN etc., while the other is born dedicatedly for the AI hardware, like HiAI [6] (HUAWEI SOCs), SNPE [5] (SnapDragon), NeuralPilot (MediaTek). In this paper, we select the TFLITE inference engine for general purposes of deployment and evaluation with three different delegates, *e.g.*, CPU, GPU, and NNAPI[2]. Specifically, at the evaluation phase, the edge server will schedule the converted TFLITE models to the targeted devices via wireless-ADB and execute the TFLITE benchmark scripts for on-device evaluation. Then the evaluated results are collected and transmitted to the web server for further analysis and visualization.

Table 1. Hardware specifications of selected mobile devices.

Brand	Chipset	CPU	GPU	Accelerator
Samsung S22	SnapDragon 8Gen1	1 × X2, 3.00 GHz 3 × A710, 2.40 GHz 4 × A510, 1.70 GHz	Adreno 730	Qualcomm AI
Oppo Find X5	Dimensity 9000	1 × X2, 3.05 GHz 3 × A710, 2.85 GHz 4 × A510, 1.80 GHz	Mali-G710	APU
Huawei P40	Kirin 990 5G	2 × A76, 2.86 GHz 2 × A76, 2.36 GHz 4 × A55, 1.95 GHz	Mali-G76	DaVinci NPU
Honor 70 Pro	Dimensity 8000	4 × A78, 2.75 GHz 4 × A55, 2.00 GHz	Mali-G610	APU
RedMi K30 Ultra	Dimensity 1000+	4 × A77, 2.20 GHz 4 × A55, 2.00 GHz	Mali-G77	APU

[2] The compatibility of NNAPI may be different across smartphone manufacturers even armed with the same SOCs, while in this work we ensure fair comparison among participated methods by evaluating them on the same devices with same drivers.

3.3 Benchmark Protocols

3.3.1 Execution Protocols

Targeted Resolution. Previous researches select 720p as the targeted resolution for all tasks, on evaluating the efficiency of SR on GPU servers. However, since there are limited hardware resources on mobile devices, large input resolution may consume too much memory footprint which will cause the evaluation application to crash. To this end, we set 480 as the targeted resolution for all tasks of our online benchmark, including $\times 2$ and $\times 4$ upscaling tasks.

Arithmetic. Previous research mainly discusses about the efficiency of SR models in FP32 format, while the INT8 and FP16[3] format are more acceptable on commodity mobile devices because the low-bit arithmetic (INT8 and FP16) can theoretically run faster and consume lower energy than FP32. Therefore, except for the FP32 arithmetic, we also incorporate FP16 and INT8 formats for evaluating SR models. Specifically, we evaluate the model in FP32, FP16, and INT8 format with CPU, GPU, and NNAPI delegates, respectively.

Statistics. To evaluate the efficiency of SR models on mobile devices, we present comprehensive on-device statistics, including the on-device latency, memory, and power consumption (measured by mW). What's more, the indirect proxies, such as parameter size, FLOPs, and the number of activations, for indicating the efficiency of SR models are also provided for reference. Based on the collected average power consumption and latency, we can calculate the average energy consumption E_i (measured by mWh per execution) of an SR model on the $i-$th setting, by simply multiplying the collected average latency Lat_i by the average power consumption P_i,

$$E_i = Lat_i \cdot P_i \tag{1}$$

4 Experiments

4.1 Implementation Details

To measure the latency on mobile devices, we use TFLite Benchmark Tool [4] with arguments *num_threads = 4* by default, and set *allow_fp16 = true* and *nnapi = true* for FP16 and INT8, respectively. We also schedule our tasks to big cores of SOCs to reduce the variation of CPU time among different runs. Limited by the space, we select two high-end smartphones of this year, *e.g.*, the Samsung S22 (SnapDragon 8Gen1) and Oppo Find X5 Pro (Mediatek Dimensity 9000) as the representative mobile device for later discussion (More details of the mobile devices can be referred to Table 1), the remainder will later be presented on our website). During the evaluation, the screen of mobile devices is turned on with the brightest light, and other modules like GPS, BLE, 4G/5G, *etc.*, are turned off to make our measurement of on-device latency, memory, and energy consumption accurate and stable.

[3] The even lower bit arithmetic, like 2/4-bit, depends on special compute hardware which is usually unavailable for commodity mobile devices.

Table 2. Efficiency comparison of different SR models in three different arithmetic, *e.g.*, FP32(CPU), FP16(GPU), INT8(NNAPI). The on-device latency, memory, and energy consumption are measured by generating upscaled images with the shape of 640×480 on the Samsung S22 smartphone. 'nan' means that the model fails to execute with this setting.

Scale	Model	#Params (K)	#FLOPs (G)	#Acts (M)	#Conv	FP32(CPU)			FP16(GPU)			INT8(NNAPI)		
						Latency (ms)	Memory (MB)	Energy (mWh)	Latency (ms)	Memory (MB)	Energy (mWh)	Latency (ms)	Memory (MB)	Energy (mWh)
×2	SRCNN [16]	57.28	17.60	29.80	3	910.0	3947.0	2.106	74.1	225.0	0.176	15.4	149.0	0.029
	FSRCNN [17]	12.75	2.02	13.52	8	69.4	255.0	0.126	25.1	278.0	0.049	126.0	211.0	0.144
	ESPCN [48]	21.28	1.63	7.68	3	81.0	533.0	0.171	9.3	229.0	0.019	1.1	153.0	0.001
	LapSRN [31]	870.11	48.74	74.34	14	904.0	1878.0	1.721	129.0	272.0	0.293	7.5	132.0	0.015
	TPSR [32]	60.78	4.67	16.90	14	207.0	598.0	0.488	17.6	219.0	0.042	108.0	308.0	0.185
	CARN-M [8]	257.03	30.56	217.19	42	1366.0	1544.0	2.873	1366.0	1841.0	2.830	16.0	157.0	0.035
	IDN [22]	550.98	45.88	133.63	31	1456.0	566.0	2.818	1205.0	713.0	2.548	23.8	127.0	0.040
	CARN [8]	964.16	74.40	172.95	33	1903.0	1560.0	4.040	324.0	2014.0	0.717	8.5	161.0	0.017
	RFDN [41]	417.21	30.52	128.52	64	1123.0	509.0	2.273	104.0	325.0	0.270	147.0	154.0	0.124
	RLFN [30]	302.28	21.94	91.04	39	673.0	483.0	1.404	66.0	288.0	0.137	222.0	174.0	0.176
	FMEN [18]	324.83	24.87	81.72	34	927.0	429.0	1.938	89.3	252.0	0.218	9.7	150.0	0.018
	EFDN [35]	272.04	19.76	93.27	59	798.0	406.0	1.670	649.0	529.0	1.406	262.0	266.0	0.377
	BSRN [37]	150.68	1.32	77.29	43	520.0	471.0	1.148	58.5	345.0	0.141	269.0	237.0	0.559
	VDSR [29]	668.23	205.28	374.48	20	5195.0	1697.0	10.846	318.0	423.0	0.842	16.7	154.0	0.038
	ECBSR [57]	593.80	45.60	83.87	18	1060.0	473.0	2.184	67.9	222.0	0.185	3.2	156.0	0.007
	IMDN [21]	694.40	53.02	141.01	46	1955.0	620.0	3.274	1406.0	972.0	2.932	660.0	342.0	1.226
	EDSR [39]	1369.86	105.49	187.70	36	2765.0	1543.0	5.602	497.0	1880.0	1.194	9.2	166.0	0.018
	RDN [59]	22123.40	1699.36	748.03	150	60124.0	3628.0	109.676	2750.0	3554.0	7.231	128.0	280.0	0.247
	RCAN [59]	15444.64	1176.58	2045.66	814	40539.0	1639.0	76.592	1979.0	2166.0	5.722	153.0	266.0	0.306
	SwinIR-light [38]	910.15	65.2	28.57	127	8588.0	745.0	15.856	Nan	nan	nan	nan	nan	nan
	SwinIR [38]	11752.49	767.00	136.09	192	38104.0	1406.0	68.694	nan	nan	nan	nan	nan	nan
	ELAN-light [56]	582.01	56.13	613.79	98	3258.0	557.0	6.414	3140.0	866.0	6.615	nan	nan	nan
	ELAN [56]	8254.09	655.00	3000.73	146	21300.0	774.0	41.222	20724.0	1959.0	40.204	nan	nan	nan
×4	SRCNN [16]	57.28	17.60	29.80	3	978.0	3967.0	0.544	75.1	257.0	0.042	15.3	155.0	0.009
	FSRCNN [17]	12.75	1.55	3.61	8	17.6	163.0	0.010	4.1	184.0	0.002	26.4	136.0	0.015
	ESPCN [48]	24.75	0.47	2.15	3	21.8	217.0	0.012	8.1	194.0	0.004	0.8	127.0	0.000
	LapSRN [31]	870.11	60.88	88.01	27	1056.0	1916.0	0.587	66.6	293.0	0.037	8.6	157.0	0.005
	TPSR [32]	61.12	1.19	5.15	15	60.8	278.0	0.034	16.8	221.0	0.009	35.2	213.0	0.020
	CARN-M [8]	368.39	10.18	69.73	42	545.0	1603.0	0.303	459.0	1701.0	0.255	4.6	156.0	0.003
	IDN [22]	550.98	15.05	34.10	31	314.0	260.0	0.174	388.0	327.0	0.216	6.1	158.0	0.003
	CARN [8]	1407.30	27.51	58.68	33	787.0	1568.0	0.437	324.0	1881.0	0.180	4.5	140.0	0.002
	ECBSR [57]	600.72	11.53	21.20	18	232.0	208.0	0.129	52.9	286.0	0.029	1.4	255.0	0.001
	IMDN [21]	715.18	13.65	35.94	46	330.0	235.0	0.183	382.0	374.0	0.212	184.0	189.0	0.102
	RFDN [41]	433.45	7.94	32.81	64	311.0	222.0	0.173	32.9	223.0	0.018	41.2	152.0	0.023
	RLFN [30]	317.22	5.77	23.44	39	152.0	250.0	0.084	19.8	228.0	0.011	37.3	179.0	0.021
	FMEN [18]	341.07	6.53	21.12	34	252.0	200.0	0.140	26.6	223.0	0.015	3.3	161.0	0.002
	EFDN [35]	272.04	4.94	23.32	59	188.0	223.0	0.104	251.0	263.0	0.139	71.0	144.0	0.039
	BSRN [37]	166.92	0.64	20.00	43	102.0	237.0	0.057	17.5	240.0	0.010	222.0	244.0	0.123
	VDSR [29]	668.23	205.28	374.48	20	5327.0	1691.0	2.971	319.0	424.0	0.177	16.9	151.0	0.009
	EDSR [39]	1517.57	38.10	67.28	37	899.0	1541.0	0.500	820.0	3106.0	0.456	5.3	4035.0	0.003
	RDN [59]	22271.11	436.56	207.36	151	13079.0	1617.0	7.325	1311.0	2408.0	0.729	29.0	251.0	0.016
	RCAN [59]	15592.35	305.87	531.78	815	7513.0	1621.0	4.202	821.0	2045.0	0.456	30.5	156.0	0.017
	SwinIR-light [38]	929.63	16.53	7.83	127	1923.0	341.0	1.070	nan	nan	nan	nan	nan	nan
	SwinIR [38]	11900.20	194.67	54.37	192	9143.0	1773.0	5.112	nan	nan	nan	nan	nan	nan
	ELAN-light [56]	601.49	14.40	164.41	98	771.0	280.0	0.429	671.0	358.0	0.373	nan	nan	nan
	ELAN [56]	8312.45	164.67	800.93	146	6075.0	477.0	3.389	5017.0	645.0	2.797	nan	nan	nan

4.2 Benchmark Results

We evaluate the state-of-the-art lightweight, traditional SR models, as well as the transformer-based models, and we also include some representative methods from recent efficient SR competitions, like AIM 2020 [54] and NTIRE 2022 [35]. The light-weight models include SRCNN [16], FSRCNN [17], ESPCN [48], LapSRN [31], TPSR [32], CARN-M/CARN [8], IDN [22], ECBSR [57], IMDN [21], RFDN [41], RLFN [30], EFDN [35], BSRN [37]. The selected traditional SR models are VDSR [29], EDSR [39], RDN [59], RCAN [59]. What's more, we also deploy and evaluate the popular transformer-based methods[4], *i.e.*, SwinIR [38] and ELAN [56], and present their performance on mobile scenarios for reference. To ensure all the selected networks are deployed and executed on mobile devices, we also rewrite some of the networks to equivalent definitions which are more mobile-friendly, *i.e.*, SwinIR [38] and ELAN [56] series. For the collected on-devices statistics, please refer to Table 2 and Table 3. Several interesting observations can be made in the following.

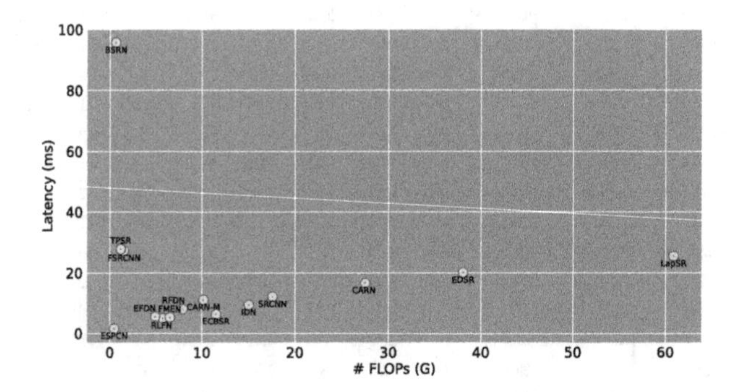

Fig. 3. Illustration of correlations between FLOPs and Latency. The latency of all models is measured on Oppo Find X5 Pro with NNAPI(INT8) delegate.

The On-device Indices are Imperative for the Evaluation of SR Models on Smartphones. Figure 3 illustrates the correlation of FLOPs with respect to inference latency[5]. Based on the observation, we further report the Spearman rank-order correlation coefficient (SROCC) values of {FLOPs, Latency} and {Activations, Latency} which are surprisingly low, 0.348 and 0.401, respectively. It leads to the conclusion that the frequently used indirect proxies, like FLOPs and Activations [54] can not faithfully reflect the real efficiency of an SR model. Thus, the on-device indices, *i.e.*, the on-device latency, and energy consumption, are imperative for evaluating the efficiency of SR models.

[4] The quantized version of transformer-based methods is currently not supported on mobile devices with TFLITE in the early version, we thus provide the performance in the format of FP32 and FP16.

[5] Similar observation can be made between activations and inference latency.

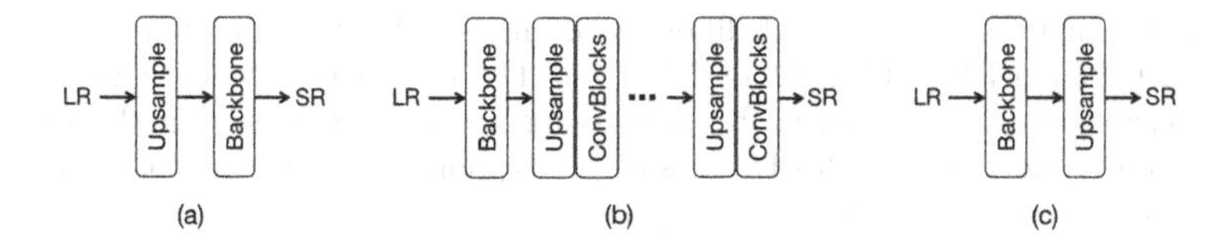

Fig. 4. Illustration of common upsampling strategies. (a) The pre-upsampling strategy. (b) Gradual post-upsampling strategy. (c) Straight post-upsampling strategy.

SR Models May Behave Quite Inconsistently in AI-specific Hardware. We notice that some of SR models behave inconsistently across different hardware, it is because the support and implementation of operators are quite different among CPU, GPU, and NNAPI(DSP or APU in our experiments). For example, the Leaky ReLU (LReLU) and Parameterized ReLU (PReLU) are used extensively on SR tasks, they can be well accelerated on CPU and GPU, while are not friendly for deployment on NPU and DSP hardware. They are not supported well and may fall back to be executed on CPU sessions, which will inevitably slow down the execution of the SR network. What's more, the group-wise convolution (as introduced in CARN-M [8] and BSRN [37]) is well supported on CPU and GPU, however, it is not supported for most AI-specific hardware (DSP and NPU). Revision of graph[6] is needed when deploying SR models with group convolution on AI-specific hardware, this can slow down the inference of SR models when the number of groups is large.

Efficiency of SR Models May be Different Across Distinct Mobile Devices. It is not surprising that even the same SR model behaves quite differently (inference latency, memory, and energy consumption) across distinct mobile devices, *e.g.*, Samsung S22 and Oppo Find X5 pro. The efficiency of an SR model on mobile devices is systematically dominated by the hardware design (different SOCs and hardware layout), system (controlling how mobile devices schedule the task) and driver (controlling how the hardware behaves), and the inference engine (determining how the SR models execute), *etc.* Thus, in this benchmark, we evaluate and compare SR models on the same mobile device with the system image, driver, and same version of TFLIE runtime, for the purpose of fair comparisons (Fig. 4).

4.3 Analysis of Architecture

Based on the benchmark results shown in Table 2 and Table 3, we further analyze the architecture design and its behaviors in mobile scenarios. We list them in the following.

The Upsampling Strategies. The pioneer attempts of SR, like SRCNN [16] and VDSR [29], follow a pre-upsampling strategy, which upscales the LR image

[6] A group-wise convolution can be rewritten with "Split-Conv-Concat" which is equivalent to the naive group-wise convolution.

Table 3. Efficiency comparison of different SR models in three different arithmetic, *e.g.*, FP32(CPU), FP16(GPU), INT8(NNAPI). The on-device latency, memory, and energy consumption are measured by generating upscaled images with the shape of 640×480 on Oppo Find X5 Pro Dimensity smartphone. 'nan' means that the model fails to execute with this setting.

Scale	Model	#Params (K)	#FLOPs (G)	#Acts (M)	#Conv	FP32(CPU)			FP16(GPU)			INT8(NNAPI)		
						Latency (ms)	Memory (MB)	Energy (mWh)	Latency (ms)	Memory (MB)	Energy (mWh)	Latency (ms)	Memory (MB)	Energy (mWh)
× 2	SRCNN [16]	57.28	17.60	29.80	3	407.0	567.0	0.466	116.0	283.0	0.111	13.1	163.0	0.009
	FSRCNN [17]	12.75	2.02	13.52	8	31.6	274.0	0.033	14.6	192.0	0.015	92.6	215.0	0.085
	ESPCN [48]	21.28	1.63	7.68	3	45.5	191.0	0.038	14.8	203.0	0.012	2.2	156.0	0.002
	LapSRN [31]	870.11	48.74	74.34	14	437.0	762.0	0.609	74.4	217.0	0.060	20.6	141.0	0.019
	TPSR [32]	60.78	4.67	16.90	14	207.0	313.0	0.166	47.0	202.0	0.036	254.0	278.0	0.195
	CARN-M [8]	257.03	30.56	217.19	42	597.0	1020.0	0.760	1245.0	1806.0	1.162	31.4	148.0	0.028
	IDN [22]	550.98	45.88	133.63	31	664.0	1080.0	0.880	1093.0	639.0	1.353	32.7	137.0	0.026
	CARN [8]	964.16	74.40	172.95	33	1058.0	1016.0	1.441	336.0	1876.0	0.314	42.7	136.0	0.029
	RFDN [41]	417.21	30.52	128.52	64	1387.0	689.0	1.179	306.0	287.0	0.255	30.6	141.0	0.025
	RLFN [30]	302.28	21.94	91.04	39	997.0	377.0	0.808	107.0	227.0	0.093	15.9	141.0	0.012
	FMEN [18]	324.83	24.87	81.72	34	365.0	401.0	0.511	113.0	216.0	0.090	16.8	155.0	0.016
	EFDN [35]	272.04	19.76	93.27	59	337.0	529.0	0.478	681.0	471.0	0.807	19.0	160.0	0.018
	BSRN [37]	150.68	1.32	77.29	43	267.0	755.0	0.389	71.0	275.0	0.062	161.0	220.0	0.172
	ECBSR [57]	593.80	45.60	83.87	18	521.0	295.0	0.611	75.7	197.0	0.071	22.2	144.0	0.019
	IMDN [21]	694.40	53.02	141.01	46	797.0	1036.0	1.019	1260.0	902.0	1.334	539.0	306.0	0.490
	VDSR [29]	668.23	205.28	374.48	20	3351.0	878.0	3.659	331.0	387.0	0.323	96.2	157.0	0.072
	EDSR [39]	1369.86	105.49	187.70	36	1344.0	504.0	1.929	294.0	1828.0	0.294	51.5	138.0	0.032
	RDN [59]	22123.40	1699.36	748.03	150	44768.0	2147.0	43.636	5556.0	2613.0	4.500	928.0	253.0	0.661
	RCAN [59]	15444.64	1176.58	2045.66	814	17200.0	2608.0	19.899	2257.0	2196.0	2.258	nan	nan	nan
	SwinIR-light [38]	910.15	65.20	28.57	127	nan	nan	nan	nan	nan	nan	nan	nan	nan
	SwinIR [38]	11752.49	767.00	136.09	192	nan	nan	nan	nan	nan	nan	nan	nan	nan
	ELAN-light [56]	582.01	56.13	613.79	98	2458.0	2223.0	3.081	2972.0	727.0	3.574	nan	nan	nan
	ELAN [56]	8254.09	655.00	3000.73	146	34388.0	8694.0	39.064	29533.0	1646.0	20.284	nan	nan	nan
× 4	SRCNN [16]	57.28	17.60	29.80	3	426.0	595.0	0.413	106.4	225.0	0.984	12.2	158.0	0.011
	FSRCNN [17]	12.75	1.55	3.61	8	17.8	146.0	0.019	5.2	157.0	0.005	27.3	140.0	0.023
	ESPCN [48]	24.75	0.47	2.15	3	21.5	209.0	0.025	30.3	218.0	0.028	1.6	161.0	0.001
	LapSR [31]	870.11	60.88	88.01	27	2113.0	2217.0	1.739	188.0	2344.0	0.214	25.5	123.0	0.016
	TPSR [32]	61.12	1.19	5.15	15	66.8	262.0	0.083	14.9	196.0	0.011	27.8	206.0	0.029
	CARN-M [8]	368.39	10.18	69.73	42	239.0	517.0	0.240	445.0	1645.0	0.472	11.2	140.0	0.010
	IDN [22]	550.98	15.05	34.10	31	253.0	236.0	0.250	250.0	287.0	0.171	9.5	129.0	0.008
	CARN [8]	1407.30	27.51	58.68	33	1418.0	1634.0	1.185	747.0	1840.0	0.469	16.7	1833.0	0.015
	RFDN [41]	433.45	7.94	32.81	64	270.0	288.0	0.225	86.6	212.0	0.062	8.2	142.0	0.005
	RLFN [30]	317.22	5.77	23.44	39	127.0	176.0	0.094	20.0	199.0	0.018	5.0	155.0	0.004
	FMEN [18]	341.07	6.53	21.12	34	235.0	203.0	0.240	245.0	196.0	0.176	5.3	130.0	0.003
	EFDN [35]	272.04	4.94	23.32	59	178.0	193.0	0.238	270.0	248.0	0.261	5.6	122.0	0.005
	BSRN [37]	166.92	0.64	20.00	43	228.0	301.0	0.227	24.5	201.0	0.025	95.7	179.0	0.072
	VDSR [29]	668.23	205.28	374.48	20	3370.0	859.0	3.723	334.0	333.0	0.313	95.2	127.0	0.055
	ECBSR [57]	600.72	11.53	21.20	18	126.0	199.0	0.158	21.2	202.0	0.024	6.5	154.0	0.006
	IMDN [21]	715.18	13.65	35.94	46	303.0	246.0	0.343	290.0	351.0	0.303	306.0	398.0	0.293
	EDSR [39]	1517.57	38.10	67.28	37	477.0	387.0	0.710	327.0	1665.0	0.427	20.1	159.0	0.019
	RDN [59]	22271.11	436.56	207.36	151	19231.0	892.0	14.447	1996.0	2264.0	1.810	239.0	258.0	0.173
	RCAN [59]	15592.35	305.87	531.78	815	14563.0	978.0	10.191	1643.0	2000.0	1.381	6119.0	511.0	4.407
	SwinIR-light [38]	929.63	16.53	7.83	127	1740.0	291.0	1.880	nan	nan	nan	nan	nan	nan
	SwinIR [38]	11900.20	194.67	54.37	192	13128.0	1618.0	8.683	nan	nan	nan	nan	nan	nan
	ELAN-light [56]	601.49	14.40	164.41	98	850.0	398.0	0.771	765.0	322.0	0.633	nan	nan	nan
	ELAN [56]	8312.45	164.67	800.93	146	8248.0	460.0	5.418	5282.0	640.0	4.094	nan	nan	nan

with bicubic interpolation, then employ CNN on the upscaled image for recovering high-frequency details. However, it introduces a large memory footprint and FLOPs which make the execution of the SR model inefficient. The rest methods employ a post-upsampling strategy, which can be also categorized into two types. One is to gradually upscaled the intermediate results, as introduced in EDSR [39], IMDN [21] , RCAN [59], LapSRN [31], and SwinIR [38], which still involving calculation on several intermediate feature with high resolution. The other is to straightly upscale (with the combination of Conv→PixelShuffle) to the results with the targeted resolution, which shows its superiority over others on latency, memory and energy consumption.

Backbone Design. The backbone dominates the computational overhead of SR networks. Previous methods, including FSRCNN [17], ESPCN [17] and ECBSR [57], cascade several convolution blocks in a plain topology which have shown their advantage on both low inference speed and energy consumption across mobile devices and compute hardware. However, the capacity of the network may be limited by the plain topology, and the heavy usage of convolution operation also constrains the scalability of the network towards low latency and energy consumption.

Another method is proposed to replace the convolution block with other operations in lower FLOPs. For example, CARN-M replaces the convolution operation of CARN with group-wise convolution, and BSRN highly employs the depth-wise convolution. This may cause operator compatibility on some mobile scenarios, *i.e.*, GPU and AI-specific hardware. Thus, as shown in the benchmark results, the latency and energy consumption of CARN-M are higher than CARN on GPU and AI-specific hardware, even with lower FLOPs. The same observation of BSRN on NNAPI with AI-specific hardware can also be made.

The recent developed efficient SR models, including EFDN, and RLFN, are highly modified from IMDN and RFDN, which are the first place winners of the AIM 2019 [55] constrained image SR challenge and the AIM 2020 [54] efficient image SR challenge on GPU server, respectively. However, as observed in the benchmark results, heavy usage of shortcut connection, element-wise operation, and associated 1×1 convolution could harm the inference speed and cause high energy consumption. Thus, the FMEN proposes an enhanced residual block (ERB) for reducing the memory access cost introduced by shortcut connection. From Table 3 and Table 2, none of the design shares consistent improvement across computing hardware, and avoiding fragmented topology are vital for designing efficient SR models on mobile devices.

We also provide the PSNR-oriented large models for intuitive comparison on mobile scenarios, including deep architecture design armed with complicated attention modules and transformer-based methods. RDN and RCAN are two representatives of the former class, reaching the depth of 150 and 814. They both need a long time and lots of energy for executing the network, and the RCAN even fails to prepare the model for execution on Oppo Find X5 Pro for $\times 2$ upscaling task. As for the transformer-based methods, we make an extensive optimization strategy to get the equivalent mobile-friendly graph[7] as the origin. One interesting observation is that ELAN can achieve even better than RCAN on ×4 upscaling task on CPU hardware, which shows opposite results against that on GPU server [56]. It is because the bandwidth of DDR and computational capacity of hardware are much lower than the GPU server, leading to the quite different compute behavior of RCAN and ELAN on mobile devices. Both of the two designs mentioned above, however, are not suitable for efficient SR model design and evaluation on mobile devices.

Activation Selection. Recently, the activation selection becomes an effective strategy of efficient SR design for boosting performance without introducing much computational complexity [35]. Though it has been successfully validated on the GPU server, it cannot be straightly applied to SR tasks on mobile scenarios. For example, IMDN and RFDN extensively use Leaky ReLU (LReLU) as the activation function in network design which may introduce neglectable FLOPs. However, LReLU involves more memory access cost than ReLU which is time-consuming when the shape of the feature is large. Thus, we make a simple experiment by replacing the LReLU in IMDN with ReLU on × 2 upscaling tasks. As can be seen from Table 4, with only a simple revision to IMDN, the average energy consumption reduced to 59.7% and 78.6% of the origins for Samsung S22 and Oppo Find X5 Pro, respectively. Thus, activation selection also plays an important role in designing and evaluating SR models in mobile scenarios.

Table 4. Efficiency comparison of IMDN with LReLU and ReLU in INT8(NNAPI) format. The on-device latency, memory, and energy consumption are measured by generating super-resolution with the shape of 640 × 480.

Device	Model	#Params (K)	#FLOPs (G)	#Acts (M)	#Conv	INT8(NNAPI)		
						Latency (ms)	Memory (MB)	Energy (mWh)
Samsung S22	IMDN-LReLU [21]	694.40	53.02	141.01	46	660.0	342.0	1.226
	IMDN-ReLU [21]	694.40	53.02	141.01	46	328.0	337.0	0.733
Oppo Find X5 Pro	IMDN-LReLU [21]	694.40	53.02	141.01	46	539.0	306.0	0.490
	IMDN-ReLU [21]	694.40	53.02	141.01	46	303.0	301.0	0.385

5 Conclusion

In this paper, we develop an online benchmark to narrow the gap between model design and efficiency evaluation on mobile devices, providing several key performance indices for reference, including hardware latency, memory, and average energy consumption. Our benchmark system consists of three components, including a web server, an edge server, and multiple mobile devices. The

[7] The transformer-based methods involve a lot of IO-intensive operations on 6D dimensional tensor, like reshape, permute, *etc.* We optimize them to work on 4D dimensional tensors which are more suitable for mobile devices. More details will be made public on our website.

benchmark system can automatically conduct a series of model conversions and graph optimization, and schedule the deployable TFLITE models to the wireless-connected mobile devices for efficiency evaluation within 15 min. The results of the evaluated model then will be fairly ranked and visualized on the website. What's more, we perform comprehensive deployment, evaluation, and comparison of existing prevailing SR model designs on multiple mid and high-end mobile devices, extensive experiments show that our benchmark can cover the majority of SR designs and effectively evaluate them on mobile devices. Although our benchmark system can automatically measure energy consumption, it can also be affected by several factors, like the environment temperature, the aging of the battery, and the version of the software system. In the future, we will further design a more robust system for energy evaluation, and excavate the design of efficient SR models for the newer architecture of mobile devices.

References

1. Caffe2. https://caffe2.ai/
2. NCNN. https://github.com/Tencent/ncnn
3. ONNX. https://onnx.ai/
4. Tflite. https://www.tensorflow.org/lite
5. SNPE (2016). https://developer.qualcomm.com/sites/default/files/docs/snpe/overview.html
6. HIAI (2017). https://developer.huawei.com/consumer/en/doc/2020315
7. Abadi, M., et al.: {TensorFlow}: a system for {Large-Scale} machine learning. In: 12th USENIX Symposium on Operating Systems Design and Implementation (OSDI 2016), pp. 265–283 (2016)
8. Ahn, N., Kang, B., Sohn, K.-A.: Fast, accurate, and lightweight super-resolution with cascading residual network. In: Ferrari, V., Hebert, M., Sminchisescu, C., Weiss, Y. (eds.) ECCV 2018. LNCS, vol. 11214, pp. 256–272. Springer, Cham (2018). https://doi.org/10.1007/978-3-030-01249-6_16
9. Chen, H., et al.: Pre-trained image processing transformer. In: Proceedings of the IEEE/CVF Conference on Computer Vision and Pattern Recognition, pp. 12299–12310 (2021)
10. Chen, T., et al.: MXNet: a flexible and efficient machine learning library for heterogeneous distributed systems. arXiv preprint arXiv:1512.01274 (2015)
11. Chu, X., Zhang, B., Ma, H., Xu, R., Li, Q.: Fast, accurate and lightweight super-resolution with neural architecture search. arXiv preprint arXiv:1901.07261 (2019)
12. Chu, X., Zhang, B., Xu, R.: Multi-objective reinforced evolution in mobile neural architecture search. In: Bartoli, A., Fusiello, A. (eds.) ECCV 2020. LNCS, vol. 12538, pp. 99–113. Springer, Cham (2020). https://doi.org/10.1007/978-3-030-66823-5_6
13. Dai, T., Cai, J., Zhang, Y., Xia, S.T., Zhang, L.: Second-order attention network for single image super-resolution. In: Proceedings of the IEEE/CVF Conference on Computer Vision and Pattern Recognition, pp. 11065–11074 (2019)
14. DeVito, Z., Ansel, J., Constable, W., Suo, M., Zhang, A., Hazelwood, K.: Using python for model inference in deep learning. arXiv preprint arXiv:2104.00254 (2021)

15. Ding, H., Pu, J., Hu, C.: Tinyneuralnetwork: an efficient deep learning model compression framework. https://github.com/alibaba/TinyNeuralNetwork (2021)
16. Dong, C., Loy, C.C., He, K., Tang, X.: Image super-resolution using deep convolutional networks. IEEE Trans. Pattern Anal. Mach. Intell. **38**(2), 295–307 (2015)
17. Dong, C., Loy, C.C., Tang, X.: Accelerating the super-resolution convolutional neural network. In: Leibe, B., Matas, J., Sebe, N., Welling, M. (eds.) ECCV 2016. LNCS, vol. 9906, pp. 391–407. Springer, Cham (2016). https://doi.org/10.1007/978-3-319-46475-6_25
18. Du, Z., Liu, D., Liu, J., Tang, J., Wu, G., Fu, L.: Fast and memory-efficient network towards efficient image super-resolution. In: Proceedings of the IEEE/CVF Conference on Computer Vision and Pattern Recognition, pp. 853–862 (2022)
19. Du, Z., Liu, J., Tang, J., Wu, G.: Anchor-based plain net for mobile image super-resolution. In: Proceedings of the IEEE/CVF Conference on Computer Vision and Pattern Recognition, pp. 2494–2502 (2021)
20. Huang, G., Liu, Z., Van Der Maaten, L., Weinberger, K.Q.: Densely connected convolutional networks. In: Proceedings of the IEEE Conference on Computer Vision and Pattern Recognition, pp. 4700–4708 (2017)
21. Hui, Z., Gao, X., Yang, Y., Wang, X.: Lightweight image super-resolution with information multi-distillation network. In: Proceedings of the 27th ACM International Conference on Multimedia, pp. 2024–2032 (2019)
22. Hui, Z., Wang, X., Gao, X.: Fast and accurate single image super-resolution via information distillation network. In: Proceedings of the IEEE Conference on Computer Vision and Pattern Recognition, pp. 723–731 (2018)
23. Ignatov, A., Timofte, R., Denna, M., Younes, A.: Real-time quantized image super-resolution on mobile NPUs, mobile AI 2021 challenge: Report. In: Proceedings of the IEEE/CVF Conference on Computer Vision and Pattern Recognition, pp. 2525–2534 (2021)
24. Ignatov, A., et al.: Ai benchmark: all about deep learning on smartphones in 2019. In: 2019 IEEE/CVF International Conference on Computer Vision Workshop (ICCVW), pp. 3617–3635. IEEE (2019)
25. Ignatov, A., et al.: PIRM challenge on perceptual image enhancement on smartphones: report. In: Leal-Taixé, L., Roth, S. (eds.) PIRM challenge on perceptual image enhancement on smartphones: Report. LNCS, vol. 11133, pp. 315–333. Springer, Cham (2019). https://doi.org/10.1007/978-3-030-11021-5_20
26. Jacob, B., et al.: Quantization and training of neural networks for efficient integer-arithmetic-only inference. In: Proceedings of the IEEE Conference on Computer Vision and Pattern Recognition, pp. 2704–2713 (2018)
27. Jia, Y., et al.: Caffe: Convolutional architecture for fast feature embedding. In: Proceedings of the 22nd ACM international conference on Multimedia, pp. 675–678 (2014)
28. Jiang, X., et al.: MNN: a universal and efficient inference engine. arXiv preprint arXiv:2002.12418 (2020)
29. Kim, J., Lee, J.K., Lee, K.M.: Accurate image super-resolution using very deep convolutional networks. In: Proceedings of the IEEE Conference on Computer Vision and Pattern Recognition, pp. 1646–1654 (2016)
30. Kong, F., et al.: Residual local feature network for efficient super-resolution. In: Proceedings of the IEEE/CVF Conference on Computer Vision and Pattern Recognition, pp. 766–776 (2022)
31. Lai, W.S., Huang, J.B., Ahuja, N., Yang, M.H.: Deep Laplacian pyramid networks for fast and accurate super-resolution. In: Proceedings of the IEEE Conference on Computer Vision and Pattern Recognition, pp. 624–632 (2017)

32. Lee, R., et al.: Journey towards tiny perceptual super-resolution. In: Vedaldi, A., Bischof, H., Brox, T., Frahm, J.-M. (eds.) ECCV 2020. LNCS, vol. 12371, pp. 85–102. Springer, Cham (2020). https://doi.org/10.1007/978-3-030-58574-7_6
33. Li, H., et al.: PAMS: quantized super-resolution via parameterized max scale. In: Vedaldi, A., Bischof, H., Brox, T., Frahm, J.-M. (eds.) ECCV 2020. LNCS, vol. 12370, pp. 564–580. Springer, Cham (2020). https://doi.org/10.1007/978-3-030-58595-2_34
34. Li, Y., Gu, S., Zhang, K., Van Gool, L., Timofte, R.: DHP: differentiable meta pruning via HyperNetworks. In: Vedaldi, A., Bischof, H., Brox, T., Frahm, J.-M. (eds.) ECCV 2020. LNCS, vol. 12353, pp. 608–624. Springer, Cham (2020). https://doi.org/10.1007/978-3-030-58598-3_36
35. Li, Y., et al.: Ntire 2022 challenge on efficient super-resolution: methods and results. In: Proceedings of the IEEE/CVF Conference on Computer Vision and Pattern Recognition, pp. 1062–1102 (2022)
36. Li, Y., et al.: Mqbench: towards reproducible and deployable model quantization benchmark. arXiv preprint arXiv:2111.03759 (2021)
37. Li, Z., et al.: Blueprint separable residual network for efficient image super-resolution. In: Proceedings of the IEEE/CVF Conference on Computer Vision and Pattern Recognition, pp. 833–843 (2022)
38. Liang, J., Cao, J., Sun, G., Zhang, K., Van Gool, L., Timofte, R.: SWINIR: image restoration using swin transformer. In: Proceedings of the IEEE/CVF International Conference on Computer Vision, pp. 1833–1844 (2021)
39. Lim, B., Son, S., Kim, H., Nah, S., Mu Lee, K.: Enhanced deep residual networks for single image super-resolution. In: Proceedings of the IEEE Conference on Computer Vision and Pattern Recognition Workshops, pp. 136–144 (2017)
40. Liu, D., Wen, B., Fan, Y., Loy, C.C., Huang, T.S.: Non-local recurrent network for image restoration. In: Advances in Neural Information Processing Systems 31 (2018)
41. Liu, J., Tang, J., Wu, G.: Residual feature distillation network for lightweight image super-resolution. In: Bartoli, A., Fusiello, A. (eds.) ECCV 2020. LNCS, vol. 12537, pp. 41–55. Springer, Cham (2020). https://doi.org/10.1007/978-3-030-67070-2_2
42. Liu, Z., et al.: SWIN transformer: Hierarchical vision transformer using shifted windows. In: Proceedings of the IEEE/CVF International Conference on Computer Vision, pp. 10012–10022 (2021)
43. Mei, Y., Fan, Y., Zhou, Y.: Image super-resolution with non-local sparse attention. In: Proceedings of the IEEE/CVF Conference on Computer Vision and Pattern Recognition, pp. 3517–3526 (2021)
44. Niu, B., et al.: Single image super-resolution via a holistic attention network. In: Vedaldi, A., Bischof, H., Brox, T., Frahm, J.-M. (eds.) ECCV 2020. LNCS, vol. 12357, pp. 191–207. Springer, Cham (2020). https://doi.org/10.1007/978-3-030-58610-2_12
45. Paszke, A., et al.: Pytorch: an imperative style, high-performance deep learning library. In: Advances in Neural Information Processing Systems, vol. 32 (2019)
46. Reddi, V.J., et al.: MLPERF mobile inference benchmark. arXiv preprint arXiv:2012.02328 (2020)
47. Reed, J., DeVito, Z., He, H., Ussery, A., Ansel, J.: torch. fx: practical program capture and transformation for deep learning in python. In: Proceedings of Machine Learning and Systems, vol. 4, pp. 638–651 (2022)
48. Shi, W., et al.: Real-time single image and video super-resolution using an efficient sub-pixel convolutional neural network. In: Proceedings of the IEEE Conference on Computer Vision and Pattern Recognition, pp. 1874–1883 (2016)

49. Tai, Y., Yang, J., Liu, X., Xu, C.: MEMNET: a persistent memory network for image restoration. In: Proceedings of the IEEE International Conference on Computer Vision, pp. 4539–4547 (2017)
50. Wang, Z., Cun, X., Bao, J., Zhou, W., Liu, J., Li, H.: UFORMER: a general u-shaped transformer for image restoration. In: Proceedings of the IEEE/CVF Conference on Computer Vision and Pattern Recognition, pp. 17683–17693 (2022)
51. Wu, H., Judd, P., Zhang, X., Isaev, M., Micikevicius, P.: Integer quantization for deep learning inference: principles and empirical evaluation. arXiv preprint arXiv:2004.09602 (2020)
52. Zamir, S.W., Arora, A., Khan, S., Hayat, M., Khan, F.S., Yang, M.H.: Restormer: efficient transformer for high-resolution image restoration. In: Proceedings of the IEEE/CVF Conference on Computer Vision and Pattern Recognition, pp. 5728–5739 (2022)
53. Zhan, Z., et al.: Achieving on-mobile real-time super-resolution with neural architecture and pruning search. In: Proceedings of the IEEE/CVF International Conference on Computer Vision, pp. 4821–4831 (2021)
54. Zhang, K., et al.: AIM 2020 challenge on efficient super-resolution: methods and results. In: Bartoli, A., Fusiello, A. (eds.) .: Aim 2020 challenge on efficient super-resolution: methods and results. LNCS, vol. 12537, pp. 5–40. Springer, Cham (2020). https://doi.org/10.1007/978-3-030-67070-2_1
55. Zhang, K., et al.: Aim 2019 challenge on constrained super-resolution: methods and results. In: 2019 IEEE/CVF International Conference on Computer Vision Workshop (ICCVW), pp. 3565–3574. IEEE (2019)
56. Zhang, X., Zeng, H., Guo, S., Zhang, L.: Efficient long-range attention network for image super-resolution. arXiv preprint arXiv:2203.06697 (2022)
57. Zhang, X., Zeng, H., Zhang, L.: Edge-oriented convolution block for real-time super resolution on mobile devices. In: Proceedings of the 29th ACM International Conference on Multimedia, pp. 4034–4043 (2021)
58. Zhang, Y., Li, K., Li, K., Wang, L., Zhong, B., Fu, Y.: Image super-resolution using very deep residual channel attention networks. In: Ferrari, V., Hebert, M., Sminchisescu, C., Weiss, Y. (eds.) ECCV 2018. LNCS, vol. 11211, pp. 294–310. Springer, Cham (2018). https://doi.org/10.1007/978-3-030-01234-2_18
59. Zhang, Y., Tian, Y., Kong, Y., Zhong, B., Fu, Y.: Residual dense network for image super-resolution. In: Proceedings of the IEEE Conference on Computer Vision and Pattern Recognition, pp. 2472–2481 (2018)

Style Adaptive Semantic Image Editing with Transformers

Edward Günther[1], Rui Gong[1(✉)], and Luc Van Gool[1,2]

[1] Computer Vision Lab, ETH Zurich, Zürich, Switzerland
gedward@student.ethz.ch, {gongr,vangool}@vision.ee.ethz.ch
[2] VISICS, ESAT/PSI, KU Leuven, Leuven, Belgium

Abstract. The goal of semantic image editing is to modify an image based on an input semantic label map, to carry out the necessary image manipulation. Existing approaches typically lack control over the style of the editing, resulting in insufficient flexibility to support the desired level of customization, *e.g.*, to turn an object into a particular style or to pick a specific instance. In this work, we propose Style Adaptive Semantic Image Editing (SASIE), where a reference image is used as an additional input about style, to guide the image manipulation process in a more adaptive manner. Moreover, we propose a new transformer-based architecture for SASIE, in which intra-/inter-image multi-head self-attention blocks transfer intra-/inter-knowledge. The content of the edited areas is synthesized according to the given semantic label, while the style of the edited areas is inherited from the reference image. Extensive experiments on multiple datasets suggest that our method is highly effective and enables customizable image manipulation.

Keywords: Semantic image editing · Transformers

1 Introduction

The task of image manipulation aims at adding, altering, and removing the instances of certain classes (*e.g.*, car, pedestrian) or semantic concepts (*e.g.*, sky, road) in the images. It has been a core problem in computer vision and graphics community for years, with a wide range of applications including data augmentation [29], visual media [30], and human-machine interaction [2].

Powered by the recent advances in generative adversarial networks (GANs) [11], semantic image editing (SIE) [13,28] has received an increasing attention and achieved impressive results. Recent works [13,28] adopt an image inpainting based strategy, where the edited image areas are synthesized conditioned on the given semantic label maps, and a deterministic mapping is learned between the two. These approaches enabled realistic editing of image "*content*";

E. Günther and R. Gong—These authors contributed equally to this work.

Supplementary Information The online version contains supplementary material available at https://doi.org/10.1007/978-3-031-25063-7_12.

L. Karlinsky et al. (Eds.): ECCV 2022 Workshops, LNCS 13802, pp. 187–203, 2023.
https://doi.org/10.1007/978-3-031-25063-7_12

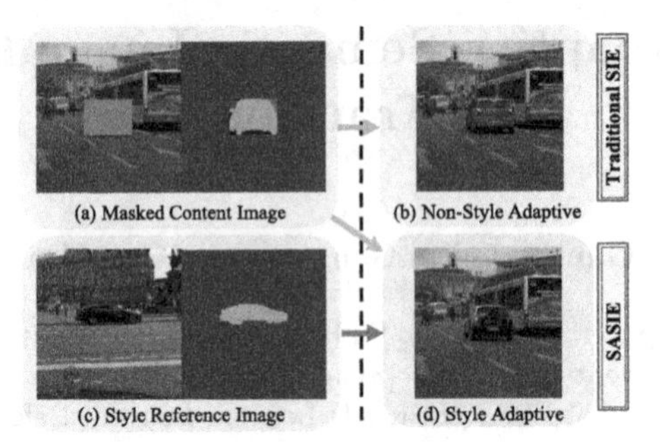

Fig. 1. Traditional SIE *vs.* SASIE. Traditional SIE only aims at synthesizing the content of the edited object. However, our SASIE not only considers the content of the edited object, but also can customize its style.

however, the control over "*style*" of the editing is not yet possible. For instance, users are able to add a "bed" to their own bedroom image with [13,28], but couldn't choose which specific bed or style of bed to add. The lack of control over "*style*" makes it cumbersome to deal with more complex scenarios, and has greatly limited the flexibility of what users can achieve with image editing systems.

In order to alleviate this limitation, based on the traditional SIE strategy in [28], we propose a novel style adaptive semantic image editing (SASIE) strategy. In our setup, besides the masked image and the semantic label maps of the masked out areas, an additional reference image is given (see Fig. 1). SAISE *synthesizes* the masked out areas according to the given semantic label maps (add, alter and remove the instances of certain classes or semantic concepts), and *stylizes* the pixels of the synthesized areas with the style of the reference image. For example, by providing an indoor scene image with the "bed" region masked out, the semantic label map denoting the region ought to contain a "bed", and the reference "bed" style image, SASIE inpaints the "bed" compatible with the semantic layout, according to the style, texture and color of the reference "bed". By providing different style reference images, the inpainted images for each follow these styles.

SASIE is challenging for two main reasons, i) *image inpainting*, *i.e.*, the synthesized pixels in the inpainted areas need to preserve the object information from the semantic map (such as object class, pose and shape), be realistic, and be consistent with the regions out of the inpainted areas; ii) *local style transfer*, *i.e.*, the style of the synthesized pixels in the inpainted areas should follow the style of the reference image. These challenges can not be resolved by naively combining traditional SIE [28] and style transfer [14] (see Sect. 3.2 and Fig. 4). As to challenge i), we follow the generative adversarial networks (GANs) framework proposed in [28], to exploit the strong ability of GANs to synthesize realistic images. In order to capture more global knowledge inside/outside the inpainted areas to improve the realism and compatibility, we introduce a transformer-based structure into the generator, instead of the popularly utilized pure convolutional structure. The intra-image multi-head self-attention block is developed to transfer knowledge inside/outside the inpainted areas. As to challenge ii), different from the traditional style transfer framework which stylizes the global image with the reference image, SASIE aims at transferring the style locally, into the inpainted area. To this end, we propose the mask-aware inter-image multi-head self-attention block, where the "key" and "value" of the self-attention block is extracted from the reference image while the "query" is from the content image.

In a nutshell, our key contributions are three-fold. 1) We propose a novel SASIE strategy for image manipulation, where the style of the edited pixels is conditioned on the reference image. 2) We propose a transformer-based framework, to synthesize the edited areas realistically, and stylize the edited areas adaptively through the intra-/inter-image multi-head self-attention blocks. 3) Extensive experiments on different benchmarks are conducted, showing the effectiveness of our method.

2 Related Work

Image Manipulation. Over the last year, image manipulation has served as an important topic in machine learning [30], computer vision [42], and graphics [39]. The amount of research on image manipulation in general, means that a complete survey is out of the scope of this work. We mainly focus on semantic-aware image manipulation [8,13,16,21,28,29,36,43,44], which allows for semantic control over the manipulation process. This area can be divided into two main categories: 1) semantic layout to image translation, 2) SIE. The former one [21,29,36] aims at translating semantic label maps to the corresponding RGB values in a pixel-to-pixel way. The latter one [13,28] aims at filling the masked region under the guidance of the semantic label map, in the image inpainting sense. Category 1) methods unavoidably change the parts outside the edited area, since they rely on the whole image synthesis. Our work falls into the category 2) SIE.

An inpainting based hierarchical framework for SIE is proposed in [13]. But [13] requires the availability of the full semantic segmentation map even for local editing. On the other hand, [28] develops a different generator and discriminator structure, able to work with the local semantic segmentation map of the edited areas. We follow this setting, due to its flexibility. However, these works can only generate a deterministic edited result once the model is trained. Instead, we propose SASIE, where the style of the edited result is conditioned on the reference image. Even though recent works [5,24] touch the problem of locally editing the image attributes based on the pretrained StyleGAN model [17–19], they still suffer from inevitably changing the parts outside the edited area.

Arbitrary Style Transfer. The arbitrary style transfer task [4,14,15,33] aims at applying the style extracted from a reference image to the content image. The first to demonstrate the effectiveness of DNNs for style transfer was [10], matching feature representations in the convolutional layers. Then, a series of works [14,28,29] have shown the advantage and validity of alignment on the normalization layer between the content image and the style reference image. However, they focus on transferring the style onto the entire content image, whereas SASIE aims at locally adapting the style of the inpainted areas. The latter needs to follow the given semantic layout, compatible with the surrounding areas in the original image. Such challenges make it non-trivial to extend the previous style transfer framework to our setting (see Fig. 4). Moreover, instead of the popularly adopted feature concatenation [10] or normalization layer assignment [14,29] for style transfer, we introduce the mask-aware inter-image multi-head self-attention block.

Vision Transformer. Recently, transformer-based methods [35] have been proven successful at different computer vision tasks, such as classification [9,25], semantic segmentation [32,37] and object detection [3]. However, few of the above works touch the cross-image transferring problem. As an exception, [7] proposes a transformer-based structure for arbitrary style transfer. However, from the task aspect, [7] still only focuses on stylizing the whole content image, rather than locally stylizing like we do. From a technical point of view, we develop a mask-aware inter-image multi-head self-attention block, for local style transfer. Our transformer decoder layer is composed of intra- and inter-image multi-head self-attention blocks, for both intra- and inter-image knowledge transfer. Besides, an inpainting based reconstruction and perceptual loss are developed, to synthesize and stylize the inpainted areas.

3 Methodology

3.1 Problem Statement

The full content image is denoted as $\mathbf{x}^t \in \mathbb{R}^{H\times W\times 3}$, where the inpainting area $\mathbf{x}^t_m \in \mathbb{R}^{H_m\times W_m\times 3}$ is masked out. Similar to traditional SIE, in our SASIE setup, we are given the non-masked area $\mathbf{x}^t_n \in \mathbb{R}^{H\times W\times 3}$, *i.e.*, $\mathbf{x}^t_n = \mathbf{x}^t \setminus \mathbf{x}^t_m$. The semantic label of the inpainting area $\mathbf{y}^t_m \in \mathbb{R}^{H_m\times W_m\times C}$ is also given, and C is the number of semantic classes. But differently, the style reference image $\mathbf{x}^s$, and the corresponding semantic label $\mathbf{y}^s$ are also provided in SASIE. The purpose of SASIE is to, 1) synthesize the masked-out area $\hat{\mathbf{x}}^t_m$, conditioned on $\mathbf{y}^t_m$, 2) transfer the style from $\mathbf{x}^s$ to $\hat{\mathbf{x}}^t_m$. Unlike SIE that only focuses on 1), SASIE considers both 1) and 2) (see Fig. 1). The final inpainted image is generated, by $\hat{\mathbf{x}}^t = \mathbf{x}^t_n \cup \hat{\mathbf{x}}^t_m$. Introducing the notation of the mask/inpainting area $M \in \{0,1\}^{H\times W}$, $\mathbf{x}^t_n$ and $\hat{\mathbf{x}}^t$ can be represented as,

$$M(h,w) = \begin{cases} 1, \text{if } \mathbf{x}^t(h,w) \in \mathbf{x}^t_m, \\ 0, \text{if } \mathbf{x}^t(h,w) \in \mathbf{x}^t_n, \end{cases} \tag{1}$$

$$\mathbf{x}^t_n = (1-M) \odot \mathbf{x}^t, \tag{2}$$

$$\hat{\mathbf{x}}^t = ((1-M) \odot \mathbf{x}^t) \cup \hat{\mathbf{x}}^t_m, \tag{3}$$

where (h,w) is the (row, column) index, and $\odot$ is element-wise multiplication.

3.2 Intra-/Inter-Image Knowledge Transfer

As analyzed in Sect. 1, the challenges of SASIE lie in i) image inpainting (*i.e.*, *intra-image knowledge transfer*), and ii) local style transfer (*i.e.*, *inter-image knowledge transfer*).

Intra-Image Knowledge Transfer. The challenge i) image inpainting includes the objects synthesis given semantic label, and the consistency inside/outside the edited areas. Since the GANs framework [13,28,29] is proven to synthesize the objects according to the given semantic label realistically. Thus, the main key point for the challenge i) is to improve the harmony, compatibility and smoothness inside/outside the edited areas. Thus, the global knowledge excavation of the whole image is important. The previous generator in GAN-based SIE framework [28] is pure convolutional structure, mainly focusing on the local knowledge extraction within the edited area. [13] excavates the global knowledge by introducing the full semantic label map, which is unavailable in SASIE. In order to transfer the knowledge between the edited and non-edited areas, we propose the generative transformer structure, which exploits the strength of the intra-image multi-head self-attention block for global knowledge extraction.

Inter-Image Knowledge Transfer. The challenge ii) local style transfer aims at stylizing the edited area locally. The previous works [14,22,29,38] focus on transferring the style of the reference image to the whole content image, instead of the instances or the objects in the content image. The naive extension [20] of the previous methods under SASIE is to stylize the whole inpainted image or the inpainted area from SIE, and then mix the stylized edited area with the non-edited area according to the mask, namely global mix and cut mix, resp. However, the naive extension is easy to cause the obvious artifacts (see Fig. 4) due to 1) inaccuracy of the mask, 2) unnatural transition between inside and outside the edited areas, and 3) lack of content preservation. In order to locally transfer the style of the reference image to the inpainted area and reduce artifacts, we further develop the inter-image multi-head self-attention block in the generative transformer structure, where the "key" and "value" in attention mechanism are from the style reference image while the "query" is from the content image.

3.3 Generative Transformer Structure for SASIE

We adopt the GANs framework [11,28,43], *i.e.*, the generator G and discriminator D pair, due to its strong ability and effectiveness for the image synthesis. We adopt the discriminator structure in [28], to utilize the semantic guidance and distinguishing ability. In order to realize the intra-/inter-image knowledge transfer in the unified way for SASIE, we propose the generative transformer structure for the generator, which is shown in Fig. 2. Our generator G is composed of three modules, i) convolutional encoder, V_c, ii) transformer decoder, F_d, iii) convolutional decoder, V_d. First, the convolutational encoder V_c is used to extract the compact feature representation. Taking the non-masked area $\mathbf{x}_n^t$, and the semantic label of the masked area $\mathbf{y}_m^t$ as the input, the compact feature $\mathbf{f}^t$ is extracted, *i.e.*, $\mathbf{f}^t = V_c(\mathbf{x}_n^t, \mathbf{y}_m^t)$. Second, the transformer decoder F_d is exploited to globally transfer the intra-image knowledge inside/outside the edited areas (see intra-image multi-head self-attention block in Sect. 3.4), and locally transfer the inter-image knowledge from the style reference image to the edited areas (see inter-image multi-head self-attention block in Sect. 3.4).

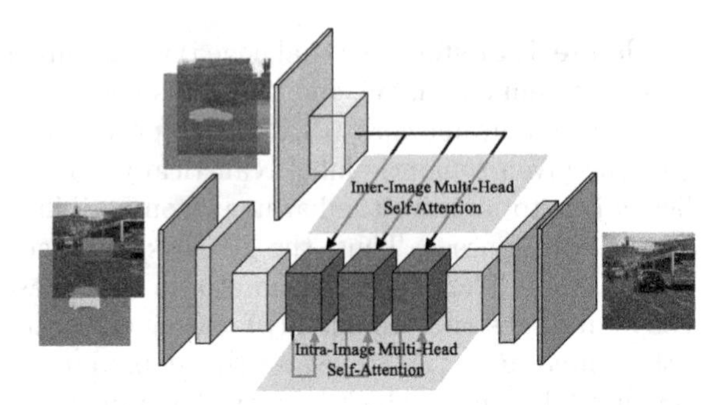

Fig. 2. Generative transformer structure. The generator is composed of i) convolutional encoder (yellow part), ii) transformer decoder (blue part) and iii) convolutional decoder (green parts). The pink part is the downsampling and linear projection for the style reference image. The transformer decoder layer in the generator unifies the intra-/inter-image multi-head self-attention block. (Color figure online)

By feeding the compact feature representation $\mathbf{f}^t$, the style reference image $\mathbf{x}^s$, and the corresponding reference image label $\mathbf{y}^s$, the edited and stylized feature representation $\tilde{\mathbf{f}}^t$ is obtained, *i.e.*, $\tilde{\mathbf{f}}^t = F_d(\mathbf{f}^t, \mathbf{x}^s, \mathbf{y}^s)$. Third, the convolutional decoder V_d is utilized to decompose the feature and generate the output. Taking the edited and stylized feature $\tilde{\mathbf{f}}^t$ as the input, the intermediate synthesized and stylized image $\tilde{\mathbf{x}}^t$ is generated, *i.e.*, $\tilde{\mathbf{x}}^t = V_d(\tilde{\mathbf{f}}^t)$. Then, only the edited area of $\tilde{\mathbf{x}}^t$ is retained, *i.e.*, $\hat{\mathbf{x}}^t_m \subseteq M \odot \tilde{\mathbf{x}}^t$. Combining with Eq. (3), the final inpainted image is generated by $\hat{\mathbf{x}}^t = M \odot \tilde{\mathbf{x}}^t + (1 - M) \odot \mathbf{x}^t$. We follow [28] for the convolutional encoder structure, and adopt the ResNet upsampling block [12,28] for the convolutional decoder. The detailed structure of the transformer decoder layer is introduced in Sect. 3.4.

3.4 Transformer Decoder Layer

As shown in Fig. 3, the transformer decoder layer is composed of intra- and inter-image multi-head self-attention blocks, which excavate the intra- and inter-image knowledge, respectively. Next, we detail intra- and inter-image multi-head self-attention blocks.

Intra-Image Multi-head Self-Attention (IAMA) Block. Given the compact feature representation $\mathbf{f}^t$ from the convolutional encoder V_c, 1×1 patches are extracted by flattening the spatial dimensions of the feature map and projecting to the embedding dimension C_t in order to obtain the input sequence $\mathbf{e}^t = \{\mathbf{e}^t_i, i = 1, ..., L\}$, where L represents the number of patches. We add learnable position embeddings $\mathbf{p}^t = \{\mathbf{p}^t_i, i = 1, ..., L\}$ to the input sequence to retain positional information, *i.e.*, $\mathbf{z}^t = \{\mathbf{z}^t_i = \mathbf{e}^t_i + \mathbf{p}^t_i, i = 1, ..., L\}$. The corresponding key $\mathbf{k}^t_j$, query $\mathbf{q}^t_j$ and, value $\mathbf{v}^t_j$ matrices of the j-th attention head can then be,

$$\mathbf{q}^t_j = \mathbf{z}^t \mathbf{w}^t_{q_j}, \mathbf{k}^t_j = \mathbf{z}^t \mathbf{w}^t_{k_j}, \mathbf{v}^t_j = \mathbf{z}^t \mathbf{w}^t_{v_j}, \tag{4}$$

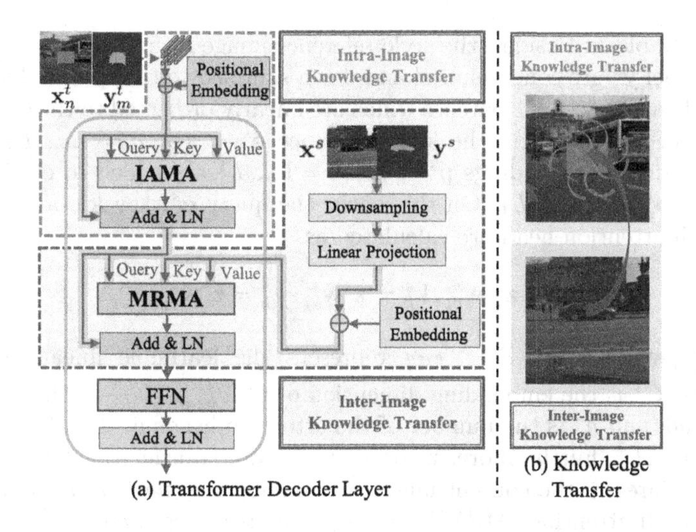

Fig. 3. Transformer decoder layer structure. The transformer decoder layer is composed of intra-image multi-head self-attention (IAMA) and inter-image multi-head self-attention (MRMA) blocks, enabling intra-/inter-image knowledge transfer.

where $\mathbf{w}^t_{q_j}, \mathbf{w}^t_{k_j}, \mathbf{w}^t_{v_j} \in \mathbb{R}^{C^t \times d^t_{head}}$ represent the learnable linear projection parameters. C^t is the embedding dimension of $\mathbf{z}^t_i$, $d^t_{head} = \frac{C^t}{h^t}$ is the dimension of each head, and h^t is the number of the attention heads in the IAMA block, *i.e.*, $j = 1, .., h^t$. On this basis, the intra-image multi-head self-attention (IAMA) block can be described as,

$$\mathrm{IAMA}(\mathbf{z}^t) = \mathrm{Concat}(\mathrm{head}^t_1, ..., \mathrm{head}^t_{h^t})\mathbf{w}^t_o, \tag{5}$$

$$\mathrm{head}^t_j = \mathrm{Attention}(\mathbf{q}^t_j, \mathbf{k}^t_j, \mathbf{v}^t_j), \tag{6}$$

$$\mathrm{Attention}(\mathbf{q}^t_j, \mathbf{k}^t_j, \mathbf{v}^t_j) = \mathrm{softmax}(\frac{\mathbf{q}^t_j {\mathbf{k}^t_j}^T}{\sqrt{d^t_{head}}})\mathbf{v}^t_j, \tag{7}$$

where Concat$(\cdot)$ is the concatenation manipulation, and $\mathbf{w}^t_o \in \mathbb{R}^{C_t \times C_t}$. Then, the output sequence $\mathbf{z}^{o_t} = \{\mathbf{z}^{o_t}_i, i = 1, ..., L\}$ is obtained by applying the residual connection and the layer normalization (LN),

$$\mathbf{z}^{o_t} = \mathrm{LN}(\mathbf{z}^t + \mathrm{IAMA}(\mathbf{z}^t)). \tag{8}$$

Inter-Image Multi-Head Self-Attention (MRMA) Block. In the intra-image multi-head self-attention block, the query $\mathbf{q}^t_j$, key $\mathbf{k}^t_j$, and value $\mathbf{v}^t_j$ are all extracted from the content image. Instead, in the inter-image multi-head self-attention block, only the query $\mathbf{q}^s_k$ is from the content image, while the key $\mathbf{k}^s_k$ and the value $\mathbf{v}^s_k$ are both from the style reference image. More specifically, taking the output from the intra-image multi-head self-attention block $\mathbf{z}^{o_t}$ as the input, $\mathbf{z}^{o_t}$ is linearly projected to the query $\mathbf{q}^s_k$, as done in the intra-image multi-head self-attention block. Besides, the style reference image $\mathbf{x}^s$ and the corresponding semantic label $\mathbf{y}^s$ are concatenated and down-sampled to the spatial dimensions of $\mathbf{f}^t$. We then extract 1×1 patches which are linearly projected to the embedding dimension of $\mathbf{e}_i^t$, obtaining the style sequence $\mathbf{e}^s = \{\mathbf{e}_i^s, i = 1, ..., L\}$. Again, learnable position embeddings $\mathbf{p}^s = \{\mathbf{p}_i^s, i = 1, .., L\}$ are added to $\mathbf{e}^s$, *i.e.*, $\mathbf{z}_s = \{\mathbf{z}_i^s = \mathbf{e}_i^s + \mathbf{p}_i^s, i = 1, ..., L\}$. On this basis, the query $\mathbf{q}_k^s$, key $\mathbf{k}_k^s$, and value $\mathbf{v}_k^s$ for the k-th attention head are calculated as,

$$\mathbf{q}_k^s = \mathbf{z}^{o_t}\mathbf{w}_{q_k}^s, \mathbf{k}_k^s = \mathbf{z}^s\mathbf{w}_{k_k}^s, \mathbf{v}_k^s = \mathbf{z}^s\mathbf{w}_{v_k}^s, \tag{9}$$

where $\mathbf{w}_{q_k}^s, \mathbf{w}_{k_k}^s, \mathbf{w}_{v_k}^s \in \mathbb{R}^{C^s \times d_{head}^s}$ represent the learnable linear projection parameters. C^s is the embedding dimension of $\mathbf{z}_i^s$, $d_{head}^s = \frac{C^s}{h^s}$ is the dimension of each head, and h^s is the number of the attention heads in the MRMA block, *i.e.*, $k = 1, .., h^s$. Furthermore, in order to locally transfer the reference style to the mask area of the content image, we develop the *mask-aware* inter-image multi-head self-attention (MRMA) block, which is represented as,

$$\mathrm{MRMA}(\mathbf{z}^{o_t}, \mathbf{z}^s, \mathbf{m}) = \mathrm{Concat}(\mathrm{head}_1^s, ..., \mathrm{head}_{h^s}^s)\mathbf{w}_o^s, \tag{10}$$

$$\mathrm{head}_k^s = \mathrm{MaskAttention}(\mathbf{q}_k^s, \mathbf{k}_k^s, \mathbf{v}_k^s, \mathbf{m}), \tag{11}$$

$$\mathrm{MaskAttention}(\mathbf{q}_k^s, \mathbf{k}_k^s, \mathbf{v}_k^s, \mathbf{m}) = \mathrm{softmax}\Big(\frac{\mathbf{q}_k^s {\mathbf{k}_k^s}^T}{\sqrt{d_{head}^s}}\Big)\mathbf{m}\mathbf{v}_k^s, \tag{12}$$

where $\mathbf{m} \in \{0, 1\}^{L\times L}$ is the masking matrix and the values of the positions corresponding to the mask area are 1, while others are 0, and $\mathbf{w}_o^s \in \mathbb{R}^{C_s \times C_s}$. Similar to the intra-image multi-head self-attention block, the output sequence of the inter-image multi-head self-attention block $\mathbf{z}^{o_s} = \{\mathbf{z}_i^{o_s}, i = 1, .., L\}$ is obtained by adding the residual connection and the layer normalization (LN),

$$\mathbf{z}^{o_s} = \mathrm{LN}(\mathbf{z}^{o_t} + \mathrm{MRMA}(\mathbf{z}^{o_t}, \mathbf{z}^s, \mathbf{m})). \tag{13}$$

Feed-Forward Network (FFN) Block. Following [9], we feed the output of the inter-image multi-head self-attention block, $\mathbf{z}^{o_s}$, into the feed-forward network (FFN). The multi-layer perceptrons (MLPs) in [9] is adopted for the FFN structure. Besides, the residual connection and the LN is also applied to obtain the output $\mathbf{z}^{o_f}$,

$$\mathbf{z}^{o_f} = \mathrm{LN}(\mathrm{FFN}(\mathbf{z}^{o_s}) + \mathbf{z}^{o_s}). \tag{14}$$

3.5 Training Loss and Objective

For the challenge i) image inpainting, following [28], we adopt the adversarial loss for the GANs framework training. Additionally, by introducing the intra-image reconstruction loss, we explicitly transfer the ground truth inpainting supervision to the mask area, to facilitate and ease the inpainting of the mask area. For the challenge ii) local style transfer, we adopt the inter-image perceptual loss.

Intra-Image Reconstruction Loss. As discussed in Sect. 3.4., the MRMA block is able to transfer the inter-image knowledge to the mask area explicitly. Besides, by replacing the style reference image and semantic label $\mathbf{x}^s, \mathbf{y}^s$ with the ground truth content image inpainting area and corresponding semantic label $\mathbf{x}^t_m, \mathbf{y}^t_m$ in the MRMA block, we aim to reconstruct the ground truth content image $\mathbf{x}^t$. In this way, the ground truth supervision $\mathbf{x}^t_m$ of the mask area is further strengthened, and eases the inpainting task. In order to measure the similarity between the synthesized image $\hat{\mathbf{x}}^t_r$ and the ground truth content image $\mathbf{x}^t$, we utilize the perceptual loss $\mathcal{L}_{rec}$, represented as,

$$\mathcal{L}_C(\hat{\mathbf{x}}^t_r, \mathbf{x}^t) = \tfrac{1}{N_l} \textstyle\sum_{p=0}^{N_l} ||\phi_p(\hat{\mathbf{x}}^t_r) - \phi_p(\mathbf{x}^t)||_1, \tag{15}$$

$$\mathcal{L}_S(\hat{\mathbf{x}}^t_r, \mathbf{x}^t) = \tfrac{1}{N_l} \textstyle\sum_{p=0}^{N_l} ||E^\phi_p(\hat{\mathbf{x}}^t_r) - E^\phi_p(\mathbf{x}^t)||_1, \tag{16}$$

$$E^\phi_p(\mathbf{x}) = \phi_p(\mathbf{x})\phi_p(\mathbf{x})^T, \tag{17}$$

$$\mathcal{L}_{rec} = \lambda_C \mathcal{L}_C(\hat{\mathbf{x}}^t_r, \mathbf{x}^t) + \lambda_S \mathcal{L}_S(\hat{\mathbf{x}}^t_r, \mathbf{x}^t), \tag{18}$$

where $\mathcal{L}_C$ and $\mathcal{L}_S$ represent the content loss and the style loss, resp. $\phi_p(\cdot)$ denotes the feature map extracted from the p-th layer of a pretrained VGG-19 model [31]. N_l is the number of the selected layers in VGG-19 model. $E^\phi_p(\cdot)$ denotes the Gram matrix. λ_C and λ_S are hyper-parameters to balance the content and style loss.

Adversarial Loss. Following [28], we utilize the Hinge Loss [23,27,28,34] $\mathcal{L}_{hinge}$ as the adversarial loss for the GANs framework training, and the Feature Matching Loss [36] $\mathcal{L}_{fm}$ for stabilization. The Feature Matching Loss is only applied to the image feature maps produced by the RGB stream of the discriminator.

Inter-Image Perceptual Loss. By feeding $\mathbf{x}^t_n, \mathbf{y}^t_m, \mathbf{x}^s, \mathbf{y}^s$ into our framework, the style adapted and semantic edited image $\hat{\mathbf{x}}^t$ is synthesized. In SASIE, the inpainted area of $\hat{\mathbf{x}}^t$ needs to preserve the content of $\mathbf{x}^t$, while the style of the inpainted area should follow the style of $\mathbf{x}^s$. Thus, we adopt the inter-image perceptual loss $\mathcal{L}_{percep}$ between $\hat{\mathbf{x}}^t$ and $\mathbf{x}^t, \mathbf{x}^s$, formulated as,

$$\mathcal{L}_{percep} = \lambda_C \mathcal{L}_C(\hat{\mathbf{x}}^t, \mathbf{x}^t) + \lambda_S \mathcal{L}_S(\hat{\mathbf{x}}^t, \mathbf{x}^s). \tag{19}$$

Total Objective. The intra-image reconstruction loss and the inter-image perceptual loss are used in the alternative way. Our above losses focus on the objects of interest in the inpainted areas. In order to further improve the smoothness of the inpainted parts around the objects of interest, we utilize an additional L1 Loss, $\mathcal{L}_{context}$, to match the context between the synthesized images $\hat{\mathbf{x}}^t_r$ and $\hat{\mathbf{x}}^t$, in the inpainted areas but around the objects of interest. Then, the total objective of our proposed approach can be denoted as,

$$\begin{aligned} \mathcal{L}_G = \delta \mathcal{L}_{rec} + (1-\delta)\mathcal{L}_{percep} + \lambda_{hinge}\mathcal{L}_{hinge} \\ + \lambda_{fm}\mathcal{L}_{fm} + \lambda_{context}\mathcal{L}_{context}, \end{aligned} \tag{20}$$

where $\delta \in \{0, 1\}$ is used to control the alternative training, *i.e.*, δ is 1 and 0 when the training iteration is odd and even, resp. λ_{hinge}, λ_{fm} and $\lambda_{context}$ are hyper-parameters to balance between different losses. The discriminator is also trained with the corresponding Hinge Loss [28], $\mathcal{L}_D$. During the training, we alternatively train the generator G and the discriminator D, with $\mathcal{L}_G$ and $\mathcal{L}_D$.

4 Experiments

4.1 Experimental Setup

First, the datasets and semantic classes involved in our experiments include Cityscapes [6] (pedestrian, rider, and car), ADE20K [40,41] (bed, picture, and lamp), CelebA [26] (eye and hair), and Flickr [28] (clouds, river, mountain, grass, sky, tree, and sea). The details of the datasets are provided in the appendix.

Then, the experimental setting of SIE and SASIE are as follows:

Traditional Semantic Image Editing (SIE). We follow the SIE setup in [28]. SIE only focuses on the image inpainting, without touching the style transfer. Thus, the key point of SIE is the intra-image knowledge transfer to inpaint the mask area realistically and compatibly. In order to prove the validity of our approach for the *intra-image knowledge transfer* (see Sect. 3.2), we conduct the experiments on Cityscapes and ADE20K benchmarks. Our framework is adjusted to the SIE by removing the inter-image multi-head self-attention block.

Style-Adaptive Semantic Image Editing (SASIE). Different from the traditional SIE, our proposed SASIE introduces the additional style reference image to stylize the inpainting areas. As anaylyzed in Sect. 3.2, the key points of SASIE are both the intra- and inter-image knowledge transfer. In order to prove the effectiveness of our approach for both *intra- and inter-image knowledge transfer*, experiments on Cityscapes, ADE20K, CelebA, and Flickr are carried out.

Next, since our method is the first approach for SASIE, we construct the baselines for SASIE based on the SOTA SIE method, SESAME, as follows:

SESAME-Concat. Following the framework in [28], and the concatenation technique for semantic image editing in [13], we concatenate the style reference image and the corresponding semantic label $\mathbf{x}^s, \mathbf{y}^s$, with the non-masked area image $\mathbf{x}^t_n$ and the semantic label of the inpainting area $\mathbf{y}^t_m$, and then feed them into the SESAME framework.

SESAME-SPADE. Following the framework in [28], and the SPADE normalization technique in [28,29], we condition the SPADE layer in [28] on the style reference image and the corresponding semantic label $\mathbf{x}^s, \mathbf{y}^s$.

4.2 Experimental Results

SIE Quantitative and Qualitative Results. As shown in Table 1, under the SIE setting, we quantitatively compare our proposed approach with the previous state-of-the-art SIE method, SESAME, on the Cityscapes and ADE20K datasets. It is shown that our method outperforms the SESAME method on

Table 1. Quantitative comparison results under traditional SIE setting, on Cityscapes and ADE20K datasets.

Models	Dataset	Setting							
		Addition				Removal			
		SSIM↑	FID↓	mIoU (%)↑	Acc (%)↑	SSIM↑	FID↓	mIoU (%)↑	Acc (%)↑
SESAME [28]	Cityscapes	0.3979	10.19	62.51	91.51	0.5940	10.80	63.19	92.90
Ours		**0.4048**	**10.12**	**62.59**	**91.53**	**0.5961**	**9.69**	**63.68**	**92.92**
SESAME [28]	ADE20K	0.2240	25.49	51.55	92.82	0.4245	20.98	50.12	92.39
Ours		**0.2706**	**22.79**	**52.10**	**93.35**	**0.4639**	**18.61**	**54.35**	**93.89**

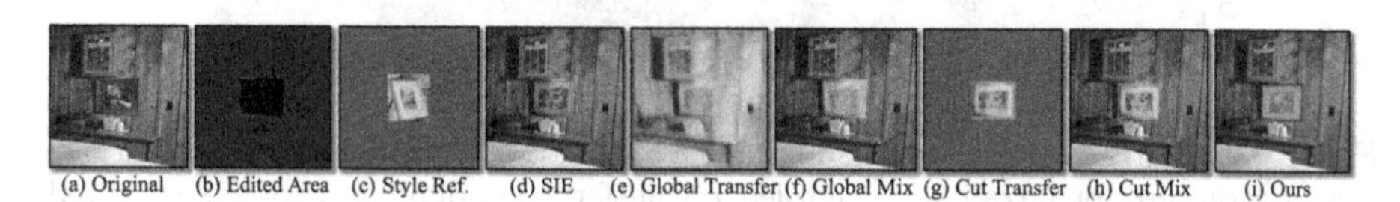

Fig. 4. Comparison with Global/Cut Mix (SIE+Style Transfer), under SASIE. (d) is the inpainted image from SIE. Global mix: (e) is obtained through the global style transfer [14] applied on (d). (f) is the mixture of (d) and (e) based on the mask in (b). Cut mix: (g) is obtained by applying [14] only to the cut inpainted area in (d). (h) is the mixture of (d) and (g) based on the mask in (b).

all metrics and different benchmarks quantitatively, proving the validity of our method for the intra-image knowledge transfer. In Fig. 5, we show the qualitative comparison results between our proposed method and the SESAME method, under the SIE setting. Benefiting from the effective intra-image knowledge transfer, it is observed that the boundary artifacts between the inside and outside of the edited areas are highly reduced. It further verifies the effectiveness of our proposed method for SIE, especially for the intra-image knowledge transfer.

SASIE Quantitative and Qualitative Results. In Table 2, the quantitative experimental results show that our proposed method outperforms baseline methods, SESAME-Concat and SESAME-SPADE, by a large margin, under the SASIE setting. It verifies the validity and strength of our proposed approach for both the intra- and inter-image knowledge transfer. From Fig. 6 and Fig. 7, it is observed that our proposed approach is able to synthesize the inpainted areas realistically, and stylize the inpainted areas according to the provided style reference image effectively. However, other baseline methods tend to directly paste the style reference object into the mask area qualitatively, losing the original semantic layout and object content information of the inpainted area. In Fig. 8, we show the qualitative results of our method on the CelebA and Flickr datasets.

Content Preservation. In order to measure the object content information preservation in the edited area quantitatively, following [1], we measure the content loss $d_c = \lambda_C \mathcal{L}_C$ and the SSIM between the inpainted image $\hat{\mathbf{x}}^t$ and the ground truth full content image $\mathbf{x}^t$. As shown in Table 2, our method preserves the semantic layout and content of the objects much better, compared with other

Fig. 5. Qualitative comparison results under SIE setting. Benefiting from our transformer structure for intra-image knowledge transfer, the boundary artifacts of edited areas are highly reduced.

Table 2. Quantitative comparison under SASIE setting, on Cityscapes and ADE20K datasets. FID, mIoU and Acc are used to measure the synthesized image quality for SASIE, while SSIM and d_c are exploited to measure the content preservation in the edited area for SASIE.

Models	Dataset									
	Cityscapes					ADE20K				
	FID↓	mIoU (%)↑	Acc (%)↑	SSIM↑	d_c ↓	FID↓	mIoU (%)↑	Acc (%)↑	SSIM↑	d_c ↓
SESAME-SPADE [28,29]	12.90	61.88	91.29	0.3464	1.07	50.47	46.73	91.60	0.1616	3.66
SESAME-Concat [13,28]	12.87	61.69	91.29	0.3432	1.01	48.05	47.07	91.95	0.1592	3.43
Ours	**11.35**	**62.41**	**91.41**	**0.3554**	**0.90**	**42.51**	**49.08**	**92.15**	**0.1860**	**2.93**

baseline methods. It further proves the effectiveness of our method for both the intra- and inter-image knowledge transfer.

User Study. We adopt the Amazon Mechanical Turk (AMT) to conduct the user study. The data for each trial is gathered from 3 participants, who are different for different trials, and there are 150 trials for Cityscapes and 145 trials for ADE20K in total. In the user study, each participant is shown the masked-out content image, the ground truth content image, the style reference image, and the inpainted images generated by different methods. Then, the participant is required to choose the one 1) with fewer artifacts, 2) whose attributes (*e.g.*, shape and pose) of the inpainted object are more similar to the ground truth content image, 3) and the inpainted object style (*e.g.*, color and texture) of which is more similar to the style reference image, at the same time. The user preference results in Table 4 show that more users prefer our synthesized images.

Ablation Study. In order to assess the effectiveness of different components and proposed losses in our method, we conduct a set of ablation studies under the SASIE setting. As shown in Table 3, each component and proposed training loss all contribute to the final performance.

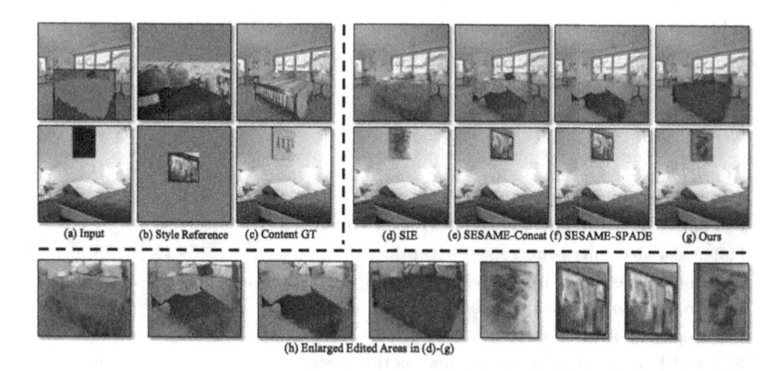

Fig. 6. Qualitative comparison results on ADE20K dataset, under SASIE. The baseline methods SESAME-Concat and SESAME-SPADE tend to directly paste the style reference object to the edited area, losing the original content information and generating artifacts (see "gray" parts of the edited area in (h)).

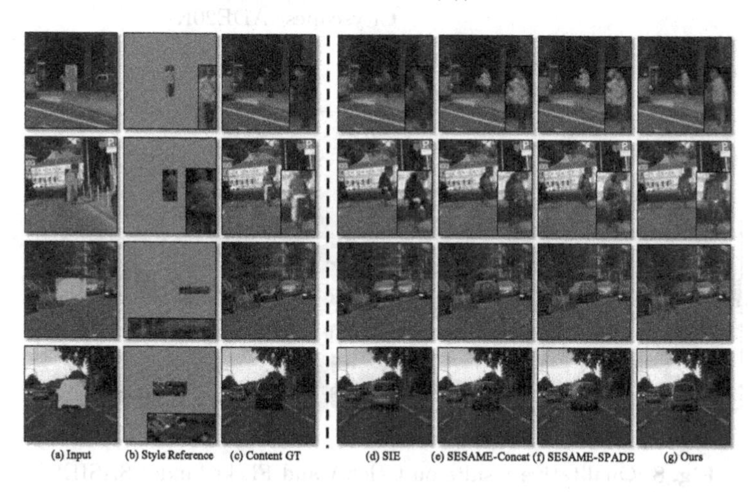

Fig. 7. Qualitative comparison results on Cityscapes dataset, under SASIE. The baseline methods SESAME-Concat and SESAME-SPADE tend to directly paste the style reference object to the edited area, losing the original content information and generating artifacts (see "gray" parts of the edited area in (e) and (f)).

Full Image Synthesis based Editing Method *vs.* Our SASIE based Method. As discussed in Sect. 2, the full image synthesis based methods, *e.g.*, the semantic layout to image translation [21], and the feature factorization [5, 24], can also edit object attributes. But since they rely on synthesizing all the pixels in the image, they unavoidably change the parts outside the edited areas. Instead, our method only locally changes the edited areas, while preserving other parts as the original image. As shown in Fig. 9, for "hair" editing, our SASIE based method only edits the hair part, while the full image synthesis based method [21] even changes the face and background.

Table 3. Ablation study. In the variants part, the upper are the variants of the structure, while the lower are the variants of the loss function.

Models	Components	Cityscapes Dataset		
		FID↓	mIoU (%)↑	Acc (%)↑
Full model	All modules *w.* all losses	**11.35**	**62.41**	**91.41**
Variants	**Vision transformer structure (ViT)** [9]	15.98	58.52	89.97
	Full model, V_c**(conv-encoder)** → V_c**(patch embedding)**	12.33	61.81	91.22
	Full model ***w/o.*** **intra-image self-attention**	14.20	61.76	91.32
	Full model, ***w/o.*** **inter-image self-attention *w.* SPADE layer**	11.68	61.85	91.37
	Full model, ***w/o.*** **inter-image self-attention *w.* Concat**	12.67	61.59	91.26
	Full model, ***w/o.*** **context loss**	11.63	62.03	91.37
	Full model, ***w/o.*** **intra-image reconstruction loss**	11.60	61.90	91.29

Table 4. User study results, under SASIE setting.

Models	User Preference (%)	
	Cityscapes	ADE20K
SESAME-SPADE	16.7	21.1
SESAME-Concat	27.8	27.6
Ours	**55.5**	**51.3**

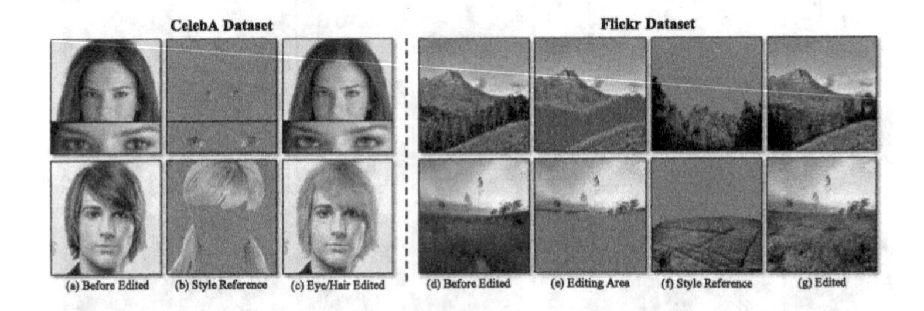

Fig. 8. Qualitative results on CelebA and Flickr, under SASIE.

Fig. 9. Comparison with the full image synthesis based method [21] for hair editing. Our SASIE based method can preserve the parts outside the edited hair area, *e.g.*, face and background.

Mixture of Traditional SIE and Style Transfer under SASIE. As shown in Fig. 4, and analyzed in Sect. 3.2, obvious artifacts, induced by naively mixing traditional SIE [28] and style transfer [14] methods under SASIE setting, indicate that the challenges i) image inpainting and ii) local style transfer of SASIE need to be resolved jointly, validating the necessity and effectiveness of our approach.

5 Conclusion

In this paper, we have proposed a novel SASIE strategy, enabling customizable image manipulation. Unlike the traditional semantic image editing strategy that only focuses on editing "*content*", our SASIE considers both "*content*" and "*style*" of the edited areas. Then we developed a transformer-based method for SASIE, facilitating the intra-/inter-image knowledge transfer. Extensive experiments on various datasets show the effectiveness of our method.

References

1. An, J., Huang, S., Song, Y., Dou, D., Liu, W., Luo, J.: ArtFlow: unbiased image style transfer via reversible neural flows. In: CVPR (2021)
2. Bau, D., et al.: Semantic photo manipulation with a generative image prior. ACM TOG **38**(4) (2019)
3. Carion, N., Massa, F., Synnaeve, G., Usunier, N., Kirillov, A., Zagoruyko, S.: End-to-end object detection with transformers. In: Vedaldi, A., Bischof, H., Brox, T., Frahm, J.-M. (eds.) ECCV 2020. LNCS, vol. 12346, pp. 213–229. Springer, Cham (2020). https://doi.org/10.1007/978-3-030-58452-8_13
4. Cheng, J., Jaiswal, A., Wu, Y., Natarajan, P., Natarajan, P.: Style-aware normalized loss for improving arbitrary style transfer. In: CVPR (2021)
5. Collins, E., Bala, R., Price, B., Susstrunk, S.: Editing in style: uncovering the local semantics of GANs. In: CVPR (2020)
6. Cordts, M., et al.: The cityscapes dataset for semantic urban scene understanding. In: CVPR (2016)
7. Deng, Y., Tang, F., Pan, X., Dong, W., Ma, C., Xu, C.: StyTr^2: unbiased image style transfer with transformers. arXiv preprint arXiv:2105.14576 (2021)
8. Dhamo, H., et al.: Semantic image manipulation using scene graphs. In: CVPR (2020)
9. Dosovitskiy, A., et al.: An image is worth 16×16 words: transformers for image recognition at scale. In: ICLR (2021)
10. Gatys, L.A., Ecker, A.S., Bethge, M.: Image style transfer using convolutional neural networks. In: CVPR (2016)
11. Goodfellow, I., et al.: Generative adversarial nets. In: NeurIPS (2014)
12. He, K., Zhang, X., Ren, S., Sun, J.: Deep residual learning for image recognition. In: CVPR (2016)
13. Hong, S., Yan, X., Huang, T.E., Lee, H.: Learning hierarchical semantic image manipulation through structured representations. In: NeurIPS (2018)
14. Huang, X., Belongie, S.: Arbitrary style transfer in real-time with adaptive instance normalization. In: ICCV (2017)
15. Huang, X., Liu, M.-Y., Belongie, S., Kautz, J.: Multimodal unsupervised image-to-image translation. In: Ferrari, V., Hebert, M., Sminchisescu, C., Weiss, Y. (eds.) ECCV 2018. LNCS, vol. 11207, pp. 179–196. Springer, Cham (2018). https://doi.org/10.1007/978-3-030-01219-9_11

16. Isola, P., Zhu, J.Y., Zhou, T., Efros, A.A.: Image-to-image translation with conditional adversarial networks. In: CVPR (2017)
17. Karras, T., et al.: Alias-free generative adversarial networks. In: NeurIPS (2021)
18. Karras, T., Laine, S., Aila, T.: A style-based generator architecture for generative adversarial networks. In: CVPR (2019)
19. Karras, T., Laine, S., Aittala, M., Hellsten, J., Lehtinen, J., Aila, T.: Analyzing and improving the image quality of StyleGAN. In: CVPR (2020)
20. Kurzman, L., Vazquez, D., Laradji, I.: Class-based styling: real-time localized style transfer with semantic segmentation. In: ICCVW (2019)
21. Lee, C.H., Liu, Z., Wu, L., Luo, P.: MaskGAN: towards diverse and interactive facial image manipulation. In: CVPR (2020)
22. Li, Y., Liu, M.-Y., Li, X., Yang, M.-H., Kautz, J.: A closed-form solution to photorealistic image stylization. In: Ferrari, V., Hebert, M., Sminchisescu, C., Weiss, Y. (eds.) ECCV 2018. LNCS, vol. 11207, pp. 468–483. Springer, Cham (2018). https://doi.org/10.1007/978-3-030-01219-9_28
23. Lim, J.H., Ye, J.C.: Geometric GAN. arXiv preprint arXiv:1705.02894 (2017)
24. Ling, H., Kreis, K., Li, D., Kim, S.W., Torralba, A., Fidler, S.: EditGAN: high-precision semantic image editing. In: NeurIPS (2021)
25. Liu, Z., et al.: Swin transformer: hierarchical vision transformer using shifted windows. In: ICCV (2021)
26. Liu, Z., Luo, P., Wang, X., Tang, X.: Deep learning face attributes in the wild. In: ICCV (2015)
27. Miyato, T., Kataoka, T., Koyama, M., Yoshida, Y.: Spectral normalization for generative adversarial networks. In: ICLR (2018)
28. Ntavelis, E., Romero, A., Kastanis, I., Van Gool, L., Timofte, R.: SESAME: semantic editing of scenes by adding, manipulating or erasing objects. In: Vedaldi, A., Bischof, H., Brox, T., Frahm, J.-M. (eds.) ECCV 2020. LNCS, vol. 12367, pp. 394–411. Springer, Cham (2020). https://doi.org/10.1007/978-3-030-58542-6_24
29. Park, T., Liu, M.Y., Wang, T.C., Zhu, J.Y.: Semantic image synthesis with spatially-adaptive normalization. In: CVPR (2019)
30. Park, T., et al.: Swapping autoencoder for deep image manipulation. In: NeurIPS (2020)
31. Simonyan, K., Zisserman, A.: Very deep convolutional networks for large-scale image recognition. In: ICLR (2015)
32. Strudel, R., Garcia, R., Laptev, I., Schmid, C.: Segmenter: transformer for semantic segmentation. In: ICCV (2021)
33. Svoboda, J., Anoosheh, A., Osendorfer, C., Masci, J.: Two-stage peer-regularized feature recombination for arbitrary image style transfer. In: CVPR (2020)
34. Tran, D., Ranganath, R., Blei, D.M.: Hierarchical implicit models and likelihood-free variational inference. In: NeurIPS (2017)
35. Vaswani, A., et al.: Attention is all you need. In: NeurIPS (2017)
36. Wang, T.C., Liu, M.Y., Zhu, J.Y., Tao, A., Kautz, J., Catanzaro, B.: High-resolution image synthesis and semantic manipulation with conditional GANs. In: CVPR (2018)
37. Xie, E., Wang, W., Yu, Z., Anandkumar, A., Alvarez, J.M., Luo, P.: SegFormer: simple and efficient design for semantic segmentation with transformers. In: NeurIPS (2021)
38. Yoo, J., Uh, Y., Chun, S., Kang, B., Ha, J.W.: Photorealistic style transfer via wavelet transforms. In: ICCV (2019)
39. Zhang, R., et al.: Real-time user-guided image colorization with learned deep priors. ACM TOG **9**(4) (2017)

40. Zhou, B., Zhao, H., Puig, X., Fidler, S., Barriuso, A., Torralba, A.: Scene parsing through ade20k dataset. In: CVPR (2017)
41. Zhou, B., et al.: Semantic understanding of scenes through the ADE20K dataset. IJCV **127**(3), 302–321 (2019)
42. Zhu, J.-Y., Krähenbühl, P., Shechtman, E., Efros, A.A.: Generative visual manipulation on the natural image manifold. In: Leibe, B., Matas, J., Sebe, N., Welling, M. (eds.) ECCV 2016. LNCS, vol. 9909, pp. 597–613. Springer, Cham (2016). https://doi.org/10.1007/978-3-319-46454-1_36
43. Zhu, J.Y., Park, T., Isola, P., Efros, A.A.: Unpaired image-to-image translation using cycle-consistent adversarial networks. In: ICCV (2017)
44. Zhu, P., Abdal, R., Qin, Y., Wonka, P.: Sean: image synthesis with semantic region-adaptive normalization. In: CVPR (2020)

Third Time's the Charm? Image and Video Editing with StyleGAN3

Yuval Alaluf[1(✉)], Or Patashnik[1], Zongze Wu[2], Asif Zamir[1], Eli Shechtman[3], Dani Lischinski[2], and Daniel Cohen-Or[1]

[1] Tel-Aviv University, Tel Aviv, Israel
yuvalalaluf@gmail.com
[2] Hebrew University of Jerusalem, Jerusalem, Israel
[3] Adobe Research, Cambridge, USA

Abstract. StyleGAN is arguably one of the most intriguing and well-studied generative models, demonstrating impressive performance in image generation, inversion, and manipulation. In this work, we explore the recent StyleGAN3 architecture, compare it to its predecessor, and investigate its unique advantages, as well as drawbacks. In particular, we demonstrate that while StyleGAN3 can be trained on unaligned data, one can still use aligned data for training, without hindering the ability to generate unaligned imagery. Next, our analysis of the disentanglement of the different latent spaces of StyleGAN3 indicates that the commonly used $\mathcal{W}/\mathcal{W}+$ spaces are more entangled than their StyleGAN2 counterparts, underscoring the benefits of using the *StyleSpace* for fine-grained editing. Considering image inversion, we observe that existing encoder-based techniques struggle when trained on unaligned data. We therefore propose an encoding scheme trained solely on aligned data, yet can still invert unaligned images. Finally, we introduce a novel video inversion and editing workflow that leverages the capabilities of a fine-tuned StyleGAN3 generator to reduce texture sticking and expand the field of view of the edited video.

Keywords: Generative Adversarial Networks · Image and video editing

1 Introduction

In recent years, Generative Adversarial Networks (GANs) [13] have revolutionized image processing. Specifically, StyleGAN generators [19–22] synthesize exceedingly realistic images, and enable editing [3,4,7,10,24,28,46], and image-to-image translation [26,27,31], particularly in well-structured domains. StyleGAN architectures are notable for their semantically-rich, disentangled, and generally well-behaved latent spaces.

Y. Alaluf and O. Patashnik—Equal contribution.

Supplementary Information The online version contains supplementary material available at https://doi.org/10.1007/978-3-031-25063-7_13.

L. Karlinsky et al. (Eds.): ECCV 2022 Workshops, LNCS 13802, pp. 204–220, 2023.
https://doi.org/10.1007/978-3-031-25063-7_13

Fig. 1. Image editing with StyleGAN3. Using the recent StyleGAN3 generator, we edit unaligned input images across various domains using off-the-shelf editing techniques. Using a trained StyleGAN3 encoder, these techniques can likewise be used to edit real images and videos.

When it comes to video processing, new challenges arise. Editing should not only be disentangled and realistic, but also temporally consistent across frames. The texture-sticking phenomenon in StyleGAN1 and StyleGAN2 [20] hinders the temporal consistency and realism of generated and manipulated videos. For example, when interpolating within the latent space, the hair and face typically do not move in unison. The recent StyleGAN3 architecture [20] is specifically designed to overcome such texture-sticking, and additionally offers translation and rotation equivariance. Naturally, these unique properties make StyleGAN3 better suited for video processing than previous style-based generators. However, significant changes introduced in StyleGAN3 architecture raise many questions and new challenges. Central among these is the disentanglement of its latent spaces and the ability to accurately invert and edit real images.

In this paper, we analyze StyleGAN3, aiming at understanding and exploring its capabilities and performance. Some of the questions we attempt to answer are: How does the disentanglement of the latent representations in StyleGAN3 compare to StyleGAN2? Do the techniques devised for identifying latent editing controls still work? Inversion is a fundamental task required for editing *real* images, and has been extensively studied in the context of StyleGAN2 [1,2,5, 6,31,32,37,45]. We therefore examine how well these existing techniques can be adapted to achieve comparable performance with StyleGAN3, and, in particular, cope with the inversion of unaligned images.

In light of the translation and rotation equivariance provided by StyleGAN3, we also examine the differences between generators trained on aligned and on unaligned data. Perhaps surprisingly, we observe that both kinds of generators are comparable in terms of their ability to generate unaligned images and control their position and rotation. However, we find that using the aligned generators is preferable for tasks such as disentangled editing and inversion. We therefore leverage aligned generators in our proposed image and video inversion and editing workflows (see Fig. 1 for sample results).

Applying the insights gained in our analysis and experiments over still images, we propose a novel workflow for inverting and editing real videos using

StyleGAN3. Notably, we leverage the capabilities of StyleGAN3 to reduce texture sticking and expand the field of view when working on a video with a cropped subject.

2 Related Work

StyleGAN2 [22] features several semantically-rich latent spaces, which have been heavily studied and exploited in the context of image manipulation [3,10,15,33, 34,36] and image inversion [1,2,8,14,29,47]. In this work, we experiment with the image manipulation techniques from Shen *et al.* [33], Wu *et al.* [40] and Patashnik *et al.* [28] in the context of StyleGAN3 [20].

To manipulate real images they must first be projected into one of the StyleGAN latent spaces. We refer the reader to the exposition in Xia *et al.* [41] for a comprehensive review of GAN inversion and the applications it enables. In this work, we leverage existing encoder-based inversion techniques [5,31,37] for achieving more accurate inversions of images, and in particular unaligned images, using StyleGAN3. We additionally employ existing generator tuning techniques [32] to achieve higher-fidelity reconstructions of a wide range facial expressions, which we find to be necessary for inverting and editing videos.

In contrast to editing images with StyleGAN, few works have addressed video-based attribute editing. Yao *et al.* [43] use a pre-trained encoder and introduce a latent space transformer to achieve consistent edits of the inverted video frames. Two concurrent works [38,42] also explore temporally consistent video editing with StyleGAN2. However, previous methods rely on facial alignment, segmentation, or a dedicated post-processing pipeline. We are the first to attempt to leverage StyleGAN3's rotation and translation equivariance to achieve accurate and consistent video editing while reducing the overhead of previous techniques, especially when working with a cropped subject.

We refer readers to the supplementary materials for additional details on StyleGAN's latent spaces, inversion techniques, and editing capabilities it offers.

3 The StyleGAN3 Architecture

To better understand the capabilities of StyleGAN3 [20], it is important to understand the overall structure and function of the different components comprising the architecture. First, as in StyleGAN [21], a simple fully-connected mapping network translates an initial latent code $z \sim \mathcal{N}(0,1)^{512}$, into an intermediate code w residing in a learned latent space $\mathcal{W}$.

Compared to StyleGAN2 [22], StyleGAN3's synthesis network is composed of a fixed number of convolutional layers (16), irrespective of the output image resolution. We denote by $(w_0, ..., w_{15})$ the set of input codes passed to these layers. In StyleGAN3, the constant 4×4 input tensor from StyleGAN2 is replaced by Fourier features, that can be rotated and translated using four parameters $(\cos\alpha,\, \sin\alpha,\, x,\, y)$, obtained from w_0 via a learned affine layer. In the remaining

Fig. 2. (a) While StyleGAN3 trained on aligned data normally generates aligned images (leftmost image), translation and in-plane rotation can be controlled by applying an explicit transformation (r, t_x, t_y) over the Fourier features. (b) Pseudo-aligning images generated by unaligned generators. While these generators normally produce unaligned images (row 1), replacing w_0 with the average latent $\overline{w}$ yields roughly aligned images (row 2).

layers, each w_i is fed into an independently learned affine layer, which yields modulation factors used to adjust the convolutional kernel weights.

In StyleGAN2, the space spanned by the outputs of these affine layers has been referred to as the *StyleSpace* [40], or simply $\mathcal{S}$. In this work, we similarly define the $\mathcal{S}$ space of StyleGAN3, with 9,894 dimensions in total for a 1024×1024 generator. Finally, as in StyleGAN2, we may define the extended space $\mathcal{W}+$ [1] where a different latent code is inserted into each layer of the synthesis network.

Since the translation and in-plane rotation of the synthesized images are given by explicit parameters obtained from w_0, the result may be easily adjusted by concatenating another transformation. We parameterize this transformation using three parameters (r, t_x, t_y), where r is the rotation angle (in degrees), and t_x, t_y are the translation parameters, and denote the resulting image by:

$$y = G(w; (r, t_x, t_y)), \tag{1}$$

where, by default, $t_x = t_y = 0$ and $r = 0$. This transformation can be applied even in a generator trained solely on aligned data, enabling it to generate rotated and translated images, see Fig. 2a.

Conversely, generators trained on unaligned data may be "coerced" to generate roughly aligned images by setting w_0 to the generator's average latent code $\overline{w}$, i.e., given by $G((\overline{w}, w_1, ..., w_{15}); (0, 0, 0))$. Figure 2b demonstrates this idea on two different StyleGAN3 generators trained on unaligned FFHQ and AFHQ datasets, respectively. Intuitively, this approximate alignment may be due to the fact that the average input pose in the training distribution is roughly aligned and centered, combined with the fact that the translation and rotation transformations in StyleGAN3 are mainly controlled by the first layer. It is important to note that equivariance is at the core of the StyleGAN3 design: translations or rotations in earlier layers are preserved across later layers and appear in the generated output.

4 Analysis

4.1 Rotation Control

As discussed above, the latent code fed into the first layer, w_0, controls the translation and rotation of the image content. However, as illustrated in Fig. 2b, w_0 affects each image in a slightly different manner. For example, the leftmost human face is slightly rotated while the face in the third column is perfectly upright. This suggests that rotation is also affected by other layers of the generator, where it is entangled with other visual attributes.

To examine the extent of this phenomenon, we perform two experiments, illustrated in Fig. 3. First, we examine a series of images $G((w^*, w_1, w^*, ..., w^*))$ that differ only in their randomly sampled w_1 latent entry (top row). It may be seen that altering w_1 affects the in-plane rotation of the face, but this change is entangled also with other attributes, such as face shape and eyes. In our second experiment, we generate a series of images $G((w_0, w_1, w^*, ..., w^*))$, where w_0 and w_1 are held fixed, while the remaining latent entries are set to a randomly sampled code w^*. It may be seen that with both w_0 and w_1 fixed, the generated images all share the same head pose. Thus, we conclude that the subsequent layers do not appear to induce any further translation or rotation, and those are determined primarily by w_0 and w_1.

Fig. 3. The roles and entanglement of w_0 and w_1. Top row: altering only w_1 affects the in-plane rotation, as well as other visual aspects, implying they are entangled in w_1. Bottom row: holding w_0 and w_1 fixed while randomly sampling the remaining latent entries demonstrates that, for all practical purposes, the first two layers determine the translation and rotation.

4.2 Disentanglement Analysis

To analyze the disentanglement of the different latent spaces of StyleGAN3, we follow Wu *et al.* [40] and compute the DCI (disentanglement/completeness/informativeness) metrics [11] of each latent space. To compute the above metrics we employ pre-trained attribute regressors for various attributes, as described by Wu *et al.* [40]. Observe that accurately computing attribute scores for unaligned images is challenging given that the attribute classifiers were trained solely on aligned images. To this end, we provide the classifiers with pseudo-aligned images, generated as described in Sect. 3.

We report the DCI metrics for the $\mathcal{Z}$, $\mathcal{W}$, and $\mathcal{S}$ spaces in Table 1 for both aligned and unaligned StyleGAN3 generators trained on the FFHQ [21] dataset. $\mathcal{S}$ achieves the highest DCI scores for both StyleGAN3 generators, as it also

Table 1. DCI metrics for StyleGAN2 and StyleGAN3. For both StyleGAN architectures, and for aligned and unaligned datasets, the DCI scores (disentanglement/completeness/informativeness) improve consistently from the initial Gaussian noise $\mathcal{Z}$, through the intermediate space $\mathcal{W}$, and to the style parameters $\mathcal{S}$, which control channel-wise statistics ($\mathcal{S} > \mathcal{W} > \mathcal{Z}$).

Generator	Space	Disent.	Compl.	Inform.
StyleGAN2	$\mathcal{Z}$	0.31	0.21	0.72
StyleGAN2	$\mathcal{W}$	0.54	0.57	0.97
StyleGAN2	$\mathcal{S}$	**0.75**	**0.87**	**0.99**
StyleGAN3 (Aligned)	$\mathcal{Z}$	0.37	0.27	0.80
StyleGAN3 (Aligned)	$\mathcal{W}$	0.47	0.43	0.94
StyleGAN3 (Aligned)	$\mathcal{S}$	**0.89**	**0.76**	**0.99**
StyleGAN3 (Unaligned)	$\mathcal{Z}$	0.36	0.26	0.80
StyleGAN3 (Unaligned)	$\mathcal{W}$	0.45	0.41	0.94
StyleGAN3 (Unaligned)	$\mathcal{S}$	**0.79**	**0.85**	**0.99**

does for StyleGAN2. Furthermore, while gaps in D and C between $\mathcal{Z}$ and $\mathcal{W}$ are smaller in StyleGAN3 than they are in StyleGAN2, the gap between $\mathcal{W}$ and $\mathcal{S}$ is larger, suggesting that using $\mathcal{S}$ for editing may be even more beneficial in StyleGAN3.

Since most StyleGAN inversion methods invert images into the $\mathcal{W}+$ latent space, it is also beneficial to examine the DCI metrics for this extended latent space. To do so, we randomly sample a set of latent codes $w \in \mathcal{W}$ and concatenate them to form latent codes in $\mathcal{W}+$. However, we find the resulting generated images are unnatural (see the supplementary materials for examples). Moreover, applying the pre-trained DCI classifiers on such images results in inaccurate attribute scores, making the computed metrics unreliable.

5 Image Editing

In this section, we examine the effectiveness of various techniques for image editing with StyleGAN3, starting with the $\mathcal{W}$ and $\mathcal{W}+$ latent spaces, and proceeding to $\mathcal{S}$.

Editing via Linear Latent Directions. Here we use InterFaceGAN [33] for finding linear directions in $\mathcal{W}$ for aligned and unaligned StyleGAN3 generators. Editing aligned images is simple and follows the approach used in StyleGAN2 [22]: given a randomly sampled latent code $w \in \mathcal{W}$, an editing direction D, and a step size δ, the edited image is generated by $G_{aligned}(w + \delta D; (0, 0, 0))$, where $G_{aligned}$ is the aligned generator.

As for unaligned images, there are two options. First, one may simply use an unaligned generator. Yet, one problem that arises in doing so is the fact that the attribute scores needed to learn these directions are obtained from

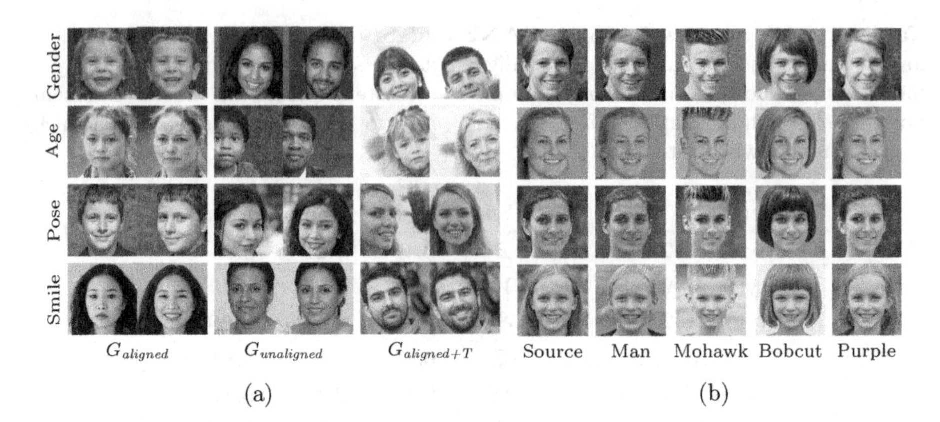

Fig. 4. (a) **Linear editing in $\mathcal{W}$.** Editing synthetic images using InterFaceGAN [33] directions in $\mathcal{W}$. Editing unaligned images can be done either using an unaligned generator $G_{unaligned}$, or using an aligned generator with an extra transformation $G_{aligned+T}$. (b) **Non-linear editing in $\mathcal{W}+$.** We edit images using the StyleCLIP mapping technique with StyleGAN3 trained on aligned faces. Even with non-linear editing paths, the edits are still entangled: local edits (e.g., expression/hairstyle) alter other attributes (e.g., background/identity).

classifiers pre-trained on aligned images. The scores produced by these classifiers on unaligned images may be inaccurate, resulting in poorly-learned directions in $\mathcal{W}$. To assist the pre-trained classifiers, we generate pseudo-aligned images by replacing w_0 with the generator's average latent code $\overline{w}$, as shown in Sect. 3. The image generated by this modified latent code is then passed to the pre-trained classifier to obtain the original latent's attribute score. Yet, another problem with using the unaligned generator is that it requires learning a separate set of directions.

A second approach to mitigate the above overhead is to generate images using the aligned generator, but apply the user-defined transformations to control the rotation and placement of the generated object. Specifically, an edited unaligned image can be synthesized by $G_{aligned}(w + \delta D; (r, t_x, t_y))$, where (r, t_x, t_y) is controlled by the user. This approach gives the added benefit that the same latent directions may be used to edit both aligned and unaligned images.

In Fig. 4a we provide editing results obtained using the three approaches above. Notably, it is possible to achieve comparable edits on unaligned images via both the aligned and the unaligned generators. We also find that linear directions found in the latent space of $G_{unaligned}$ to be generally more entangled than those found in the latent space of $G_{aligned}$. This is most notable in the "smile" direction found in $G_{unaligned}$, see the last row. We attribute this entanglement to two factors: (1) the pseudo-aligned images may still be out-of-domain with respect to the classifier trained on aligned images, resulting in less accurate attribute scores; and (2) the linear editing directions make it more challenging to attain disentangled editing.

Given the insight that unaligned images may be edited using a single aligned generator and the fact that the aligned generator produces higher quality images (as shown in StyleGAN3 [20]), we focus our subsequent analysis on the aligned generator.

Editing via Non-linear Latent Paths. Various works have demonstrated that editing images via non-linear latent paths typically results in more faithful, disentangled edits [3,16]. Following these works, we now explore learning non-linear latent editing paths within the $\mathcal{W}+$ latent space using the StyleCLIP mapper technique [28]. As shown in Fig. 4b, the resulting edits are still entangled. For example, the image background typically changes across the different edits, even for local edits such as "angry". These results lead us to explore whether editing within the $\mathcal{S}$ space of StyleGAN3 achieves latent edits that are more disentangled than those achievable with $\mathcal{W}$ and $\mathcal{W}+$.

Editing via Latent Directions in $\mathcal{S}$ Recall that our DCI analysis (Table 1) indicates that the $\mathcal{S}$ space is more disentangled and complete than the $\mathcal{W}$ latent spaces. Here, we examine whether this finding extends to the editing quality of these spaces, particularly in terms of editing disentanglement. To this end, we find global linear editing directions in $\mathcal{S}$ using StyleCLIP [28].

Fig. 5. Editing in $\mathcal{S}$. We edit synthetic images using the StyleCLIP [28] global directions technique using StyleGAN3 generators trained on the FFHQ [21], AFHQv2 [9,20], and Landscapes HQ [35] datasets.

Figure 5 demonstrates that, in the domain of human faces, editing in $\mathcal{S}$ results in disentangled edits for both the aligned and unaligned StyleGAN3 images. Particularly, notice how the image backgrounds are much better preserved compared to the $\mathcal{W}$-based editing. Further, observe that the face identity is well-preserved for unrelated edits and that local edits, such as those changing hairstyle and expression, do not alter unrelated image regions (e.g., expression is consistent across the "gender", "hi-top fade", and "tanned" edits). Notably, this disentanglement also holds for other domains such as animal faces (AFHQv2 [9]) and landscapes (Landscapes HQ [35]). When editing animals, the fur color, pose, and backgrounds are well-preserved under the various edits. Additionally, altering the landscapes preserves key contents of the original image, such as the lake (top) or the road (bottom).

6 StyleGAN3 Inversion

In this section, we address the task of inverting a pre-trained StyleGAN3 generator G. In other words, given a target image x, we seek a latent code $\hat{w}$ that optimally reconstructs it:

$$\hat{w} = \arg\min_{w} \mathcal{L}\left(x, G(w; (r, t_x, t_y))\right), \tag{2}$$

where $\mathcal{L}$ is the L_2 or LPIPS [44] reconstruction loss.

Motivated by the goal of employing StyleGAN3 for editing real videos, solving the inversion task via a learned encoder (as opposed to latent vector optimization) may assist in achieving better temporal consistency due to its natural smoothness and bias for learning lower frequency representations [30,38].

More formally, we seek to train an encoder E over a large set of images $\{x_i\}_{i=1}^{N}$ for minimizing the objective:

$$\sum_{i=1}^{N} \mathcal{L}(x_i, G(E(x_i))), \tag{3}$$

where $E(x)$ encodes an input image x into a latent code w.

As discussed in Sect. 3, having obtained the latent code $w = E(x)$, an additional transformation may be passed to the generator to control the translation and rotation of the reconstructed image $y = G(w; (r, t_x, t_y))$. Finally, some latent manipulation f may be applied over this latent code to obtain an edited image

$$y_{edit} = G(f(w); (r, t_x, t_y)). \tag{4}$$

6.1 Designing the Encoder Network

To enable the encoding and editing of aligned and unaligned images (such as those found in a video sequence), our inversion scheme must support the generation of both input types. A natural first attempt at doing so is to design an encoder trained on both types of images, paired with an unaligned generator. Specifically, one can employ the training schemes of existing StyleGAN2 encoders [5,31,37] to minimize the objective given in Eq. (3) for unaligned inputs. Yet, we find that such a training scheme struggles in capturing the high variability of the unaligned facial images, resulting in poor reconstructions, see supplementary materials for an ablation study of such a design.

Encoding Unaligned Images. We instead choose to leverage an aligned StyleGAN3 generator and design an encoder trained solely on *aligned* images. As previously shown, this scheme can then be used for editing and synthesizing both aligned and unaligned imagery. In this formulation, the encoder no longer needs to correctly capture the highly-variable placement and pose of the unaligned input images. This in turn simplifies the encoder's training objective, allowing it to instead focus on faithfully capturing the input identity and other image features.

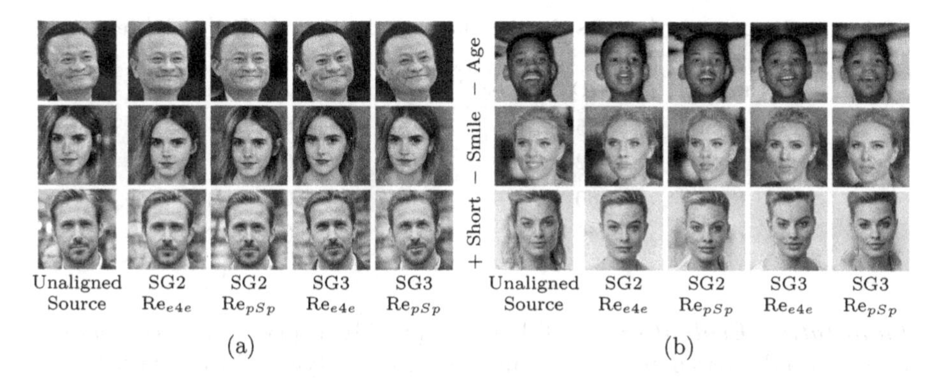

Fig. 6. (a) **Reconstruction quality comparison** between ReStyle [5] encoders trained for inverting StyleGAN2 and StyleGAN3 generators. (b) **Editing comparison.** We perform various edits [28,33] over codes obtained by each method.

Given the encoder trained to reconstruct aligned images, we are left with the question of how to extend this encoding scheme to support the encoding and editing of unaligned images at inference time. Assume we have a given unaligned image $x_{unaligned}$. We begin by using an off-the-shelf facial detector [23] to detect and align the image, resulting in an aligned version of the input, denoted by $x_{aligned}$. We then predict the translation (t_x, t_y) and rotation r between $x_{aligned}$ and $x_{unaligned}$ by detecting and aligning the eyes in the two images. We refer the reader to the supplementary for details on computing these parameters.

Finally, the inversion and reconstruction of $x_{unaligned}$ are given by:

$$w_{aligned} = E(x_{aligned}) \tag{5}$$

$$y_{unaligned} = G(w_{aligned}; (r, t_x, t_y)). \tag{6}$$

Observe that while the encoder receives the aligned image $x_{aligned}$, the reconstruction is able to capture the placement and rotation of the unaligned input through the use of the transformation (r, t_x, t_y). As such, our inversion scheme, although trained solely on aligned images, is able to faithfully encode *both* aligned and unaligned images by leveraging the unique design of StyleGAN3. Please see the supplementary for additional details on our inversion scheme.

6.2 Inverting Images into StyleGAN3

Qualitative Evaluation. As shown in Fig. 6a, our StyleGAN3 encoders attain visually comparable results to their StyleGAN2 counterparts. Observe that with StyleGAN3 we are able to faithfully reproduce the input position, even when given aligned inputs, by using our landmark-based predicted transformations.

Table 2. Reconstruction results measured on the CelebA-HQ [18,25] test set.

Method	↑ ID	↑ MS-SSIM	↓ LPIPS	↓ L_2	Time (s)
SG2 ReStyle$_{pSp}$	0.66	0.79	0.13	0.03	0.37
SG2 ReStyle$_{e4e}$	0.52	0.74	0.19	0.04	0.37
SG3 ReStyle$_{pSp}$	0.60	0.77	0.17	0.03	0.52
SG3 ReStyle$_{e4e}$	0.49	0.70	0.22	0.06	0.52

Quantitative Evaluation. In Table 2 we provide a quantitative comparison between encoder-based inversion techniques for both StyleGAN2 and StyleGAN3 generators on the human facial domain. Since StyleGAN2 is limited to encoding aligned images, we perform our evaluation on the CelebA-HQ [18,25] test set. In addition to the inference time required by each inversion technique, we report the L_2 distance, the LPIPS [44] distance, identity similarity [17], and MS-SSIM [39] score between the reconstructions and their sources. Our StyleGAN3 encoders reach a slightly worse performance compared to the StyleGAN2 encoders. We believe the higher difficulty in inverting StyleGAN3 is in part due to its less well-behaved $\mathcal{W}+$ latent space. This is also supported by our experiment in Sect. 4.2, where we observed that the quality of the generated images quickly deteriorates as we move away from the $\mathcal{W}$ space. We believe this quick collapse of the latent space contributes to the challenge of training inversion encoders for StyleGAN3.

Editability via Latent Space Manipulation. We now turn to evaluating the *editability* of our ReStyle encoders for StyleGAN3. As illustrated in Fig. 6b, our Restyle$_{e4e}$ encoder achieves realistic and meaningful edits while preserving the input identity. This is in contrast to Restyle$_{pSp}$, which despite achieving high-quality reconstructions, yields visibly less editable inversions. Notably, observe that in StyleGAN3, the gap in editing quality achieved by ReStyle$_{e4e}$ compared to ReStyle$_{pSp}$ is much larger compared to in StyleGAN2. This is most evident in artifacts along the hair in the last row and hints at the increased importance of inverting into well-behaved latent regions.

7 Inverting and Editing Videos with StyleGAN3

Our evaluations in the previous section point to StyleGAN2 as still being the default choice when operating over images. We now extend our inversion method to encoding and editing videos. In doing so, we hope to better understand where the design of StyleGAN3 may provide users an advantage over its predecessor. This extension introduces two central challenges. First, the reconstructed and edited video frames should be *temporally consistent*, which may be difficult to attain when inverting each frame independently. Second, individual frames of a human face video often feature more challenging facial expressions than those found in still image training sets, (e.g., closed eyes or a mouth open mid-speech). These challenges must be addressed, regardless of the architecture. Using StyleGAN3 is appealing because it reduces texture sticking and inherently handles

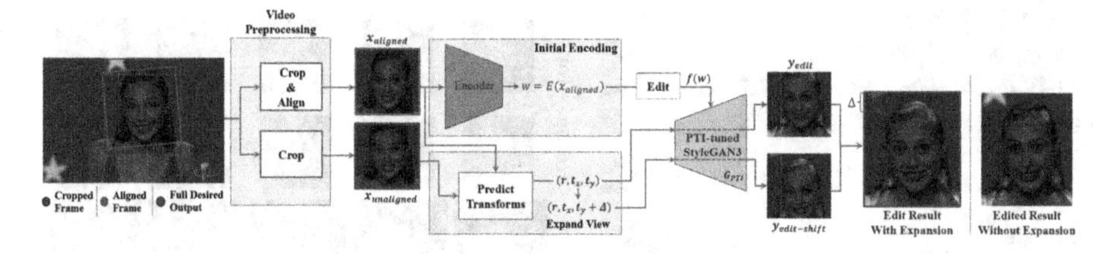

Fig. 7. An overview of our video editing workflow. The region containing the face is first cropped (red frame) and aligned (green frame) using an off-the-shelf detector [23]. Note that the top of the head is cropped out. The aligned frame is inverted by our encoder to a latent code $w = E(x_{aligned})$, and undergoes some manipulation f (e.g., hair color), yielding an edited code $f(w)$. In parallel, we compute the transformation (r, t_x, t_y) between $x_{aligned}$ and $x_{unaligned}$. Given $f(w)$ and (r, t_x, t_y), the edited frame is generated by $y_{edit} = G_{PTI}(f(w); (r, t_x, t_y))$. To fix the over-cropping, we generate an additional *shifted* image by translating the Fourier features by some value Δ (marked in orange). The two generated images are then merged into an extended frame comprising the entire edited head shown on the right. In the rightmost image, we show the edited result obtained *without* expansion. Observe that the edited result is not uniform along the entirety of the hair region when pasted back to the original context. (Color figure online)

varying face positions and rotations. Additionally, as we demonstrate below, StyleGAN3 may be leveraged to increase the field of view, resulting in wide-view reconstructed and edited videos, rather than close-ups of an individual. This even enables the faithful edit of attributes that partially "spill out" of the frame. Below, we describe our end-to-end video encoding and editing pipeline, summarized in Fig. 7.

Video Preprocessing. Given an input video, we begin by cropping each video frame to be compatible with the input head size expected by StyleGAN3. For achieving a stable video that looks as if it was captured from a non-moving camera, we crop a fixed bounding box across all video frames (as illustrated by the red bounding box in Fig. 7). We denote the resulting cropped images by $\{x_{i,unaligned}\}_{i=1}^{N}$. To invert each frame using our encoder, we additionally align each frame (green box in Fig. 7), yielding images $\{x_{i,aligned}\}_{i=1}^{N}$. We also compute the transformation $(r_i, t_{x,i}, t_{y,i})$ between each $(x_{i,aligned}, x_{i,unaligned})$ pair.

Initial Video Encoding. We use our trained ReStyle$_{e4e}$ encoder E to obtain the initial frame inversions $w_i = E(x_{i,aligned})$, whose unaligned reconstructions are given by $y_i = G(w_i; (r_i, t_{x,i}, t_{y,i}))$. We can additionally apply some manipulation f to obtain an edited version of the input frame: $y_{i,edit} = G(f(w_i); (r_i, t_{x,i}, t_{y,i}))$.

Latent Vector Smoothing. Inverting each frame independently may result in inconsistencies between successive reconstructed frames. This may be caused by the pre-processing alignment, the encoder network itself, or by the manipulation applied on the inverted latent codes. To mitigate temporal discontinuities, we temporally smooth the inverted, edited latent codes $f(w_i)$ and the

Fig. 8. (a) **Video editing results.** Results of our video editing pipeline with StyleGAN3. (b) Video reconstruction results obtained using our FOV technique. In row 2 we provide the unaligned reconstructions. In row 3 we provide the expanded video reconstruction. Additional results in the supplementary materials.

predicted transformation matrix T_i applied on the Fourier features and derived by $(r_i, t_{x,i}, t_{y,i})$, using a weighted moving average:

$$w_{i,smooth} = \sum_{j=i-2}^{i+2} \mu_j f(w_j) \quad \text{and} \quad T_{i,smooth} = \sum_{j=i-2}^{i+2} \mu_j T_j \tag{7}$$

where $[\mu_{i-2}, \mu_{i-1}, \mu_i, \mu_{i+1}, \mu_{i+2}] = \frac{1}{4}[0.25, 0.75, 1, 0.75, 0.5]$ for all i. We find that this smoothing operation improves temporal coherence without harming the reconstruction quality.

Pivotal Tuning for Improved Reconstructions. To further improve the frame reconstructions, we adopt the pivotal tuning inversion (PTI) method [32]. Specifically, the initial inversions are used for fine-tuning the weights of the StyleGAN3 generator to achieve better reconstructions of the input frames [38]. Note, while the encoder network is trained to reconstruct aligned images, we perform the PTI fine-tuning using the original *unaligned* images. That is, when performing PTI, losses are computed between the $x_{i,unaligned}$ images and their refined reconstructions given by:

$$y_{i,\text{PTI}} = G_{\text{PTI}}(w_i; (r_i, t_{x,i}, t_{y,i})), \tag{8}$$

where G_{PTI} is the PTI-modified generator.

Bringing It Together. Having obtained the smoothed edited latent codes, the corresponding smoothed transformations, and fine-tuned generator, we can generate the unified edited video. The final edited i-th frame is given by:

$$y_{i,final} = G_{\text{PTI}}(w_{i,smooth}; T_{i,smooth}). \tag{9}$$

We provide reconstruction and editing results in Fig. 8a and in the supplementary materials. In addition, by training StyleGAN-NADA [12] on G_{PTI} for a given video and text prompt, we can generate edited videos in various styles. We refer the reader to the supplementary results along with additional details.

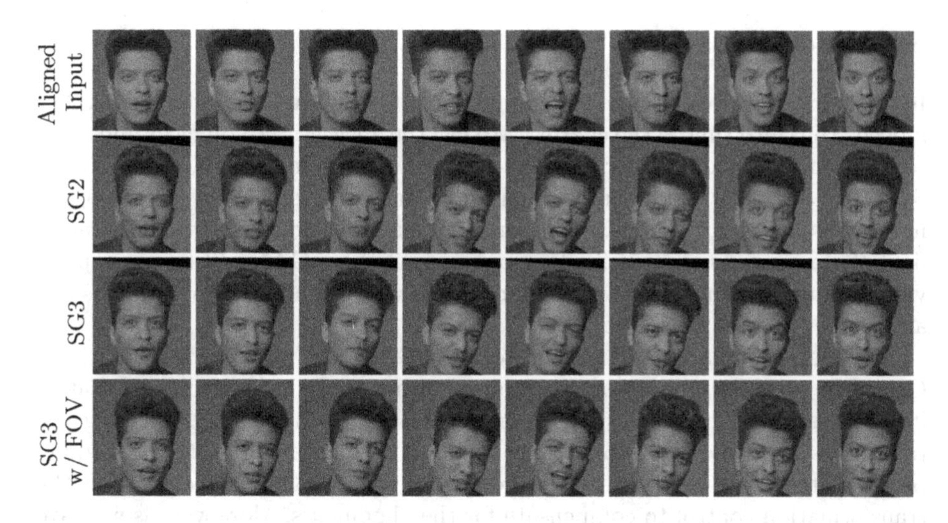

Fig. 9. Video editing comparison. Results of inverting and editing a video with StyleGAN2 (SG2) and StyleGAN3 (SG3). For StyleGAN3, we show results with and without our FOV expansion technique.

Expanding the Field-of-View. We now describe how we can expand the field-of-view (FOV) of the video reconstruction. Denote by Δ the desired expansion (illustrated in the original frame of Fig. 7). To expand the FOV, we construct a transformation matrix T_{Δ}. For example, for a vertical expansion of the frame, we define a transform matrix corresponding to a vertical shift derived from the parameters $(0, 0, \Delta)$. For each input frame, we then generate two images: $y = G_{\text{PTI}}(w_{i,smooth}; T_{i,smooth})$ and $y_{shift} = G_{\text{PTI}}(w_{i,smooth}; T_{\Delta} \cdot T_{i,smooth})$. Finally, we adjoin to y the added non-overlapping parts from y_{shift}, obtaining the wider output frame. Results of such an expansion are shown in Fig. 8b where we demonstrate the ability to reconstruct the entirety of the individual's head.

Notice, images generated by StyleGAN2 are aligned, and as such, attributes we may wish to edit may overflow outside the frame. Since the images are aligned, we must project the edited frame back to the original context. Doing so, we may obtain a mismatch between regions within the generated image boundary (which were edited) and those outside the boundary (which were untouched).

This idea is demonstrated in Fig. 9. Here, we use Tzaban *et al.* [38] to edit a video with StyleGAN2 and use our encoder and FOV expansion technique for editing a video with StyleGAN3. As can be seen, the man's hair overflows outside of the original aligned image. Therefore with StyleGAN2, when pasting the edited image back into its original context, we get a non-uniform edit along the hair. In StyleGAN3, however, our FOV expansion technique allows editing the desired attribute in its entirety, resulting in a full, coherent edit.

8 Conclusions

In this work, we have explored the competence of StyleGAN3, wondering whether indeed "the third time is the charm". We feel the answer is still unclear and more research may be required for a definite answer. On the one hand, the ability of StyleGAN3 to control the translation and rotation of generated images opens new intriguing opportunities. The prominent example explored is the generative field-of-view expansion, which allows one to apply StyleGAN editing on cropped video frames in a more consistent manner, alleviating the need for cumbersome and challenging seamless stitching.

On the other hand, the benefits of StyleGAN3 do come with limitations. Generally speaking, its latent space is somewhat more entangled than that of its predecessors. This makes the inversion task more challenging, affecting the robustness of frame inversions along a video. We have shown that this may be alleviated by applying the inversion on aligned images and exploiting the transformation control to compensate for the alignments. Moreover, as we have shown, training the encoder solely on aligned images does not introduce additional overheads, and even gains higher-quality synthesis.

We have naturally focused on facial images and videos. More research is required to investigate the power of StyleGAN3 for other domains. In particular, avoiding texture-sticking may be significant in videos of outdoor scenes containing high-frequency textures, like foliage or water streams. Another intriguing direction is to consider an encoder architecture that mirrors the StyleGAN3 generator, and might also employ 1×1 convolutions and Fourier features.

References

1. Abdal, R., Qin, Y., Wonka, P.: Image2stylegan: how to embed images into the stylegan latent space? In: Proceedings of the IEEE International Conference on Computer Vision, pp. 4432–4441 (2019)
2. Abdal, R., Qin, Y., Wonka, P.: Image2stylegan++: how to edit the embedded images? In: Proceedings of the IEEE/CVF Conference on Computer Vision and Pattern Recognition. pp. 8296–8305 (2020)
3. Abdal, R., Zhu, P., Mitra, N., Wonka, P.: Styleflow: attribute-conditioned exploration of stylegan-generated images using conditional continuous normalizing flows (2020)
4. Alaluf, Y., Patashnik, O., Cohen-Or, D.: Only a matter of style: age transformation using a style-based regression model. ACM Trans. Graph. **40**(4) (2021). https://doi.org/10.1145/3450626.3459805
5. Alaluf, Y., Patashnik, O., Cohen-Or, D.: Restyle: a residual-based stylegan encoder via iterative refinement. In: Proceedings of the IEEE/CVF International Conference on Computer Vision (ICCV), October 2021
6. Alaluf, Y., Tov, O., Mokady, R., Gal, R., Bermano, A.H.: Hyperstyle: Stylegan inversion with hypernetworks for real image editing (2021)
7. Bau, D., et al.: Paint by word (2021)
8. Bau, D., et al.: Semantic photo manipulation with a generative image prior **38**(4) (2019). https://doi.org/10.1145/3306346.3323023, https://doi.org/10.1145/3306346.3323023

9. Choi, Y., Uh, Y., Yoo, J., Ha, J.W.: Stargan v2: diverse image synthesis for multiple domains (2020)
10. Collins, E., Bala, R., Price, B., Süsstrunk, S.: Editing in style: uncovering the local semantics of GANs. In: 2020 IEEE/CVF Conference on Computer Vision and Pattern Recognition (CVPR), pp. 5770–5779 (2020)
11. Eastwood, C., Williams, C.K.: A framework for the quantitative evaluation of disentangled representations. In: International Conference on Learning Representations (2018)
12. Gal, R., Patashnik, O., Maron, H., Chechik, G., Cohen-Or, D.: Stylegan-nada: clip-guided domain adaptation of image generators (2021)
13. Goodfellow, I.J., et al.: Generative adversarial nets. In: Proceedings of the 27th International Conference on Neural Information Processing Systems - Volume 2. NIPS 2014, Cambridge, MA, USA, pp. 2672–2680. MIT Press (2014)
14. Guan, S., Tai, Y., Ni, B., Zhu, F., Huang, F., Yang, X.: Collaborative learning for faster stylegan embedding. arXiv preprint arXiv:2007.01758 (2020)
15. Härkönen, E., Hertzmann, A., Lehtinen, J., Paris, S.: Ganspace: discovering interpretable GAN controls. arXiv preprint arXiv:2004.02546 (2020)
16. Hou, X., Zhang, X., Shen, L., Lai, Z., Wan, J.: Guidedstyle: Attribute knowledge guided style manipulation for semantic face editing (2020)
17. Huang, Y., et al.: Curricularface: adaptive curriculum learning loss for deep face recognition. In: Proceedings of the IEEE/CVF Conference on Computer Vision and Pattern Recognition, pp. 5901–5910 (2020)
18. Karras, T., Aila, T., Laine, S., Lehtinen, J.: Progressive growing of gans for improved quality, stability, and variation. arXiv preprint arXiv:1710.10196 (2017)
19. Karras, T., Aittala, M., Hellsten, J., Laine, S., Lehtinen, J., Aila, T.: Training generative adversarial networks with limited data (2020)
20. Karras, T., et al.: Alias-free generative adversarial networks. CoRR abs/2106.12423 (2021)
21. Karras, T., Laine, S., Aila, T.: A style-based generator architecture for generative adversarial networks. In: Proceedings of the IEEE Conference on Computer Vision and Pattern Recognition, pp. 4401–4410 (2019)
22. Karras, T., Laine, S., Aittala, M., Hellsten, J., Lehtinen, J., Aila, T.: Analyzing and improving the image quality of stylegan. In: Proceedings of the IEEE/CVF Conference on Computer Vision and Pattern Recognition, pp. 8110–8119 (2020)
23. King, D.E.: DLIB-ML: a machine learning toolkit. J. Mach. Learn. Res. **10**, 1755–1758 (2009)
24. Ling, H., Kreis, K., Li, D., Kim, S.W., Torralba, A., Fidler, S.: Editgan: high-precision semantic image editing. In: Advances in Neural Information Processing Systems (NeurIPS) (2021)
25. Liu, Z., Luo, P., Wang, X., Tang, X.: Deep learning face attributes in the wild (2015)
26. Menon, S., Damian, A., Hu, S., Ravi, N., Rudin, C.: Pulse: self-supervised photo upsampling via latent space exploration of generative models. In: Proceedings of the IEEE/CVF Conference on Computer Vision and Pattern Recognition, pp. 2437–2445 (2020)
27. Park, T., et al.: Swapping autoencoder for deep image manipulation. arXiv preprint arXiv:2007.00653 (2020)
28. Patashnik, O., Wu, Z., Shechtman, E., Cohen-Or, D., Lischinski, D.: Styleclip: text-driven manipulation of stylegan imagery (2021)

29. Pidhorskyi, S., Adjeroh, D.A., Doretto, G.: Adversarial latent autoencoders. In: Proceedings of the IEEE/CVF Conference on Computer Vision and Pattern Recognition, pp. 14104–14113 (2020)
30. Rahaman, N., et al: On the spectral bias of neural networks. In: International Conference on Machine Learning, pp. 5301–5310. PMLR (2019)
31. Richardson, E., et al.: Encoding in style: a stylegan encoder for image-to-image translation. In: Proceedings of the IEEE/CVF Conference on Computer Vision and Pattern Recognition (2021)
32. Roich, D., Mokady, R., Bermano, A.H., Cohen-Or, D.: Pivotal tuning for latent-based editing of real images. arXiv preprint arXiv:2106.05744 (2021)
33. Shen, Y., Gu, J., Tang, X., Zhou, B.: Interpreting the latent space of GANs for semantic face editing. In: Proceedings of the IEEE/CVF Conference on Computer Vision and Pattern Recognition, pp. 9243–9252 (2020)
34. Shen, Y., Zhou, B.: Closed-form factorization of latent semantics in GANs. arXiv preprint arXiv:2007.06600 (2020)
35. Skorokhodov, I., Sotnikov, G., Elhoseiny, M.: Aligning latent and image spaces to connect the unconnectable. arXiv preprint arXiv:2104.06954 (2021)
36. Tewari, A., et al.: Stylerig: rigging stylegan for 3D control over portrait images. arXiv preprint arXiv:2004.00121 (2020)
37. Tov, O., Alaluf, Y., Nitzan, Y., Patashnik, O., Cohen-Or, D.: Designing an encoder for stylegan image manipulation (2021)
38. Tzaban, R., Mokady, R., Gal, R., Bermano, A.H., Cohen-Or, D.: Stitch it in time: gan-based facial editing of real videos (2022)
39. Wang, Z., Simoncelli, E.P., Bovik, A.C.: Multiscale structural similarity for image quality assessment. In: The Thrity-Seventh Asilomar Conference on Signals, Systems & Computers, vol. 2, pp. 1398–1402. IEEE (2003)
40. Wu, Z., Lischinski, D., Shechtman, E.: Stylespace analysis: disentangled controls for stylegan image generation. In: Proceedings of the IEEE/CVF Conference on Computer Vision and Pattern Recognition, pp. 12863–12872 (2021)
41. Xia, W., Zhang, Y., Yang, Y., Xue, J.H., Zhou, B., Yang, M.H.: Gan inversion: a survey (2021)
42. Xu, Y., AlBahar, B., Huang, J.B.: Temporally consistent semantic video editing. arXiv e-prints pp. arXiv-2206 (2022)
43. Yao, X., Newson, A., Gousseau, Y., Hellier, P.: A latent transformer for disentangled face editing in images and videos. In: Proceedings of the IEEE/CVF International Conference on Computer Vision, pp. 13789–13798 (2021)
44. Zhang, R., Isola, P., Efros, A.A., Shechtman, E., Wang, O.: The unreasonable effectiveness of deep features as a perceptual metric. In: CVPR (2018)
45. Zhu, J., Shen, Y., Zhao, D., Zhou, B.: In-domain GAN inversion for real image editing. arXiv preprint arXiv:2004.00049 (2020)
46. Zhu, P., Abdal, R., Femiani, J., Wonka, P.: Barbershop: GAN-based image compositing using segmentation masks. ACM Trans. Graph. **40**(6) (2021). https://doi.org/10.1145/3478513.3480537
47. Zhu, P., Abdal, R., Qin, Y., Wonka, P.: Improved stylegan embedding: where are the good latents? ArXiv abs/2012.09036 (2020)

CNSNet: A Cleanness-Navigated-Shadow Network for Shadow Removal

Qianhao Yu, Naishan Zheng, Jie Huang, and Feng Zhao(✉)

University of Science and Technology of China, Hefei, China
{nbyqh,nszheng,hj0117}@mail.ustc.edu.cn, fzhao956@ustc.edu.cn

Abstract. The key to shadow removal is recovering the contents of the shadow regions with the guidance of the non-shadow regions. Due to the inadequate long-range modeling, the CNN-based approaches cannot thoroughly investigate the information from the non-shadow regions. To solve this problem, we propose a novel cleanness-navigated-shadow network (CNSNet), with a shadow-oriented adaptive normalization (SOAN) module and a shadow-aware aggregation with transformer (SAAT) module based on the shadow mask. Under the guidance of the shadow mask, the SOAN module formulates the statistics from the non-shadow region and adaptively applies them to the shadow region for region-wise restoration. The SAAT module utilizes the shadow mask to precisely guide the restoration of each shadowed pixel by considering the highly relevant pixels from the shadow-free regions for global pixel-wise restoration. Extensive experiments on three benchmark datasets (ISTD, ISTD+, and SRD) show that our method achieves superior de-shadowing performance.

Keywords: Shadow removal · Shadow-aware aggregation · Shadow-oriented adaptive normalization

1 Introduction

"Where there is light, there is shadow." The shadows, which are prevalent in nature images, often appear when objects partially or completely hinder the light sources. However, undesirable shadows not only fail to satisfy the human perception requirements, but also degrade the performance of the subsequent computer vision tasks [4,20,27,31,51], such as object detection, segmentation, and tracking. To improve the human perception and machine perception, it is essential to apply the shadow removal to recover the contents of the shadow regions with the guidance of the shadow-free regions.

As a long-standing computer vision problem, shadow removal has drawn much attention. Existing approaches can be roughly classified into 2 categories:

F. Zhao—Corresponding author.

Q. Yu and N. Zheng—Co-first authors contributed equally.

L. Karlinsky et al. (Eds.): ECCV 2022 Workshops, LNCS 13802, pp. 221–238, 2023.
https://doi.org/10.1007/978-3-031-25063-7_14

model-based and learning-based techniques. The traditional model-based methods largely depend on the handcrafted priors, e.g., image gradients [9,13], illumination [32,44,50], and regions [15,37]. Due to the limitations of such priors, these algorithms exhibit poor performance when applied to diverse shadow scenes.

Benefiting from the large-scale datasets and the strong learning ability of deep convolutional neural networks (CNNs) [6,29], the learning-based methods have provided superior results over the conventional approaches. For instance, Le *et al.* [21,22] proposed a two-stage network to formulate a linear shadow illumination model to acquire shadow-free images via shadow mattes. DHAN [5] applies the dilated convolution to aggregate the multi-context features and attentions hierarchically for artifact-free images. Fu *et al.* [10] reformulated the shadow removal as multi-exposure fusion on the multiple estimated overexposed images. However, the convolution operation hinders the CNN-based methods from establishing the long-range pixel dependencies between non-shadow and shadow regions. Hence, these methods cannot fully investigate the information from the shadow-free regions to restore each pixel of the shadow regions. Recently, Chen *et al.* [3] explored the potential context relationships between shadow and non-shadow regions, transferring the contextual information from shadow-free patches to shadow patches. Nonetheless, the patch-wise transferring manner is hampered by inaccurate information transformation and complex matching.

Due to the effectiveness of the long-range modeling [36,41], the transformer has recently achieved widespread dominance in many computer vision tasks [1,7,25,42,48]. Intuitively, the property of transformer can be utilized to recover the shadow region by establishing the relationship from all the pixels in the non-shadow region to the shadow region. However, the transformer constructs the interaction between all the pixels for recovering the shadow region. Since the features of the shadow region are corrupted while the shadow-free region in the same image has reasonable visibility, only the connection from the pixels with high relevance in the non-shadow region to the shadow region should be formed.

To address the aforementioned issues, we propose a cleanness-navigated-shadow network for shadow removal, namely CNSNet. With the guidance of the shadow mask, the proposed CNSNet investigates the characteristics between the shadow region and the non-shadow region, and establishes the connection from the highly relevant pixels in the non-shadow regions to the shadow region. It comprises three distinct components: a shadow-oriented adaptive normalization (SOAN) module, a shadow-aware aggregation with transformer (SAAT) module, and a soft-region mask predictor. Specifically, for the statistical shift between the shadow-free and shadow regions, the SOAN module performs region-wise restoration by extracting the mean and variance from the shadow-free region and adaptively applying them to the shadow region under the guidance of the shadow mask. This guarantees the statistical consistency between the two regions in a region-wise manner. Furthermore, to build the connection from the non-shadow region to the shadow region, we design the SAAT module with the guidance of the shadow mask. However, the hard shadow mask separates the two regions absolutely, causing the loss of information transmission in the transformer. The soft-region mask predictor is introduced to measure the correlation between the two regions. Therefore, under the guidance of the soft mask, the SAAT module constructs the connection from the pixels with a high correlation in the non-shadow region to the shadow region for pixel-wise restoration.

In summary, our contributions in this work are as follows:

- We propose a cleanness-navigated-shadow network (CNSNet) for shadow removal, which investigates the relationship from the shadow-free region to the shadow region under the guidance of the shadow mask.
- We design a SOAN module to extract the statistics (e.g., mean and variance) from the shadow-free region and adaptively apply them to the shadow region for region-wise restoration. Besides, a SAAT module is introduced to take the pixels with high correlation in the non-shadow region into account to recover the shadow region for pixel-wise restoration.
- Extensive experiments on the public ISTD, ISTD+, and SRD datasets demonstrate that our CNSNet not only achieves competitive results over existing state-of-the-art methods, but also maintains the balance of network parameters, efficiency, and performance.

2 Related Work

2.1 Shadow Removal

Early studies on shadow removal typically make use of various hand-crafted prior information, such as image gradients [9,13], illumination properties [32,44,50], region characteristics [15,37], and user interactions [12,13,43]. For example, Finlayson *et al.* [8,9] applied the gradient consistency-based manipulation to recover the shadow images. Shor *et al.* [32] utilized the areas around the shadow edges to estimate the parameters of affine transformations from the shadow to non-shadow regions.

Recently, due to the appearance of large-scale datasets, CNNs have greatly improved the shadow removal performance and gradually become the mainstream of this task [3,5,10,11,16,21,22,29,38,53]. For instance, DSC [16] creates a direction-aware spatial attention module and aggregates both global and context information. Zhu *et al.* [53] implemented shadow removal from the perspective of invertible neural networks, and proposed the BMNet with much fewer network parameters and less computational cost. Moreover, generative adversarial networks (GANs) have been widely used in de-shadowing [6,17,19,26,35,39,49], building on the bidirectional guidance of shadow generation, detection, and removal. ST-CGAN [39] collaboratively detects and removes shadows with the architecture of stacked conditional GANs. G2R [26] employs the shadow generators to synthesize numerous pseudo shadow pairs for joint training. RIS-GAN [49] utilizes the explored relationship among the negative residual images, the inverse illumination maps, and the shadows. Besides, DC-ShadowGAN [19] and Mask-ShadowGAN [17] exploit adversarial learning and mask-guided cycle consistency constraints and apply unsupervised learning with unpaired datasets.

2.2 Region-Wise Information

In recent years, regional information has drawn much attention from researchers in low-level computer vision tasks [3,24,40,46,47,52], especially in works related to segmentation or fusion. For example, Ling *et al.* [24] introduced a region-aware module to develop the visual style from the background and apply it in the foreground, reinterpreting the image harmonization as a style transfer problem. Yu *et al.* [47] proposed a region normalization, which standardizes the features in different regions during the inpainting network training. DSNet [40] further combines the deformable convolution with the regional mechanism and dynamically uses region-wise normalization methods for better image inpainting.

In the shadow removal task, previous works have primarily focused on pairing features from the shadow and non-shadow regions. Guo *et al.* [14,15] computed the illumination ratios by randomly sampling pair patches from both sides of the shadow boundary. In [46], shadow detection is treated as a shadow region labeling problem to train a region classifier, and then applies pairs of shadow regions and neighboring shadow-free regions to achieve regional relighting. On the other hand, CANet [3] removes shadows by transferring the contextual information of non-shadow regions to shadow regions in a patch-level way.

2.3 Vision Transformer

Recently, due to the success of transformer-based models in the field of NLP [36], transformer and its variants have widely exhibited outstanding performance in low-level computer vision tasks (e.g., image restoration, enhancement, super-resolution, and dehazing) [1,7,25,33,38,42,45,48]. Unlike CNNs, transformer-based network structures are naturally adept at capturing long-range dependencies through the global self-attention. Vision transformer (ViT) [7] is the pioneer in implementing a pure transformer architecture by treating images as token sequences via path-wise linear embedding. For example, IPT [1] utilizes typical transformer blocks to train on images with multi-heads and multi-tails for various tasks. Uformer [42] is a hybrid structure consisting of UNet [30] and transformer for image restoration, with inserted depth-wise convolution in the feed-forward network. Similar to Uformer, Restormer [48] changes self-attention from spatial dimension to channels, aiming to reduce the computational complexity. Swin transformer [25] separates tokens into windows and performs self-attention within a window to maintain the linear computational cost.

3 Method

Intuitively, objects from shadow regions and non-shadow regions exist in similar contexts except for illumination, to some extent, making it possible to allow the non-shadow regions to guide the shadow regions. On the basis of this, we elaborate in Sect. 3.1 on the overview of our proposed cleanness-navigated-shadow network (CNSNet), which is a composite CNN-transformer framework. With the

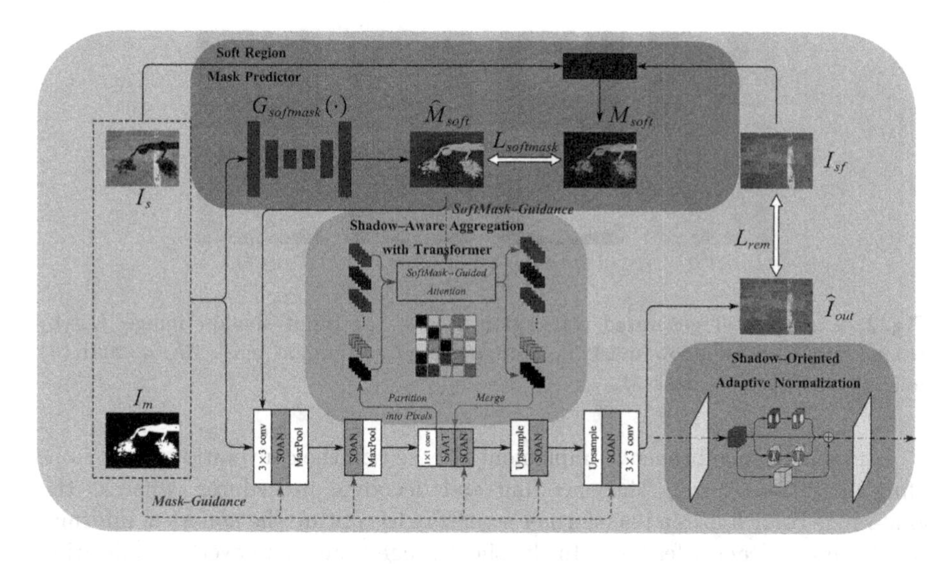

Fig. 1. Illustration of our proposed cleanness-navigated-shadow network (CNSNet) for shadow removal. It involves three key elements: soft-region mask predictor (green box), shadow-oriented adaptive normalization (SOAN) module (orange box), and shadow-aware aggregation with transformer (SAAT) module (purple box). First, the predictor takes in a shadow image and its corresponding shadow mask to obtain a soft-region mask. Then, both hard and soft masks are concatenated with the input image, entering the UNet-like network to produce the shadow-free results. Note that the guidance (dotted arrows) of both hard and soft masks is applied in the region-wise SOAN and pixel-wise SAAT modules, respectively. (Color figure online)

input shadow image and the corresponding shadow mask, CNSNet consists of three key parts: soft-region mask predictor, shadow-oriented adaptive normalization (SOAN) module, and shadow-aware aggregation with transformer (SAAT) module (see Sects. 3.2–3.4 for more details).

3.1 Cleanness-Navigated-Shadow Network

Traditional networks for image enhancement adopt convolutional operations in the hidden layer. However, most simple convolutional operations focus more on the surrounding pixels and only have a small receptive field to extract local information, which may be inadequate to recover the entire images. Specifically, in the shadow removal task, this local information mainly comes from the regions of the same nature (shadow or non-shadow), while ignoring the association and mutual influence of the shadow and non-shadow regions to a considerable extent.

To address this critical issue, we propose the CNSNet with complementary short-range and long-range communications, fully leveraging the regional information. As illustrated in Fig. 1, our CNSNet is an end-to-end designed framework, including both encoder and decoder procedures during the training process. Besides, the soft-region masks are intermediately produced as supplementary auxiliary information for the network training.

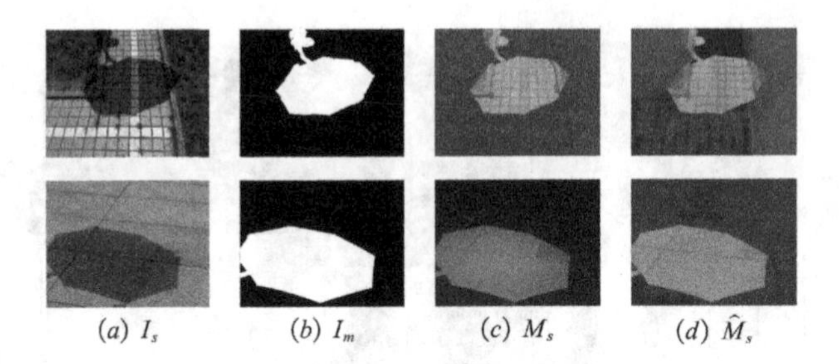

Fig. 2. Examples of generated soft-region masks. (a) Input shadow image I_s, (b) ground-truth hard-shadow mask I_m, (c) expected soft-region mask M_{soft}, and (d) generated soft-region mask $\hat{M}_{soft}$.

The short-range branch is implemented based on the convolutional and sampling operations during the encoding and decoding procedures, whereas the long-range branch uses a transformer structure to capture the non-local information from the deepest features. In the short-range branch, a novel normalization method SOAN is designed to utilize the non-shadow regional statistics as affine function parameters for the shadow regions after regional instance normalization, thereby roughly ensuring the region-wise statistical consistency. In the long-range branch, the corresponding soft-region mask acquired from the predictor is used to direct the transformer to restore each pixel by taking all the pixels with high relevance into account for the global pixel-wise restoration.

Finally, we employ the pixel-wise L_1 distance between our shadow removal outputs $\hat{I}_{out}$ and the ground-truth shadow-free images I_{sf} as a loss function $\mathcal{L}_{rem}$ for shadow removal:

$$\mathcal{L}_{rem} = ||\hat{I}_{out} - I_{sf}||_1. \tag{1}$$

3.2 Soft-Region Mask Predictor

Referring to [19], we first compute the difference between the input shadow image I_s and the corresponding shadow-free image I_{sf} to obtain the expected soft-region mask M_{soft}, and apply the function $F(I_s, I_{sf})$ on the difference:

$$M_{soft} = F(I_s, I_{sf}) = \frac{1}{3} \sum_{c \in \{R,G,B\}} |N(I_{s_c} - I_{sf_c})|, \tag{2}$$

where $N(\cdot)$ is a normalization function on the channel dimension defined as $N(I) = (I - I_{\min}) / (I_{\max} - I_{\min})$. Here, $I_{\min}$ and $I_{\max}$ are the minimum and maximum values of I, respectively. Note that the values of M_{soft} are in the range of $[0, 1]$. Figure 2 shows some examples of generated soft-region masks.

The network architecture of $G_{softmask}(\cdot)$ employs a traditional UNet [30] structure, combining the shadow image I_s and the hard shadow mask I_m as inputs to generate a soft-region mask $\hat{M}_{soft}$. As we can see, utilizing $G_{softmask}(\cdot)$ can produce a high-quality soft-region mask close to the reference. In other words, the soft-region mask predictor seeks to learn the regional correlation of

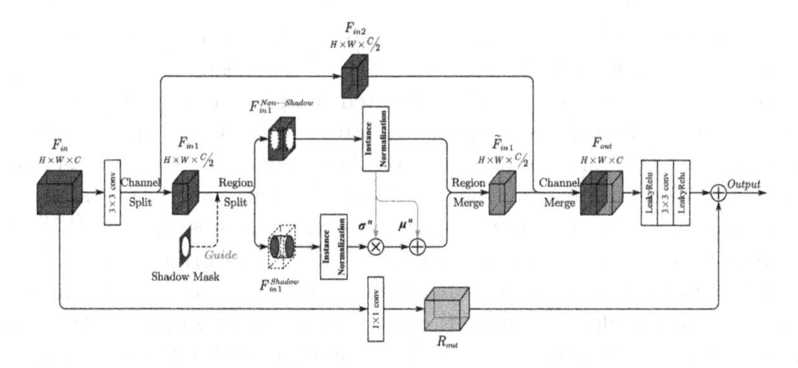

Fig. 3. Illustration of our proposed shadow-oriented adaptive normalization (SOAN) module. Taking the input features F_{in} and the corresponding resized shadow mask M_{in} as the priors, the features are then split across the channel dimensions, and half of them (F_{in1}) performs regional instance normalization, while the other half (F_{in2}) keeps the context information at the same time. Finally, the output is obtained by concatenating the processed features F_{out} with the residual features R_{out}.

the shadow and the non-shadow in a fuzzy number-based manner. We explicitly integrate the learned soft-region mask information into the transformer to guide it for better restoration of every shadowed pixel.

During the training phase, we set the predictor to obtain the soft-region mask $\hat{M}_{soft}$ through the L_1 distance loss by:

$$\mathcal{L}_{soft} = ||\hat{M}_{soft} - M_{soft}||_1 = ||G_{softmask}(I_s, I_m) - F(I_s, I_{sf})||_1. \tag{3}$$

3.3 Shadow-Oriented Adaptive Normalization (SOAN)

Here, we suppose a n_{total}-pixel image with shadows, containing n_{shadow} shadow pixels and n_{non} non-shadow pixels. The mean and variance of the two regions are recorded as μ_{shadow}, μ_{non}, σ_{shadow}, and σ_{non}, while μ_{total} and σ_{total} represent the statistics of the entire image. Their detailed relationships are as follows:

$$n_{total} = n_{shadow} + n_{non}, \tag{4}$$

$$\mu_{total} = \frac{n_{shadow}}{n_{total}} \cdot \mu_{shadow} + \frac{n_{non}}{n_{total}} \cdot \mu_{non}, \tag{5}$$

$$\sigma_{total}^2 = \frac{n_{shadow}}{n_{total}} \cdot \left(\sigma_{shadow}^2 + \mu_{shadow}^2\right) + \frac{n_{non}}{n_{total}} \cdot \left(\sigma_{non}^2 + \mu_{non}^2\right) - \mu_{total}^2. \tag{6}$$

Due to the common sense that the values of RGB channels in the shadow region are generally much lower than those in the shadow-free regions, we can observe through the above formulas that both μ_{total} and σ_{total} have a large shift compared to μ_{shadow}, μ_{non}, σ_{shadow}, and σ_{non}. Thus, the conventional normalization technique (e.g., BN [18] or IN [2,34]) on the entire image is not competent to overcome this difficulty. In addition, although RN [47] separately standardizes features based on different regions, where the partition processing is so absolute to ignore any semantic relationship between regions.

Therefore, we design the shadow-oriented adaptive normalization (SOAN) module. While maintaining the original features, our SOAN utilizes the mean and variance of the non-shadow areas to adaptively assist the recovery of the shadow areas, roughly ensuring the consistency of statistics in the two regions.

As shown in Fig. 3, the SOAN block takes the features $F_{in} \in \mathbb{R}^{H \times W \times C}$ as inputs and the rescaled shadow masks $M_{in} \in \mathbb{R}^{H \times W}$ as prior guidance. H, W, and C individually denote the height, width, and channels of the current feature maps. Firstly, we divide the convolutional features into two parts on the channel dimension, i.e., $F_{in1}, F_{in2} \in \mathbb{R}^{H \times W \times C/2}$. As for F_{in1}, we further split it into two regions: R^{shadow} (shadow regions) and R^{non} (non-shadow regions) in a spatial-wise manner according to M_{in} as below:

$$F_{in1} = F_{in1}^{Shadow} \cup F_{in1}^{Non-Shadow}. \tag{7}$$

The two regions are separately standardized by IN [2,34] and then re-merge together. Unlike RN [47], the IN is used in this case without the learnable affine parameters and the normalized features of shadow regions are affined with the learned scale and bias from non-shadow regions. Specifically, the normalized value of the shadow pixel p located in (h, w, c) can be computed by:

$$\tilde{p}_{h,w,c} = \frac{p_{h,w,c} - \mu_c^s}{\sigma_c^s} \cdot \sigma_c^n + \mu_c^n, \tag{8}$$

where $p_{h,w,c}$ and $\tilde{p}_{h,w,c}$ are the initiation and standardization of the pixel value, μ_c^s and σ_c^s are the channel-wise mean and variance of the shadow features, while μ_c^n and σ_c^n represent the statistics of the non-shadow regions, calculated by:

$$\mu_c^{Region} = \frac{1}{Num^{Region}} \sum_{p_{h,w,c} \in R^{Region}} p_{h,w,c}, \tag{9}$$

$$\sigma_c^{Region} = \sqrt{\frac{1}{Num^{Region}} \sum_{p_{h,w,c} \in R^{Region}} \left(p_{h,w,c} - \mu_c^{Region}\right)^2 + \epsilon}. \tag{10}$$

Then, F_{in2} re-concatenates with the normalized $\tilde{F}_{in1}$ on the channel dimension, which keeps the context information at the meantime. After that, the SOAN module output $F_{out} \in \mathbb{R}^{H \times W \times C}$ is integrated through the convolution layers and finally adds with the residual features $R_{out} \in \mathbb{R}^{H \times W \times C}$.

In comparison to other normalization methods shown in Table 4, our SOAN is significantly better than single BN [18] and IN [2,34], further proving the rationality of our aforementioned analysis.

3.4 Shadow-Aware Aggregation with Transformer (SAAT)

In a variety of image enhancement tasks, traditional transformers [7,36] can extract non-local information via image patches. However, in general architectures, the attention mechanisms focus on all the patches, which may bring in

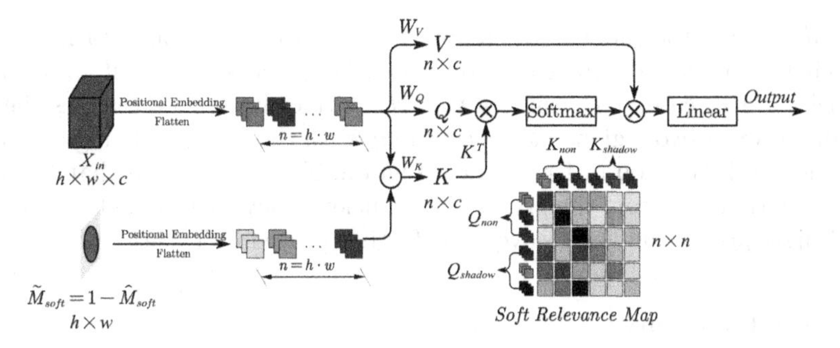

Fig. 4. Illustration of our shadow-aware aggregation with transformer (SAAT) module with a single head of the transformer layer. The difference from the traditional transformer structure is that we introduce the guidance of a soft-region mask, which multiplies with the input features when calculating the key vectors to acquire the soft relevance map. Based on this map, SAAT produces the outputs in a pixel-wise manner.

worthless information. Taking our shadow removal task as an example, to restore a pixel in the shadow regions, the long-range attention may be captured from pixels of both shadow regions and non-shadow regions, while the shadow patches are frequently ineffective. Hence, the inaccurate information brought by the traditional transformer will interfere with the subsequent shadow removal.

To this end, we propose a shadow-aware aggregation with transformer (SAAT) module, which improves the performance of this special task by utilizing soft shadow mask-guided attention. The SAAT module consists of two cascading transformer layers, including the multi-head self-attention (MSA) modules and the feed-forward networks (FFNs). Figure 4 shows the transformer layer with a single head. Given the input feature maps $X_{in} \in \mathbb{R}^{h\times w\times c}$ and associated soft-region mask $\tilde{M}_{soft} = 1 - \hat{M}_{soft} \in \mathbb{R}^{h\times w}$ acquired from the soft-region mask predictor, we partition X_{in} and $\tilde{M}_{soft}$ in a pixel-wise manner, where each pixel is an input token and its channels are token embeddings. Totally, there are $n = h \times w$ feature patches. Every token executes positional encoding by $\tilde{x} = x + pe$, $\tilde{m}_{soft} = m_{soft} + pe$, where pe is the positional embedding. Then, we flatten $\tilde{X}, \tilde{M}_{soft} \in \mathbb{R}^{n\times c}$ into 1D vectors and perform the following computation:

$$\tilde{X} = [\tilde{x}_1, \tilde{x}_2, ..., \tilde{x}_n], \tilde{M}_{soft} = [\tilde{m}_{soft1}, \tilde{m}_{soft2}, ..., \tilde{m}_{soft\ n}], \tag{11}$$

$$Q = \tilde{X}W_q,\ K = \left(\tilde{X} \cdot \tilde{M}_{soft}\right) W_k,\ V = \tilde{X}W_v, \tag{12}$$

where $W_q, W_k, W_v \in \mathbb{R}^{c\times d}$ represent the linear learnable matrices, and $Q, K, V \in \mathbb{R}^{n\times d}$ are the query, key, and value features, respectively. Here, the soft-region masks affect the values of key features to build better connections between different regions. Following that, we obtain the attention score map $A \in \mathbb{R}^{n\times n}$ and the final output features $\tilde{Y} \in \mathbb{R}^{(h\times w)\times d}$ as follows:

$$Attention\,(Q, K, V) = softmax\left(\frac{QK^T}{\sqrt{c}}\right) V = AV = [\tilde{y}_1, \tilde{y}_2, ..., \tilde{y}_n], \tag{13}$$

$$\tilde{Y} = FFN\left([\tilde{y}_1, \tilde{y}_2, ..., \tilde{y}_n]\right), \tag{14}$$

where d denotes the number of channels in self-attention computation, which is equal to c in our design to simplify and keep the same input and output channels. In this case, our design of the transformer layer characterizes the correlation between two regions more accurately with the help of soft-region masks, avoiding the distraction of irrelevant attention. Thus, we ensure that the long-range attentions are from the pixels with sufficient relevance and help to produce global high-quality shadow recovery, as further proved in Table 4 of Sect. 4.4.

3.5 Loss Functions

Following the previous works [3,5,10,21,22,26,53], except for the $\mathcal{L}_{rem}$ and $\mathcal{L}_{soft}$ mentioned above in Sect. 3.1 and Sect. 3.2, we also use a perceptual loss $\mathcal{L}_{per}$ and a gradient loss $\mathcal{L}_{grad}$ based on Poison image editing [28].

Here, $\mathcal{L}_{per}$ is the perceptual-consistency loss that aims to preserve the image structure with semantic measures and low-level details in multiple contexts. We estimate the feature differences in pre-trained VGG19 networks between the ground-truth shadow-free image I_{gt} and our shadow-removed image $\hat{I}_{out}$ as follows:

$$\mathcal{L}_{per} = \sum_{k=1}^{5} w_k ||VGG_k\left(\hat{I}_{out}\right) - VGG_k\left(I_{gt}\right)||_2, \tag{15}$$

where $VGG_k(\cdot)$ outputs the multi-scale features of the k-th intermediate layers, and w_1, w_2, w_3, w_4, and w_5 are set to 1/32, 1/16, 1/8, 1/4, and 1 in this work.

In addition, $\mathcal{L}_{grad}$ is proposed by Fu *et al.* in [10], purposing to reduce the gradient domain along the shadow boundary:

$$\mathcal{L}_{grad} = \left(1 - \tilde{M}_{in}\right) \cdot MSE\left(\nabla\hat{I}_{out}, \nabla I_{in}\right) + \tilde{M}_{in} \cdot MSE\left(\nabla\hat{I}_{out}, \nabla I_{gt}\right), \tag{16}$$

where I_{in}, I_{gt}, $\hat{I}_{out}$, and $\tilde{M}_{in}$ respectively represent the initial shadow images, the ground-truth shadow-free images, our shadow-removed results, and the shadow masks dilated with 7 pixels, and ∇ denotes the Laplacian gradient operator. It minimizes the gradient domain differences between $\hat{I}_{out}$ and I_{gt}, while maintaining the gradient domain of non-shadow regions between $\hat{I}_{out}$ and I_{in}.

In summary, the total loss function of our CNSNet is a weighted sum of the four components described above, which is calculated by:

$$\mathcal{L}_{total} = \lambda_1\mathcal{L}_{rem} + \lambda_2\mathcal{L}_{soft} + \lambda_3\mathcal{L}_{per} + \lambda_4\mathcal{L}_{grad}, \tag{17}$$

where λ_1, λ_2, λ_3, and λ_4 are hyperparameters to balance different loss terms and are respectively set to 10.0, 5.0, 1.0, and 1.0 in our experiments.

4 Experiments

4.1 Datasets and Evaluation Measurements

Benchmark Datasets. We utilize three representative public datasets: ISTD, adjusted ISTD (ISTD+), and SRD, to train and evaluate our proposed model.

Both the ISTD and ISTD+ datasets contain 1870 image triplets of shadow images, shadow-free images, and shadow masks, which have 1330 training triplets and 540 testing triplets. Moreover, due to the color mismatch in ISTD, the ISTD+ has reduced the color inconsistency using an image augmentation method. On the other hand, the SRD dataset consists of 2680 training and 408 testing pairs of shadow and shadow-free images without shadow masks. We additionally use the detection results of DHAN [5] for SRD shadow masks.

Implementation Details. Our proposed method is implemented in PyTorch with a single GPU (NVIDIA GeForce GTX 3090). In the experiments, we employ the Adam optimizer to train our network for over 200 epochs with a batch size of 8 and the input patch size of 256 × 256. The initial learning rate is set to 1e-3 and gradually decreases with a dynamic decay strategy. As for the data augmentation, we randomly adopt the cropping, rotating, and flipping operations during the training, to circumvent the overfitting problem.

Evaluation Metrics. Following the previous works, we use the root mean square error (RMSE) in the LAB color space between the shadow-removed result and its ground truth to evaluate the performance. Note that the RMSE value is actually calculated by the mean absolute error (MAE) in this task. The values of RMSE are calculated at each pixel of the shadow region, non-shadow region, as well as the whole image, and a lower value indicates better performance. Furthermore, to verify the effectiveness of our algorithm more comprehensively, we additionally assess the experimental results with the peak signal-to-noise ratio (PSNR) and the structural similarity (SSIM). The higher, the better.

Table 1. Quantitative shadow removal results of our network compared to state-of-the-art shadow removal methods on the ISTD dataset. The outcomes of the methods marked with an asterisk "*" are referenced from their original papers.

Method	RMSE			PSNR (dB)			SSIM		
	S	NS	ALL	S	NS	ALL	S	NS	ALL
Input Image	32.11	6.83	10.97	22.40	27.30	20.56	0.936	0.975	0.892
ST-CGAN [39]	9.55	6.13	6.69	33.73	29.50	27.43	0.981	0.957	0.928
DSC [16]	8.50	<u>5.13</u>	<u>5.68</u>	34.64	<u>31.22</u>	28.97	0.983	0.968	0.943
DHAN [5]	<u>7.55</u>	5.39	5.74	<u>35.52</u>	31.01	<u>29.08</u>	<u>0.988</u>	0.969	<u>0.953</u>
RIS-GAN* [49]	8.99	6.33	6.95	—					
DC-SNet [19]	10.57	5.83	6.60	31.68	28.98	26.37	0.976	0.957	0.921
CANet* [3]	8.86	6.07	6.15	—					
Fu *et al.* [10]	7.77	5.57	5.93	34.71	28.60	27.19	0.975	0.880	0.845
BMNet [53]	7.89	5.30	5.73	34.58	30.85	28.70	<u>0.988</u>	<u>0.970</u>	0.950
CNSNet (Ours)	**6.56**	**4.23**	**4.61**	**36.67**	**32.15**	**30.29**	**0.991**	**0.979**	**0.965**

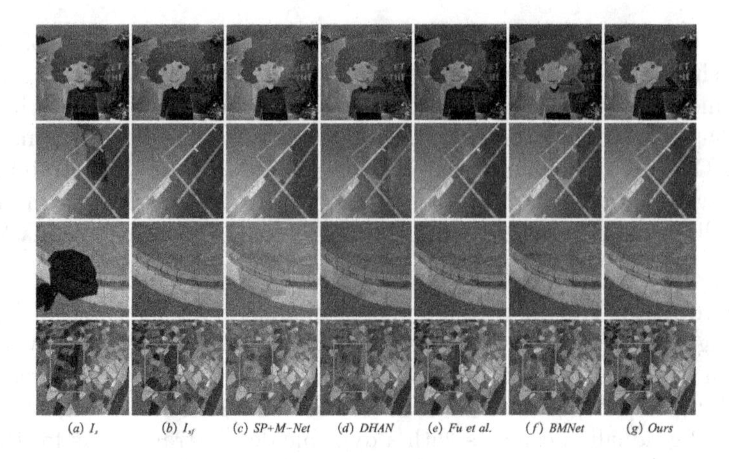

Fig. 5. Visual comparison results of shadow removal on the ISTD dataset. (a) Input shadow images, (b) corresponding ground-truth shadow-free images, and (c)-(g) results from SP+M-Net [21,23], DHAN [5], Fu *et al.* [10], BMNet [53], and our CNSNet.

4.2 Shadow Removal Evaluation on ISTD Dataset

As shown in Tables 1 and 2, we report the quantitative shadow removal results of our CNSNet on the ISTD and ISTD+ datasets, and compare it with recent state-of-the-art (SOTA) algorithms, including ST-CGAN [39], DSC [16], DHAN [5], RIS-GAN [49], DC-ShadowNet [19], G2R [26], CANet [3], Fu *et al.* [10], and BMNet [53]. In addition, we compare the network parameters (Param.) and floating point operations (FLOPs) in Table 2, where the values of Param. and FLOPs are directly referred from [53]. For the sake of fairness of comparison, these statistics are calculated from the de-shadowing results with a resolution of 256 × 256, presented by the authors or directly acquired from the original papers. In the following tables, S, NS, and ALL indicate the shadow region, non-shadow region, and entire image, respectively. Note that RMSE is calculated by averaging the RMSE over all the pixels in certain regions of the whole testing set, not per image. The first row (input image) shows the metrics of the original corresponding pair images of shadow and shadow-free, as a blank-control group. The best and the second-place values for each metric are respectively highlighted in **bold** and underlined.

From Tables 1 and 2, we can observe that our CNSNet achieves the best shadow removal performance than other SOTA methods by a large margin. The visualization results of the shadow removal comparison are displayed in Fig. 5, which further justifies the effectiveness of our method. We produce a better visual restoration effect with fewer artifacts and boundary traces between the shadow and non-shadow regions. Specifically, although BMNet [53] has slightly fewer network parameters and FLOPs, the values of RMSE, PSNR, SSIM in our algorithm are significantly improved by 1.12, 1.59 dB, 0.015 in entire images, 1.33, 2.09 dB, 0.003 in shadow regions and 1.07, 1.30 dB, 0.009 in non-shadow regions. Besides, on the ISTD+ dataset, our method also has a great improvement, outperforming the BMNet [53] by 0.10, 0.11, 0.11 lower RMSE values and 0.25 dB, 0.35 dB, 0.29 dB higher PSNR values in shadow regions, non-shadow regions, and whole images, respectively. From the overall view, we effectively ensure the good balance of network parameters, efficiency, and performance.

4.3 Shadow Removal Evaluation on SRD Dataset

In Table 3, we further report the shadow removal results on the SRD dataset. The compared baseline methods include DSC [16], DHAN [5], RIS-GAN [49], DC-ShadowNet [19], CANet [3], Fu *et al.* [10], and BMNet [53]. Our CNSNet still presents a competitive de-shadowing performance by decreasing the total RMSE value from 4.46 to 4.29, and increasing the PSNR value of the shadow regions from 35.05 dB to 35.10 dB.

4.4 Ablation Studies

In this subsection, we conduct several ablation studies on the ISTD dataset to demonstrate the contribution of each essential component in our framework. The notations used are listed as follows:

- "Ours (default)": Taking the final results as the default control group;
- "Ours w/o $\mathcal{L}$": Removing the certain loss term;
- "Ours w/o SOAN/SAAT": Removing the SOAN/SAAT module;
- "$SOAN_{BN}$": Utilizing BN in SOAN instead of regional IN;
- "$SOAN_{IN}$": Utilizing direct IN in SOAN instead of regional IN; and
- "$SAAT_{hardmask}$": Utilizing the guidance of hard masks in SAAT instead of soft masks.

In Table 4, we first justify the effects of the loss functions. It can be seen that $\mathcal{L}_{soft}$ enables the network to acquire more accurate soft-region masks for better

Table 2. Quantitative shadow removal results of our network compared to state-of-the-art shadow removal methods on the ISTD+ dataset.

Method	RMSE			PSNR (dB)			Param	FLOPs
	S	NS	ALL	S	NS	ALL	(M:10^6)	(G:10^9)
Input Image	39.04	2.52	8.50	20.83	37.34	20.45	—	
DSC [16]	7.54	3.16	3.88	35.97	35.76	32.05	22.30	123.47
DHAN [5]	11.30	7.17	7.85	32.91	27.14	25.65	21.75	262.87
DC-SNet [19]	10.43	3.68	4.78	32.00	33.53	28.76	21.16	105.00
G2R [26]	7.41	3.03	3.74	35.76	35.54	31.88	22.76	113.87
Fu *et al.* [10]	6.58	3.83	4.28	36.04	31.15	29.44	143.01	160.32
BMNet [53]	<u>5.70</u>	<u>2.58</u>	<u>3.09</u>	<u>37.85</u>	<u>37.39</u>	<u>33.91</u>	**0.37**	**10.99**
CNSNet (Ours)	**5.60**	**2.47**	**2.98**	**38.10**	**37.74**	**34.20**	<u>1.17</u>	<u>17.67</u>

Table 3. Quantitative shadow removal results of our network compared to state-of-the-art shadow removal methods on the SRD dataset.

Method	RMSE			PSNR (dB)			SSIM		
	S	NS	ALL	S	NS	ALL	S	NS	ALL
Input Image	39.31	4.54	14.11	18.96	31.44	18.19	0.871	0.975	0.829
DSC [16]	9.31	<u>3.46</u>	5.07	32.20	34.90	29.87	0.969	<u>0.984</u>	0.943
DHAN [5]	7.77	3.49	4.67	33.83	35.02	30.72	0.980	<u>0.984</u>	<u>0.957</u>
RIS-GAN* [49]	8.22	6.05	6.78	—					
DC-SNet [19]	8.28	3.71	4.97	33.40	34.93	30.55	0.974	0.983	0.947
CANet* [3]	7.82	5.88	5.98	—					
Fu *et al.* [10]	8.93	5.26	6.27	32.43	30.83	27.96	0.968	0.950	0.901
BMNet* [53]	**6.61**	3.61	<u>4.46</u>	<u>35.05</u>	**36.02**	**31.69**	<u>0.981</u>	0.982	0.956
CNSNet (Ours)	<u>6.92</u>	**3.29**	**4.29**	**35.10**	<u>35.69</u>	**31.69**	**0.982**	**0.986**	**0.959**

Table 4. Ablation studies on choosing the loss functions and variants of the two key modules in our proposed CNSNet on the ISTD dataset.

Method	RMSE			PSNR (dB)		
	S	NS	ALL	S	NS	ALL
Ours (default)	**6.56**	4.23	4.61	**36.67**	32.15	30.29
Ours w/o $\mathcal{L}_{soft}$	7.54	4.66	5.13	35.38	31.51	29.48
Ours w/o $\mathcal{L}_{grad}$	6.91	3.97	4.45	36.56	33.15	30.91
Ours w/o $\mathcal{L}_{per}$	7.05	4.26	4.71	36.35	32.54	30.27
Ours w/o SOAN	7.54	4.43	4.94	35.39	31.80	29.75
$SOAN_{BN}$	7.23	4.24	4.73	36.43	32.30	30.18
$SOAN_{IN}$	7.17	4.07	4.58	35.85	33.12	30.53
Ours w/o SAAT	7.49	5.04	5.44	36.09	30.08	28.70
$SAAT_{HardMask}$	6.95	4.01	4.49	36.32	33.31	30.91

pixel-to-pixel connection, while $\mathcal{L}_{grad}$ helps to balance the difference between the shadow and non-shadow regions for smooth recovery on the shadow boundaries, due to the dilated masks. Then, we investigate the performance gain brought by our SOAN module compared to other normalization methods (i.e., BN [18] and IN [2,34]). Obviously, applying the SOAN block can maintain the statistical consistency of deep features between the two regions, thereby improving the de-shadowing quality of shadow regions by a large margin. Finally, in the SAAT module, we verify the superior performance of the generated soft-region masks over the hard masks, solving the problem of lost information transmission caused by absolute regional separation.

5 Conclusions

In this paper, we develop a cleanness-navigated-shadow network (CNSNet) to achieve shadow removal via the short-range and long-range modeling. Our CNSNet exploits the auxiliary guidance of shadow masks to thoroughly investigate the regional information in both region-wise and pixel-wise ways through two novel modules, i.e., shadow-oriented adaptive normalization (SOAN) and shadow-aware aggregation with transformer (SAAT). The SOAN module keeps the statistical consistency by applying the information from the shadow-free region to the shadow region, while the SAAT module builds up the pixel-to-pixel connection between the two regions. Comprehensive experimental results have demonstrated the efficacy and superiority of our method, maintaining the balance of network complexity and performance at the meanwhile.

Acknowledgments. This work was supported by the Anhui Provincial Natural Science Foundation under Grant 2108085UD12. We acknowledge the support of GPU cluster built by MCC Lab of Information Science and Technology Institution, USTC.

References

1. Chen, H., et al.: Pre-trained image processing transformer. In: Proceedings of the IEEE/CVF Conference on Computer Vision and Pattern Recognition, pp. 12299–12310 (2021)
2. Chen, L., Lu, X., Zhang, J., Chu, X., Chen, C.: Hinet: half instance normalization network for image restoration. In: Proceedings of the IEEE/CVF Conference on Computer Vision and Pattern Recognition, pp. 182–192 (2021)
3. Chen, Z., Long, C., Zhang, L., Xiao, C.: CaNet: a context-aware network for shadow removal. In: Proceedings of the IEEE/CVF International Conference on Computer Vision, pp. 4743–4752 (2021)
4. Cucchiara, R., Grana, C., Piccardi, M., Prati, A.: Detecting moving objects, ghosts, and shadows in video streams. IEEE Trans. Pattern Anal. Mach. Intell. **25**(10), 1337–1342 (2003)
5. Cun, X., Pun, C.M., Shi, C.: Towards ghost-free shadow removal via dual hierarchical aggregation network and shadow matting GAN. In: Proceedings of the AAAI Conference on Artificial Intelligence, vol. 34, pp. 10680–10687 (2020)
6. Ding, B., Long, C., Zhang, L., Xiao, C.: ARGAN: attentive recurrent generative adversarial network for shadow detection and removal. In: Proceedings of the IEEE/CVF International Conference on Computer Vision, pp. 10213–10222 (2019)
7. Dosovitskiy, A., et al.: An image is worth 16x16 words: transformers for image recognition at scale. arXiv preprint arXiv:2010.11929 (2020)
8. Finlayson, G.D., Drew, M.S., Lu, C.: Entropy minimization for shadow removal. Int. J. Comput. Vision **85**(1), 35–57 (2009)
9. Finlayson, G.D., Hordley, S.D., Lu, C., Drew, M.S.: On the removal of shadows from images. IEEE Trans. Pattern Anal. Mach. Intell. **28**(1), 59–68 (2005)
10. Fu, L., et al.: Auto-exposure fusion for single-image shadow removal. In: Proceedings of the IEEE/CVF Conference on Computer Vision and Pattern Recognition, pp. 10571–10580 (2021)

11. Gao, J., Zheng, Q., Guo, Y.: Towards real-world shadow removal with a shadow simulation method and a two-stage framework. In: Proceedings of the IEEE/CVF Conference on Computer Vision and Pattern Recognition, pp. 599–608 (2022)
12. Gong, H., Cosker, D.: Interactive removal and ground truth for difficult shadow scenes. J. Opt. Soc. Am. A **33**(9), 1798–1811 (2016)
13. Gryka, M., Terry, M., Brostow, G.J.: Learning to remove soft shadows. ACM Trans. Graphics **34**(5), 1–15 (2015)
14. Guo, R., Dai, Q., Hoiem, D.: Single-image shadow detection and removal using paired regions. In: Proceedings of the IEEE Conference on Computer Vision and Pattern Recognition, pp. 2033–2040 (2011)
15. Guo, R., Dai, Q., Hoiem, D.: Paired regions for shadow detection and removal. IEEE Trans. Pattern Anal. Mach. Intell. **35**(12), 2956–2967 (2012)
16. Hu, X., Fu, C.W., Zhu, L., Qin, J., Heng, P.A.: Direction-aware spatial context features for shadow detection and removal. IEEE Trans. Pattern Anal. Mach. Intell. **42**(11), 2795–2808 (2019)
17. Hu, X., Jiang, Y., Fu, C.W., Heng, P.A.: Mask-ShadowGAN: learning to remove shadows from unpaired data. In: Proceedings of the IEEE/CVF International Conference on Computer Vision, pp. 2472–2481 (2019)
18. Ioffe, S., Szegedy, C.: Batch normalization: Accelerating deep network training by reducing internal covariate shift. In: Proceedings of the International Conference on Machine Learning, pp. 448–456 (2015)
19. Jin, Y., Sharma, A., Tan, R.T.: DC-ShadowNet: single-image hard and soft shadow removal using unsupervised domain-classifier guided network. In: Proceedings of the IEEE/CVF International Conference on Computer Vision, pp. 5027–5036 (2021)
20. Jung, C.R.: Efficient background subtraction and shadow removal for monochromatic video sequences. IEEE Trans. Multimedia **11**(3), 571–577 (2009)
21. Le, H., Samaras, D.: Shadow removal via shadow image decomposition. In: Proceedings of the IEEE/CVF International Conference on Computer Vision, pp. 8578–8587 (2019)
22. Le, H., Samaras, D.: From shadow segmentation to shadow removal. In: Proceedings of the European Conference on Computer Vision, pp. 264–281 (2020)
23. Le, H., Samaras, D.: Physics-based shadow image decomposition for shadow removal. IEEE Trans. Pattern Anal. Mach. Intell. **01**, 1–1 (2021)
24. Ling, J., Xue, H., Song, L., Xie, R., Gu, X.: Region-aware adaptive instance normalization for image harmonization. In: Proceedings of the IEEE/CVF Conference on Computer Vision and Pattern Recognition, pp. 9361–9370 (2021)
25. Liu, Z., et al.: Swin transformer: hierarchical vision transformer using shifted windows. In: Proceedings of the IEEE/CVF International Conference on Computer Vision, pp. 10012–10022 (2021)
26. Liu, Z., Yin, H., Wu, X., Wu, Z., Mi, Y., Wang, S.: From shadow generation to shadow removal. In: Proceedings of the IEEE/CVF Conference on Computer Vision and Pattern Recognition, pp. 4927–4936 (2021)
27. Nadimi, S., Bhanu, B.: Physical models for moving shadow and object detection in video. IEEE Trans. Pattern Anal. Mach. Intell. **26**(8), 1079–1087 (2004)
28. Pérez, P., Gangnet, M., Blake, A.: Poisson image editing. In: ACM SIGGRAPH, pp. 313–318 (2003)
29. Qu, L., Tian, J., He, S., Tang, Y., Lau, R.W.: DeshadowNet: a multi-context embedding deep network for shadow removal. In: Proceedings of the IEEE Conference on Computer Vision and Pattern Recognition, pp. 4067–4075 (2017)

30. Ronneberger, O., Fischer, P., Brox, T.: U-Net: convolutional networks for biomedical image segmentation. In: Proceedings of the International Conference on Medical Image Computing and Computer-Assisted Intervention, pp. 234–241 (2015)
31. Sanin, A., Sanderson, C., Lovell, B.C.: Improved shadow removal for robust person tracking in surveillance scenarios. In: Proceedings of the 20th International Conference on Pattern Recognition, pp. 141–144 (2010)
32. Shor, Y., Lischinski, D.: The shadow meets the mask: Pyramid-based shadow removal. In: Computer Graphics Forum, vol. 27, pp. 577–586 (2008)
33. Song, Y., He, Z., Qian, H., Du, X.: Vision transformers for single image dehazing. arXiv preprint arXiv:2204.03883 (2022)
34. Ulyanov, D., Vedaldi, A., Lempitsky, V.: Instance normalization: The missing ingredient for fast stylization. arXiv preprint arXiv:1607.08022 (2016)
35. Vasluianu, F.A., Romero, A., Van Gool, L., Timofte, R.: Shadow removal with paired and unpaired learning. In: Proceedings of the IEEE/CVF Conference on Computer Vision and Pattern Recognition, pp. 826–835 (2021)
36. Vaswani, A., et al.: Attention is all you need. In: Advances in Neural Information Processing Systems, vol. 30 (2017)
37. Vicente, T.F.Y., Hoai, M., Samaras, D.: Leave-one-out kernel optimization for shadow detection and removal. IEEE Trans. Pattern Anal. Mach. Intell. **40**(3), 682–695 (2017)
38. Wan, J., Yin, H., Wu, Z., Wu, X., Liu, Z., Wang, S.: CRFormer: a cross-region transformer for shadow removal. arXiv preprint arXiv:2207.01600 (2022)
39. Wang, J., Li, X., Yang, J.: Stacked conditional generative adversarial networks for jointly learning shadow detection and shadow removal. In: Proceedings of the IEEE Conference on Computer Vision and Pattern Recognition, pp. 1788–1797 (2018)
40. Wang, N., Zhang, Y., Zhang, L.: Dynamic selection network for image inpainting. IEEE Trans. Image Process. **30**, 1784–1798 (2021)
41. Wang, X., Girshick, R., Gupta, A., He, K.: Non-local neural networks. In: Proceedings of the IEEE Conference on Computer Vision and Pattern Recognition, pp. 7794–7803 (2018)
42. Wang, Z., Cun, X., Bao, J., Zhou, W., Liu, J., Li, H.: Uformer: a general u-shaped transformer for image restoration. In: Proceedings of the IEEE/CVF Conference on Computer Vision and Pattern Recognition, pp. 17683–17693 (2022)
43. Wen, C.L., Hsieh, C.H., Chen, B.Y., Ouhyoung, M.: Example-based multiple local color transfer by strokes. In: Computer Graphics Forum, vol. 27, pp. 1765–1772 (2008)
44. Xiao, C., She, R., Xiao, D., Ma, K.L.: Fast shadow removal using adaptive multi-scale illumination transfer. In: Computer Graphics Forum, vol. 32, pp. 207–218 (2013)
45. Xu, X., Wang, R., Fu, C.W., Jia, J.: SNR-aware low-light image enhancement. In: Proceedings of the IEEE/CVF Conference on Computer Vision and Pattern Recognition, pp. 17714–17724 (2022)
46. Yarlagadda, S.K., Zhu, F.: A reflectance based method for shadow detection and removal. In: Proceedings of the IEEE Southwest Symposium on Image Analysis and Interpretation, pp. 9–12 (2018)
47. Yu, T., et al.: Region normalization for image inpainting. In: Proceedings of the AAAI Conference on Artificial Intelligence, vol. 34, pp. 12733–12740 (2020)
48. Zamir, S.W., Arora, A., Khan, S., Hayat, M., Khan, F.S., Yang, M.H.: Restormer: Efficient transformer for high-resolution image restoration. In: Proceedings of the IEEE/CVF Conference on Computer Vision and Pattern Recognition, pp. 5728–5739 (2022)

49. Zhang, L., Long, C., Zhang, X., Xiao, C.: RIS-GAN: explore residual and illumination with generative adversarial networks for shadow removal. In: Proceedings of the AAAI Conference on Artificial Intelligence, vol. 34, pp. 12829–12836 (2020)
50. Zhang, L., Zhang, Q., Xiao, C.: Shadow remover: image shadow removal based on illumination recovering optimization. IEEE Trans. Image Process. **24**(11), 4623–4636 (2015)
51. Zhang, W., Zhao, X., Morvan, J.M., Chen, L.: Improving shadow suppression for illumination robust face recognition. IEEE Trans. Pattern Anal. Mach. Intell. **41**(3), 611–624 (2018)
52. Zhu, P., Abdal, R., Qin, Y., Wonka, P.: Sean: image synthesis with semantic region-adaptive normalization. In: Proceedings of the IEEE/CVF Conference on Computer Vision and Pattern Recognition, pp. 5104–5113 (2020)
53. Zhu, Y., Huang, J., Fu, X., Zhao, F., Sun, Q., Zha, Z.J.: Bijective mapping network for shadow removal. In: Proceedings of the IEEE/CVF Conference on Computer Vision and Pattern Recognition, pp. 5627–5636 (2022)

Unifying Conditional and Unconditional Semantic Image Synthesis with OCO-GAN

Marlène Careil[1,2](✉), Stéphane Lathuilière[2], Camille Couprie[1], and Jakob Verbeek[1]

[1] Meta AI, New York, USA
marlenec@fb.com
[2] LTCI, Télécom Paris, Institut Polytechnique de Paris, Palaiseau, France

Abstract. Generative image models have been extensively studied in recent years. In the unconditional setting, they model the marginal distribution from unlabelled images. To allow for more control, image synthesis can be conditioned on semantic segmentation maps that instruct the generator the position of objects in the image. While these two tasks are intimately related, they are generally studied in isolation. We propose OCO-GAN, for Optionally COnditioned GAN, which addresses both tasks in a unified manner, with a shared image synthesis network that can be conditioned either on semantic maps or directly on latents. Trained adversarially in an end-to-end approach with a shared discriminator, we are able to leverage the synergy between both tasks. We experiment with Cityscapes, COCO-Stuff, ADE20K datasets in a limited data, semi-supervised and full data regime and obtain excellent performance, improving over existing hybrid models that can generate both with and without conditioning in all settings. Moreover, our results are competitive or better than state-of-the art specialised unconditional and conditional models.

1 Introduction

Remarkable progress has been made in modeling complex data distributions with deep generative models, allowing considerable improvement in generative models for images [3,11,16,31,39], videos [14] and other contents [26,42]. In particular, GANs [9,17] stand out for their compelling sample quality, and are able to produce near-photo realistic images in restricted domains such as human faces [16] or object centric images [3,34].

Despite these advances, modelling more complex indoor or outdoor scenes with many objects remains challenging [5]. To move beyond this limitation, and allow for more control on the generation process, conditional image generation based on semantic segmentation maps or labeled bounding boxes has been investigated, see *e.g.* [30,35–37]. Moreover, semantically conditioned image generation, or "semantic image synthesis" for short, enables users with precise control

Supplementary Information The online version contains supplementary material available at https://doi.org/10.1007/978-3-031-25063-7_15.

L. Karlinsky et al. (Eds.): ECCV 2022 Workshops, LNCS 13802, pp. 239–255, 2023.
https://doi.org/10.1007/978-3-031-25063-7_15

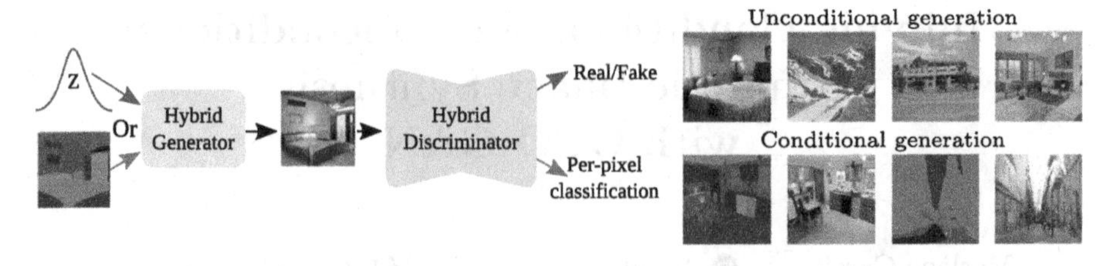

Fig. 1. Our OCO-GAN uses a shared synthesis network that is either conditioned on a semantic segmentation map, or on latent variables only for unconditional generation. The shared discriminator provides a training signal by classifying entire images as real or fake, and uses an additional per-pixel classification loss for conditionally generated images. We show image generation examples with OCO-GAN trained on ADE20K.

over what content should be generated where in automated content generation. Yet, training semantic image generation models requires images annotated with segmentation maps, which are expensive to acquire due to their detailed nature.

Unconditional and conditional image generation are closely related, and indeed to some extent semantics can be recovered from unconditionally trained GANs, see *e.g.* [2,40]. Despite their relatedness, however, both tasks are generally studied in isolation, even though they could potentially mutually benefit each other. Unlabeled images do not require expensive annotation, and jointly training an unconditional generator may help conditional generation since parameter-sharing acts as a form of regularization. In particular in semi-supervised settings where only few images are annotated, jointly training a single hybrid model can help conditional generation. Conversely, annotated images can provide per-pixel loss terms to train the network, and help the unconditional network to output images with better structure and higher levels of detail. This supervision, can be especially useful in settings with small training datasets where unconditional generation is challenging.

The few works that address both problems together, take an existing conditional or unconditional model, and add a module to address the other task. For example, by taking a pre-trained unconditional GAN, and learning an inference network that takes a semantic map as input and which produces the GAN latents that reconstruct the corresponding training image [32]. Another approach is to learn a semantic map generating network, which can be combined with a conditional semantic synthesis network to perform unconditional generation [1]. Such stage-wise approaches, however, do not fully leverage the potential of learning both tasks simultaneously. In the former approach the unconditional GAN is pre-trained without taking advantage of the semantic segmentation data, while the latter does not use unlabeled images to train the image synthesis network.

To address limited data and semi-supervised settings, we introduce OCO-GAN, for Optionally COnditioned GAN, which enables both unconditional generation as well as conditional semantic image synthesis in a single framework, where most of the parameters between the two tasks are shared. We also propose a single end-to-end training process that does not require stage-wise training like previous hybrid models resulting in performance gain for unconditional and semantic image synthesis. See Fig. 1 for an illustration. Building off StyleMap-GAN [20], our model uses a style-based synthesis network with locally defined styles. The synthesis network is shared across both tasks, and is either conditioned on a semantic map or on latent variables. To train the model, we use a discriminator that is also shared between both tasks. A real/fake classification head for the entire image is used to train for unconditional generation, and a per-pixel classification loss [36] is used for conditional generation.

We perform experiments on the Cityscapes, COCO-Stuff and ADE20K datasets for the two addressed settings: for the limited data regime we train the models on small subsets of the original datasets, for the semi-supervised setting we train from the full datasets but use labels only for a subset of the images. To complete our evaluation, we also report results of experiments using the full datasets.

In summary, our contributions are the following:

- We propose OCO-GAN, a unified style-based model capable of both unconditional image generation as well as conditional semantic image generation.
- We introduce an end-to-end training procedure that combines conditional and unconditional losses using a shared discriminator network.
- Our experiments show that OCO-GAN improves over existing state-of-the-art hybrid approaches, while also yielding results that are competitive or better than those of state-of-the-art specialized methods for either conditional or unconditional generation in the three addressed settings.

2 Related Work

GANs for Image Generation Without Spatial Guidance. GANs have been widely used in computer vision applications, due to their excellent performance at synthesizing (near-)photo-realistic images. State-of-the-art unconditional generative models include StyleGAN [16,18,19] and BigGAN [3]. StyleGAN2 [16] is composed of two main components: a mapping network that produces the style of an image, and a synthesis network that incorporates the style at different feature levels through convolutional layers to generate the image. BigGAN uses a convolutional upsampling architecture, and conditions on class labels by using class-specific gains and biases in the BatchNorm layers. Unconditional GANs, such as StyleGAN2, obtain very good results when trained on datasets with limited diversity, such as ones containing only human faces, or a single class of objects such as cars. For more diverse datasets, *e.g.*ImageNet, class-conditional GANs such as BigGAN or StyleGAN-XL [34] allow for better generation quality by relying on manually defined labels for training and generation. StyleMapGAN [20] adds locality to the styles that are used in StyleGAN for affine modulation of the features in the layers of the image synthesis network. In our work we build upon the StyleMapGAN architecture, as the spatiality of the styles fits well with the semantic image generation task.

GANs for Semantic Image Synthesis. Conditional GANs for semantic image synthesis, a task consisting of generating photo-realistic images from semantic segmentation masks, have considerably developed over the last few years and offer more control over the generation process. Various models with different architectures and training techniques have been proposed. Pix2pix [13] is one of the first works, and proposes to use a U-Net [33] generator with a patch-based discriminator. SPADE [29] introduces spatially adaptive normalization layers that modulate the intermediate features with the input labels, which efficiently propagate semantic condition through the generator, and uses a multi-scale patch-based discriminator that evaluates the image/label pairs at different scales. CC-FPSE [23] also modifies the generator to produce conditional spatially-varying convolutional kernels and adopts a feature-pyramid discriminator. SC-GAN [41] divides the generator into two tasks: semantic encoding and style rendering. These semantic vector generator is trained with a regression loss, while adversarial and perceptual losses are used to train the style rendering generator. CollageGAN [22] aims at high resolution semantic image synthesis by developing a conditional version of StyleGAN2 and using class-specific generators, trained on additional datasets, which enable it to have better image quality for small objects. OASIS [36] adopts a new type of discriminator: a U-Net which is trained to predict the segmentation labels of pixels in real images and an additional "fake" label for pixels in generated images. In contrast to previous methods that are trained with perceptual and ℓ_2 reconstruction losses, OASIS demonstrates good performance when using only the adversarial loss, showing better diversity in the generated images without sacrificing image quality. In our work we build upon the state-of-the-art OASIS architecture, extending it to enable both conditional and unconditional generation.

Hybrid Conditional-Unconditional GANs. Almost all GANs in the literature are trained for either conditional or unconditional image synthesis. Existing hybrid approaches combine conditional and unconditional blocks in stage-wise training procedures. For example they train an encoder on top of a pre-trained unconditional GAN to perform conditional generation or image editing [27,32,45–47]. Among them, pixel2style2pixel (PSP) [32] is a network built for various tasks of image-to-image translation, that produces style vectors in the $W+$ latent space of a pre-trained StyleGAN2 generator. Semantic Bottleneck GAN (SB-GAN) [1] treats the semantic map as a latent variable to enable unconditional image synthesis. They separately train SPADE for conditional image generation, and a second unconditional model that can generate semantic segmentation maps. They then fine-tuning these two networks in an end-to-end approach, using Gumbel-Softmax approximation to differentiate through the discrete segmentation maps. They obtain better results than ProGAN [15] and BigGAN [3] for unconditional generation, and SPADE for conditional synthesis on the Cityscapes and ADE-indoor datasets. In contrast to these approaches, our OCO-GAN is based on a single end-to-end architecture, which takes semantic maps or latents as conditioning input to the shared generator, and uses a shared discriminator with a per-pixel loss and entire image classification loss to obtain the training signal for conditional and unconditional generation respectively. Training our hybrid architecture in a single training process with mixed batches of conditional and unconditional samples allows us to better leverage the synergy

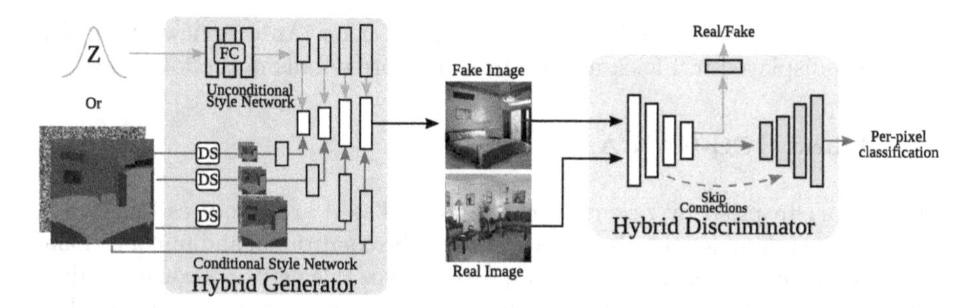

Fig. 2. Overview of OCO-GAN which enables conditional (red path) and unconditional (blue path) generation. Our hybrid generator takes alternatively a random vector, transformed into style maps in our unconditional style network in blue, or the concatenation of a segmentation map with a random map, going through downsampling (DS) layers into our conditional style network in red. (Color figure online)

between both tasks, as most of the network parameters are in the shared parts of the architecture, and are consistently trained by both losses. Recently, [12] introduced Product-Of-Experts GANs, able to synthesize images from multiple modalities, including text, segmentation maps and sketches, by using a product of experts to model a hierarchical conditional latent space and training with a multimodal discriminator. **Learning GANs from limited data.** Improving GANs by discriminator augmentation has recently been explored in several works to regularize the discriminator in low-data regimes [17,38,43,44]. For example, [17] uses an adaptive augmentation method (ADA), applying augmentations on real and synthetic images with a certain probability before they enter the discriminator. Specifically, the probability of augmentation is set proportionally to the fraction of real images receiving positive discriminator output. With this approach, the results of StyleGAN2 can often be matched using an order of magnitude less data. Another approach is explored in [24], which combines self-supervised and semi-supervised training. A feature extractor is learned with two heads for rotation and class label prediction respectively. It labels remaining data with a few-shot classifier, and trains a conditional GAN. These augmentation techniques are orthogonal to our hybrid OCO-GAN approach, and we show experimentally that ADA can be successfully applied to OCO-GAN.

3 Method

In this section we present OCO-GAN, which can generate images both in an unconditional manner, as well as conditioned on semantic segmentation maps. For training we leverage datasets composed of RGB images $\mathbf{x}_n \in \mathbb{R}^{H\times W\times 3}, n \in \{1,\dots,N\}$. For part or all of the images, we have corresponding segmentation maps $\mathbf{s}_n \in \{0,1\}^{H\times W\times C}$, represented with one-hot encoding across C classes. OCO-GAN consists of a shared synthesis network which takes input from either a conditional or unconditional style network. To train our hybrid model, we take an adversarial approach and use a shared discriminator network with separate branches for conditional and unconditional generation. An overview of the full pipeline is displayed in Fig. 2, and we provide more details in the following.

3.1 Hybrid OCO-GAN Architecture

Generator. Inspired by recent success of style-based architectures [16,19,34], we rely on a style space that is shared between the conditional and unconditional generation tasks. The generator is composed of two style networks corresponding to the conditional and unconditional cases. Their style representations are then processed by the shared image synthesis network. Sharing parameters between the conditional and unconditional tasks helps the unconditional branch to benefit from the available annotations in the limited data setting, while taking advantage of the large available quantity of unlabeled images in the partially annotated setting. We adopt a style representation with spatial dimensions, which captures more local variation and is better suited for semantic image synthesis.

The **unconditional style network** S_U is inspired from StyleMapGAN [20] which introduces spatial dimensions into style-based architectures. It consists of an MLP mapping network that transforms the latent variable $\mathbf{z} \sim \mathcal{N}(\mathbf{0}, \mathbf{I})$ into a style map, which is transformed at different resolutions by a stack of convolutional and up-sampling layers, to match the size of feature maps in the synthesis network.

The **conditional style network** S_C takes a segmentation map as input and produces spatial styles at different resolutions to feed into the synthesis network. To allow for diverse generations for a given semantic map, we follow OASIS [36] and concatenate a 3D noise tensor $\mathbf{Z} \in \mathbb{R}^{H \times W \times 64}$ to the input semantic map along the channel dimension. The noise tensor is obtained by spatially replicating a noise vector sampled from a unit Gaussian of dimension 64. We resize the extended input semantic map and apply several convolutional layers, to produce the style maps at matching resolutions for the layers of the synthesis network.

The styles from the conditional or unconditional style network are then incorporated in the shared **synthesis network** S_G through modulated convolutional layers in every residual block. The activations in the synthesis network are first normalized by subtracting and dividing by the mean and standard deviation computed per channel and per image. The styles are then convolved to generate gains and biases that are respectively multiplied with and added to the normalized feature maps [19]. We progressively upsample the features through transposed convolutions in residual blocks.

Discriminator. For our discriminator D, we use a U-Net [33] encoder-decoder architecture with skip connections, based on the one from OASIS [36]. The decoder outputs a per-pixel classification map across C+1 classes, representing the C labels in the dataset and an additional "fake" label. The task of the discriminator consists in segmenting real images using supervision from the ground truth label maps and classifying pixels in generated images as fake. The generator, on the other hand, aims to make the discriminator recognize generated pixels as belonging to the corresponding class in the ground truth label map.

Our discriminator differs from the one in OASIS in several respects. First, we add a classification head to the output of the encoder that consists of convolutional and fully connected layers, and which is used to map the entire input image to a real/fake classification score. The parameters in the discriminator's encoder are shared between both tasks to favor synergy during training. Parameter sharing is particularly useful in the limited data and partially annotated settings where each task can benefit from the other. Second, unlike OASIS, we don't use spectral normalization in the encoder part, but we find it beneficial to keep it in the decoder network. Third, in the decoder, before applying the residual upsampling blocks, we transform the output of the encoder by an *atrous* spatial pyramid pooling [6] module which has multiple *atrous* convolutions in parallel at different rates. As it increases the spacing between the convolution kernel elements, *atrous* convolutions allow to increase receptive fields while maintaining spatial dimension, making them well suited for segmentation tasks [7].

Details on the network architectures are given in the supplementary material.

3.2 Training

We train our models for both tasks jointly using the two discriminator branches, and use mixed batches of conditional and unconditional samples in training.

The unconditional discriminator branch classifies entire images as real or fake, and is trained using the standard binary cross-entropy loss:

$$\mathcal{L}_D^u = -\mathbb{E}_{\mathbf{x}}[\log D_{enc}(\mathbf{x})] - \mathbb{E}_{\mathbf{z}}[\log(1 - D_{enc}(S_{G,U}(\mathbf{z}))], \tag{1}$$

where $S_{G,U}(\mathbf{z}) = S_G(S_U(\mathbf{z}))$, the expectation over $\mathbf{x}$ is with respect to the empirical distribution of training images, and the expectation over $\mathbf{z}$ with respect to its unit Gaussian prior. We use the non-saturating GAN loss [9] to train the unconditional branch of the generator network:

$$\mathcal{L}_G^u = -\mathbb{E}_z[\log D_{enc}(S_{G,U}(\mathbf{z}))]. \tag{2}$$

In preliminary experiments, we observed best performance when this loss is applied only for unconditional generation.

For the conditional discriminator branch, we rely on the multi-class cross-entropy loss of OASIS [36], which aims to classify pixels from real images according to their ground-truth class, and pixels in generated images as fake:

$$\begin{aligned}\mathcal{L}_D^c = & -\mathbb{E}_{(\mathbf{x},\mathbf{s})}\left[\sum_{k=1}^{C}\alpha_k\sum_{h,w}^{H\times W}\mathbf{s}_{h,w,k}\log D_{dec}(\mathbf{x})_{h,w,k}\right] \\ & -\mathbb{E}_{(\mathbf{z},\mathbf{s})}\left[\sum_{h,w}^{H\times W}\log D_{dec}(S_{G,C}(\mathbf{z},\mathbf{s}))_{h,w,C+1}\right],\end{aligned} \tag{3}$$

where the α_k are class balancing terms, set to the inverse class frequency, expectation over $(\mathbf{x},\mathbf{s})$ is w.r.t.the empirical distribution, and over $(\mathbf{z},\mathbf{s})$ is w.r.t.the empirical distribution of segmentation maps $\mathbf{s}$ in the train set and the prior over $\mathbf{z}$. As for the generator, we use the non-saturating version of Eq. (3):

$$\mathcal{L}_G^c = -\mathbb{E}_{(\mathbf{z},\mathbf{s})}\left[\sum_{k,h,w=1}^{C,H,W} \alpha_k \mathbf{s}_{h,w,k} \log D_{dec}(S_{G,C}(\mathbf{z},\mathbf{s}))_{h,w,C+1}\right],$$

which aims to make the discriminator mistake generated pixels for each class with real pixels of that class.

To further improve the training of our models, we use additional regularization terms on both discriminator branches. For the unconditional branch, we add the R1 regularization loss [25] also used in StyleGAN2, that penalizes gradients of the discriminator on real images with high Euclidean norm. Like in StyleGAN2, we perform this regularization every 16 minibatches. For the conditional branch, we add the LabelMix loss of [36], which mixes a real training image with a generated one using the segmentation map of the real image. The loss aims to minimize differences in the class logits computed from the mixed image and the mixed logits computed from the real and generated images.

4 Experiments

We first describe our experimental setup, and then present quantitative and qualitative results in different settings, followed by ablation studies.

4.1 Experimental Setup

Datasets. To validate our models we use the Cityscapes [8], COCO-Stuff [4], ADE20K [45] and CelebA-HQ [21] datasets, and use the standard validation and training sets used in the GAN literature [5,29,36]. We resize images and segmentation maps to 256×256, except for Cityscapes where it is 256×512. Besides experiments on the full datasets, we consider the following two settings.

In the *limited data setting*, we use smaller subsets of the four datasets to test models. This is interesting as obtaining semantic maps for training is time consuming. As in [1], we use Cityscapes5K to refer to the subset of the 25K images in the full dataset for which fine semantic segmentation maps are available. For COCO-Stuff we randomly selected a subset of 12K images (10% of all images), which we refer to as COCO-Stuff12K. ADE-Indoor is a subset of ADE20K used in [1] consisting of 4,377 training and 433 validation samples of indoor scenes. Moreover, we sampled a random subset of 4K images (20% of all data), which we denote ADE4K. For CelebA-HQ dataset, we train models using 3k images.

In a *partially annotated setting*, we test to what extent using a large number of unlabeled images is beneficial in cases where limited labeled training images are available. Here we train with the full datasets, but use only a small portion of the semantic maps. We used only 500 labeled images for Cityscapes and CelebA-HQ datasets, and 2K labels for ADE20K and COCO-Stuff datasets. Due space limitations, we report results on CelebA-HQ in the supplementary material.

Table 1. Results in the **limited data setting**. In italic: Results taken from original papers. In bold: Best result among unconditional, conditional, and hybrid models, resp.

	Cityscapes5K			ADE-Indoor			ADE4K			COCO-Stuff12K		
	↓FID	**↓CFID**	**↑mIoU**	↓FID	↓CFID	↑mIoU	↓FID	↓CFID	↑mIoU	↓FID	↓CFID	↑mIoU
StyleGAN2 [16]	79.6			87.3			95.7			77.4		
StyleMapGAN [20]	**70.4**			**74.8**			**87.8**			**65.2**		
BigGAN [3]	80.5			109.6			112.7			82.2		
SPADE [30]		*71.8*	*62.3*		**50.3**	44.7		**43.4**	34.3		28.3	33.8
OASIS [36]		**47.7**	**69.3**		50.8	**57.5**		49.9	**36.1**		**25.6**	**37.6**
PSP [32]	79.6	118.7	19.3	87.3	148.7	35.6	95.7	157.7	3.2	77.4	225.1	3.1
SB-GAN [1]	*65.5*	*60.4*	n/a	*85.3*	*48.2*	n/a	101.5	49.3	24.9	99.6	52.6	27.2
OCO-GAN (Ours)	**57.4**	**41.6**	**69.4**	**69.0**	**47.9**	**58.6**	**80.6**	**43.8**	**36.1**	**51.5**	**25.6**	**40.3**

We will release our data spits along with our code and trained models upon publication to facilitate reproduction of our results.

Evaluation Metrics. To assess image quality we report the standard **FID** metric [10] for both unconditional and conditional generation. We use **CFID** to denote the FID metric computed for **C**onditionally generated images. To assess the consistency of semantic image synthesis with the corresponding segmentation map, we report the **mIoU** metric, calculated using the same segmentation networks as [36]. We report metrics averaged over five sets of samples, and provide standard deviations and more evaluation details in the supplementary material.

Baselines. We compare to two hybrid models, capable of both conditional and unconditional image generation. To our knowledge, they are the only existing methods that handle both types of generation. SB-GAN [1] was evaluated on Cityscapes5K, Cityscapes and ADE-Indoor, and we train it on the other datasets using the settings provided by the authors for ADE-Indoor. PSP [32] trains an encoder on top of a (fixed) pre-trained StyleGAN2 generator, and was only evaluated on human face and animal face datasets. As unconditional models we include StyleGAN2 [16], StyleMAPGAN [20], and BigGAN [3]. For the class-conditional BigGAN, we use a single trivial class label for all images to obtain an unconditional model, as in [1,5,24,28]. For semantic image synthesis we compare our method to the influential SPADE approach [30], and the recent OASIS [36] model which reports the best performance on the (full) datasets we consider.

Where possible we report results from the original papers for the baselines (in italic). In other cases we train models using code released by the authors. In the partially annotated case, we train SB-GAN and PSP with the same labelled and unlabelled images as OCO-GAN. For PSP method, we use a pretrained StyleGAN2 on all the unconditional dataset and train the encoder using the limited amount of labelled data, while for SB-GAN, we train semantic synthesizer and SPADE generator with limited labeled data and use all the dataset for training the unconditional discriminator in the finetuning training phase.

4.2 Main Results

Limited Data Setting. In Table 1 we evaluate OCO-GAN in limited data regime in terms of FID, CFID and mIoU metrics, along with state-of-the-art

Fig. 3. Unconditional (left, first three columns) and conditional (right, last three columns) generations from models in the **limited data setting**, trained on ADE-Indoor (top three rows) and COCO-Stuff12K (bottom three rows). SB-GAN unconditional images for ADE-Indoor taken from original paper.

baselines. On both Cityscapes and ADE-Indoor, SB-GAN outperforms StyleGAN2 and BigGAN unconditional models, as well as SPADE on conditional generation, showing that this is a strong baseline for hybrid generation.

OCO-GAN consistently improves the hybrid PSP and SB-GAN baselines on all metrics and datasets. We observe poor CFID and mIoU values for PSP on all the datasets, showing that this two-stage approach fails to generate realistic images that adhere to the semantic conditioning. OCO-GAN also outperforms the unconditional baselines StyleGAN2, StyleMapGAN and BigGAN on all the datasets, with an average improvement of 9.9 points of FID in comparison to the best baseline. It shows that the unconditional generation benefits from the semantic supervision signal during training. When it comes to conditional generation, OCO-GAN matches or improves the CFID and mIoU scores of the state-of-the-art semantic synthesis model SPADE and OASIS. CFID is improved by 6.1 points on Cityscapes5K, while we notice an improvement of 2.4 points on

Table 2. Results in the **Partially annotated setting**. In bold: Best result among unconditional, conditional, and hybrid models, resp.

	Cityscapes			ADE20K			COCO-Stuff		
	↓FID	↓CFID	↑mIoU	↓FID	↓CFID	↑mIoU	↓FID	↓CFID	↑mIoU
StyleGAN2 [16]	49.1			43.1			37.0		
StyleMapGAN [20]	**48.8**			**39.6**			**33.0**		
BigGAN [3]	67.8			81.0			70.5		
SPADE [30]		83.7	52.1		**43.4**	34.3		**42.0**	**28.6**
OASIS [36]		**77.4**	**63.5**		49.9	**36.1**		77.2	23.5
SB-GAN [1]	96.1	93.1	50.2	82.6	45.5	31.0	103.5	57.6	19.2
PSP [32]	**49.1**	111.8	13.8	43.1	147.2	2.8	**37.0**	168.1	2.4
OCO-GAN	49.6	**49.3**	**65.2**	**40.4**	**40.1**	**37.6**	44.4	**36.8**	**32.1**

Fig. 4. Semantic synthesis on Cityscapes25K in the **partially annotated setting**.

ADE-Indoor. On ADE4K, SPADE is slightly better than OCO-GAN in CFID by 0.4, but has a worse mIoU value by 1.8. In the case of COCO-Stuff12K dataset, OCO-GAN is on par with OASIS in CFID, but improves mIoU by 2.7.

In Fig. 3 we present samples for the ADE-Indoor and COCO-Stuff12K datasets. For the unconditional case, we show samples depicting similar scenes to aid comparison. In these more diverse datasets, OCO-GAN is able to produce images of higher quality containing sharper and more precise details, in both conditional and unconditional settings. See for example the unconditional samples depicting kitchens and dining rooms for ADE-Indoor, or the conditional giraffe sample for COCO-Stuff12K, in particular the legs and torso of the giraffe, as well as the trees in the background in the same image.

Partially Annotated Setting. We present our results in the partially annotated setting in Table 2. Also in this setting, OCO-GAN considerably outperforms hybrid baselines in most cases. Indeed, PSP only works well for unconditional generation, as it is based on a pretrained StyleGAN2 generator (trained on the full unlabeled dataset in this setting), but performs poorly for semantic synthesis due to the limited data to learn the style inference network. SB-GAN has worse metrics than OCO-GAN for both types of generation. While our model is competitive with unconditional baselines in terms of FID, we observe a

Table 3. Results on the **full datasets**. Italic: Results taken from original papers. On Cityscapes, SB-GAN uses a segmentation net to train with additional segmentations. In bold: Best result among unconditional, conditional, and hybrid models, resp.

	Cityscapes			ADE20K			COCO-Stuff			Inf. time secs.
	↓FID	↓CFID	↑mIoU	↓FID	↓CFID	↑mIoU	↓FID	↓CFID	↑mIoU	↓Unc./Cond.
StyleGAN2 [16]	49.1			43.1			37.0			-/0.02
StyleMapGAN [20]	**48.8**			**39.6**			**33.0**			-/0.06
BigGAN	67.8			81.0			70.5			-/0.01
SPADE [30]		*71.8*	*62.3*		*33.9*	*38.5*		*22.6*	*37.4*	0.03/-
OASIS [36]		***47.7***	***69.3***		***28.3***	***48.8***		***17.0***	***44.1***	0.01/-
SB-GAN [1]	*63.0*	*54.1*	n/a	62.8	35.0	34.5	89.6	42.6	19.3	0.07/0.03
PSP [32]	49.1	99.2	15.6	43.1	146.9	2.8	**37.0**	211.6	2.1	0.02/0.12
OCO-GAN	**48.8**	**41.5**	**69.1**	**38.6**	**30.4**	**47.1**	38.5	**20.9**	**42.1**	0.05/0.06

Fig. 5. Unconditional (left) and cond. (right) samples from ADE20K models in the **full data setting**.

substantial improvements in conditional generation when comparing to SPADE and OASIS. Indeed, for Cityscapes, ADE20K and COCO-Stuff, OCO-GAN is respectively 21.6, 3.3 and 3.0 points of CFID lower, and 9.9, 1.5 and 3.5 points of mIoU higher than the best conditional model. Where in the limited data setting above unconditional generation benefits from joint training, in the current setting conditional generation improves by jointly training our hybrid architecture. In Fig. 4, we show conditional samples of Cityscapes. We observe sharper details and better quality images for OCO-GAN compared to OASIS and SB-GAN, especially when looking at cars and vegetation.

Full Data Setting. We provide results on the full datasets in Table 3. As in the limited data and partially annotated experiments, OCO-GAN outperforms the hybrid baselines, SB-GAN and PSP in all cases, except on COCO-Stuff where PSP has slightly better FID. Although in this setting there is less benefit of addressing both tasks with a single unified model, our results are still competitive with specialized unconditional and conditional generation models. For the unconditional task, OCO-GAN achieves the best FID values on Cityscapes and

Table 4. Ablation of architectures and training modes. We evaluate models in the limited data setting with Cityscapes5K and partially annotated setting with Cityscapes25K (with 500 annotated images). Italics: results from original paper.

	Cityscapes5K			Cityscapes25K		
	↓FID	↓CFID	↑mIoU	↓FID	↓CFID	↑mIoU
Unconditional training						
StyleGAN2	79.6			49.1		
StyleMapGAN	70.4			**48.8**		
OCO-GAN - uncond. only	82.0			50.5		
Conditional training						
OASIS		*47.7*	*69.3*		77.4	63.5
OCO-GAN - cond. only		43.7	**70.3**		76.8	57.7
Hybrid training						
OCO-GAN	**57.4**	**41.6**	69.4	49.8	**55.8**	**65.2**

ADE20K, but obtains a higher FID than StyleGAN2 on COCO-Stuff. For conditional generation, OCO-GAN obtains slightly worse scores than OASIS (except for CFID on Cityscapes, where OCO-GAN is the best), but outperforms SPADE.

We present unconditional and conditional synthesized samples of models trained on ADE20k in Fig. 5. We notice the improvement of image quality from SB-GAN to OCO-GAN. Indeed, the difference is particularly striking when looking at the building in the second row for semantic image synthesis, which is better rendered by OCO-GAN. We additionally report inference time and comment on the results in section C.7 of Supplementary Material.

4.3 Ablation Studies

Impact of Hybrid Training and Architecture. In Table 4, we conduct an ablation on Cityscapes5K in the limited data regime and Cityscapes25K in the partially annotated setting to separate out the effect of the OCO-GAN architecture and that of the joint training of the unconditional and conditional branches.

For unconditional generation, we see the benefit of hybrid training in the unconditional generation results of OCO-GAN in the *limited data regime*. We obtain 82.0 FID by training OCO-GAN for unconditional generation only, which improves to 57.4 FID when jointly training both (un)conditional branches. In the partially annotated setting we do not observe this dramatic improvement, as in this case many more unlabeled images are available during training.

For conditional generation, when we only train the conditional branch, OCO-GAN improves over OASIS on both datasets and metrics, in particular for the partially annotated setting (CityScapes25K) where there are only 500 labeled images. Hybrid training improves conditional generation in all cases, except in mIoU for Cityscapes5K, where the mIoU degrades by 0.9 points, from 70.3 to 69.4, to the level of OASIS (69.3). In the partially annotated case the CFID is very strongly by 19.3 points from 76.8 to 57.5 CFID, demonstrating that OCO-GAN effectively leverages the many unlabeled images available in this setting.

Table 5. Effect of Adaptive Data Augmentation (ADA) on FID for StyleGAN2 and OCO-GAN (trained in hybrid mode) for the datasets of the limited data setting.

	City5K	ADE-Ind	ADE4K	CCS12K
StyleGAN2	79.6	87.3	95.7	77.4
same w/ ADA	60.6	63.9	77.8	**48.9**
OCO-GAN	57.4	69.0	80.6	51.5
same w/ ADA	**48.0**	**58.8**	**67.2**	50.0

Moreover, we ran two additional experiments on Cityscapes5K. In the first, we start training the unconditional part only, then fix it and train the conditional branch. In the second, we proceed in the reverse order. In the first case we get a FID of 73.8 and CFID of 126.4 while in the second case we have FID 95.3 and CFID 50.2, which is to compare with a FID 57.4 and CFID 41.6 in the joint training. These results show the benefit of our joint training approach, as compared to stage-wise training, as done *e.g.*in PSP [32].

Comparison with Adaptive Data Augmentation. Data augmentation can boost the performance of generative models, especially when dealing with small datasets, see *e.g.* [17,38,43,44]. To assess to what extent such methods are complementary to our hybrid approach, we perform experiments with Adaptive Data Augmentation (ADA) [17] in the limited data setting. The results in Table 5 show that both StyleGAN2 and OCO-GAN benefit from ADA for all datasets. We observe an improvement of 9.4, 10.2 and 13.4 points FID when adding ADA to OCO-GAN on Cityscapes5K, ADE-Indoor and ADE4K respectively, while for the larger COCO-Stuff-12K the improvement is limited to 1.5 points.

Ablation of Regularisation Terms to Train Our Models. We show in Table 5 of Supplementary Material the effects of regularization during training.

5 Conclusion

We propose OCO-GAN, a model capable of synthesizing images with and without conditioning on semantic maps. We combine a shared style-based synthesis network, containing most of the network parameters, with conditional and unconditional style generating networks. We train the model using a U-Net discriminator with a full-image real-fake classification branch, and a per-pixel branch that classifies pixels across semantic classes and an additional "fake" label.

We validate our model on the Cityscapes, ADE20K, and COCO-Stuff datasets, using three settings: a limited data setting, a semi-supervised setting, and the full datasets. In all settings our approach outperforms existing hybrid models that can generate both with and without spatial semantic guidance, and is competitive with specialized conditional or unconditional state-of-the-art baselines. In limited data regime our method improves over state-of-the-art unconditional models, by leveraging the additional training signal provided by the labeled images. In the semi-supervised setting we improve over state-of-the-art conditional models, by using the large pool of unlabeled images to regularize the training.

References

1. Azadi, S., Tschannen, M., Tzeng, E., Gelly, S., Darrell, T., Lucic, M.: Semantic bottleneck scene generation. arXiv preprint arXiv:1911.11357 (2019)
2. Bau, D., et al.: GAN dissection: visualizing and understanding generative adversarial networks. In: ICLR (2019)
3. Brock, A., Donahue, J., Simonyan, K.: Large scale GAN training for high fidelity natural image synthesis. In: ICLR (2019)
4. Caesar, H., Uijlings, J., Ferrari, V.: COCO-Stuff: thing and stuff classes in context. In: CVPR (2018)
5. Casanova, A., Careil, M., Verbeek, J., Drozdzal, M., Romero-Soriano, A.: Instance-conditioned GAN. In: NeurIPS (2021)
6. Chen, L., Papandreou, G., Schroff, F., Adam, H.: Rethinking atrous convolution for semantic image segmentation. In: CVPR (2017)
7. Chen, L.C., Zhu, Y., Papandreou, G., Schroff, F., Adam, H.: Encoder-decoder with atrous separable convolution for semantic image segmentation. In: CVPR (2018)
8. Cordts, M., et al.: The Cityscapes dataset for semantic urban scene understanding. In: CVPR (2016)
9. Goodfellow, I., et al.: Generative adversarial nets. In: NeurIPS (2014)
10. Heusel, M., Ramsauer, H., Unterthiner, T., Nessler, B., Hochreiter, S.: GANs trained by a two time-scale update rule converge to a local Nash equilibrium. In: NeurIPS (2017)
11. Ho, J., Jain, A., Abbeel, P.: Denoising diffusion probabilistic models. In: NeurIPS (2020)
12. Huang, X., Mallya, A., Wang, T.C., Liu, M.Y.: Multimodal conditional image synthesis with product-of-experts GANs. arXiv preprint arXiv:2112.05130 (2021)
13. Isola, P., Zhu, J., Zhou, T., Efros, A.A.: Image-to-image translation with conditional adversarial networks. In: CVPR (2017)
14. Jacob Walker, Ali Razavi, A.v.: Predicting video with VQVAE. arXiv preprint arXiv:2103.01950 (2021)
15. Karras, T., Aila, T., Laine, S., Lehtinen, J.: Progressive growing of GANSs for improved quality, stability, and variation. In: ICLR (2018)
16. Karras, T., Laine, S., Aittala, M., Hellsten, J., Lehtinen, J., Aila, T.: Analyzing and improving the image quality of stylegan. In: CVPR (2020)
17. Karras, T., Aittala, M., Hellsten, J., Laine, S., Lehtinen, J., Aila, T.: Training generative adversarial networks with limited data. In: NeurIPS (2020)
18. Karras, T., et al.: Alias-free generative adversarial networks. In: NeurIPS (2021)
19. Karras, T., Laine, S., Aila, T.: A style-based generator architecture for generative adversarial networks. In: CVPR (2019)
20. Kim, H., Choi, Y., Kim, J., Yoo, S., Uh, Y.: Exploiting spatial dimensions of latent in GAN for real-time image editing. In: CVPR (2021)
21. Lee, C.H., Liu, Z., Wu, L., Luo, P.: MaskGAN: towards diverse and interactive facial image manipulation. In: CVPR (2020)
22. Li, Y., Li, Y., Lu, J., Shechtman, E., Lee, Y.J., Singh, K.K.: Collaging class-specific GANs for semantic image synthesis. In: ICCV (2021)

23. Liu, X., Yin, G., Shao, J., Wang, X.: Learning to predict layout-to-image conditional convolutions for semantic image synthesis. In: NeurIPS (2019)
24. Lucic, M., Tschannen, M., Ritter, M., Zhai, X., Bachem, O., Gelly, S.: High-fidelity image generation with fewer labels. In: ICML (2019)
25. Mescheder, L., Nowozin, S., Geiger, A.: Which training methods for gans do actually converge? In: International Conference on Machine Learning (ICML) (2018)
26. Nash, C., Ganin, Y., Eslami, S., Battaglia, P.: PolyGen: an autoregressive generative model of 3D meshes. In: ICML (2020)
27. Nguyen, A., Clune, J., Bengio, Y., Dosovitskiy, A., Yosinski, J.: Plug and play generative networks: conditional iterative generation of images in latent space. In: CVPR (2017)
28. Noroozi, M.: Self-labeled conditional GANs. arXiv preprint arXiv:2012.02162 (2020)
29. Park, J., Florence, P., Straub, J., Newcombe, R., Lovegrove, S.: DeepSDF: learning continuous signed distance functions for shape representation. In: CVPR (2019)
30. Park, T., Liu, M.Y., Wang, T.C., Zhu, J.Y.: Semantic image synthesis with spatially-adaptive normalization. In: CVPR (2019)
31. Ramesh, A., et al.: Zero-shot text-to-image generation. In: Meila, M., Zhang, T. (eds.) Proceedings of the 38th International Conference on Machine Learning. Proceedings of Machine Learning Research, vol. 139, pp. 8821–8831. PMLR, 18–24 July 2021
32. Richardson, E., et al.: Encoding in style: a StyleGAN encoder for image-to-image translation. In: CVPR (2021)
33. Ronneberger, O., Fischer, P., Brox, T.: U-net: Convolutional networks for biomedical image segmentation. In: Medical Image Computing and Computer-Assisted Intervention (2015)
34. Sauer, A., Schwarz, K., Geiger, A.: StyleGAN-XL: Scaling StyleGAN to large diverse datasets. arXiv preprint arXiv:2202.00273 (2022)
35. Sun, W., Wu, T.: Learning layout and style reconfigurable GANs for controllable image synthesis. IEEE Trans. Pattern Anal. Mach. Intell. 1–1 (2021). https://doi.org/10.1109/TPAMI.2021.3078577
36. Sushko, V., Schönfeld, E., Zhang, D., Gall, J., Schiele, B., Khoreva, A.: You only need adversarial supervision for semantic image synthesis. In: ICLR (2021)
37. Sylvain, T., Zhang, P., Bengio, Y., Hjelm, R.D., Sharma, S.: Object-centric image generation from layouts. CoRR abs/2003.07449 (2020)
38. Tran, N.T., Tran, V.H., Nguyen, N.B., Nguyen, T.K., Cheung, N.M.: On data augmentation for GAN training. arXiv preprint arXiv:2006.05338 (2020)
39. Vahdat, A., Kautz, J.: NVAE: a deep hierarchical variational autoencoder. arXiv preprint arXiv:2007.03898 (2020)
40. Voynov, A., Babenko, A.: Unsupervised discovery of interpretable directions in the GAN latent space. In: ICML (2020)
41. Wang, Y., Qi, L., Chen, Y.C., Zhang, X., Jia, J.: Image synthesis via semantic composition. In: ICCV (2021)
42. Yang, S., et al.: Statistical parametric speech synthesis using generative adversarial networks under a multi-task learning framework. arXiv preprint arXiv:1707.01670 (2017)
43. Zhao, S., Liu, Z., Lin, J., Zhu, J.Y., Han, S.: Differentiable augmentation for data-efficient GAN training. In: NeurIPS (2020)
44. Zhao, Z., Singh, S., Lee, H., Zhang, Z., Odena, A., Zhang, H.: Improved consistency regularization for GANs. arXiv preprint arXiv:2002.04724 (2020)

45. Zhou, B., Zhao, H., Puig, X., Fidler, S., Barriuso, A., Torralba, A.: Scene parsing through ADE20K dataset. In: CVPR (2017)
46. Zhu, J., Shen, Y., Zhao, D., Zhou, B.: In-domain GAN inversion for real image editing. In: Vedaldi, A., Bischof, H., Brox, T., Frahm, J.-M. (eds.) ECCV 2020. LNCS, vol. 12362, pp. 592–608. Springer, Cham (2020). https://doi.org/10.1007/978-3-030-58520-4_35
47. Zhu, J.-Y., Krähenbühl, P., Shechtman, E., Efros, A.A.: Generative visual manipulation on the natural image manifold. In: Leibe, B., Matas, J., Sebe, N., Welling, M. (eds.) ECCV 2016. LNCS, vol. 9909, pp. 597–613. Springer, Cham (2016). https://doi.org/10.1007/978-3-319-46454-1_36

Efficient Image Super-Resolution Using Vast-Receptive-Field Attention

Lin Zhou[1], Haoming Cai[1], Jinjin Gu[2,3], Zheyuan Li[1], Yingqi Liu[1], Xiangyu Chen[1,2,4], Yu Qiao[1,2], and Chao Dong[1,2(✉)]

[1] ShenZhen Key Lab of Computer Vision and Pattern Recognition, SIAT-SenseTime Joint Lab, Shenzhen Institutes of Advanced Technology, Chinese Academy of Sciences, Shenzhen, China
{zy.li3,yq.liu3,yu.qiao,chao.dong}@siat.ac.cn
[2] Shanghai AI Laboratory, Shanghai, China
[3] The University of Sydney, Sydney, Australia
jinjin.gu@sydney.edu.au
[4] University of Macau, Zhuhai, China

Abstract. The attention mechanism plays a pivotal role in designing advanced super-resolution (SR) networks. In this work, we design an efficient SR network by improving the attention mechanism. We start from a simple pixel attention module and gradually modify it to achieve better super-resolution performance with reduced parameters. The specific approaches include: (1) increasing the receptive field of the attention branch, (2) replacing large dense convolution kernels with depthwise separable convolutions, and (3) introducing pixel normalization. These approaches paint a clear evolutionary roadmap for the design of attention mechanisms. Based on these observations, we propose VapSR, the Vast-receptive-field Pixel attention network. Experiments demonstrate the superior performance of VapSR. VapSR outperforms the present lightweight networks with even fewer parameters. And the light version of VapSR can use only 21.68% and 28.18% parameters of IMDB and RFDN to achieve similar performances to those networks. The code and models are available at https://github.com/zhoumumu/VapSR.

Keywords: Image super-resolution · Deep convolution network · Attention mechanism

1 Introduction

Single image Super-Resolution (SISR) is a fundamental low-level vision problem that aims at recovering a high-resolution (HR) image from its low-resolution (LR) observations. SISR has attracted increasing attention in both the research community and industry. Since SRCNN [13] introduced deep learning into SR, deep networks have become the de facto approach for advanced SR algorithms

L. Zhou and H. Cai—Equal Contributions.

L. Karlinsky et al. (Eds.): ECCV 2022 Workshops, LNCS 13802, pp. 256–272, 2023.
https://doi.org/10.1007/978-3-031-25063-7_16

due to their ease of use and high performance. However, deep SR networks rely on a large number of parameters that can provide sufficiently complex capacity to map LR images to HR images. These parameters and high computation costs limit the application of SR networks. The design of SR networks with efficiency as the primary goal has gradually become an important issue.

Among the numerous SR networks, the studies related to the attention mechanism have achieved a lot of success. The channel attention brought by RCAN [54] makes it practical to train very deep high-performance SR networks. PAN [56] has achieved good progress in designing a lightweight SR network using pixel attention. After image processing entered the Transformer era, the application of the attention mechanism underwent great changes. Vision Transformers [15] rely on attention mechanisms to achieve excellent performance. Many works have proved that introducing large receptive fields and local windows [9,44] in the attention branch improves the SR effect. However, many advanced design ideas have not been verified in designing the attention mechanism for convolutional lightweight SR networks. In this paper, we start from a basic pixel attention module to explore better attention mechanisms designed for efficient SR.

The first effort we made in this paper was to introduce the large receptive field design into the attention mechanism. This is in line with other recent design trends using large kernel sizes [16], as well as the design principles of transformers [9,36,44]. We show the advantages of using large kernel convolutions in the attention branch. Secondly, we use depthwise separable convolution to split dense large convolution kernels. A large receptive field is achieved in the attention branch using a depthwise and a depthwise dilated convolution. We also replace the 3×3 convolutions in the backbone network with 1×1 convolutions to reduce the number of parameters. Thirdly, we present a novel pixel normalization that can make the training less prone to crashing.

Along the above footprints, we demonstrate a novel path to an efficient SR architecture called VapSR (VAst-receptive-field Pixel attention network). Compared with the current state-of-the-art algorithms, the proposed VapSR reduces a lot of parameters while improving the SR effect. For example, compared with the champion of the NTIRE2022 efficient SR competition [27], VapSR achieves an improvement on PSNR by more than 0.1dB with 185K fewer parameters. Our experiments demonstrate the effectiveness of the proposed method.

2 Related Work

Deep Networks for SR. Since SRCNN [13] was proposed as the pioneering work for employing a three-layer convolutional neural network for the SR task, numerous methods [14,24,30,37,43,45] have been proposed to achieve better performance. FSRCNN [14] proposes a pipeline that upsamples features at the end of the network, which boosts the performance while keeping the model lightweight. VDSR [24] introduces skip connections for residual learning to increase the depth of the SR network. DRCN [25] and DRRN [45] both adopt the recursive structure to improve the reconstruction performance. SRDenseNet [47] and RDN [55] prove that the dense connection is beneficial to improving the capacity of SR models. RCAN [54] employs a channel attention scheme to bring the attention mechanism to the SR methods. SAN [11] proposes a second-order attention module, which brings further performance improvement. SwinIR [36] proposed to utilize Swin Transformer [39] for the SR task. HAT [9] refreshes state-of-the-art performance through hybrid attention schemes and a pre-training strategy.

Attention Schemes for SR. The attention mechanism can be interpreted as a way to bias the allocation of available resources towards the most informative parts of an input signal. There are approximately four attention schemes: channel attention, spatial attention, combined channel-spatial attention and self-attention mechanism. RCAN [54], inspired by SENet [18], reweights the channel-wise features according to their respective weight responses. SelNet [10] and PAN [56] employ the spatial attention mechanism, which calculates the weight for each element. Regarding combined channel-spatial attention, HAN [42] additionally proposes a spatial attention module via 3D convolutional operation. The self-attention mechanism was adopted from natural language processing to model the long-range dependence [50]. IPT [7] is the first Transformer-based SR method based on the ViT [15]. It relies on a large model scale (over 115.5M parameters). SwinIR [36] calculates the window-based self-attention to save the computations. HAT [9] further proposes multiple attention schemes to improve the window-based self-attention and introduce channel-wise attention to SR Transformer.

Efficient SR Models. Efficient SR design aims to reduce model complexity and latency for SR networks [2,8,20,21,28,35,38]. CARN [2] employs the combination of group convolution and 1×1 convolution to save computations. After IDN [21] proposes the residual feature distillation structure, there is a series of works [20,35,38] following this micro-architecture design. IMDN [20] improves IDN via an information multi-distillation block by using a channel splitting strategy. RFDN [38] rethinks the channel splitting operation and introduces a new equivalent architecture of the progressive refinement module. In NTIRE 2022 Efficient SR Challenge [34], RLFN [28] won the championship in the runtime track by ditching the multi-branch design of RFDN for faster parallel computing and introducing a contrastive loss for better performance. BSRN [35] won the first place in the model complexity track by replacing the standard convolution with a well-designed depth-wise separable convolution to save computations and utilizing two effective attention schemes to enhance the model ability.

Large Kernel Design. CNNs used to be the common choice for computer vision tasks. However, CNNs have been greatly challenged by Transformers recently [6,15,31,39], and Transformer-based methods have also shown leading performances on the SR task [7,9,32,36]. In Transformer, self-attention is designed to be either global [7,15] or local, both accompanied by larger kernels [9,36,39]. Thus, information can be gathered from a large region. Inspired by this characteristic of Transformer, a series of works have been proposed to design better CNNs [12,16,40,48]. ConvMixer [48] utilizes large kernel convolutions to build the model and achieve the competitive performance to the ViT [15]. ConvNeXt

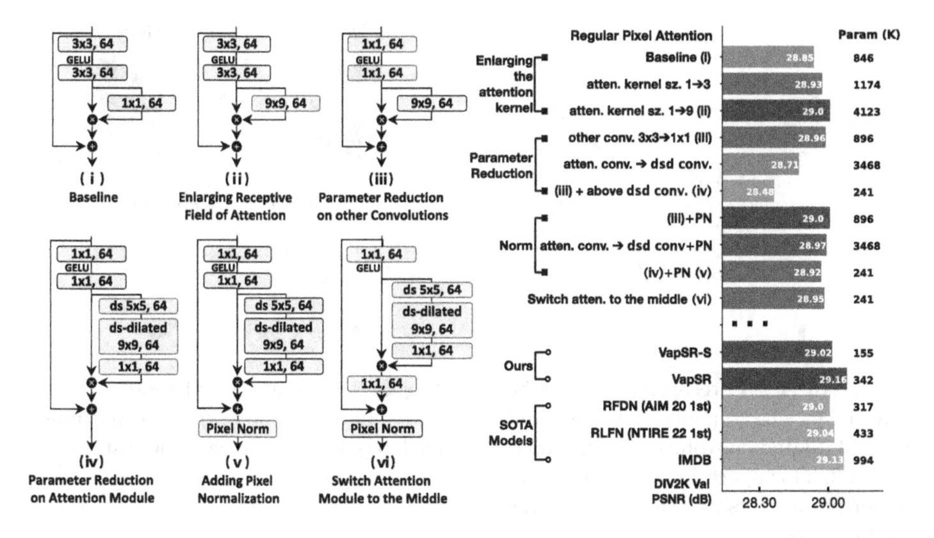

Fig. 1. The evolutionary design roadmap of the proposed method. The figures on the left are the key architectural milestones. The plot on the right shows the PSNR performance on DIV2K and the corresponding parameters. Every evolution and modification of the main design stages are marked with red box on the left and described with the text on the right. And we omit some micro designs in this plot but they will be elaborated lately in Sect. 5.3. (Color figure online)

[40] proves that well-designed CNN with large kernel convolution can obtain similar performance to Swin Transformer [39]. RepLKNet [12] scales up the filter kernel size to 31×31 and outperforms the state-of-the-art Transformer-based methods. VAN [16] conducts an analysis of the visual attention and proposes the large kernel attention based on the depth-wise convolution.

3 Motivation

The attention mechanism has been proven effective in SR networks. In particular, an efficient SR model PAN [56] using pixel attention achieves good performance while greatly reducing the number of parameters. Pixel attention performs an attention operation on each element of the features. Compared with channel attention and spatial attention, pixel attention is a more general form of attention operation and thus provides a good baseline for our further exploration.

Inspired by recent advances in self-attention [50] and vision transformers [15], we believe that there is still considerable room for improvement even for the attention mechanism based on convolutional operations. In this section, we show the process of improving SR network attention through three design criteria in pixel attention. First, we show the advantages of using large kernel convolutions in the attention branch. Then we use well-designed depth-wise separable convolutions to reduce the huge computational burden brought by large kernel convolutions. We demonstrate the potential of this network topology design for

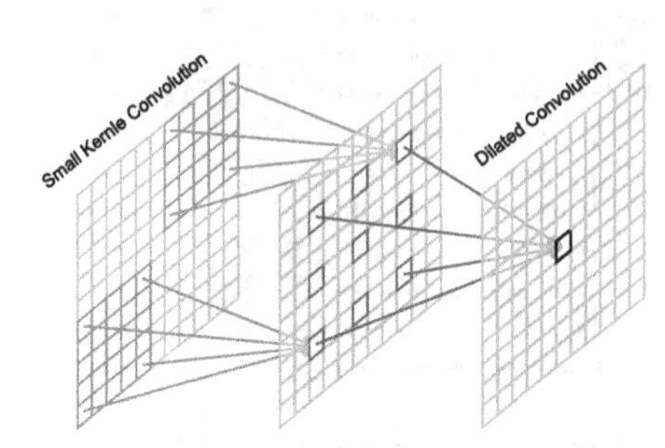

Fig. 2. An 11×11 receptive field can be replaced by a 5×5 small convolution and a 3×3 dilated convolution with a dilation of 3. This operation saves the number of parameters and achieves a large receptive field.

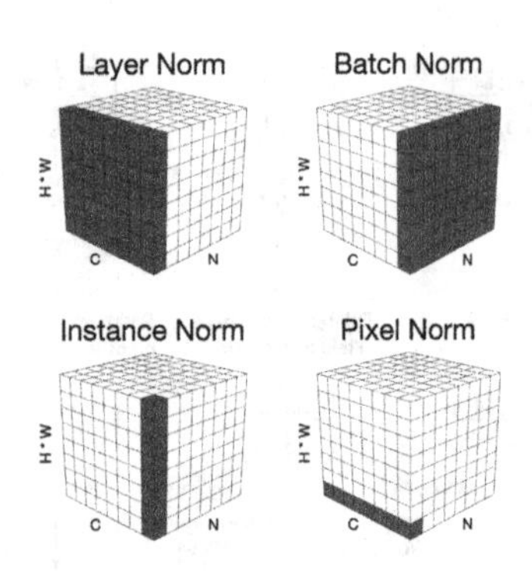

Fig. 3. Different normalization methods. The cubes in purple are normalized by the same mean and variance.

efficient SR. Finally, inspired by vision Transformers, we introduce a pixel-wise normalization operation in the convolutional network to train SR networks with complex attention efficiently and stably. We demonstrate a solid roadmap toward our improved network design.

3.1 Large Kernel in Visual Attention

We start with the building blocks of one of the most common SR networks with a basic pixel attention operation. This block is shown in Fig. 1(i). In general, the main operations that provide receptive fields in SR networks are the two 3×3 convolutions in the residual block. Whereas the attention branch contains only one 1×1 convolution. However, inspired by the vision Transformers, we are able to improve the performance by increasing the receptive field of the attention branch. We enlarge the kernel size in the attention branch of the baseline block to 3 and 9 to study the effect of enlarging the attention kernel, respectively. This modification is shown in Fig. 1(ii). Its performance variation is shown by the first 3 experiments in the roadmap on the right of Fig. 1. It can be seen that although it brings tonnes of additional parameters, enlarging the kernel size in attention brings about 0.15 dB of performance improvement. After showing that performing large kernel convolutions in attention can bring benefits, we continue to explore this basis.

3.2 Parameter Reduction

Enlarging kernel size brings a large number of parameters. We then try to remove relatively unimportant parts of the network as much as possible and reduce parameters accordingly. The above kernel enlarging strategy provides an architecture that relies on one large kernel convolution in attention and 3×3

convolutions outside the attention. The good news is that dense convolution kernels are often not the best choice for large kernel sizes. We can reduce the parameters of the network by implementing a more sparse large kernel convolution.

Depthwise separable convolution is a classic, intuitive solution for large kernel convolution parameter reduction. Depthwise separable convolution splits a dense convolution operation into spacial depthwise convolutions and point-size convolutions in channels. The depthwise convolution is in the form of a group convolution that assigns only one kernel for each feature channel. The pointwise convolution is a 1×1 convolution for channel fusion. Inspired by [16], the depthwise convolution can be further decomposed. Taking a convolution of size 11×11 as an example, we can convert it into a 5×5 normal convolution and a 3×3 dilated convolution with a dilation of 3 while keeping its equivalent receptive field size unchanged, as shown in Fig. 2. This design reduces the number of parameters as much as possible while keeping the receptive field even larger. In the fourth experiment on the right of Fig. 1, we find a performance drop of only 0.04dB after replacing the large attention kernel with a set of depthwise separable convolutions. This operation saves about 3,200K parameters.

In addition to the above solution, compared with the large kernel attention operation, the receptive field brought by the rest 3×3 convolutions in the original backbone is no longer important. We replace the two 3×3 convolutions with 1×1 convolutions and find a parameter reduction of 655K. However, the performance drops dramatically after this modification. By combining the depthwise separable convolution attention and the 1×1 body convolutions, we can further reduce the parameter to 241K. The implementation of such architecture is shown in Fig. 1(iv). We argue that with good training of this network, we might be able to achieve high performance with such small parameters.

3.3 Pixel Normalization for Stable Attention Training

Due to the introduction of element-wise multiplication in the attention mechanism, the training stability is greatly reduced. At a small learning rate, the network cannot converge well, and increasing the learning rate will cause the network to return abnormal gradients, causing training collapse. The above parameter reduction solution produces such a difficult-to-train network that it suffers a performance drop of about 0.5 dB due to the training issue.

We find that this training problem is partly due to internal covariate shift [22] phenomena. For a network with attention layers, its multiplication makes the degree of shift more difficult to control. To solve this problem, we introduce a pixel normalization layer to normalize the shifted layer distribution to a standard normal distribution. The difference between pixel normalization and other normalization methods is shown in Fig. 3. Given a feature tensor that can be formulated as $x \in \mathbb{R}^{HW \times C}$, ($H$, W and C are the height, width and feature dimension), x can be viewed as HW feature vectors, and each vector belongs to a pixel position. We represent the feature vector of the ith pixel with $x^i \in \mathbb{R}^C$. The mean and variance of x^i are:

$$\mu^i = \frac{1}{C}\sum_{j=1}^{C} x_j^i, \quad \sigma^i = \frac{1}{C}\sum_{j=1}^{C}(x_j^i - \mu^i)^2. \tag{1}$$

The output of the pixel normalization can be formulated as

$$\tilde{x}^i = \frac{x^i - \mu^i}{\sqrt{\sigma^i + \epsilon}} \odot \gamma + \beta, \tag{2}$$

where γ and β represent the parameter vectors for scaling and shifting and have the same dimensions as C. Different from the other normalization methods in the existing literature, pixel normalization calculates the mean and variance of the features of different pixels and normalizes them separately. In other words, pixel normalization's shifting and scaling operations are spatially inhomogeneous.

The boost from using pixel normalization is huge. According to Fig. 1, using pixel normalization on the reduced parameter model yields excellent results close to using a large dense kernel in attention. When equipped with the above two practices that reduce parameters, we can reduce the number of network parameters to 241K on the basis of outperforming the baseline. At this point, we get the novel architecture shown in Fig. 1(v).

3.4 Discussion

The generated network design correlates with some existing models in several respects. Firstly, in vision transformers, layer normalization has been proven important for the good performance of Transformers [15,36]. The Transformer networks [36] usually reshape the feature map $C \times H \times W$ into shape $HW \times C$ and then perform layer normalization. However, the layer normalization at this time is no longer consistent with the layer norm originally used for convolution networks [4], but is equivalent to the pixel norm described in this paper when the token size is 1×1. The original layer normalization will introduce parameter numbers consistent with the element number in the feature tensor; thus, models built with it can no longer handle arbitrary resolutions. The pixel normalization we describe is the most flexible and efficient of the many other normalization layers available. Note that although the equivalent method of our pixel normalization has already appeared in the vision transformers, its successful use in convolution networks has not yet been witnessed.

We also found that our findings are very similar to a concurrent work, the large kernel attention (LKA) [16]. The main difference is that LKA uses two 1×1 convolutional layers as projection layers and places the attention layer in the middle of the convolutional layers. This design is also very similar to Transformer's use of the attention mechanism. Figure 1 shows the result when using a similar approach to LKA. It can be seen that the changing layer order brings a performance improvement of 0.03dB. At this point, our method is shown in the Fig. 1 (vi). We show that the proposed VapSR and VapSR-S built using this building block use fewer parameters while achieving SR performance higher than the existing methods.

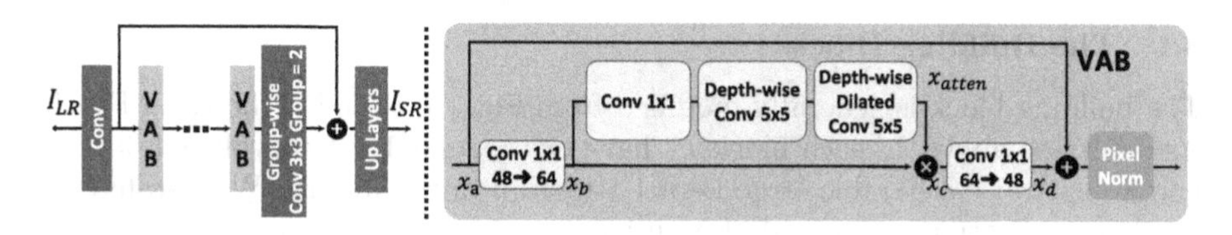

Fig. 4. The architecture of the proposed VapSR. The rightmost block is the detailed illustration of the proposed main block VAB. I_{LR} and I_{SR} are the corresponding input low-resolution image and the output image of the network.

4 Network Architecture

Based on the building block discussed above, we build a novel SR network called VapSR (VAst-receptive-field Pixel attention network). The architecture is illustrated in Fig. 4. The high-level design of the proposed network follows the common design of deep SR networks. VapSR contains three modules: (1) feature extraction, (2) nonlinear mapping, and (3) reconstruction.

Given a low-resolution image I_{LR}, the feature extraction contains a convolution layer with a kernel size of 3×3 to extract features from I_{LR}.

$$x_0 = f_{ext}(I_{LR}), \tag{3}$$

and x_0 is the extracted higher dimensional feature maps.

In the nonlinear mapping stage, x_0 is then fed into a stack of the building blocks to enhance the feature representations. We denote the building blocks as $f_{VAB}(\cdot)$, and this process can be formulated as

$$x_n = f^n_{VAB}(f^{n-1}_{VAB}(...f^0_{VAB}(x_0)...), \tag{4}$$

where x_n represents the output feature map of the nth VAB. At the end of the nonlinear mapping stage, we add a 3×3 convolution layer $f_{ref}(\cdot)$ after the building blocks and perform a residual connection with x_0:

$$x_{map} = f_{ref}(x_n) + x_0. \tag{5}$$

At last, we utilize the reconstruction module to upsample the features to the HR size. Here we obtain:

$$I_{SR} = f_{rec}(x_{map}), \tag{6}$$

where $f_{rec}(\cdot)$ is the reconstruction module, and I_{SR} is the final result of the network. Our reconstruction module contains two $\times 2$ pixel-shuffle layers to implement a $\times 4$ upsampling scale, which brings a consistent promotion compared to a single $\times 4$ pixel-shuffle layer. There are convolution layers before both pixel-shuffle layers, and the number of channels can be adjusted as required.

4.1 The Building Blocks

The building block is generally modified from Fig. 1 (vi) and is shown in Fig. 4. As described above, in each block we have two regular 1×1 convolution layers, and a depthwise separable large kernel attention in the middle. We also have a pixel normalization at the end of each block.

Given input feature x_a, the first 1×1 convolution layer projects x_a to x_b and expands the number of channels from 48 to 64. We perform GELU activation [17] to x_b, and this is the only activation in the building block. In the attention module, we firstly use a 1×1 point-wise convolution for channel fusion. Then we use a depth-wise convolution with a kernel size of 5 and a depth-wise dilation convolution with a kernel size of 5 and a dilation of 3. The combination of these two convolution layers is able to implement a receptive field of 17. The feature x_{atten} generated by the attention branch is the same size as the original feature x_b through reasonable padding. The attention is implemented using an element-wise product as $x_c = x_{atten} \odot x_b$. Then, another 1×1 convolution layer projects x_c to x_d and shrinks the number of channels back to 48. At last, the pixel normalization is performed on $x_d + x_a$.

5 Experiments

5.1 Experimental Setup

Datasets and Evaluation Metrics. The training images consist of 2650 images from Flickr2K [37] and 800 images from DIV2K [1] train. We evaluate our models on widely used benchmark datasets - Set5 [5], Set14 [53], BSD100 [41], and Urban100 [19]. The commonly used data augmentation methods are applied in the training dataset. Specifically, We use the random combination of random rotation by 90°, 180°, 270° and flipping horizontally for data augmentation. The average peak-signal-to-noise ratio (PSNR) and the structural similarity [23,51] (SSIM) on the luminance (Y) channel are used as the evaluation metrics.

Implementation Details. We implement two models, VapSR and VapSR-S. VapSR $\times 4$ consists of 21 VABs and VapSR-S (also for the $\times 4$ scale) is the light version of VapSR with 11 VABs. And we configure the input and output feature to 32 channels instead of 48 channels for VapSR-S. Both of them have two $\times 2$ pixel-shuffle layers and two convolution layers. We make minor adjustment on the Up-Layers and the number of blocks for $\times 2$ and $\times 3$ scale.

Training Details. The model is trained using the Adam optimizer [26] with $\beta_1 = 0.9$ and $\beta_2 = 0.99$. Notably, using $\beta_2 = 0.99$ stead of commonly used $\beta_2 = 0.999$ can bring better performance for our proposed model design. The learning rate is set to 1×10^{-3} during the whole 1×10^6 training iterations. And we set a smaller learning rate specially for the $\times 2$ scale. The weight of the exponential moving average (EMA) [3] is set to 0.999. Only the L1 loss is used to optimize the model. For VapSR, the mini-batch is set to 64, and the input patch size is set to 48×48. We enlarge the minibatch size to 192, and the patch size to 64×64 for VapSR-S.

Table 1. Quantitative comparison with state-of-the-art methods on benchmark datasets. The best and second-best performance are in red and blue colors, respectively. 'Multi-Adds' is calculated with a 1280 × 720 GT image.

Method	Scale	Params[K]	Multi-Adds[G]	Set5 PSNR/SSIM	Set14 PSNR/SSIM	B100 PSNR/SSIM	Urban100 PSNR/SSIM
Bicubic	×2	–	–	33.66/0.9299	30.24/0.8688	29.56/0.8431	26.88/0.8403
SRCNN [13]		8	52.7	36.66/0.9542	32.45/0.9067	31.36/0.8879	29.50/0.8946
FSRCNN [14]		13	6.0	37.00/0.9558	32.63/0.9088	31.53/0.8920	29.88/0.9020
VDSR [24]		666	612.6	37.53/0.9587	33.03/0.9124	31.90/0.8960	30.76/0.9140
LapSRN [29]		251	29.9	37.52/0.9591	32.99/0.9124	31.80/0.8952	30.41/0.9103
DRRN [45]		298	6,796.9	37.74/0.9591	33.23/0.9136	32.05/0.8973	31.23/0.9188
MemNet [46]		678	2,662.4	37.78/0.9597	33.28/0.9142	32.08/0.8978	31.31/0.9195
IDN [21]		553	124.6	37.83/0.9600	33.30/0.9148	32.08/0.8985	31.27/0.9196
CARN [2]		1592	222.8	37.76/0.9590	33.52/0.9166	32.09/0.8978	31.92/0.9256
IMDN [20]		694	158.8	38.00/0.9605	33.63/0.9177	32.19/0.8996	32.17/0.9283
PAN [56]		261	70.5	38.00/0.9605	33.59/0.9181	32.18/0.8997	32.01/0.9273
LAPAR-A [33]		548	171.0	38.01/0.9605	33.62/0.9183	32.19/0.8999	32.10/0.9283
RFDN [38]		534	95.0	38.05/0.9606	33.68/0.9184	32.16/0.8994	32.12/0.9278
RLFN [27]		527	115.4	38.07/0.9607	33.72/0.9187	32.22/0.9000	32.33/0.9299
BSRN [35]		332	73.0	38.10/0.9610	33.74/0.9193	32.24/0.9006	32.34/0.9303
VapSR (ours)		329	74.0	38.08/0.9612	33.77/0.9195	32.27/0.9011	32.45/0.9316
Bicubic	×3	–	–	30.39/0.8682	27.55/0.7742	27.21/0.7385	24.46/0.7349
SRCNN [13]		8	52.7	32.75/0.9090	29.30/0.8215	28.41/0.7863	26.24/0.7989
FSRCNN [14]		13	5.0	33.18/0.9140	29.37/0.8240	28.53/0.7910	26.43/0.8080
VDSR [24]		666	612.6	33.66/0.9213	29.77/0.8314	28.82/0.7976	27.14/0.8279
LapSRN [29]		502	149.4	33.81/0.9220	29.79/0.8325	28.82/0.7980	27.07/0.8275
DRRN [45]		298	6,796,9	34.03/0.9244	29.96/0.8349	28.95/0.8004	27.53/0.8378
MemNet [46]		678	2,662.4	34.09/0.9248	30.00/0.8350	28.96/0.8001	27.56/0.8376
IDN [21]		553	56.3	34.11/0.9253	29.99/0.8354	28.95/0.8013	27.42/0.8359
CARN [2]		1592	118.8	34.29/0.9255	30.29/0.8407	29.06/0.8034	28.06/0.8493
IMDN [20]		703	71.5	34.36/0.9270	30.32/0.8417	29.09/0.8046	28.17/0.8519
PAN [56]		261	39.0	34.40/0.9271	30.36/0.8423	29.11/0.8050	28.11/0.8511
LAPAR-A [33]		544	114.0	34.36/0.9267	30.34/0.8421	29.11/0.8054	28.15/0.8523
RFDN [38]		541	42.2	34.41/0.9273	30.34/0.8420	29.09/0.8050	28.21/0.8525
BSRN [35]		340	33.3	34.46/0.9277	30.47/0.8449	29.18/0.8068	28.39/0.8567
VapSR (ours)		337	33.6	34.52/0.9284	30.53/0.8452	29.19/0.8077	28.43/0.8583
Bicubic	×4	–	–	28.42/0.8104	26.00/0.7027	25.96/0.6675	23.14/0.6577
SRCNN [13]		8	52.7	30.48/0.8626	27.50/0.7513	26.90/0.7101	24.52/0.7221
FSRCNN [14]		13	4.6	30.72/0.8660	27.61/0.7550	26.98/0.7150	24.62/0.7280
VDSR [24]		666	612.6	31.35/0.8838	28.01/0.7674	27.29/0.7251	25.18/0.7524
LapSRN [29]		813	149.4	31.54/0.8852	28.09/0.7700	27.32/0.7275	25.21/0.7562
DRRN [45]		298	6,796.9	31.68/0.8888	28.21/0.7720	27.38/0.7284	25.44/0.7638
MemNet [46]		678	2,662.4	31.74/0.8893	28.26/0.7723	27.40/0.7281	25.50/0.7630
IDN [21]		553	32.3	31.82/0.8903	28.25/0.7730	27.41/0.7297	25.41/0.7632
CARN [2]		1592	90.9	32.13/0.8937	28.60/0.7806	27.58/0.7349	26.07/0.7837
IMDN [20]		715	40.9	32.21/0.8948	28.58/0.7811	27.56/0.7353	26.04/0.7838
PAN [56]		272	28.2	32.13/0.8948	28.61/0.7822	27.59/0.7363	26.11/0.7854
LAPAR-A [33]		659	94.0	32.15/0.8944	28.61/0.7818	27.61/0.7366	26.14/0.7871
RFDN [38]		550	23.9	32.24/0.8952	28.61/0.7819	27.57/0.7360	26.11/0.7858
RLFN [27]		527	29.8	32.24/0.8952	28.62/0.7813	27.60/0.7364	26.17/0.7877
BSRN-S [35]		156	8.3	32.16/0.8949	28.62/0.7823	27.58/0.7365	26.08/0.7849
VapSR-S (ours)		155	9.0	32.14/0.8951	28.64/0.7826	27.60/0.7373	26.05/0.7852
BSRN [35]		352	19.4	32.35/0.8966	28.73/0.7847	27.65/0.7387	26.27/0.7908
VapSR(ours)		342	19.5	32.38/0.8978	28.77/0.7852	27.68/0.7398	26.35/0.7941

Table 2. Quantitative comparison of Micro designs subordinate components in the structure. The model (vi) here is the same one shown in the roadmap Fig. 1

Exp. Idx.	Group conv.	Inverted bottleneck	Deeper up. block	Parameters [K]	DIV2K PSNR
Model (vi)	–	–	–	241.1	28.95
Model (vi)+	✓	–	–	222.7	28.92
Model (vi)++	✓	✓	–	152.2	28.84
Model (vii)	✓	✓	✓	156.0	28.86

5.2 Comparison with State-of-the-Art Methods

We compare the proposed VapSR with exsisting common lightweight SR approaches with upscaling foctor $\times 2$, $\times 3$ and $\times 4$, including SRCNN [13], FSRCNN [14], VDSR [24], LapSRN [29], DRRN [45], MemNet [46], IDN [21], CARN [2], IMDB [20], PAN [56], LAPAR-A [33], RFDN [38], RLFN [27], and BSRN [35]. Table 1 shows the quantitative comparison results for different upscale factors. We also provide the number of parameters and Multi-Adds calculated on the 1280×720 output. Benefit from the simple yet efficient structure, the proposed VapSR achieves state-of-the-art performance with remarkably few parameters. Specifically, our VapSR $\times 4$ uses 21.68% and 28.18% parameters of RFDN $\times 4$ and IMDN $\times 4$, while obtains average 0.187 dB improvement on four evaluation datasets. Moreover, the proposed VapSR-S using the fewest parameters to achieve competitive performance to BSRN-S [35], which is the winner of the model complexity sub-track in NTIRE 2022 Challenge on Efficient Super-Resolution [34]. Interestingly, our structure has relative advantages on metric SSIM than PSNR as well. The results on SSIM can keep on top of the competing models even though the PSNR is slightly lower.

Figure 5 shows the qualitative comparison of the proposed method. Our approach can reconstruct stripes and line patterns more accurately than the existing methods, reflecting the advantage of the proposed method on metric SSIM as we mentioned above. We take the image "img_092" for example. Most of the existing methods generate noticeable artifacts and blurry effects, while our method produces accurate lines. For the details of the buildings in "img_011", "img_062" and "img_074", VapSR could also make reconstruction with fewer artifacts.

5.3 Ablation Study on Micro Design

In this section, we conduct ablation studies on the micro designs involved in our final method. The micro designs have four parts: subordinate components in the architecture, normalization type, attention layers' sequence, and receptive field. The following ablation study applies a smaller batch size and fewer training iterations to make time consumption more acceptable.

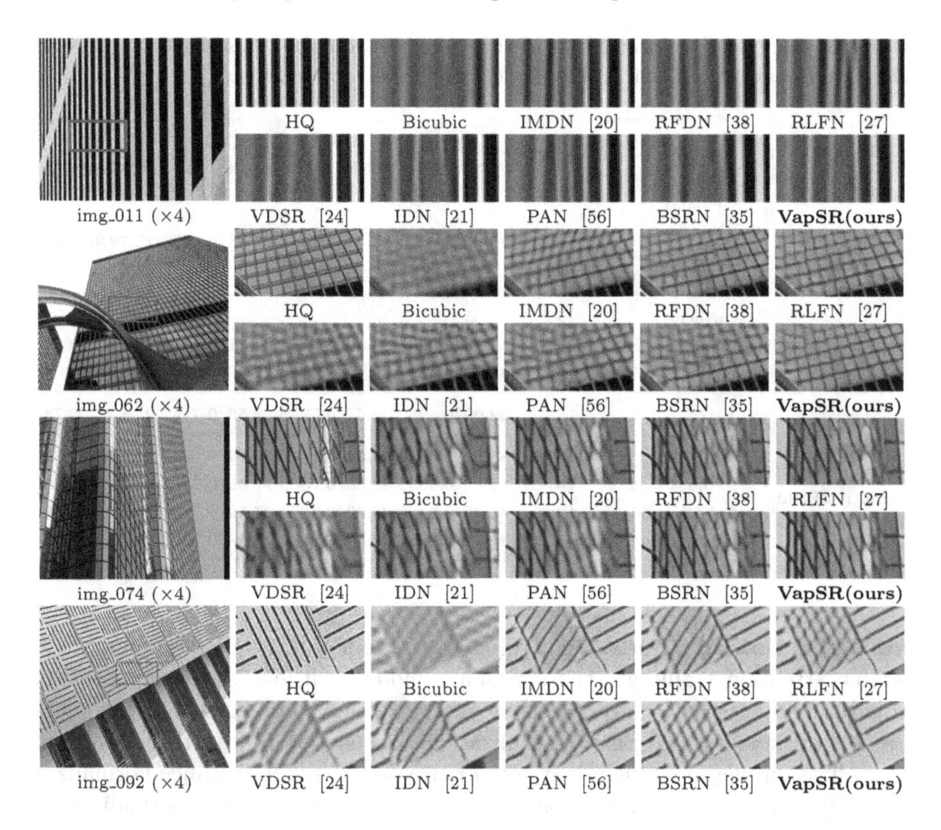

Fig. 5. Visual comparison about image SR (×4) in some challenging cases.

Subordinate Components in Structure. We make some detailed designs for the network to achieve further efficiency, as shown in Table 2. These changes are all small but essential tricks and could shed light on designing a high-performance SR network. The first promoted trick is replacing the regular convolution layer with a group-wise convolution layer at the end of the non-linear mapping module We find that it's a harmless way to reduce the model size while maintaining performance. The second promoted trick is the inverted bottleneck architecture. We implement this architecture by letting the two 1×1 convolution layers at both ends of the VAB expand and then shrink the channels, as explained in Sect. 4.1. This trick reduces about 30% of model parameters while maintaining its PSNR performance in an acceptable region. The third one is using deeper upsampling layers. We implement two convolution layers instead of a single one. This trick brings performance improvements while introducing acceptable extra parameters. We apply these tricks to the model (vi) shown in Fig. 1 step by step, and finally, we get the model (vii).

Table 3. Quantitative comparison of three ablation variables based on model (vii). The performance values surpassing model (vii) are in violet.

Variable	Method	Params[K]	Set5 PSNR/SSIM	Set14 PSNR/SSIM	B100 PSNR/SSIM	Urban100 PSNR/SSIM
-	model (vii){ PixelNorm 5 - 7 - 1 k = 7, block = 10}	156	32.00/0.8929	28.52/0.7797	27.53/0.7343	25.79/0.7768
Normalization	BatchNorm	156	-	-	-	-
	InstanceNorm	156	31.78/0.8897	28.33/0.7764	27.41/0.7328	25.15/0.7646
	GroupNorm	156	31.93/0.8912	28.40/0.7766	27.46/0.7323	25.62/0.7693
Order	1 - 5 - 7	156	32.00/0.8930	28.52/0.7798	27.53/0.7347	25.80/0.7773
Receptive Field	k = 5, block = 11	152	32.01/0.8933	28.52/0.7801	27.54/0.7349	25.83/0.7782
	k = 5, block = 12	164	32.02/0.8932	28.55/0.7805	27.56/0.7356	25.90/0.7800
	k = 9, block = 9	161	31.96/0.8925	28.49/0.7789	27.50/0.7337	25.77/0.7761
	k = 11, block = 8	166	31.93/0.8917	28.47/0.7785	27.49/0.7333	25.72/0.7745

Normalization Methods. According to the experiments of designing stage "stable attention training", those primary models without any normalization would finally crash at a large learning rate $1 \times e^{-3}$. To demonstrate the effectiveness of the described pixel normalization, we compare it with three common normalization methods, including batch normalization [22], instance normalization [49] and group normalization [52]. Since the standard layer normalization [4] raises the network's parameters to 2.7M, it's outside the scope of this ablation experiment. The validation curves are shown in Fig. 6(b) and the results are listed in the Table 3. It can be observed that batch normalization does not help in preventing training crashes. Group normalization and instance normalization decrease PSNR and SSIM by a large margin, and the training are unstable or well-converged. In contrast, our pixel normalization can achieve a steady training curve and the best performance growth.

The Sequence of Layers in Attention. We conducted the experiment of adjusting the sequence of the attention layers. [35] proved that putting the point-wise convolution before the depth-wise convolution achieves better performance in the SR task. Hence, we rearrange the order of the attention layers denoting as 1-5-7, which represents putting 1×1 point-wise convolution in front, next the 5×5 depth-wise convolution and then 7×7 depth-wise dilation convolution. The experiment show that order 1-5-7 performs slightly better than 5-7-1 which the model (vii) take, which verified the conclusion mentioned above again. Moreover, we also conducted experiments to prove that the influence of two depth-wise convolution layers' positions is relatively smaller and inconsistent. So we adopted the order 1-5-7 as the final structure.

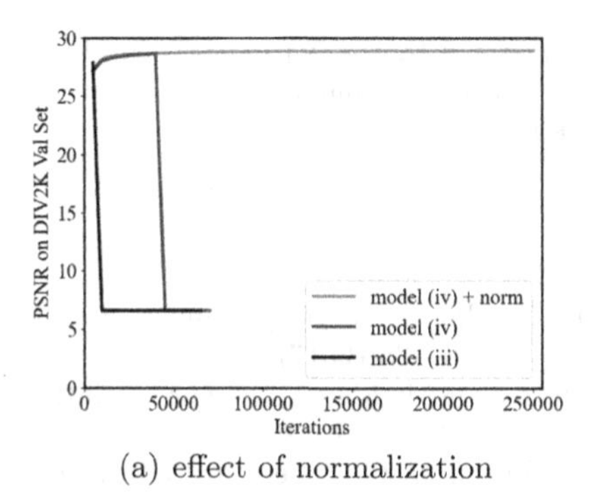

(a) effect of normalization

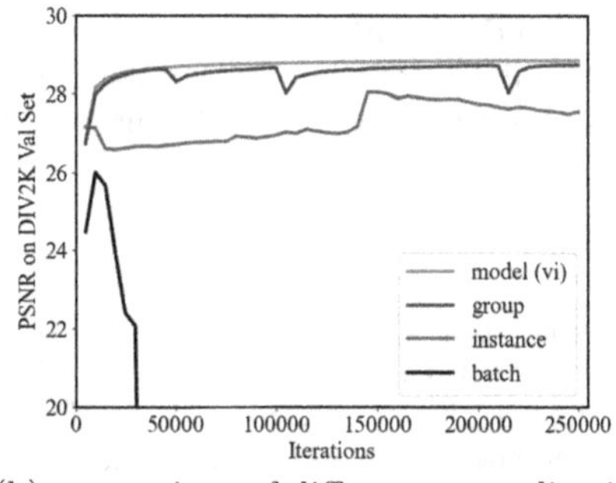

(b) comparison of different normalization

Fig. 6. Validation curves with or without normalization under the setting of large learning rate. Plot (a) shows lack of normalization causes the models crash early. In plot (b), the effect of different normalization from good to bad are as followed: pixel normalization (red), group normalization (green), instance normalization (purple) and batch normalization (blue). (Color figure online)

Attention Layers' Receptive Field. We explored the effect of attention layers with different receptive fields, which are mainly dominated by the dilated convolution layer. The field of a 7×7 convolution with dilation 3 is equivalent to 19. We modify the kernel size of this layer to 5, 9, and 11 separately to change the field and adjust the number of blocks to keep the networks' parameters close to or slightly more than the model (vii). As shown in Table 3, increasing the kernel size to 9 and 11 causes an apparent successive performance drop. While the smaller kernel size of 5 can obtain performance promotion, even when we keep the parameters less than model(vii) with only 11 blocks. Therefore we choose the structure with a kernel size of 5 finally. With the retrofit of order and receptive field, the eventual structure of VapSR is settled now, as illustrated in Fig. 4.

6 Conclusions

This work proposes a lightweight convolutional neural network called VapSR to achieve efficient image super-resolution. The experiments demonstrate that our VapSR can achieve state-of-the-art performance with concise structure and fewer parameters. Starting from the motivation of improving the attention mechanism, we first verified the advantages of using large kernel convolutions on the SR task. Then we successfully apply the efficient depth-wise separable convolution to reduce the model size. Thirdly, the proposed pixel normalization makes it possible to train this architecture steadily, and we prove it superior to other normalizations. We detailed the design process in the form of a roadmap. It reveals how we squeeze the complexity of the model while keeping the performance step by step clearly and eventually leading to VapSR.

Acknowledgements. This work is partially supported by the National Natural Science Foundation of China (61906184, U1913210), and the Shanghai Committee of Science and Technology, China (Grant No. 21DZ1100100).

References

1. Agustsson, E., Timofte, R.: NTIRE 2017 challenge on single image super-resolution: Dataset and study. In: CVPRW, pp. 126–135 (2017)
2. Ahn, N., Kang, B., Sohn, K.A.: Fast, accurate, and lightweight super-resolution with cascading residual network. In: ECCV, pp. 252–268 (2018)
3. Athiwaratkun, B., Finzi, M., Izmailov, P., Wilson, A.G.: There are many consistent explanations of unlabeled data: why you should average. arXiv preprint arXiv:1806.05594 (2018)
4. Ba, J.L., Kiros, J.R., Hinton, G.E.: Layer normalization. arXiv preprint arXiv:1607.06450 (2016)
5. Bevilacqua, M., Roumy, A., Guillemot, C., Alberi-Morel, M.L.: Low-complexity single-image super-resolution based on nonnegative neighbor embedding. In: BMVC, pp. 135.1–135.10 (2012)
6. Carion, N., Massa, F., Synnaeve, G., Usunier, N., Kirillov, A., Zagoruyko, S.: End-to-end object detection with transformers. In: Vedaldi, A., Bischof, H., Brox, T., Frahm, J.-M. (eds.) ECCV 2020. LNCS, vol. 12346, pp. 213–229. Springer, Cham (2020). https://doi.org/10.1007/978-3-030-58452-8_13
7. Chen, H., et al.: Pre-trained image processing transformer. In: CVPR, pp. 12299–12310 (2021)
8. Chen, H., Gu, J., Zhang, Z.: Attention in attention network for image super-resolution. arXiv preprint arXiv:2104.09497 (2021)
9. Chen, X., Wang, X., Zhou, J., Dong, C.: Activating more pixels in image super-resolution transformer. arXiv preprint arXiv:2205.04437 (2022)
10. Choi, J.S., Kim, M.: A deep convolutional neural network with selection units for super-resolution. In: CVPRW, pp. 154–160 (2017)
11. Dai, T., Cai, J., Zhang, Y., Xia, S.T., Zhang, L.: Second-order attention network for single image super-resolution. In: CVPR, pp. 11065–11074 (2019)
12. Ding, X., Zhang, X., Han, J., Ding, G.: Scaling up your kernels to 31×31: revisiting large kernel design in CNNs. In: CVPR, pp. 11963–11975 (2022)
13. Dong, C., Loy, C.C., He, K., Tang, X.: Learning a deep convolutional network for image super-resolution. In: Fleet, D., Pajdla, T., Schiele, B., Tuytelaars, T. (eds.) ECCV 2014. LNCS, vol. 8692, pp. 184–199. Springer, Cham (2014). https://doi.org/10.1007/978-3-319-10593-2_13
14. Dong, C., Loy, C.C., Tang, X.: Accelerating the super-resolution convolutional neural network. In: Leibe, B., Matas, J., Sebe, N., Welling, M. (eds.) ECCV 2016. LNCS, vol. 9906, pp. 391–407. Springer, Cham (2016). https://doi.org/10.1007/978-3-319-46475-6_25
15. Dosovitskiy, A., et al.: An image is worth 16×16 words: transformers for image recognition at scale. arXiv preprint arXiv:2010.11929 (2020)
16. Guo, M.H., Lu, C.Z., Liu, Z.N., Cheng, M.M., Hu, S.M.: Visual attention network. arXiv preprint arXiv:2202.09741 (2022)
17. Hendrycks, D., Gimpel, K.: Gaussian error linear units (GELUs). arXiv preprint arXiv:1606.08415 (2016)
18. Hu, J., Shen, L., Sun, G.: Squeeze-and-excitation networks. In: CVPR, pp. 7132–7141 (2018)

19. Huang, J.B., Singh, A., Ahuja, N.: Single image super-resolution from transformed self-exemplars. In: CVPR, pp. 5197–5206 (2015)
20. Hui, Z., Gao, X., Yang, Y., Wang, X.: Lightweight image super-resolution with information multi-distillation network. In: ACM Multimedia, pp. 2024–2032 (2019)
21. Hui, Z., Wang, X., Gao, X.: Fast and accurate single image super-resolution via information distillation network. In: CVPR, pp. 723–731 (2018)
22. Ioffe, S., Szegedy, C.: Batch normalization: accelerating deep network training by reducing internal covariate shift. In: ICML, pp. 448–456 (2015)
23. Jinjin, G., Haoming, C., Haoyu, C., Xiaoxing, Y., Ren, J.S., Chao, D.: PIPAL: a large-scale image quality assessment dataset for perceptual image restoration. In: Vedaldi, A., Bischof, H., Brox, T., Frahm, J.-M. (eds.) ECCV 2020. LNCS, vol. 12356, pp. 633–651. Springer, Cham (2020). https://doi.org/10.1007/978-3-030-58621-8_37
24. Kim, J., Lee, J.K., Lee, K.M.: Accurate image super-resolution using very deep convolutional networks. In: CVPR, pp. 1646–1654 (2016)
25. Kim, J., Lee, J.K., Lee, K.M.: Deeply-recursive convolutional network for image super-resolution. In: CVPR, pp. 1637–1645 (2016)
26. Kingma, D.P., Ba, J.: Adam: a method for stochastic optimization. arXiv preprint arXiv:1412.6980 (2014)
27. Kong, F., et al.: Residual local feature network for efficient super-resolution. In: CVPR, pp. 766–776 (2022)
28. Kong, F., et al.: Residual local feature network for efficient super-resolution. In: CVPRW, pp. 766–776 (2022)
29. Lai, W.S., Huang, J.B., Ahuja, N., Yang, M.H.: Deep laplacian pyramid networks for fast and accurate super-resolution. In: CVPR, pp. 624–632 (2017)
30. Ledig, C., et al.: Photo-realistic single image super-resolution using a generative adversarial network. In: CVPR, pp. 4681–4690 (2017)
31. Li, K., et al.: Uniformer: unifying convolution and self-attention for visual recognition. arXiv preprint arXiv:2201.09450 (2022)
32. Li, W., Lu, X., Lu, J., Zhang, X., Jia, J.: On efficient transformer and image pre-training for low-level vision. arXiv preprint arXiv:2112.10175 (2021)
33. Li, W., Zhou, K., Qi, L., Jiang, N., Lu, J., Jia, J.: LAPAR: linearly-assembled pixel-adaptive regression network for single image super-resolution and beyond. In: NIPS, vol. 33, pp. 20343–20355 (2020)
34. Li, Y., et al.: NTIRE 2022 challenge on efficient super-resolution: methods and results. In: CVPR, pp. 1062–1102 (2022)
35. Li, Z., et al.: Blueprint separable residual network for efficient image super-resolution. In: CVPR, pp. 833–843 (2022)
36. Liang, J., Cao, J., Sun, G., Zhang, K., Van Gool, L., Timofte, R.: SwinIR: image restoration using swin transformer. In: ICCV, pp. 1833–1844 (2021)
37. Lim, B., Son, S., Kim, H., Nah, S., Mu Lee, K.: Enhanced deep residual networks for single image super-resolution. In: CVPR, pp. 136–144 (2017)
38. Liu, J., Tang, J., Wu, G.: Residual feature distillation network for lightweight image super-resolution. In: Bartoli, A., Fusiello, A. (eds.) ECCV 2020. LNCS, vol. 12537, pp. 41–55. Springer, Cham (2020). https://doi.org/10.1007/978-3-030-67070-2_2
39. Liu, Z., et al.: Swin transformer: hierarchical vision transformer using shifted windows. arXiv preprint arXiv:2103.14030 (2021)
40. Liu, Z., Mao, H., Wu, C.Y., Feichtenhofer, C., Darrell, T., Xie, S.: A convnet for the 2020s. In: CVPR, pp. 11976–11986 (2022)

41. Martin, D., Fowlkes, C., Tal, D., Malik, J.: A database of human segmented natural images and its application to evaluating segmentation algorithms and measuring ecological statistics. In: ICCV, vol. 2, pp. 416–423 (2001)
42. Niu, B., et al.: Single image super-resolution via a holistic attention network. In: Vedaldi, A., Bischof, H., Brox, T., Frahm, J.-M. (eds.) ECCV 2020. LNCS, vol. 12357, pp. 191–207. Springer, Cham (2020). https://doi.org/10.1007/978-3-030-58610-2_12
43. Qian, G., et al.: Rethinking the pipeline of demosaicing, denoising and super-resolution. In: ICCP (2022)
44. Shi, S., Gu, J., Xie, L., Wang, X., Yang, Y., Dong, C.: Rethinking alignment in video super-resolution transformers. arXiv preprint arXiv:2207.08494 (2022)
45. Tai, Y., Yang, J., Liu, X.: Image super-resolution via deep recursive residual network. In: CVPR, pp. 3147–3155 (2017)
46. Tai, Y., Yang, J., Liu, X., Xu, C.: MemNet: a persistent memory network for image restoration. In: ICCV, pp. 4539–4547 (2017)
47. Tong, T., Li, G., Liu, X., Gao, Q.: Image super-resolution using dense skip connections. In: ICCV, pp. 4799–4807 (2017)
48. Trockman, A., Kolter, J.Z.: Patches are all you need? arXiv preprint arXiv:2201.09792 (2022)
49. Ulyanov, D., Vedaldi, A., Lempitsky, V.: Instance normalization: the missing ingredient for fast stylization. arXiv preprint arXiv:1607.08022 (2016)
50. Vaswani, A., et al.: Attention is all you need. In: NIPS, vol. 30, pp. 5998–6008 (2017)
51. Wang, Z., Bovik, A.C., Sheikh, H.R., Simoncelli, E.P.: Image quality assessment: from error visibility to structural similarity. TIP **13**(4), 600–612 (2004)
52. Wu, Y., He, K.: Group normalization. In: ECCV, pp. 3–19 (2018)
53. Zeyde, R., Elad, M., Protter, M.: On single image scale-up using sparse-representations. In: Boissonnat, J.-D., et al. (eds.) Curves and Surfaces 2010. LNCS, vol. 6920, pp. 711–730. Springer, Heidelberg (2012). https://doi.org/10.1007/978-3-642-27413-8_47
54. Zhang, Y., Li, K., Li, K., Wang, L., Zhong, B., Fu, Y.: Image super-resolution using very deep residual channel attention networks. In: ECCV, pp. 286–301 (2018)
55. Zhang, Y., Tian, Y., Kong, Y., Zhong, B., Fu, Y.: Residual dense network for image super-resolution. In: CVPR, pp. 2472–2481 (2018)
56. Zhao, H., Kong, X., He, J., Qiao, Yu., Dong, C.: Efficient image super-resolution using pixel attention. In: Bartoli, A., Fusiello, A. (eds.) ECCV 2020. LNCS, vol. 12537, pp. 56–72. Springer, Cham (2020). https://doi.org/10.1007/978-3-030-67070-2_3

Unsupervised Scene Sketch to Photo Synthesis

Jiayun Wang[1(✉)], Sangryul Jeon[1], Stella X. Yu[1], Xi Zhang[2], Himanshu Arora[2], and Yu Lou[2]

[1] UC Berkeley/ICSI, Berkeley, CA, USA
{peterwg,srjeon,stellayu}@berkeley.edu
[2] Amazon, Seattle, WA, USA
{xizhn,arorah,ylou}@amazon.com

Abstract. Sketches make an intuitive and powerful visual expression as they are fast executed freehand drawings. We present a method for synthesizing realistic photos from scene sketches. Without the need for sketch and photo pairs, our framework directly learns from readily available large-scale photo datasets in an unsupervised manner. To this end, we introduce a standardization module that provides *pseudo* sketch-photo pairs during training by converting photos and sketches to a standardized domain, i.e. the edge map. The reduced domain gap between sketch and photo also allows us to disentangle them into two components: holistic scene structures and low-level visual styles such as color and texture. Taking this advantage, we synthesize a photo-realistic image by combining the structure of a sketch and the visual style of a reference photo. Extensive experimental results on perceptual similarity metrics and human perceptual studies show the proposed method could generate realistic photos with high fidelity from scene sketches and outperform state-of-the-art photo synthesis baselines. We also demonstrate that our framework facilitates a controllable manipulation of photo synthesis by editing strokes of corresponding sketches, delivering more fine-grained details than previous approaches that rely on region-level editing.

Keywords: Sketch · Scene sketch · Photo synthesis · Unsupervised learning

1 Introduction

Sketching is an intuitive way to represent visual signals. With a few sparse strokes, humans could understand and envision a photo from a sketch. Additionally, unlike photos which are rich in color and texture, sketches are easily editable as strokes are easy to modify. We aim to synthesize photos that preserve the structure of scene sketches while delivering the low-level visual style of reference photos.

Unlike previous works [15,24,32] that synthesize photos from categorical object-level sketches, our goal in which scene-level sketches are used as input

L. Karlinsky et al. (Eds.): ECCV 2022 Workshops, LNCS 13802, pp. 273–289, 2023.
https://doi.org/10.1007/978-3-031-25063-7_17

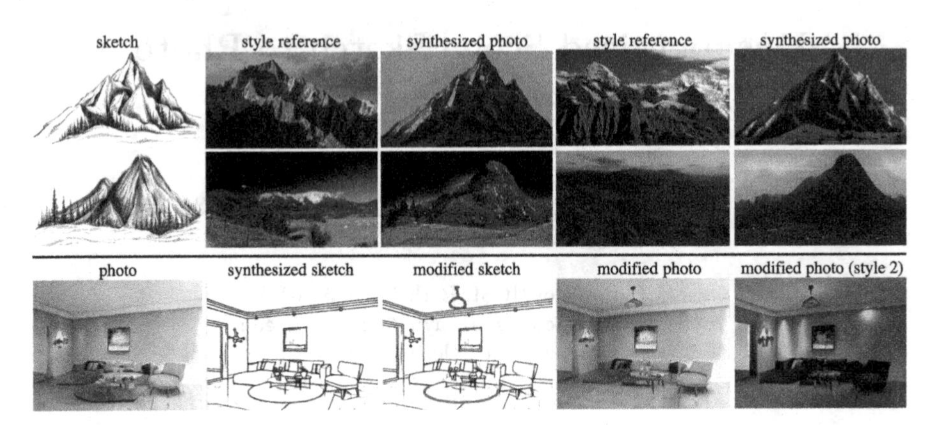

Fig. 1. *Upper:* Given a sketch and a style reference photo, our method is capable of transferring low-level visual styles of the reference while preserving the content structure of the sketch. We show synthesis results with different references. *Lower:* Given an arbitrary photo, users could easily and interactively edit it by adding or removing strokes on the synthesized sketch. (Color figure online)

poses additional challenges due to **1) Lack of data.** There is no training data available for our task due to the complexity of scene sketches. Not only the insufficient amount of scene sketches, but the lack of paired scene sketch-image datasets make supervised learning from one modality to another intractable. **2) Complexity of scene sketches.** A scene sketch usually contains many objects of diverse semantic categories with complicated spatial organization and occlusions. Isolating objects, synthesizing object photos and combining them together [7] do not work well and are hard to generalize. For one, detecting objects from sketches is hard due to the sparse structure. For another, one may encounter objects that do not belong to seen categories, and the composition could also make the synthesized photo unrealistic.

We propose to alleviate these issues via **1)** a standardization module, and **2)** disentangled representation learning.

For the lack of data, we propose a standardization module, where input images are converted to a standardized domain, edge maps. Edge maps can be considered as *synthetic sketches* due to the high similarity to real sketches. With the standardization, readily-available large-scale photo datasets could be used for training by converting them to edge maps. Additionally, during inference, sketches of various individual styles are also standardized such that the gap between training and inference is narrowed.

For the complexity of scene sketches, we learn disentangled holistic content and low-level style representations from photos and sketches by encouraging only content representations of photo-sketch pairs to be similar. As a definition, content representations encode holistic semantic and geometric structures of a sketch or photo. Style representations encode the low-level visual information such as color and texture. A sketch could depict similar contents as a photo, but contain no color or texture information. By factorizing out colors and textures, the model

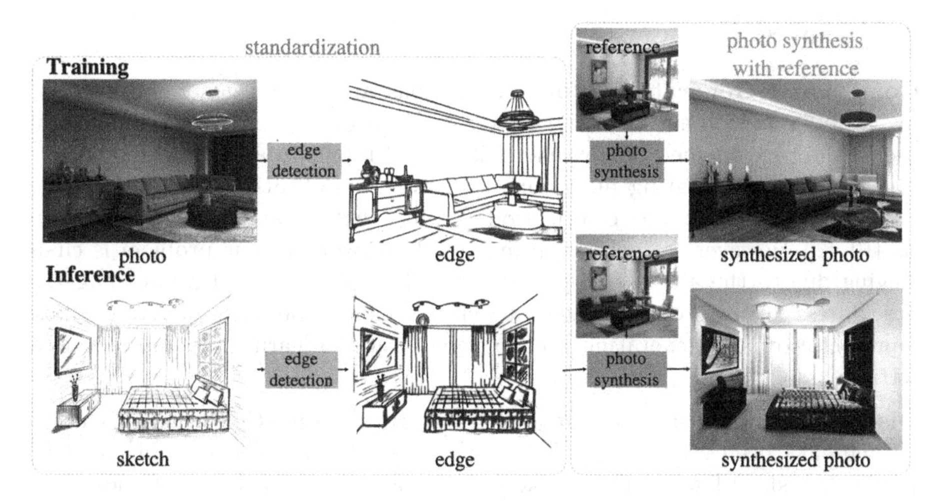

Fig. 2. Our method consists of two components, standardization and photo synthesis. **Left:** The standardization module converts photos or sketches into a standardized domain, edge maps, to reduce the domain gap between training and inference. **Right:** From the standardized edge map, the photo synthesis module generates a photo with a similar style as the given reference image.

could directly learn from large-scale photos for scene structures and transfer the knowledge to sketches. Additionally, combining the content representation of a sketch and a style representation of a reference photo could decode a realistic photo. The decoded photo should depict similar contents as the sketch and shares a similar style with the reference photo. This is the underlying mechanics of the proposed reference-guided scene sketch to photo synthesis approach. Note that the disentangled representations have been studied previously for photos [28,34] and we extend the concept to sketches.

As exemplified in Fig. 1, not only photo synthesis from scene sketch, our model can promote also controllable photo editing by allowing users to directly modify strokes of a corresponding sketch. The process is easy and fast as strokes are easy and flexible to modify, compared with photo editing from segmentation maps proposed by previous works [15,22,26,28]. Specifically, the standardization module first converts a photo to a sketch. Users could modify strokes of the sketch and synthesize a newly edited photo with our model. Additionally, the style of the photo could also be modified with another reference photo as guidance.

We summarize our contribution as follows: **1)** We propose an unsupervised scene sketch to photo synthesis framework. We introduce a standardization module that converts arbitrary photos to standardized edge maps, enabling a vast amount of real photos to be utilized during training. **2)** Our framework facilitates controllable manipulation of photo synthesis through editing scene sketches with more plausibility and simplicity than previous approaches. **3)** Technically, we propose novel designs for scene sketch to photo synthesis, including shared content representations to enable knowledge transfer from photos to sketches and model fine-tuning with sketch-reference-photo triplets for improved performance.

2 Related Work

Conditional Generative Models. Previous approaches generated realistic images by conditioning generative adversarial networks [9] on a given input from users. More recent methods extended it to multi-domain and multi-modal setting [4,13,23], facilitating numerous downstream applications including image inpainting [14,29], photo colorization [20,40], texture and geometry synthesis [10,42]. However, naively adopting this framework to our problem is challenging due to the absence of paired data where sketches and photos aligned. We address this by projecting arbitrary sketches and photo into the intermediate representation and generating pseudo paired data to learn in an unsupervised setting.

Disentanglement of Content and Style Representations. The disentanglement has been studied [31,44] prior to the surge of deep learning models, where they show low-level style like texture can be modeled as statistics of an image. Deep generative models [16,21,28,34] also achieved success in photo style transfer by the disentanglement. We extend the disentanglement idea to sketches and show its application in photo synthesis.

Sketch to Photo Synthesis. Following a seminal work, SketchGAN [3], several efforts has been made on synthesizing photos [8,24,37] or reconstructing 3D shapes [5,35,36] from sketches. They however mainly focused on categorical single-object sketches without substantial background clutters, and thus have difficulties when encountered with complicated scene-level sketches.

Scene sketch to photo synthesis is limited by lack of the data. SketchyScene [45] is the only scene dataset with object segmentation and corresponding cartoon images. However, their sketch is manually composited from multiple object sketches with reference to a cartoon image. The composite sketch has a large domain gap to real scene sketches with reference to a real scene. Their composition idea greatly impacts how researchers solve the photo synthesis. [7] detect objects of composite sketches and generate individual photos as well as a background image and combine them together. Holistic scene structures are ignored and the photo composition leads to artifacts and unrealism. We learn holistic scene structures from massive photo datasets and transfer the knowledge to sketches.

Deep Image Editing. By the favor of powerful generative models [17], previous works edited photos by modifying the extracted latent vector. Typically they sampled the desired latent vector from a fixed distribution according to a user's semantic control [43], or let a user spatially annotate the region-based semantic layout [27,28]. DeepFaceDrawing [2] enables user to sketch progressive for face image synthesis. Our work differs in that we allow users to directly edit strokes of a complicated scene sketch, thus enabling much more fine-grained editing.

3 Methods

As illustrated in Fig. 2, our framework mainly consists of two components: domain standardization and reference-based photo synthesis. For standardiza-

Fig. 3. The standardization module converts photos and sketches to a standardized domain, edge maps. After the standardization, edges of photos and sketches share higher similarity, which makes the domain gap between training and evaluation narrower. Within the test set, edges of sketches with different individual styles also a share higher similarity, making the intra-sketch-set discrepancy smaller.

tion (details in Sect. 3.1), input photos and sketches are converted to standardized edge maps, which bypass the lack of data issue. The second part is reference-guided photo synthesis (details in Sect. 3.2), where synthesized photos are generated based on input sketches and style reference photos.

3.1 Domain Standardization

Due to the lack of paired sketch-photo datasets, it is intractable for supervised models to synthesize photos from sketches. We adopt a similar idea as [35], where they converted inputs to a standardized domain, and showed learning from such domain has better performance compared to directly using unprocessed inputs.

As shown in Fig. 2**L**, the standardization can be considered as data prepossessing and is different for training and inference. During training, we collect a large scale photo dataset of a specific category, e.g., indoor scenes. Each photo is converted to a standardized edge map for later use with an off-the-shelf deep-learning-based edge detector [30]. During inference, unlike the training, the input is a sketch. We use the same edge detector to convert it to the edge map for later use. Figure 3 depicts examples of photo, sketches and their corresponding edges. The standardized edge maps have small domain discrepancies. In addition to narrowing the domain gap between the training and test data, the standardization module during inference could narrow the gap of individual sketching styles (e.g., stroke width), which was also similarly shown in [35]. Given that edge maps serve as a proxy for real sketches, we slightly abuse the wording of *synthetic sketches* (or omitted as sketches) hereinafter as they may refer to standardized edge maps.

3.2 Reference-Guided Photo Synthesis

Previous works [28,34] show that photos can be encoded to two disentangled representations: content and style representations. We extend the concept to

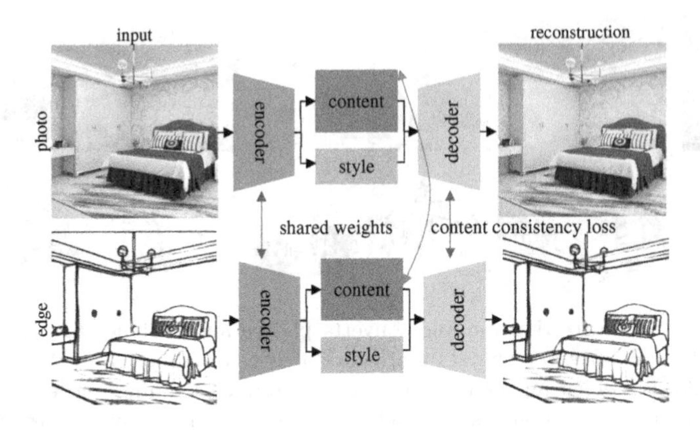

Fig. 4. *Disentangled representation encoding* is the first stage of the sketch-to-photo synthesis module. For each photo, we generate a standardized edge map and form an image pair. Each image of the pair is encoded as content and style representations by the encoder. We add content consistency loss to make content representations of the photo and the edge to be similar. The representations are then decoded to a reconstructed image by the decoder. The network learns the representations through the auto-encoding process. For the performance of sketch to photo synthesis later, both photos and their corresponding standardized edges are fed to the network for auto-encoding.

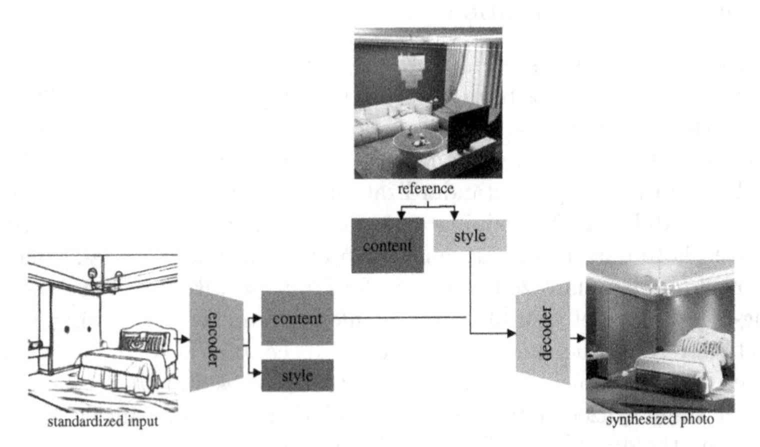

Fig. 5. *Fine-tuning with sketch-reference-photo triplets* is the second stage of the sketch-to-photo synthesis module. The input is a standardized edge map and a reference photo. The model is pre-trained in the representation encoding phase. Both the edge map and the reference photo are encoded by the network for content and style representations. The content and representations are fed to the decoder to reconstruct the synthesized photo.

sketches and show that they can be encoded to disentangled representations. Preserving content representation while replacing the sketch style with a real photo style representation could generate a realistic synthesized photo.

The module is trained in two stages. **1)** Disentangled representation encoding stage learns content and style representations from images via auto-encoding. **2)** We further fine-tune the model with sketch-reference-photo triplets, with regularization loss to guarantee the synthesizing quality. Our model is inspired by and based on previous arts on disentangled representation learning [28] and style transfer [34], with novel designs for the goal of scene sketch to photo synthesis.

Disentangled Representation Encoding. Figure 4 depicts the pipeline of the disentangled representation encoding stage. Denote a pair of input images and its corresponding edge as $\{\mathbf{x}, \mathbf{x}'\}$, the encoder as E, decoder as G, and discriminator as D. The encoder encodes input pairs $\{\mathbf{x}, \mathbf{x}'\}$ to two representation pairs, content $\{c_{\mathbf{x}}, c_{\mathbf{x}'}\}$ and style $\{s_{\mathbf{x}}, s_{\mathbf{x}'}\}$, i.e., $E(\{\mathbf{x}, \mathbf{x}'\}) = \{\{c_{\mathbf{x}}, c_{\mathbf{x}'}\}, \{s_{\mathbf{x}}, s_{\mathbf{x}'}\}\}$. From the encoded representations, the decoder reconstructs a photo $G(c_{\mathbf{x}}, s_{\mathbf{x}})$ and its edge $G(c_{\mathbf{x}'}, s_{\mathbf{x}'})$. The auto-encoder ensures the reconstructed image pair is similar to the input image pair by the following reconstruction loss in ℓ_1-norm:

$$\mathcal{L}_{\text{rec}_1} = \mathrm{E}_{\mathbf{x} \sim \mathbf{X}, \mathbf{x}' \sim \mathbf{X}'}[|\mathbf{x} - G(c_{\mathbf{x}}, s_{\mathbf{x}})| + |\mathbf{x}' - G(c_{\mathbf{x}'}, s_{\mathbf{x}'})|] \tag{1}$$

Since the photo and the edge depict the same content, we ask their content representations to be similar in ℓ_1-norm:

$$\mathcal{L}_{\text{content}} = \mathrm{E}_{\mathbf{x} \sim \mathbf{X}, \mathbf{x}' \sim \mathbf{X}'}[|c_{\mathbf{x}} - c_{\mathbf{x}'}|] \tag{2}$$

Further, the adversarial GAN loss [9] is required to train discriminator G for realistic reconstructions:

$$\mathcal{L}_{\text{GAN}_1} = \mathrm{E}_{\mathbf{x} \sim \mathbf{X}, \mathbf{x}' \sim \mathbf{X}'}[-\log D(G(c_{\mathbf{x}}, s_{\mathbf{x}})) - \log D(G(c_{\mathbf{x}'}, s_{\mathbf{x}'}))] \tag{3}$$

The final loss is $\mathcal{L}_{\text{rec}_1} + \theta \mathcal{L}_{\text{content}} + \alpha \mathcal{L}_{\text{GAN}_1}$, where θ, α are both set to be 0.5.

Fine-Tuning with Sketch-Reference-Photo Triplets. Figure 5 depicts the pipeline of the fine-tuning stage. Denote the sketch, reference photo and output synthesized photo as $\mathbf{x}^{\mathbf{k}}, \mathbf{x}^{\mathbf{r}}, \mathbf{x}^{\mathbf{o}}$, respectively. With the pre-trained model from the previous representation learning stage, the encoder is able to encode content and style representations of sketches and photos. The output image is generated by the decoder from the content representation of the sketch $c_{\mathbf{x}^{\mathbf{k}}}$, and the style representation of the reference $s_{\mathbf{x}^{\mathbf{r}}}$:

$$\mathbf{x}^{\mathbf{o}} = G(c_{\mathbf{x}^{\mathbf{k}}}, s_{\mathbf{x}^{\mathbf{r}}}) \tag{4}$$

As the model has been pre-trained in the previous stage for encoding content and style representations, the model has a good starting point for synthesizing photos from sketches. To ensure the output image has similar content as the sketch and a similar style as the reference, however, we enforce the following regularization loss on content and style representations in ℓ_1-norm:

$$\mathcal{L}_{\text{reg}} = \mathrm{E}_{\mathbf{x}^{\mathbf{k}} \sim \mathbf{X}^{\mathbf{k}}, \mathbf{x}^{\mathbf{r}} \sim \mathbf{X}^{\mathbf{r}}, \mathbf{x}^{\mathbf{o}} \sim \mathbf{G}(\mathbf{c}_{\mathbf{x}^{\mathbf{k}}}, \mathbf{s}_{\mathbf{x}^{\mathbf{r}}})}[|c_{\mathbf{x}^{\mathbf{o}}} - c_{\mathbf{x}^{\mathbf{k}}}| + |s_{\mathbf{x}^{\mathbf{o}}} - s_{\mathbf{x}^{\mathbf{r}}}|] \tag{5}$$

Additionally, the adversarial GAN loss is required:

$$\mathcal{L}_{\text{GAN}_2} = \mathrm{E}_{\mathbf{x}^{\mathbf{k}} \sim \mathbf{X}^{\mathbf{k}}, \mathbf{x}^{\mathbf{r}} \sim \mathbf{X}^{\mathbf{r}}}[-\log D(G(c_{\mathbf{x}^{\mathbf{k}}}, s_{\mathbf{x}^{\mathbf{r}}}))] \tag{6}$$

The final loss is $\mathcal{L}_{\text{reg}} + \beta \mathcal{L}_{\text{GAN}_2}$, where β is set to be 0.5 in the work.

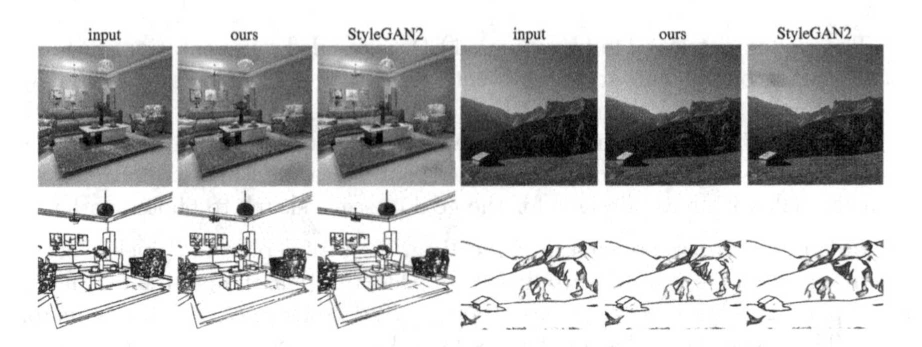

Fig. 6. The reconstruction results of our method and StyleGAN2 [34]. Images are projected into embedding spaces for ours and StyleGAN2 [34]. Both photos and standardized edges are fed to the network for reconstruction. The high faithfulness in reconstruction demonstrates that the learned content and style representations are effective.

Table 1. **(a)** Reconstruction performance measured in LPIPS (↓) [41]. Images are projected into embedding spaces for ours and StyleGAN2 [34]. We reconstruct photos and edges with a similar performance as StyleGAN2 [34], demonstrating the disentanglement to content and style representations is effective. **(b)** Reference-guided sketch to photo synthesis performance measured in FID (↓) [12]. Our method outperforms other baseline methods in all three categories.

(a)

input	method	indoor	church	mountain	mean
photo	ours	**0.254**	**0.214**	**0.221**	**0.229**
	StyleGAN2	0.256	0.220	0.224	0.233
edge	ours	**0.180**	**0.166**	**0.171**	**0.172**
	StyleGAN2	0.161	0.188	0.173	0.174

(b)

FID (↓)	indoor	church	mountain	mean
ours	**105.5**	**48.7**	**73.8**	**76.0**
SAE [28]	107.7	52.4	74.1	78.1
ObjSketch [24]	136.5	62.1	95.4	98.0
SpliceViT [33]	204.2	119.7	140.7	154.9
DTP [18]	205.2	124.2	143.5	157.6
Style2Paints [39]	254.2	217.3	247.7	239.7

4 Experimental Results

4.1 Network Architectures and Training Details

Network Architectures. Images are fed to the encoder to obtain content and style representations. First, images go through 4 down-sampling residual blocks [11] to obtain an intermediate representation. The intermediate representation is fed to another convolution layer to obtain the content representation with a spatial size of 16×16. The intermediate representation is also fed to another two convolution layers to obtain a style representation/vector dimension of 2048. The decoder consists of 4 up-sampling residual blocks. The style representation is injected to the decoder convolution layers with weight modulation techniques described in StyleGAN2 [34]. The discriminator is the same as that of StyleGAN2.

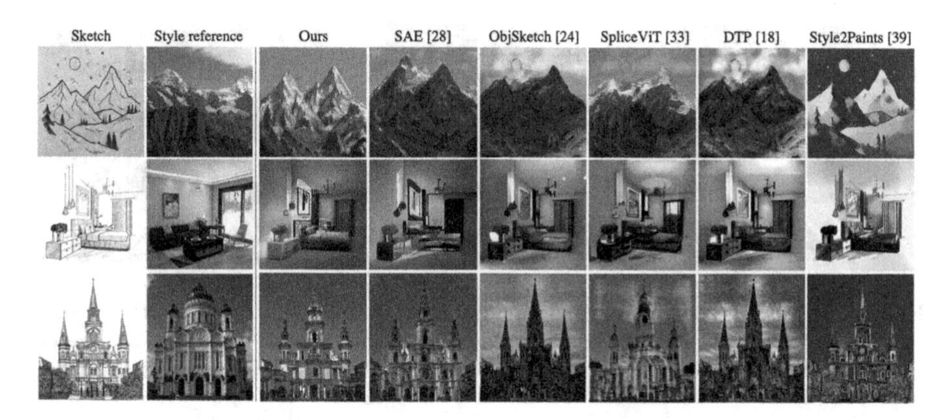

Fig. 7. Various baseline photo syntheses from sketches with style guidance. Note that SpliceViT [33] and DTP [18] are designed for test-time optimization and are not trained on the full dataset, making them disadvantageous to other methods. All other methods are trained on the same dataset with a similar iteration as the proposed method. Style2Paints is designed to synthesize painting, not realistic photos. Our model synthesizes photos that share a similar content as the sketch and a similar visual style as the style photo reference.

Hyper-Parameters and Training Schedules. For representation encoding, the initial learning rate is 2e−3. We use Adam optimizer [19] with $\beta = (0, 0.99)$. For fine-tuning, we start from the previously pre-trained model. The training schedule stays the same with the initial learning rate being 4e−4. The entire training time for the 3D-front indoor scene dataset is 7 days on 4 V100 GPUs.

Baselines. We follow the released code and the same settings of all baseline methods and retrain on datasets used in the paper. Specifically, some baselines [18,24,28,33] only work on photos, but not sketches. We use a gray-scale images as a proxy to ensure the photo synthesis quality. Specifically, we first train a sketch to gray-scale photo model using the same setting as step 1 of [24], where the input to the model is a standardized sketch. The generated gray-scale photo is then used to train a gray-scale to color photo model with the same setting of the baseline methods. SpliceViT [33] and DTP [18] are designed for test-time optimization and are not trained on the entire dataset. All other baseline methods are trained on the same dataset as the proposed method with a similar iteration.

4.2 Datasets

We train on the following scene photo datasets: **1) 3D-Front Indoor Scene** [6] consists of 14,761 training and 5,479 validation photos. They are rendered with Blender from synthetic indoor scenes including bedrooms and living rooms. Photos are resized to 286 and randomly cropped to 256 during training. **2) LSUN Church** [38] consists of 126,227 photos of outdoor churches. We randomly sam-

Fig. 8. The indoor scene, church and mountain sketch to photo synthesis with different references. We synthesize high-fidelity scene photos with similar content as the sketch and similar style as the reference photos.

ple 25,255 photos as the validation set. Photos are resized to 286 and randomly cropped to 256 during training. **3) GeoPose3K Mountain Landscape** [1] has 3,114 mountain landscape photos. 623 photos are randomly sampled for validation. Training photos are resized to 572 and randomly cropped.

For evaluation, we collect a **Scene Sketch Evaluation Set**. For each category (indoor scenes, mountain and church), we collect 50 sketches from the Internet, respectively. The sketches are collected with an intention to cover various sketching styles, e.g. different levels of line width, geometric distortion, use of shading, etc.

4.3 Representation Encoding

With effective learned representation, the model could reconstruct photos or sketches with high quality. We evaluate reconstruction performance in LPIPS [41].

Table 1**a** reports the LPIPS distance of reconstructed and input photos and synthetic sketches of our stage 1 model and StyleGAN2 [34]. Figure 6 depicts several examples of the input and reconstruction. Our representation encoding model has a slightly better reconstruction performance compared to StyleGAN2, indicating the learned content and style representations are adequate and ready for further fine-tuning with sketch-reference-photo pairs.

4.4 Photo Synthesis

We evaluate the photo synthesis performance of our method and baselines in terms of photo-realism. We calculate the Fréchet inception distance (FID) [12] between the synthesized photo set and the training photo set for each category (Table 1**b**). Our method outperforms other baselines under the FID metric. Figure 7 depicts synthesis results of our method and baselines. Note that Splice-ViT [33] and DTP [18] designed for test-time optimization and was not trained on the full dataset, making it disadvantageous to other methods. Style2Paints is designed to synthesizing painting, not realistic photos. We however include it

Table 2. A human perceptual study of the synthesized photos. **(a)** The fooling rate of our synthesized model over real photos measures the realism of the generation. **(b)** User preference on which method synthesizes photos that depicts more similar content to the sketch. **(c)** User preference on which method synthesizes photos that depicts more similar visual style to the reference photo. Compared with [28], we have a higher fooling rate over real photos, better content and style matching preference rate.

(a) Fooling rate (↑)

(%)	indoor scene	church	mountain	mean
ours	**25.00**	**44.3**	**48.9**	**39.4**
SAE [28]	10.0	6.6	20.0	12.2

(b) Content matching (↑)

(%)	indoor scene	church	mountain	mean
ours	**80.1**	**92.1**	**75.0**	**82.4**
SAE [28]	19.9	7.9	25.0	17.6

(c) Style matching (↑)

(%)	indoor scene	church	mountain	mean
ours	**61.9**	**90.9**	**71.0**	**74.6**
SAE [28]	38.1	9.1	29.0	25.4

as it is one of the few works that study synthesizing from scene sketches. Our synthesis result outperforms all other methods, with SAE [28] being the second. As for if the content of the output photo matches with the input sketch or if the style matches with the reference photo, we provide human perceptual evaluation in Sect. 4.5.

We also provide more visualization of our synthesis results of indoor scenes, churches and mountains in Fig. 8.

4.5 Human Perceptual Study

We conduct a human perceptual study to evaluate the realism of synthesized photos, and if synthesized photos match contents and styles as desired. We only evaluate our method and SAE [28], the second best-performing synthesis method, due to limited resources.

We create a survey consisting of three parts: photorealism, content matching with sketches and style matching with reference photos. As guidance to the participants, we state our research purpose at the beginning of the survey. For each part, a detailed description and an example question with answers and explanations are provided for the participant's reference. The order of our results, baseline results, and real images are randomly shuffled in the survey to minimize the potential bias from the participant. Each part consists of 13 questions, with one question being a *bait question* with an obvious answer. The bait question is designed to check if the participant is paying attention and if the answers are reliable. There are in total 51 participants, with 1 being ruled out due to failing one of the bait questions. Thus we finally collect 1,950 valid human judgments.

To evaluate the photorealism, we randomly select synthesized photos of ours and SAE evenly from three categories. Both methods use the same input sketch and reference photo. For each synthesized photo, we use Google's search by image feature to find the most similar real photo and ask participants which one they think looks more like a real photo. We then calculate the percentage of participants being fooled. Note that the fooling rate of random guessing is 50%. Table 2a reports the fooling rate of our method and SAE. Ours is 27% higher than SAE. Specifically, for churches and mountains, ours achieves a fooling rate over 44%: the generated photos are almost indistinguishable from real photos.

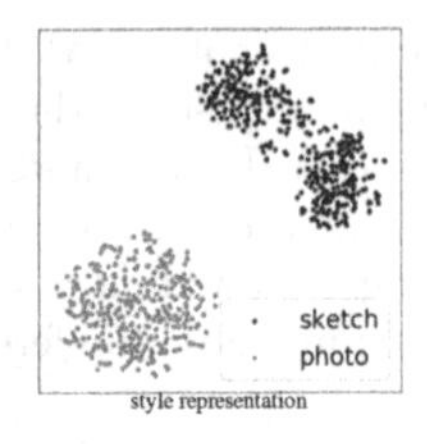

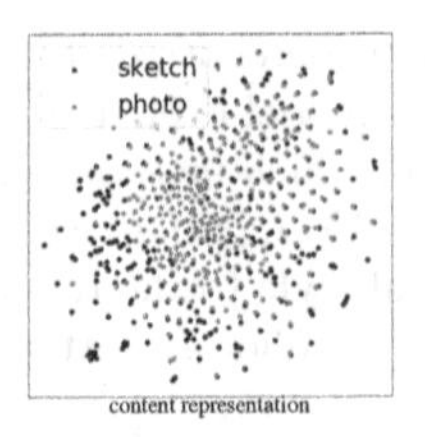

Fig. 9. The style representations of sketches and photos are well separated, while the content representations of sketches and photos are tangled together. We visualize learned content and style representations of sketches and photos with T-SNE [25]. The results show that sketches and photos share the content space and it is appropriate to train on photos and transfer knowledge to sketches.

Fig. 10. Sketch to photo synthesis with combined style representations of two references. We encode style representations from two photos, e.g. a winter photo and a summer photo. By increasing the weight of the summer image and decreasing that of the winter image, the synthesized photo from the sketch gradually changes from winter appearance to summer appearance.

To evaluate if the synthesized photos match the content of the input sketch, we show participants an input sketch and two synthesized results from our method and SAE, and ask them to pick one that has the most similar content as the sketch. Table 2**b** reports the preference rate of ours over SAE. We achieve 82% on average preference rate, well outperforming the baseline.

To evaluate if the synthesized photos match the style of the reference photo, we show participants a reference photo and two synthesized results from our method and SAE, and ask them to pick one that has the most similar style to the sketch. Table 2**c** reports the preference rate of ours over SAE. We achieve a 75% average preference rate, well outperforming the baseline.

4.6 Photo Editing Through Sketch

As depicted in Fig. 11, given an input photo, we convert it to a standardized edge map (where we refer as sketch for simplicity). Users could add and remove strokes to edit the photo. We also show the possibility of sequential editing in the figure. We evaluate the photo editing performance for the indoor scene validation dataset, and the FID [12] of edited images to the training set is 69.2. One limitation is that the content in the unmodified region of a given photo may not be well preserved as the edited photo is solely generated from the edge map.

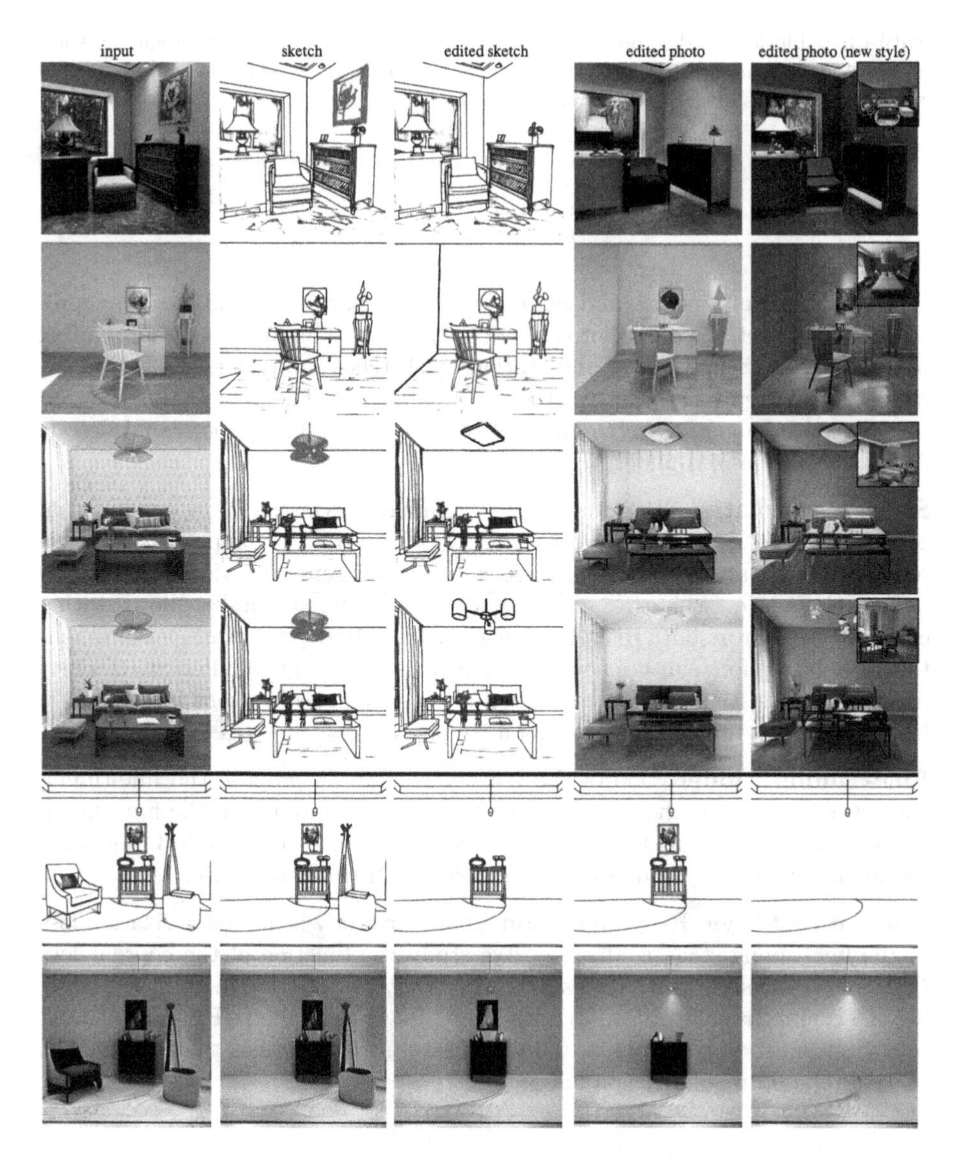

Fig. 11. Photo editing and style transfer via sketches. *Upper:* Given an input image, we first convert it to a standardized edge map. We then add or remove strokes in the edge map and convert it back to a photo. The visual style of the photo could also be changed with a reference photo (top right). *Lower:* Sequential editing by gradually removing strokes. (Color figure online)

4.7 Analysis and Ablation Studies

Analysis of Style Representations. We visualize the learned content and style representations of photos and sketches using T-SNE [25] in Fig. 9: style

Table 3. Ablation studies on the fine-tuning stage, content and style regularization loss for indoor scenes in FID (↓) [12] distance. Having both stage 2 fine-tuning and the regularization loss gives the best result.

no fine-tune	fine-tune+style loss	fine-tune+content loss	fine-tune+all loss
107.9	107.0	106.1	**105.5**

representations of sketches and photos are well separated, while content representations of sketches and photos are not separable. This verifies the grounding of the method: the content representations of sketches and photos can be shared, while the style representations for the two are different. Thus, combining the content representation of a sketch and style representation of a photo could decode a realistic synthesized photo.

Style Interpolation. We study if the reference style can be a combination of style of two different reference images $\mathbf{x}^{\mathbf{r_1}}$ and $\mathbf{x}^{\mathbf{r_2}}$. Suppose their style representations are $\mathbf{s}_{\mathbf{x}^{\mathbf{r_1}}}$ and $\mathbf{s}_{\mathbf{x}^{\mathbf{r_2}}}$. The combined representation $s_{\text{combined}} = \gamma\mathbf{s}_{\mathbf{x}^{\mathbf{r_1}}} + (1-\gamma)\mathbf{s}_{\mathbf{x}^{\mathbf{r_2}}}$, where $\gamma \in [0,1]$. By adjusting γ, we synthesize photos with a combined style from both reference images. Figure 10 depicts examples of mountain sketch to photo synthesis with combined styles from two different reference images. By adjusting γ, the synthesized photos have a continuous interpolation from winter to summer, and afternoon to dusk.

Fine-Tuning Model. One of the novelty is that we propose the fine-tuning with sketch-reference-photo triplets for the task. We evaluate if the fine-tuning is necessary by removing the fine-tuning stage. As reported in Table 3, removing the model fine-tuning leads to 2.4 worse results in the FID metric.

Content and Style Regularization Loss. We study if the regularization loss at the fine-tuning stage is effective. We study the function of the content loss ($|c_{\mathbf{x}^{\mathbf{o}}} - c_{\mathbf{x}^{\mathbf{k}}}|$) and style loss ($|s_{\mathbf{x}^{\mathbf{o}}} - s_{\mathbf{x}^{\mathbf{r}}}|$) respectively. As reported in Table 3, removing the content regularization loss leads to 1.5 worse results in FID metric, and removing the style loss leads to 0.6 worse results. This verifies the effectiveness of the proposed regularization loss.

5 Summary

We propose a reference-guided framework for photo synthesis from scene sketches. We first convert all input photos and sketches to standardized edge maps, allowing the model to learn in unsupervised setting without the need of real sketches or sketch-photo pairs. Sequentially, the standardized input and reference image are disentangled into content and style components to synthesize new hybrid image that preserves the content of standardized input while transferring the style of reference image. Extensive experiments demonstrate that our method can generate and edit a realistic photo from a user's scene sketch with a reference photo as style guidance, surpassing the previous approaches on three benchmarks.

A major insight of this work is that, we learn to synthesize scene structures directly from the vast amount of readily-available photos, rather than synthesizing and combining individual objects. Rather than worrying about the acclimated errors from sketch-based object detection, photo synthesis and spatial combination for the final output, we treat the scene sketches as a whole and learn the holistic structures for photo synthesis.

One limitation is that the deep-learning based standardization step could eliminate strokes that reflect the details of the scene, or misinterpret the strokes as textures. Future work could study a sketch-to-edge standardization process that preserves higher fidelity of the sketch. Another limitation lies in the sketch-based photo editing - the unchanged regions of a given photo may not be well preserved. This is due to the model takes sketch as the only input. Future work could improve the performance by taking the original photo into consideration.

Acknowledgements. This research was supported, in part, by BAIR-Amazon Commons and AWS. We thank Yubei Chen for helpful discussions. We thank Tian Qin for providing some scene sketches used in the study. We thank Li Tang, Lu Yuan, Martin Zhai, Xingchen Liu, Karl Hillesland, Amin Kheradmand, Nasim Souly, Charlotte Wang, Valerie Moss and other anonymous participants in our human perceptual study.

References

1. Brejcha, J., Čadík, M.: GeoPose3K: mountain landscape dataset for camera pose estimation in outdoor environments. Image Vis. Comput. **66**, 1–14 (2017)
2. Chen, S.Y., Su, W., Gao, L., Xia, S., Fu, H.: DeepFaceDrawing: deep generation of face images from sketches. ACM Trans. Graph. (TOG) **39**(4), 72-1 (2020)
3. Chen, W., Hays, J.: SketchyGAN: towards diverse and realistic sketch to image synthesis. In: Proceedings of the IEEE Conference on Computer Vision and Pattern Recognition, pp. 9416–9425 (2018)
4. Choi, Y., Uh, Y., Yoo, J., Ha, J.W.: StarGAN v2: diverse image synthesis for multiple domains. In: Proceedings of the IEEE/CVF Conference on Computer Vision and Pattern Recognition, pp. 8188–8197 (2020)
5. Delanoy, J., Aubry, M., Isola, P., Efros, A.A., Bousseau, A.: 3D sketching using multi-view deep volumetric prediction. Proc. ACM Comput. Graph. Interact. Tech. **1**(1), 1–22 (2018)
6. Fu, H., et al.: 3D-front: 3D furnished rooms with layouts and semantics. In: Proceedings of the IEEE/CVF International Conference on Computer Vision, pp. 10933–10942 (2021)
7. Gao, C., Liu, Q., Xu, Q., Wang, L., Liu, J., Zou, C.: SketchyCOCO: image generation from freehand scene sketches. In: Proceedings of the IEEE/CVF Conference on Computer Vision and Pattern Recognition, pp. 5174–5183 (2020)
8. Ghosh, A., et al.: Interactive sketch & fill: multiclass sketch-to-image translation. In: Proceedings of the IEEE/CVF International Conference on Computer Vision, pp. 1171–1180 (2019)
9. Goodfellow, I., et al.: Generative adversarial nets. Adv. Neural Inf. Process. Syst. **27** (2014)

10. Guérin, É., et al.: Interactive example-based terrain authoring with conditional generative adversarial networks. ACM Trans. Graph. (TOG) **36**(6), 1–13 (2017)
11. He, K., Zhang, X., Ren, S., Sun, J.: Deep residual learning for image recognition. In: Proceedings of the IEEE Conference on Computer Vision and Pattern Recognition, pp. 770–778 (2016)
12. Heusel, M., Ramsauer, H., Unterthiner, T., Nessler, B., Hochreiter, S.: GANs trained by a two time-scale update rule converge to a local nash equilibrium. Adv. Neural Inf. Process. Syst. **30** (2017)
13. Huang, X., Liu, M.Y., Belongie, S., Kautz, J.: Multimodal unsupervised image-to-image translation. In: Proceedings of the European Conference on Computer Vision (ECCV), pp. 172–189 (2018)
14. Iizuka, S., Simo-Serra, E., Ishikawa, H.: Globally and locally consistent image completion. ACM Trans. Graph. (ToG) **36**(4), 1–14 (2017)
15. Isola, P., Zhu, J.Y., Zhou, T., Efros, A.A.: Image-to-image translation with conditional adversarial networks. In: Proceedings of the IEEE Conference on Computer Vision and Pattern Recognition, pp. 1125–1134 (2017)
16. Karras, T., Laine, S., Aila, T.: A style-based generator architecture for generative adversarial networks. In: Proceedings of the IEEE/CVF Conference on Computer Vision and Pattern Recognition, pp. 4401–4410 (2019)
17. Karras, T., Laine, S., Aittala, M., Hellsten, J., Lehtinen, J., Aila, T.: Analyzing and improving the image quality of StyleGAN. In: Proceedings of the IEEE/CVF Conference on Computer Vision and Pattern Recognition, pp. 8110–8119 (2020)
18. Kim, S., Kim, S., Kim, S.: Deep translation prior: test-time training for photorealistic style transfer. arXiv preprint arXiv:2112.06150 (2021)
19. Kingma, D.P., Ba, J.: Adam: a method for stochastic optimization. In: ICLR (Poster) (2015)
20. Larsson, G., Maire, M., Shakhnarovich, G.: Learning representations for automatic colorization. In: Leibe, B., Matas, J., Sebe, N., Welling, M. (eds.) ECCV 2016. LNCS, vol. 9908, pp. 577–593. Springer, Cham (2016). https://doi.org/10.1007/978-3-319-46493-0_35
21. Lee, H.Y., et al.: DRIT++: diverse image-to-image translation via disentangled representations. Int. J. Comput. Vision **128**(10), 2402–2417 (2020)
22. Ling, H., Kreis, K., Li, D., Kim, S.W., Torralba, A., Fidler, S.: EditGAN: high-precision semantic image editing. Adv. Neural Inf. Process. Syst. **34**, 16331–16345 (2021)
23. Liu, M.Y., et al.: Few-shot unsupervised image-to-image translation. In: Proceedings of the IEEE/CVF International Conference on Computer Vision, pp. 10551–10560 (2019)
24. Liu, R., Yu, Q., Yu, S.X.: Unsupervised sketch to photo synthesis. In: Vedaldi, A., Bischof, H., Brox, T., Frahm, J.-M. (eds.) ECCV 2020. LNCS, vol. 12348, pp. 36–52. Springer, Cham (2020). https://doi.org/10.1007/978-3-030-58580-8_3
25. Van der Maaten, L., Hinton, G.: Visualizing data using t-SNE. J. Mach. Learn. Res. **9**(11), 2579–2605 (2008)
26. Meng, C., He, Y., Song, Y., Song, J., Wu, J., Zhu, J.Y., Ermon, S.: SDEdit: guided image synthesis and editing with stochastic differential equations. In: International Conference on Learning Representations (2021)
27. Park, T., Liu, M.Y., Wang, T.C., Zhu, J.Y.: Semantic image synthesis with spatially-adaptive normalization. In: Proceedings of the IEEE/CVF Conference on Computer Vision and Pattern Recognition, pp. 2337–2346 (2019)
28. Park, T., et al.: Swapping autoencoder for deep image manipulation. Adv. Neural. Inf. Process. Syst. **33**, 7198–7211 (2020)

29. Pathak, D., Krahenbuhl, P., Donahue, J., Darrell, T., Efros, A.A.: Context encoders: feature learning by inpainting. In: Proceedings of the IEEE Conference on Computer Vision and Pattern Recognition, pp. 2536–2544 (2016)
30. Poma, X.S., Riba, E., Sappa, A.: Dense extreme inception network: towards a robust CNN model for edge detection. In: Proceedings of the IEEE/CVF Winter Conference on Applications of Computer Vision, pp. 1923–1932 (2020)
31. Portilla, J., Simoncelli, E.P.: A parametric texture model based on joint statistics of complex wavelet coefficients. Int. J. Comput. Vision **40**(1), 49–70 (2000)
32. Richardson, E., et al.: Encoding in style: a styleGAN encoder for image-to-image translation. In: Proceedings of the IEEE/CVF Conference on Computer Vision and Pattern Recognition, pp. 2287–2296 (2021)
33. Tumanyan, N., Bar-Tal, O., Bagon, S., Dekel, T.: Splicing ViT features for semantic appearance transfer. arXiv preprint arXiv:2201.00424 (2022)
34. Viazovetskyi, Y., Ivashkin, V., Kashin, E.: StyleGAN2 distillation for feed-forward image manipulation. In: Vedaldi, A., Bischof, H., Brox, T., Frahm, J.-M. (eds.) ECCV 2020. LNCS, vol. 12367, pp. 170–186. Springer, Cham (2020). https://doi.org/10.1007/978-3-030-58542-6_11
35. Wang, J., Lin, J., Yu, Q., Liu, R., Chen, Y., Yu, S.X.: 3D shape reconstruction from free-hand sketches. arXiv preprint arXiv:2006.09694 (2020)
36. Wang, L., Qian, C., Wang, J., Fang, Y.: Unsupervised learning of 3D model reconstruction from hand-drawn sketches. In: Proceedings of the 26th ACM International Conference on Multimedia, pp. 1820–1828 (2018)
37. Xiang, X., Liu, D., Yang, X., Zhu, Y., Shen, X., Allebach, J.P.: Adversarial open domain adaptation for sketch-to-photo synthesis. In: Proceedings of the IEEE/CVF Winter Conference on Applications of Computer Vision, pp. 1434–1444 (2022)
38. Yu, F., Seff, A., Zhang, Y., Song, S., Funkhouser, T., Xiao, J.: LSUN: construction of a large-scale image dataset using deep learning with humans in the loop. arXiv preprint arXiv:1506.03365 (2015)
39. Zhang, L., Li, C., Simo-Serra, E., Ji, Y., Wong, T.T., Liu, C.: User-guided line art flat filling with split filling mechanism. In: IEEE/CVF Conference on Computer Vision and Pattern Recognition (CVPR) (2021)
40. Zhang, R., Isola, P., Efros, A.A.: Colorful image colorization. In: Leibe, B., Matas, J., Sebe, N., Welling, M. (eds.) ECCV 2016. LNCS, vol. 9907, pp. 649–666. Springer, Cham (2016). https://doi.org/10.1007/978-3-319-46487-9_40
41. Zhang, R., Isola, P., Efros, A.A., Shechtman, E., Wang, O.: The unreasonable effectiveness of deep features as a perceptual metric. In: Proceedings of the IEEE Conference on Computer Vision and Pattern Recognition, pp. 586–595 (2018)
42. Zhou, Y., Zhu, Z., Bai, X., Lischinski, D., Cohen-Or, D., Huang, H.: Non-stationary texture synthesis by adversarial expansion. arXiv preprint arXiv:1805.04487 (2018)
43. Zhu, J.-Y., Krähenbühl, P., Shechtman, E., Efros, A.A.: Generative visual manipulation on the natural image manifold. In: Leibe, B., Matas, J., Sebe, N., Welling, M. (eds.) ECCV 2016. LNCS, vol. 9909, pp. 597–613. Springer, Cham (2016). https://doi.org/10.1007/978-3-319-46454-1_36
44. Zhu, S.C., Wu, Y., Mumford, D.: Filters, random fields and maximum entropy (frame): towards a unified theory for texture modeling. Int. J. Comput. Vision **27**(2), 107–126 (1998)
45. Zou, C., et al.: SketchyScene: richly-annotated scene sketches. In: Proceedings of the European Conference on Computer Vision (ECCV), pp. 421–436 (2018)

U-shape Transformer for Underwater Image Enhancement

Lintao Peng, Chunli Zhu, and Liheng Bian(✉)

Beijing Institute of Technology, Beijing, China
bian@bit.edu.cn

Abstract. The light absorption and scattering of underwater impurities lead to poor underwater imaging quality. The existing data-driven based underwater image enhancement (UIE) techniques suffer from the lack of a large-scale dataset containing various underwater scenes and high-fidelity reference images. Besides, the inconsistent attenuation in different color channels and space areas is not fully considered for boosted enhancement. In this work, we constructed a large-scale underwater image (LSUI) dataset including 4279 image pairs, and reported an U-shape Transformer network where the transformer model is for the first time introduced to the UIE task. The U-shape Transformer is integrated with a channel-wise multi-scale feature fusion transformer (CMSFFT) module and a spatial-wise global feature modeling transformer (SGFMT) module specially designed for UIE task, which reinforce the network's attention to the color channels and space areas with more serious attenuation. Meanwhile, in order to further improve the contrast and saturation, a novel loss function combining RGB, LAB and LCH color spaces is designed following the human vision principle. The extensive experiments on available datasets validate the state-of-the-art performance of the reported technique with more than 2dB superiority. The dataset and demo code are available on https://bianlab.github.io/codes.html.

Keywords: Underwater image enhancement · Transformer · Multi-color space loss function · Underwater image dataset

1 Introduction

Underwater Image Enhancement (UIE) technology [42,50] is essential for obtaining underwater images and investigating the underwater environment, which has wide applications in ocean exploration, biology, archaeology, underwater robots [20] and among other fields. However, underwater images frequently have problematic issues, such as color casts, color artifacts and blurred details [43]. Those issues could be explained by the strong absorption and scattering effects on light, which are caused by dissolved impurities and suspended matter in the medium (water). Therefore, UIE-related innovations are of great significance for improving the visual quality and merit of images in accurately understanding the underwater world.

L. Karlinsky et al. (Eds.): ECCV 2022 Workshops, LNCS 13802, pp. 290–307, 2023.
https://doi.org/10.1007/978-3-031-25063-7_18

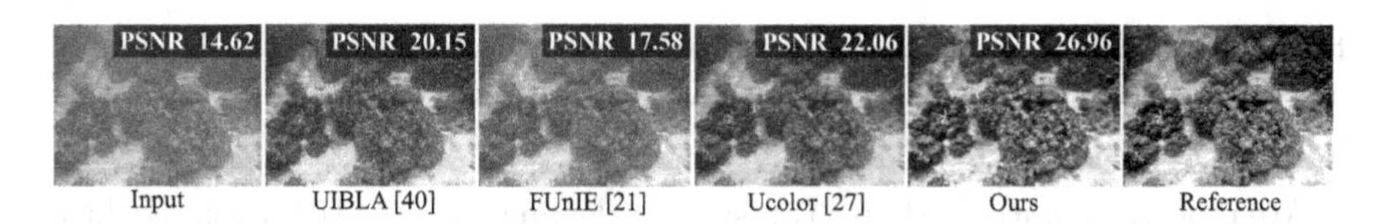

Fig. 1. Compared with the existing UIE methods, the image produced by our U-shape Transformer has the highest PSNR [23] score and best visual quality.

In general, the existing UIE methods could be categorized into three types, which are physical model-based, visual prior-based and data-driven methods, respectively. Among them, visual prior-based UIE methods [2,12,14,18,19,24] mainly concentrated on improving the visual quality of underwater images by modifying pixel values from the perspectives of contrast, brightness and saturation. Nevertheless, the ignorance of the physical degradation process limits the improvement of enhancement quality. In addition, physical-model based UIE methods [5,8,9,13,16,25,29,39,48] mainly focus on the accurate estimation of medium transmission. With the estimated medium transmission and other key underwater imaging parameters such as the homogeneous background light, a clean image can be obtained by reversing a physical underwater imaging model. However, the performance of physical model-based UIE is restricted to complicated and diverse real-world underwater scenes. That is because (1) *model hypothesis is not always plausible with complicated and dynamic underwater environment*; (2) *evaluating multiple parameters simultaneously is challenging*. More recently, as to the data-driven methods [1,10,11,15,20,26,28,30,31,43–45,49], which could be regarded as deep learning technologies in UIE domain, exhibit impressive performance on UIE task. However, the existing underwater datasets more-or-less have the disadvantages, such as a small number of images, few underwater scenes, or even not real-world scenarios, which limits the performance of the data-driven UIE method. Besides, the inconsistent attenuation of the underwater images in different color channels and space areas has not been unified in one framework.

In this work, we first built a large scale underwater image (LSUI) dataset, which covers more abundant underwater scenes (water types, lighting conditions and target categories) and better visual quality reference images than existing underwater datasets [1,28,31,34]. The dataset contains 4279 real-world underwater images, and the corresponding clear images are generated as comparison references. We also provide the semantic segmentation map and medium transmission map for each image. Furthermore, with the prior knowledge that the attenuation of different color channels and space areas in underwater images is inconsistent, we designed a channel-wise multi-scale feature fusion transformer (CMSFFT) module and a spatial-wise global feature modeling transformer (SGFMT) module based on the attention mechanism, and embedded them in our U-shape Transformer which is designed based on [21]. Moreover, according to [26,32], we designed a multi-color space loss function including

RGB, LAB and LCH color space. Figure 1 shows the result of our UIE method and some comparison UIE methods, and the main contributions of this paper can be summarized as follows:

- We reported a novel U-shape Transformer dealing with the UIE task, in which the designed CMSFFT and SGFMT modules based on transformer enables to effectively remove color artifacts and casts.
- We designed a novel multi-color space loss function combing the RGB, LCH and LAB color-space features, which further improves the contrast and saturation of enhanced images.
- We released a large-scale dataset containing 4279 real underwater images and the corresponding high-quality reference images, semantic segmentation maps, and medium transmission maps, which facilitates further development of UIE techniques.

2 Related Work

2.1 Data-driven UIE Methods

As we mentioned the pros and cons of physical model-based and visual prior-based UIE methods in Sect. 1, this part concerns only data-driven UIE methods.

Current data-driven UIE methods can be divided into two main technical routes, (1) *designing an end-to-end module;* (2) *utilizing deep models directly to estimate physical parameters, and then restore the clean image based on the degradation model.* To alleviate the need for real-world underwater paired training data, Li et al. [31] proposed a WaterGAN to generate underwater-like images from in-air images and depth maps in an unsupervised manner, in which the generated dataset is further used to train the WaterGAN. Moreover, [30] exhibited a weakly supervised underwater color transmission model based on CycleGAN [56]. Benefiting from the adversarial network architecture and multiple loss functions, that network can be trained using unpaired underwater images, which refines the adaptability of the network model to underwater scenes. However, images in the training dataset used by the above methods are not matched real underwater images, which leads to limited enhancement effects of the above methods in diverse real-world underwater scenes. Recently, Li et al. [28] proposed a gated fusion network named WaterNet, which uses gamma-corrected images, contrast-improved images, and white-balanced images as the inputs to enhance underwater images. Yang et al. [51] proposed a conditional generative adversarial network (cGAN) to improve the perceptual quality of underwater images.

The methods mentioned above usually use existing deep neural networks for general purposes directly on UIE tasks and neglect the unique characteristics of underwater imaging. For example, [30] directly used the CycleGAN [56] network structure, and [28] adopted a simple multi-scale convolutional network. Other models such as UGAN [10],WaterGAN [31] and cGAN [51], still inherited the disadvantage of GAN-based models, which produces unstable enhancement results. In addition, Ucolor [26] combined the underwater physical imaging model and designed a medium transmission guided model to reinforce the network's response to areas with more severe quality degradation, which could improve the visual quality of the network output to a certain extent. However, physical models sometimes failed with varied underwater environments.

From above, our proposed network aims at generating high visual quality underwater images by designing a novel multi-color space loss function and properly accounting the inconsistent attenuation characteristics of underwater images in different color channels and space areas.

2.2 Underwater Image Datasets

The sophisticated and dynamic underwater environment results in extreme difficulties in the collection of matched underwater image training data in real-world underwater scenes. Present datasets can be classified into two types, they are (1) Non-reference datasets. Liu et al. [34] proposed the RUIE dataset, which encompasses varied underwater lighting, depth of field, blurriness and color cast scenes. Akkaynak et al. [1] published a non-reference underwater dataset with a standard color comparison chart. Those datasets, however, cannot be used for end-to-end training for lacking matched clear reference underwater images. (2) Full-reference datasets. Li et al. [31] presented an unsupervised network dubbed WaterGAN to produce underwater-like images using in-air images and depth maps. Similarly, Fabbri et al. [10] used CycleGAN to generate distorted images from clean underwater images based on weakly-supervised distribution transfer. However, these methods rely heavily on training samples, which is easy to produce artifacts that are out of reality and unnatural. Li et al. [28] constructed a real UIE benchmark UIEB, including 890 image pairs, in which reference images were hand-crafted using the existing optimal UIE methods. Although those images are authentic and reliable, the number, content and coverage of underwater scenes are limited. In contrast, our LSUI dataset contains 4279 real-world underwater images with more abundant underwater scenes (water types, lighting conditions and target categories) than existing underwater datasets [1,28,31,34], and the corresponding clear images are generated as comparison references. We also provide the semantic segmentation map and medium transmission map for each raw underwater image.

2.3 Transformers

Although CNN-based UIE methods [10,20,26,28,45] achieved significant improvement compared with traditional UIE methods. There are still two aspects that limit its further promotion (1) *uniform convolution kernel is not able to characterize the inconsistent attenuation of underwater images in different color channels and spatial regions;* (2) *the CNN architecture concerns more on local features, while ineffective for long-dependent and global feature modeling.*

Recently, transformer [46] has gained more and more attention, its content-based interactions between image content and attention weights can be interpreted as spatially varying convolution, and the self-attention mechanism is efficient at modeling long-distance dependencies and global features. Benefiting from these advantages, transformers have shown outstanding performance in several vision tasks [7,35,54,55]. Compared with previous CNN-based UIE networks, our CMSFFT and SGFMT modules designed based on the transformer can guide the network to pay more attention to the more serious attenuated color channels and spatial areas. Moreover, by combining CNN with transformer, we achieve better performance with a relatively small amount of parameters.

3 Proposed Dataset and Method

3.1 LSUI Dataset

Data Collection. We have collected 8018 underwater images, some of them are collected by ourselves, and some of them are sourced from existing public datasets [1,10,28,31,34]. Real underwater images with rich water scenes, water types, lighting conditions and target categories, are selected to the extent possible, for further generating clear reference images.

Reference Image Generation. The reference images were selected with two round subjective and objective evaluations. In the first round, we firstly use 18 existing optimal UIE methods [2,5,8,9,11–13,20,25,29–31,36,39,40,44,45,49] to process the collected underwater images successively, and a set with $18 * 8018$ images is generated for the next-step optimal reference dataset selection. To reduce the number of images that need to be selected manually, non-reference metrics UIQM [38] and UCIQE [52] are adopted to score all generated images with equal weights. Then, the top-three reference images of each original one form a set with the size $3 * 8018$. Considering individual differences, 20 volunteers with image processing experience were invited to rate images according to 5 most important judgments (contrast; saturation; color correction effects; artifacts degree; over or under-enhancement degree) of UIE tasks with a score from 0–10, where the higher score represents the more contentedness. And the total score of each reference picture is 100 ($5 * 20$) after normalizing each score to 0–1. The top-one reference image of each raw underwater image was chosen with the highest summation value.

After the first round, some of the generated reference images still have problems such as blur, color cast and noise. So in the second round, we invited volunteers to vote on each reference picture again to select its existing problems and determine the corresponding optimization method, and then use appropriate image enhancement methods [33,53,54] to process it. Next, all volunteers were invited to conduct another round of voting to remove image pairs that more than half of the volunteers were dissatisfied with. To improve the utility of the LSUI dataset, we also hand-labeled a segmentation map and generated

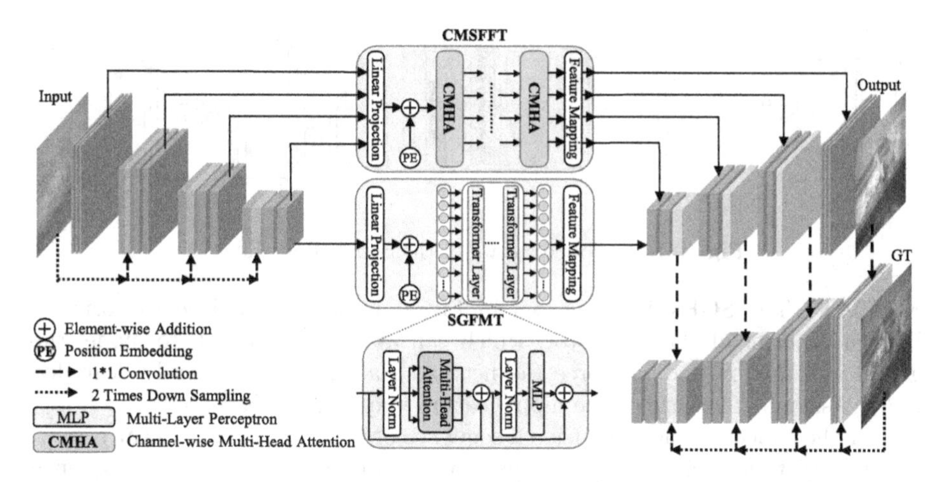

Fig. 2. The network structure of U-shape Transformer. CMSFFT and SGFMT modules specially designed for UIE tasks reinforce the network's attention to the more severely attenuated color channels and spatial regions. The multi-scale connections of the generator and the discriminator make the gradient flow freely between the generator and the discriminator, therefore making the training process more stable.

a medium transmission map for each image. Eventually, our LSUI dataset contains 4279 images and the corresponding high-quality reference images, semantic segmentation maps, and medium transmission maps for each image.

3.2 U-shape Transformer

Overall Architecture. The overall architecture of the U-shape Transformer is shown as Fig. 2, which includes a generator and a discriminator.

In the generator (1) Encoding: Except for being directly input to the network, the original image will be downsampled three times respectively. Then after 1*1 convolution, the three scale feature maps are input into the corresponding scale convolution block. The outputs of four convolutional blocks are the inputs of the CMSFFT and SGFMT; (2) Decoding: After feature remapping, the output of the SGFMT is directly sent to the first convolutional block. Meanwhile, four convolutional blocks with varied scales will receive the four outputs from CMSFFT.

In the discriminator, the input of the four convolutional blocks includes: the feature map output by its own upper layer, the feature map of the corresponding size from the decoding part and the feature map generated by $1 * 1$ convolution after downsampling to the corresponding size using the reference image. With the described multi-scale connections, the gradient flow can flow freely on multiple scales between the generator and the discriminator, such that a stable training process could be obtained, details of the generated images could be enriched. The detailed structure of SGFMT and CMSFFT in the network will be described in the following two subsections.

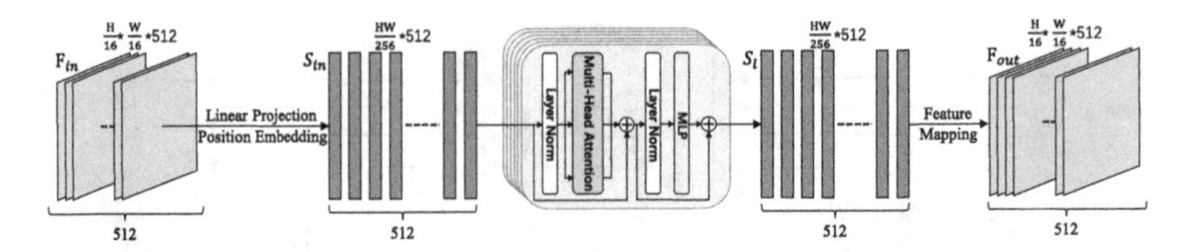

Fig. 3. Data flow diagram of the SGFMT module.

SGFMT. The SGFMT (as shown in Fig. 3) is used to replace the original bottleneck layer of the generator, which can assist the network to model the global information and reinforce the network's attention on severely degraded parts. Assuming the size of the input feature map is $F_{in} \in \mathbb{R}^{\frac{H}{16} * \frac{W}{16} * C}$.

For the expected one-dimensional sequence of the transformer, linear projection is used to stretch the two-dimensional feature map into a feature sequence $S_{in} \in \mathbb{R}^{\frac{HW}{256} * C}$. For preserving the valued position information of each region, learnable position embedding is merged directly, which can be expressed as,

$$S_{in} = W * F_{in} + \mathrm{PE}, \tag{1}$$

where $W * F_i$ represents a linear projection operation, PE represents a position embedding operation.

Then we input the feature sequence S_{in} to the transformer block, which contains 4 standard transformer layers [46]. Each transformer layer contains a multi-head attention block (MHA) and a feed-forward network (FFN). The FFN includes a normalization layer and a fully connected layer. The output of the l-th($l \in [1, 2, \ldots, l]$) layer in the transformer block can be calculated by,

$$S_l^{'} = \mathrm{MHA}(\mathrm{LN}(S_{l-1})) + S_{l-1}, \quad S_l = \mathrm{FFN}(\mathrm{LN}(S_l^{'})) + S_l^{'}, \tag{2}$$

where LN represents layer normalization, and S_l represents the output sequence of the l-th layer in the transformer block. The output feature sequence of the last transformer block is $S_l \in \mathbb{R}^{\frac{HW}{256} * C}$, which is restored to the feature map of $F_{out} \in \mathbb{R}^{\frac{H}{16} * \frac{W}{16} * C}$ after feature remapping.

CMSFFT. To reinforce the network's attention on the more serious attenuation color channels, inspired by [47], we designed the CMSFFT block to replace the skip connection of the original generator's encoding-decoding architecture (Fig. 4), which consists of the following three parts.

Multi-scale Feature Encoding. The inputs of CMSFFT are the feature maps $F_i \in \mathbb{R}^{\frac{H}{2^i} * \frac{W}{2^i} * C_i} (i = 0, 1, 2, 3)$ with different scales. Differs from the linear projection in Vit [6] which is applied directly on the partitioned original image, we use convolution kernels with related filter size $\frac{P}{2^i} * \frac{P}{2^i} (i = 0, 1, 2, 3)$ and step size $\frac{P}{2^i} (i = 0, 1, 2, 3)$, to conduct linear projection on feature maps with varied scales. In this work, P is set as 32. After that, four feature sequence

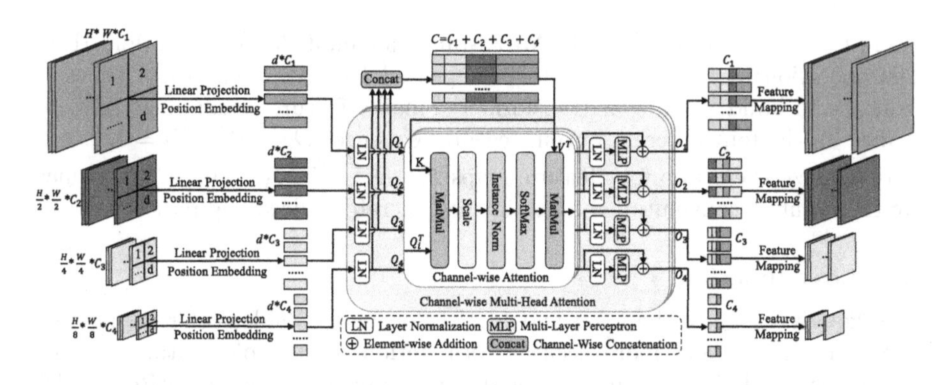

Fig. 4. Detailed structure of the CMSFFT module.

$S_i \in \mathbb{R}^{d*\ C_i}(i = 1, 2, 3, 4)$ could be obtained, where $d \in \frac{HW}{P^2}$. Those four convolution kernels divide feature maps into the same number of blocks, while the number of channels $C_i (i = 1, 2, 3, 4)$ remains unchanged. Then, four query vectors $Q_i \in \mathbb{R}^{d*\ C_i}(i = 1, 2, 3, 4)$, $K \in \mathbb{R}^{d*\ C}$ and $V \in \mathbb{R}^{d*\ C}$ can be obtained by Eq. (3).

$$Q_i = S_i W_{Q_i}, \quad K = SW_K, \quad V = SW_V, \tag{3}$$

where $W_{Q_i} \in \mathbb{R}^{d*\ C_i}(i = 1, 2, 3, 4)$, $W_K \in \mathbb{R}^{d*\ C}$ and $W_V \in \mathbb{R}^{d*\ C}$ stands for learnable weight matrices; S is generated by concatenating $S_i \in \mathbb{R}^{d*\ C_i}(i = 1, 2, 3, 4)$ via the channel dimension, where $C = C_1 + C_2 + C_3 + C_4$. In this work, C_1, C_2, C_3, and C_4 are set as 64, 128, 256, 512, respectively.

Channel-Wise Multi-head Attention (CMHA). The CMHA block has six inputs, which are $K \in \mathbb{R}^{d*\ C}$, $V \in \mathbb{R}^{d*\ C}$ and $Q_i \in \mathbb{R}^{d*\ C_i}(i = 1, 2, 3, 4)$. The output of channel-wise attention $\mathrm{CA_i} \in \mathbb{R}^{C_i*\ d}(i = 1, 2, 3, 4)$ could be obtained by,

$$\mathrm{CA}_i = \mathrm{SoftMax}(\mathrm{IN}(\frac{Q_i^T K}{\sqrt[2]{C}}))V^T, \tag{4}$$

where IN represents the instance normalization operation. This attention operation performs along the channel-axis instead of the classical patch-axis [6], which can guide the network to pay attention to channels with more severe image quality degradation. In addition, IN is used on the similarity maps to assist the gradient flow spreads smoothly.

The output of the i-th CMHA layer can be expressed as,

$$\mathrm{CMHA_i} = (\mathrm{CA_i}^1 + \mathrm{CA_i}^2 + ..., +\mathrm{CA_i}^N)/N + Q_i, \tag{5}$$

where N is the number of heads, which is set as 4 in our implementation.

Feed-Forward Network (FFN). Similar to the forward propagation of [6], the FFN output can be expressed as,

$$O_i = \mathrm{CMHA_i} + \mathrm{MLP}(\mathrm{LN}(\mathrm{CMHA_i})), \tag{6}$$

where $O_i \in \mathbb{R}^{d* C_i} (i = 1, 2, 3, 4)$; MLP stands for multi-layer perception. Here, The operation in Eq. (6) needs to be repeated l (l=4 in this work) times in sequence to build the l-layer transformer. Finally, feature remappings are performed on the four different output feature sequences $O_i \in \mathbb{R}^{C_i* d} (i = 1, 2, 3, 4)$ to reorganize them into four feature maps $F_i \in \mathbb{R}^{\frac{H}{2^i}*\frac{W}{2^i}*C_i} (i = 0, 1, 2, 3)$, which are the input of convolutional block in the generator's decoding part.

3.3 Loss Function

To take advantage of the LAB and LCH color spaces' wider color gamut representation range and more accurate description of the color saturation and brightness, we designed a multi-color space loss function combining RGB, LAB and LCH color spaces to train our network. The image from RGB space is firstly converted to LAB and LCH space, and reads,

$$\begin{aligned} L^{G(x)}, A^{G(x)}, B^{G(x)} &= \mathrm{RGB2LAB(G(x))}, \quad \mathrm{L^y, A^y, B^y = RGB2LAB(y)} \\ L^{G(x)}, C^{G(x)}, H^{G(x)} &= \mathrm{RGB2LCH(G(x))}, \quad \mathrm{L^y, C^y, H^y = RGB2LCH(y)} \end{aligned} \tag{7}$$

where x, y and $G(x)$ represents the original inputs, the reference image, and the clear image output by the generator, respectively.

Loss functions in the LAB and LCH space are written as Eq. (8) and Eq. (9).

$$\begin{aligned} &Loss_{LAB}(G(x), y) = E_{x,y}[(L^{y} - L^{G(x)})^2 - \\ &\sum_{i=1}^{n} Q(A_i^y) log(Q(A_i^{G(x)})) - \sum_{i=1}^{n} Q(B_i^y) log(Q(B_i^{G(x)}))], \end{aligned} \tag{8}$$

$$\begin{aligned} &Loss_{LCH}(G(x), y) = E_{x,y}[-\sum_{i=1}^{n} Q(L_i^y) log(Q(L_i^{G(x)})) \\ &+ (C^y - C^{G(x)})^2 + (H^y - H^{G(x)})^2], \end{aligned} \tag{9}$$

where Q stands for the quantization operator.

L_2 loss in the RGB color space $Loss_{RGB}$ and the perceptual loss $Loss_{per}$ [22], as well as $Loss_{LAB}$ and $Loss_{LCH}$ are the four loss functions for the generator.

Besides, standard GAN loss function is introduced for minimizing the loss between generated and reference pictures, and written as,

$$L_{GAN}(G, D) = E_y[log D(y)] + E_x[log(1 - D(G(x)))], \tag{10}$$

where D represents the discriminator. D aims at maximizing $L_{GAN}(G, D)$, to accurately distinguish the generated image from the reference image. And the goal of generator G is to minimize the loss between generated pictures and reference pictures. Then, the final loss function is expressed as,

$$\begin{aligned} G^* =& arg \min_G \max_D L_{GAN}(G, D) + \alpha Loss_{LAB}(G(x), y) \\ &+ \beta Loss_{LCH}(G(x), y) + \gamma Loss_{RGB}(G(x), y) + \mu Loss_{per}(G(x), y), \end{aligned} \tag{11}$$

where $\alpha, \beta, \gamma, \mu$ are hyperparameters, which are set as 0.001, 1, 0.1, 100, respectively, with numerous experiments.

4 Experiments

4.1 Experiment Settings

Benchmarks. The LSUI dataset was randomly divided as Train-L (3879 images) and Test-L400 (400 images) for training and testing, respectively. Besides Train-L, the second training set Train-U contains 800 pairs of underwater images from UIEB [28] and 1,250 synthetic underwater images from [27]; the third training set Train-E contains the paired training images in the EUVP [20] dataset. Testing datasets are categorized into two types (1) full-reference testing dataset: Test-L400 and Test-U90 (remaining 90 pairs in UIEB); (2) non-reference testing dataset: Test-U60 and SQUID. Here, Test-U60 includes 60 non-reference images in UIEB; 16 pictures from SQUID [1] forms the second non-reference testing dataset.

Compared Methods. We compare U-shape Transformer with 10 UIE methods to verify our performance superiority. It includes two physical-based models (UIBLA [39], UDCP [9]), three visual prior-based methods (Fusion [2], retinex based [12], RGHS [18]), and five data-driven methods (WaterNet [28], FUnIE [20], UGAN [10], UIE-DAL [45], Ucolor [26]).

Evaluation Metrics. For the testing dataset with reference images, we conducted full-reference evaluations using PSNR [23] and SSIM [17] metrics. For images in the non-reference testing dataset, non-reference evaluation metrics UCIQE [52] and UIQM [38] are employed, in which higher UCIQE or UIQM score suggests better human visual perception. For UCIQE and UIQM cannot accurately measure the performance in some cases [3,28], we also conducted a survey following [26], which results are stated as "perception score (PS)". PS ranges from 1–5, with higher scores indicating higher image quality. Moreover, NIQE [37], which lower value represents a higher visual quality, is also adopted as the metrics.

4.2 Dataset Evaluation

The effectiveness of LSUI is evaluated by retraining the compared methods (U-net [41], UGAN [10] and U-shape Transformer) on Train-L, Train-U and Train-E. The trained network was tested on Test-L400 and Test-U90. As shown in Table 1, the model trained on our dataset is the best of PSNR and SSIM. It could be explained that LSUI contains richer underwater scenes and better visual quality reference images than existing underwater image datasets, which could improve the enhancement and generalization ability of the tested network.

4.3 Network Architecture Evaluation

Full-Reference Evaluation. The Test-L400 and Test-U90 datasets were used for evaluation. We retrained the 5 open-sourced deep learning-based UIE methods on our dataset. And the statistical results and visual comparisons are summarized in Table 2 and Fig. 5. We also provide the running time (image size is

Table 1. Dataset evaluation results. The highest PSNR and SSIM scores are marked in **bold**.

Methods	Training Data	Test-U90		TestL-400	
		PSNR	SSIM	PSNR	SSIM
U-net [41]	Train-U	17.07	0.76	19.19	0.79
	Train-E	17.46	0.76	19.45	0.78
	Ours	**20.14**	**0.81**	**20.89**	**0.82**
UGAN [10]	Train-U	20.71	0.82	19.89	0.79
	Train-E	20.72	0.82	19.82	0.78
	Ours	**21.56**	**0.83**	**21.74**	**0.84**
Ours	Train-U	21.25	0.84	22.87	0.85
	Train-E	21.75	0.86	23.01	0.87
	Ours	**22.91**	**0.91**	**24.16**	**0.93**

256*256) of all UIE methods in Table 2, as well as the FLOPs and parameter amount of each data-driven UIE method.

As in Table 2, our U-shape Transformer demonstrates the best performance on both PSNR and SSIM metrics with relatively few parameters, FLOPs, and running time. The potential limitations of the performance of the 5 data-driven methods are analyzed as follows. The strength of FUnIE [20] lies in achieving fast, lightweight,and fewer parameter models, while naturally limits its scalability on complex and distorted testing samples. UGAN [10] and UIE-DAL [45] did not consider the inconsistent characteristics of the underwater images. Ucolor's media transmission map prior can not effectively represent the attenuation of each area, and simply introducing the concept of multi-color space into the network's encoder part cannot effectively take advantage of it, which causes unsatisfactory results in terms of contrast, brightness, and detailed textures.

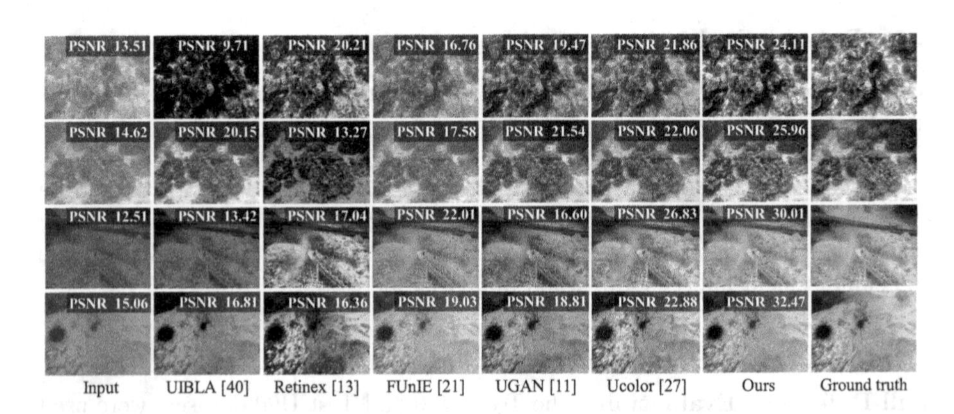

Fig. 5. Visual comparison of enhancement results sampled from the Test-L400(LSUI) and Test-U90(UIEB [28]) dataset. We regard the reference picture as ground truth (GT) to calculate PSNR.

Table 2. Quantitative comparison among different UIE methods on the full-reference testing set. The highest scores of PSNR and SSIM are marked in **bold**, and all UIE methods are tested on a PC with an INTEL(R) I5-10500 CPU, 16.0GB RAM, a NVIDIA GEFORCE RTX 1660 SUPER.

Methods	Test-L400		Test-U90		FLOPs↓	#param.↓	time↓
	PSNR↑	SSIM↑	PSNR↑	SSIM↑			
UIBLA [39]	13.54	0.71	15.78	0.73	×	×	42.13 s
UDCP [9]	11.89	0.59	13.81	0.69	×	×	30.82 s
Fusion [2]	17.48	0.79	19.04	0.82	×	×	6.58 s
Retinex based [12]	13.89	0.74	14.01	0.72	×	×	1.06 s
RGHS [18]	14.21	0.78	14.57	0.79	×	×	8.92 s
WaterNet [28]	17.73	0.82	19.81	0.86	193.7G	24.81M	0.61 s
FUnIE [20]	19.37	0.84	19.45	0.85	10.23G	7.019M	0.09 s
UGAN [10]	19.79	0.78	20.68	0.84	38.97G	57.17M	0.05 s
UIE-DAL [45]	17.45	0.79	16.37	0.78	29.32G	18.82M	0.07 s
Ucolor [26]	22.91	0.89	20.78	0.87	443.85G	157.4M	2.75 s
Ours	**24.16**	**0.93**	**22.91**	**0.91**	66.2G	65.6M	0.07 s

The visual comparisons shown in Fig. 5 reveal that enhancement results of our method are the closest to the reference image, which has fewer color artifacts and high-fidelity object areas. Five selected methods tend to produce color artifacts that deviated from the original color of the object. Among the methods, UIBLA [39] exhibits severe color casts. Retinex based [12] could improve the image contrast to a certain extent, but cannot remove the color casts and color artifacts effectively. The enhancement result of FUnLE [20] is yellowish and reddish overall. Although UGAN [10] and Ucolor [28] could provide relatively good color appearance, they are often affected by local over-enhancement, and there are still some color casts in the result.

Non-reference Evaluation. The Test-U60 and SQUID datasets were utilized for the non-reference evaluation, in which statistical results and visual comparisons are shown in Table 3 and Fig. 6 (a).

As in Table 3, our method achieved the highest scores on PS and NIQE metrics, which confirmed the initial idea to contemplate the human eye's color perception and better generalization ability to varied real-world underwater scenes. Note that UCIQE and UIQM of all deep learning-based UIE methods are weaker than physical model-based or visual prior-based, also reported in [26]. Those two metrics are of valuable reference, but cannot as absolute justifications [3,28], for they are non-sensitive to color artifacts & casts and biased to some features.

As in Fig. 6 (a), enhancement results of our method have the highest PS value, which index reflects the visual quality. Among the methods involved in comparison, results of the UIBLA [39] and FUnIE [20] have a certain degree of color cast. Retinex based [12] method introduces artifacts and unnatural colors. UGAN [10] and UIE-DAL [45] have the issue of local over-enhancement and

Table 3. Quantitative comparison among different UIE methods on the non-reference testing set. The highest scores are marked in **bold**.

Methods	Test-U60				SQUID			
	PS↑	UIQM↑	UCIQE↑	NIQE↓	PS↑	UIQM↑	UCIQE↑	NIQE↓
input	1.46	0.82	0.45	7.16	1.23	0.81	0.43	4.93
UIBLA [39]	2.18	1.21	0.60	6.13	2.45	0.96	0.52	4.43
UDCP [9]	2.01	1.03	0.57	5.94	2.57	1.13	0.51	4.47
Fusion [2]	2.12	**1.23**	0.61	4.96	2.89	**1.29**	0.61	5.01
Retinex based [12]	2.04	0.94	0.69	4.95	2.33	1.01	0.66	4.86
RGHS [18]	2.45	0.66	0.71	4.82	2.67	0.82	**0.73**	4.54
WaterNet [28]	3.23	0.92	0.51	6.03	2.72	0.98	0.51	4.75
FUnIE [20]	3.12	1.03	0.54	6.12	2.65	0.98	0.51	4.67
UGAN [10]	3.64	0.86	0.57	6.74	2.79	0.90	0.58	4.56
UIE-DAL [45]	2.03	0.72	0.54	4.99	2.21	0.79	0.57	4.88
Ucolor [26]	3.71	0.84	0.53	6.21	2.82	0.82	0.51	4.32
Ours	**3.91**	0.85	**0.73**	**4.74**	**3.23**	0.89	0.67	**4.24**

color artifacts, which main reason is they ignore the inconsistent attenuation characteristics of the underwater images in the different space areas and the color channels. Although Ucolor [26] introduces the transmission medium prior to reinforce the network's attention on the spatial area with severe attenuation, it still ignores the inconsistent attenuation characteristics in different color channels, which results in the problem of overall color cast. In our method, the reported CMSFFT and SGFMT modules could reinforce the network's attention to the color channels and spatial regions with serious attenuation, therefore obtaining high visual quality enhancement results without artifacts and color casts.

4.4 Ablation Study

To prove the effectiveness of each component, we conduct a series of ablation studies on the Test-L400 and Test-U90. Four factors are considered including the CMSFFT, the SGFMT, the multi-scale gradient flow mechanism (MSG), and the multi-color space loss function (MCSL).

Experiments are all trained by Train-L. Statistical results are shown in Table 4, in which baseline model (BL) refers to [21], full models is the complete U-shape Transformer. In Table 4, our full model achieves the best quantitative performance on the two testing dataset, which reflects the effectiveness of the combination of CMSFFT, SGFMT, MSG, and MCSL modules.

As in Fig. 6 (b), the enhancement result of the full model has the highest PSNR and best visual quality. The results of BL+MSG have less noise and artifacts than the BL module because the MSG mechanism helps to reconstruct

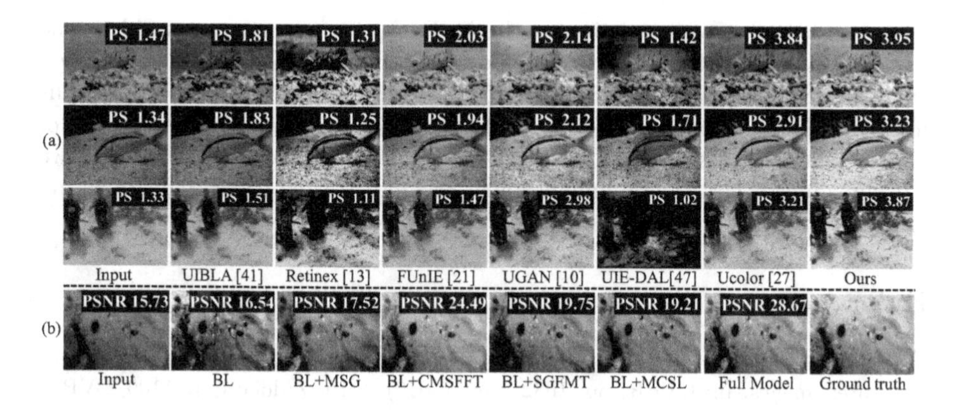

Fig. 6. (a) Visual comparison of the non-reference evaluation sampled from the Test-U60(UIEB [28]) dataset. (b) Visual comparison of the ablation study sampled from the Test-L400 dataset.

Table 4. Statistical results of ablation study on the Test-L400 and the Test-U90. The highest scores are marked in **bold**.

Models	Test-L400		Test-U90	
	PSNR	SSIM	PSNR	SSIM
BL	19.34	0.79	19.36	0.81
BL+CMSFFT	22.47	0.88	21.72	0.86
BL+SGFMT	21.78	0.86	21.36	0.87
BL+MSG	20.11	0.82	21.24	0.85
BL+MCSL	21.51	0.82	20.16	0.81
Full Model	**24.16**	**0.93**	**22.91**	**0.91**

local details. Thanks to the multi-color space loss function, the overall color of BL+MCSL's result is close to the reference image. The unevenly distributed visualization and artifacts in local areas of BL+MCSL are due to the lack of efficient attention guidance. Although the enhanced results of BL+CMSFFT and BL+SGFMT are evenly distributed, the overall color is not accurate. The investigated four modules have their particular functionality in the enhancement process, which integration could improve the overall performance of our network.

5 Conclusions

In this work, we released the LSUI dataset which is the largest real-world underwater dataset with high-fidelity reference images. Besides, we reported an U-shape Transformer network for state-of-the-art enhancement. The network's CMSFFT and SGFMT modules could solve the inconsistent attenuation issue of underwater images in different color channels and space regions, which has not been considered among existing methods. Extensive experiments validate the superior ability of the network to remove color artifacts and casts. Combined with the multi-color space loss function, the contrast and saturation of output images are further improved. Nevertheless, it is impossible to collect images of all the complicated scenes such as deep-ocean low-light scenarios. Therefore, we will introduce other general enhancement techniques such as low-light boosting [4] for future work.

References

1. Akkaynak, D., Treibitz, T.: Sea-thru: a method for removing water from underwater images. In: CVPR, pp. 1682–1691 (2019). https://doi.org/10.1109/CVPR.2019.00178
2. Ancuti, C., Ancuti, C.O., Haber, T., Bekaert, P.: enhancing underwater images and videos by fusion. In: CVPR, pp. 81–88 (2012). https://doi.org/10.1109/CVPR.2012.6247661
3. Berman, D., Levy, D., Avidan, S., Treibitz, T.: Underwater single image color restoration using haze-lines and a new quantitative dataset. IEEE TPAMI **43**(8), 2822–2837 (2021). https://doi.org/10.1109/TPAMI.2020.2977624
4. Chen, C., Chen, Q., Xu, J., Koltun, V.: Learning to see in the dark. In: CVPR, pp. 3291–3300 (2018). https://doi.org/10.1109/CVPR.2018.00347
5. Chiang, J.Y., Chen, Y.C.: Underwater image enhancement by wavelength compensation and dehazing. IEEE TIP **21**(4), 1756–1769 (2012). https://doi.org/10.1109/TIP.2011.2179666
6. Dosovitskiy, A., et al.: An image is worth 16x16 words: Transformers for image recognition at scale. (2021) ArXiv abs/2010.11929
7. Dosovitskiy, A., et al.: An image is worth 16x16 words: transformers for image recognition at scale. (2020) arXiv preprint arXiv:2010.11929
8. Drews, P.L., Nascimento, E.R., Botelho, S.S., Montenegro Campos, M.F.: Underwater depth estimation and image restoration based on single images. IEEE Comput. Graph. Appl. **36**(2), 24–35 (2016). https://doi.org/10.1109/MCG.2016.26
9. Drews Jr, P., do Nascimento, E., Moraes, F., Botelho, S., Campos, M.: Transmission estimation in underwater single images. In: ICCV workshops, pp. 825–830 (2013). https://doi.org/10.1109/ICCVW.2013.113
10. Fabbri, C., Islam, M.J., Sattar, J.: Enhancing underwater imagery using generative adversarial networks. ICRA, pp. 7159–7165 (2018)
11. Fu, X., Fan, Z., Ling, M., Huang, Y., Ding, X.: Two-step approach for single underwater image enhancement. In: ISPACS, pp. 789–794 (2017). https://doi.org/10.1109/ISPACS.2017.8266583
12. Fu, X., Zhuang, P., Huang, Y., Liao, Y., Zhang, X.P., Ding, X.: A retinex-based enhancing approach for single underwater image. In: ICIP, pp. 4572–4576 (2014). https://doi.org/10.1109/ICIP.2014.7025927
13. Galdran, A., Pardo, D., Picón, A., Alvarez-Gila, A.: Autom. Red-Channel Underwater Image Restor. JVCIR **26**, 132–145 (2015)
14. Ghani, A.S.A., Isa, N.A.M.: Underwater image quality enhancement through composition of dual-intensity images and rayleigh-stretching. In: ICCE, pp. 219–220 (2014). https://doi.org/10.1109/ICCE-Berlin.2014.7034265
15. Guo, Y., Li, H., Zhuang, P.: Underwater image enhancement using a multiscale dense generative adversarial network. IEEE J. Oceanic Eng. **45**(3), 862–870 (2019)

16. He, K., Sun, J., Tang, X.: Single image haze removal using dark channel prior. In: CVPR, pp. 1956–1963 (2009). https://doi.org/10.1109/CVPR.2009.5206515
17. Horé, A., Ziou, D.: Image quality metrics: PSNR vs. SSIM. In: ICPR, pp. 2366–2369 (2010). https://doi.org/10.1109/ICPR.2010.579
18. Huang, D., Wang, Y., Song, W., Sequeira, J., Mavromatis, S.: Shallow-Water Image Enhancement Using Relative Global Histogram Stretching Based on Adaptive Parameter Acquisition. In: Schoeffmann, K., et al. (eds.) MMM 2018. LNCS, vol. 10704, pp. 453–465. Springer, Cham (2018). https://doi.org/10.1007/978-3-319-73603-7_37
19. Iqbal, K., Odetayo, M., James, A., Salam, R.A., Talib, A.Z.H.: Enhancing the low quality images using unsupervised colour correction method. In: IEEE Int. Conf. Syst. Man. Cybern, pp. 1703–1709 (2010). https://doi.org/10.1109/ICSMC.2010.5642311
20. Islam, M.J., Xia, Y., Sattar, J.: Fast underwater image enhancement for improved visual perception. IEEE Robot. Autom. Lett. **5**(2), 3227–3234 (2020). https://doi.org/10.1109/LRA.2020.2974710
21. Isola, P., Zhu, J.Y., Zhou, T., Efros, A.A.: Image-to-image translation with conditional adversarial networks. In: CVPR, pp. 1125–1134 (2017)
22. Johnson, J., Alahi, A., Fei-Fei, L.: Perceptual Losses for Real-Time Style Transfer and Super-Resolution. In: Leibe, B., Matas, J., Sebe, N., Welling, M. (eds.) ECCV 2016. LNCS, vol. 9906, pp. 694–711. Springer, Cham (2016). https://doi.org/10.1007/978-3-319-46475-6_43
23. Korhonen, J., You, J.: Peak signal-to-noise ratio revisited: is simple beautiful? In: QoMEX, pp. 37–38. IEEE (2012)
24. Li, C.Y., Guo, J.C., Cong, R.M., Pang, Y.W., Wang, B.: Underwater image enhancement by dehazing with minimum information loss and histogram distribution prior. IEEE TIP **25**(12), 5664–5677 (2016). https://doi.org/10.1109/TIP.2016.2612882
25. Li, C.Y., Guo, J.C., Cong, R.M., Pang, Y.W., Wang, B.: Underwater image enhancement by dehazing with minimum information loss and histogram distribution prior. IEEE TIP **25**(12), 5664–5677 (2016). https://doi.org/10.1109/TIP.2016.2612882
26. Li, C., Anwar, S., Hou, J., Cong, R., Guo, C., Ren, W.: Underwater image enhancement via medium transmission-guided multi-color space embedding. IEEE TIP **30**, 4985–5000 (2021)
27. Li, C., Anwar, S., Porikli, F.: Underwater scene prior inspired deep underwater image and video enhancement. Pattern Recognit. **98**, 107038 (2020)
28. Li, C., Guo, C., Ren, W., Cong, R., Hou, J., Kwong, S., Tao, D.: An underwater image enhancement benchmark dataset and beyond. IEEE TIP **29**, 4376–4389 (2020). https://doi.org/10.1109/TIP.2019.2955241
29. Li, C., Guo, J., Chen, S., Tang, Y., Pang, Y., Wang, J.: Underwater image restoration based on minimum information loss principle and optical properties of underwater imaging. In: ICIP, pp. 1993–1997 (2016). https://doi.org/10.1109/ICIP.2016.7532707
30. Li, C., Guo, J., Guo, C.: Emerging from water: underwater image color correction based on weakly supervised color transfer. IEEE Signal Process. Lett. **25**(3), 323–327 (2018)
31. Li, J., Skinner, K.A., Eustice, R.M., Johnson-Roberson, M.: Watergan: unsupervised generative network to enable real-time color correction of monocular underwater images. IEEE Robot. Autom. Lett. **3**(1), 387–394 (2017)

32. Li, X., Li, A.: An improved image enhancement method based on lab color space retinex algorithm. In: Li, C., Yu, H., Pan, Z., Pu, Y. (eds.) International Society for Optics and Photonics, SPIE (2019) ICGIP. vol. 11069, pp. 756–765. https://doi.org/10.1117/12.2524449
33. Liang, J., Cao, J., Sun, G., Zhang, K., Van Gool, L., Timofte, R.: Swinir: image restoration using swin transformer. In: ICCV Workshops, pp. 1833–1844 (2021)
34. Liu, R., Fan, X., Zhu, M., Hou, M., Luo, Z.: Real-world underwater enhancement: Challenges, benchmarks, and solutions under natural light. IEEE Trans. Circuits Syst. Video Technol. **30**, 4861–4875 (2020)
35. Liu, Z., et al.: Swin transformer: hierarchical vision transformer using shifted windows. (2021) arXiv preprint arXiv:2103.14030
36. Ma, Z., Oh, C.: A wavelet-based dual-stream network for underwater image enhancement. (2021) arXiv preprint arXiv:2202.08758
37. Mittal, A., Soundararajan, R., Bovik, A.C.: Making a completely blind image quality analyzer. IEEE Signal Process. Lett. **20**(3), 209–212 (2013). https://doi.org/10.1109/LSP.2012.2227726
38. Panetta, K., Gao, C., Agaian, S.: Human-visual-system-inspired underwater image quality measures. IEEE J. Ocean. Eng. **41**(3), 541–551 (2016). https://doi.org/10.1109/JOE.2015.2469915
39. Peng, Y.T., Cosman, P.C.: Underwater image restoration based on image blurriness and light absorption. IEEE TIP **26**(4), 1579–1594 (2017)
40. Qi, Q., Li, K., Zheng, H., Gao, X., Hou, G., Sun, K.: Sguie-net: Semantic attention guided underwater image enhancement with multi-scale perception. (2022) arXiv preprint arXiv:2201.02832
41. Ronneberger, O., Fischer, P., Brox, T.: U-Net: Convolutional Networks for Biomedical Image Segmentation. In: Navab, N., Hornegger, J., Wells, W.M., Frangi, A.F. (eds.) MICCAI 2015. LNCS, vol. 9351, pp. 234–241. Springer, Cham (2015). https://doi.org/10.1007/978-3-319-24574-4_28
42. Sahu, P., Gupta, N., Sharma, N.: A survey on underwater image enhancement techniques. IJCA, **87**(13) (2014)
43. Schettini, R., Corchs, S.: Underwater image processing: State of the art of restoration and image enhancement methods. EURASIP. J. Adv. Signal Process. **2010**, 1–14 (2010)
44. Song, W., Wang, Y., Huang, D., Tjondronegoro, D.: A Rapid Scene Depth Estimation Model Based on Underwater Light Attenuation Prior for Underwater Image Restoration. In: Hong, R., Cheng, W.-H., Yamasaki, T., Wang, M., Ngo, C.-W. (eds.) PCM 2018. LNCS, vol. 11164, pp. 678–688. Springer, Cham (2018). https://doi.org/10.1007/978-3-030-00776-8_62
45. Uplavikar, P.M., Wu, Z., Wang, Z.: All-in-one underwater image enhancement using domain-adversarial learning. In: CVPR Workshops, pp. 1–8 (2019)
46. Vaswani, A., et al.: Attention is all you need. In: NIPS, pp. 5998–6008 (2017)
47. Wang, H., Cao, P., Wang, J., Zaiane, O.R.: Uctransnet: rethinking the skip connections in u-net from a channel-wise perspective with transformer (2021)
48. Wang, Y., Liu, H., Chau, L.P.: Single underwater image restoration using adaptive attenuation-curve prior. In: IEEE Trans. Circuits. Syst. I. Regul. Pap. 65(3), 992–1002 (2018). https://doi.org/10.1109/TCSI.2017.2751671
49. Yang, H.Y., Chen, P.Y., Huang, C.C., Zhuang, Y.Z., Shiau, Y.H.: Low complexity underwater image enhancement based on dark channel prior. In: IBICA, pp. 17–20 (2011). https://doi.org/10.1109/IBICA.2011.9
50. Yang, M., Hu, J., Li, C., Rohde, G., Du, Y., Hu, K.: An in-depth survey of underwater image enhancement and restoration. IEEE Access. **7**, 123638–123657 (2019)

51. Yang, M., Hu, K., Du, Y., Wei, Z., Sheng, Z., Hu, J.: Underwater image enhancement based on conditional generative adversarial network. Signal Process., Image Commun. 81, 115723 (2020)
52. Yang, M., Sowmya, A.: An underwater color image quality evaluation metric. IEEE TIP **24**(12), 6062–6071 (2015). https://doi.org/10.1109/TIP.2015.2491020
53. Ye, T., et al.: Perceiving and modeling density is all you need for image dehazing (2021)
54. Zamir, S.W., Arora, A., Khan, S., Hayat, M., Khan, F.S., Yang, M.H.: Restormer: efficient transformer for high-resolution image restoration (2021)
55. Zhao, H., Jiang, L., Jia, J., Torr, P.H., Koltun, V.: Point transformer. In: ICCV, pp. 16259–16268 (2021)
56. Zhu, J.Y., Park, T., Isola, P., Efros, A.A.: Unpaired image-to-image translation using cycle-consistent adversarial networks. In: ICCV, pp. 2242–2251 (2017). https://doi.org/10.1109/ICCV.2017.244

Hybrid Transformer Based Feature Fusion for Self-Supervised Monocular Depth Estimation

Snehal Singh Tomar(✉), Maitreya Suin, and A. N. Rajagopalan

Indian Institute Of Technology Madras, Chennai, India
snehal@smail.iitm.ac.in, raju@ee.iitm.ac.in

Abstract. With an unprecedented increase in the number of agents and systems that aim to navigate the real world using visual cues and the rising impetus for 3D Vision Models, the importance of depth estimation is hard to understate. While supervised methods remain the gold standard in the domain, the copious amount of paired stereo data required to train such models makes them impractical. Most State of the Art (SOTA) works in the self-supervised and unsupervised domain employ a ResNet-based encoder architecture to predict disparity maps from a given input image which are eventually used alongside a camera pose estimator to predict depth without direct supervision. The fully convolutional nature of ResNets makes them susceptible to capturing per-pixel local information only, which is suboptimal for depth prediction. Our key insight for doing away with this bottleneck is to use Vision Transformers, which employ self-attention to capture the global contextual information present in an input image. Our model fuses per-pixel local information learned using two fully convolutional depth encoders with global contextual information learned by a transformer encoder at different scales. It does so using a mask-guided multi-stream convolution in the feature space to achieve state-of-the-art performance on most standard benchmarks.

1 Introduction

The ability of humans to perceive 3D geometry in the complex environment around them is an indispensable sensory endowment. All agents seeking human-like performance in real-world navigation must learn to emulate this capability. Depth estimation is a crucial component of understanding 3D geometry. Traditional Computer Vision techniques estimate depth by calculating disparity over stereo image pairs with known correspondences using prior knowledge of intrinsic camera parameters. Most modern-day autonomous agents synergize image data with information from other sensor modalities such as LiDARs, RGB-D

S. S. Tomar and M. Suin—Equal contribution.

Supplementary Information The online version contains supplementary material available at https://doi.org/10.1007/978-3-031-25063-7_19.

L. Karlinsky et al. (Eds.): ECCV 2022 Workshops, LNCS 13802, pp. 308–326, 2023.
https://doi.org/10.1007/978-3-031-25063-7_19

cameras, short-range RADARs, and Ultrasonic Sensors for accurate depth estimation. Acquiring such data is expensive in terms of capital and human effort. Monocular Depth Estimation is the task of predicting the depth information of a scene using a single RGB image as input. Although supervised methods ([1,29,33]) have presented the best results for the task, getting ground truth depth data to make them work is challenging. SOTA self-supervised and unsupervised methods ([16,26,28]) that perform Monocular Depth Estimation are inspired from [61] and make use of fully convolutional architectures for predicting disparity from a given image. They also have a separate network for predicting camera pose, the predictions of which are clubbed with the depth predictions and prior knowledge of intrinsic camera parameters to warp a given frame in a monocular video sequence onto another temporally consistent frame. These temporally consistent frames thus, serve as a surrogate for direct depth map supervision to perform this ill-posed task.

A series of recent supervised and unsupervised works have attempted to provide richer representations to their encoder-decoder architectures for depth estimation. They have mainly focused on fusing their latent representations with representations learned for other downstream vision tasks such as semantic segmentation ([7,20,30]) or using geometric priors ([37,38]) to improve their depth map predictions. Most self-supervised approaches deploy variants of ResNet [22] for predicting the depth of a scene. However, due to the limited receptive field of convolutional operation, they struggle to model the global context and capture the long-range relationship between different regions Inspired by the recent success of vision transformers [7], our endeavor in this work is to explore the utility of transformer and its interaction with traditional CNN architectures for depth prediction. Although [41] recently deployed transformer for various supervised image-to-image tasks, including depth prediction, it is relatively unexplored for self-supervised monocular depth estimation. We validate that a vision transformer can be extremely helpful for depth estimation due to its effective long-range modeling ability with a global receptive field. However, a transformer alone is insufficient for accurately predicting the finer local details like sharp depth boundaries. To address this, we design a hybrid-transformer architecture, where we deploy parallel 'local' encoder branches parallel to a 'global' transformer for capturing different abstractions of the scene. The local branches specialize in extracting neighboring features around every pixel at different spatial resolutions. In the end, complementary information from these three branches is fused using a multi-scale fusion block. Our key contributions are:

1. We propose an Encoder architecture consisting of a Vision Transformer (ViT) based module, a Low-Resolution Local (LRL), and a High-Resolution Local (HRL) convolutional module. The ViT based module captures global contextual information, whereas the LRL and HRL modules capture local information at low and high resolutions, respectively.
2. We propose a mask-guided multi-scale fusion block to combine complementary information from our encoder's convolutional and transformer-based elements at different scales and resolutions in the latent space.

3. We illustrate with our ablations, that depth encoder architectures that fuse global, local, HR, and LR features perform better than architectures which use them in isolation.
4. Extensive experiments on multiple datasets ([15,16,19,26]) demonstrate the strong performance of our approach with interpretable operations.

2 Related Works

Self-Supervised Monocular Depth Estimation: The advent of deep learning architectures in Computer Vision made Monocular Depth Estimation a reality. Monocular Depth Estimation had been attempted by several works ([12,13]) primarily in a supervised fashion only until [61] presented a reliable Structure from Motion (SfM) inspired self-supervised framework. A category of literature ([16,19,26,28]) has built upon the foundation laid by [61] by making important improvements to the Depth and Pose estimation networks. Amongst these, [16] is a seminal work that makes several important contributions, such as a robust architecture based on ResNet18 [22], multi-scale depth estimation, and per-pixel minimum reprojection loss.

Representation Learning for Augmenting Depth Estimation: Both supervised and self-supervised Monocular Depth Estimation methods have recently started to look beyond ground-truth depth maps and monocular video, respectively, as a supervision signal for their models without resorting to multimodal sensor fusion. They learn rich data representations for related downstream tasks such as semantic segmentation or classification and fuse them with the representations learned for depth estimation. Representations learnt for semantic segmentation has been used extensively by such models ([5,30,43,62]). Features learned for Monocular Depth Estimation usually aid Semantic Segmentation models ([23,51]). The methods described in this section differ from methods that use additional supervision in the form of stereo images or data from additional sensor modalities.

Feature Fusion in Latent Space. Much research has gone into finding optimal fusion techniques for combining latent representations learned via different architectures, latent representations learned for different tasks, and latent representations learned over different data modalities. [58] proposes *Feature Attention Fusion* and *Multi-Scale Feature Fusion Dense Pyramid* for Monocular Depth Estimation. [24] proposes combining representations learnt via both 2D and 3D convolutions. [54] explores latent space feature fusion for depth map aggregation. [59] fuses latent space features corresponding to global and local attributes for image retrieval. The success of such approaches has motivated our method, which learns from both global and local receptive fields for the current task. We describe our fusion module in Sect. 3.1.

Vision Transformers. Transformer Models, originally proposed for Natural Language tasks [49] have revolutionized the way sequence-to-sequence problems are attempted. Vision Transformers (ViTs) ([11,48] , [10]) have adopted the idea by splitting images into patches and learning their representations as a sequence. A transformer architecture can extract information from an entire image in a single step. The recent past has seen a steep surge in the utility of ViTs for standard vision tasks such as image recognition [10], segmentation ([3,18,56,57]), object detection [21], restoration [52], and inpainting [9] . Many recent methods have applied self-attention within local regions [34,52] to limit the computational requirement. In our work, we adopt the transposed attention-based transformer design of [60] for its efficient operation.

Although convolutional networks excel at processing local details, they lack the required adaptability and contextual information. Correlation among different regions should result in better depth estimation accuracy for objects of different shapes and scales. Recently, [26] embedded a self-attention layer within a CNN. But, it limits the potential as it was applied only on the CNN-generated feature maps of the lowest resolution. Although [41] utilized a pure transformer architecture for depth prediction, it was trained using GT supervision. The behavior of transformers and their relation with CNN is rather unexplored in the self-supervised domain. A pure transformer architecture potentially lacks the required finesse in processing the local details, such as object boundaries. Thus, we design a hybrid transformer architecture that exploits the long-range modeling capability of the transformer while maintaining the local pixel-level accuracy using parallel convolutional branches operating at different resolutions. A detailed description of the architecture is given in the following section.

3 Method

Our goal is to generate the per-pixel depth map of a single frame at inference time. To learn such a function, usually, image-to-image CNNs are trained in a supervised manner where the supervision of the ground-truth depth map is available while training. But, practically, creating such annotated datasets for a large number of scenes is costly. Instead, we follow the self-supervised framework of [16], which only requires short image sequences of a scene captured by a moving camera. In such an approach, the primary supervision comes from the task of novel view synthesis. A depth-prediction network is trained to predict the depth of a scene. A separate pose-estimation is required while training that produces the relative camera pose given two consecutive frames. Given the depth map and the relative pose, the appearance of the target frame seen from a different camera pose is synthesized. As the networks are trained to perform view synthesis using the predicted depth map of the current scene, it inherently learns to extract the underlying scene structure. Thus, the whole framework can be trained in a self-supervised manner without requiring GT annotations. Note that the pose estimation is only needed for training to perform view-synthesis. The depth estimation network alone is sufficient for predicting the depth at inference.

Our overall approach is shown in Fig. 1. For the depth-estimation encoder, we first deploy a global attention-based branch focusing on long-range information modeling. Next, to capture the local details better, we deploy two local high-resolution (HRL) and low-resolution (LRL) branches, focusing on either overall

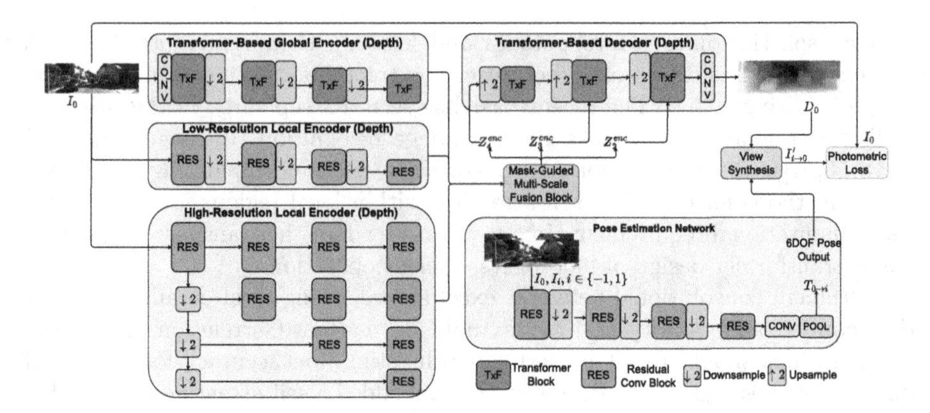

Fig. 1. An overview of our hybrid-transformer based self-supervised depth prediction framework. Three parallel encoder branches are shown on the left. The information is fused using the mask-guided multi-scale fusion block, and finally the depth decoder estimates the depth map. The pose network is derived from ResNet18 [22] architecture.

scene content (LRL) or sharper pixel details (HRL). Encoder features from these three encoder branches are fused using a multi-scale fusion block, which utilizes a spatial masking operation to extract the most useful information. Next, a transformer-based decoder module uses the fused encoded information to predict the depth map of the scene. We use a simpler ResNet-based pose network as we found it adequate for the relatively easier pose estimation task. In the following section, we first describe the structure of the depth and pose estimation networks, followed by a discussion of the training strategies.

3.1 Depth Estimation Network

The depth estimation network has an encoder-decoder framework. We deploy three parallel encoders for capturing local as well as global information. First, we will discuss the individual encoder branches, followed by the description of the information fusion module and depth prediction in the decoder.

Transformer-Based Global Encoder: Let $I \in \mathbb{R}^{3\times H\times W}$ be the input frame. We first apply a single convolutional layer in the global branch to generate the initial shallow feature embeddings $X \in \mathbb{R}^{C\times H\times W}$. Next, X is passed through four encoder levels with an increasing number of feature maps and decreasing spatial resolution. For downsampling, we apply pixel-unshuffle operation [46]. Each level of the encoder contains multiple Vision-Transformer blocks. The operation of one such block is shown in Fig. 2 (a). Given the input tensor $\hat{X} \in \mathbb{R}^{\hat{C}\times\hat{H}\times\hat{W}}$ at any particular level, we first generate three different embeddings: query ($Q \in \mathbb{R}^{d_q\times\hat{H}\times\hat{W}}$), key ($K \in \mathbb{R}^{d_q\times\hat{H}\times\hat{W}}$), and value ($V \in \mathbb{R}^{d_v\times\hat{H}\times\hat{W}}$). We apply standard convolutional operations on X to generate Q, K, V. In the original transformer architecture [49], the correlations between all possible elements were calculated using $Q^T K$, which will result in a huge $\hat{H}\hat{W} \times \hat{H}\hat{W}$ matrix for images. Instead, we follow the design of [60] and generate a transposed-attention map QK^T significantly reducing the memory requirement. Next, each pixel value of the output feature map Y is updated using global information as

$$Y = \hat{X} + \text{conv}_{1\times1}(V\ \text{softmax}(QK^T)) \tag{1}$$

Similar to the original design of [49], we divide the number of channels into multiple 'heads' and process those in parallel.

In a typical transformer architecture, the self-attention operation is followed by feed-forward and normalization layers. Motivated by [8], we deploy a gated feed-forward function. The gating mechanism controls the information passed on in the hierarchy. Given the feature map Y, a gating function can be generated as $G = \psi(\text{conv}(LN(Y)))$, where ψ is a non-linear function, such as sigmoid, GELU, etc., LN represents a standard layer-normalization operation [49]. Given G, the output of the gated-feedforward layer can be expressed as

$$Y^{TG} = Y + G \otimes \text{conv}(LN(Y)) \tag{2}$$

Ultimately, we will have four different feature representations from four different levels $Y_e^{TG} \in \mathbb{R}^{C*e \times \frac{H}{2^{(e-1)}} \times \frac{H}{2^{(e-1)}}}, e \in \{1, 2, 3, 4\}$.

Low-resolution Local Encoder (LRL): We introduce two convolutional branches to compensate for the deficiencies of fine-grained local information in the global transformer branch. In contrast to methods like [26], which embed the self-attention layer into the traditional CNN, our global branch encodes the images independently, which can fully exert the advantages of the transformer. To complement this operation, first, we deploy a ResNet-based low-resolution local branch. We use the first four levels of a pre-trained ResNet18 architecture. Given the input image I, it gradually convolves and downsamples the feature maps. At four different levels of the encoder, we will again have feature maps $Y_e^{LRL} \in \mathbb{R}^{C*e \times \frac{H}{2^{(e-1)}} \times \frac{H}{2^{(e-1)}}}, e \in \{1, 2, 3, 4\}$. Y_e^{LRL} can be expressed as

$$Y_e^{LRL} = \text{conv}(\text{down}(Y_{e-1}^{LRL})) + \text{down}(Y_{e-1}^{LRL}) \tag{3}$$

The final and deepest feature map of the LRL encoder, i.e., Y_4^{LRL} has the lowest spatial resolution. Thus, it loses much of the finer pixel details but is able to capture the overall scene content better with a large number of feature maps.

High-resolution Local Encoder (HRL): For the high-resolution counterpart, we maintain the spatial resolution of the feature maps at multiple levels. Specifically, we deploy four parallel streams, where the spatial resolution is fixed for a particular stream. But, different streams have different resolutions. This design is derived from [50]. It allows us to generate deep, high-resolution feature maps throughout the entire encoding process.

Let $(Y0)_e^{HRL}$ be the input for the eth stream. The output Y_e^{HRL} of the eth stream can be expressed as

$$Y_e^{HRL} = \text{conv}((Y0)_e^{HRL}) + (Y0)_e^{HRL} \tag{4}$$

where $Y_e^{HRL} \in \mathbb{R}^{C*e \times \frac{H}{2^{(e-1)}} \times \frac{H}{2^{(e-1)}}}$. Unlike Eq. 3, we do not use any downsampling operation inside a stream. Thus, for the higher resolution streams, Y_e^{HRL} is much more deeper and expressive than Y_e^{LRL}. It helps in predicting highly accurate local depth boundaries. On the contrary, for lower spatial resolution, Y_e^{LRL} is more robust than Y_e^{HRL} due to the large number of feature maps and captures the overall scene content better.

Fusion Module: At the end of the encoder, we will have $Y_e^{TG}, Y_e^{LRL}, Y_e^{HRL}$, for each level $e \in \{1, 2, 3, 4\}$. To fuse the complementary information from these three branches, we design an atrous convolution-based fusion block (Fig. 2 (b)) that extracts multi-scale information from the feature maps. In most existing works, the final prediction is performed at multiple scales. The coarser outputs usually contain the global structure but may lack intricate local details, which are present in the finer predictions. Motivated by this observation, we propose a multi-convolution-based fusion block that aims to extract and fuse features at multiple scales. We deploy multiple convolution layers with different dilation rates in parallel. Convolution with dilation 1 will better capture the immediate local information, whereas a higher dilation rate will capture the coarser but more global information. We concatenate these feature representations and fuse them using a 1×1 convolutional layer. Our experimental results demonstrate that such operations in the feature domain significantly help the depth estimation network capture a better representation of the scene and boost the accuracy.

We use four parallel convolution layers with different dilation rates (1, 3, 5, 7) to extract information at multiple scales. Formally, for input $Y \in \mathbb{R}^{C \times H \times W}$, multiple atrous convolutions are applied over the input feature map as follows:

$$Y^r[j] = \sum_k Y[j + r \cdot k]W[k] \tag{5}$$

where k is the size of the filter W and the dilation rate r determines the stride. Standard convolution is a special case in which rate $r = 1$. Next, we concatenate these outputs and pass through a 1×1 convolution layer to maintain the number of feature maps. The operation can be expressed as

$$Z = \text{conv}_{1\times1}(\text{concat}(Y^1, Y^3, Y^5, Y^7)) \tag{6}$$

It allows the network to capture multi-scale abstractions of the scene.

Before extracting multi-scale information from $Y_e^{TG}, Y_e^{LRL}, Y_e^{HRL}$ using Eq. 6, we emphasize the most useful feature locations for the local convolutional branches using a spatial masking technique. Given, $Y_e^{LRL} \in \mathbb{R}^{\hat{C}\times\hat{H}\times\hat{W}}$, we first generate a spatial mask $M^{LRL} \in \mathbb{R}^{\hat{C}\times\hat{H}\times\hat{W}}$. Next, we elementwise multiply the feature map Y_e^{LRL} with the corresponding spatial mask as

$$M^{LRL} = \text{sigmoid}(\text{conv}(Y_e^{LRL})$$
$$Y_e^{\prime LRL} = Y_e^{LRL} \odot M^{LRL}$$

This process allows us to enhance the crucial and helpful features from the local branches while suppressing the redundant ones. The same is applied to generate $Y_e^{\prime HRL}$ as well. Next, we pass each of these enhanced feature maps ($Y_e^{\prime LRL}, Y_e^{\prime HRL}$ and Y_e^{TG}) through multiple atrous-convolution layers (Eqs. 5 and 6) to generate the updated feature maps $Z_e^{LRL}, Z_e^{HRL}, Z_e^{TG}$ with multi-scale information. Next, we concatenate a pair of feature sets for a particular level and pass through an 1×1 convolutional layer to maintain the number of channels. This process is repeated twice to fuse information from three branches. Formally, the final output of encoder for level e can be expressed as

$$Z_e^{enc} = \text{conv}_{1\times1}(\text{concat}(\text{conv}_{1\times1}(\text{concat}(Z_e^{TG}, Z_e^{LRL})), Z_e^{HRL})) \quad (7)$$

Note that, information from all three branches can be concatenated and fused at one go, but it increases the computational load due to high number of feature channels.

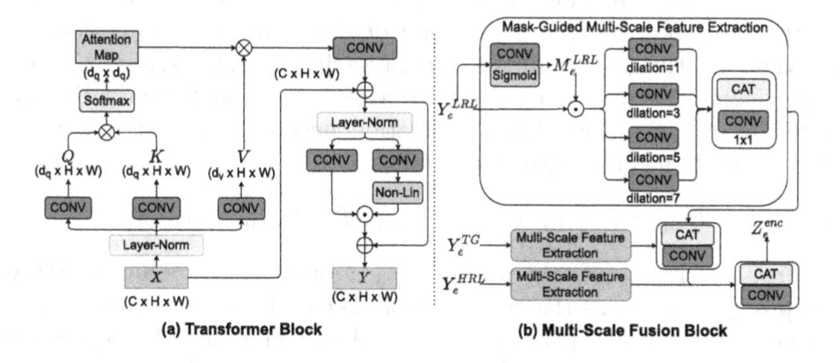

Fig. 2. An overview of the transformer block and the multi-scale fusion module.

Decoder. We aim to extract a robust set of feature representations in the encoder. Information from different branches is merged in the proposed fusion block. Finally, the input to the decoder is Z_4^{enc}. This is further refined in the first decoder level and upsampled at the end. We resort to a purely transformer-based decoder as it excels in modeling the dependencies between different feature maps and locations. The core operations are the same as Eqs. 1 and 2. The upsampled decoder feature is concatenated with the encoder feature of the same level Z_3^{enc}, constituting the input to the next decoder level. This process is repeated for all levels. At the end, we pass the feature map through a convolution followed by a sigmoid layer to estimate the final disparity map. Note that, while training, we predict the disparity map at every decoder level following the multi-scale prediction approach of [16]. Thus, we simply add one convolution and sigmoid layer at every decoder level for coarser predictions. During inference, we discard these layers and only take the full-scale output at the end.

3.2 Pose Estimation Network

The input to the pose estimation network is the target view I_0 concatenated with one of the source views I_i, $i \in \{-1, +1\}$. It predicts the relative pose $T_{0 \to i}$ between the two views. For pose estimation, we follow [16] and use a ResNet18-based network. It consists of a series of residual convolution and downsampling blocks derived from ResNet18. In the end, we use a 1×1 convolutional layer and a 6-channel spatial average pooling operation. The output is the set of 6 DoF transformations between $T_{0 \to i}$ (corresponding to 3 Euler angles and 3-D translation). The translation is parametrized in Euclidean coordinates $\{x, y, z\}$ and rotation uses Euler angles $\{\alpha, \beta, \gamma\}$. This process is repeated for each source view producing independent transformations.

3.3 Training Strategies

Following [16], we also use the assumption that the world is static and the view change is only caused by moving the camera. So, we should be able to synthesize the target view I_0 given the depth map of the scene, pixel intensities of the source view, relative pose, and the camera intrinsics (K). Usually, K is known for a particular dataset. Given the relative pose $T_{0 \to i}$ and the depth map D_0 predicted by the pose and depth estimation networks, respectively, the view synthesis operation can be expressed as

$$I'_{i \to 0} = I_i < \text{proj}_{cam}(\text{proj}_{world}(I_0, D_0, T_{0 \to i}), K) > \tag{8}$$

where proj_{world} represents the inverse projection of image points to 3D world points, proj_{cam} represents the projection of 3D coordinates to image points, $<>$ represents the sampling operation. We use the differentiable bilinear sampling mechanism of [25] that linearly interpolates the values of the 4-pixel neighbors (top-left, top-right, bottom-left, and bottom-right) to calculate $I'_{i \to 0}$.

Next, we utilize two main regularizing function for training the network. We employ photometric loss to calculate the difference between I_0 and $I'_{i \to 0}$. Following [16,26], the loss function can be defined as

$$R_p^i = \frac{\alpha}{2}(1 - \text{SSIM}(I_0, I'_{i \to 0})) + (1 - \alpha)||I_0 - I'_{i \to 0}||_1 \tag{9}$$

where SSIM represents Structural Similarity [53] and $\alpha = 0.85$. Moreover, to promote the smoothness of the generated depth map, we use the widely used edge-aware smoothness during training

$$R_s = |\partial_x d_0| e^{-|\partial_x I_0|} + |\partial_y d_0| e^{-|\partial_y I_0|} \tag{10}$$

We also utilize the minimum photometric error, masking of the stationary pixels and multi-scale losses introduced in [16]. The final loss is computed as the weighted sum of the minimum reprojection loss and the smoothness term

$$R_{final} = \min_i R_p^i + \beta R_s \tag{11}$$

4 Experiments

4.1 Dataset Details

The KITTI dataset [15] is an accepted standard dataset for the task of Monocular Depth Estimation. It's Eigen Split [12] for Monocular Depth Estimation comprises 22,600 training images, 888 validation images, and 697 test images. Our model uses images of dimensions $640 \times 192 \times 3$ for training and those of dimensions $1280 \times 384 \times 3$ for evaluation. We apply the same preprocessing techniques as [16].

4.2 Training and Implementation Details

We build upon the standard framework of Monodepth2 [16]. All presented models were trained on the KITTI [15] dataset. Our model was trained using 4 NVIDIA RTX 3090 GPUs with a batch size of 16 for 20 epochs using Adam [27] optimizer at an initial learning rate of 10^{-4} with a geometric progression based decay schedule. In the following sections, we present quantitative and qualitative results to back our findings, along with several ablation studies to corroborate the importance of each element of our proposed method and the synergy between them.

4.3 Quantitative Analysis

Keeping in line with standard practice, we evaluate the efficacy of our model using Absolute Relative Error (Abs Rel), Square Relative Error (Sq Rel), Root Mean Squared Error (RMSE), Log RMSE, and the standard accuracy thresholds viz. $\delta < 1.25$, $\delta < 1.25^2$, and $\delta < 1.25^3$. Amongst these, RMSE is the most stringent error measure since it is the RMS value of the absolute pixel-wise error observed between a predicted depth map and the ground truth, whereas others except Log RMSE are relative to a certain degree. In Table 1, we have reported the quantitative scores of our model and existing supervised, semantic-guided, and unsupervised depth estimation networks. Our model achieves SOTA performance on KITTI Eigen split [12] with respect to RMSE and also beats existing methods on most other metrics. For completeness, we also evaluate on a subset of images from Make3D [44,45] and HR-WSI [55] datasets. We compare our approach with two recent SOTA works - [19,26] with open-source implementation. All the models were trained on KITTI and then tested on a subset of images from [44,55]. As can be observed in Table 1, our approach achieves superior performance on diverse range of scenes from other datasets, as well.

Table 1. Quantitative Results for our model's performance with respect to SOTA. The best performing entries have been highlighted in bold. The second best entries have been underlined. Methods indicated with $*$ and † require semantic data and GT depth map, respectively. Our model performs best on five and second best on one of the seven metrics. (↓: lower is better, ↑: higher is better.)

Data	Method	Abs Rel ↓	Sq Rel ↓	RMSE ↓	RMSE log ↓	$\delta < 1.25$ ↑	$\delta < 1.25^2$ ↑	$\delta < 1.25^3$ ↑
[15]	Zhou *et al.* [61]	0.208	1.768	6.856	0.283	0.678	0.885	0.957
	GeoNet [40]†	0.164	1.303	6.090	0.247	0.765	0.919	0.968
	Struct2Depth [4]*	0.141	1.026	5.291	0.215	0.816	0.945	0.979
	Mahjourian [36]	0.163	1.240	6.220	0.250	0.762	0.916	0.968
	CC [42]	0.140	1.070	5.326	0.217	0.826	0.941	0.975
	DF-Net [63]	0.150	1.124	5.507	0.223	0.806	0.933	0.973
	Li *et al.* [32]	0.150	1.127	5.564	0.229	0.823	0.936	0.974
	Pilzer *et al.* [39]	0.142	1.231	5.785	0.239	0.795	0.924	0.968
	EPC++ [35]	0.141	1.029	5.350	0.216	0.816	0.941	0.976
	Bian *et al.* [2]	0.137	1.089	5.439	0.217	0.830	0.942	0.975
	GLNet [6]	0.135	1.070	5.230	0.210	0.841	0.948	0.980
	Gordon *et al.* [17]*	0.128	0.959	5.230	0.212	0.845	0.947	0.976
	Monodepth2 [16]	0.115	0.882	4.701	0.19	0.879	0.961	0.982
	Packnet-SfM [19]	0.107	0.802	4.538	0.186	**0.889**	0.962	0.981
	Semantics [47]*	0.126	0.835	4.937	0.199	0.844	0.953	0.982
	SGDepth [28]	0.117	0.907	4.844	0.196	0.875	0.958	0.98
	Jonston *et al.* [26]	**0.106**	0.861	4.699	**0.185**	**0.889**	0.962	0.982
	Gao *et al.* [14]	0.112	0.866	4.693	0.189	0.881	0.961	**0.983**
	Lee *et al.* [31]	0.112	0.777	4.772	0.191	0.872	0.959	0.982
	Ours	0.112	**0.75**	**4.528**	0.187	0.881	**0.963**	**0.983**
[44]	Johnston *et al.* [26]	0.711	15.868	27.934	1.978	**0.115**	0.216	0.312
	Packnet-SfM [19]	0.725	17.102	28.352	2.037	0.096	0.194	0.287
	Ours	**0.601**	**14.391**	**26.016**	**1.519**	0.104	**0.227**	**0.357**
[55]	Jonston *et al.* [26]	0.393	1.839	1.102	0.274	0.115	0.199	0.254
	Packnet-SfM [19]	1.304	6.04	3.861	1.150	**0.121**	**0.241**	**0.356**
	Ours	**0.188**	**0.512**	**0.972**	**0.262**	0.078	0.160	0.238

4.4 Qualitative Analysis

We have visualized the depth map predicted by our approach and SOTA Packnet-SfM [19], Johnston *et al.* [26], Monodepth2 [16] in Fig. 3. e have used the models trained on KITTI dataset [15] and directly tested on DDAD [19], Make3D [44,45], HR-WSI [55] datasets (Fig. 4). As highlighted in the figures, existing works struggle to accurately predict the depth maps for complex regions or objects and their boundaries. For example, for the 4th column, Fig. 3, existing works often fail to distinguish between the vehicle boundary and the background. Similarly, for the 4th row, Fig. 4, they struggle to predict the depth of the highlighted tree region properly. In comparison, our approach is able to gather useful global information from the other regions of the image (e.g., other green tree regions), which helps in accurately identifying an object appearing in front of a complex background. For the 1st row, Fig. 4, our approach produces the local boundaries of the highlighted road sign more accurately, demonstrating its superior ability in processing local scene information, as well.

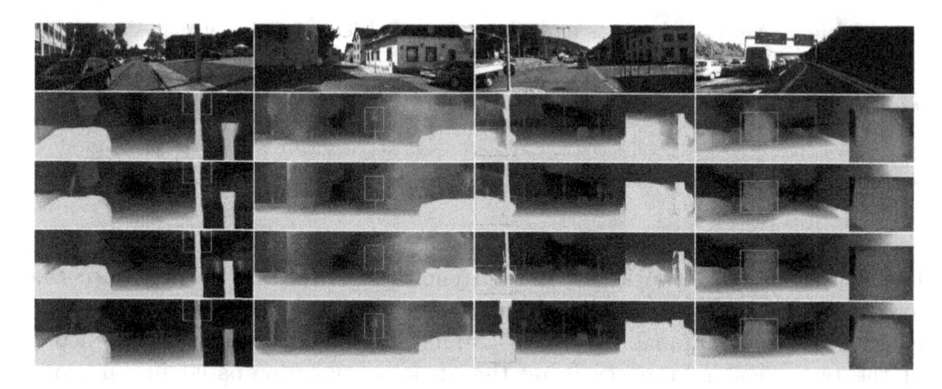

Fig. 3. Qualitative comparisons on selected test images from the Eigen split [12] benchmark of the KITTI dataset [15]. From top: Input, Packnet-SfM [19], Monodepth2 [16], Jonston *et al.* [26] and our outpus.

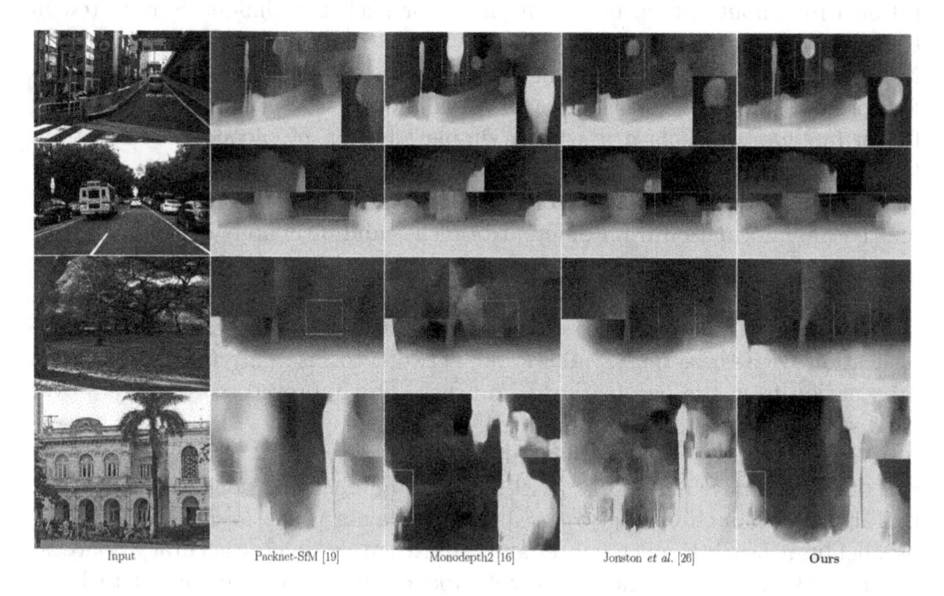

Fig. 4. Qualitative comparisons on selected test images from the DDAD [19] (Rows 1,2), Make3D [44,45] (Row 3), HR-WSI [55] (Row 4) datasets.

4.5 Ablation Studies

We perform the following experiments, as reported in Table 2 on the KITTI dataset [15]. Net1: Backbone vision transformer encoder-decoder architecture. Net2: Net1 + low-resolution local encoder branch. Net3: Net2 + high-resolution local encoder branch. Net4: We replace the atrous convolution based fusion block in Net3 with simple addition of features from the three encoder branches. Net5: Net3 + spatial masking in the atrous-fusion module.

Table 2. Network Analysis on [15].

	LRL	TG	HRL	Sum	Atr.	Mask	Abs Rel	Sq Rel	RMSE	R. log
Net1		✓					0.123	0.937	4.917	0.201
Net2	✓	✓			✓		0.113	0.791	4.637	0.188
Net3	✓	✓	✓		✓		0.11	0.754	4.602	0.184
Net4	✓	✓	✓	✓			0.113	0.807	4.641	0.188
Net5	✓	✓	✓		✓	✓	0.112	0.75	4.528	0.187

Table 3. Efficiency analysis on [15].

	Time	Params	Abs Rel	Sq Rel	RMSE	R. log
[26]	**0.04 s**	38.3 M	**0.106**	0.861	4.699	0.185
[19]	1.91 s	129 M	0.107	0.802	4.538	0.186
CNN_L	0.20 s	91 M	0.115	0.826	4.658	0.188
TxF_L	0.17 s	82.5 M	0.12	0.858	4.843	0.198
Ours_S	0.21 s	**32.7M**	0.112	0.81	4.561	**0.184**
Ours_L	0.23 s	67.4 M	0.112	**0.75**	**4.528**	0.187

The strong performance of our baseline (Net1) shows the utility of transformer-based networks for the current task. For Net2, we introduce the low-resolution local encoder branch in parallel, which processes the image using consecutive convolutional and downsampling operations. We fuse the output of the two encoder branches using the atrous-convolution-based fusion module without any masking operation. The inclusion of the LRL branch improves the performance of the baseline network. For Net3, we deploy another parallel encoder branch that maintains the high-resolution and deep feature representation throughout the encoder. The need for such a technique is reflected in the improved quantitative score of Net3. Overall, the significant improvement of Net2 and Net3 over Net1 shows the complementary behavior of the global transformer and local convolutional branches. To analyze the utility of the proposed fusion module, we replace it with a simple addition of encoder features from the three parallel branches in Net4. The inferior performance of Net4 compared to Net3 demonstrates the need for an adaptive fusion module that can extract multi-scale information from different branches and ultimately capture a better representation of the scene. For Net5, our final model, we introduce the spatial masking operation in the fusion module. Every mask element lies between 0 and 1, representing the importance of different feature positions. Elementwise multiplication with these masks allows our network to fuse only the most helpful information from the parallel local branches. The improved score of Net5 over Net3 strengthens our argument.

We have reported the inference time (per image) and the number of parameters of two recent SOTA works ([19,26]) and different ablations of our network in Table 3. Although our network is slightly heavier than [26] due to the three parallel branches, it is much lighter and faster than [19], whose accuracy is closest to our work. To analyze the role of network complexity for the current task, we train two large baseline networks: TxF_L and CNN_L, with a higher number of parameters than our final model (Ours_L). TxF_L is a pure transformer network based on [60], and CNN_L is a pure CNN network based on [16]. These baselines' performance is much inferior to our model, despite having similar or more parameters and runtime. Next, we train a lightweight version of our network (Ours_S) that achieves comparable performance to our final model (Ours_L) while having the least number of parameters among all the baselines and SOTA works. These experiments demonstrate that the performance improvement is primarily due to the adaptive fusion of multi-scale information from the three branches, which is the key contribution of our work.

We further visualize the spatial mask of Net4 in Fig. 5 for an intuitive understanding. The masks for the low-resolution and high-resolution local branches

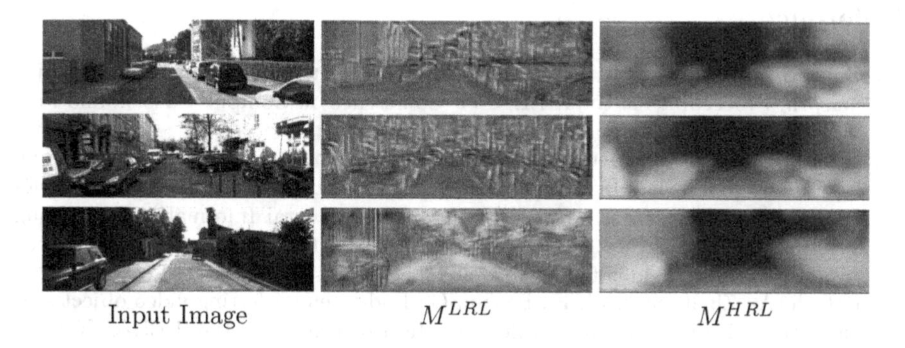

Fig. 5. Visualization of the spatial mask (M) for different encoder branches.

(i.e., M^{LRL} and M^{HRL}) are shown in Columns 2 and 3, respectively. The spatial masks visually correlate with crucial spatial locations, for example, different objects and structures. Also, the difference between M^{LRL} and M^{HRL} demonstrates the distinct representations learned by the two branches. Interestingly, M^{LRL} is uniformly distributed throughout the scene while highlighting the relevant parts. This correlates with our intuition that LRL primarily gathers high-level information from the entire frame. In contrast, M^{HRL} is more focused on foreground objects where the depth variation is significant. We argue that HRL excels in handling challenging regions with high-depth variations. Such adaptive ability can be considered crucial to the observed performance improvement of our hybrid network.

5 Conclusion

We have presented a hybrid transformer-based framework for self-supervised monocular depth estimation, achieving improvements over prior arts. Most existing works use a convolutional architecture that fails to satisfactorily model the long-range dependencies between different regions of an image. Due to successive downsampling operations, such designs often fail to preserve the detailed pixel information. On the other hand, using a pure transformer architecture might lack fine-grained local depth information and boundaries. To incorporate the strength of CNNs with the transformer model, we introduce a hybrid framework that extracts helpful information from three encoder branches operating at different spatial resolutions. Complimentary information from the branches is fused adaptively using spatial masks. It would be interesting to extend such a hybrid framework to other image-to-image tasks, such as image enhancement, which would be addressed in future works.

References

1. Bhat, S.F., Alhashim, I., Wonka, P.: AdaBins: depth estimation using adaptive bins. In: Proceedings of the IEEE/CVF Conference on Computer Vision and Pattern Recognition (CVPR), pp. 4009–4018 (2021)
2. Bian, J., et al.: Unsupervised scale-consistent depth and ego-motion learning from monocular video. In: Wallach, H., Larochelle, H., Beygelzimer, A., d' Alché-Buc, F., Fox, E., Garnett, R. (eds.) Advances in Neural Information Processing Systems, vol. 32. Curran Associates, Inc. (2019). https://proceedings.neurips.cc/paper/2019/file/6364d3f0f495b6ab9dcf8d3b5c6e0b01-Paper.pdf
3. Botach, A., Zheltonozhskii, E., Baskin, C.: End-to-end referring video object segmentation with multimodal transformers. In: Proceedings of the IEEE/CVF Conference on Computer Vision and Pattern Recognition (CVPR), pp. 4985–4995 (2022)
4. Casser, V., Pirk, S., Mahjourian, R., Angelova, A.: Depth prediction without the sensors: leveraging structure for unsupervised learning from monocular videos. CoRR abs/1811.06152 (2018). http://arxiv.org/abs/1811.06152
5. Chen, P.Y., Liu, A.H., Liu, Y.C., Wang, Y.C.F.: Towards scene understanding: unsupervised monocular depth estimation with semantic-aware representation. In: 2019 IEEE/CVF Conference on Computer Vision and Pattern Recognition (CVPR), pp. 2619–2627 (2019). https://doi.org/10.1109/CVPR.2019.00273
6. Chen, Y., Schmid, C., Sminchisescu, C.: Self-supervised learning with geometric constraints in monocular video: connecting flow, depth, and camera. In: Proceedings of the IEEE/CVF International Conference on Computer Vision (ICCV) (2019)
7. Choi, J., Jung, D., Lee, D., Kim, C.: SAFENet: self-supervised monocular depth estimation with semantic-aware feature extraction (2020). https://doi.org/10.48550/ARXIV.2010.02893. https://arxiv.org/abs/2010.02893
8. Dauphin, Y.N., Fan, A., Auli, M., Grangier, D.: Language modeling with gated convolutional networks. In: International Conference on Machine Learning, pp. 933–941. PMLR (2017)
9. Dong, Q., Cao, C., Fu, Y.: Incremental transformer structure enhanced image inpainting with masking positional encoding. In: Proceedings of the IEEE/CVF Conference on Computer Vision and Pattern Recognition (CVPR), pp. 11358–11368 (2022)
10. Dosovitskiy, A., et al.: An image is worth 16x16 words: transformers for image recognition at scale. In: International Conference on Learning Representations (2021). https://openreview.net/forum?id=YicbFdNTTy
11. Dosovitskiy, A., et al.: An image is worth 16x16 words: transformers for image recognition at scale. arXiv preprint arXiv:2010.11929 (2020)
12. Eigen, D., Puhrsch, C., Fergus, R.: Depth map prediction from a single image using a multi-scale deep network. In: Ghahramani, Z., Welling, M., Cortes, C., Lawrence, N., Weinberger, K. (eds.) Advances in Neural Information Processing Systems, vol. 27. Curran Associates, Inc. (2014). https://proceedings.neurips.cc/paper/2014/file/7bccfde7714a1ebadf06c5f4cea752c1-Paper.pdf
13. Fu, H., Gong, M., Wang, C., Batmanghelich, K., Tao, D.: Deep ordinal regression network for monocular depth estimation. In: Proceedings of the IEEE Conference on Computer Vision and Pattern Recognition (CVPR) (2018)
14. Gao, F., Yu, J., Shen, H., Wang, Y., Yang, H.: Attentional separation-and-aggregation network for self-supervised depth-pose learning in dynamic scenes. CoRR abs/2011.09369 (2020). https://arxiv.org/abs/2011.09369

15. Geiger, A., Lenz, P., Urtasun, R.: Are we ready for autonomous driving? the KITTI vision benchmark suite. In: 2012 IEEE Conference on Computer Vision and Pattern Recognition, pp. 3354–3361 (2012). https://doi.org/10.1109/CVPR.2012.6248074
16. Godard, C., Mac Aodha, O., Firman, M., Brostow, G.J.: Digging into self-supervised monocular depth prediction (2019)
17. Gordon, A., Li, H., Jonschkowski, R., Angelova, A.: Depth from videos in the wild: Unsupervised monocular depth learning from unknown cameras. In: Proceedings of the IEEE/CVF International Conference on Computer Vision (ICCV) (2019)
18. Gu, J., et al.: Multi-scale high-resolution vision transformer for semantic segmentation. In: Proceedings of the IEEE/CVF Conference on Computer Vision and Pattern Recognition (CVPR), pp. 12094–12103 (2022)
19. Guizilini, V., Ambrus, R., Pillai, S., Raventos, A., Gaidon, A.: 3D packing for self-supervised monocular depth estimation. In: Proceedings of the IEEE/CVF Conference on Computer Vision and Pattern Recognition (CVPR) (2020)
20. Guizilini, V., Hou, R., Li, J., Ambrus, R., Gaidon, A.: Semantically-guided representation learning for self-supervised monocular depth. CoRR abs/2002.12319 (2020). https://arxiv.org/abs/2002.12319
21. He, C., Li, R., Li, S., Zhang, L.: Voxel set transformer: a set-to-set approach to 3D object detection from point clouds. In: Proceedings of the IEEE/CVF Conference on Computer Vision and Pattern Recognition (CVPR), pp. 8417–8427 (2022)
22. He, K., Zhang, X., Ren, S., Sun, J.: Deep residual learning for image recognition. In: Proceedings of the IEEE Conference on Computer Vision and Pattern Recognition (CVPR) (2016)
23. He, L., Lu, J., Wang, G., Song, S., Zhou, J.: SOSD-Net: joint semantic object segmentation and depth estimation from monocular images. Neurocomputing **440**, 251–263 (2021). https://doi.org/10.1016/j.neucom.2021.01.126. https://www.sciencedirect.com/science/article/pii/S0925231221002344
24. Huynh, L., Nguyen, P., Matas, J., Rahtu, E., Heikkilä, J.: Boosting monocular depth estimation with lightweight 3D point fusion. CoRR abs/2012.10296 (2020). https://arxiv.org/abs/2012.10296
25. Jaderberg, M., Simonyan, K., Zisserman, A., et al.: Spatial transformer networks. In: Advances in Neural Information Processing Systems 28 (2015)
26. Johnston, A., Carneiro, G.: Self-supervised monocular trained depth estimation using self-attention and discrete disparity volume. In: Proceedings of the IEEE/CVF Conference on Computer Vision and Pattern Recognition (CVPR) (2020)
27. Kingma, D.P., Ba, J.: Adam: a method for stochastic optimization (2014). https://doi.org/10.48550/ARXIV.1412.6980. https://arxiv.org/abs/1412.6980
28. Klingner, M., Termöhlen, J.-A., Mikolajczyk, J., Fingscheidt, T.: Self-supervised monocular depth estimation: solving the dynamic object problem by semantic guidance. In: Vedaldi, A., Bischof, H., Brox, T., Frahm, J.-M. (eds.) ECCV 2020. LNCS, vol. 12365, pp. 582–600. Springer, Cham (2020). https://doi.org/10.1007/978-3-030-58565-5_35
29. Kopf, J., Rong, X., Huang, J.B.: Robust consistent video depth estimation. In: Proceedings of the IEEE/CVF Conference on Computer Vision and Pattern Recognition (CVPR), pp. 1611–1621 (2021)
30. Kumar, V.R., Klingner, M., Yogamani, S., Milz, S., Fingscheidt, T., Mader, P.: SynDistNet: self-supervised monocular fisheye camera distance estimation synergized with semantic segmentation for autonomous driving. In: Proceedings of the IEEE/CVF Winter Conference on Applications of Computer Vision (WACV), pp. 61–71 (2021)

31. Lee, S., Im, S., Lin, S., Kweon, I.S.: Learning monocular depth in dynamic scenes via instance-aware projection consistency. Proceed. AAAI Conf. Artif. Intell. **35**(3), 1863–1872 (2021). https://ojs.aaai.org/index.php/AAAI/article/view/16281
32. Li, S., Xue, F., Wang, X., Yan, Z., Zha, H.: Sequential adversarial learning for self-supervised deep visual odometry. In: Proceedings of the IEEE/CVF International Conference on Computer Vision (ICCV) (2019)
33. Li, Z.: Monocular depth estimation toolbox. https://github.com/zhyever/Monocular-Depth-Estimation-Toolbox (2022)
34. Liang, J., Cao, J., Sun, G., Zhang, K., Van Gool, L., Timofte, R.: SwinIR: image restoration using swin transformer. In: Proceedings of the IEEE/CVF International Conference on Computer Vision, pp. 1833–1844 (2021)
35. Luo, C., et al.: Every pixel counts ++: joint learning of geometry and motion with 3D holistic understanding. IEEE Trans. Pattern Anal. Mach. Intell. **42**(10), 2624–2641 (2020). https://doi.org/10.1109/TPAMI.2019.2930258
36. Mahjourian, R., Wicke, M., Angelova, A.: Unsupervised learning of depth and ego-motion from monocular video using 3D geometric constraints. In: Proceedings of the IEEE Conference on Computer Vision and Pattern Recognition (CVPR) (2018)
37. Naderi, T., Sadovnik, A., Hayward, J., Qi, H.: Monocular depth estimation with adaptive geometric attention. In: 2022 IEEE/CVF Winter Conference on Applications of Computer Vision (WACV), pp. 617–627 (2022). https://doi.org/10.1109/WACV51458.2022.00069
38. Patil, V., Sakaridis, C., Liniger, A., Van Gool, L.: P3Depth: monocular depth estimation with a piecewise planarity prior (2022). https://doi.org/10.48550/ARXIV.2204.02091. https://arxiv.org/abs/2204.02091
39. Pilzer, A., Lathuiliere, S., Sebe, N., Ricci, E.: Refine and distill: exploiting cycle-inconsistency and knowledge distillation for unsupervised monocular depth estimation. In: Proceedings of the IEEE/CVF Conference on Computer Vision and Pattern Recognition (CVPR) (2019)
40. Qi, X., Liao, R., Liu, Z., Urtasun, R., Jia, J.: GeoNet: geometric neural network for joint depth and surface normal estimation. In: Proceedings of the IEEE Conference on Computer Vision and Pattern Recognition (CVPR) (2018)
41. Ranftl, R., Bochkovskiy, A., Koltun, V.: Vision transformers for dense prediction. In: Proceedings of the IEEE/CVF International Conference on Computer Vision, pp. 12179–12188 (2021)
42. Ranjan, A., Jampani, V., Kim, K., Sun, D., Wulff, J., Black, M.J.: Adversarial collaboration: joint unsupervised learning of depth, camera motion, optical flow and motion segmentation. CoRR abs/1805.09806 (2018). http://arxiv.org/abs/1805.09806
43. Saeedan, F., Roth, S.: Boosting monocular depth with panoptic segmentation maps. In: Proceedings of the IEEE/CVF Winter Conference on Applications of Computer Vision (WACV), pp. 3853–3862 (2021)
44. Saxena, A., Chung, S., Ng, A.: Learning depth from single monocular images. In: Weiss, Y., Schölkopf, B., Platt, J. (eds.) Advances in Neural Information Processing Systems, vol. 18. MIT Press (2005). https://proceedings.neurips.cc/paper/2005/file/17d8da815fa21c57af9829fb0a869602-Paper.pdf
45. Saxena, A., Sun, M., Ng, A.Y.: Make3D: Learning 3D scene structure from a single still image. IEEE Trans. Pattern Anal. Mach. Intell. **31**(5), 824–840 (2009). https://doi.org/10.1109/TPAMI.2008.132

46. Shi, W., et al.: Real-time single image and video super-resolution using an efficient sub-pixel convolutional neural network. In: Proceedings of the IEEE Conference on Computer Vision and Pattern Recognition, pp. 1874–1883 (2016)
47. Tosi, F., Aleotti, F., Ramirez, P.Z., Poggi, M., Salti, S., Stefano, L.D., Mattoccia, S.: Distilled semantics for comprehensive scene understanding from videos. In: Proceedings of the IEEE/CVF Conference on Computer Vision and Pattern Recognition (CVPR) (2020)
48. Touvron, H., Cord, M., Douze, M., Massa, F., Sablayrolles, A., Jégou, H.: Training data-efficient image transformers & distillation through attention. In: International Conference on Machine Learning, pp. 10347–10357. PMLR (2021)
49. Vaswani, A., et al.: Attention is all you need. In: Advances in Neural Information Processing Systems 30 (2017)
50. Wang, J., et al.: Deep high-resolution representation learning for visual recognition. IEEE Trans. Pattern Anal. Mach. Intell. **43**(10), 3349–3364 (2020)
51. Wang, L., Zhang, J., Wang, O., Lin, Z., Lu, H.: SDC-Depth: semantic divide-and-conquer network for monocular depth estimation. In: 2020 IEEE/CVF Conference on Computer Vision and Pattern Recognition (CVPR), pp. 538–547 (2020). https://doi.org/10.1109/CVPR42600.2020.00062
52. Wang, Z., Cun, X., Bao, J., Zhou, W., Liu, J., Li, H.: Uformer: A general u-shaped transformer for image restoration. In: Proceedings of the IEEE/CVF Conference on Computer Vision and Pattern Recognition. pp. 17683–17693 (2022)
53. Wang, Z., Bovik, A.C., Sheikh, H.R., Simoncelli, E.P.: Image quality assessment: from error visibility to structural similarity. IEEE Trans. Image Process. **13**(4), 600–612 (2004)
54. Weder, S., Schönberger, J.L., Pollefeys, M., Oswald, M.R.: NeuralFusion: online depth fusion in latent space. In: 2021 IEEE/CVF Conference on Computer Vision and Pattern Recognition (CVPR), pp. 3161–3171 (2021). https://doi.org/10.1109/CVPR46437.2021.00318
55. Xian, K., Zhang, J., Wang, O., Mai, L., Lin, Z., Cao, Z.: Structure-guided ranking loss for single image depth prediction. In: The IEEE/CVF Conference on Computer Vision and Pattern Recognition (CVPR) (2020)
56. Xu, J., et al.: GroupViT: semantic segmentation emerges from text supervision. In: Proceedings of the IEEE/CVF Conference on Computer Vision and Pattern Recognition (CVPR), pp. 18134–18144 (2022)
57. Xu, L., Ouyang, W., Bennamoun, M., Boussaid, F., Xu, D.: Multi-class token transformer for weakly supervised semantic segmentation. In: Proceedings of the IEEE/CVF Conference on Computer Vision and Pattern Recognition (CVPR), pp. 4310–4319 (2022)
58. Xu, X., Chen, Z., Yin, F.: Monocular depth estimation with multi-scale feature fusion. IEEE Signal Process. Lett. **28**, 678–682 (2021). https://doi.org/10.1109/LSP.2021.3067498
59. Yang, M., et al.: DOLG: single-stage image retrieval with deep orthogonal fusion of local and global features. In: Proceedings of the IEEE/CVF International Conference on Computer Vision (ICCV), pp. 11772–11781 (2021)
60. Zamir, S.W., Arora, A., Khan, S., Hayat, M., Khan, F.S., Yang, M.H.: Restormer: efficient transformer for high-resolution image restoration. In: Proceedings of the IEEE/CVF Conference on Computer Vision and Pattern Recognition, pp. 5728–5739 (2022)
61. Zhou, T., Brown, M., Snavely, N., Lowe, D.G.: Unsupervised learning of depth and ego-motion from video. In: Proceedings of the IEEE Conference on Computer Vision and Pattern Recognition (CVPR) (2017)

62. Zhu, S., Brazil, G., Liu, X.: The edge of depth: explicit constraints between segmentation and depth. In: 2020 IEEE/CVF Conference on Computer Vision and Pattern Recognition (CVPR), pp. 13113–13122 (2020). https://doi.org/10.1109/CVPR42600.2020.01313
63. Zou, Y., Luo, Z., Huang, J.-B.: DF-Net: unsupervised joint learning of depth and flow using cross-task consistency. In: Ferrari, V., Hebert, M., Sminchisescu, C., Weiss, Y. (eds.) ECCV 2018. LNCS, vol. 11209, pp. 38–55. Springer, Cham (2018). https://doi.org/10.1007/978-3-030-01228-1_3

Towards Real-World Video Deblurring by Exploring Blur Formation Process

Mingdeng Cao[1], Zhihang Zhong[2], Yanbo Fan[3], Jiahao Wang[1], Yong Zhang[3], Jue Wang[3], Yujiu Yang[1(✉)], and Yinqiang Zheng[2(✉)]

[1] Tsinghua Shenzhen International Graduate School, Tsinghua University, Beijing, China
yang.yujiu@sz.tsinghua.edu.cn
[2] The University of Tokyo, Tokyo, Japan
yqzheng@ai.u-tokyo.ac.jp
[3] Tencent AI Lab, Shenzhen, China

Abstract. This paper aims at exploring how to synthesize close-to-real blurs that existing video deblurring models trained on them can generalize well to real-world blurry videos. In recent years, deep learning-based approaches have achieved promising success on video deblurring task. However, the models trained on existing synthetic datasets still suffer from generalization problems over real-world blurry scenarios with undesired artifacts. The factors accounting for the failure remain unknown. Therefore, we revisit the classical blur synthesis pipeline and figure out the possible reasons, including shooting parameters, blur formation space, and image signal processor (ISP). To analyze the effects of these potential factors, we first collect an ultra-high frame-rate (940 FPS) RAW video dataset as the data basis to synthesize various kinds of blurs. Then we propose a novel realistic blur synthesis pipeline termed as RAW-Blur by leveraging blur formation cues. Through numerous experiments, we demonstrate that synthesizing blurs in the RAW space and adopting the same ISP as the real-world testing data can effectively eliminate the negative effects of synthetic data. Furthermore, the shooting parameters of the synthesized blurry video, *e.g.*, exposure time and frame-rate play significant roles in improving the performance of deblurring models. Impressively, the models trained on the blurry data synthesized by the proposed RAW-Blur pipeline can obtain more than 5dB PSNR gain against those trained on the existing synthetic blur datasets. We believe the novel realistic synthesis pipeline and the corresponding RAW video dataset can help the community to easily construct customized blur datasets to improve real-world video deblurring performance largely, instead of laboriously collecting real data pairs.

Keywords: Video deblurring · Real-world deblurring · Synthetic blurs · RAW signal processing

---

Supplementary Information The online version contains supplementary material available at https://doi.org/10.1007/978-3-031-25063-7_20.

L. Karlinsky et al. (Eds.): ECCV 2022 Workshops, LNCS 13802, pp. 327–343, 2023.
https://doi.org/10.1007/978-3-031-25063-7_20

Fig. 1. The real-world deblurring results of DBN [27] and EDVR [30]. The blurry videos come from the testing subset in the real-world video deblurring dataset BSD [35], and the models trained on the popular synthetic datasets, including GORRO [18], DVD [27], REDS [17], and HFR-DVD [14], cannot obtain satisfying results.

1 Introduction

Video deblurring aims at restoring the latent sharp frame from the blurry input frames and has received considerable research attention. In which, deep neural networks (DNN)-based deblurring approaches [3,14,18,27,30,36,38] are in the leading positions in recent years. In order to benchmark the video deblurring performance and facilitate the development of DNN-based models, large-scale and high-quality datasets are required. Therefore, developing an effective and convenient pipeline to create blurry-sharp pairs is indispensable.

To construct realistic blur datasets, most recently, researchers [23,35] designed specific imaging systems that can record the blur-sharp pairs simultaneously. However, these systems require extremely high physical precision to align the blurry and sharp image acquisition cameras. Otherwise, the collected pairs may suffer from the misalignment problem easily. Meanwhile, these specific imaging systems are not flexible and can only capture sharp-blurry pairs with some fixed settings, which further restricts real-world deblurring performance.

Compared to these real-world blur acquisition systems, synthesizing blurs is more convenient and flexible in constructing large-scale datasets for DNN-based model training. A classical approach is to convolve the sharp image with predefined motion blur kernels [10,11,25]. Yet, the synthesized blurs are unnatural since the bur kernels can only simulate limited uniform camera motions. A more realistic and popular way is to imitate the real-world blur formation process, and some researchers have proposed various large-scale synthetic deblurring datasets, such as GOPRO [18], DVD [27], REDS [17], and HFR-DVD [14]. These datasets usually capture high-frame-rate sharp videos and then average several consecutive sharp frames to generate blurry frames blindly. Unfortunately, these blur synthesis pipelines are still quite different from the real-world blur formation process, including blur formation space, image signal processing, noise level, *etc.*. We test some models trained on these popular synthetic datasets on the real-world blurry scenes in [35] shown in Fig. 1. We can see that these models cannot effectively remove the blurs and may even destroy the image contents since the real-world blurs have a completely different blur formation process compared to existing synthetic blur pipelines. Thus developing a convenient pipeline is urgently needed to help the community synthesize realistic blurs to improve real-world video deblurring performance.

To move beyond aforementioned drawbacks of existing video deblurring datasets, in this work, 1) we aim at figuring out the intrinsic reasons why the models trained on existing video deblurring datasets cannot generalize well to the real-world blurry scenes, 2) and how to eliminate this gap on the data side. To this end, we firstly revisit existing blur synthesis pipelines by comparing them to the real-world blur formation process in detail. We find that the frame-rate of the sharp videos, blur formation space, noise, image signal processor, and blurry video parameters are the main factors attributing to the gap. Secondly, we collect a series of ultra high frame-rate RAW sharp videos so that we can synthesize various blurs to analyze and validate the factors effectively. Meanwhile, based on these RAW videos, we propose a novel pipeline termed RAW-Blur to synthesize realistic blurs for video deblurring. Thus existing models trained on these data can remove the blurs in the real world. Our RAW-Blur pipeline consists of three parts. Firstly, we should ensure the synthesized blurry video parameters (*e.g.*, exposure time and frame-rate) match the real blurry scene as closely as possible. Secondly, we average sharp frames in RAW space and add the reduced noise during the average process to simulate the real blur formation process. Last, the synthesized RAW blurs are transformed into RGB counterparts by a specific ISP close to the one adopted by the captured real blurry video. The whole pipeline is very close to the natural blur formation process, enabling handling blurs in the real world. Thus the models trained on the blur data synthesized by RAW-Blur can remove the real-world blurs shown in Fig. 1(e).

Our contributions can be summarized as follows[1]:

- We revisit existing blur synthesis pipeline, and figure out the intrinsic reasons for the performance gap between synthetic and real blurs.
- We propose a novel realistic blur synthesis pipeline RAW-Blur, which can synthesize close-to-real blurs in RAW space. Meanwhile, we contribute a ultra high frame-rate RAW video dataset to help the community construct their customized blur datasets.
- The quantitative and qualitative experimental results on the real-world blurry scenarios show the excellent real-world video deblurring performance of the proposed pipeline against existing blur synthesis methods.

2 Related Works

2.1 Deep Deblurring

Recently, deep learning-based image and video deblurring approaches have achieved significant progress through more efficient spatial and temporal modeling.

[1] This work was partially done when Mingdeng was a intern in Tencent AI Lab.

As for image deblurring, which tries to restore the clear image through effective spatial modeling, researchers usually design some specific architectures to enable a large receptive field. Nah *et al.* [18] proposed a multi-scale network to restore the clear images in a coarse-to-fine strategy. Based on this, Tao et al. [29] further shared the parameters of the restoration network among all scales, which reduces the parameters largely. Meanwhile, [28,33] utilized the hierarchical multi-patch network to deal with blurry images, achieving 40$\times$ faster runtime compared to previous multi-scale-based methods. Yuan et al. [32] also utilized deformable convolution [4] with optical flow guided training for spatially variant deconvolution.

Compared to image deblurring, video deblurring (or multi-image deblurring) requires additional efficient temporal modeling to obtain satisfying results. Inspired by the encoder-decoder network, Su *et al.* [27] designed a deep video deblurring network that directly outputs the central clear frame by concatenating consecutive frames as the model's input. To exploit the temporal information better, researchers adopt 3D convolution and recurrent neural network architecture for temporal modeling. However, the 3D-CNN-based model [34] requires huge computational costs, and simple RNN-based models [9,19,35] still are inefficient for temporal modeling. Based on these models, some temporal alignment and aggregation methods are proposed to model the temporal variations. Wang et al. [30] utilized deformable convolution, and Zhou *et al.* [38] adopted dynamic convolution to align neighboring frames, achieving better performance with moderate parameters. Meanwhile, optical flow is further adopted to align multiple frames to aggregate complementary information in [14,21,26].

2.2 Motion Blur Datasets

In the early days, the research community applied blur kernels to synthesized blurry images, which usually convolve the sharp images with predefined uniform blur kernels and add Gaussian noise. Since assuming a uniform blur over the image is not realistic even for planar scenes, Levin *et al.* [13] locked the tripod's Z-axis rotation handle but release the X and Y handles to create eight spatially invariant blur kernels. The maximum collected kernel size is up to 41×41. Then, Lai et al. [12] utilized the algorithm in [24] to synthesize 4 larger blur kernels, ranging from 51×51 to 101×101 by randomly sampling 6D camera trajectories. More recently, Shen et al. [25] synthesized 20,000 motion blur kernels with sizes ranging from 13×13 to 27×27 based on [2]'s random 3D camera trajectories to generate a blurred face dataset. However, images with uniform blur are different from real captured cases which usually have spatially varying blurs.

The blurring process can be simulated by the integration of sharp images during the shutter time. To generate non-uniform blurs and reduce the gap between synthetic and real-world blurry images, researchers resort to average consecutive frames from high frame-rate videos to synthesize blur-sharp image/video pairs. Su *et al.* simulated a motion blur dataset, dubbed as DVD [27], at 30FPS by sub-sampling every eighth frame of 240FPS sharp videos from various devices, including iPhone 6s, GoPro Hero 4 Black, and Canon 7D. The blurry images are synthesized by averaging 7 nearby frames. To alleviate the ghosting artifacts (*e.g.*, unnatural spikes or steps in the blur trajectory) caused by adjacent exposures, *i.e.*, duty cycle, Su *et al.* compute optical flow between adjacent frames and generate extra 10 evenly spaced inter-frame images before averaging. Meanwhile, Nah *et al.* proposed a motion blur dataset GOPRO [18], using 240FPS videos from GOPRO4 Hero Black camera. Instead of averaging consecutive frames directly in RGB space, GOPRO [18] applies an inverse gamma function to the images and then averages them to avoid the effects caused by nonlinear CRFs. Then, Nah *et al.* create a large dataset REDS [17] by incorporating the experiences from DVD [27] and GOPRO [18]. The difference is that REDS [17] directly uses the off-the-shelf video-frame-interpolation model [20] to interpolate the videos from 120FPS to 1920FPS before averaging. Nevertheless, the current deep learning-based models trained on previous synthetic datasets still have poor generalization on real-world samples.

Recently, some researchers have started to shoot the real-world blur-sharp image pairs directly by designing a precise beam-splitter acquisition system. Specifically, two cameras are physically aligned to receive the light split in half by the beam splitter. With one camera using a long exposure and the other using a short exposure, the acquisition system can take both sharp and blurry image pairs of the same scenarios simultaneously. Based on the above devices, Rim *et al.* and Zhong *et al.* propose the first real-world motion blur image dataset RealBlur [23] and video dataset BSD [35], respectively. Zhong *et al.* further propose a dataset BS-RSCD [37] for the joint rolling shutter deblurring and correction task. However, real-world datasets have very high precision requirements for the beam-splitter acquisition system, otherwise it is very easy to cause image misalignment. For shooting videos with different devices or different exposure parameters, it is cumbersome and risky to use such equipment.

Although these datasets promote the development of the deblurring significantly, the models trained on these datasets still cannot generalize well to real-world blurry scenarios.

3 Blur Formation Process Revisit

3.1 Real-World Blur Formation

We simplify the typical digital camera imaging process into two key steps: image signal acquisition (RAW data) and RGB rendering with ISP. As shown in Fig. 2(a), during the signal acquisition, the input lights are captured and transformed into digital signals by the camera sensor with a series of camera settings, *e.g.*, exposure time and frame-rate. During this process, the blurs occur when relative movements arise during the exposure time. This real blur accumulation process can be formulated as:

$$B_{real} = \int_0^{\tau} S(t)dt, \tag{1}$$

where B_{real} and $S(t)$ are the blurry RAW frame and the signal captured by the camera sensor at time t, respectively. τ is the exposure time. After that, these signals (RAW image) are rendered into RGB format with camera ISP, including white balance, demosaicing, gamma correction, color correction, *etc.*.

3.2 Existing Blur Synthesis Pipeline

Existing synthetic blur datasets, *e.g.*, GOPRO [18], DVD [27], REDS [17], and HFR-DVD [14], are constructed in a blind way that discretizes the continuous blur accumulation process (described in Eq. 1) by averaging consecutive frames of sharp videos. In practice, researchers first record the high frame-rate observed (in RGB format) videos directly, then DVD and HFR-DVD average these observed frames to synthesize blurs:

$$B_{syn} = \frac{1}{M}\sum_{i=1}^{M} \hat{S}[i], \tag{2}$$

where B_{syn} is the synthesized blurry frame, M and $\hat{S}[i]$ are the number of sampled frames, and the $i-$th observed sharp frame during the exposure time. These synthesized blurs cannot generalize well in real-world blurs since the RGB space is non-linear, and averaging in this space may generate different blur patterns against real-world blurs. GOPRO and REDS further adopt the inverse CRF to transform the observed sharp frames into linear space and then synthesize blurs [18]:

$$B_{syn} = g(\frac{1}{M}\sum_{i=1}^{M} g^{-1}(\hat{S}[i])), \tag{3}$$

where g is the non-linear CRF. In this way, the blur synthesis process is closer to the natural blur accumulation process. Yet, they still suffer from domain gaps since the CRF is unknown and hard to estimate accurately. Meanwhile, DVD and REDS utilize frame interpolation to increase video continuity; thus, more frames interpolated during inter-frame time generate more continuous blur. To be convenient, we summarise the detailed configurations and the critical operations in blur synthesis of these datasets in Table 1.

Table 1. The detailed configurations of existing synthetic video deblurring datasets

Datasets	GOPRO [18]	DVD [27]	REDS [17]	HFR-DVD [14]
Sharp FPS	240	240	120	1000
Blurry FPS	≈25	≈30	24	25
Frame Interpolation		✓	✓	
Inverse CRF Calibration	✓		✓	

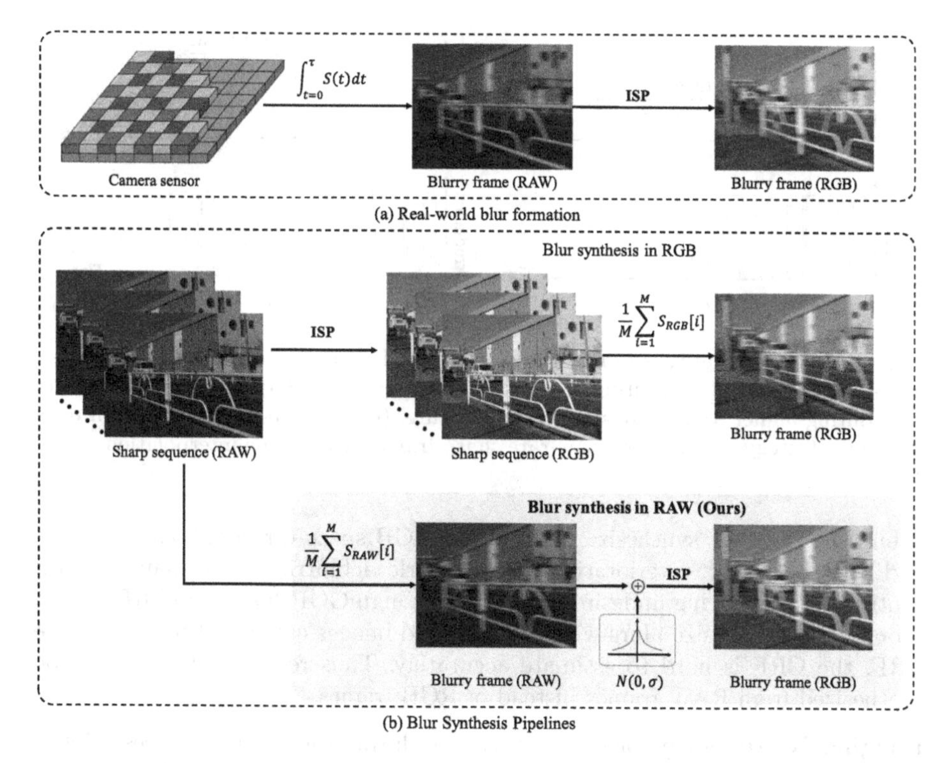

Fig. 2. Real-world and synthetic blur formation processes. Our pipeline directly synthesize the blurs in RAW space and further add the noise to simulate the real blurs.

3.3 Synthetic Blur Analysis

In the following, we analyze the intrinsic reasons that the models trained on these synthetic datasets cannot generalize well on real-world blurry scenes by comparing synthetic blurs to real ones.

High Frame-rate Sharp Video. To simulate the real blur formation process (shown in Fig. 3(a)), short exposure time and high frame-rate are required to acquire sharp videos. However, existing synthetic datasets usually capture sharp videos at 120 FPS (*e.g.*, REDS) or 240 FPS (*e.g.*, GOPRO, DVD), which own large inter-frame time (shown in Fig. 3(c)), resulting in synthesizing unreal blurs that are discontinuous and suffer from undesired spikes. Accordingly, DVD and REDS try to alleviate this problem by further increasing frames during the inter-frame time with frame interpolation techniques (shown in Fig 3(d)). Nevertheless, the interpolated videos suffer from artifacts due to the occlusions [1]. Thus higher frame-rate sharp videos with short inter-frame time are needed for more continuous and close-to-real blurs synthesis.

Blur Formation Space. We should reclaim that the real-world blurs are formatted as RAW signals rather than the observed RGB images. Existing synthetic

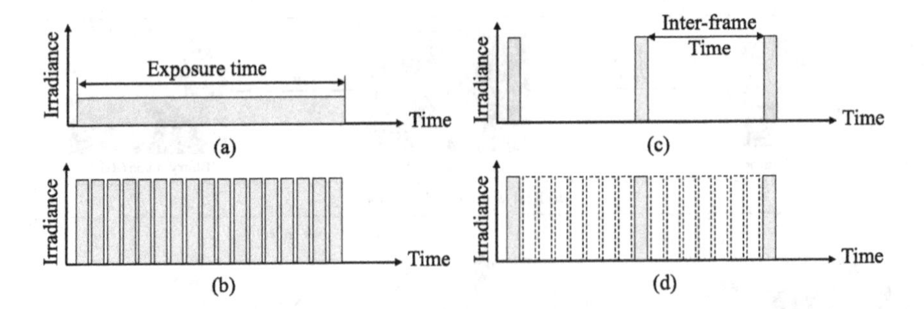

Fig. 3. Exposure strategies of real-world blurry and sharp videos acquisition process. **(a)** Long exposure time during real-world blur formation. **(b)** Very short exposure time during higher frame-rate sharp video capture. **(c)** Lower frame-rate sharp frames exposure strategy. **(d)** Lower frame-rate sharp frames with frame interpolation

deblurring datasets synthesize blurs in the RGB space directly, *e.g.*, DVD [27] and HFR-DVD [14], deteriorating the real-world deblurring performance significantly and introducing undesired artifacts. Though GOPRO [18] and REDS [17] proposed to synthesize blurs with the observed images calibrated by the inverse CRF, the CRF is hard to estimate accurately. Thus realistic blurs should be synthesized from RAW frames instead of RGB frames.

Imaging Noise. Some noises are inevitable during the imaging process [7,16], especially when blurs occur, where the exposure time is relatively long in low-light conditions. However, when we adopt the pipeline described in Sect. 3.2 to synthesize blurs, the noise tends to be eliminated for the following reasons. Firstly, the captured sharp videos are of different levels of imaging noise because of the short exposure time compared to blurry videos. Secondly, averaging multiple frames can reduce the noise largely (please refer to the proof in the provided supplementary materials). Yet, existing blur synthesis pipelines ignore the noise, widening the gap between these synthetic and real-world blurs. Thus, it is reasonable to consider additional noise to generate more realistic blurs.

Image Signal Process (ISP). Camera ISPs transform the captured RAW data into RGB images. In this process, different camera manufacturers develop their own processing pipelines based on the traditional ISP algorithms. Therefore, the blurs synthesized from the sharp frames captured by one camera may not generalize well to the blurry scenes recorded by the cameras adopting another ISPs. Thus we should further consider the impacts of different ISPs in removing real-world blurs.

Video Parameter Setting. As for video deblurring, modeling on the temporal variation is significant. The exposure time and frame-rate of the synthesized blurry videos are two critical factors. Specifically, the exposure time highly relates to the blurring degrees, and the frame-rate is essential for temporal modeling. Thus we can consider that the difference in terms of these two factors between synthetic and real-world blurry videos is also one of the reasons why the real-world video deblurring can not be resolved well by existing synthetic blur datasets [14,17,18,27].

4 RAW-Blur Dataset Construction

In this section, we elaborate on synthesizing a blur dataset for real-world video deblurring considering the factors described in Sect. 3. We propose a novel synthesis pipeline called RAW-Blur, which can generalize well in real blurry videos. In brief, we first collected an ultra-high frame-rate RAW video dataset named UHFRaw. Then, we synthesize realistic blurs based on the proposed UHFRaw.

4.1 Ultra-High Frame-rate RAW Video Dataset

We first collect a high frame-rate raw video dataset termed as UHFRaw. The videos are captured at a ultra-high frame-rate with the RAW format. Specifically, we collect 45 videos of resolution 812×620 at 940 FPS totally, by a BITRAN CS-700C camera. Each video is about 3 s long and contains about 2900 frames. Due to the high-frequency exposure strategy, the exposure time is approximately 1.01 ms and inter-frame time is nearly zero. Therefore, the captured raw frames are sharp and continuous enough to synthesize realistic blurs without frame interpolation and inverse CRF estimation. The scenes in the captured videos are mainly street scenes, where various vehicles, pedestrians, *etc.*., are included.

We also note that a RAW dataset named Deblur-RAW is also proposed for image deblurring in [15]. Compared to our UHFRaw, they only record the RAW videos at a low frame-rate (30 FPS) and average 3–5 frames to synthesize blurs. Therefore, the synthesized blurs are unreal and unnatural, further introducing undesired artifacts (please refer to our supplementary material).

4.2 RAW-Blur Synthesis Pipeline

Our blur synthesis pipeline for real-world video deblurring is shown in Fig. 2, which contains three main parts. First, we should set the exposure time and the frame-rate of the synthesized video. Then, we synthesize the RAW blurry images by averaging consecutive RAW sharp frames, and further add the reduced noise to the blurry RAW frame, making it closer to the real blurry scenes. Last, the noisy blurry RAW frames are rendered into RGB format by the Camera ISP.

Blurry Video Parameter Setting. As our aim is to synthesize a realistic dataset for video deblurring, we should consider the frame-rate of the video and exposure time. We denote the time of each blurry frame (inverse of frame-rate) and its exposure time as T-ms and τ-ms. Thus the dead-time between consecutive frames is $(T - \tau)$-ms, and each frame owns the duty cycle $\frac{\tau}{T}$. Since our UHFRaw the exposure time of each frame in UHFRaw is nearly 1-ms, thus T and τ are approximately equal to the number of frames. We can easily synthesize blurry videos with various video parameters by changing these two hyper-parameters. Our experiments in Sect. 5 demonstrate the impact of the synthesized video parameters for real-world video deblurring.

RAW-Blur Synthesis. Instead of averaging consecutive frames in the RGB space adopted by existing blur synthesis methods, we average the sharp high frame-rate RAW frames directly to synthesize more realistic blurs blindly (the detailed comparison is shown Fig. 2(b)). Meanwhile, as we analyzed in Sect. 3.3, the noise in the sharp frames is corrupted significantly when we adopt averaging consecutive frames strategy to simulate the blurring process. To synthesize close-to-real blurs in the wild, we add the reduced Gaussian noise into the synthesized RAW blurry frame. Thus the whole RAW blur synthesis pipeline can be formulated as:

$$B_{raw} = \frac{1}{\tau}\sum_{i=0}^{\tau} S_{raw}[i] + N(0,\sigma), \tag{4}$$

where B_{raw} is the synthesized RAW blurry frame, and $S_{raw}[i]$ is the sharp RAW frame at index i. $N(0,\sigma)$ is the noise distribution with expectation 0 and variance σ. In our pipeline, σ consists of the signal-independent and dependent parts, *i.e.*, the classical Poisson-Gaussian noise model.

Rendered with ISP. After synthesizing the RAW blurry frames with noise, we should translate them into RGB frames with the camera ISP for model training. Since each manufacturer designs its own specific and nonpublic ISP for processing the RAW images, the blurry videos rendered by different ISPs own different image properties, like brightness, color constancy. As a result, the models trained with the blurs rendered by one ISP cannot generalize on the blurs rendered by another. Therefore, we propose to apply the same ISP of the captured real-world blurry videos to transform the synthesized raw blurry frames into RGB blurry frames as far as possible.

Compared to the existing blur synthesis pipeline, our RAW-Blur synthesis pipeline has the following advantages that can generate close-to-real blurs: 1) RAW-Blur considers the effects of blurry video parameters for real-world video deblurring. 2) The ultra-high frame-rate sharp frames ensure the temporal continuity to synthesize natural and real blurs, avoiding the artifacts introduced by the interpolation operation. 3) Synthesizing blurs in RAW space and adding reduced noise is closest to the real blur formation process, while existing blur synthesis methods suffer from the non-linear property in RGB space. 4) The effect of camera ISP is further considered to improve existing models' real-world video deblurring performance.

5 Experiments

5.1 Experimental Setting

Datasets. We compare the proposed realistic blur synthesis pipeline to the popular synthetic datasets, including GOPRO, DVD, REDS, and HFR-DVD. Meanwhile, to evaluate the real-world video deblurring performance quantitatively, we conduct validation on the recently proposed real-world video deblurring dataset BSD [35]. There are three subsets with different exposure times for sharp and blurry image pairs in BSD, *i.e.*, 1 ms–8 ms, 2 ms–16 ms, and 3 ms–24 ms, respectively. In the following, T and τ are set as 33 and 11 in our synthesized datasets without special instructions.

Implementation Details. We adopt a classical DNN-based video deblurring model DBN [27] and a state-of-the-art model EDVR [30] without any modifications to validate the effectiveness of the proposed blur synthesis pipeline on deblurring real-world blurry videos. During training, we randomly crop the input frames into 256×256 and randomly flip them both horizontally and vertically for the data augmentation. The initial learning rates are 4×10^{-4} and 10^{-4} for DBN and EDVR, respectively. ADAM [6] optimizer is employed to optimize the model's parameters. We train the models with 1000 epochs totally, and the learning rate decays 10 times at 700-, 850-, and 920-th epoch. For evaluation, both PSNR and SSIM [31] are adopted to evaluate the deblurring performance quantitatively. All codes to reproduce the results will be made public.

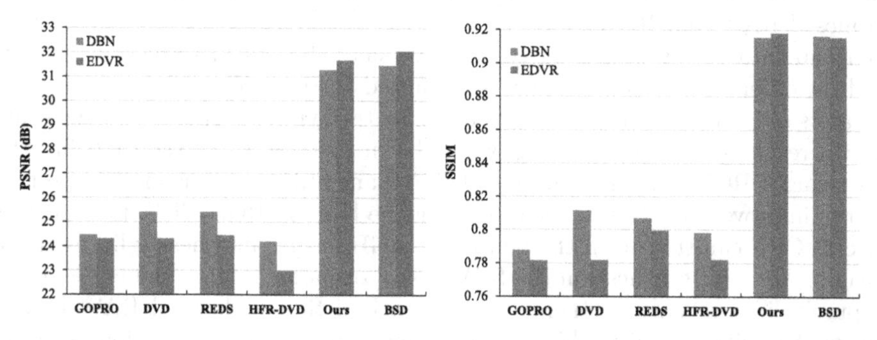

Fig. 4. The performance (*i.e.*, PSNR and SSIM) comparison of the models trained on different synthetic datasets and real-world datasets in the testing subset of BSD. We can see that both DBN and EDVR obtain much higher PSNR when trained on our synthetic dataset and nearly the same PSNR as the models trained with BSD. This demonstrates our synthesis pipeline is closer to the real-world blur formation process. We further provide the PSNR and SSIM metrics in our supplementary materials.

5.2 Comparison to Existing Synthetic Datasets

The quantitative comparison between the dataset synthesized by our RAW-Blur pipeline and existing synthetic datasets is shown in Fig. 4. We see that when trained on existing synthetic datasets, both DBN and EDVR obtain very low evaluation metrics since the blurs in these synthetic datasets deviate from real-world blurry scenes. On the contrary, these two models trained on our synthesized dataset achieve higher PSNR and SSIM with a large margin against other synthetic datasets. Besides, the metrics are much closer to those of the models trained on the training subset of BSD[2]. This reveals that synthesizing blurs in RAW space with noise is closer to the real blur formation process.

[2] Note that the dataset synthesized by RAW-Blur only contains 45 videos with 3800 frames in total, which is 35% less than the training data in BSD [35].

Meanwhile, we further visualize the deblurred results of EDVR trained on different datasets in Fig. 5. We see that EDVR fails in dealing with real-world blurry scenes when trained on existing synthetic datasets, and our RAW blur pipeline can restore friendly visual results. These quantitative and qualitative results demonstrate the effectiveness of our blur synthesis pipeline, which can be used to remove real-world blurs.

5.3 Analysis of RAW-Blur Synthesis Pipeline

In the following, we analyze the effects of the key components mentioned in Setc. 3.3 for real-world video deblurring by conducting numerous comparative experiments corresponding to each factor.

Blur Synthesis Space. We first conduct experiments of the blur synthesis space for real-world video deblurring, including RGB space adopted in DVD [27] and HFR-DVD, RGB-CRF space adopted in GOPRO [18] and REDS [17], and the proposed RAW and RAW-Noise (RAW with noise) space. Thus we synthesize four sub-datasets by averaging consecutive sharp frames in these four spaces, and the difference of each strategy is visualized in Fig. 6. The newly synthesized datasets are with the same setting except for the synthesis space, and the evaluation results of different strategies are shown in Table 2. We see that directly averaging in RGB space gets the lowest PSNR and SSIM, and the model's performance improves largely when synthesizing the blurs in RGB-CRF, for that the inverse CRF can transform the non-linear RGB images into nearly linear RGB images. And blur synthesis in the RAW space surpasses it since the RAW space is purely linear. When further adding noise into the synthesized RAW blurry images to imitate realistic blurs, we further improve EDVR's performance in terms of 0.3dB higher PSNR.

(a) Blurry Frame (b) GOPRO (c) DVD (d) REDS
(e) HFR-DVD (f) Ours (g) BSD (h) Sharp

Fig. 5. Visual comparison of EDVR [30] on the real-world BSD testing subset. The model trained on the blur dataset synthesized by RAW-Blur shows a strong competitive edge against existing synthetic datasets. More results can be found in the supplementary materials.

These results demonstrate the importance of the synthesis space, and simply averaging frames cannot reflect the real-world blur formation process. Therefore, we recommend synthesizing realistic blurs in RAW space and adding noise.

Ultra High Frame-rate Sharp Videos. We then validate the effectiveness of the high frame-rate sharp videos. To do so, we further synthesize the blurs with frame interpolation in RGB space since there are nearly no RAW video interpolation techniques. Specifically, we first temporally downsample the frame-rate 8 times to generate sharp videos at about 120 FPS. Then we adopt a state-of-the-art video interpolation model RIFE [8] to interpolate the low frame-rate videos into 940 FPS. The quantitative results on the BSD are shown in Tab. 2. We see that the model trained on the interpolated data achieved 1dB lower PSNR than the one without interpolation. These results ravels that the interpolation may introduce some artifacts, and the synthesized blurs still own the domain gap compared to the real-world accumulated blurs. Thus, we highly recommend capturing native high FPS videos to synthesize realistic blurry frames.

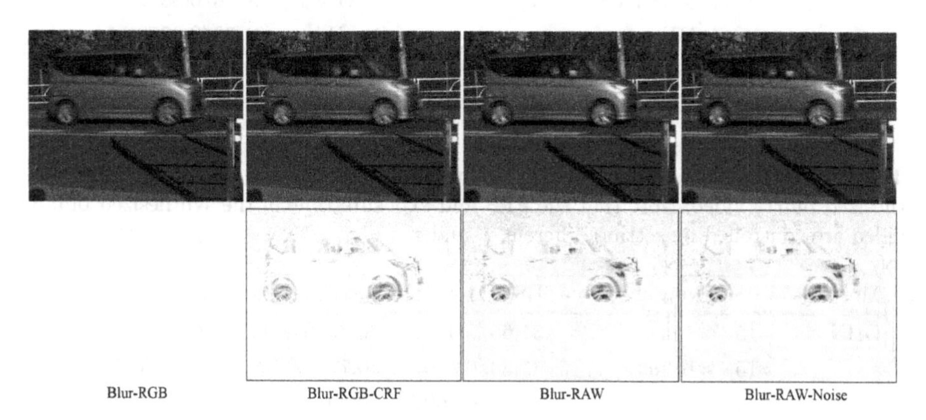

Fig. 6. The visualization of the synthesized blurs in different space and the difference between the blurs synthesized in RGB space and RGB-CRF, RAW, RAW-Noise space. We see that these data synthesised in different spaces differ significantly mainly in the blurry areas, *e.g.*, the blurry car.

Different ISPs. We also adopt a free RAW image processing software named DarkTable [5] to render the synthesized RAW blurry frames to analyze the effect of different ISPs, shown in Tab. 2. Both DBN and EDVR suffer from 1.0dB and 1.7dB PSNR drop compared to the models trained over the frames rendered by RawPy [22]. This reveals that the blurs rendered by different ISPs differ greatly, resulting in the models trained with the frames processed by one specific ISP cannot generalize well on the frames processed by another. Thus, we highly recommend adopting the same ISP of the real-world blurry scenes to render the synthesized RAW blurry frames for model training.

Table 2. Effects of synthesis space, video frame interpolation and ISP. The performance (PSNR/SSIM) are evaluated in the testing subset of BSD [35]. Note that the frames in BSD are rendered by RawPy [22]

Methods	Synthesis Space	Interpolation	ISP	BSD (2–16 ms)
DBN	RGB		RawPy	29.10/0.8925
	RGB	✓	RawPy	28.03/0.8695
	RGB-CRF		RawPy	30.67/0.9003
	RGB-CRF	✓	RawPy	28.19/0.8521
	RAW		RawPy	30.86/0.9025
	RAW		DarkTable	29.86/0.8887
	RAW-Noise		RawPy	31.28/0.9156
EDVR	RGB		RawPy	29.77/0.8893
	RGB	✓	RawPy	27.91/0.8739
	RGB-CRF		RawPy	31.06/0.9059
	RGB-CRF	✓	RawPy	29.68/0.8812
	RAW		RawPy	31.39/0.9072
	RAW		DarkTable	29.68/0.8891
	RAW-Noise		RawPy	31.68/0.9192

Table 3. The quantitative evaluation results of different video parameters of the synthesized blurry video. Note that the FPS and exposure time of the synthesized blurry video are controlled by setting different T and τ.

Methods	FPS	Exposure Time	BSD (1–8 ms)	BSD (2–16 ms)	BSD (3–24 ms)
DBN	≈15	≈8 ms	31.65/0.9151	30.72/0.8983	30.05/0.8971
	≈15	≈16 ms	31.33/0.9105	30.77/0.9018	30.60/0.9113
	≈15	≈24 ms	30.70/0.9007	29.69/0.8816	30.07/0.8999
	≈28	≈11 ms	31.22/0.9094	30.87/0.9019	30.58/0.9083
	≈43	≈11 ms	29.37/0.8912	29.12/0.8839	30.26/0.9048
EDVR	≈15	≈8ms	32.02/0.9213	31.62/0.9129	31.05/0.9153
	≈15	≈16 ms	32.04/0.9200	31.76/0.9129	31.81/0.9274
	≈15	≈24 ms	30.67/0.9072	31.04/0.9047	31.60/0.9235
	≈28	≈11ms	29.85/0.9043	31.62/0.9109	28.91/0.9053
	≈43	≈11 ms	27.49/0.8555	27.22/0.8443	28.62/0.8763

Exposure Time and FPS. We further study the effects of synthesized blurry video parameters for video deblurring, *e.g.*, exposure time and FPS. To achieve so, we further synthesize five new deblurring datasets with different exposure times and the FPS (by changing T and τ). Table 3 demonstrate that both exposure time and FPS play significant roles in video deblurring for the following reasons. The exposure time of the blurry video reflect the degree of the blurs to some extent. Meanwhile, frame-rate affects the temporal modeling of the deblurring models. As a result, we should synthesize the blurry video for model training with closer video parameters of real-world blurry videos to achieve better deblurring performance.

6 Conclusion and Limitation

In this paper, we explore the real-world video deblurring task by synthesizing realistic blurs in the RAW space with noise. A novel blur synthesis pipeline RAW-Blur and a corresponding ultra high frame-rate RAW video dataset UHFRaw are proposed. With numerous experimental results, existing video deblurring models trained with the blur dataset synthesized by RAW-Blur can significantly improve real-world video deblurring performance compared to existing blur synthesis pipelines. Meanwhile, we highly recommend synthesizing realistic blurs in RAW space with ultra high FPS RAW frames and considering the effects of ISP and blurry video parameter settings. However, our pipeline cannot be applied to the existing synthetic datasets since they do not provide the RAW sources. One feasible solution is to invert the RGB frames into RAW images and use our RAW-Blur synthesis pipeline.

Acknowledgement. This work was supported partially by the Major Research Plan of the National Natural Science Foundation of China (Grant No. 61991450), the Shenzhen Key Laboratory of Marine IntelliSense and Computation (under Contract ZDSYS20200811142605016), and JSPS KAKENHI (Grant Number 22H00529).

References

1. Bao, W., Lai, W.S., Zhang, X., Gao, Z., Yang, M.H.: Memc-net: Motion estimation and motion compensation driven neural network for video interpolation and enhancement. IEEE TPAMI **43**(3), 933–948 (2019)
2. Boracchi, G., Foi, A.: Modeling the performance of image restoration from motion blur. IEEE TIP **21**(8), 3502–3517 (2012)
3. Cao, M., Fan, Y., Zhang, Y., Wang, J., Yang, Y.: Vdtr: Video deblurring with transformer. arXiv preprint arXiv:2204.08023 (2022)
4. Dai, J., et al.: Deformable convolutional networks. In: ICCV, pp. 764–773 (2017)
5. Darktable: Darktable is an open source photography workflow application and raw developer. https://www.darktable.org/ (2022)
6. Diederik, K., Jimmy, B., et al.: Adam: A method for stochastic optimization. In: ICLR, pp. 273–297 (2014)
7. Foi, A., Trimeche, M., Katkovnik, V., Egiazarian, K.: Practical poissonian-gaussian noise modeling and fitting for single-image raw-data. IEEE TIP **17**(10), 1737–1754 (2008)
8. Huang, Z., Zhang, T., Heng, W., Shi, B., Zhou, S.: Rife: Real-time intermediate flow estimation for video frame interpolation. arXiv preprint arXiv:2011.06294 (2020)
9. Hyun Kim, T., Mu Lee, K., Scholkopf, B., Hirsch, M.: Online video deblurring via dynamic temporal blending network. In: ICCV, pp. 4038–4047 (2017)

10. Kaufman, A., Fattal, R.: Deblurring using analysis-synthesis networks pair. In: CVPR, pp. 5811–5820 (2020)
11. Kupyn, O., Budzan, V., Mykhailych, M., Mishkin, D., Matas, J.: Deblurgan: Blind motion deblurring using conditional adversarial networks. In: CVPR, pp. 8183–8192 (2018)
12. Lai, W.S., Huang, J.B., Hu, Z., Ahuja, N., Yang, M.H.: A comparative study for single image blind deblurring. In: CVPR, pp. 1701–1709 (2016)
13. Levin, A., Weiss, Y., Durand, F., Freeman, W.T.: Understanding and evaluating blind deconvolution algorithms. In: CVPR, pp. 1964–1971 (2009)
14. Li, D., et al.: Arvo: Learning all-range volumetric correspondence for video deblurring. In: CVPR, pp. 7721–7731 (2021)
15. Liang, C.H., Chen, Y.A., Liu, Y.C., Hsu, W.: Raw image deblurring. IEEE TMM (2020)
16. Liu, C., Freeman, W.T., Szeliski, R., Kang, S.B.: Noise estimation from a single image. In: CVPR, vol. 1, pp. 901–908 (2006)
17. Nah, S., et al.: Ntire 2019 challenge on video deblurring and super-resolution: Dataset and study. In: CVPR Workshops (2019)
18. Nah, S., Hyun Kim, T., Mu Lee, K.: Deep multi-scale convolutional neural network for dynamic scene deblurring. In: CVPR, pp. 3883–3891 (2017)
19. Nah, S., Son, S., Lee, K.M.: Recurrent neural networks with intra-frame iterations for video deblurring. In: CVPR, pp. 8102–8111 (2019)
20. Niklaus, S., Mai, L., Liu, F.: Video frame interpolation via adaptive separable convolution. In: ICCV, pp. 261–270 (2017)
21. Pan, J., Bai, H., Tang, J.: Cascaded deep video deblurring using temporal sharpness prior. In: CVPR, pp. 3043–3051 (2020)
22. Riechert, M.: Raw image processing for python, a wrapper for libraw. https://github.com/charlespwd/project-title (2014)
23. Rim, J., Lee, H., Won, J., Cho, S.: Real-world blur dataset for learning and benchmarking deblurring algorithms. In: Vedaldi, A., Bischof, H., Brox, T., Frahm, J.-M. (eds.) ECCV 2020. LNCS, vol. 12370, pp. 184–201. Springer, Cham (2020). https://doi.org/10.1007/978-3-030-58595-2_12
24. Schmidt, U., Rother, C., Nowozin, S., Jancsary, J., Roth, S.: Discriminative non-blind deblurring. In: CVPR, pp. 604–611 (2013)
25. Shen, Z., Lai, W.S., Xu, T., Kautz, J., Yang, M.H.: Deep semantic face deblurring. In: CVPR, pp. 8260–8269 (2018)
26. Son, H., Lee, J., Lee, J., Cho, S., Lee, S.: Recurrent video deblurring with blur-invariant motion estimation and pixel volumes. ACM TOG **40**(5), 1–18 (2021)
27. Su, S., Delbracio, M., Wang, J., Sapiro, G., Heidrich, W., Wang, O.: Deep video deblurring for hand-held cameras. In: CVPR, pp. 1279–1288 (2017)
28. Suin, M., Purohit, K., Rajagopalan, A.: Spatially-attentive patch-hierarchical network for adaptive motion deblurring. In: CVPR, pp. 3606–3615 (2020)
29. Tao, X., Gao, H., Shen, X., Wang, J., Jia, J.: Scale-recurrent network for deep image deblurring. In: CVPR, pp. 8174–8182 (2018)
30. Wang, X., Chan, K.C., Yu, K., Dong, C., Change Loy, C.: Edvr: Video restoration with enhanced deformable convolutional networks. In: CVPR Workshops (2019)
31. Wang, Z., Bovik, A.C., Sheikh, H.R., Simoncelli, E.P.: Image quality assessment: from error visibility to structural similarity. IEEE TIP **13**(4), 600–612 (2004)
32. Yuan, Y., Su, W., Ma, D.: Efficient dynamic scene deblurring using spatially variant deconvolution network with optical flow guided training. In: CVPR, pp. 3555–3564 (2020)

33. Zhang, H., Dai, Y., Li, H., Koniusz, P.: Deep stacked hierarchical multi-patch network for image deblurring. In: CVPR, pp. 5978–5986 (2019)
34. Zhang, K., Luo, W., Zhong, Y., Ma, L., Liu, W., Li, H.: Adversarial spatio-temporal learning for video deblurring. IEEE TIP **28**(1), 291–301 (2018)
35. Zhong, Z., Gao, Y., Zheng, Y., Zheng, B.: Efficient spatio-temporal recurrent neural network for video deblurring. In: Vedaldi, A., Bischof, H., Brox, T., Frahm, J.-M. (eds.) ECCV 2020. LNCS, vol. 12351, pp. 191–207. Springer, Cham (2020). https://doi.org/10.1007/978-3-030-58539-6_12
36. Zhong, Z., Sun, X., Wu, Z., Zheng, Y., Lin, S., Sato, I.: Animation from blur: Multi-modal blur decomposition with motion guidance. In: ECCV (2022). https://doi.org/10.1007/978-3-031-19800-7_35
37. Zhong, Z., Zheng, Y., Sato, I.: Towards rolling shutter correction and deblurring in dynamic scenes. In: CVPR, pp. 9219–9228 (2021)
38. Zhou, S., Zhang, J., Pan, J., Xie, H., Zuo, W., Ren, J.: Spatio-temporal filter adaptive network for video deblurring. In: ICCV, pp. 2482–2491 (2019)

Unified Transformer Network for Multi-Weather Image Restoration

Ashutosh Kulkarni[(✉)], Shruti S. Phutke, and Subrahmanyam Murala

CVPR Lab, Indian Institute of Technology Ropar, Punjab, India
{ashutosh.20eez0008,2018eez0019,subbumurala}@iitrpr.ac.in

Abstract. Vision based applications routinely involve restoration as a preprocessing step, making it impossible to have separate architectures for different types of weather restoration. But, most of the existing methods focus on weather specific application. Further, the methods for multi-weather image restoration have high computational constraints. To overcome these limitations, we propose a compact transformer based network, with 4.5M parameters ($1/10^{th}$ of the existing method) for unified (simultaneous) removal of rain, snow and hazy effect with *single set of trained parameters*. We propose two parallel streams to handle the degradations: First, original resolution transformer stream (ORTS) focuses mainly on extracting fine level features through original scales of the inputs. Second, multi-level feature aggregation stream (MFAS) learns different sizes of the weather degradations. Further, it also uses coarse outputs from the first stream and utilizes edge boosting skip connections (EBSC) for propagating crucial edge details essential for image restoration. Finally, we present a memory replay training approach for generalization of the proposed network on multi-weather degraded scenarios. Substantial experiments on synthetic as well as real-world images, along with extensive ablation studies, demonstrate that the proposed method performs competitively with the existing methods for multi-weather image restoration. The code is provided at https://github.com/AshutoshKulkarni4998/UMWTransformer.

Keywords: Compact transformer network · Memory replay training · Multi-weather restoration

1 Introduction

Advancements in industrial applications have led to escalated utilization of various computer vision based applications such as object recognition [6], semantic segmentation [38], scene analysis [14], depth estimation [12], *etc.* These applications mostly rely on visual sensors for effective functioning. But, adverse weather conditions like rain, snow, and haze considerably degrade visual quality of the images captured. In light of this, it is imperative to develop an algorithm that can restore images taken in any of the aforementioned weather conditions.

Supplementary Information The online version contains supplementary material available at https://doi.org/10.1007/978-3-031-25063-7_21.

L. Karlinsky et al. (Eds.): ECCV 2022 Workshops, LNCS 13802, pp. 344–360, 2023.
https://doi.org/10.1007/978-3-031-25063-7_21

For rain removal, the existing hand-crafted methods [17,24,27] make use of frequency domain representation [17], sparse coding [27] and Gaussian mixture models [24]. The advancements in deep learning and its generalizability has inspired the researchers to develop deep convolutional neural networks (CNNs) [1,8,16,23,30,33,41,43] to overcome the limitations of hand-crafted based algorithms. These methods make use of gated convolutions [1], rain-density labels [43], diverse receptive fields [8], multiple scales [16,41], *etc.*. Further, the emergence of the transformer networks has lead to an increased research towards its applicability on rain removal task [36,40].

For snow removal, from the existing methods [3,4,20,26,35,37,39,46], hand-crafted based methods make use of multi-guided filter [46], hierarchical image decompositon[35], and deep learning based methods make use of partial convolutions [3], dual-tree complex wavelet representation [4], *etc.*

For haze removal, hand-crafted based methods [13,31] make use of dark channel prior [13] along with morphological operations for further refinement of dark channel prior transmission maps [31] and deep learning based methods use gated convolutions [1], multi-scale features [44], unpaired adversarial learning [45], *etc.* But there exist very few methods that aim at restoring multi-weather degraded images, *e.g.* [21] targets removal of rain with veiling effect through a network consiting of encoders for estimating rain streak, transmission map, and atmospheric light. [3,4] attempt at removing snow streaks and veiling effect through a single network, and [22] attempts at removing rain streaks, veiling effect, raindrops and snow streaks through three encoder streams and one generalized decoder.

According to the research reviewed above, the problem of weather degraded image restoration has been addressed in a variety of ways. However, many of these methods are based on weather-specific applications, which lead to undesirable results when weather degradation in the input images vary. In order to cope with various types of weather, a cascaded network comprised of several weather-specific networks may be utilized. But, if any of the stages produce incorrect results, the final result will also be impacted. In addition to this, there is a computational burden associated with such cascading of various weather network types. Further, existing methods for multi-weather image restoration [3,5,21,22,34] either consist of task-specific encoders [3,22], or utilize additional priors [21] *i.e.,* rain-streak maps, transmission maps, atmospheric light, *etc.* which increases the data requirement. Since, multi-weather degraded image restoration is a crucial preprocessing step for various computer vision applications to be applied in adverse weather conditions, there is a dire need of a generalized method that can handle various types of weather degradations, with plausible perceptual quality and less computational complexity. Overall, an efficient method for processing multi-weather features on a generalized scale is required.

Considering these requirements, we propose a unified approach with feature processing blocks that are designed for handling frequent rain and snow streaks, as well as hazy effects in images. Instead of having separate encoders for weather specific task [22], the proposed method consists of single trainable network which can handle multiple weather degradations. The proposed network amalgamates fine feature extraction through original scale and multi-scale weather feature extraction, along with edge detail enhancement. The major contributions are:

- A compact unified transformer network is proposed to remove multiple types of weather degradations present in an image. The proposed network has only **4.5M** trainable parameters, which are comparatively less *i.e.*, $\mathbf{1/10^{th}}$ of the state-of-the-art method for multi-weather image restoration.
- A novel parallel two stream transformer based architecture is proposed, where the first stream acts towards extracting crucial fine level features from the input by maintaining the resolution intact. Second stream works towards extraction of multi-scale weather degradations and utilization of first stream outputs for effective image restoration.
- Further, we propose utilization of memory replay training for effective multi-weather image restoration. *To the best of author's knowledge, this is the first attempt of using memory replay strategy for multi-weather image restoration.*

Extensive experiments on various weather degraded synthetic datasets and real-world images demonstrate that the proposed method performs competitively with the existing state-of-the-art methods.

2 Related Work

Rain Removal: Inverse recovery of rain-free image from a rain degraded image is an ill-posed problem, and various hand-crafted and deep learning based approaches have been proposed to solve this problem. The hand-crafted prior based methods include Gaussian mixture model [24] and discriminative sparse coding [27], *etc.* Afterwards, the emergence of deep learning networks brought advancements in rain removal methods, and the researchers proposed various approaches towards the task of rain removal from images. Zhang and Patel [43] proposed a density-label aware classification and rain removal method using a multi-stream dense network. In [1], authors adopted the technique of smoothed dilation to avoid grid artifacts caused by traditional dilated convolutions. In [30], an approach which takes advantage of recursive computations is proposed that leverages intra-stage recurrent computations resulting in reduction of the network parameters while maintaining deraining performance. Zamir *et al.* [41] proposed a multi-stage approach which progressively restores a degraded image by injecting supervision at each stage, while processing different scales of input.

Snow Removal: Snow removal from single images is even more challenging task due to the non-transparency property of snow, which results in occluded background region and loss of details. For recovering snow-free images from snow degraded images, hand-crafted prior [35,46] and deep learning based methods [3,4,26] were proposed. In [46], the authors analyzed the difference between snow streaks and clear background edges, and applied multi-guided filter to remove snowflakes. Eigen *et al.* [11] trained a model making use of convolutional neural networks for snow removal. Further, in deep learning approaches, Liu *et al.*. [26] made the initial attempt towards learning-based snow removal. The methods

[3,4] that mainly focused on snow with veil removal will be discussed ahead in separate subsection for multi-weather image restoration.

Haze Removal: Initial attempts were directed towards image dehazing using hand-crafted priors [13]. He *et al.* [13] proposed a baseline haze relevant prior to get the image coarse-level depth information for de-hazing. However, it fails in sky regions and exhibits halo effect near complicated edge structures. Researchers [9,45] have developed convolutional neural networks to calculate the transmission map for image dehazing. A boosted decoder was proposed by Dong *et al.* [9], where only reconstruction error calculated using ground truth is used as supervision for gradually obtaining the haze-free image. Zhao *et al.* [45] proposed a weakly supervised two stage framework which utilizes unpaired adversarial learning. Recently, Shin *et al.* [32] proposed a dual-supervised triple convolutional network for image dehazing. Jia *et al.*[15] proposed a meta-attention based network for restoration of hazy images. Liu *et al.* [25] proposed a multi-branch feature extraction based method for integrating all characteristic information and reconstructing the haze-free image. Chen *et al.* [7] proposed generalization of a pre-trained network on synthetic data to adapt on real-world images. Li *et al.* [19] proposed an unsupervised learning based approach with compact multi-scale feature attention and multi-frequency representations.

Multi-Weather Image Restoration: The literature explained above mainly focuses on restoration of single type of weather degradation. Now-a-days, there has been an escalation in the research towards restoration of multiple types of weather degradations simultaneously. In [21], authors have proposed a two stage network where first stage calculates image priors and second stage recovers the background details. In [22], a multi-weather image restoration approach is proposed where authors leverage neural architecture search for processing the features extracted from the task-specific encoders. Chen *et al.* [3] proposed a three stage network addressing veiling effect, snow-streak size and transparency. Furthermore, [4] applied dual-tree wavelet transform based hierarchical decomposition paradigm for representing complex snow shape along with a complex wavelet loss. *The well known rain, snow and haze formation models are provided in the supplementary material.*

Transformers for Computer Vision: Due to the proven superiority of the Transformers over CNNs in capturing long-range dependencies due to self-attention, they have been vastly used for several applications. Initial work for computer vision using transformers was done in the form of Vision Transformers (ViT) [10] for visual recognition. It uses flattened patches of images for training the Transformer network. The transformers are also utilized for low-level vision applications. Using the image processing transformer, [2] illustrated how pre-training on large datasets can lead to improved performance for low-level applications. Uformer [36] utilized a U-Net like structure using transformers for image restoration problems *i.e.*, deraining, deblurring and denoising. Recently, Restoration Transformer (Restormer) [40] was introduced for image restoration, which makes the vanilla transformer efficient by making changes in the basic building blocks of transformer.

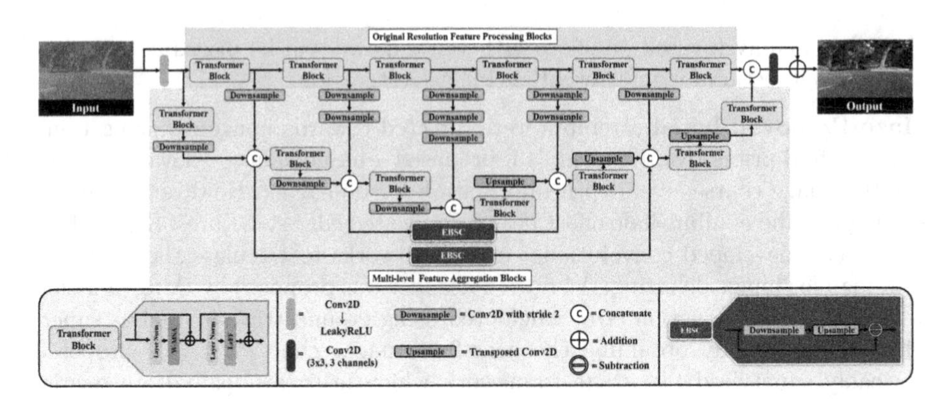

Fig. 1. Overview of the proposed network for multi-weather image restoration.

3 Proposed Method

Network Architecture: Initial level features are extracted through convolution layer and these features are then passed to original resolution transformer stream (ORTS) and multi-level feature aggregation stream (MFAS) to learn multiple scales and densities of weather degradations along with preserving finer details in the input image. In the final stage, features from both the streams are fused and passed through output convolution layer to get the residual output of dimensions $H \times W \times 3$, which is added with the input image to get the restored image. The overview of the proposed transformer network for multi-weather image restoration is provided in Fig. 1.

3.1 Original Resolution Transformer Stream

Image restoration is a challenging task and recovery of fine level details is an essential step. Most of the finer edge details are present in the original input image. Following this, for fine level feature extraction, we make use of the original scale of inputs in the proposed original resolution transformer stream (ORTS). The input is passed through a series of transformer blocks to extract high level as well as low level features. Due to its efficiency and lesser computational complexity than vanilla transformers, we have used transformer block from [36]. But unlike [36,40] which only use encoder-decoder like network and perform transformer operations on different scales for restoration, we have utilized capability of transformer block to capture contextual information even on original resolution of the input features. Operation of transformer block is explained ahead.

Transformer Block: As image restoration is an essential pre-processing step for various surveillance applications, it is expected to be time and computationally efficient. To address this, inspired by [36], we have used a window transformer block, which utilizes both self attention and convolution operator in transformer to capture long range dependencies and local context respectively. The main transformer block consists of two core designs:

Table 1. Quantitative results comparison of the proposed method with the existing state-of-the-art single image rain removal methods on *Outdoor-Rain* dataset. The sequence defines the order of testing. We use image dehazing method [9] for dehazing the images. Red and Blue values denote best and second best performance respectively.

Methods	Publication	Sequence	PSNR	SSIM
GCANet [1]	WACV-19	Derain + Dehaze	16.89	0.69
PReNet [30]	CVPR-19	Derain + Dehaze	16.77	0.67
MPRNet [41]	CVPR-21	Dehaze + Derain	17.87	0.74
Restormer [40]	CVPR-22	Dehaze + Derain	18.37	0.77
HRR [21]	CVPR-19	-	21.56	0.84
NAS [22]	CVPR-20	-	24.71	0.89
TransWeather [34]	CVPR-22	-	27.96	0.95
KDMW [5]	CVPR-22	-	26.88	0.93
Ours	-	-	27.49	0.94

(i) Window-based Multi-head Self-Attention (WMSA): Self-attention within non-overlapping local windows of dimensions $M \times M$ is performed to reduce the computational cost.
(ii) Feed-Forward Network (FF): Depthwise convolutional block is added to the feed forward network in the transformer block to overcome the issue of lesser local context. Operations of the transformer block can be represented as:

$$F_l' = WMSA(LN(F_{l-1})) + F_{l-1} \ ; \ F_l = FF(LN(F_l')) + F_l' \tag{1}$$

where F_l' and F_l are the outputs of $WMSA$ and FF blocks respectively. LN is layer normalization. The heads of queries($\mathbf{Q}$), keys ($\mathbf{K}$) and values ($\mathbf{V}$) have same dimensions and attention is calculated as:

$$Attention(\mathbf{Q}, \mathbf{K}, \mathbf{V}) = softmax(\mathbf{Q}\mathbf{K}^T)\mathbf{V} \tag{2}$$

The effectiveness of the ORTS is explained in the ablation study section. *More details about ORTS are given in supplementary material.*

3.2 Multi-level Feature Aggregation Stream

Rain streaks and snow degradations in an image have various sizes. Thus, learning different sizes of such degradations is essential for effective restoration of weather degraded images. Further, different scales of input features represent different densities of haze, *i.e.,* smaller scales represent densely accumulated haze and larger scales represent evenly distributed haze. To this context, we propose a multi-level feature aggregation stream (MFAS) for learning different scales and densities of the degradations present in the image. As the capturing window size is kept constant across different scales of the input features, the stream is able to learn different sizes and densities of the degradations. The

Table 2. Quantitative results comparison of the proposed method with existing state-of-the-art methods on *SRRS* dataset for snow with veil removal.

Methods	Eigen *et al.*[11]	DeSnowNet[26]	JSTASR[3]	NAS[22]	ASR[4]	Ours
Publication	ICCV-13	TIP-18	ECCV-20	CVPR-20	ICCV-21	-
PSNR	17.36	20.38	25.82	24.98	27.78	29.7
SSIM	0.66	0.84	0.89	0.88	0.92	0.95

Table 3. Quantitative results comparison of the proposed method with existing state-of-the-art methods on *SOTS* dataset for haze removal.

Methods	Publication	PSNR	SSIM
DCP [13]	TPAMI-10	17.54	0.848
GCANet [1]	WACV-19	26.20	0.930
RefineDNet [45]	TIP-21	28.82	0.953
MSBDN [9]	CVPR-20	33.79	0.984
USID [19]	TMM-22	23.89	0.919
TSDNet [25]	TII-22	24.24	0.959
MADN [15]	TII-21	28.13	0.957
PSD [7]	CVPR-21	26.33	0.942
TransWeather [34]	CVPR-22	33.56	0.978
KDMWR [5]	CVPR-22	33.95	0.980
Ours	-	33.87	0.986

ORTS makes use of original scales of inputs. Though original scale of input is essential for extracting fine level details, it does not have access to multiple scales of the weather degradations. So, for further utilizing intermediate outputs of the ORTS, they are fused in this stream for further processing and accumulating additional information.

The rain and snow degradations tend to overlap on the useful edge features, and haze degradation fades the edge details in an image. Therefore, in order to enhance the edge features, instead of using direct skip connections between encoder and decoder features like [36], we have used Edge Boosting Skip Connections (EBSC). Both, encoder features and coarse outputs from ORTS facilitate EBSC. The output of the EBSC block can be equated as:

$$EBSC_{out} = EBSC_{in} - \uparrow (\downarrow (EBSC_{in})) \tag{3}$$

Here, $\uparrow$ represents upsampling with transposed convolution and $\downarrow$ represents downsampling with strided convolution. With such adaptation, essential edge features are propagated through the skip connection which helps in effective multi-weather image restoration. The features from the MFAS are then fused with the ORTS features at last and passed through output convolution layer for obtaining residual image, which is added with the input image to get the restored output. The effectiveness of MFAS and EBSC is explained in the ablation study.

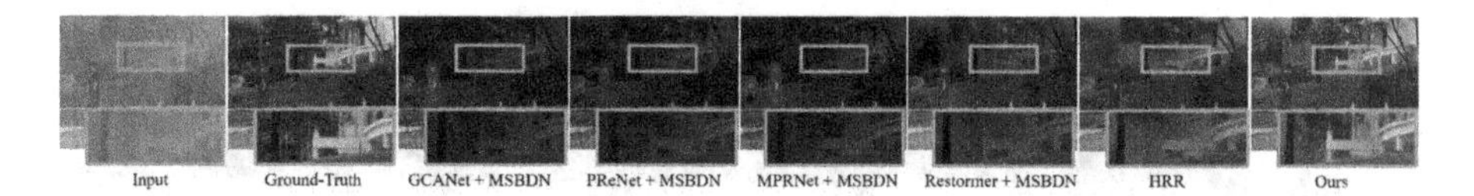

Fig. 2. Qualitative results comparison of the proposed method with GCANet [1], PReNet [30], MPRNet [41], Restormer [40], HRR [21] on Outdoor-Rain dataset for rain with veil removal. MSBDN[9] is used for dehazing the images of existing deraining methods [1,30,40,41].

Fig. 3. Qualitative results comparison of the proposed method with Eigen *et al.*[11], JSTASR [3] and ASR [4] on SRRS dataset for snow with veil removal.

3.3 Training Strategy

The learning strategy of any unified multi-weather image restoration algorithm should be robust enough to restore multiple types of weather degraded images with a single set of trained parameters. The learning can be approached as:
(i) Sequential Learning: Multiple tasks are learned in a sequential manner *i.e.*, the model is trained on the first task and then the same model is used for fine-tuning on further tasks. But, this training strategy imposes risk of catastrophic forgetting *i.e.*, being able to perform better on the task in hand but has reduced performance on the previously learned tasks.
(ii) Combined Training: A solution to the problem of catastrophic forgetting [42] can be approached by combining the datasets of all the tasks and training on a unified dataset. But, dataset for each task has different physical properties and learning of all the tasks in combined manner is ineffective as the trained model does not reach peak performance for each task.
(iii) Memory Replay Training: Compared to other training strategies, memory replay training is an effective training strategy solution as it learns different weather conditions without the problem of catastrophic forgetting (faced in sequential training) and ineffective generalization (faced in combined training). To avoid the limitations imposed by above two training strategies, we have trained the proposed network with memory replay training strategy. In this method, the model is trained on the first task T_1. The same trained model is used for training on the second task T_2, and data samples from the dataset of T_1 are also used along with dataset for T_2. In this work, we have trained our network for 3 tasks, *i.e.*, rain + veil removal, snow + veil removal and haze removal. With such training, the model learns all types of weather restoration and gives peak performance for each task. Further analysis on memory replay training is provided in the ablation study section.

Fig. 4. Qualitative results comparison of the proposed method with DCP [13], GCANet [1], RefineDNet [45], USID [19], MADN[15], TSDNet [25], MSBDN [9] and PSD [7] on SOTS dataset for haze removal.

Fig. 5. Qualitative results comparison of the proposed method with PReNet [30], GCANet [1], MPRNet [41], Restormer [40], HRR [21] on real-world rainy image.

3.4 Loss Functions

Loss functions play a vital role while training the network. $\mathbb{L}_1$ loss has been utilized for achieving lesser per pixel difference. Edge loss ($\mathbb{L}_{Edge}$) has been optimized in order to preserve the edge details while training. Y-channel loss ($\mathbb{L}_Y$) is also optimized for additional monitoring while training the network because the Y channel has most of the rain and snow streak information. Perceptual loss ($\mathbb{L}_P$) is used for achieving perceptual similarity between ground truth and generated (restored) image. The total loss ($\mathbb{L}_{Total}$) is calculated as weighted sum of all of the above mentioned losses with appropriate weightage (λ_{loss}) given to each loss as :

$$\mathbb{L}_{Total} = \lambda_1 \mathbb{L}_1 + \lambda_{Edge} \mathbb{L}_{Edge} + \lambda_Y \mathbb{L}_Y + \lambda_P \mathbb{L}_P \quad (4)$$

We assign $\lambda_1 = 5$, $\lambda_{Edge} = 1$, $\lambda_Y = 1$ and $\lambda_P = 10$ which are obtained and verified experimentally. *Detailed equations are provided in the supplementary material.*

4 Experimental Analysis

4.1 Datasets

The datasets used for training and testing the proposed method are:

- **Outdoor-Rain Dataset** [21]: This dataset contains images which are degraded by rain with veiling effect. We have used 9000 training and 1500 testing images.

Fig. 6. Qualitative results comparison of the proposed method with DeSnowNet [26], NAS [22], JSTASR [3] and ASR [4] on real world snowy image. As evaluation codes of the methods DeSnowNet [26] and NAS [22] are not publically available, we use the results provided in ASR [4] (*Please zoom in for clearer view*).

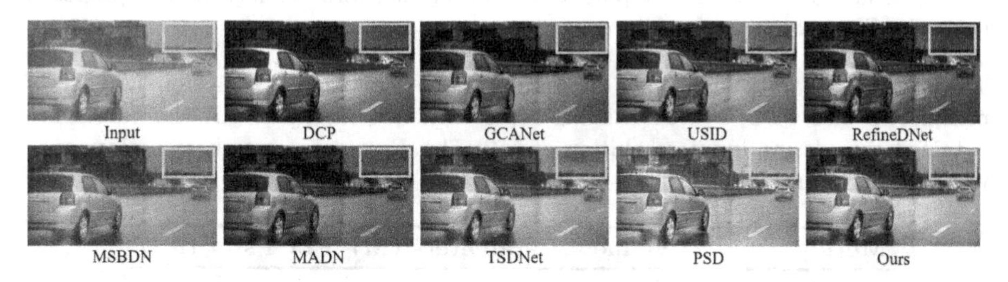

Fig. 7. Qualitative results comparison of the proposed method with DCP [13], GCANet [1], USID [19], RefineDNet [45], MSBDN[9], MADN [15], TSDNet [25] and PSD [7] on real world hazy image.

- **SRRS Dataset** [3]: This dataset contains images which are degraded by snow with veiling effect. We have used 9000 training and 2000 testing images.
- **SOTS Dataset** [18]: This dataset contains images which are haze degraded with different levels of haze degradations. We have used 9000 training and 500 testing images containing outdoor-haze scenarios.

4.2 Training Details

Input images are first cropped into patches of 256×256. While training, the ADAM optimizer is used and initial learning rate is kept at 2×10^{-4} which is varied with cosine annealing strategy. The proposed network is implemented using Pytorch library, and trained on NVIDIA-DGX station with 2.2 GHz processor, Intel Xeon E5-2698, NVIDIA Tesla V100 16 GB GPU.

4.3 Quantitative and Qualitative Analysis

Quantitative Evaluation: Analysis in comparison with the existing methods on Outdoor-Rain dataset consisting of rain with veiling effect is given in Table 1. As most of the existing methods for rain removal have task specific application, for fair comparison on Outdoor-Rain dataset, we test the existing rain streak removal methods along with image dehazing method [9] with different orders of execution *i.e.*, deraining first then dehazing and vice a versa. Comparison with the existing image de-snowing methods on SRRS dataset, which consist

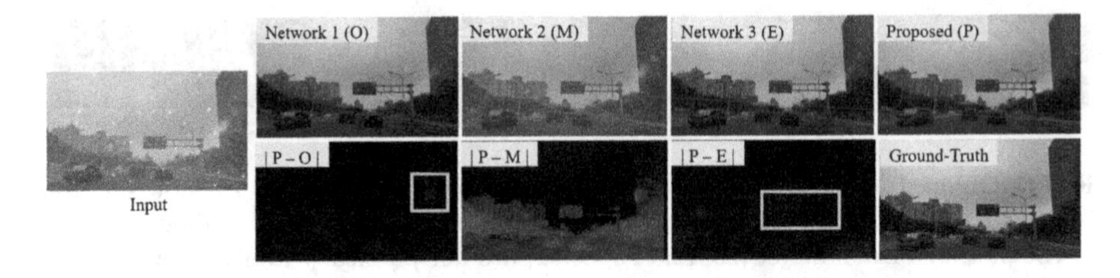

Fig. 8. Qualitative results comparison of various network settings of the proposed network (*Please zoom in for clearer view).*

Table 4. Comparative analysis on computational complexity based on number of parameters and FLOPs with existing state-of-the-art methods. All the values are obtained considering input resolution of 512×512. Red represents best performance value. (Color figure online)

Method	Venue	Parameters (M)	FLOPs ($\times 10^{11}$)
GCANet [1]	WACV-19	0.70	1.48
PReNet [30]	CVPR-19	0.16	3.55
MPRNet [41]	CVPR-21	3.67	5.64
HRR [21]	CVPR-19	40.63	15.87
MSBDN [9]	CVPR-20	31.35	3.32
RefineDNet [45]	TIP-21	65.78	6.03
Restormer [40]	CVPR-22	26.12	5.34
Ours	-	4.5	0.99

of snow with veiling effect, is given in Table 2. Comparison with the existing image de-hazing methods on SOTS dataset is given in Table 3. Peak signal-to-noise ratio (PSNR), which is evaluated on luminance channel, and structural similarity index measure (SSIM) are used as evaluation metrics. It is observed that the proposed method outperforms the existing methods on rain + veil, snow + veil and haze removal. It is worth to note that the proposed method does not use extra priors for training of the network *i.e.,* transmission maps [21], ground-truth rain maps [21], *etc.* Also, it is worth to note that the proposed method has only **4.5M** trainable parameters, which are significantly less than existing methods. We compare the performance in terms of number of parameters and number of FLOPs (Floating Point Operations) in Table 4.

Qualitative Evaluation: Qualitative comparison of the proposed method with existing state-of-the-art methods on Outdoor-Rain dataset is given in Fig. 2, on SRRS dataset is given in Fig. 3 and SOTS dataset is given in Fig. 4. In the research of weather degraded image restoration, we know that there is significant gap between synthetic and real data. Thus, we have evaluated the proposed method on real-world rainy, snowy and hazy images, and compared with the

Table 5. Quantitative evaluation on various settings of the proposed method on SRRS (✓ and × denote inclusion and exclusion of a module respectively).

Network Setting	ORTS	MFAS	EBSC	PSNR	SSIM
Network 1	×	✓	✓	27.19	0.90
Network 2	✓	×	×	25.76	0.87
Network 3	✓	✓	×	28.04	0.93
Ours	✓	✓	✓	**29.71**	**0.95**

Table 6. Quantitative results of various training configurations (RV = Rain + Veil, SV = Snow + Veil, H = Haze)

Datasets→		**Outdoor-Rain**	**SRRS**	**SOTS**
Training Methods↓		**PSNR / SSIM**	**PSNR / SSIM**	**PSNR / SSIM**
Sequential	RV→	27.63 / 0.95	18.32 / 0.83	17.90 / 0.86
	RV→SV	24.91 / 0.91	28.87 / 0.92	25.44 / 0.93
	RV→SV→H	22.95 / 0.87	26.79 / 0.90	30.12 / 0.95
Combined (RV + SV + H)		25.83 / 0.92	28.97 / 0.93	30.89 / 0.96
Memory Replay		**27.49 / 0.94**	**29.71 / 0.95**	**31.41 / 0.98**

state-of-the-art methods in Fig. 5, Fig. 6 and Fig. 7 respectively. It can be seen from the qualitative results, our method is able to completely remove the rain streaks, snow streaks and hazy effect from the degraded images and retain the perceptual quality when compared to existing state-of-the-art methods. *More qualitative results are provided in the supplementary material.*

4.4 Ablation Study

In this section, we analyse the contribution of each module in the proposed network. *SRRS* dataset for snow with veil removal is used to evaluate the contribution of individual element discussed ahead.

Contribution of ORTS: We evaluate the performance of the proposed network with and without ORTS. As seen from the output of Network 1 (O) in Fig. 8, and the absolute difference (|P - O|) between the output of proposed method (P), it is verified that the proposed ORTS is effective in extracting finer details from the input features. Also, refer the PSNR values of Network 1 and the proposed network from Table 5 for quantitative verification.

Contribution of MFAS: We carry out analysis with and without MFAS for scrutinizing its effectiveness. As observed from the output of Network 2 (M) in Fig. 8, and the absolute difference (|P - M|) between the output of proposed method (P), it is verified that the proposed MFAS is effective in learning features with various sizes and densities of rain, snow and hazy degradations streaks

Table 7. Ablation study on effect of different loss functions on training of the proposed network. Values are calculated with *SRRS* dataset.

Loss Function	PSNR	SSIM
$\mathbb{L}_1$	25.36	0.88
$\mathbb{L}_1 + \mathbb{L}_{Edge}$	26.89	0.90
$\mathbb{L}_1 + \mathbb{L}_{Edge} + \mathbb{L}_Y$	27.53	0.92
Proposed Method: $\mathbb{L}_1 + \mathbb{L}_{Edge} + \mathbb{L}_Y + \mathbb{L}_P$	29.71	0.95

Table 8. Ablation study on effect of different number of modules associated with each stream. Values in gray cells represent the number of modules in the proposed network. PSNR and SSIM are calculated using *SRRS* dataset. (Color figure online)

Settings	Number of Modules	PSNR	SSIM
Transformer Blocks in ORTS	4	27.31	0.91
	5	28.89	0.93
	6	**29.71**	**0.95**
	7	29.65	0.94
Downsampling Levels in MFAS	2	28.73	0.93
	3	**29.71**	**0.95**
	4	29.01	0.94

which facilitates complete restoration of the images. Quantitative verification is also done in Table 5.

Contribution of EBSC: Upon visual analysis of output of Network 3 (E) from Fig. 8, and the absolute difference (|P - E|) between the output of proposed method (P), it is verified that the proposed EBSC is effective in extracting and propagating crucial edge information towards the decoder side. Further, we have scrutinized the effectiveness of EBSC quantitatively in Table 5.

Contribution of Memory Replay Training: We train the proposed network with different training settings as conveyed in Table 6. As seen from the results, the memory replay training has been proven most effective for training the proposed method and yields best results out of all the training strategies.

Effect of Different Loss Functions: We have utilized different loss functions while training the proposed network. The effect of different loss functions is analysed in Table 7. As seen from the results, the combination of different loss functions performs in a better manner.

Effect of Different Number of Modules: We have analysed the effect of different number of modules present in each stream and conveyed the results through Table 8. As seen from the results, further increasing the number of

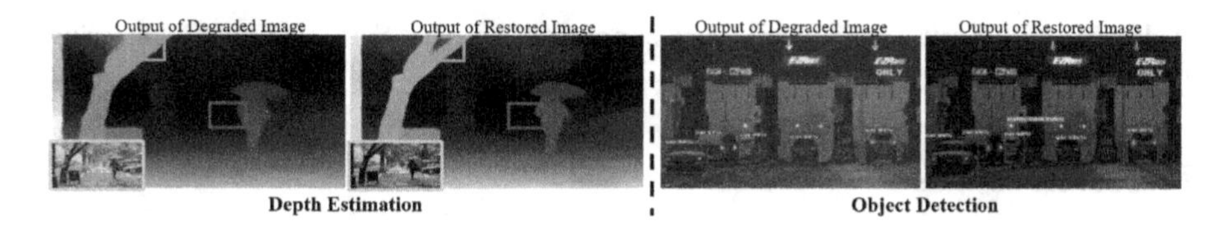

Fig. 9. Application of the proposed method on depth estimation and object detection.

transformer blocks in ORTS improves the results with a small margin quantitatively, but no noticeable difference is observed in qualitative results. This also increases the computational complexity of the network. Further increase in the downsampler levels in Multi-level Feature Aggregation Stream (MFAS) decreases the performance of the network, as reconstruction of low-resolution data becomes a challenging task after certain number of downsamples. *More analysis on ablation study is given in the supplementary material.*

5 Real World Application

We analyse our method for real-world computer vision applications such as depth estimation and object detection. For this, the weather degraded image is passed through state-of-the-art depth estimation method [28] and object detection method [29]. In order to see the effect of restoration, we restore the degraded image through our method and pass the output through the depth estimation and object detection networks. As seen from Fig. 9, predictions of the restored images are semantically more accurate than in the weather degraded images. This validates the applicability of the proposed method on various computer vision applications in real-world weather degraded scenarios.

6 Conclusion

In this paper, we proposed a unified transformer network for multiple weather degraded image restoration. In that, we propose a parallel two stream network, where original resolution transformer stream focuses on extracting finer details from original resolution of the input features. Multi-level feature aggregation stream learns different sizes and densities of the weather degradations through various scales of the inputs. Further, it also makes use of coarse outputs from the previous stream and utilizes edge boosting skip connections for propagating crucial edge details. Also, we experimented different training settings for multi-weather image restoration and found memory replay training to be the best working. We discussed the applicability of the proposed method for real world computer vision applications. Substantial experiments and ablation study on a number of synthetic datasets and real world weather degraded images verify the superior performance of the proposed method over the state-of-the-art methods for multi-weather image restoration.

Acknowledgement. This work was supported by the Science and Engineering Research Board (DST-SERB), India, under Grant ECR/2018/001538.

References

1. Chen, D., et al.: Gated context aggregation network for image dehazing and deraining. In: 2019 IEEE Winter Conference On Applications Of Computer Vision (WACV), pp. 1375–1383. IEEE (2019)
2. Chen, H., et al.: Pre-trained image processing transformer. In: Proceedings of the IEEE/CVF Conference on Computer Vision and Pattern Recognition, pp. 12299–12310 (2021)
3. Chen, W.-T., Fang, H.-Y., Ding, J.-J., Tsai, C.-C., Kuo, S.-Y.: JSTASR: joint size and transparency-aware snow removal algorithm based on modified partial convolution and veiling effect removal. In: Vedaldi, A., Bischof, H., Brox, T., Frahm, J.-M. (eds.) ECCV 2020. LNCS, vol. 12366, pp. 754–770. Springer, Cham (2020). https://doi.org/10.1007/978-3-030-58589-1_45
4. Chen, W.T., et al.: All snow removed: Single image desnowing algorithm using hierarchical dual-tree complex wavelet representation and contradict channel loss. In: Proceedings of the IEEE/CVF International Conference on Computer Vision, pp. 4196–4205 (2021)
5. Chen, W.T., Huang, Z.K., Tsai, C.C., Yang, H.H., Ding, J.J., Kuo, S.Y.: Learning multiple adverse weather removal via two-stage knowledge learning and multi-contrastive regularization: Toward a unified model. In: Proceedings of the IEEE/CVF Conference on Computer Vision and Pattern Recognition, pp. 17653–17662 (2022)
6. Chen, Y., Bai, W., Huang, Q., Xiao, J.: Efficient motion symbol detection and multi-kernel learning for aer object recognition. IEEE Transactions on Cognitive and Developmental Systems (2021). https://doi.org/10.1109/TCDS.2021.3122131
7. Chen, Z., Wang, Y., Yang, Y., Liu, D.: Psd: Principled synthetic-to-real dehazing guided by physical priors. In: Proceedings of the IEEE/CVF Conference on Computer Vision and Pattern Recognition, pp. 7180–7189 (2021)
8. Deng, S., et al.: Detail-recovery image deraining via context aggregation networks. In: Proceedings of the IEEE/CVF Conference On Computer Vision And Pattern Recognition, pp. 14560–14569 (2020)
9. Dong, H., et al.: Multi-scale boosted dehazing network with dense feature fusion. In: Proceedings of the IEEE/CVF Conference On Computer Vision And Pattern Recognition, pp. 2157–2167 (2020)
10. Dosovitskiy, A., et al.: An image is worth 16×16 words: Transformers for image recognition at scale. arXiv preprint arXiv:2010.11929 (2020)
11. Eigen, D., Krishnan, D., Fergus, R.: Restoring an image taken through a window covered with dirt or rain. In: Proceedings of the IEEE International Conference On Computer Vision, pp. 633–640 (2013)
12. Godard, C., Mac Aodha, O., Firman, M., Brostow, G.J.: Digging into self-supervised monocular depth estimation. In: Proceedings of the IEEE/CVF International Conference on Computer Vision, pp. 3828–3838 (2019)
13. He, K., Sun, J., Tang, X.: Single image haze removal using dark channel prior. IEEE Trans. Pattern Anal. Mach. Intell. **33**(12), 2341–2353 (2010)
14. Itti, L., Koch, C., Niebur, E.: A model of saliency-based visual attention for rapid scene analysis. IEEE Trans. Pattern Anal. Mach. Intell. **20**(11), 1254–1259 (1998)

15. Jia, T., Li, J., Zhuo, L., Li, G.: Effective meta-attention dehazing networks for vision-based outdoor industrial systems. IEEE Trans. Industr. Inf. **18**(3), 1511–1520 (2022). https://doi.org/10.1109/TII.2021.3059020
16. Jiang, K., et al.: Multi-scale progressive fusion network for single image deraining. In: Proceedings of the IEEE/CVF Conference On Computer Vision And Pattern Recognition, pp. 8346–8355 (2020)
17. Kang, L.W., Lin, C.W., Fu, Y.H.: Automatic single-image-based rain streaks removal via image decomposition. IEEE Trans. Image Process. **21**(4), 1742–1755 (2011)
18. Li, B., et al.: Benchmarking single-image dehazing and beyond. IEEE Trans. Image Process. **28**(1), 492–505 (2019). https://doi.org/10.1109/TIP.2018.2867951
19. Li, J., Li, Y., Zhuo, L., Kuang, L., Yu, T.: Usid-net: Unsupervised single image dehazing network via disentangled representations. IEEE Trans. Multimedia (2022). https://doi.org/10.1109/TMM.2022.3163554
20. Li, M., Cao, X., Zhao, Q., Zhang, L., Gao, C., Meng, D.: Video rain/snow removal by transformed online multiscale convolutional sparse coding. arXiv preprint arXiv:1909.06148 (2019)
21. Li, R., Cheong, L.F., Tan, R.T.: Heavy rain image restoration: Integrating physics model and conditional adversarial learning. In: Proceedings of the IEEE/CVF Conference on Computer Vision and Pattern Recognition, pp. 1633–1642 (2019)
22. Li, R., Tan, R.T., Cheong, L.F.: All in one bad weather removal using architectural search. In: Proceedings of the IEEE/CVF Conference on Computer Vision and Pattern Recognition, pp. 3175–3185 (2020)
23. Li, X., Wu, J., Lin, Z., Liu, H., Zha, H.: Recurrent squeeze-and-excitation context aggregation net for single image deraining. In: Ferrari, V., Hebert, M., Sminchisescu, C., Weiss, Y. (eds.) ECCV 2018. LNCS, vol. 11211, pp. 262–277. Springer, Cham (2018). https://doi.org/10.1007/978-3-030-01234-2_16
24. Li, Y., Tan, R.T., Guo, X., Lu, J., Brown, M.S.: Rain streak removal using layer priors. In: Proceedings of the IEEE Conference On Computer Vision And Pattern Recognition. pp. 2736–2744 (2016)
25. Liu, R.W., Guo, Y., Lu, Y., Chui, K.T., Gupta, B.B.: Deep network-enabled haze visibility enhancement for visual iot-driven intelligent transportation systems. IEEE Trans. Industrial Inform. (2022). https://doi.org/10.1109/TII.2022.3170594
26. Liu, Y.F., Jaw, D.W., Huang, S.C., Hwang, J.N.: Desnownet: Context-aware deep network for snow removal. IEEE Trans. Image Process. **27**(6), 3064–3073 (2018). https://doi.org/10.1109/TIP.2018.2806202
27. Luo, Y., Xu, Y., Ji, H.: Removing rain from a single image via discriminative sparse coding. In: Proceedings of the IEEE International Conference on Computer Vision, pp. 3397–3405 (2015)
28. Ranftl, R., Bochkovskiy, A., Koltun, V.: Vision transformers for dense prediction. In: Proceedings of the IEEE/CVF International Conference on Computer Vision, pp. 12179–12188 (2021)
29. Redmon, J., Farhadi, A.: Yolov3: An incremental improvement. arXiv preprint arXiv:1804.02767 (2018)
30. Ren, D., Zuo, W., Hu, Q., Zhu, P., Meng, D.: Progressive image deraining networks: A better and simpler baseline. In: Proceedings of the IEEE/CVF Conference on Computer Vision and Pattern Recognition, pp. 3937–3946 (2019)
31. Salazar-Colores, S., Cabal-Yepez, E., Ramos-Arreguin, J.M., Botella, G., Ledesma-Carrillo, L.M., Ledesma, S.: A fast image dehazing algorithm using morphological reconstruction. IEEE Trans. Image Process. **28**(5), 2357–2366 (2018)

32. Shin, J., Park, H., Paik, J.: Region-based dehazing via dual-supervised triple-convolutional network. IEEE Trans. Multimedia **24**, 245–260 (2022). https://doi.org/10.1109/TMM.2021.3050053
33. Su, Z., Zhang, Y., Shi, J., Zhang, X.P.: Recurrent network knowledge distillation for image rain removal. IEEE Transactions on Cognitive and Developmental Systems (2021). https://doi.org/10.1109/TCDS.2021.3131045
34. Valanarasu, J.M.J., Yasarla, R., Patel, V.M.: Transweather: Transformer-based restoration of images degraded by adverse weather conditions. In: Proceedings of the IEEE/CVF Conference on Computer Vision and Pattern Recognition, pp. 2353–2363 (2022)
35. Wang, Y., Liu, S., Chen, C., Zeng, B.: A hierarchical approach for rain or snow removing in a single color image. IEEE Trans. Image Process. **26**(8), 3936–3950 (2017)
36. Wang, Z., Cun, X., Bao, J., Zhou, W., Liu, J., Li, H.: Uformer: A general u-shaped transformer for image restoration. In: Proceedings of the IEEE/CVF Conference on Computer Vision and Pattern Recognition (CVPR), pp. 17683–17693 (June 2022)
37. Xu, J., Zhao, W., Liu, P., Tang, X.: An improved guidance image based method to remove rain and snow in a single image. Comput. Inform. Sci. **5**(3), 49 (2012)
38. Yi, D., Fang, H., Hua, Y., Su, J., Quddus, M., Han, J.: Improving synthetic to realistic semantic segmentation with parallel generative ensembles for autonomous urban driving. IEEE Trans. Cognitive Developm. Syst. (2021). https://doi.org/10.1109/TCDS.2021.3117925
39. Yu, S., et al.: Content-adaptive rain and snow removal algorithms for single image. In: International Symposium on Neural Networks. pp. 439–448. Springer (2014). https://doi.org/10.1007/978-3-319-12436-0_49
40. Zamir, S.W., Arora, A., Khan, S., Hayat, M., Khan, F.S., Yang, M.H.: Restormer: Efficient transformer for high-resolution image restoration. In: Proceedings of the IEEE/CVF Conference on Computer Vision and Pattern Recognition (CVPR), pp. 5728–5739 (June 2022)
41. Zamir, S.W., et al.: Multi-stage progressive image restoration. In: Proceedings of the IEEE/CVF Conference on Computer Vision and Pattern Recognition, pp. 14821–14831 (2021)
42. Zhai, M., Chen, L., Tung, F., He, J., Nawhal, M., Mori, G.: Lifelong gan: Continual learning for conditional image generation. In: Proceedings of the IEEE/CVF International Conference on Computer Vision, pp. 2759–2768 (2019)
43. Zhang, H., Patel, V.M.: Density-aware single image de-raining using a multi-stream dense network. In: Proceedings of the IEEE Conference On Computer Vision And Pattern Recognition, pp. 695–704 (2018)
44. Zhang, J., Tao, D.: Famed-net: A fast and accurate multi-scale end-to-end dehazing network. IEEE Trans. Image Process. **29**, 72–84 (2019)
45. Zhao, S., Zhang, L., Shen, Y., Zhou, Y.: Refinednet: A weakly supervised refinement framework for single image dehazing. IEEE Trans. Image Process. **30**, 3391–3404 (2021)
46. Zheng, X., Liao, Y., Guo, W., Fu, X., Ding, X.: Single-image-based rain and snow removal using multi-guided filter. In: Lee, M., Hirose, A., Hou, Z.-G., Kil, R.M. (eds.) ICONIP 2013. LNCS, vol. 8228, pp. 258–265. Springer, Heidelberg (2013). https://doi.org/10.1007/978-3-642-42051-1_33

DSR: Towards Drone Image Super-Resolution

Xiaoyu Lin([✉]), Baran Ozaydin, Vidit Vidit, Majed El Helou, and Sabine Süsstrunk

School of Computer and Communication Sciences, EPFL, Lausanne, Switzerland
{xiaoyu.lin,baran.ozaydin,vidit.vidit, majed.elhelou,sabine.susstrun}@epfl.ch

Abstract. Despite achieving remarkable progress in recent years, single-image super-resolution methods are developed with several limitations. Specifically, they are trained on fixed content domains with certain degradations (whether synthetic or real). The priors they learn are prone to overfitting the training configuration. Therefore, the generalization to novel domains such as drone top view data, and across altitudes, is currently unknown. Nonetheless, pairing drones with proper image super-resolution is of great value. It would enable drones to fly higher covering larger fields of view, while maintaining a high image quality.

To answer these questions and pave the way towards drone image super-resolution, we explore this application with particular focus on the single-image case. We propose a novel drone image dataset, with scenes captured at low and high resolutions, and across a span of altitudes. Our results show that off-the-shelf state-of-the-art networks witness a significant drop in performance on this different domain. We additionally show that simple fine-tuning, and incorporating altitude awareness into the network's architecture, both improve the reconstruction performance (Our code and data are available at https://github.com/IVRL/DSR).

Keywords: Image super-resolution · Drone imaging · Super-resolution dataset

1 Introduction

The task of single-image super-resolution (SR) is the reconstruction of a high-resolution (HR) image from a single low-resolution (LR) capture. Super-resolution is widely studied in the literature and has witnessed significant progress with deep learning approaches. Multiple HR images can correspond to a single LR observation, making the SR task an ill-posed problem. In other words, SR methods are bound to learn an image distribution prior to reconstruct an *expected* HR image. With the power of deep networks to implicitly learn prior distributions, SR performance improved and it opened up new possibilities to push the limits of imaging systems.

Supplementary Information The online version contains supplementary material available at https://doi.org/10.1007/978-3-031-25063-7_22.

L. Karlinsky et al. (Eds.): ECCV 2022 Workshops, LNCS 13802, pp. 361–377, 2023.
https://doi.org/10.1007/978-3-031-25063-7_22

Fig. 1. Sample images from our DSR dataset. The drone takes off at ground level, moves to a predetermined set of altitudes, and captures images at each of these altitudes.

Although deep-learning-based SR methods pushed forward the state-of-the-art performance, multiple constraints were incorporated along the way. The models are trained with certain assumptions on the degradation model that are learned during training. Furthermore, the underlying datasets are human-captured and thus suffer from a content bias. Namely, these datasets are typically captured along the horizontal direction and at a human-level height. Thus, the methods trained on this domain do not generalize well to different data such as top view imaging.

Top view images constitute, however, most of what a drone would observe in the majority of drone applications. Drones can significantly benefit from SR solutions as they would enable the drone to fly at higher altitudes, survey wider areas, and preserve good image quality. However, because of the necessary reliance on data priors, it is unclear whether available SR solutions would perform well on drone images, and whether a domain gap exists across altitudes.

In an effort towards enabling drone image super-resolution, we create the **Drone Super-Resolution**, **DSR**, dataset. It is the first dataset of drone images containing low-resolution and high-resolution drone-based image pairs, with varying altitudes. We obtain our high-resolution targets with optical zoom, and we collect 10 versions of each scene for 10 different chosen altitudes, Fig. 1. With our DSR benchmark, we evaluate the performance of available SR methods. We observe lower performance on our drone data, and a difference in performance between altitudes. We show in the frequency domain that a domain gap also exists for different altitudes of drone capture. We finally propose a solution that exploits altitude meta data. We show that this altitude-aware approach, similarly to fine-tuning, outperforms previous SR methods, even degradation-adaptive ones. Our contributions can be summarized as follows:

- We introduce a novel dataset, DSR, for drone image super-resolution. Our dataset provides pairs of low-resolution and high-resolution images. Furthermore, we capture each scene at 10 different altitudes. We also provide a burst of 7 low-resolution images and RAW data to encourage future research.
- We analyze domain variability, its effect on regular data vs. top view data, and also across altitudes for the drone data. We benchmark nine off-the-shelf super-resolution methods on DSR.
- We show that an altitude-aware architecture, as well as fine-tuning, can outperform previous methods.

2 Related Work

Drone Imaging. Various novel applications rely on drones and drone imaging. The research can be divided into two categories. The first category comprises the methods designed for the drones themselves (e.g. for localization or autonomous navigation), and the second are methods developed with drone data for downstream tasks (e.g. surveillance or traffic control).

Computer vision methods already play an essential role already within drones, for instance, camera calibration and image matching [39] for drones' automated flight and stabilization, including pose estimation and obstacle detection [2]. Many more industrial applications exist that exploit images and videos captured by drones. Some drone-based image or video datasets were developed for specific tasks such as traffic and crowd detection [30], notably for surveillance [25]. Other datasets were proposed for the object detection task [15,38]. The UAVDT [10] benchmark forms a database for object detection and tracking using drone images. UAVDT also contains some real-world challenges, such as varying weather conditions, flying altitudes, and drone views. These datasets are used to evaluate available approaches and develop novel methods, especially for learning-based solutions. However, the publicly available datasets focus on object detection and other related computer vision tasks. Here, we present DSR, which is a drone dataset dedicated to the super-resolution task. Rather than sampling video frames [10], we opt for the more time-consuming approach of capturing one image at a time in order to obtain the highest quality possible. DSR consists of paired low-resolution and high-resolution images, across varying altitudes, and includes burst sequences and RAW data.

Super-Resolution. SR methods are based on edge model assumptions [8], priors on gradient distributions [32], or learn from examples as with deep neural networks which hold the state-of-the-art reconstruction performance. These networks exploit deep architectures with residual learning and attention mechanisms [45], or train over the frequency domain using wavelet transforms for memory and compute optimization [48]. A variety of other solutions were developed to exploit perceptual losses [17] or GAN models [21,36] to improve the qualitative performance of the SR networks. However, the SR networks face generalization issues as highlighted in [31], notably in their learned priors [13].

Table 1. The final number of valid low-resolution and high-resolution image pairs in our DSR dataset for each drone capture altitude (from 10 m to 140 m with respect to the ground take-off level). The dataset is split into non-overlapping training, validation, and test sets. We note that image pairs are evenly distributed across the capture altitudes and do not create data distribution imbalance.

Data split	# Valid pairs at different altitudes (m)										Total pairs
	10	20	30	40	50	70	80	100	120	140	
Training	484	513	508	505	509	521	528	530	537	540	5175
Validation	256	262	258	259	262	266	266	265	261	267	2622
Test	228	239	235	248	248	259	270	270	270	265	2532

Despite attempts to account for degradation changes [42], the generalization still suffers because of the need to estimate blur kernels (an estimation which is, itself, prone to over-fitting). Recently, two datasets were built using optical zoom with varying focal lengths to obtain paired low-resolution and high-resolution image pairs [7,44]. These datasets all include similar content from a unique domain. This is due to the fact that their images were captured from a human-perspective, with generally uniform height and a camera axis along the horizontal direction. We study with DSR the application of SR to a different domain with a different image prior, for drone images captured from a top view and at varying altitudes. We study the performance of available methods on this novel domain, and propose solutions to improve their results. We hope DSR will foster future research towards developing drone image super-resolution.

3 DSR Dataset

We create our DSR dataset by capturing image pairs of the same scene with different focal length values. We take photos at ten different altitudes for each scene. Our altitude values form the set {10, 20, 30, 40, 50, 70, 80, 100, 120, 140} m, which is chosen in such a way as to have good sampling across altitudes, various altitudes that are multiples of each other, and to span a large range of altitudes that are relevant for drone applications [10]. The dataset is split into three sets for training, validation, and testing, while ensuring that no overlaps exist between any of the sets (Table 1). We choose a large size of validation and test sets relative to the training set (one to two ratio), to emphasize the quality of the empirical evaluation. We also show in Table 1 that the final number of low-resolution and high-resolution pairs is evenly distributed across altitudes and does not cause any data imbalance.

3.1 Data Acquisition

Localization. We use a DJI Mavic 3 drone for collecting the images of our DSR dataset. For controlling the altitude, the vertical hovering accuracy range of the drone when using GNSS (GPS+Galileo+BeiDou) is ±0.5 m. We exploit this information as an altitude guide and regularly refine our estimates with a measurement from the onboard barometer, which is measuring atmospheric pressure change relative to the take-off ground position and can provide improved accuracy. We additionally rely on GNSS for horizontal positioning to ensure no overlap between acquired images, based on the drone's geographic location.

Cameras. The drone comes pre-equipped with two cameras that are very closely mounted. The first is a 4/3-in CMOS sensor Hasselblad L2D-20c camera (referred to hereinafter as "Hasselblad camera"). The second is a 1/2-in CMOS sensor Tele camera (referred to hereinafter as "Tele camera"). The focal length of the Hasselblad camera is 24 mm, and that of the Tele camera is 162 mm. The Hasselblad camera captures images of 5280×3956 pixel resolution, and the Tele camera captures images of 4000×3000 pixel resolution [9]. The Tele camera has a larger focal length, meaning the images captured by the Tele camera form an optically zoomed version of the ones captured with the Hasselblad camera at the same position. Therefore, images from the Tele camera are used to generate the high-resolution ground-truth target data and matched images from the Hasselblad camera are used to generate their low-resolution counterpart.

Capture Settings. The drone's Tele camera has a fixed f-number, optimized and set by the manufacturer to 4.4. This aperture size yields a good trade-off between depth of field and exposure. A small aperture size is advantageous for super-resolution data acquisition [7,44], as it extends the depth of field and hence the range of depth values in the captured scene that are only minimally blurred. However, exposure is also critical for a drone because of the inherent instability in such applications. As the drone is exposed to both its own vibrations as well as to the wind, a large enough aperture size enables a reduction in exposure time and hence a reduction in motion blur.

Acquisition Details. We match the exposure value between the Tele camera and the Hasselblad camera in real-time during capture onboard the drone. The Hasselblad camera settings (f-number and shutter speed) are adjusted on the drone to match the exposure value of the Tele camera capture. We thus begin by shooting with the Tele camera, then the Hasselblad camera. With the latter, we shoot a burst of 7 consecutive images. We include this burst capture for future research, as multi-frame super-resolution has the potential to achieve better results than single image super-resolution [5,11,20] despite its practical inconvenience. For each scene, i.e. horizontal geographic location, we capture at each of $\{10, 20, 30, 40, 50, 70, 80, 100, 120, 140\}$ m altitudes a Tele camera image and 7 Hasselblad images, while preserving the exposure value between paired captures. With this configuration, we follow the convention in BurstSR [5] and capture 200 scenes in total, which we split into non-overlapping train, validation, and test sets consisting of 100, 50, and 50 scenes, respectively. Every *scene* is the equivalent of 8 captures $\times$ 10 altitudes. After creating 180×180 resolution patches [5], this yields a total of 5175, 2622, and 2532 image pairs for training, validation, and test sets. Our patches are extracted from the LR images in a sliding window manner. To avoid overlap among patches, we extract 180×180 patches from LR images with a stride of 180 pixels.

3.2 Image Registration

For RAW data, we first pack each 2×2 block in the raw Bayer color filter array along channel dimensions, to obtain 4-channel images. In all following data processing steps, we obtain the parameters of homography transformations based on the RGB image and apply the same transforms to both RGB and RAW images for consistency. For burst sequences, we obtain the parameters of transforms based on the first burst frame. It is worth noting that although we focus in this paper on RGB data and single-image (first frame) SR for their practicality in real-world applications, we also provide burst sequences and RAW data in DSR to advance future research on drone SR.

Since the images captured by two cameras have different fields of view (FOV), we first crop the matched FOV from each LR image in the burst sequence. In data acquisition, we use the two adjacent cameras to capture our HR and LR images. The vibration of a flying drone causes more severe displacements than, for instance, human held capture or tripods. Thus, simply cropping around the center of the LR image [7,44] cannot always achieve FOV matching. Similar to BurstSR [5], we estimate the homography between the LR and HR paired images using SIFT [26] and RANSAC [14] to perform FOV matching. However, in our dataset, the can drone can move subtly not only horizontally but also vertically, which leads to small variations in the corresponding FOV on the LR image. Although the worst variations are only a few pixels wide, they would complicate downstream data possessing steps by causing a varying scale factor, albeit by an almost negligible magnitude. To overcome this problem, we resize the FOV to a fixed resolution (720×540). Therefore, the scale factor of our DSR dataset is always constant at exactly $\times 50/9$. We choose the nearest-neighbor interpolation method to resize the matched FOV. This choice is based on the fact that the majority of pixel values are not even affected because the worst dimension variations are only a few pixels wide within the 720×540 resolution, and most images do not have any variation. A detailed comparison including other resizing methods is provided in the supplementary material.

Exposure. Despite matching the same exposure value across captures, we still found some color difference in certain image pairs. To improve the accuracy of our image pair registration, we exploit the pixel-wise color transfer algorithm of RealSR [7]. Additionally, we apply histogram matching and evaluate both methods on our dataset (see supplementary material for further evaluation results). We thus apply color transfer and histogram matching for color correction on our paired images, to minimize any potential variations.

3.3 Alignment Analysis

A certain degree of misalignment cannot be avoided in super-resolution data acquisition, especially when working with drones. To best counter this problem, we carefully design external and internal control procedures.

External Control Procedure. We improve our alignment as well as possible by controlling external conditions. We acquire data in mild weather conditions

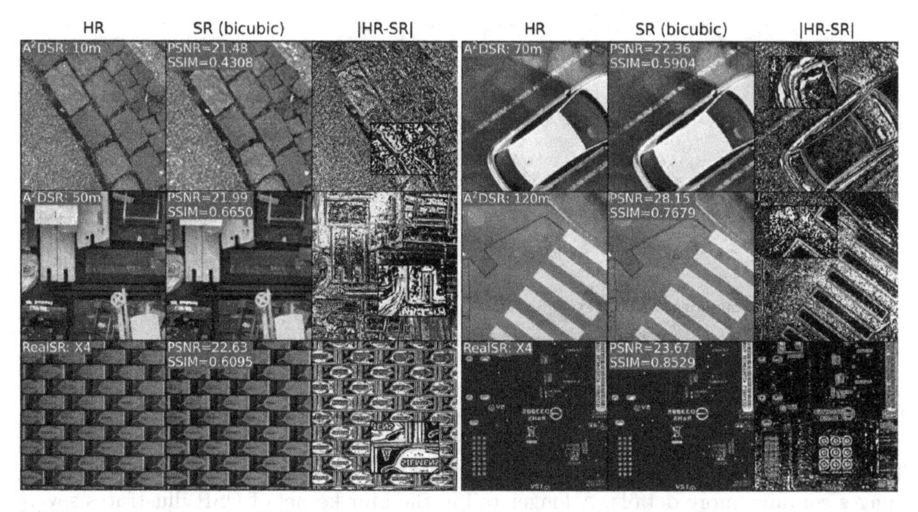

(a) Examples from the DSR dataset at $10m$ and $50m$, and the RealSR dataset.

(b) Examples from the DSR dataset at $70m$ and $120m$, and the RealSR dataset.

Fig. 2. We analyze the registration on the higher resolution for more precision. The HR image is compared with the bicubic upsampled LR. We note that the absolute error between the two is around edges, which are sharper in the HR image. We can see in the second row of (b), that the errors on symmetric surfaces (yellow rectangles) are similarly symmetric and follow the same contour shape. This illustrates the alignment quality, as any misalignment would cause shape distortions between the image and error shape contours. For reference, the bottom row (a, b) shows a similar analysis on RealSR images [7]. (Color figure online)

with minimal wind and clouds, to reduce wind-induced movement and avoid moving cloud shadows in the scene. We carefully stabilize the drone at each altitude before capturing data. Furthermore, we look for static scenes and avoid moving vehicles or mobile objects. And lastly, we reduce the occurrences of vegetation especially trees with light leaves that can be moved by the wind.

Internal Control Procedure. For each patch, we reiterate the homography estimation between the LR patch and the corresponding region in the HR image to perform more refined local alignment. We align the LR image within the HR counterpart (rather than the reverse) for more precise alignment [5]. As a further control step, we remove all HR and LR pairs that have a normalized cross correlation lower than 0.9. The final numbers of *valid* image pairs for each set at each altitude are listed in Table 1, and are balanced across all altitudes.

We visualize the absolute error map between the HR image and a bicubic upsampling of its LR image counterpart. The resulting mean spectral error value is visualized in Fig. 2. We can note that the absolute error over symmetric surfaces is consistently symmetric following the same contour shape, indicating a good alignment. The figure also shows in the last row a similar example from the RealSR [7] that contains similar errors despite the use of a tripod during

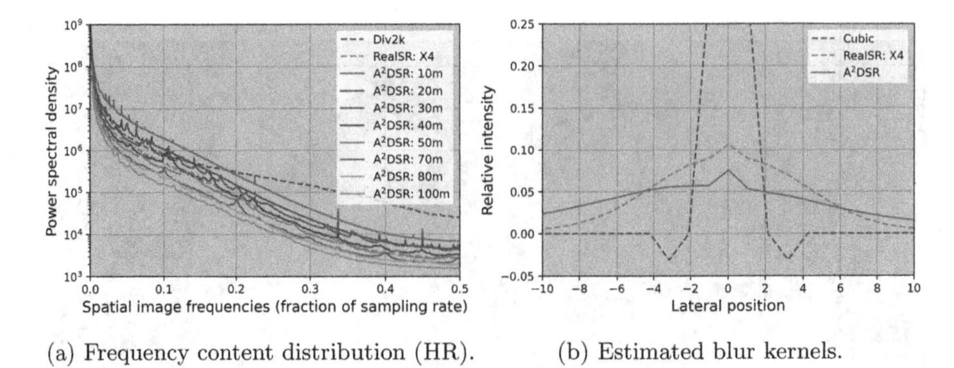

(a) Frequency content distribution (HR). (b) Estimated blur kernels.

Fig. 3. (a) Average PSD in different datasets and altitudes, and (b) blur kernel estimation. DSR has similar spatial frequency characteristics as other real-world datasets. The curve of lower altitude lies above the curve of higher altitude, the corresponding images contains more details. A longer tail in the blur kernel of DSR illustrates severe blur and thus harder-to-perform super-resolution.

acquisition. These errors are due to the fact that LR images do not contain the sharp details of their HR counterparts, resulting in differences along edges [44].

3.4 Data Analysis

Our DSR dataset contains image pairs captured at different altitudes. We analyze in this section, in the frequency domain, the potential domain gap across altitudes. We plot in Fig. 3a the average power spectral density (PSD) of our test set HR images for a set of altitudes and for two other datasets; the Div2K dataset [1] and a real-world SR dataset RealSR [6,7]. To enforce FOV invariance across altitudes for our DSR dataset, we crop the center from higher altitudes. The PSD curves of DSR are closer to the curve of RealSR [6,7], while that of the Div2K dataset [1] is more flat with a slower drop along high frequencies. Div2K indeed contains high definition images with more high frequencies. What is most interesting to observe in the variation in our DSR dataset. We can see that the PSD at lower altitudes lies above the PSD of higher altitude images. This clearly highlights the existence of a domain gap, in terms of frequency content, across altitudes. As SR networks learn to predict conditional frequency bands [13], it is important to note any differences in that domain.

We also estimate the blur kernel between our LR and HR observations. We use the blur kernel estimation proposed in [47]. We estimate the blur kernel on the test set of DSR, and on the RealSR dataset [6,7] with scale factor of 4 for reference. We also plot the widely used MATLAB bicubic kernel. The results are visualized in Fig. 3b. From this visualization, it is evident that the bicubic kernel significantly differs from the blur kernels of real-world datasets, as noted in [47]. The blur kernel estimated from DSR is similar to the in RealSR but is even more different than the bicubic kernel. Its wider shape indicates stronger blur, and a more challenging SR reconstruction task on our DSR dataset.

Table 2. PSNR (dB)/SSIM results of various SR methods on commonly used public datasets. We highlight the best and the second best results on each dataset in red and blue, respectively.

Method	Public datasets				
	Div2K	Set5	Set14	B100	Urban100
Bicubic	26.56/.9593	26.37/.7504	24.30/.7065	24.62/.6013	21.82/.7722
EDSR [23]	28.27/.9796	28.91/.8581	26.07/.7870	25.80/.6657	23.76/.8793
RDN [46]	28.20/.9791	28.73/.8540	26.06/.7841	25.75/.6626	23.62/.8732
RCAN [45]	28.27/.9795	28.78/.8575	26.11/.7867	25.81/.6668	23.82/.8814
ESRGAN [36]	26.35/.9666	26.66/.7976	24.53/.7186	24.01/.5890	22.25/.8372
BSRNet [41]	27.22/.9651	27.17/.7924	24.99/.7309	25.16/.6285	22.75/.8269
SwinIR [22]	28.45/.9809	28.79/.8630	26.10/.7930	25.88/.6727	24.09/.8946
NLSN [29]	28.31/.9796	28.91/.8607	26.15/.7891	25.86/.6689	24.00/.8862
DASR [34]	27.97/.9775	28.25/.8447	25.83/.7769	25.64/.6590	23.36/.8633

4 Evaluating SR on Drone Data

We benchmark state-of-the-art methods on DSR. We evaluate the performance on each altitude to explore its effect. To look for potential domain gaps, and to give reference results, we also benchmark on commonly used public datasets. And, for consistent comparisons, all datasets are fixed to the same scale factor. We then evaluate fine-tuning and our proposed altitude-aware solutions.

4.1 Pretrained Off-the-Shelf SR Methods

We use bicubic interpolation and eight state-of-the-art SR networks for benchmarking results: EDSR [23], RDN [46], RCAN [45], ESRGAN [36], SwinIR [22], BSRNet [41], NLSN [29] and DASR [34]. We exclusively use the trained networks provided by the corresponding authors with the scale factor of 4. Out of these SR networks, ESRGAN, SwinIR and DASR were trained on both the Div2K [1] and Flickr2K [23,33] datasets. BSRNet was trained on the Div2K [1], Flickr2K [23,33], WED [27], and FFHQ [18] datasets. The rest of the networks were trained on the Div2K [1] dataset. To address the real-world scenario, SwinIR, NLSN and DASR exploit a designed degradation model applied to HR images when generating the LR counterparts. The rest of the networks only apply the standard degradation model (i.e., the MATLAB bicubic downsampling).

We first benchmark on five commonly used public datasets: Div2K [1], Set5 [3], Set14 [40], B100 [28], and Urban100 [16]. The LR images are generated with MATLAB's bicubic downsampling with DSR's scale factor (i.e. $\times 50/9$). Since the SR networks have a scale factor of 4, we upscale the outputs of the networks to the target size using bicubic interpolation for consistent evaluation. Following the method originally presented in [45], we evaluate the performance of SR solutions using PSNR and SSIM [37] on the Y channel in the transformed

Table 3. PSNR (dB)/SSIM results of SR methods on our DSR test set. †These SR networks are trained with more complex but practical degradation models considering blur, downsampling and noise. We highlight the best and the second best results at each altitude in red and blue, respectively.

Method	DSR dataset				
	10 m	20 m	30 m	40 m	50 m
Bicubic	23.24/.7156	23.28/.7112	23.46/.7118	23.94/.7271	23.93/.7341
EDSR	23.03/.7155	23.12/.7138	23.29/.7151	23.73/.7289	23.75/.7362
RDN	23.04/.7157	23.14/.7139	23.30/.7153	23.75/.7291	23.76/.7364
RCAN	23.01/.7154	23.12/.7137	23.28/.7150	23.72/.7290	23.73/.7361
ESRGAN	22.84/.7125	22.96/.7111	23.09/.7123	23.50/.7260	23.53/.7331
BSRNet	22.54/.7099	22.68/.7103	22.89/.7126	23.39/.7272	23.36/.7343
SwinIR†	22.97/.7150	23.08/.7133	23.24/.7146	23.68/.7288	23.69/.7357
NLSN†	23.02/.7155	23.13/.7138	23.30/.7153	23.74/.7292	23.75/.7363
DASR†	23.12/.7166	23.20/.7142	23.37/.7156	23.82/.7297	23.82/.7369
Method	DSR dataset				
	70 m	80 m	100 m	120 m	140 m
Bicubic	24.13/.7456	24.31/.7486	24.39/.7574	24.88/.7741	24.79/.7777
EDSR	23.95/.7484	24.12/.7506	24.22/.7603	24.65/.7764	24.59/.7802
RDN	23.96/.7486	24.13/.7508	24.24/.7604	24.67/.7766	24.61/.7804
RCAN	23.93/.7483	24.10/.7505	24.21/.7602	24.64/.7763	24.58/.7802
ESRGAN	23.73/.7454	23.89/.7476	24.00/.7572	24.39/.7731	24.33/.7767
BSRNet	23.55/.7467	23.72/.7491	23.85/.7596	24.31/.7761	24.25/.7802
SwinIR†	23.88/.7478	24.05/.7501	24.16/.7598	24.58/.7759	24.52/.7798
NLSN†	23.95/.7485	24.12/.7507	24.23/.7604	24.66/.7766	24.60/.7803
DASR†	24.03/.7489	24.20/.7512	24.30/.7608	24.76/.7771	24.68/.7807

YCbCr space. The results are shown in Table 2, and give a reference indication on the performances of these methods on public datasets. With the same setup, we benchmark the aforementioned SR methods on our DSR test set. To quantitatively explore the effects of altitude, we evaluate separately on each altitude. The results are shown in Table 3. Comparing the results of the eight benchmarked SR networks on public datasets and our DSR, we see that the SR networks suffer a performance drop on DSR. Only the Urban100 shows lower PSNR and SSIM results, as this dataset contains a large number of edges and high frequency urban images. Such a drop in performance is commonly observed in real-world scenarios [31,34], and the degradation model of DSR is challenging. It is also interesting to observe the variation in performance across altitudes, and that the ranking of different SR methods actually changes between datasets.

Table 4. PSNR (dB)/SSIM results of the pretrained and fine-tuned FCNN and SwinIR networks on our DSR test set. ‡The network is pretrained on Div2K. *The network is fine-tuned on DSR using all altitudes. We highlight the best and the second best results in red and blue, respectively.

Method		10 m	20 m	30 m	40 m	50 m
FCNN	Pretrain‡	23.11/.7174	23.19/.7150	23.38/.7164	23.83/.7305	23.83/.7378
	Fine-tuned*	23.45/.7191	23.50/.7126	23.68/.7131	24.14/.7292	24.11/.7350
	Altitude-aware	23.52/.7104	23.54/.7028	23.70/.7051	24.17/.7235	24.15/.7293
SwinIR	Pretrained‡	23.07/.7171	23.17/.7151	23.34/.7165	23.78/.7305	23.79/.7375
	Fine-tuned*	23.67/.7219	23.73/.7162	23.92/.7172	24.36/.7350	24.35/.7421
	Altitude-aware	23.74/.7204	23.76/.7147	23.95/.7169	24.40/.7350	24.38/.7420
Method		70 m	80 m	100 m	120 m	140 m
FCNN	Pretrained‡	24.04/.7500	24.22/.7525	24.32/.7619	24.79/.7782	24.71/.7818
	Fine-tuned*	24.32/.7475	24.51/.7503	24.59/.7587	25.07/.7758	25.00/.7802
	Altitude-aware	24.34/.7428	24.52/.7465	24.58/.7561	25.06/.7738	24.98/.7782
SwinIR	Pretrained‡	24.00/.7497	24.16/.7520	24.27/.7615	24.71/.7778	24.64/.7815
	Fine-tuned*	24.61/.7570	24.78/.7601	24.88/.7684	25.36/.7847	25.28/.7897
	Altitude-aware	24.62/.7572	24.78/.7604	24.89/.7698	25.38/.7866	25.29/.7920

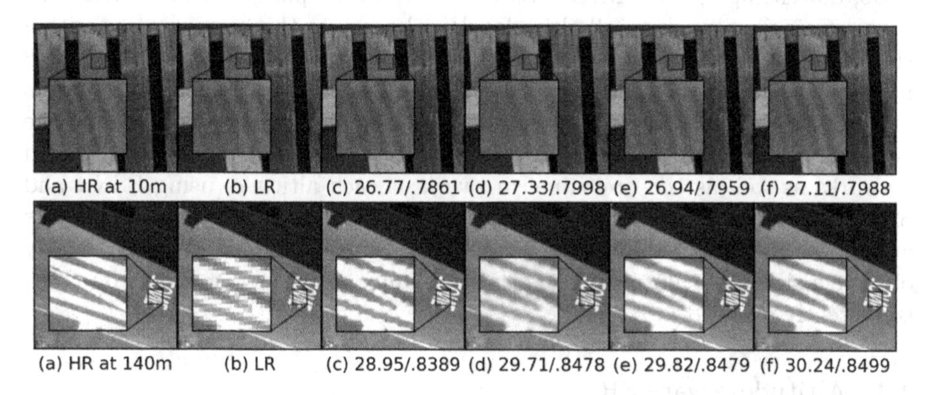

Fig. 4. Qualitative and quantitative (PSNR (dB)/SSIM) results of pretrained and fine-tuned SR networks on the DSR test images. Per row: (a) HR ground-truth, (b) LR counterpart, (c) Output of SwinIR pretrained on Div2K, (d) Output of SwinIR fine-tuned on DSR 10 m data, (e) Output of SwinIR fine-tuned on DSR 140 m data, (f) Output of SwinIR fine-tuned on all DSR altitudes (+0.34 dB and +1.29 dB improvement over the baseline model).

4.2 Fine-Tuned SR Methods

We further evaluate the performance of learning-based networks on DSR at various altitudes with a retraining step. We fine-tune the SwinIR [22] network on DSR as it achieves the best performance on the public datasets. To fit the scale factor, we add an upsampling layer followed by a convolutional layer with

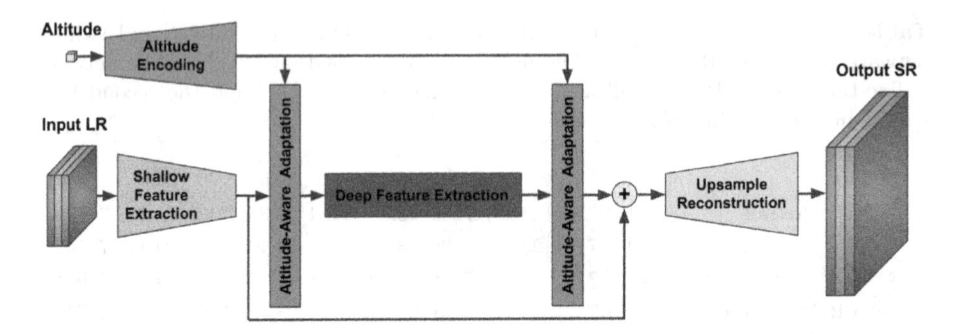

Fig. 5. The architecture of our proposed altitude-aware SwinIR (AASwinIR) for drone SR. DAB from DASR [34] is used for altitude-aware feature adaptation. The altitude is first enmbedded with an encoding step, and it is then used to modify internal feature extraction. A similar modification is made over FCNN, as described in the text.

LeakyReLU activation at the end of the upsampling block of the original SwinIR. Following the setup in SwinIR [22], We use the $\ell 1$ loss with the ADAM [19] optimizer for fine-tuning. We first pretrain the network on Div2K synthetic image pairs, downsampled with MATLAB bicubic downsampling. Then, we fine-tune the pretrained network on all altitudes. We also repeat this configuration with a simple and fully convolutional network we refer to as FCNN [31]. It contains 8 hidden layers, each with 128 channels. We use ReLU activations between layers and only learn the residual. The detailed training procedure is presented in the supplementary material. Finally, we evaluate the performance of those fine-tuned networks as well as the pretrained network at each altitude, using PSNR and SSIM [37] on the Y channel in the transformed YCbCr space. Numerical results are shown in Table 4. Qualitative results are given in Fig. 4. We can clearly observe a significant improvement in performance with our fine-tuning on DSR, consistently over +0.6 dB.

4.3 Altitude-Aware SR

The domain gaps between images captured at different altitudes (see Fig. 3a), motivates us to build a method for incorporating altitude information, which we can obtain as meta data. Inspired by internal feature manipulation techniques [12,24] and DASR [34], we use a degradation-aware convolutional block (DAB) for altitude awareness. The core of DAB is the degradation-aware layer (DAL), which predicts a kernel of depth-wise convolution and attention weights for channel-wise adaptation conditioned on the encoded altitude. Specifically, the encoded altitude is fed into two full-connected (FC) layers and then reshaped to a convolutional kernel. Next, the input image feature is convolved (depth-wise) with the obtained kernel. The resulting feature is sent to a 1×1 convolutional layer. The encoded altitude is also fed into the other two FC layers, followed by a sigmoid activation to generate weights for channel-wise attention. The input image feature is processed by a channel attention operation using the generated

Table 5. PSNR (dB)/SSIM results of the pretrained and fine-tuned altitude-aware SwinIR (AASwinIR) on our DSR test set. ‡The network is pretrained on Div2K. The network fine-tuned with altitude information consistently yields better performance than the standard SwinIR. Without altitude information, it achieves similar results as the standard SwinIR. We highlight the best and the second best results at each altitude in red and blue, respectively.

Method	10 m	20 m	30 m	40 m	50 m
AASwinIR Div2K‡	23.06/.7168	23.16/.7148	23.32/.7161	23.76/.7301	23.77/.7371
Fine-tuned SwinIR	23.67/.7219	23.73/.7162	23.92/.7172	24.36/.7350	24.35/.7421
AASwinIR w/o altitude	23.65/.7209	23.73/.7164	23.91/.7170	24.36/.7340	24.34/.7408
AASwinIR	23.74/.7204	23.76/.7147	23.95/.7169	24.40/.7350	24.38/.7420
Method	70 m	80 m	100 m	120 m	140 m
AASwinIR Div2K‡	23.97/.7492	24.14/.7515	24.25/.7610	24.68/.7772	24.62/.7811
Fine-tuned SwinIR	24.61/.7570	24.78/.7601	24.88/.7684	25.36/.7847	25.28/.7897
AASwinIR w/o altitude	24.58/.7552	24.74/.7581	24.85/.7667	25.35/.7832	25.26/.7882
AASwinIR	24.62/.7572	24.78/.7604	24.89/.7698	25.38/.7866	25.29/.7920

weights. The results from depth-wise convolution and channel-wise attention are summed together to obtain the output of DAL. The DAB consists of two DALs, with each DAL followed by a 3×3 convolutional layer.

In our altitude-aware SwinIR (AA-SwinIR), the shallow and deep image features of SwinIR [22], along with the encoded altitude information, act as an input for DABs. We use two FC layers to encode the altitude. The detailed architecture of our AA-SwinIR is shown in Fig. 5. We also modify the FCNN network to incorporate altitude information by adding one AAL between each standard convolutional layer to build our altitude-aware FCNN network (AA-FCNN).

We first pretrain our altitude-aware SwinIR network on Div2K using the same setup presented in Sect. 4.2. Since Div2K does not contain altitude information, we set all altitudes to be 1. We fine-tune this network on DSR along with the altitude information. Additionally, we normalize the altitude values by a factor of 80. The detailed training setup is presented in the supplementary material. The quantitative results of the altitude-aware AA-SwinIR pretrained on Div2K, and the fine-tuned version with altitude information are reported in Table 5. We also report the results where the altitude is frozen at training and inference to a default value (i.e. feed in a constant 1).

Comparing the results in Table 4 and Table 5, our fine-tuning and altitude-aware networks consistently improve the performance of SwinIR across altitudes, to a further extent with the altitude information. The improvement with the altitude-aware approach is largest at the altitudes where the fine-tuned SwinIR struggles the most (i.e. lower altitudes). However, both our fine-tuned and altitude-aware methods significantly boost the performance, up to nearly +0.7 dB. Sample qualitative results are illustrated in columns (e) and (f) in Fig. 6. The quantitative results of AASwinIR *without* altitude information are

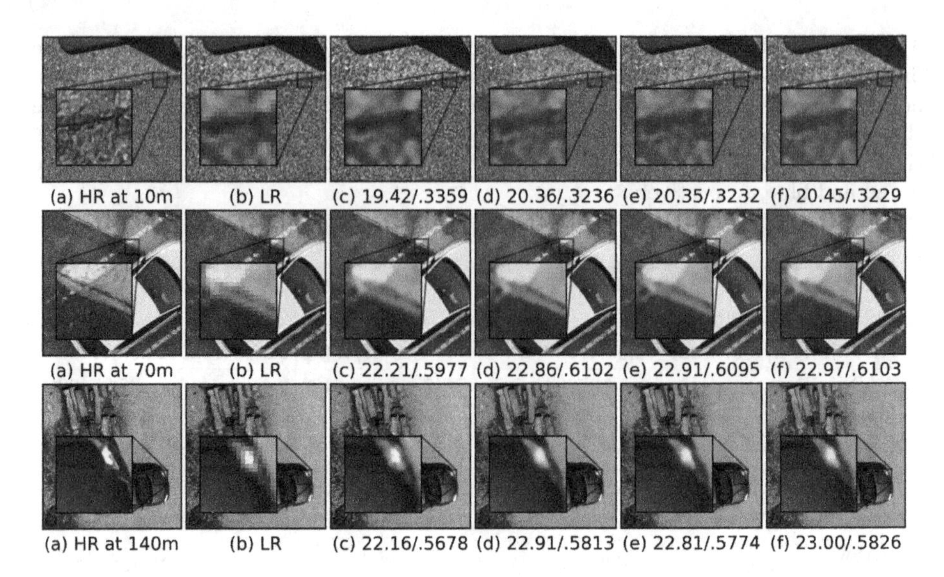

Fig. 6. Qualitative and quantitative (PSNR (dB)/SSIM) results of pretrained and fine-tuned altitude-aware SwinIR network on the DSR test images. (a) HR ground-truth, (b) LR counterpart, (c) SR output of the pretrained altitude-aware AASwinIR network on Div2K, (d) SR output of SwinIR fine-tuned on all ten DSR altitudes (the same as (f) in Fig. 4), (e) SR output of AASwinIR fine-tuned on DSR without feeding altitude information, (f) SR output of the fine-tuned AASwinIR on DSR with correct altitude information.

similar to the standard fine-tuned SwinIR, which reveals that altitude information (and not the architectural change) contributes to the improvement. We show corresponding qualitative examples in Fig. 6 for this altitude-ablation study.

5 Conclusion and Future Directions

We create the first drone super-resolution dataset (DSR) with paired low-resolution and optical zoom high-resolution images. The paired data are captured for each scene at ten different altitudes. We show that off-the-shelf SR methods cannot be readily applied to drone images without a drop in performance, due to a domain gap in the learned priors. Furthermore, we show that a domain gap in terms of frequency content also emerges across varying altitudes. We show that fine-tuning previous methods on drone data improves the results, and even more so with our proposed altitude-aware architecture.

To foster future research towards drone super-resolution, our dataset also includes burst sequences of seven images for each of our LR captures. Future work involves extension to multi-image super-resolution, as opposed to single-image super-resolution, as well as onboard-drone efficient super-resolution solutions [4, 35,43]. We also carefully selected the capture altitudes such that they mutually form different multiples of each other. This can be exploited to pursue cross-altitude super-resolution methods that do not rely on pixel-level matching, but rather on learning distribution transformation.

References

1. Agustsson, E., Timofte, R.: NTIRE 2017 challenge on single image super-resolution: dataset and study. In: Proceedings of the IEEE Conference on Computer Vision and Pattern Recognition Workshops, pp. 126–135 (2017)
2. Al-Kaff, A., Martin, D., Garcia, F., de la Escalera, A., Armingol, J.M.: Survey of computer vision algorithms and applications for unmanned aerial vehicles. Expert Syst. Appl. **92**, 447–463 (2018)
3. Bevilacqua, M., Roumy, A., Guillemot, C., Alberi-Morel, M.L.: Low-complexity single-image super-resolution based on nonnegative neighbor embedding (2012)
4. Bhardwaj, K., et al.: Collapsible linear blocks for super-efficient super resolution. In: Proceedings of Machine Learning and Systems, vol. 4, pp. 529–547 (2022)
5. Bhat, G., Danelljan, M., Van Gool, L., Timofte, R.: Deep burst super-resolution. In: Proceedings of the IEEE/CVF Conference on Computer Vision and Pattern Recognition, pp. 9209–9218 (2021)
6. Cai, J., Gu, S., Timofte, R., Zhang, L.: NTIRE 2019 challenge on real image super-resolution: methods and results. In: Proceedings of the IEEE Conference on Computer Vision and Pattern Recognition Workshops (2019)
7. Cai, J., Zeng, H., Yong, H., Cao, Z., Zhang, L.: Toward real-world single image super-resolution: a new benchmark and a new model. In: Proceedings of the IEEE International Conference on Computer Vision (2019)
8. Chan, T.M., Zhang, J., Pu, J., Huang, H.: Neighbor embedding based super-resolution algorithm through edge detection and feature selection. Pattern Recogn. Lett. **30**(5), 494–502 (2009)
9. DJI: DJI Mavic 3 - user manual v1.6 (2022). https://dl.djicdn.com/downloads/DJI_Mavic_3/20220531/DJI_Mavic_3_User_Manual_v1.6_en.pdf
10. Du, D., et al.: The unmanned aerial vehicle benchmark: object detection and tracking. In: Ferrari, V., Hebert, M., Sminchisescu, C., Weiss, Y. (eds.) ECCV 2018. LNCS, vol. 11214, pp. 375–391. Springer, Cham (2018). https://doi.org/10.1007/978-3-030-01249-6_23
11. Dudhane, A., Zamir, S.W., Khan, S., Khan, F., Yang, M.H.: Burst image restoration and enhancement. arXiv preprint arXiv:2110.03680 (2021)
12. El Helou, M., Süsstrunk, S.: Blind universal Bayesian image denoising with Gaussian noise level learning. IEEE Trans. Image Process. **29**, 4885–4897 (2020)
13. El Helou, M., Zhou, R., Süsstrunk, S.: Stochastic frequency masking to improve super-resolution and denoising networks. In: Vedaldi, A., Bischof, H., Brox, T., Frahm, J.-M. (eds.) ECCV 2020. LNCS, vol. 12361, pp. 749–766. Springer, Cham (2020). https://doi.org/10.1007/978-3-030-58517-4_44
14. Fischler, M.A., Bolles, R.C.: Random sample consensus: a paradigm for model fitting with applications to image analysis and automated cartography. Commun. ACM **24**(6), 381–395 (1981)
15. Hsieh, M.R., Lin, Y.L., Hsu, W.H.: Drone-based object counting by spatially regularized regional proposal network. In: Proceedings of the IEEE International Conference on Computer Vision, pp. 4145–4153 (2017)

16. Huang, J.B., Singh, A., Ahuja, N.: Single image super-resolution from transformed self-exemplars. In: Proceedings of the IEEE Conference on Computer Vision and Pattern Recognition, pp. 5197–5206 (2015)
17. Johnson, J., Alahi, A., Fei-Fei, L.: Perceptual losses for real-time style transfer and super-resolution. In: Leibe, B., Matas, J., Sebe, N., Welling, M. (eds.) ECCV 2016. LNCS, vol. 9906, pp. 694–711. Springer, Cham (2016). https://doi.org/10.1007/978-3-319-46475-6_43
18. Karras, T., Laine, S., Aila, T.: A style-based generator architecture for generative adversarial networks. In: Proceedings of the IEEE/CVF Conference on Computer Vision and Pattern Recognition, pp. 4401–4410 (2019)
19. Kingma, D.P., Ba, J.: Adam: a method for stochastic optimization. arXiv preprint arXiv:1412.6980 (2014)
20. Lecouat, B., Ponce, J., Mairal, J.: Lucas-kanade reloaded: end-to-end super-resolution from raw image bursts. In: Proceedings of the IEEE/CVF International Conference on Computer Vision, pp. 2370–2379 (2021)
21. Ledig, C., et al.: Photo-realistic single image super-resolution using a generative adversarial network. In: Proceedings of the IEEE/CVF Conference on Computer Vision and Pattern Recognition (2017)
22. Liang, J., Cao, J., Sun, G., Zhang, K., Van Gool, L., Timofte, R.: SwinIR: image restoration using swin transformer. In: Proceedings of the IEEE/CVF International Conference on Computer Vision, pp. 1833–1844 (2021)
23. Lim, B., Son, S., Kim, H., Nah, S., Mu Lee, K.: Enhanced deep residual networks for single image super-resolution. In: Proceedings of the IEEE Conference on Computer Vision and Pattern Recognition Workshops, pp. 136–144 (2017)
24. Lin, X., Bhattacharjee, D., El Helou, M., Süsstrunk, S.: Fidelity estimation improves noisy-image classification with pretrained networks. IEEE Sig. Process. Lett. **28**, 1719–1723 (2021)
25. Liu, K., Mattyus, G.: Fast multiclass vehicle detection on aerial images. IEEE Geosci. Remote Sens. Lett. **12**(9), 1938–1942 (2015)
26. Lowe, D.G.: Object recognition from local scale-invariant features. In: Proceedings of IEEE International Conference on Computer Vision, vol. 2, pp. 1150–1157. IEEE (1999)
27. Ma, K., et al.: Waterloo exploration database: new challenges for image quality assessment models. IEEE Trans. Image Process. **26**(2), 1004–1016 (2016)
28. Martin, D., Fowlkes, C., Tal, D., Malik, J.: A database of human segmented natural images and its application to evaluating segmentation algorithms and measuring ecological statistics. In: Proceedings of the IEEE International Conference on Computer Vision, vol. 2, pp. 416–423. IEEE (2001)
29. Mei, Y., Fan, Y., Zhou, Y.: Image super-resolution with non-local sparse attention. In: Proceedings of the IEEE/CVF Conference on Computer Vision and Pattern Recognition, pp. 3517–3526 (6 2021)
30. Oh, S., et al.: A large-scale benchmark dataset for event recognition in surveillance video. In: CVPR 2011, pp. 3153–3160. IEEE (2011)
31. Shocher, A., Cohen, N., Irani, M.: "zero-shot" super-resolution using deep internal learning. In: Proceedings of the IEEE Conference on Computer Vision and Pattern Recognition, pp. 3118–3126 (2018)
32. Sun, J., Xu, Z., Shum, H.Y.: Image super-resolution using gradient profile prior. In: Proceedings of the IEEE/CVF Conference on Computer Vision and Pattern Recognition (2008)

33. Timofte, R., Agustsson, E., Van Gool, L., Yang, M.H., Zhang, L.: NTIRE 2017 challenge on single image super-resolution: methods and results. In: Proceedings of the IEEE Conference on Computer Vision and Pattern Recognition Workshops, pp. 114–125 (2017)
34. Wang, L., et al.: Unsupervised degradation representation learning for blind super-resolution. In: Proceedings of the IEEE/CVF Conference on Computer Vision and Pattern Recognition, pp. 10581–10590 (2021)
35. Wang, X., Dong, C., Shan, Y.: RepSR: training efficient VGG-style super-resolution networks with structural re-parameterization and batch normalization. arXiv preprint arXiv:2205.05671 (2022)
36. Wang, X., et al.: ESRGAN: enhanced super-resolution generative adversarial networks. In: Leal-Taixé, L., Roth, S. (eds.) ECCV 2018. LNCS, vol. 11133, pp. 63–79. Springer, Cham (2019). https://doi.org/10.1007/978-3-030-11021-5_5
37. Wang, Z., Bovik, A.C., Sheikh, H.R., Simoncelli, E.P.: Image quality assessment: from error visibility to structural similarity. IEEE Trans. Image Process. **13**(4), 600–612 (2004)
38. Xu, X., et al.: DAC-SDC low power object detection challenge for UAV applications. IEEE Trans. Pattern Anal. Mach. Intell. **43**(2), 392–403 (2019)
39. Xu, Y., Ou, J., He, H., Zhang, X., Mills, J.: Mosaicking of unmanned aerial vehicle imagery in the absence of camera poses. Remote Sens. **8**(3), 204 (2016)
40. Zeyde, R., Elad, M., Protter, M.: On single image scale-up using sparse-representations. In: Boissonnat, J.-D., et al. (eds.) Curves and Surfaces 2010. LNCS, vol. 6920, pp. 711–730. Springer, Heidelberg (2012). https://doi.org/10.1007/978-3-642-27413-8_47
41. Zhang, K., Liang, J., Van Gool, L., Timofte, R.: Designing a practical degradation model for deep blind image super-resolution. In: IEEE International Conference on Computer Vision, pp. 4791–4800 (2021)
42. Zhang, K., Zuo, W., Zhang, L.: Learning a single convolutional super-resolution network for multiple degradations. In: Proceedings of the IEEE/CVF Conference on Computer Vision and Pattern Recognition (2018)
43. Zhang, X., Zeng, H., Zhang, L.: Edge-oriented convolution block for real-time super resolution on mobile devices. In: Proceedings of the 29th ACM International Conference on Multimedia, pp. 4034–4043 (2021)
44. Zhang, X., Chen, Q., Ng, R., Koltun, V.: Zoom to learn, learn to zoom. In: Proceedings of the IEEE/CVF Conference on Computer Vision and Pattern Recognition, pp. 3762–3770 (2019)
45. Zhang, Y., Li, K., Li, K., Wang, L., Zhong, B., Fu, Y.: Image super-resolution using very deep residual channel attention networks. In: Ferrari, V., Hebert, M., Sminchisescu, C., Weiss, Y. (eds.) ECCV 2018. LNCS, vol. 11211, pp. 294–310. Springer, Cham (2018). https://doi.org/10.1007/978-3-030-01234-2_18
46. Zhang, Y., Tian, Y., Kong, Y., Zhong, B., Fu, Y.: Residual dense network for image super-resolution. In: Proceedings of the IEEE Conference on Computer Vision and Pattern Recognition, pp. 2472–2481 (2018)
47. Zhou, R., El Helou, M., Sage, D., Laroche, T., Seitz, A., Süsstrunk, S.: W2S: microscopy data with joint denoising and super-resolution for widefield to SIM mapping. In: Bartoli, A., Fusiello, A. (eds.) ECCV 2020. LNCS, vol. 12535, pp. 474–491. Springer, Cham (2020). https://doi.org/10.1007/978-3-030-66415-2_31
48. Zhou, R., Lahoud, F., El Helou, M., Süsstrunk, S.: A comparative study on wavelets and residuals in deep super resolution. In: Electronic Imaging (2019)

CEN-HDR: Computationally Efficient Neural Network for Real-Time High Dynamic Range Imaging

Steven Tel(✉), Barthélémy Heyrman, and Dominique Ginhac

ImViA EA7535, University Burgundy Franche-Comté, 21078 Dijon, France
{steven.tel,barthelemy.heyrman,dginhac}@u-bourgogne.fr

Abstract. High dynamic range (HDR) imaging is still a challenging task in modern digital photography. Recent research proposes solutions that provide high-quality acquisition but at the cost of a very large number of operations and a slow inference time that prevent the implementation of these solutions on lightweight real-time systems. In this paper, we propose CEN-HDR, a new computationally efficient neural network by providing a novel architecture based on a light attention mechanism and sub-pixel convolution operations for real-time HDR imaging. We also provide an efficient training scheme by applying network compression using knowledge distillation. We performed extensive qualitative and quantitative comparisons to show that our approach produces competitive results in image quality while being faster than state-of-the-art solutions, allowing it to be practically deployed under real-time constraints. Experimental results show our method obtains a score of 43.04 μ-PSNR on the *Kalantari2017* dataset with a framerate of 33 FPS using a Macbook M1 NPU. The proposed network will be available at https://github.com/steven-tel/CEN-HDR

Keywords: High dynamic range imaging · Efficient computational photography

1 Introduction

In the last decades, applications based on computer vision have become increasingly important in everyday life. Currently, many research works are conducted to propose more reliable algorithms in areas such as object detection, action recognition, or scene understanding. However, the accuracy of these algorithms depends largely on the quality of the acquired images. Most standard cameras are unable to faithfully reproduce the illuminations range of a natural scene, as the limitations of their sensors generate a loss of structural or textural information in under-exposed and over-exposed regions of the acquired scene. To tackle this challenge, sensors with a higher dynamic range (HDR) have been proposed [17,26] to capture more intensity levels of the scene illumination, but

L. Karlinsky et al. (Eds.): ECCV 2022 Workshops, LNCS 13802, pp. 378–394, 2023.
https://doi.org/10.1007/978-3-031-25063-7_23

these solutions are expensive, preventing high dynamic range acquisition from being readily available.

Towards making HDR imaging practical and accessible, software solutions were proposed, based on the emergence of deep learning in computer vision applications. They acquire one Low Dynamic Range (LDR) image and try to expends its dynamic range thanks to a generative adversarial network [8,12,37]. Although these methods produce images with a higher illumination range, they have limitations in extending the dynamic of the input image to the dynamic of the acquired scene. A more effective approach is to acquire multiple LDR images with different exposure times and merge them into one final HDR image. Traditional computer vision algorithms [3] allow the acquisition of good-quality static scenes when there is no camera or object motion between images with different exposure times. However, in a lot of use cases, images are captured in a rapid sequence from a hand-held device resulting in inevitable misalignments between low dynamic range shots. Therefore scenes with motions introduce new challenges as ghost-like artifacts for large motion regions or loss of details in occluded regions. Following recent advances in the field of deep learning, several methods based on Convolutional Neural Network (CNN) were proposed to spatially align input frames to a reference one before merging them into a final HDR image.

State-of-the-art deep learning solutions for multi-frame merging HDR [13,34] tend to be based on a previously proposed method [30] and add additional processing to increase the accuracy of the HDR merging system. As a consequence, the computational cost and execution time are significantly increased, preventing these solutions to be used in lightweight systems and/or in real-time applications. Then, the primary goal of HDR imaging software solutions, which was to make HDR imaging more widely available compared to hardware solutions, is therefore not being achieved at all. Therefore, in this paper, we propose a Computationally Efficient neural Network for High Dynamic Range imaging (CEN-HDR). CEN-HDR is based on an encoder-decoder neural network architecture for generating ghost-free HDR images from scenes with large foreground and camera movements. Unlike previously published solutions, we decided to develop a new approach keeping in mind the constraint of inference time and computational cost.

1. We propose CEN-HDR a novel efficient convolutional neural network based on a new attention mechanism and sub-pixel convolution that overcomes ghost-like artifacts and occluded regions while keeping a low computational cost, allowing our solution to be implemented in real-time on a lightweight system.
2. We demonstrate the efficiency of network compression for the realization of CEN-HDR by applying a knowledge distillation scheme.
3. We perform extensive experiments to determine the best trade-off between accuracy and inference cost with the main objective to demonstrate the relevance of CEN-HDR.

2 Related Works

We briefly summarize existing HDR merging approaches into two categories: deep learning-based architectures and efficient learning-based architectures. In the first category, the proposed methods aim to achieve better quality in HDR imaging without taking inference cost into account. Approaches belonging to the second category seek to optimize the compromise between the quality of the generated images and the computation cost. This leads to the use of new operators in the proposed deep learning architectures.

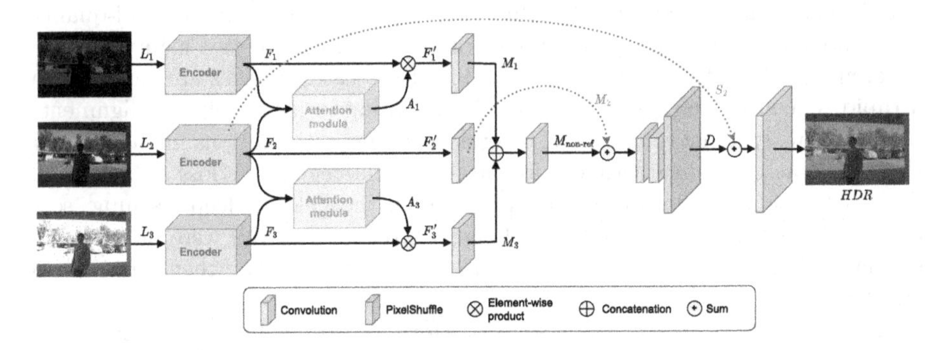

Fig. 1. Architecture of the proposed CEN-HDR solution. The spatial size of input features is divided by 2 at the encoding step. The attention module allows registering non-reference features to the reference ones. The full spatial size is recovered thanks to the pixel shuffle operation.

2.1 Deep Learning Based HDR Merging

Using multiple input images for HDR generation leads to the need to align the features of the LDR images to the reference image. The first common method for feature registration is by computing the motion between inputs features using an optical flow algorithm. Multiple studies [7,22,31] used Liu [11] optical flow algorithm. In *Kalantari et al.* [7], the input images are aligned by selecting the image with better pixels as a reference and computing the optical flow between this reference and other input LDR images. Then, the warped images are fed to a supervised convolutional neural network (CNN) to merge them into an HDR image. However, since the optical flow algorithm initially assumes that the input images have the same exposure time, trying to warp the different exposures with occluded regions can result in artifacts in the final HDR image. To address this issue, *DeepHDR* [28] proposes an image translation network able to hallucinate information in the occluded regions without the need for optical flow. Moreover, many solutions have been developed to correct the ghost effect introduced by the misalignment of the input images. In *AHDRNet* [30] an attention module is proposed to emphasize the alignment of the input images to the reference image. Input images are then merged using several dilated residual dense blocks. The high performance of the AHDRNet network led it to be used as a base network for other methods. For example, *ADNet* [13] follows the same main architecture that AHDRNet but adds a Pyramidal alignment module based on deformable convolution allowing a better representation of the extracted features but in the counterpart of a larger number of operations.

2.2 Efficient Learning-Based HDR Merging Architectures

To our best knowledge, the first architecture which aims to be efficient was proposed by *Prabhakar et al.* [21] by processing low-resolution images and upscaling the result to the original full resolution thanks to a bilateral guided upsampling module. Recently, the HDR community tends to focus more on efficient HDR image generation [20] and no longer only aims at improving the image quality but also at significantly limiting the number of processing operations. This results in efficient solutions such as *GSANet* [9] that propose efficient ways to process gamma projections of input images with spatial and channel attention blocks to increase image quality while limiting the number of parameters. *Yu et al.* [36] introduce a multi-frequency lightweight encoding module to extract features and a progressive dilated u-shape block for features merging. Moreover, the different standard convolution operations are replaced by depth-wise separable convolution operations firstly proposed in [2], they are composed of a depth-wise convolution followed by a pointwise convolution which allows more efficient use of model parameters. Another efficient method, proposed by *Yan et al.* [33] is a lightweight network based on an u-net [23] like encoder-decoder architecture, allowing for spatially reduced processed features. While these solutions focus on the number of performed operations, their inference time still remains too long to be considered as real-time solutions.

3 Proposed Method

We consider three LDR images $I_i \in \mathbb{R}^{3\times H\times W}$ with their respective exposure times t_i as inputs. The generated HDR image is spatially aligned with the central LDR frame I_2 selected as the reference image. To make our solution more robust to exposure difference between inputs, the respective projection of each LDR input frame into the HDR domain is calculated using the gamma encoding function described in Eq. 1, following previous works [13,18,30]:

$$H_i = \frac{I_i^\gamma}{t_i}, \quad \gamma = 2,2 \tag{1}$$

where $H_i \in \mathbb{R}^{3\times H\times W}$ is the gamma-projected input. Then, each input is concatenated with their respective gamma-projection to obtain $L_i \in \mathbb{R}^{6\times H\times W}$:

$$L_i = I_i \oplus H_i \tag{2}$$

where $\oplus$ represents the concatenation operation. L_i will then be fed to our proposed merging network whose architecture is detailed in Fig. 1.

3.1 Feature Encoding

Using high-resolution images as inputs presents an additional challenge in the design of a real-time HDR merging network. To solve such a problem, previous works [29,33] propose to use a U-net [23] like architecture to reduce the spatial size of the features processed by the merging network. However, a too large reduction of the spatial dimensions causes the extraction of coarse features that degrade the final result. So, we decide to limit the spatial reduction to 2 by using an encoder block composed of 2 sequential convolutions as described in Eq. 3.

$$F_i = conv_{E_1}(conv_{E_2}(L_i)) \tag{3}$$

where $F_i \in \mathbb{R}^{32\times\frac{H}{2}\times\frac{W}{2}}$ is the features map extracted from the encoder for each LDR input. $conv_{E_1}$ and $conv_{E_2}$ are 3×3-convolution layers extracting respectively 16 and 32 features map. The spatial size is divided by 2 setting a stride of 2 for $conv_2$.

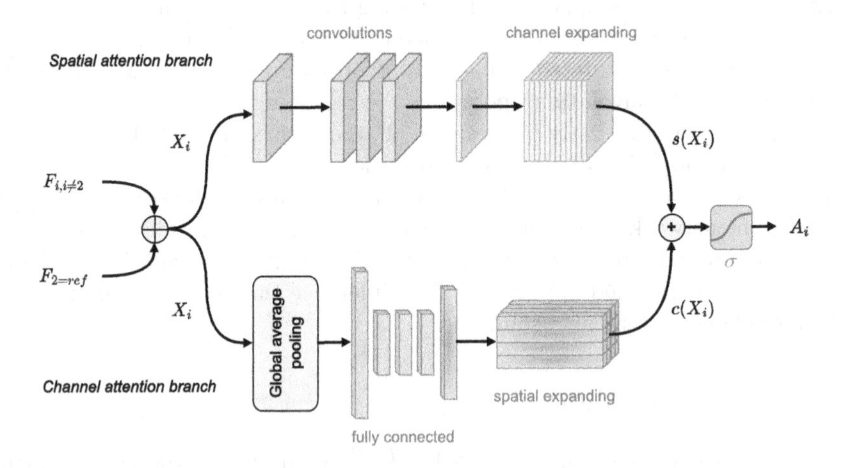

Fig. 2. Illustration of the attention module composed of 2 branches, respectively responsible for spatial attention and channel attention. A sigmoid activation function is used to keep the value between 0 and 1.

3.2 Attention Module

The final generated HDR image must be aligned with the reference image. To address this requirement [30] demonstrates the effectiveness of using a spatial attention module after the encoding step. Since then, many spatial and channel attention modules have been proposed in the literature which can be integrated into networks to improve their performance. According to the inference cost study done in Table 5, we propose the *Spatial-Channel Reference Attention Module* (SCRAM) a slightly modified version of the Bottleneck Attention Module (BAM) proposed in [19]. Indeed while BAM aims to generate a mask of its input feature maps, in our case we want to generate attention maps from the concatenation of reference and non-reference features, these attention maps are then applied to non-reference features only resulting in a reduction of the number of feature maps in the proposed attention module. Moreover, batch normalization is not applied in SCRAM. The detailed structure of SCRAM is illustrated in Fig. 2.

The non-reference features $F_{i \neq 2}$ are concatenated with the features of the reference image:

$$X_i = F_i \oplus F_{2=ref}, \quad i \neq 2 \tag{4}$$

where $X_i \in \mathbb{R}^{64 \times \frac{H}{2} \times \frac{W}{2}}$ is the input of SCRAM and $\oplus$ is the concatenation operator.

Following [19], SCRAM is composed of two branches respectively responsible for the spatial and channel features alignment of the non-reference images to the reference ones:

$$A_i = \sigma(s(X_i) + c(X_i)), \quad i \neq 2 \tag{5}$$

where s is the spatial attention branch and c is the channel attention branch. The sum of produced features by each branch passes through σ, a sigmoid activation function to keep the output values between 0 and 1.

Spatial Attention: The objective of the spatial branch is to produce an attention map allowing to keep the most relevant information for the spatial alignment of the non-reference images to the reference one. To limit the computation, we first reduce the number of features map by 3 using a pointwise convolution. With the objective to extract more global features while keeping the same computation, we then make the receptive field larger by employing 3 dilated convolutions [35] layers with a factor of dilatation set to 2. The final attention map of size $(1, H, W)$ is produced by using a pointwise convolution and then expanded across the channel dimension to obtain $A_S \in \mathbb{R}^{32 \times H \times W}$.

Channel Attention: This branch aims to perform a channel-wise feature recalibration. We first squeeze the spatial dimension by applying a global average pooling which sums out the spatial information to obtain the features vector of size $(64, 1, 1)$. A multilayer perceptron with three hidden layers is then used in purpose to estimate cross-channel attention. The last activation size is set to 32 to fit the number of channels of the non-reference features F_i. Finally, the resulting vector map is spatially expanded to obtain the final feature map $A_C \in \mathbb{R}^{32 \times H \times W}$.

The Attention features A_i are then used to weight the non-reference features F_i:

$$F_i' = F_i \otimes A_i, \quad i \neq 2 \tag{6}$$

where $\otimes$ is the element-wise product and F_i' is the aligned non-reference features. For the reference features we set $F_2' = F_2$.

3.3 Features Merging

While most of the computation is usually done in the merging block [13,18,30], we propose a novel efficient feature merging block that first focuses on merging the non-references features. Each feature map F_i' produced by the encoder goes through a convolution layer:

$$M_i = conv_{M_1}(F_i') \tag{7}$$

where $conv_{M_1}$ is a 3×3-convolution producing $M_i \in \mathbb{R}^{64 \times \frac{H}{2} \times \frac{W}{2}}$. Then we focus on merging the non-reference features maps by concatenating them and feeding the result features in a convolution layer:

$$M_{\text{non-ref}} = conv_{M_2}(M_1 \oplus M_3) \tag{8}$$

where $conv_{M_2}$ is a 3×3-convolution and $M_{\text{non-ref}} \in \mathbb{R}^{64 \times \frac{H}{2} \times \frac{W}{2}}$ is the non-reference merged features.

As we emphasize the reference features throughout all our network, here we merge our reference features with the non-reference $M_{non-ref}$ only by adding them together to limit the number of features map processed later:

$$M = conv_{M_4}(conv_{M_3}(M_2 + M_{non-ref})) \tag{9}$$

where $conv_{M_3}$ and $conv_{M_4}$ are 3×3-convolutions producing each 64 features map and $M \in \mathbb{R}^{64 \times \frac{H}{2} \times \frac{W}{2}}$ contains features from all LDR input images.

3.4 Features Decoding

The role of the decoder is to produce the final HDR image from the features produced by the merger block. At the encoding stage, we divided the spatial dimensions by 2. While the original spatial size is usually recovered using bilinear upsampling or transposed convolution [14] operation, we propose to use the pixel shuffle operation first proposed in [25], it is presented as an efficient sub-pixel convolution with a stride of $1/r$ where r is the upscale factor. In our case, we set $r = 2$. As illustrated in Fig. 3, the pixel shuffle layer rearranges elements in a tensor of shape $(C \times r^2, H, W)$ to a tensor of shape $(C, H \times r, W \times r)$:

$$D = PixelShuffle(M) \tag{10}$$

where $D \in \mathbb{R}^{16 \times H \times W}$ is the resulting upscaled features.

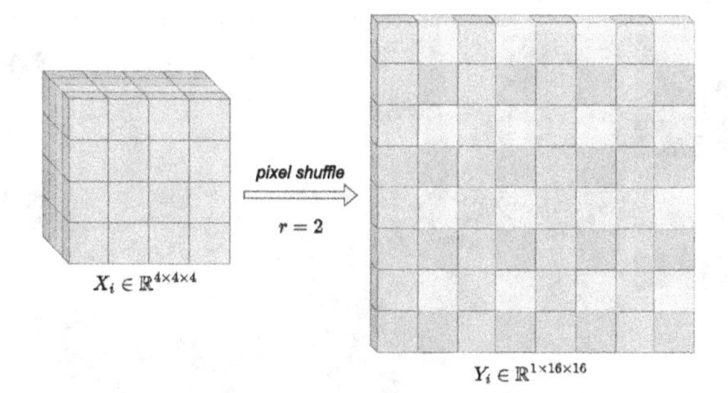

Fig. 3. Illustration of the pixel rearrangement by the pixel shuffle layer for an upscale factor set to $r = 2$ and an input shape of $(4, 4, 4)$. In the proposed solution, the input shape is $(64, \frac{H}{2}, \frac{W}{2})$, the produced output size is $(16, H, W)$.

The final HDR image is obtained by following the Eq. 11.

$$HDR = \sigma(conv_D(D + S_2)) \tag{11}$$

where S_2 is the reference features extracted by the first convolution layer of our network $conv_{E_1}$ to stabilize the training of our network. Finally, we generate the final HDR image using a 3×3-convolution layer, followed by a Sigmoid activation function.

4 Experimental Settings

4.1 Datasets

The CEN-HDR network has been trained using the dataset provided by [7] composed of 74 training samples and 15 test samples. Each sample represents the acquisition of a dynamic scene caused by large foreground or camera motions and is composed of three input LDR images (with EV of -2.00, 0.00, +2.00 or -3.00, 0.00, +3.00) and a reference HDR image aligned with the medium exposure image. The network has also been separately trained and tested using the dataset from the NTIRE [20] dataset, where the 3 LDR images are synthetically generated from the HDR images provided by [4]. The dataset is composed of 1500 training samples, 60 validation samples, and 201 testing samples. The ground-truth images for the testing sample are not provided.

4.2 Loss Function

Following previous works [7,30], the images have been mapped from the linear HDR domain to the LDR domain before evaluating the loss function. In order to train the network, the tone-mapping function has to be differentiable around zero, so, the μ-law function is defined as follows:

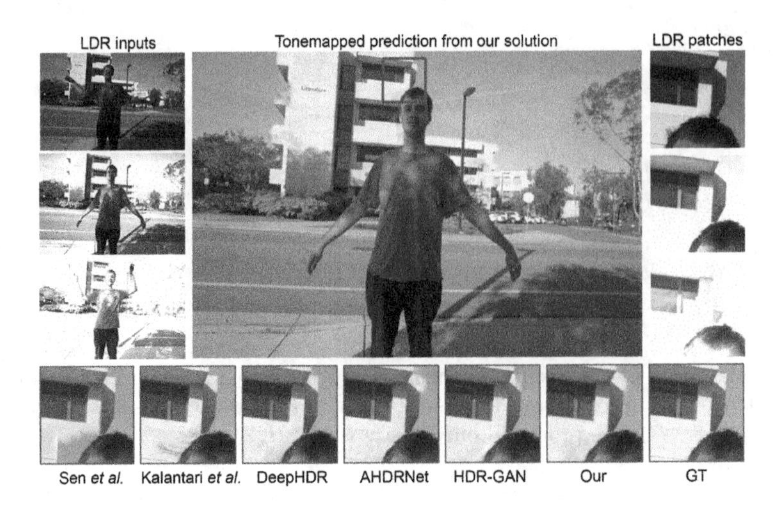

Fig. 4. Qualitative comparison of the proposed CEN-HDR solution with other HDR merging methods. The cropped patch demonstrates that the proposed efficient network has equivalent capabilities to the state-of-the-art methods to correct the ghost effect due to the large movement in the scene.

$$T(H) = \frac{log(1 + \mu H)}{log(1 + \mu)}, \quad \mu = 5000 \tag{12}$$

where H is the linear HDR image and μ the amount of compression.

To make an efficient network, a network compression method defined as knowledge distillation proposed in [5] has been used. By using knowledge distillation, we assume that the capacity of a large network is not fully exploited, so the objective is to transfer the knowledge of this large teacher network to our lighter network as described in the Eq. 13.

$$\mathcal{L} = \alpha \times \mathcal{L}(T, T_{GT}) + (1 - \alpha) \times \mathcal{L}(T, T_{Teacher}) \tag{13}$$

where $\mathcal{L}$ is the L_1 loss function. T is the tone mapped prediction of our network, T_{GT} the tone mapped ground truth provided in the dataset and $T_{Teacher}$ the tone mapped prediction of the large teacher network. we use the HDR-GAN [18] model as the teacher. α is a trade-off parameter set to 0.2. Moreover, the method proposed by [7] to produce the training dataset focuses mainly on foreground motion. It does not allow the generation of reliable ground-truth images for chaotic motions in the background, such as the movement of tree leaves due to wind, which results in a ground truth image with blurred features that do not reflect reality. This has the effect of producing a greater error when the predicted image contains sharper features than in the ground truth image. Using also a predicted image from a teacher model allows for dealing with this data misalignment. The comparison of the performance obtained between training done with knowledge distillation and without is made in Table. 1.

4.3 Implementation Details

The CEN-HDR network has been trained using cropped patches of size 256×256 pixels with a stride of 128 and evaluated on the full-resolution test images. During training, random augmentations are applied on the cropped patch such as horizontal symmetry and rotation of 90, 180, or 270°C. Training has been done using the Adam optimizer with a batch size of 8. The learning rate is initially set to 10^{-4}, keep fixed for 80 epochs, and decreased by 0.8 every 20 epochs after. The training lasts for 500 epochs.

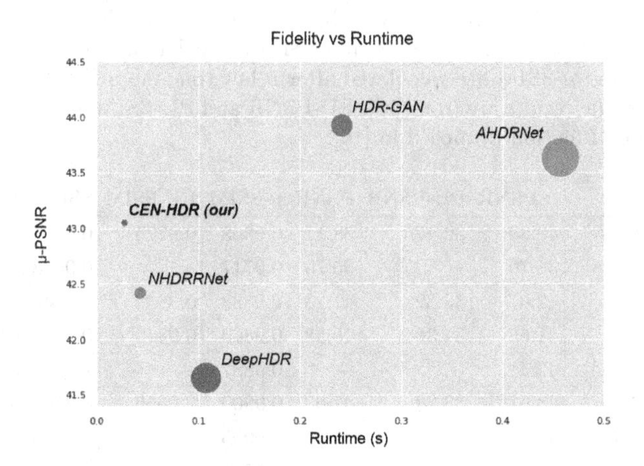

Fig. 5. Comparison of the proposed CEN-HDR solution with other HDR merging methods. The X-axis represents the mean runtime using the M1 NPU with an input size of 1280×720 pixels. The Y-axis is the fidelity score on the test images from [7] dataset. The best solutions tend to be in the upper left corner. The radius of circles represents the number of operations, the smaller the better.

5 Experimental Results

5.1 Fidelity Performance

In Table 2 and Fig. 4, the proposed CEN-HDR solution is compared against seven lightweight state-of-the-art methods: [6,24] are based on input patch registration methods. [7] is based on a sequential CNN, the inputs need first to be aligned thanks to an optical flow algorithm. For [28], the background of each LDR input is aligned by homography before being fed to an encoder-decoder-based CNN. [32] proposes an encoder-decoder architecture with a non-local attention module. [30] is a CNN based on an attention block for features registration and on multiple dilated residual dense blocks for merging. [18] is the first GAN-based approach for HDR merging with a deep supervised HDR method. Quantitative evaluation is done using objective metrics. The standard peak signal-to-noise

Table 1. Comparison of the proposed CEN-HDR architecture performances with and without knowledge distillation. In the first case, the network is trained with the HDR ground-truth proposed in [7]. In the second case, we also use the prediction of HDR-GAN [18] as label (Eq. 13). In both cases, the network is trained for 500 epochs using the l_1 criterion.

Training method	μ-PSNR	PSNR	μ-SSIM	SSIM	HDR-VDP2
w/o knowledge distillation	40.8983	40.0298	0.9772	0.9926	62.17
with knowledge distillation	43.0470	40.5335	0.9908	0.9956	64.34

Table 2. Quantitative comparison with lightweight state-of-the-art methods on the Kalatanri2017 [7] test samples. PSNR and SSIM are calculated in the linear domain while μ-PSNR and μ-SSIM are calculated after μ-law tone mapping (Eq. 12). For compared methods, the results are from [18]. PU-PSNR and PU-SSIM are calculated applying the encoding function proposed in [16].

Method	μ-PSNR	PU-PSNR	PSNR	μ-SSIM	PU-SSIM	SSIM	HDR-VDP2
Sen et al. [24]	40.80	32.47	38.11	0.9808	0.9775	0.9721	59.38
Hu et al. [6]	35.79	–	30.76	0.9717	–	0.9503	57.05
Kalantari et al. [7]	42.67	33.82	41.23	0.9888	0.9832	0.9846	65.05
DeepHDR [28]	41.65	31.36	40.88	0.9860	0.9815	0.9858	64.90
NHDRRNet [32]	42.41	–	–	0.9887	–	–	61.21
AHDRNet [30]	43.61	33.94	41.03	0.9900	0.9855	0.9702	64.61
HDRGAN [18]	43.92	34.04	41.57	0.9905	0.9851	0.9865	65.45
CEN-HDR(our)	43.05	33.23	40.53	0.9908	0.9821	0.9856	64.34

ratio (PSNR) and structural similarity (SSIM) are computed both directly in the linear domain and after tone mapping by applying the μ-law function (Eq. 12). The HDR-VDP2 [15] metric predicts the quality degradation with the respect to the reference image. We set the diagonal display size to 24 inches and the viewing distance to 0.5 m. In addition, PU-PSNR and PU-SSIM are calculated by applying the encoding function proposed in [16] with a peak luminance set to 4000.

In Fig. 6, we present the results for 3 test scenes of the NTIRE [20] dataset. While the network can produce images with a high dynamic range, we notice that the high sensor noise present in the input images is not fully corrected in the dark areas of the output HDR images. Moreover, the motion blur introduced in the dataset produces less sharp characteristics in the output image.

In Table 6 we study the effect of the attention module on the performance of proposed HDR deghosting architecture. SCRAM-C and SCRAM-S respectively correspond to the SCRAM module composed only of the channel attention branch and the spatial attention branch. The proposed SCRAM allows to achieve a similar quality as the spatial attention module proposed in AHDR-Net[28], while Table 5 shows that the inference cost of SCRAM is lower.

5.2 Efficiency Comparison

As we want to propose an efficient HDR generation method, in Table 3 we compare the computation cost and the inference time of our network with state-of-the-art HDR networks that achieve similar performance on quality metrics.

The number of operations and parameters are measured using the script provided by the NTIRE [20] challenge. To evaluate runtimes, all the compared networks are executed on the Neural Processing Unit (NPU) of a MacBook Pro (2021) powered with an M1 chip. The time shown is the average for 500 inference runs after a warm-up of 50 runs. The input size is set to 1280 × 720 pixels. The gamma-projection of LDR inputs (Eq. 1) and the tone mapping of the HDR output are included in the inference time measurement. Note that the background alignment of inputs frame using homography for [28] is not included.

Table 3. Inference cost comparison of the proposed CEN-HDR solution against state-of-the-art lightweight deep learning-based methods. The number of operations and parameters are measured using the script provided by the NTIRE [20] challenge. To measure the inference time, all the compared networks are executed on an M1 NPU. The input size is set to 1280 × 720 pixels.

Method	Num. of params	Num. of op. (GMAccs)	Runtime(s)	FPS
DeepHDR [28]	14618755	843.16	0.1075	9.30
AHDRNet [30]	1441283	1334.95	0.4571	2.18
NHDRRNet [32]	7672649	166.11	0.0431	23.20
HDR-GAN [18]	2631011	479.78	0.2414	4.14
CEN-HDR(our)	282883	78.36	0.0277	36.38

Table 4 compares the number of parameters and operations of the proposed CEN-HDR solution with recent efficient methods [9,33,36]. The input size is set to 1900 × 1060 pixels corresponding to the size of the inputs from the dataset proposed by [20]. For compared methods [9,33,36] the measurements are provided by the NTIRE [20] challenge. We could not compare the inference time of the CEN-HDR solution with these three architectures as they were recently proposed and their implementation is not yet available.

Figure 5 compares the trade-off between fidelity to the ground truth label and runtime of the proposed CEN-HDR solution with other HDR merging methods. The X-axis represents the mean runtime using an M1 NPU with an input size of 1280×720 pixels. The Y-axis is the fidelity score on the test images from [7] dataset. The best solutions tend to be in the upper left corner. The radius of circles represents the number of operations. Our solution is shown as the best solution for real-time HDR merging with a high-fidelity score.

Table 4. Inference cost comparison of the proposed CEN-HDR solution versus recent efficient merging networks. The number of operations and parameters for [9,33,36] and our solution are computed following the method described in [20].The input size is set to 1900×1060 pixels.

Method	Num. of params	Num. of op. (GMAccs)
GSANet [9]	80650	199.39
Yan et al. [33]	18899000	156.12
Yu et al. [36]	1013250	199.88
CEN-HDR(our)	282883	128.78

Table 5. Inference cost comparison of attention modules. Spatial and Channel attention modules are studied by feeding a tensor of size $(1, \frac{H}{4}, \frac{W}{4})$ corresponding to the concatenation of the reference and non-reference tensors after the encoding step.

Method	Attention type		params.	GMAccs	Runtime(s)
	Spatial	Channel			
AHDRNet attention [30]	✓		55392	20.772	0.0085
EPSANet [38]	✓		42560	15.768	0.0111
SK attention [10]		✓	125984	43.104	0.0155
Double attention [1]	✓	✓	33216	12.456	0.0101
CBAM [27]	✓	✓	22689	7.525	0.0734
BAM [19]	✓	✓	17348	5.008	0.0060

Table 6. Effect of the attention module on the performance of proposed HDR deghosting network. SCRAM-C and SCRAM-S respectively correspond to the SCRAM module composed only of the channel attention branch and the spatial attention branch.

Method	μ-PSNR	PSNR	μ-SSIM	SSIM
Without attention module	42.12	39.95	0.9850	0.9823
AHDRNet [30] attention module	42.94	40.49	0.9903	0.9852
SCRAM-C	42.32	40.14	0.9854	0.9829
SCRAM-S	42.89	40.41	0.9884	0.9835
SCRAM	43.05	40.53	0.9908	0.9856

Fig. 6. Qualitative results of the proposed CEN-HDR solution on samples from the NTIRE [20] challenge dataset. The ground truth images are not provided. We notice that the high sensor noise present in the input images is not fully corrected in the dark areas of the output HDR images. Moreover, the motion blur introduced in the dataset produces less sharp characteristics in the output image.

6 Conclusions

In this paper, we propose CEN-HDR, a novel computationally efficient HDR merging network able to correct the ghost effect caused by large object motions in the scene and camera motion. The proposed lightweight network architecture effectively succeeds in generating real-time HDR images with a dynamic range close to that of the original scene. By integrating the knowledge distillation methods in our training scheme, we demonstrate that the majority of the representation capabilities of a large HDR merging network can be transferred into a lighter network, opening the door to real-time HDR embedded systems.

References

1. Chen, Y., Kalantidis, Y., Li, J., Yan, S., Feng, J.: a^2-nets: double attention networks (2018). https://doi.org/10.48550/ARXIV.1810.11579. https://arxiv.org/abs/1810.11579
2. Chollet, F.: Xception: deep learning with depthwise separable convolutions (2016). https://doi.org/10.48550/ARXIV.1610.02357. https://arxiv.org/abs/1610.02357
3. Debevec, P.E., Malik, J.: Recovering high dynamic range radiance maps from photographs. In: Proceedings of the 24th Annual Conference on Computer Graphics and Interactive Techniques, pp. 369–378. SIGGRAPH 1997, ACM Press/Addison-Wesley Publishing Co., USA (1997). https://doi.org/10.1145/258734.258884. https://doi.org/10.1145/258734.258884
4. Froehlich, J., Grandinetti, S., Eberhardt, B., Walter, S., Schilling, A., Brendel, H.: Creating cinematic wide gamut HDR-video for the evaluation of tone mapping operators and HDR-displays. In: Proceedings SPIE 9023 (2014). https://doi.org/10.1117/12.2040003
5. Hinton, G., Vinyals, O., Dean, J.: Distilling the knowledge in a neural network (2015). https://doi.org/10.48550/ARXIV.1503.02531. https://arxiv.org/abs/1503.02531
6. Hu, J., Gallo, O., Pulli, K., Sun, X.: HDR deghosting: how to deal with saturation? In: 2013 IEEE Conference on Computer Vision and Pattern Recognition, pp. 1163–1170 (2013). https://doi.org/10.1109/CVPR.2013.154
7. Kalantari, N.K., Ramamoorthi, R.: Deep high dynamic range imaging of dynamic scenes. ACM Trans. Graph. (Proceedings of SIGGRAPH 2017) **36**(4), 1–12 (2017)
8. Khan, Z., Khanna, M., Raman, S.: FHDR: HDR image reconstruction from a single LDR image using feedback network (2019)
9. Li, F., et al.: Gamma-enhanced spatial attention network for efficient high dynamic range imaging. In: Proceedings of the IEEE/CVF Conference on Computer Vision and Pattern Recognition (CVPR) Workshops, pp. 1032–1040 (2022)
10. Li, X., Wang, W., Hu, X., Yang, J.: Selective kernel networks (2019). https://doi.org/10.48550/ARXIV.1903.06586. https://arxiv.org/abs/1903.06586
11. Liu, C.: Beyond pixels: exploring new representations and applications for motion analysis, Ph. D. thesis, MIT, USA (2009) aAI0822221
12. Liu, Y.L., et al.: Single-image HDR reconstruction by learning to reverse the camera pipeline (2020)
13. Liu, Z., et al.: Adnet: attention-guided deformable convolutional network for high dynamic range imaging (2021)

14. Long, J., Shelhamer, E., Darrell, T.: Fully convolutional networks for semantic segmentation. In: Proceedings of the IEEE Conference on Computer Vision and Pattern Recognition (CVPR) (2015)
15. Mantiuk, R., Kim, K.J., Rempel, A.G., Heidrich, W.: HDR-VDP-2: a calibrated visual metric for visibility and quality predictions in all luminance conditions. In: ACM SIGGRAPH 2011 Papers. SIGGRAPH 2011, Association for Computing Machinery, New York, NY, USA (2011). https://doi.org/10.1145/1964921.1964935. https://doi.org/10.1145/1964921.1964935
16. Mantiuk, R.K., Azimi, M.: Pu21: a novel perceptually uniform encoding for adapting existing quality metrics for HDR. In: 2021 Picture Coding Symposium (PCS), pp. 1–5 (2021). https://doi.org/10.1109/PCS50896.2021.9477471
17. Nayar, S., Mitsunaga, T.: High dynamic range imaging: spatially varying pixel exposures. In: Proceedings IEEE Conference on Computer Vision and Pattern Recognition, CVPR 2000 (Cat. No.PR00662), vol. 1, pp. 472–479 (2000). https://doi.org/10.1109/CVPR.2000.855857
18. Niu, Y., Wu, J., Liu, W., Guo, W., Lau, R.W.H.: HDR-GAN: HDR image reconstruction from multi-exposed LDR images with large motions. IEEE Trans. Image Process. **30**, 3885–3896 (2021). https://doi.org/10.1109/TIP.2021.3064433
19. Park, J., Woo, S., Lee, J.Y., Kweon, I.S.: Bam: bottleneck attention module (2018). https://doi.org/10.48550/ARXIV.1807.06514. https://arxiv.org/abs/1807.06514
20. Pérez-Pellitero, E., et al.: NTIRE 2022 challenge on high dynamic range imaging: methods and results. In: Proceedings of the IEEE/CVF Conference on Computer Vision and Pattern Recognition (CVPR) Workshops, pp. 1009–1023 (2022)
21. Prabhakar, K.R., Agrawal, S., Singh, D.K., Ashwath, B., Babu, R.V.: Towards practical and efficient high-resolution HDR Deghosting with CNN. In: Vedaldi, A., Bischof, H., Brox, T., Frahm, J.-M. (eds.) ECCV 2020. LNCS, vol. 12366, pp. 497–513. Springer, Cham (2020). https://doi.org/10.1007/978-3-030-58589-1_30
22. Prabhakar, K.R., Agrawal, S., Babu, R.V.: Self-gated memory recurrent network for efficient scalable HDR deghosting. CoRR abs/2112.13050 (2021). https://arxiv.org/abs/2112.13050
23. Ronneberger, O., Fischer, P., Brox, T.: U-net: convolutional networks for biomedical image segmentation. CoRR abs/1505.04597 (2015). https://arxiv.org/abs/1505.04597
24. Sen, P., Kalantari, N.K., Yaesoubi, M., Darabi, S., Goldman, D.B., Shechtman, E.: Robust patch-based HDR reconstruction of dynamic scenes. ACM Trans. Graphics (TOG) (Proceedings of SIGGRAPH Asia 2012) **31**(6), 1–11 (2012)
25. Shi, W., et al.: Real-time single image and video super-resolution using an efficient sub-pixel convolutional neural network (2016)
26. Tumblin, J., Agrawal, A., Raskar, R.: Why i want a gradient camera. In: 2005 IEEE Computer Society Conference on Computer Vision and Pattern Recognition (CVPR2005), vol. 1, pp. 103–110 (2005). https://doi.org/10.1109/CVPR.2005.374
27. Woo, S., Park, J., Lee, J.Y., Kweon, I.S.: CBAM: convolutional block attention module (2018). https://doi.org/10.48550/ARXIV.1807.06521. https://arxiv.org/abs/1807.06521
28. Wu, S., Xu, J., Tai, Y., Tang, C.: End-to-end deep HDR imaging with large foreground motions. CoRR abs/1711.08937 (2017). https://arxiv.org/abs/1711.08937
29. Wu, S., Xu, J., Tai, Y.-W., Tang, C.-K.: Deep high dynamic range imaging with large foreground motions. In: Ferrari, V., Hebert, M., Sminchisescu, C., Weiss, Y. (eds.) ECCV 2018. LNCS, vol. 11206, pp. 120–135. Springer, Cham (2018). https://doi.org/10.1007/978-3-030-01216-8_8

30. Yan, Q., et al.: Attention-guided network for ghost-free high dynamic range imaging. In: IEEE Conference on Computer Vision and Pattern Recognition (CVPR), pp. 1751–1760 (2019)
31. Yan, Q., et al.: Multi-scale dense networks for deep high dynamic range imaging. In: 2019 IEEE Winter Conference on Applications of Computer Vision (WACV), pp. 41–50 (2019). https://doi.org/10.1109/WACV.2019.00012
32. Yan, Q., et al.: Deep HDR imaging via a non-local network. IEEE Trans. Image Process. **29**, 4308–4322 (2020). https://doi.org/10.1109/TIP.2020.2971346
33. Yan, Q., et al.: A lightweight network for high dynamic range imaging. In: Proceedings of the IEEE/CVF Conference on Computer Vision and Pattern Recognition (CVPR) Workshops, pp. 824–832 (2022)
34. Ye, Q., Xiao, J., Lam, K., Okatani, T.: Progressive and selective fusion network for high dynamic range imaging. CoRR abs/2108.08585 (2021). https://arxiv.org/abs/2108.08585
35. Yu, F., Koltun, V.: Multi-scale context aggregation by dilated convolutions (2015). https://doi.org/10.48550/ARXIV.1511.07122. https://arxiv.org/abs/1511.07122
36. Yu, G., Zhang, J., Ma, Z., Wang, H.: Efficient progressive high dynamic range image restoration via attention and alignment network (2022). https://doi.org/10.48550/ARXIV.2204.09213. https://arxiv.org/abs/2204.09213
37. Zeng, H., Cai, J., Li, L., Cao, Z., Zhang, L.: Learning image-adaptive 3D lookup tables for high performance photo enhancement in real-time. In: IEEE Transactions on Pattern Analysis and Machine Intelligence, p. 1 (2020). https://doi.org/10.1109/TPAMI.2020.3026740
38. Zhang, H., Zu, K., Lu, J., Zou, Y., Meng, D.: EPSANet: an efficient pyramid squeeze attention block on convolutional neural network (2021). https://doi.org/10.48550/ARXIV.2105.14447. https://arxiv.org/abs/2105.14447

Image Super-Resolution with Deep Variational Autoencoders

Darius Chira[1(✉)], Ilian Haralampiev[1], Ole Winther[1,2,3], Andrea Dittadi[1,4], and Valentin Liévin[1]

[1] Technical University of Denmark, Copenhagen, Denmark
darius.chira14@gmail.com
[2] University of Copenhagen, Copenhagen, Denmark
[3] Rigshospitalet, Copenhagen University Hospital, Copenhagen, Denmark
[4] Max Planck Institute for Intelligent Systems, Tübingen, Germany

Abstract. Image super-resolution (SR) techniques are used to generate a high-resolution image from a low-resolution image. Until now, deep generative models such as autoregressive models and Generative Adversarial Networks (GANs) have proven to be effective at modelling high-resolution images. VAE-based models have often been criticised for their feeble generative performance, but with new advancements such as VDVAE, there is now strong evidence that deep VAEs have the potential to outperform current state-of-the-art models for high-resolution image generation. In this paper, we introduce VDVAE-SR, a new model that aims to exploit the most recent deep VAE methodologies to improve upon the results of similar models. VDVAE-SR tackles image super-resolution using transfer learning on pretrained VDVAEs. The presented model is competitive with other state-of-the-art models, having comparable results on image quality metrics.

Keywords: VDVAE · SR · Single-image super-resolution · Deep variational autoencoders · Transfer learning

1 Introduction

Single Image Super-Resolution (SISR) consists in producing a high-resolution image from its low-resolution counterpart. Image super-resolution has long been considered one of the most arduous challenges in image processing. This is yet another computer vision task that was transformed by the deep learning revolution and has potential applications including but not limited to medical imaging, security, computer graphics, and surveillance.

D. Chira and I. Haralampiev—Equal contribution, alphabetical order.
A. Dittadi and V. Liévin—Equal advising, alphabetical order.

Supplementary Information The online version contains supplementary material available at https://doi.org/10.1007/978-3-031-25063-7_24.

L. Karlinsky et al. (Eds.): ECCV 2022 Workshops, LNCS 13802, pp. 395–411, 2023.
https://doi.org/10.1007/978-3-031-25063-7_24

Deep generative models have been shown to excel at image generation. This is particularly true for autoregressive models [5,33–35,42] and Generative Adversarial Networks (GAN) [4,12,19,47], whereas Variational Autoencoders (VAE) [21,37] have long been thought to be unable to produce high-quality samples. However, recent improvements in VAE design, such as using a hierarchy of latent variables and increasing depth [6,23,28,43] have demonstrated that deep VAEs can compete with both GANs and autoregressive models for high-resolution image generation. The current state-of-the-art VAE is the Very Deep Variational Autoencoder (VDVAE) [6] which successfully scales to 78 stochastic layers, whereas previous work only experimented with up to 40 layers [43].

Since the VAE is an unconditional generative model, in order to perform image super-resolution it has to be turned into a *conditional* generative model which generates data depending on additional conditioning data. This can be achieved by using the framework of Conditional Variational Autoencoders (CVAE) [39], where the prior is conditioned on an additional random variable and parameterized by a neural network. In this work, we introduce **VDVAE-SR**, a VDVAE conditioned on low-resolution images by adding a new component that we call LR-encoder as it resembles the encoder of the original VDVAE. This component is connected to the decoder, passing information on each layer in the top-down path both to the prior and the approximate posterior. During training, the latent distributions of the low- and high-resolution images are matched using the KL divergence term in the evidence lower bound (ELBO). The learned information is used in generative mode, where only the low-resolution image is included in the model.

A drawback of deep models such as the VDVAE is that they require a large amount of computing and training time. One way to compensate in that regard is to apply transfer learning and utilize a pre-trained model in order to speed up the process. However, this is not always straightforward in practice as presenting a pre-trained model with new data could lead to exploding gradients. This is particularly relevant for deep variational autoencoders as they are prone to unstable training and can be sensitive to hyperparameters changes. We show that using transfer learning for such a model is possible, and we describe the methods to do so, by making only certain parts of the network trainable and using gate parameters to stabilise the process.

We fine-tune a VDVAE model pretrained on FFHQ 256×256 [19] using DIV2K [1], a common dataset in the image super-resolution literature [7,26,32,44]. We evaluate the fine-tuned model on a number of common datasets in the literature of single image super-resolution: Set5 [3], Set14 [45], Urban100 [16], BSD100 [29], and Manga109 [30]. Following previous work [24,32,44], we test our approach both quantitatively, in terms of PSNR and SSIM metrics, and qualitatively, by visually inspecting the generated images, and compare our results against three state-of-the-art super-resolution methods: EDSR [26], ESRGAN [44] and RFANet [27]. We investigate the role of the sampling temperature, which controls the variance of samples at each stochastic layer in VDVAEs, and show results generated with low and high temperatures. By sampling with a lower temperature, the model achieves quantitative scores better than ESRGAN, but slightly lower than EDSR. At the same time, qualitatively, when sampling with a higher temperature, the images look sharper and less blurred than those generated by the EDSR model. Even though, in general, ESRGAN generates sharper images, it is prone to produce more artifacts as well. We believe that our proposed method shows a good compromise between visual artifacts and image sharpness.

We summarize our contributions as follows:

1. We propose VDVAE-SR, an adaptation of very deep VAEs (VDVAEs) for the task of single image super-resolution. VDVAE-SR introduces an additional component that we call LR-encoder, which takes the low-resolution image as input, while its output is used to condition the prior.
2. We show how to utilize transfer learning and achieve stable training in order to take advantage of a VDVAE model already pre-trained on 32 V100 GPUs for 2.5 weeks.
3. We present competitive qualitative and quantitative results compared to state-of-the-art methods on popular test datasets for 4x upscaling.

2 Related Work

One of the first successes in image super-resolution is the SR-CNN [10], which is based on a three-layer CNN structure and uses a bicubic interpolated low-resolution image as input to the network. Later, with the proposal of residual neural networks (ResNets) [13], which provide fast training and better performance for deep architectures, numerous works have adapted ResNets-based models to the task of super-resolution, such as SR-ResNet [24] and SR-DenseNet [41]. One of the frequently used CNN-based super-resolution models in comparative studies is EDSR [26], where the authors use ResNets without batch normalization in the residual block, achieving impressive results and getting first place on the NTIRE2017 Super-Resolution Challenge.

In terms of GAN-based image super-resolution models, several methods have gained a lot of popularity starting with SRGAN [24] where the authors argue that most popular metrics (PSNR, SSIM) do not necessarily reflect perceptually better SR results and that is why they use an extensive mean opinion score (MOS) for evaluating perceptual quality. With that in mind, SRGAN introduces a perceptual loss different from previous work, based on adversarial as well as content loss. Another method, ESRGAN [44], builds upon SRGAN by improving the network architecture removing all batch normalization layers and introducing a new Residual in Residual Dense Block (RRDB). In addition, an enhanced discriminator is used based on Relativistic GAN [18] and the features before the activation loss are used to improve perceptual loss.

A recent work that uses VAEs for image super-resolution is the srVAE [11], which consists of a VAE with three latent variables, one of them being a down-scaled version of the original image. This work shows impressive generative performance in terms of FID score when tested on ImageNet-32 and CIFAR-10, but no quantitative results of their super-resolution model are reported. Another recent work that uses a VAE-based model for image super-resolution is VarSR [17]. This work focuses on very low-resolution images (8×8) and shows better results compared to some popular super-resolution methods.

Deep VAEs such as [6,23,28,43] adapt their architecture from Ladder VAEs (LVAE) [40], which introduce a novel top-down inference model and achieve stable training with multiple stochastic layers. A method that improved upon the LVAE is the Bidirectional-Inference VAE (BIVA) [28] adding a deterministic top-down path in the generative model and applying a bidirectional inference network. These modifications solved the variable collapse issue of the LVAE which may occur when the architecture consists of a very deep hierarchy of stochastic latent variables. Recently, NVAE [43] reported further improvements by using normalizing flows in order to allow for more expressive distributions and thus outperform the state-of-the-art among non-autoregressive and VAE models. Finally, the VDVAE model [6] demonstrated that the number of stochastic layers matters greatly for performance, achieving better results than previous VAE-based models and some autoregressive ones, having the potential to outperform those as well.

Denoising Diffusion Probabilistic Models (DDPM) [14] are the latest addition to the family of probabilistic generative models. DDPMs define a diffusion process that progressively turns the input image into noise, and learn to synthesize images by inverting that process. DDPMs and variations thereof excel at high-resolution image generation [8,31] and have been successfully applied to the task of single image super-resolution [15,25].

3 Preliminaries

In this section, we define variational autoencoders (VAEs) and conditional VAEs (CVAEs), for which we derive the evidence lower bound. We then introduce the VDVAE using the VAE framework.

3.1 Variational Autoencoders

The Variational Autoencoder [21] is a generative model built on probabilistic principles. It consists of a joint model $p_\theta(\mathbf{x}, \mathbf{z}) = p_\theta(\mathbf{x}|\mathbf{z})p_\theta(\mathbf{z})$ parameterized by θ and an approximate posterior $q_\phi(\mathbf{z}|\mathbf{x})$ parameterized by ϕ. All models are implemented using neural networks. During generation, the latent variable $\mathbf{z}$ is sampled from the prior and the observation variable $\mathbf{x}$ is sampled from the observation model following $\mathbf{z} \sim p_\theta(\mathbf{z}), \mathbf{x} \sim p_\theta(\mathbf{x}|\mathbf{z})$.

VAE models are optimized with stochastic gradient ascent to maximize the marginal likelihood:

$$p_\theta(\mathbf{x}) = \int_{\mathbf{z}} p_\theta(\mathbf{x}|\mathbf{z})p_\theta(\mathbf{z})d\mathbf{z} \tag{1}$$

In practice, $p_\theta(\mathbf{x})$ is intractable of the integration over $\mathbf{z}$, which makes the posterior $p_\theta(\mathbf{z}|\mathbf{x})$ also intractable. Variational Inference (VI) solves the intractability of $p_\theta(\mathbf{z}|\mathbf{x})$ using an approximate posterior $q_\phi(\mathbf{z}|\mathbf{x})$. The resulting objective function, the evidence lower bound (ELBO), is further derived using Jensen's inequality and expressed as:

$$\mathcal{L}(\mathbf{x}; \theta, \phi) = E_{q_\phi(\mathbf{z}|\mathbf{x})}\left[log\frac{p_\theta(\mathbf{x}, \mathbf{z})}{q_\phi(\mathbf{z}|\mathbf{x})}\right] \leq \log p_\theta(\mathbf{x}) \; . \tag{2}$$

3.2 Conditional Variational Autoencoders

In order to generate specific data as in the case of image super-resolution, where we need to generate a high-resolution image from its low-resolution counterpart, the Conditional Variational Autoencoders (CVAE) can be used.

Similar to the VAE, the CVAE is also built on probabilistic principles. CVAE is optimized to maximize the marginal probability similar to Eq. (2) but this time conditioned on a random variable which could be for example a low-resolution image $\mathbf{y}$:

$$p_\theta(\mathbf{x}|\mathbf{y}) = \int_{\mathbf{z}} p_\theta(\mathbf{x}|\mathbf{y},\mathbf{z})p(\mathbf{z}|\mathbf{y})d\mathbf{z} \tag{3}$$

The posterior of the latent variables is:

$$p_\theta(\mathbf{z}|\mathbf{x},\mathbf{y}) = \frac{p_\theta(\mathbf{x}|\mathbf{z},\mathbf{y})p_\theta(\mathbf{z}|\mathbf{y})}{p_\theta(\mathbf{x}|\mathbf{y})} \tag{4}$$

where again $p_\theta(\mathbf{x}|\mathbf{y})$ is intractable and needs to be approximated using a variational distribution $q_\phi(\mathbf{z}|\mathbf{x},\mathbf{y}) \approx p_\theta(\mathbf{z}|\mathbf{x},\mathbf{y})$. The conditional ELBO for the CVAE can be derived again using Jensen's inequality, resulting in:

$$\mathcal{L}(\mathbf{x},\mathbf{y};\theta,\phi) = E_{q_\phi(\mathbf{z}|\mathbf{x},\mathbf{y})}\left[\log \frac{p_\theta(\mathbf{x},\mathbf{z},\mathbf{y})}{q_\phi(\mathbf{z}|\mathbf{x},\mathbf{y})}\right] \leq \log p_\theta(\mathbf{x}|\mathbf{y}) \ . \tag{5}$$

3.3 Very Deep Variational Autoencoder (VDVAE)

The VDVAE [6] consists of a hierarchy of layers of latent variables conditionally dependent on each other. This results in a more flexible prior and posterior compared to a simple diagonal Gaussian prior which could be too limiting. An iterative interaction between "bottom-up" and "top-down" layers is achieved through parameter sharing between the inference and generative models in each layer. The prior and the approximate posterior for a model with K stochastic layers factorize as:

$$p_\theta(\mathbf{z}) = p_\theta(\mathbf{z}_0)p_\theta(\mathbf{z}_1|\mathbf{z}_0)...p_\theta(\mathbf{z}_K|\mathbf{z}_{<K}) \tag{6}$$

$$q_\phi(\mathbf{z}|\mathbf{x}) = q_\phi(\mathbf{z}_0|\mathbf{x})q_\phi(\mathbf{z}_1|\mathbf{z}_0,\mathbf{x})...q_\phi(\mathbf{z}_K|\mathbf{z}_{<K},\mathbf{x}) \tag{7}$$

where $p_\theta(\mathbf{z}_0)$ is a diagonal Gaussian distribution $\mathcal{N}(\mathbf{z}_0 \mid \mathbf{0}, \mathbf{I})$ and the latent variable group $\mathbf{z}_0$ is at the top layer that corresponds to small number of latent variables at low resolution. Intuitively, $\mathbf{z}_K$ is at the bottom of the network having a larger number of latent variables at high resolution.

The VDVAE architecture is composed of blocks of two types: the residual blocks (bottom-up path) and the top-down blocks (Fig. 1). The top-down blocks are also residual and handle two tasks: processing the information flowing through the decoder and handling the stochasticity. Each top-down block of index $j > 0$ handles the distributions $q_\phi(\mathbf{z}_j|\mathbf{z}_{j-1},\mathbf{x})$ and $p_\theta(\mathbf{z}_j|\mathbf{z}_{j-1})$. Top-down blocks are composed sequentially. Therefore, we can define $\mathbf{h}_j$ as the input to the top-down block of index j, where $\mathbf{h}_j$ is a function of the samples $\mathbf{z}_{<j}$. This allows us to express the VDVAE model as:

$$p_\theta(\mathbf{z}) = p_\theta(\mathbf{z}_0) \prod_{j=1}^{K} p_\theta(\mathbf{z}_j|\mathbf{h}_j), \qquad q_\phi(\mathbf{z}|\mathbf{x}) = q_\phi(\mathbf{z}_0|\mathbf{x}) \prod_{j=1}^{K} q_\phi(\mathbf{z}_j|\mathbf{h}_j, \mathbf{x}) . \tag{8}$$

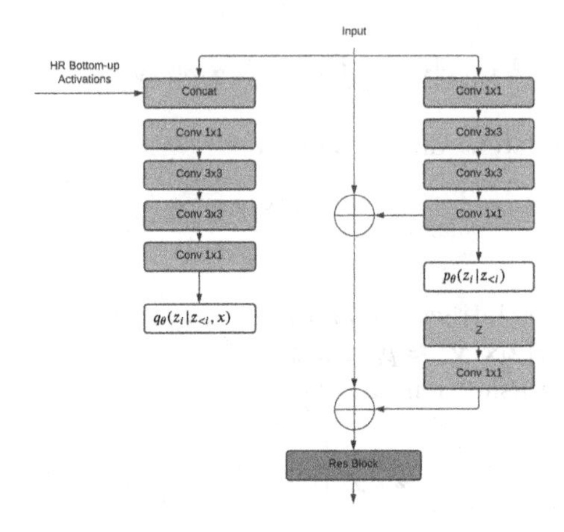

Fig. 1. Top-down block of the VDVAE [6].

4 VDVAE-SR

In this section, we introduce the proposed VDVAE-SR model. We provide an overview of the model architecture, after which we detail the conditional prior network and its integration with the VDVAE model.

4.1 LR-encoder

The dependency on the lower-resolution image $\mathbf{y}$ is implemented using the encoder of a lower-resolution VDVAE of depth $K' < K$, which we call LR-encoder. The LR-encoder maps the lower-resolution image to latent space, providing one activation $\mathbf{g}_j$ for each layer $j \in [0, K']$. Each activation $\mathbf{g}_j$ is defined as the output of the bottom-up residual block of index j.

4.2 Conditional Prior

The top-down path, or decoder, of the VDVAE is modified to depend on $\mathbf{y}$ using the LR-encoder activations $\mathbf{g}_0, \ldots, \mathbf{g}_{K'}$. This results in a conditional prior

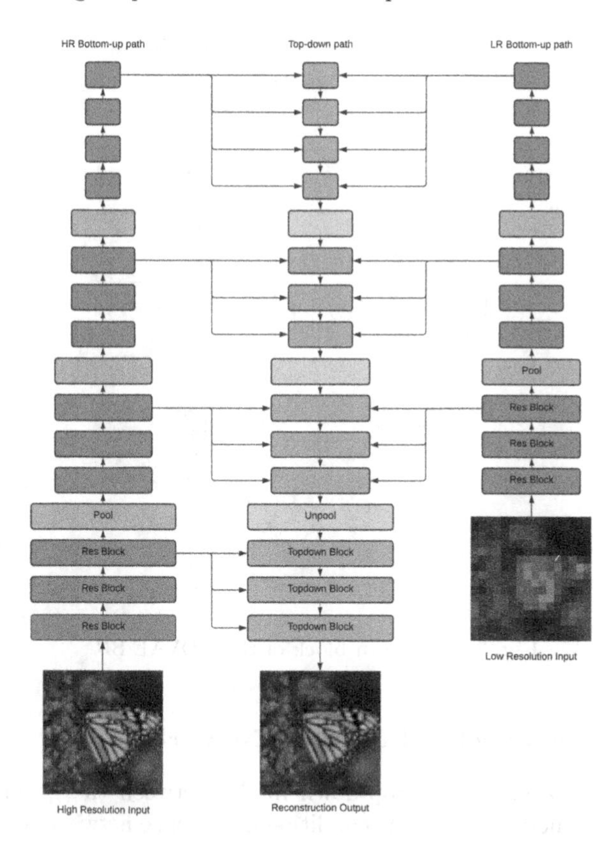

Fig. 2. Network Architecture of the proposed VDVAE-SR Model.

$p_\theta(\mathbf{z}|\mathbf{y})$ that maps the low-resolution image $\mathbf{y}$ to a distribution over the latent variables $\mathbf{z}$ (Fig. 2).

The architecture of the VDVAE-SR is identical to the one of the VDVAE, except for two alterations:

1. The input to each top-down block (see Fig. 3) is defined as:

$$\tilde{\mathbf{h}}_j = \begin{cases} \mathbf{g}_j & \text{if } j = 0 \\ \mathbf{h}_j + \alpha_j \mathbf{g}_j & \text{if } j \in [1, K'] \\ \mathbf{h}_j & \text{otherwise} \end{cases} \tag{9}$$

where $\alpha_1, \ldots, \alpha_{K'}$ are scalar gate parameters initialized to zero [2].

2. The top layer is conditioned on $\mathbf{y}$ such that

$$p_\theta(\mathbf{z}_0|\tilde{\mathbf{h}}) = \mathcal{N}\left(\mathbf{z}_0 \,|\, \mu_\theta(\tilde{\mathbf{h}}_0),\, \sigma_\theta(\tilde{\mathbf{h}}_0)\right) , \tag{10}$$

where μ_θ and σ_θ are linear layers mapping the output of the top-most LR-encoder layer to the parameter-space of $p_\theta(\mathbf{z}_0|\mathbf{y})$.

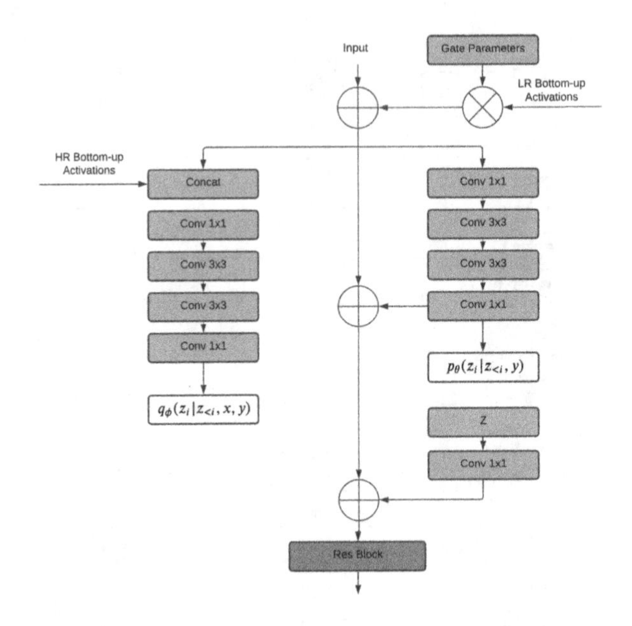

Fig. 3. Top-down block of the VDVAE-SR.

4.3 Generative Model and Inference Network

Because of the sharing of the top-down model between the generative model and the inference network [40], the conditional inference network naturally arises from the alteration of the prior, without further modification. Using the activations $\tilde{\mathbf{h}}_0, \ldots, \tilde{\mathbf{h}}_K$ and the definition of the VDVAE given in Eq. (8), we define the VDVAE-SR as:

$$p_\theta(\mathbf{z}|\mathbf{y}) = \prod_{j=0}^{K} p_\theta(\mathbf{z}_j|\tilde{\mathbf{h}}_j), \qquad q_\phi(\mathbf{z}|\mathbf{y}, \mathbf{x}) = q_\phi(\mathbf{z}_0|\mathbf{x}) \prod_{j=1}^{K} q_\phi(\mathbf{z}_j|\tilde{\mathbf{h}}_j, \mathbf{x}) \,. \tag{11}$$

5 Experiments

5.1 Datasets

Training Dataset. We train our models on the DIV2K dataset, introduced by [1]. The DIV2K dataset consists of 800 RGB high-definition high-resolution images for training, 100 images for validation, and 100 for testing. The dataset contains a variety of diverse pictures, including different types of shot such as portrait, scenery, and object shots.

Test Datasets. We test our method on popular benchmarking datasets commonly used in single-image super resolution: Set5 [3], Set14 [45], Urban100 [16], BSD100 [29], and Manga109 [30]. Having multiple test datasets gives a better understanding of the strengths and shortcomings of our model, since these datasets contain different types of pictures: BSD100, Set5, and Set14 mostly consist of natural images with a broad range of styles, while the focus on Urban100 is mainly on buildings and urban scenes, and Manga109 consists of drawings of Japanese manga.

5.2 Implementation Details

Since it takes about 2.5 weeks to train a VDVAE model on FFHQ 256×256 on 32 NVIDIA V100 GPUs, we choose to rely on pretrained VDVAEs and adapt them to the super-resolution task. We use a pretrained VDVAE with a stochastic depth of 62 layers. Our method, VDVAE-SR, includes the original VDVAE encoder and decoder, which we initialize with the weights from the pretrained model. We then freeze the encoder, allow fine-tuning of the decoder, and train the LR-encoder from scratch. We optimize the model end-to-end for 100,000 steps using the Adam optimizer [20] with a learning rate of $5 \cdot 10^{-4}$ and batch size of 1 on one NVIDIA V100 GPU.

When using transfer learning, it was observed that the model suffered from exploding gradients if the new information from the LR-encoder was introduced in an uncontrolled manner. Introducing gate parameters similar to the approach in [2] significantly improved training stability.

5.3 Evaluation

In terms of evaluation metrics, we use the traditional PSNR and SSIM quality measures, both widely used as metrics for image restoration tasks. While PSNR (Peak Signal to Noise Ratio) is calculated based on the mean squared error of the pixel-to-pixel difference, the SSIM (Structural Similarity Method) is considered to have a closer correlation with human perception by calculating distortion levels based on comparisons of structure, luminance, and contrast. Additional to the traditional PSNR and SSIM metrics, we evaluate the produced images using the DISTS [9] score, which has showed evidence that the metric matches closer to human perception. We quantitatively evaluate different super-resolution methods by applying them to low-resolution images and computing the PSNR, SSIM and DISTS metrics using the super-resolution output and the reference high-resolution image. For the PSNR and SSIM, all pictures are converted from RGB to YCbCr and the metrics are computed on the Y channel (luma component) of the pictures. The reason for this is that it has been observed (e.g., in [36]) that the results of evaluating on the luminosity channel in the YCbCr color space, rather than on the usual RGB representation, are closer to the actual perceived structural noise of the image. We thus adopt the same approach, following prior work. Finally, note that the YCbCr space is used during the testing phase exclusively, while the training and validation are still performed in the RGB color space.

5.4 Results

Quantitative Results. We compare our method to three other super-resolution methods, namely EDSR, ESRGAN and RFANet, based on their official implementation. The quantitative results on PSNR and SSIM are shown in Table 1, where EDSR performs best on both metrics, with our method ($t = 0.1$) closely following on second place.

Table 1. Evaluation metrics using PSNR and SSIM on the Y channel and DISTS. The number next to VDVAE-SR (our method) denotes the temperature used for sampling. The best scores are represented in **bold**, while the second best results are underlined.

Dataset	EDSR			ESRGAN			RFANet			VDVAE-SR 0.1			VDVAE-SR 0.8		
	PSNR	SSIM	DISTS	PSNR	SSIM	DISTS	PSNR	SSIM	DISTS	PSNR	SSIM	DISTS	PSNR	SSIM	DISTS
Set5	31.97	0.902	0.121	30.39	0.864	**0.078**	**32.53**	**0.908**	0.119	31.48	0.886	0.123	30.51	0.869	0.108
Set14	**28.33**	**0.800**	0.097	26.20	0.720	**0.064**	27.33	0.774	0.092	27.99	0.776	0.105	27.62	0.761	0.097
BSDS100	**28.46**	**0.781**	0.158	25.87	0.690	**0.094**	27.04	0.758	0.154	28.05	0.752	0.169	27.69	0.738	0.152
Manga109	**30.85**	**0.918**	**0.009**	28.77	0.870	0.010	21.09	0.739	0.015	29.92	0.904	0.013	29.55	0.896	0.011
Urban100	26.02	0.798	0.029	24.36	0.748	0.024	**26.89**	**0.823**	**0.023**	25.36	0.759	0.037	25.15	0.750	0.034

As first discussed in [24], the PSNR and SSIM scores tend to favor smoother images, this being attributed to the nature of how these metrics are calculated, which is in contrast to human visual perception. This is confirmed by the obtained scores of our method using different temperatures as decreasing the variance produces more averaged-out images and thus higher scores. Based on the DISTS metric, ESRGAN performs best on three datasets. Our method with higher temperature follows on second place on the Set5 and BSDS100 datasets.

Qualitative Results. Figures 4 to 6 show a visual comparison of two pictures from BSD100 dataset between the original HR image, Bicubic, EDSR, ESRGAN, RFANet and our method with both 0.1 and 0.8 temperatures.

It can be observed that the points made in the quantitative section still stand, as EDSR, having the best PSNR and SSIM scores, has a smoother and blurrier look, and our model with 0.1 temperature looks closer to it. As for the model with 0.8 temperature, it introduces more details compared to EDSR. It is still blurrier than the outputs of ESRGAN but has fewer artifacts and it is able to reproduce some details without introducing any generative noise. As for the RFANet, the images are still blurrier, but having more visual similarities with our method than EDSR. For this reason in most metrics it gets a better score, but visually it still does not generate highly detailed features.

In Figs. 4 to 5 it can be observed that ESRGAN produces some artifacts on the bull's head and the person's hand, while our model retains the structure of the objects. In Fig. 6 we can again see how the eye of the bird has a different shape and a more averaged-out look in the case of the EDSR, and even more drastic shape change in the case of the ESRGAN, while our models keep the rounder shape, while not averaging out the outer colors as much.

Fig. 4. SR output comparison between multiple models for a picture (image 376043) of the BSD100 DataSet.

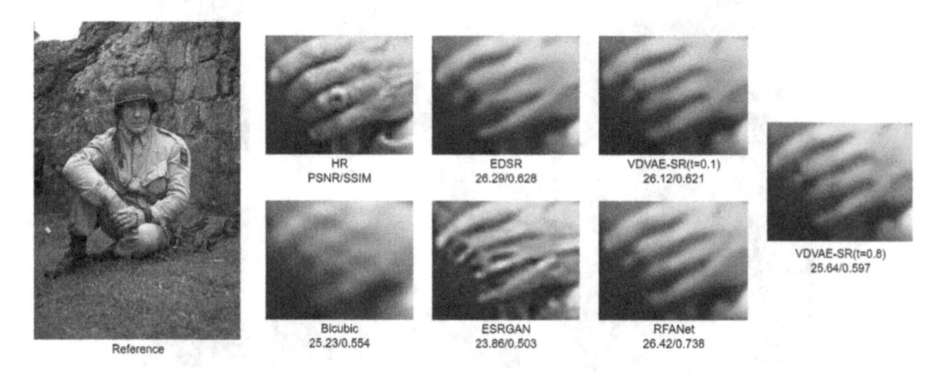

Fig. 5. SR output comparison between multiple models for a picture (image 38092) of the BSD100 DataSet.

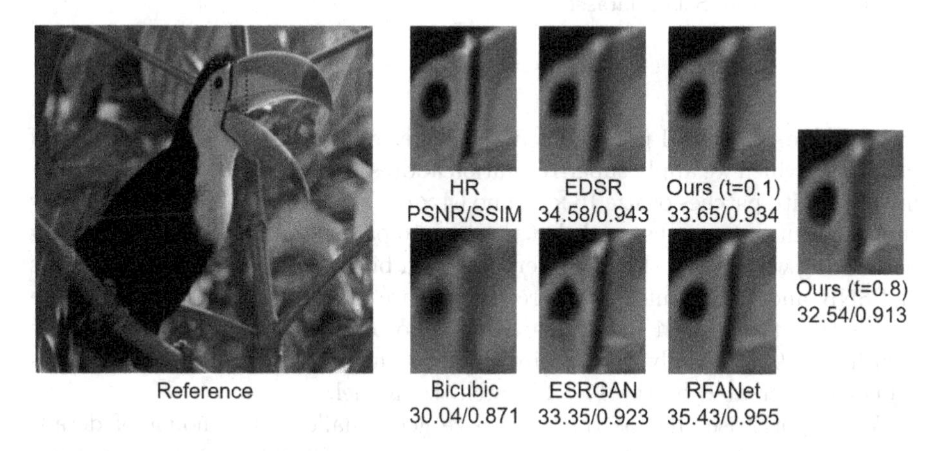

Fig. 6. SR output comparison between multiple models for the bird picture of the Set5 DataSet.

Temperature. The "temperature" parameter t, taking values between 0 and 1, is used in VDVAE when sampling from prior in generative mode, often resulting in higher-quality samples when lowered as observed in previous work [22,43]. Reducing the temperature results in reducing the variance of the Gaussian distributions in the prior and so achieving more regularity in the generated samples. Figure 7 shows examples of samples with different temperatures. We can observe how samples taken with a lower temperature look smoother, whereas those taken with a higher temperature have more details but also more artifacts. We corroborate this quantitatively in Fig. 8, which shows that the PSNR and SSIM scores (for Set5 and Set14) both decrease as the sampling temperature is increased. This agrees with our qualitative observations, as PSNR and SSIM measures are usually higher for images that are more averaged out and contain less noise.

Fig. 7. Prior sampling difference with varying temperature values for 256×256 images (comic picture from Set14 dataset).

Patch Size. A crucial parameter in our super-resolution method is the size of patches to which we apply super-resolution addressed also in [38,46]. After experimenting with patches of size 16×16 and 64×64 (i.e., 64×64 and 256×256 after super-resolution), we observed that the 16×16 patch size models were generally performing worse than their counterparts with bigger patch sizes, both in terms of PSNR and SSIM, and in a perceptual sense as the models fail to recreate details that the 64×64 patch models have no problem with. This can also be seen in Fig. 9, especially on the bird's eye, as the general shape and sharpness cannot be recreated by the 16×16 patch size model.

We hypothesise that as the patch size gets smaller, the amount of details found in a patch becomes lesser, and the models will not be able to recreate those details anymore based on context, as the patches will start to look more similar to each other and generic.

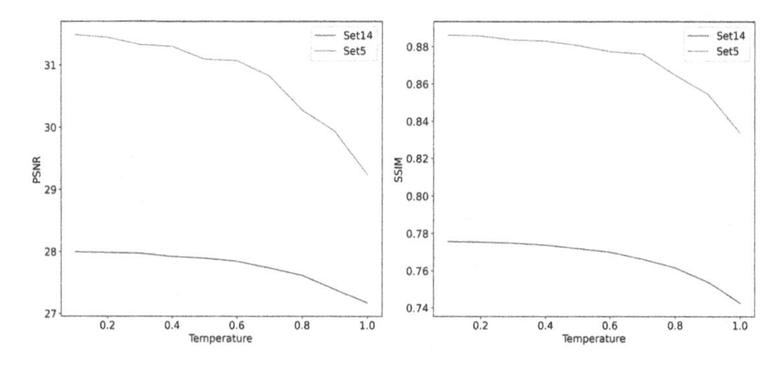

Fig. 8. PSNR and SSIM scores of Prior samples with varying temperature values for Set5 and Set14 datasets.

Fig. 9. 16×16 and 64×64 patch size model outputs for a Set5 bird image.

Activations only in Posterior. As another ablation study, we investigated the scenario where the activations from the LR-encoder are passed only in the posterior part of the top-down block as shown on Fig. 11. Doing only this, the network does not get enough information during the learning phase, only being able to generate more global features of the images, without any fine details as observed in Fig. 10.

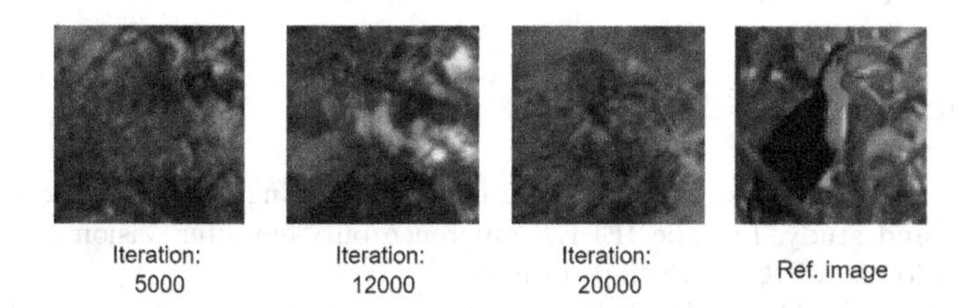

Fig. 10. The first three images are test samples taken during the training process, while the forth image is the reference.

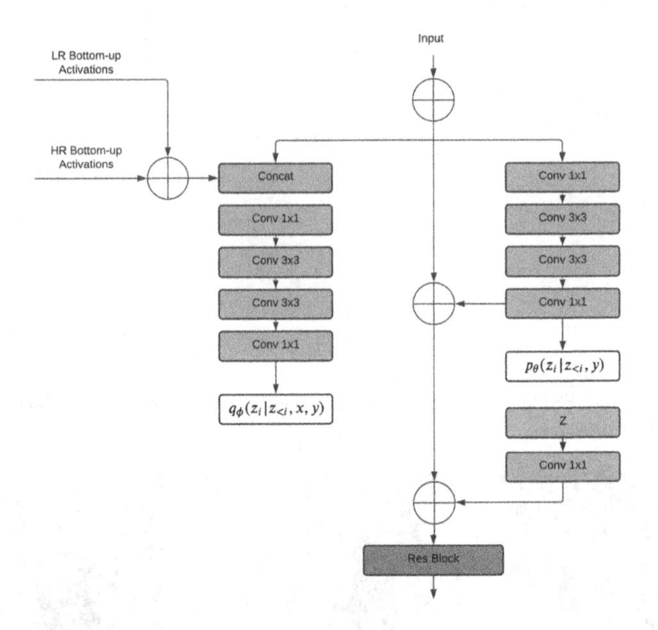

Fig. 11. Top-down block adding activations in posterior.

6 Conclusions

In this paper, we investigated the use of Very Deep Variational Autoencoders (VDVAE) for the purpose of generating super-resolution (SR) images. After the introduction of the proposed VDVAE-SR model, and based on the results presented, we conclude that the introduced model and its quantitative and qualitative results are satisfying as they are comparable to other popular methods, generating images that compensate between image sharpness and visual artifacts. As being part of the scarce family of VAE-based models for image super-resolution and the first to our knowledge that uses a deep hierarchical architecture, we believe that our proposed method still has a lot of space for building upon, to improve the results even further, as multiple modifications such as changes to training time, layer architecture, or the use of more flexible distributions can be investigated in the future.

References

1. Agustsson, E., Timofte, R.: Ntire 2017 challenge on single image super-resolution: dataset and study. In: The IEEE Conference on Computer Vision and Pattern Recognition (CVPR) Workshops (Jul 2017)
2. Bachlechner, T., Majumder, B.P., Mao, H.H., Cottrell, G.W., McAuley, J.: Rezero is all you need: Fast convergence at large depth. arXiv preprint arXiv:2003.04887 (2020)

3. Bevilacqua, M., Roumy, A., Guillemot, C., line Alberi Morel, M.: Low-complexity single-image super-resolution based on nonnegative neighbor embedding. In: Proceedings of the British Machine Vision Conference, pp. 135.1-135.10. BMVA Press (2012). https://doi.org/10.5244/C.26.135
4. Brock, A., Donahue, J., Simonyan, K.: Large scale GAN training for high fidelity natural image synthesis. arXiv preprint arXiv:1809.11096 (2018)
5. Chen, X., Mishra, N., Rohaninejad, M., Abbeel, P.: PixelSNAIL: an improved autoregressive generative model. In: Dy, J., Krause, A. (eds.) Proceedings of the 35th International Conference on Machine Learning. Proceedings of Machine Learning Research. vol. 80, pp. 864–872. PMLR (10–15 Jul 2018). https://proceedings.mlr.press/v80/chen18h.html
6. Child, R.: Very deep VAEs generalize autoregressive models and can outperform them on images. arXiv preprint arXiv:2011.10650 (2020)
7. Dai, T., Cai, J., Zhang, Y., Xia, S.T., Zhang, L.: Second-order attention network for single image super-resolution. In: Proceedings of the IEEE/CVF Conference on Computer Vision and Pattern Recognition (CVPR) (June 2019)
8. Dhariwal, P., Nichol, A.: Diffusion models beat GANs on image synthesis. arXiv preprint arXiv:2105.05233 (2021)
9. Ding, K., Ma, K., Wang, S., Simoncelli, E.P.: Image quality assessment: unifying structure and texture similarity. arXiv preprint arXiv:2004.07728 (2020)
10. Dong, C., Loy, C.C., He, K., Tang, X.: Image super-resolution using deep convolutional networks. IEEE Trans. Pattern Anal. Mach. Intell. **38**(2), 295–307 (2015)
11. Gatopoulos, I., Stol, M., Tomczak, J.M.: Super-resolution variational autoencoders. arXiv preprint arXiv:2006.05218 (2020)
12. Goodfellow, I., et al.: Generative adversarial nets. In: Advances in Neural Information Processing Systems **27** (2014)
13. He, K., Zhang, X., Ren, S., Sun, J.: Deep residual learning for image recognition. In: Proceedings of the IEEE Conference on Computer Vision and Pattern Recognition, pp. 770–778 (2016)
14. Ho, J., Jain, A., Abbeel, P.: Denoising diffusion probabilistic models. Adv. Neural Inf. Process. Syst. **33**, 6840–6851 (2020)
15. Ho, J., Saharia, C., Chan, W., Fleet, D.J., Norouzi, M., Salimans, T.: Cascaded diffusion models for high fidelity image generation. J. Mach. Learn. Res. **23**(47), 1–33 (2022)
16. Huang, J.B., Singh, A., Ahuja, N.: Single image super-resolution from transformed self-exemplars. In: Proceedings of the IEEE Conference on Computer Vision and Pattern Recognition (CVPR) (Jun 2015)
17. Hyun, S., Heo, J.-P.: VarSR: variational super-resolution network for very low resolution images. In: Vedaldi, A., Bischof, H., Brox, T., Frahm, J.-M. (eds.) ECCV 2020. LNCS, vol. 12368, pp. 431–447. Springer, Cham (2020). https://doi.org/10.1007/978-3-030-58592-1_26
18. Jolicoeur-Martineau, A.: The relativistic discriminator: a key element missing from standard GAN. arXiv preprint arXiv:1807.00734 (2018)
19. Karras, T., Laine, S., Aila, T.: A style-based generator architecture for generative adversarial networks. In: Proceedings of the IEEE/CVF Conference on Computer Vision and Pattern Recognition, pp. 4401–4410 (2019)
20. Kingma, D.P., Ba, J.: Adam: A method for stochastic optimization. arXiv preprint arXiv:1412.6980 (2014)
21. Kingma, D.P., Welling, M.: Auto-encoding variational bayes. arXiv preprint arXiv:1312.6114 (2013)

22. Kingma, D.P., Dhariwal, P.: Glow: Generative flow with invertible 1×1 convolutions. In: Advances in Neural Information Processing Systems **31** (2018)
23. Kingma, D.P., Salimans, T., Jozefowicz, R., Chen, X., Sutskever, I., Welling, M.: Improved variational inference with inverse autoregressive flow. In: Advances in Neural Information Processing Systems **29** (2016)
24. Ledig, C., et al.: Photo-realistic single image super-resolution using a generative adversarial network. In: Proceedings of the IEEE Conference on Computer Vision and Pattern Recognition,pp. 4681–4690 (2017)
25. Li, H., et al.: SRDiff: single image super-resolution with diffusion probabilistic models. Neurocomputing **479**, 47–59 (2022)
26. Lim, B., Son, S., Kim, H., Nah, S., Mu Lee, K.: Enhanced deep residual networks for single image super-resolution. In: Proceedings of the IEEE Conference on Computer Vision and Pattern Recognition Workshops, pp. 136–144 (2017)
27. Liu, J., Zhang, W., Tang, Y., Tang, J., Wu, G.: Residual feature aggregation network for image super-resolution. In: 2020 IEEE/CVF Conference on Computer Vision and Pattern Recognition (CVPR), pp. 2356–2365 (2020). https://doi.org/10.1109/CVPR42600.2020.00243
28. Maaløe, L., Fraccaro, M., Liévin, V., Winther, O.: BIVA: a very deep hierarchy of latent variables for generative modeling. arXiv preprint arXiv:1902.02102 (2019)
29. Martin, D., Fowlkes, C., Tal, D., Malik, J.: A database of human segmented natural images and its application to evaluating segmentation algorithms and measuring ecological statistics. In: Proceedings Eighth IEEE International Conference on Computer Vision. ICCV 2001. vol. 2, pp. 416–423 (2001)
30. Matsui, Y., et al.: Sketch-based manga retrieval using manga109 dataset. Multimedia Tools Appl. **76**(20), 21811–21838 (2016)
31. Nichol, A.Q., Dhariwal, P.: Improved denoising diffusion probabilistic models. In: Meila, M., Zhang, T. (eds.) Proceedings of the 38th International Conference on Machine Learning. Proceedings of Machine Learning Research, vol. 139, pp. 8162–8171. PMLR (18–24 Jul 2021). https://proceedings.mlr.press/v139/nichol21a.html
32. Niu, B., et al.: Single image super-resolution via a holistic attention network. In: Vedaldi, A., Bischof, H., Brox, T., Frahm, J.-M. (eds.) ECCV 2020. LNCS, vol. 12357, pp. 191–207. Springer, Cham (2020). https://doi.org/10.1007/978-3-030-58610-2_12
33. van den Oord, A., Kalchbrenner, N., Espeholt, L., kavukcuoglu, k., Vinyals, O., Graves, A.: Conditional image generation with pixelCNN decoders. In: Lee, D., Sugiyama, M., Luxburg, U., Guyon, I., Garnett, R. (eds.) Advances in Neural Information Processing Systems. vol. 29. Curran Associates, Inc. (2016). https://proceedings.neurips.cc/paper/2016/file/b1301141feffabac455e1f90a7de2054-Paper.pdf
34. Oord, A.V., Kalchbrenner, N., Kavukcuoglu, K.: Pixel recurrent neural networks. In: Balcan, M.F., Weinberger, K.Q. (eds.) Proceedings of The 33rd International Conference on Machine Learning. Proceedings of Machine Learning Research, vol. 48, pp. 1747–1756. PMLR, New York, USA (20–22 Jun 2016). https://proceedings.mlr.press/v48/oord16.html
35. Parmar, N., et al.: Image transformer. In: Dy, J., Krause, A. (eds.) Proceedings of the 35th International Conference on Machine Learning. Proceedings of Machine Learning Research, vol. 80, pp. 4055–4064. PMLR (10–15 Jul 2018). https://proceedings.mlr.press/v80/parmar18a.html
36. Pisharoty, N., Jadhav, M., Dandawate, Y.: Performance evaluation of structural similarity index metric in different colorspaces for HVS based assessment of quality of colour images. Int. J. Eng. Technol. **5**, 1555–1562 (2013)

37. Rezende, D.J., Mohamed, S., Wierstra, D.: Stochastic backpropagation and approximate inference in deep generative models. In: International Conference on Machine Learning, pp. 1278–1286. PMLR (2014)
38. Sajjadi, M.S., Scholkopf, B., Hirsch, M.: EnhanceNet: single image super-resolution through automated texture synthesis. In: Proceedings of the IEEE International Conference on Computer Vision, pp. 4491–4500 (2017)
39. Sohn, K., Lee, H., Yan, X.: Learning structured output representation using deep conditional generative models. In: Cortes, C., Lawrence, N., Lee, D., Sugiyama, M., Garnett, R. (eds.) Advances in Neural Information Processing Systems. vol. 28. Curran Associates, Inc. (2015). https://proceedings.neurips.cc/paper/2015/file/8d55a249e6baa5c06772297520da2051-Paper.pdf
40. Sønderby, C.K., Raiko, T., Maaløe, L., Sønderby, S.K., Winther, O.: Ladder variational autoencoders. arXiv preprint arXiv:1602.02282 (2016)
41. Tong, T., Li, G., Liu, X., Gao, Q.: Image super-resolution using dense skip connections. In: Proceedings of the IEEE International Conference on Computer Vision, pp. 4799–4807 (2017)
42. Uria, B., Côté, M.A., Gregor, K., Murray, I., Larochelle, H.: Neural autoregressive distribution estimation. J. Mach. Learn. Res. **17**(1), 7184–7220 (2016)
43. Vahdat, A., Kautz, J.: NVAE: a deep hierarchical variational autoencoder. arXiv preprint arXiv:2007.03898 (2020)
44. Wang, X., et al.: ESRGAN: enhanced super-resolution generative adversarial networks. In: Proceedings of the European conference on computer vision (ECCV) workshops, pp. 0–0 (2018)
45. Zeyde, R., Elad, M., Protter, M.: On single image scale-up using sparse-representations. In: Boissonnat, J.-D. (ed.) Curves and Surfaces 2010. LNCS, vol. 6920, pp. 711–730. Springer, Heidelberg (2012). https://doi.org/10.1007/978-3-642-27413-8_47
46. Zhang, Y., Tian, Y., Kong, Y., Zhong, B., Fu, Y.: Residual dense network for image super-resolution. In: Proceedings of the IEEE Conference on Computer Vision and Pattern Recognition, pp. 2472–2481 (2018)
47. Zhu, J.Y., Park, T., Isola, P., Efros, A.A.: Unpaired image-to-image translation using cycle-consistent adversarial networks. In: Proceedings of the IEEE International Conference on Computer Vision, pp. 2223–2232 (2017)

Light Field Angular Super-Resolution via Dense Correspondence Field Reconstruction

Yu Mo, Yingqian Wang(✉), Longguang Wang, Jungang Yang, and Wei An

National University of Defense Technology, Changsha, China
wangyingqian16@nudt.edu.cn

Abstract. Light field (LF) angular super-resolution (SR) aims at reconstructing a densely sampled LF from a sparsely sampled one. To achieve accurate angular SR, it is important but challenging to incorporate the complementary information among input views, especially when dealing with large disparities. In this paper, we propose to reconstruct dense correspondence field among different views for LF angular SR. According to the LF geometry structure, we first capture correspondences along the horizontal and vertical axes of input views with a global receptive field. We then incorporate the linear structure prior among angular viewpoints to reconstruct a dense correspondence field. With the reconstructed dense correspondence field, the relationship between each target view and the input views is constructed. Next, we develop a view projection approach to project input views to the target positions. Moreover, a projection loss is introduced to preserve the LF parallax structure. Extensive experiments demonstrate that our proposed network can recover accurate details and preserve LF parallax structure. Comparative results show the advantage of our method over state-of-the-art methods on synthetic and real-world datasets.

Keywords: Light field reconstruction · Angular super-resolution · Correspondence field

1 Introduction

With the popularity of the commercial light field (LF) cameras, LF imaging has attracted more and more attention in academia and industry. The LF, denoted by a simplified 4D function $L(x, y, u, v)$, can simultaneously record both the directions and radiance of light rays. The additional directional information in LF images allows a wider range of vision applications, such as depth estimation [1–3], post-capture refocusing [4–6], super-resolution [7–9] and so on. However, due to limited sensor resolution, there is a trade-off between spatial resolution and angular resolution in an LF [10], which limits the performance of algorithms in these vision applications.

One feasible way to relieve such a trade-off is to perform LF angular SR, i.e. synthesize intermediate views from a sparse set of input views. Traditional

L. Karlinsky et al. (Eds.): ECCV 2022 Workshops, LNCS 13802, pp. 412–428, 2023.
https://doi.org/10.1007/978-3-031-25063-7_25

methods [11,12] treated view synthesis as the approximation of 4D LF function [13], and incorporated the sparsity prior in the Fourier domain to reconstruct a densely-sampled LF. Recently, deep learning-based methods [14–22] have been successfully applied to LF angular SR, and achieved promising performance. Among them, some methods [14–16] synthesize novel views using the local LF information learned from neighboring views, epipolar plane images (EPIs) or EPI volumes by multiple convolutional layers. However, the performance of these methods is limited, since the misalignment among different views impedes the incorporation of complementary information.

To better incorporate the complementary information, recent methods [18–20] estimated disparities and performed view alignment, i.e., to warp input views to the target positions. However, due to the limited receptive field of convolutional filters [23], it is difficult for networks to capture correspondence with large disparities [15]. As a result, these methods are prone to produce blurriness and artifacts on regions with large disparities. Moreover, synthesizing each view independently by sub-pixel warping is computationally expensive.

In this paper, we achieve LF angular SR by reconstructing a dense correspondence field. According to the structural property of LF, we capture correspondences among input views with a global receptive field. Moreover, based on the fact that the disparities of in-between views have a linear relationship with their angular viewpoint positions, we propose to reconstruct a dense correspondence field from input views. Note that, the dense correspondence field contains the correspondence cues which can be used to project the input views to the target positions. In addition to the reconstruction loss that is commonly used, we introduce a projection loss to regularize the generated correspondence field.

The main contributions of our network are summarized as follows:

- We propose an LF angular SR method by modeling the relationships among input views and in-between views as a correspondence field.
- We incorporate the LF structure prior by designing a dense correspondence field reconstruction approach and handle large disparities in LF angular SR.
- Comprehensive experiments on both synthetic and real-world datasets demonstrate the effectiveness and superiority of the proposed method as compared to state-of-the-art methods.

The rest of this paper is organized as follows: Sect. 2 reviews the related literature. In Sect. 3, our method is introduced in details. Experimental results on synthetic and real-world datasets are presented in Sect. 4. Finally, conclusions are drawn in Sect. 5.

2 Related Works

The aim of LF angular SR (also termed as LF reconstruction) is to reconstruct a dense-sampled LF from a sparsely-sampled one. Recent angular SR methods can be roughly classified to two categorizes: non-depth based methods and depth based methods. A brief review is presented in this section.

2.1 Non-Depth Based Methods

Many researchers have tried to reconstruct densely-sampled LF with signal processing techniques. Among them, Shi et al. [11] proposed an LF reconstruction method based on the sparsity in the continuous Fourier spectrum. In their method, only the boundary and diagonal viewpoints were used to synthesize novel views. Vagharshakyan et al. [12] developed an effective reconstruction method based on the concepts of LF sparsification. The shearlet transform was used as the sparse transform and a restoration technique was proposed for LF reconstruction.

Due to the recent success of deep learning techniques in LF image processing [7,24–28], learning-based LF angular SR methods have been proposed. Yoon et al. [29] proposed an LFCNN to use the neighboring input views to synthesize novel views. Yeung et al. [30] proposed a spatial-angular alternating convolution network to reconstruct a densely-sampled LF. Wu et al. [31] developed a "blur-restoration-deblur" framework for LF angular SR. Liu et al. [32] proposed a network based on the multi-angular epipolar geometry structure to explore the rich angular information for LF reconstruction. Wang et al. [14] built a Pseudo 4DCNN to synthesize dense LFs from a sparse set of input views. In their subsequent work [33], they proposed an EPI structure preserving loss function to improve the reconstruction performance. Meng et al. [16] developed a deep high-dimensional dense residual network with 4D convolutions for LF reconstruction. Wang et al. [22] proposed a general disentangling mechanism and developed a DistgASR network for LF angular SR. However, it is difficult for non-depth based methods to incorporate the complementary information without alignment among different views, resulting in limited performance.

2.2 Depth Based Methods

To handle the incorporation of complementary information, some works [17,19, 20,34,35] first estimate the depth of a scene and then warp the input views to synthesize the novel view images. For example, Wanner et al. [34] estimated the depth maps using EPIs and generated novel views of a scene in a variational framework. Shi et al. [20] combined the pixel-based reconstruction module and the feature-based reconstruction module with the estimated depth maps to construct an end-to-end learning framework. Meng et al. [17] first employed two DenseNets to compute the scene depth and a warping confidence map, and then used them to synthesize target views. Wu et al. [35] proposed a CNN-based network to evaluate sheared EPIs, where the sheared value is correlated with the depth. Jin et al. [19] generated novel views by warping the input views based on the estimated depth maps and explored the spatial-angular relations among the warped views to reconstruct a high angular resolution LF.

However, it is challenging to achieve accurate depth estimation, especially for LFs with large baselines and irregular sampling patterns. To deal with the problem, Mildenhall et al. [36] used the plane sweep volumes (PSV) to handle the irregularly sampled images and applied a 3D CNN to synthesize a local

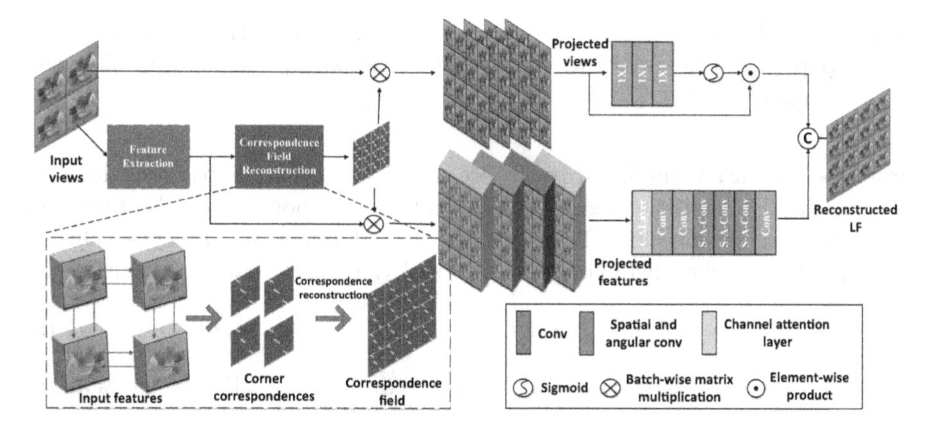

Fig. 1. An overview of our proposed network.

LF. Kalantari et al. [18] split the angular SR task into two stages including depth estimation and color prediction, and employed a sequential network to model these two components. Jin et al. [37] proposed a confidence-based blending strategy to handle the occluded regions and developed a flexible reconstruction network to reconstruct densely-sampled LFs. However, the performance of depth-based methods largely depends on the accuracy of the estimated depth map, which is often inaccurate in occluded and textureless regions.

3 Method

3.1 Light Field Representation

In this paper, we aim to synthesize the in-between views from a sparsely-sampled LF. Following the two-plane parametrization [38], we represent an LF by using a 4D function, including 2D angular dimensions (u, v) and 2D spatial dimensions (x, y). For a densely-sampled LF, we denote it as $L(H, W, N, N)$, which represents $N \times N$ SAIs of spatial dimension $H \times W$. Given $n \times n$ input views with the spatial resolution of $H \times W$, we can reconstruct the densely-sampled LF by:

$$L(H, W, N, N) = f(L_{in}(H, W, n, n), \theta), \tag{1}$$

where θ denotes the network parameter and $L_{in}(H, W, n, n)$ denotes the input LF. Specifically, $N = \beta \times (n - 1) + 1$, where β denotes the upsampling factor in the angular dimension.

Epipolar Plane Image (EPI). EPI is often used for LF image processing. An EPI can be obtained by stacking the views with a fixed angular and a fixed spatial coordinate. Therefore, an EPI contains the position relationship (i.e. oriented lines) of the same object in different angular views. According to the inherent geometry of LF, the slopes of oriented lines in an EPI reflect the disparity values. In this paper, we use an EPI to visualize the relationship among multiple views in LF, as shown in Fig. 2.

Spatial-Angular Relationship. There is a strong relationship among different views in an LF. That is, disparities between the corresponding pixels of any two adjacent views are the same. Therefore, under the Lambertian and non-occlusion assumption, the relationship of view (p, q) and view (m, n) can be formulated by

$$L(x, y, p, q) = L(x + d(m - p), y + d(n - q), m, n), \tag{2}$$

where (x, y) is the spatial coordinates, d is the disparity of adjacent views at the pixel (x, y).

Based on the theory, given the disparities of the views at both ends, we can calculate the disparities of in-between views.

3.2 Proposed Network

In this section, we describe the proposed network in details. Similar to existing LF reconstruction methods [14,32,33], our method only processes the Y channel images in the YCbCr color space. Figure 1 shows our framework.

Given a sparsely sampled LF, we first extract features from each view using our feature extraction module. Then we build the initial correspondence among the features of input views. Afterwards, we develop a correspondence reconstruction module to calculate the relationships among in-between views from the initial correspondence maps. The relationships implicitly encode the disparity information and help to keep the LF structure during the reconstruction process. Finally, the reconstructed correspondence field is used to project the input views and their features to the target positions. The projected views and features are subsequently fed to a refinement sub-network to restore texture details. Due to the occlusion, some pixels are visible in part of the projected views. Therefore, we use an attention mechanism to highlight valid pixels which are more likely to be un-occluded in projected views.

Feature Extraction. Since features with rich context information are important for correspondence estimation, we apply an atrous spatial pyramid pooling (ASPP) module [39] to enlarge the receptive field and learn multi-scale features. Specifically, the ASPP module is constructed by three dilated convolutions parallelly with dilation rates of 1, 4, 8, respectively. We chain the ASPP module with a residual block and repeat this structure for two times to extract high-dimensional features. The extracted high-dimensional features are then used to build initial correspondence among input views.

Initial Correspondence Construction. Given the extracted features of each input view, we build the initial correspondence along both horizontal and vertical axes of input views.

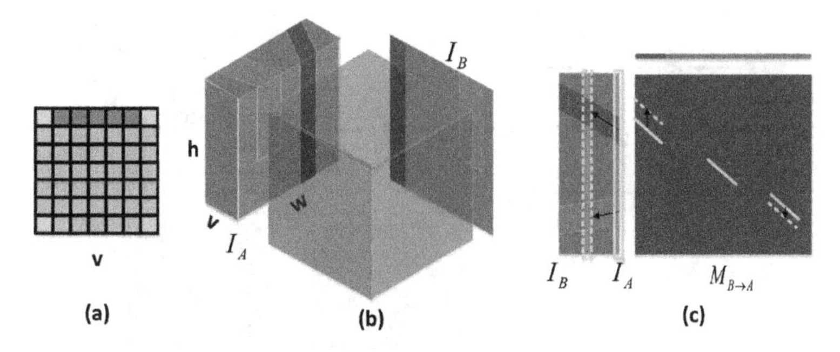

Fig. 2. A toy example to illustrate the correlations between different views and their correspondence maps. (a) The stacked views on the same angular axis. (b) The correspondence volume between view I_B and its left counterparts along the horizontal angular axis. (c) Variations of the correspondence map $M_{B\rightarrow A}$ for different views. The yellow solid box represents the view I_A. When the view moves from I_A to I_B in the stacked volume, the responses in correspondence map move towards the diagonal where the disparities are zero.

Taking the horizontal initial correspondence, two feature maps $F_A, F_B \in \mathbb{R}^{H\times W\times C}$ of view I_A and I_B are first fed to 1×1 convolutions to produce a query feature map $Q \in \mathbb{R}^{H\times W\times C}$ and a key feature map $K \in \mathbb{R}^{H\times W\times C}$, respectively. Then the key feature map is reshaped to $\mathbb{R}^{H\times C\times W}$ to perform batch-wise matrix multiplication with Q, and a softmax layer is used to generate a correspondence map $M_{B\rightarrow A} \in \mathbb{R}^{H\times W\times W}$. The response in the correspondence map represents similarity along the epipolar line between the view I_A and I_B. In the same way, F_A and F_B can be exchanged to produce $M_{A\rightarrow B} \in \mathbb{R}^{H\times W\times W}$.

Note that, $M_{A\rightarrow B}(i, j, w)$ represents the contribution of position (i, w) in view I_A to position (i, j) in view I_B. The disparity information are implicitly encoded into the correspondence map.

In this work, we take 2×2 corner views as input. Therefore, by computing the correspondence maps of each two adjacent views, we can obtain eight initial correspondence maps (i.e. corner correspondences).

Dense Correspondence Field Reconstruction. According to the spatial-angular relationship in LFs, we analyze the correlations between the angular positions of views and their correspondence maps, as shown in Fig. 2. We stack the views along the angular dimension V and show the variations of their correspondence maps. For views between I_A and I_B, their correspondence maps have a strong relationship with $M_{B\rightarrow A}$. Specifically, the points below the diagonal in $M_{B\rightarrow A}$ need to move up to produce the in-between correspondence maps. In contrast, the points above the diagonal in $M_{B\rightarrow A}$ need to move down to produce the in-between correspondence maps. Moreover, the moving distance in correspondence map is linearly related to the relative angular position between the input view and the target view. Note that, the points on the diagonal in the correspondence map represent points with zero disparities.

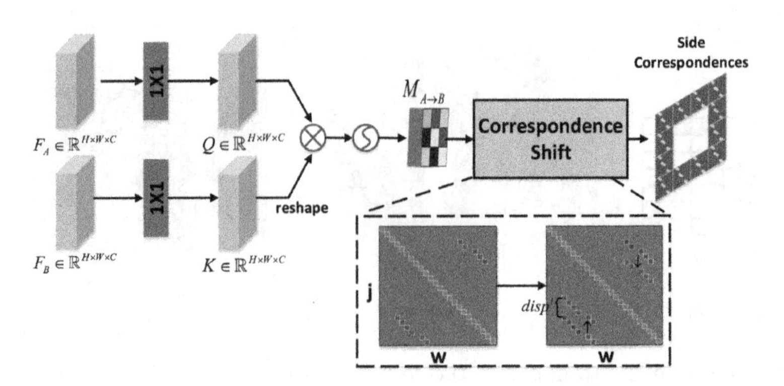

Fig. 3. An illustration of the correspondence shifting operation. According to the intrinsic geometry of LF, shifting the points vertically in corner correspondence maps can generate side correspondence maps.

Based on the above analyses, we can generate in-between correspondence maps by shifting the points in corner sorrespondence maps. Specifically, given a correspondence map $M \in \mathbb{R}^{H\times W\times W}$, each slice of correspondence map (e.g. $M_{A\to B}(i,:,:)$) represents the dependency between corresponding rows (i.e. $I_A(i,:)$ and $I_B(i,:)$). The correspondence shift operation is performed in the latter two dimensions $M^i(j,w) \in \mathbb{R}^{W\times W}$. The disparity of each row in $M^i(j,w)$ can be regressed as:

$$\overset{\wedge}{D^i} = \sum_{w=0}^{W-1} w \times M^i(:,w). \tag{3}$$

Suppose we want to generate s images between the two input views, the disparity between any two of the adjacent views can be obtained by:

$$disp^i = \frac{\overset{\wedge}{D^i}}{s+1}, \tag{4}$$

where $disp^i$ denotes the baseline distance of the i-th row, and the distance between view (m,n) and view (m,k) is computed as:

$$disp^i(k) = disp^i \times \mid k - n \mid. \tag{5}$$

Using the calculated disparities, we can generate side correspondences from the initial corner correspondences, as shown in Fig. 3.

Since there are many views in an LF, for a more accurate representation, we change the subscript of the correspondence map to the specific angular position. For example, we denote the correspondence map from view $(1,1)$ to view $(1,7)$ by $M_{(1,1)\to(1,7)}$.

For example, given two correspondence maps $M_{(1,1)\to(1,7)}$ and $M_{(1,1)\to(7,1)}$, the correspondence maps for view $(1,k)$ and $(m,1)$, $k,m = 2,3,\ldots,6$ can be generated:

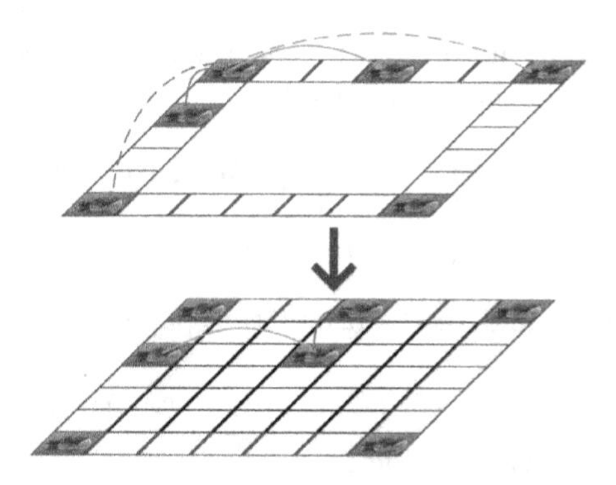

Fig. 4. An illustration of synthesizing densely-sampled views via a reconstructed correspondence field. Red squares denote projected views from input views using the side correspondences. Then these projected views can project to all the target views via a dense correspondence field. (Color figure online)

$$M_{(1,1)\rightarrow(1,k)} = f^S(M_{(1,1)\rightarrow(1,7)}, disp(k)), \quad (6)$$

$$M_{(1,1)\rightarrow(m,1)} = f^S(M_{(1,1)\rightarrow(7,1)}, disp(m)), \quad (7)$$

where f^S denotes the aforementioned shifting operator.

According to the generated side correspondences, we can produce the side views by performing view projection (see subsection **View Projection**). Similar correspondences reconstruction approach on these side views can generate the central correspondence maps.

Consequently, we can generate all views in an LF by using a dense correspondence field. In our implementation, we perform correspondence ensemble by using all the input views to repetitively generate the correspondence map of each view, and the final correspondence field is produced by concatenating all the generated correspondence maps.

In summary, our correspondence reconstruction approach provides a simple and effective approach to reconstruct a dense correspondence field for an LF and can be applied to synthesize in-between views efficiently.

View Projection. Based on the reconstructed correspondence field, the in-between views can be generated by the corresponding correspondence map, as shown in Fig. 4. To be specific, given the input view I_A and correspondence map $M_{A\rightarrow B}$, view I_B can be generated by taking a batch-wise matrix multiplication, i.e.

$$I'_{A\rightarrow B} = M_{A\rightarrow B} \otimes I_A, \quad (8)$$

Without considering occlusion and boundary regions, the generated image $I'_{A\rightarrow B}$ should be the same as I_B. This process can be seen as a projection process, so we also call M a projection map.

However, in practice, there are usually some occluded regions in most scenes. The occluded pixels are usually assigned with small weights in the correspondence map, because they can not find their correspondence pixels. To perform occlusion detection, valid masks can be generated by the correspondence map. Therefore, we can obtain the valid mask by:

$$Val^A(i,w) = \begin{cases} 1, & if \sum_{j\epsilon[1,W]} M_{A\to B}(i,j,w) > \tau \\ 0, & otherwise \end{cases}, \tag{9}$$

where τ is a threshold and set to 0.1 in this paper. Valid projection can be written as:

$$I'_{B\to A} = Val^A \odot (M_{B\to A} \otimes I_B), \tag{10}$$

where $\odot$ is the element-wise product.

Refinement Sub-network. After obtaining the dense correspondence field, we can project the input views to the target positions. However, due to the occlusion and non-Lambertian effect, some pixels cannot find their correspondences in the input views. As a result, these correspondence maps inevitably contain errors, leading to distortions in the projection views.

To mitigate this problem, we project not only the input views but also their features to the target positions. In order to fuse projection views effectively, closer values are assigned with larger weights by an attention map constructed by three 1×1 convolutions and a sigmoid layer. In summary, the fused projection view L^{fu}_{view} can be defined as:

$$L^{fu}_{view} = A_{view} \odot L^{p}_{view}, \tag{11}$$

where L^{p}_{view} is the concatenation of the projected views on the target positions, $A_{view}\epsilon R^{N^2\times H\times W}$ is the attention map.

Compared to the view projection, the feature projection can help to recover better high-frequency details. We first use a channel attention layer to model the interdependencies across feature channels. Then we exploit several sequential separable convolutions [30] to learn the LF geometry structure, which save many computational burden compared to 4D convolution.

Loss Function. Two loss functions are used to train our network. One is L_1 norm based loss function, which measures the difference between the reconstructed views and their corresponding groundtruths.

$$\ell_d = \frac{1}{N^2}\sum_{k=1}^{N^2} \| L^k_{RE} - L^k_{GT} \|_1, \tag{12}$$

where L^k_{RE} is the k-th reconstructed view in the $N\times N$ novel reconstructed views and L^k_{GT} is the corresponding groundtruth.

The other one is the projection loss, which is used to regularize our correspondence field to capture accurate correspondence in input views.

$$\ell_p = \frac{1}{2n^2}\sum_{k,m\epsilon n^2} \| Val^k \odot L^k_{in} - Val^k \odot M_{m\to k} \otimes L^m_{in} \|_1, \tag{13}$$

where L_{in}^{k} and L_{in}^{m} are the k-th and m-th input view, respectively. Val^{k} is the corresponding valid mask, $M_{m \to k}$ is the projection map from view m to view k and n^2 is the number of input views. Note that, the value of m and k cannot be equal and the absolute difference between them cannot be equal to 3.

In summary, the overall loss is defined as:

$$\ell = \ell_d + \lambda \ell_p, \tag{14}$$

where λ is empirically set to 0.05.

3.3 Training Details

Similar to most LF reconstruction methods [14,19,20,37], our training data contained both synthetic and real-world LFs. For training datasets, 20 synthetic scenes from [40] and [41], and 100 real-world scenes from [18] and [42] were used. For test datasets, 2 scenes from the *HCInew* dataset [40], 8 scenes from the *HCIold* dataset [41], 30 scenes from the *30scenes* [18], 25 scenes from the *Occlusions* category in [42] and 15 scenes from the *Reflective* category in [42] were used. Note that, these datasets contain multiple types of scenes for evaluating LF reconstruction methods. Specifically, the *HCI* dataset contain scenes with large disparities, the *Occlusions* contain scenes with occlusion areas, and the *Reflective* contain scenes with non-Lambertian surfaces.

For a broader comparison, we followed most existing works [14,19,20,30,37] to perform $2 \times 2 \to 7 \times 7$ angular SR and only process the luminance Y channel in the YCbCr color space. During training, we angularly cropped the central 7×7 views and extracted their 2×2 corner views as inputs. Each view was cropped into patches of spatial size 64×64 to generate training samples.

The proposed network was implemented in Pytorch and optimized using the Adam method [43] with $\beta_1 = 0.9$ and $\beta_2 = 0.999$. The batch size was set to 6. The learning rate was initialized to 5×10^{-4}, and reduced by a factor of 2 for every 15 epochs. The training was stopped after 60 epochs. Our experiments were conducted on a PC with two Nvidia RTX 2080Ti GPUs.

4 Experimental Results

In this section, we compare the proposed method with state-of-the-art methods, including ShearedEPI [35], LFEPICNN [31], P4DCNN [14], Kalantari et al. [18], Yeung et al. [30], LFASR-geo [19] and FS-GAF [37]. The peak signal-to-noise ratio (PSNR) and the structural similarity (SSIM) were used to evaluate these algorithms numerically, averaged on all synthesized views.

4.1 Synthetic Scenes

We evaluate our method on synthetic scenes (i.e. HCI [40,41]) to test the performance on scenes with large disparities. Table 1 shows the quantitative comparisons of our method with other state-of-the-art methods. From this table, we

Fig. 5. Visual comparisons achieved by different methods on synthetic scenes.

observe that our network (named CFR) achieves the highest PSNR and SSIM on most scenes. On average, our method outperforms the top-performing method FS-GAF [37] by nearly 1 dB in terms of PSNR. For qualitative comparisons, we select three challenging scenes (*Bicycle* from *HCInew* [40], *Buddha2* and *MonasRoom* from *HCIold* [41]) and provide reconstructed central views, error maps and extracted EPIs in Fig. 5. Since the input views are sparse (only 2×2), EPI-based methods ShearedEPI [35] and LFEPICNN [31], cannot recover the clear linear structure from only 2 rows or columns of pixels, resulting in severe aliasing effects in reconstructed views. Yeung et al. [30] does not work well on scenes with large disparities because of their limited receptive field. Depth-based method Kalantari et al. [18] performs not well on the HCI dataset with large disparities because they used a shallow network to fuse the warped views, which is

Table 1. Quantitative comparisons of state-of-the-art methods for $2 \times 2 \rightarrow 7 \times 7$ LF angular SR on the HCI dataset. The PSNR/SSIM are the average value of all the scenes of a dataset. The best results are in red and the second best results are in blue.

Light field	Method						
	LFEPICNN	ShearedEPI	Kalantari et al.	Yeung et al.	LFASR-geo	FS-GAF	Ours
bicycle	26.17/0.762	30.84/0.924	32.37/0.935	32.92/0.945	34.03/0.954	34.14/0.958	34.40/0.959
herbs	26.86/0.694	30.80/0.831	31.70/0.847	31.05/0.836	32.76/0.882	34.47/0.942	33.18/0.888
buddha	32.86/0.916	42.91/0.986	42.47/0.985	44.03/0.988	45.65/0.991	46.26/0.992	46.22/0.992
buddha2	32.63/0.902	38.03/0.966	39.51/0.969	40.61/0.973	41.48/0.975	41.49/0.976	41.83/0.979
horses	27.85/0.799	34.52/0.941	36.14/0.952	36.21/0.957	37.26/0.968	36.91/0.955	37.34/0.969
maria	35.41/0.934	38.27/0.954	40.27/0.969	41.15/0.972	42.71/0.983	42.16/0.981	42.75/0.984
medieval	31.81/0.819	32.21/0.877	32.76/0.882	33.03/0.894	34.02/0.913	32.76/0.895	35.67/0.929
monasRoom	35.45/0.946	41.06/0.983	43.09/0.985	44.92/0.989	45.88/0.990	45.29/0.983	46.20/0.991
papillon	34.55/0.936	41.42/0.981	43.04/0.983	44.73/0.986	45.51/0.987	43.13/0.979	45.53/0.988
stillLife	21.64/0.550	24.63/0.792	24.78/0.797	24.14/0.771	25.67/0.854	27.40/0.871	31.49/0.927
Average	31.15/0.841	35.47/0.924	36.62/0.930	37.28/0.931	38.49/0.950	38.41/0.953	39.46/0.961

Table 2. Quantitative comparisons of state-of-the-art methods for $2 \times 2 \rightarrow 7 \times 7$ LF angular SR on Lytro dataset. The PSNR/SSIM are the average value of all the scenes of a dataset. The best results are in red and the second best results are in blue.

Dataset	Method						
	LFEPICNN	ShearedEPI	P4DCNN	Yeung et al.	LFASR-geo	FS-GAF	Ours
30scenes [18]	33.66/0.918	39.17/0.975	38.22/0.970	42.77/0.986	42.53/0.985	42.75/0.986	42.68/0.986
Occlusions [42]	32.72/0.924	34.41/0.955	35.42/0.962	38.88/0.980	38.36/0.977	38.51/0.979	38.60/0.980
Reflective [42]	34.76/0.930	36.38/0.944	35.96/0.942	38.33/0.960	38.20/0.955	38.35/0.957	39.65/0.965
Average	33.71/0.924	36.65/0.958	36.53/0.958	39.99/0.975	39.70/0.972	39.87/0.974	40.31/0.977

difficult to capture long distance correspondences. Moreover, depth-based methods LFASR-geo [19] and FS-GAF [37] achieve better result based on a finely designed disparity estimation network and a complex refinement network. However, these depth-based method can not estimate accurate depth maps especially in complex structure such as the foreground bike in *Bicycle*. As a result, these depth-based methods produce some ghosting artifacts near the object boundaries. It can be observed from the error maps and close-up images that our method achieves a better perceptual quality and reconstruct novel views which are closer to the groundtruth.

4.2 Real-World Scenes

In this section, we conduct experiments on real-world scenes *30scenes* [18], *Occlusions* [42] and *Reflective* [42], which contain complex occlusions and reflective surfaces. Table 2 shows the quantitative comparisons under task $2 \times 2 \rightarrow 7 \times 7$ angular SR. As can be seen from the table, our method achieves comparable performance with Yeung et al. [30] at *30scenes* and *Occlusions* datasets, and the best performance on the *Reflective* dataset. Figure 6 compares the visual performance on three real-world scenes achieved by different methods, respectively. For *30scenes_ Cars* and *Reflective_ 11_ eslf*, there are some non-Lambertian surfaces,

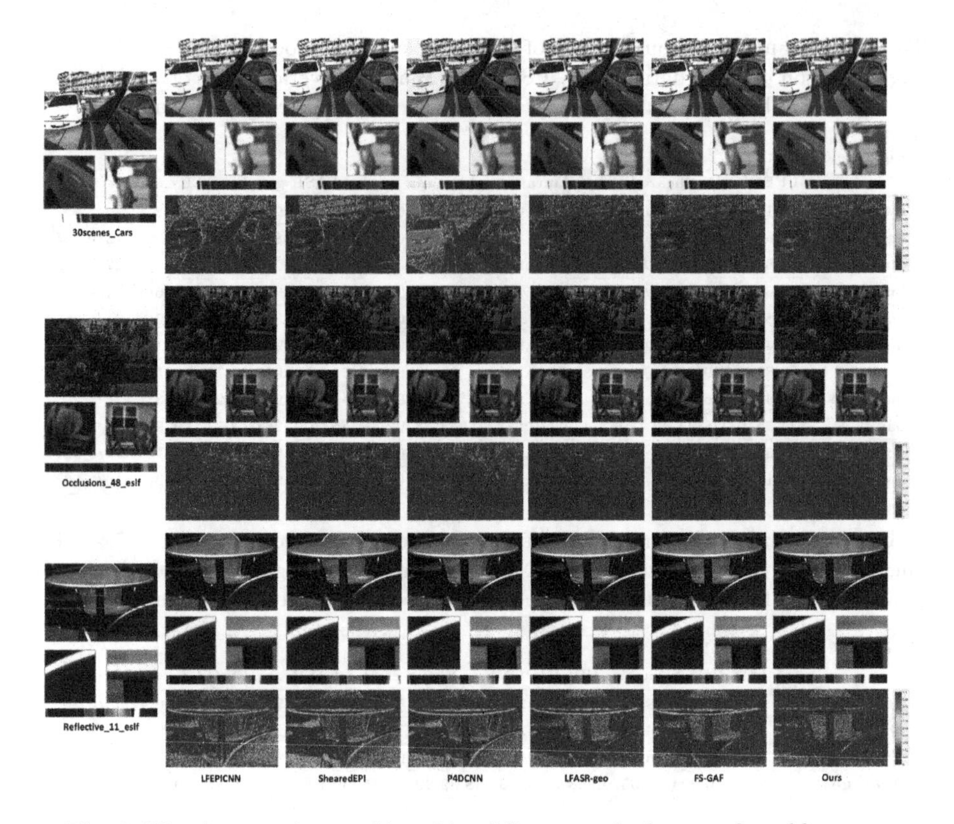

Fig. 6. Visual comparisons achieved by different methods on real-world scenes.

such as a refractive car and a reflective desk. Due to the difficulty of disparity estimation on these surfaces, depth-based methods (LFASR-geo [19] and FS-GAF [37]) cannot to reconstruct non-Lambertian surfaces accurately and introduce some blurring and ghosting artifacts. Moreover, as shown in the extracted EPIs, EPI-based methods can not recover a clear structure from a sparse input. For the second case shown in scene *Occlusion_48_eslf*, complex occlusions exist in many regions. ShearedEPI [35] and LFEPICNN [31] fail to reconstruct the complex structure in the background. LFASR-geo [19] and FS-GAF [37] estimate a proper occlusion structure and achieve good visual results on scene *Occlusions_48_eslf*. In contrast, our method achieves a closer approximation to the groundtruth.

4.3 Visualization of Correspondence Maps

In this section, we visualize the correspondence maps generated by our method in Fig. 7. In this case, the disparities between input views are about 20 pixels. For the correspondence map between input views, the values represent the degree of similarity of correspondence points. In Fig. 7(a), we can clearly see that the projection weights between horizontal views are large (close to red), which means

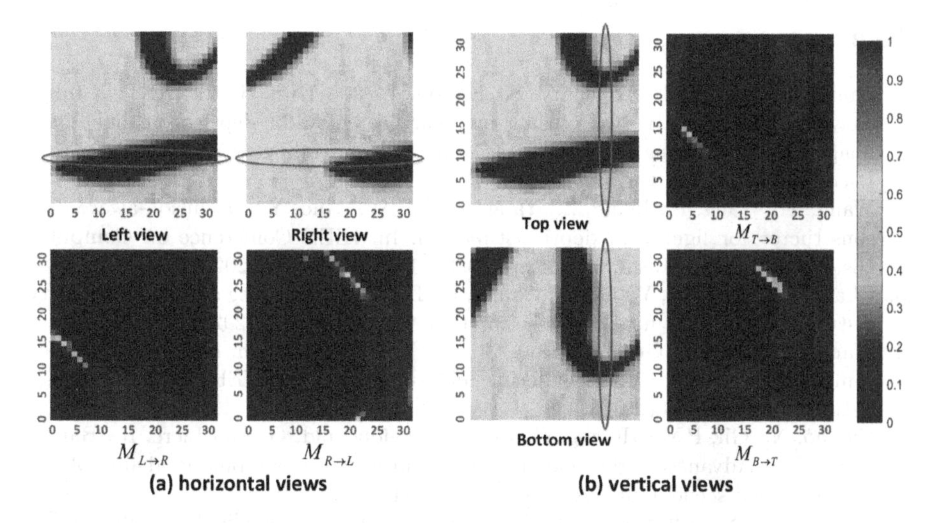

Fig. 7. Correspondence maps between the horizontal views and vertical views on a scene with large disparity. The correspondence map is the projection weight of the horizontal axis circled by a red ellipse. The axes of view and the corresponding axes in the correspondence map is marked with the same color. (Color figure online)

reliable correspondences can be captured by our module. In contrast, the projection weights in the correspondence map between top view and bottom view, as shown in Fig. 7(b), are only about 0.5. That is because, the circled vertical axis has many textureless areas, which are difficult to match. These relatively small weights are consistent with our intuition.

5 Conclusion

In this paper, we propose an end-to-end trainable method to synthesize in-between views from sparse input views by reconstructing a dense correspondence field. We model the epipolar property and capture the correspondences in a sparse-sampled LF. Based on the linear relation along the angular viewpoint position, we introduce a correspondence reconstruction approach to generate a correspondences field, which contains the correspondence information from the input views to the target view positions. Then, we design two sub-refinement networks to refine the projected views and features, respectively. Moreover, a projection loss is introduced to promote the preservation of the LF parallax structure. The evaluations on synthetic and real-world datasets with challenging scenes have demonstrated the superiority of the proposed network.

References

1. Shin, C., Jeon, H.G., Yoon, Y., So Kweon, I., Joo Kim, S.: EPINET: a fully-convolutional neural network using epipolar geometry for depth from light field images. In: Proceedings of the IEEE Conference on Computer Vision and Pattern Recognition, pp. 4748–4757 (2018)
2. Wang, Y., Wang, L., Liang, Z., Yang, J., An, W., Guo, Y.: Occlusion-aware cost constructor for light field depth estimation. In: IEEE Conference on Computer Vision and Pattern Recognition, pp. 19809–19818 (CVPR) (2022)
3. Chao, W., Wang, X., Wang, Y., Chang, L., Duan, F.: Learning sub-pixel disparity distribution for light field depth estimation. arXiv preprint (2022)
4. Wang, Y., Yang, J., Guo, Y., Xiao, C., An, W.: Selective light field refocusing for camera arrays using bokeh rendering and superresolution. IEEE Signal Process. Lett. **26**(1), 204–208 (2018)
5. Viganò, N., Gil, P.M., Herzog, C., de la Rochefoucauld, O., van Liere, R., Batenburg, K.J.: Advanced light-field refocusing through tomographic modeling of the photographed scene. Opt. Express **27**(6), 7834–7856 (2019)
6. Jayaweera, S.S., Edussooriya, C.U., Wijenayake, C., Agathoklis, P., Bruton, L.T.: Multi-volumetric refocusing of light fields. IEEE Signal Process. Lett. **28**, 31–35 (2020)
7. Wang, Y., Wang, L., Yang, J., An, W., Yu, J., Guo, Y.: Spatial-angular interaction for light field image super-resolution. In: Vedaldi, A., Bischof, H., Brox, T., Frahm, J.-M. (eds.) ECCV 2020. LNCS, vol. 12368, pp. 290–308. Springer, Cham (2020). https://doi.org/10.1007/978-3-030-58592-1_18
8. Cheng, Z., Xiong, Z., Chen, C., Liu, D., Zha, Z.J.: Light field super-resolution with zero-shot learning. In: IEEE Conference on Computer Vision and Pattern Recognition, pp. 10010–10019 (2021)
9. Wang, Y., Liang, Z., Wang, L., Yang, J., An, W., Guo, Y.: Learning a degradation-adaptive network for light field image super-resolution. arXiv preprint arXiv:2206.06214 (2022)
10. Zhu, H., Guo, M., Li, H., Wang, Q., Robles-Kelly, A.: Revisiting spatio-angular trade-off in light field cameras and extended applications in super-resolution. IEEE Trans. Visual Comput. Graphics **27**(6), 3019–3033 (2019)
11. Shi, L., Hassanieh, H., Davis, A., Katabi, D., Durand, F.: Light field reconstruction using sparsity in the continuous Fourier domain. ACM Trans. Graph. **34**(1), 1–13 (2014)
12. Vagharshakyan, S., Bregovic, R., Gotchev, A.: Light field reconstruction using shearlet transform. IEEE Trans. Pattern Anal. Mach. Intell. **40**(1), 133–147 (2017)
13. Wu, G., et al.: Light field image processing: an overview. IEEE J. Sel. Top. Signal Process. **11**(7), 926–954 (2017)
14. Wang, Y., Liu, F., Wang, Z., Hou, G., Sun, Z., Tan, T.: End-to-end view synthesis for light field imaging with pseudo 4DCNN. In: European Conference on Computer Vision, pp. 333–348 (2018)
15. Wu, G., Wang, Y., Liu, Y., Fang, L., Chai, T.: Spatial-angular attention network for light field reconstruction. IEEE Trans. Image Process. **30**, 8999–9013 (2021)
16. Meng, N., So, H.K.H., Sun, X., Lam, E.: High-dimensional dense residual convolutional neural network for light field reconstruction. IEEE Trans. Pattern Anal. Mach. Intell. **43**, 873–886 (2019)
17. Meng, N., Li, K., Liu, J., Lam, E.Y.: Light field view synthesis via aperture disparity and warping confidence map. IEEE Trans. Image Process. **30**, 3908–3921 (2021)

18. Kalantari, N.K., Wang, T.C., Ramamoorthi, R.: Learning-based view synthesis for light field cameras. ACM Trans. Graph. **35**(6), 1–10 (2016)
19. Jin, J., Hou, J., Yuan, H., Kwong, S.: Learning light field angular super-resolution via a geometry-aware network. In: AAAI Conference on Artificial Intelligence (2020)
20. Shi, J., Jiang, X., Guillemot, C.: Learning fused pixel and feature-based view reconstructions for light fields. In: IEEE Conference on Computer Vision and Pattern Recognition, pp. 2555–2564 (2020)
21. Wu, G., Liu, Y., Fang, L., Chai, T.: Revisiting light field rendering with deep anti-aliasing neural network. IEEE Trans. Pattern Anal. Mach. Intell. **44**, 5430–5444 (2021)
22. Wang, Y., Wang, L., Wu, G., Yang, J., An, W., Yu, J., Guo, Y.: Disentangling light fields for super-resolution and disparity estimation. IEEE Trans. Pattern Anal. Mach. Intell. **45**, 425–443 (2022)
23. Long, J., Ning, Z., Darrell, T.: Do convnets learn correspondence? Adv. Neural Inf. Process. Syst. **27**, 1601–1609 (2014)
24. Wang, Y., Liu, F., Zhang, K., Hou, G., Sun, Z., Tan, T.: LFNet: a novel bidirectional recurrent convolutional neural network for light-field image super-resolution. IEEE Trans. Image Process. **27**(9), 4274–4286 (2018)
25. Yoon, Y., Jeon, H.G., Yoo, D., Lee, J.Y., Kweon, I.S.: Light-field image super-resolution using convolutional neural network. IEEE Signal Process. Lett. **24**(6), 848–852 (2017)
26. Zhang, S., Lin, Y., Sheng, H.: Residual networks for light field image super-resolution. In: IEEE Conference on Computer Vision and Pattern Recognition, pp. 11046–11055 (2019)
27. Meng, N., Wu, X., Liu, J., Lam, E.Y.: High-order residual network for light field super-resolution. In: AAAI Conference on Artificial Intelligence (2020)
28. Jin, J., Hou, J., Chen, J., Kwong, S.: Light field spatial super-resolution via deep combinatorial geometry embedding and structural consistency regularization. In: IEEE Conference on Computer Vision and Pattern Recognition, pp. 2260–2269 (2020)
29. Yoon, Y., Jeon, H.G., Yoo, D., Lee, J.Y., So Kweon, I.: Learning a deep convolutional network for light-field image super-resolution. In: IEEE International Conference on Computer Vision Workshops, pp. 24–32 (2015)
30. Wing Fung Yeung, H., Hou, J., Chen, J., Ying Chung, Y., Chen, X.: Fast light field reconstruction with deep coarse-to-fine modeling of spatial-angular clues. In: European Conference on Computer Vision, pp. 137–152 (2018)
31. Wu, G., Zhao, M., Wang, L., Dai, Q., Chai, T., Liu, Y.: Light field reconstruction using deep convolutional network on EPI. In: IEEE Conference on Computer Vision and Pattern Recognition, pp. 6319–6327 (2017)
32. Liu, D., Huang, Y., Wu, Q., Ma, R., An, P.: Multi-angular epipolar geometry based light field angular reconstruction network. IEEE Trans. Comput. Imaging **6**, 1507–1522 (2020)
33. Wang, Y., Liu, F., Zhang, K., Wang, Z., Sun, Z., Tan, T.: High-fidelity view synthesis for light field imaging with extended pseudo 4DCNN. IEEE Trans. Comput. Imaging **6**, 830–842 (2020)
34. Wanner, S., Goldluecke, B.: Variational light field analysis for disparity estimation and super-resolution. IEEE Trans. Pattern Anal. Mach. Intell. **36**(3), 606–619 (2014)
35. Wu, G., Liu, Y., Dai, Q., Chai, T.: Learning sheared epi structure for light field reconstruction. IEEE Trans. Image Process. **28**(7), 3261–3273 (2019)

36. Mildenhall, B., et al.: Local light field fusion: practical view synthesis with prescriptive sampling guidelines. ACM Trans. Graph. **38**(4), 1–14 (2019)
37. Jin, J., Hou, J., Chen, J., Zeng, H., Kwong, S., Yu, J.: Deep coarse-to-fine dense light field reconstruction with flexible sampling and geometry-aware fusion. IEEE Trans. Pattern Anal. Mach. Intell. **44**, 1819–1836 (2020)
38. Levoy, M., Hanrahan, P.: Light field rendering. In: Proceedings of the 23rd Annual Conference on Computer Graphics and Interactive Techniques, pp. 31–42 (1996)
39. Wang, L., et al.: Learning parallax attention for stereo image super-resolution. In: IEEE Conference on Computer Vision and Pattern Recognition (2019)
40. Honauer, K., Johannsen, O., Kondermann, D., Goldluecke, B.: A dataset and evaluation methodology for depth estimation on 4D light fields. In: Lai, S.-H., Lepetit, V., Nishino, K., Sato, Y. (eds.) ACCV 2016. LNCS, vol. 10113, pp. 19–34. Springer, Cham (2017). https://doi.org/10.1007/978-3-319-54187-7_2
41. Wanner, S., Meister, S., Goldluecke, B.: Datasets and benchmarks for densely sampled 4D light fields. In: Vision, Modelling and Visualization, pp. 225–226. Citeseer (2013)
42. Raj, A.S., Lowney, M., Shah, R., Wetzstein, G.: Stanford lytro light field archive (2016)
43. Kingma, D.P., Ba, J.: Adam: a method for stochastic optimization. In: International Conference on Learning and Representation (2015)

Adaptive Mask-Based Pyramid Network for Realistic Bokeh Rendering

Konstantinos Georgiadis[2], Albert Saà-Garriga[1], Mehmet Kerim Yucel[1](✉), Anastasios Drosou[2], and Bruno Manganelli[1]

[1] Samsung Research UK, Staines-upon-Thames, UK
mehmet.yucel@samsung.com
[2] Centre for Research and Technology Hellas (CERTH), Information Technologies Institute, Thessaloniki, Greece

Abstract. Bokeh effect highlights an object (or any part of the image) while blurring the rest of the image, and creates a visually pleasant artistic effect. Due to the sensor-based limitations on mobile devices, machine learning (ML) based bokeh rendering has gained attention as a reliable alternative. In this paper, we focus on several improvements in ML-based bokeh rendering; i) on-device performance with high-resolution images, ii) ability to guide bokeh generation with user-editable masks and iii) ability to produce varying blur strength. To this end, we propose Adaptive Mask-based Pyramid Network (AMPN), which is formed of a Mask-Guided Bokeh Generator (MGBG) block and a Laplacian Pyramid Refinement (LPR) block. MGBG consists of two lightweight networks stacked to each other to generate the bokeh effect, and LPR refines and upsamples the output of MGBG to produce the high-resolution bokeh image. We achieve i) via our lightweight, mobile-friendly design choices, ii) via the stacked-network design of MGBG and the weakly-supervised mask prediction scheme and iii) via manually or automatically editing the intensity values of the mask that guide the bokeh generation. In addition to these features, our results show that AMPN produces competitive or better results compared to existing methods on the EBB! dataset, while being faster and smaller than the alternatives.

Keywords: Bokeh rendering · Image refocusing · Laplacian pyramid

1 Introduction

Bokeh effect is one of the fundamental photography techniques, where an object (or a region) in the image is effectively highlighted by blurring out the rest of the image. Traditionally, such an effect is achieved via focusing the camera on an area and taking the photo using a wide aperture lens. Although it is achievable with appropriate cameras with fast lenses with large apertures, or even with mobile devices with stereo setups, bokeh rendering may not be feasible for all mobile devices due to sensory/hardware constraints. Synthetically generating

K. Georgiadis, A. Saà-Garriga and M. K. Yucel—The authors have contributed equally.

L. Karlinsky et al. (Eds.): ECCV 2022 Workshops, LNCS 13802, pp. 429–444, 2023.
https://doi.org/10.1007/978-3-031-25063-7_26

the bokeh effect through machine-learning (ML) based methods are therefore viable alternatives, especially for setups without adequate hardware.

There has been a rising interest in synthetic bokeh rendering, especially following the release of EBB! dataset [8] and bokeh rendering challenges [9,11]. Starting with earlier methods focusing on portrait images [24,25], later methods leveraged advances in image-to-image methods and generative models [3–5,8,17,18,20–22,27,33]. Depth/disparity [4,8,21,27] and saliency maps [9,21], as well as segmentation maps [34] have found use as the principal guiding component for bokeh rendering, however, such methods require additional ground-truth information during training and require additional components in inference time. Conversely, methods not using any guidance at all [3,5,18,22] are essentially learning an inflexible mapping, where in-focus areas will be implicitly learned by the network and can not be changed by the user. Furthermore, ML-based methods essentially fit to the blur strength of the training data and can generate only fixed bokeh styles [20], leading to limited user interaction.

Thinking from a mobile use case standpoint, we focus on three areas of improvement; i) fast on-device runtimes with high-resolution images, ii) ability to guide bokeh generation with user-editable masks without additional compute in inference time and iii) ability to render bokeh effect with varying blur strengths. There are fast, on-device methods for bokeh rendering [3–5,8], but none of them meets the criteria ii) and iii). There are methods leveraging external guidance for bokeh rendering (i.e. criteria ii)) [4,8,9,21,27,34], but they require additional processing to generate such guidance (i.e. depth models, saliency models, etc). A recent work [20] focuses on the ability to generate varying blur strengths (criteria iii)), but their solution is not tailored for on-device performance and therefore not suitable for mobile use cases.

In light of these above mentioned criteria, we propose Adaptive Mask-based Pyramid Network (AMPN) for mobile-friendly, realistic bokeh rendering. AMPN is formed of two main building blocks; i) Mask-Guided Bokeh Generator (MGBG) and ii) Laplacian Pyramid Refinement (LPR). MGBG consists of two stacked networks, where the first one is responsible for mask prediction and the other one is for bokeh generation, guided by the predicted mask. LPR essentially refines the low-resolution bokeh output of MGBG and upsamples it to the input resolution. Using lightweight designs in both blocks lets us operate efficiently, whereas LPR lets us produce high fidelity, high-resolution outputs without the need of third party solutions for upsampling/super-resolution (meeting criteria i). Furthermore, if a user is providing its mask, we can completely remove the mask prediction network during inference, making our pipeline even more compact.

Having strong user-guidance often requires strong supervision (i.e. ground-truths) in (guidance) mask prediction, which may not be available in every dataset. Our pipeline, on the other hand, learns mask prediction in a weakly supervised way due to the stacked network paradigm we adopt in MGBG (meeting criteria ii)). We train using wide/shallow depth-of-field images using existing datasets without mask ground-truth. The level of mask guidance we achieve is visually pleasant (see Fig. 1), it empowers users with greater flexibility (i.e. mask

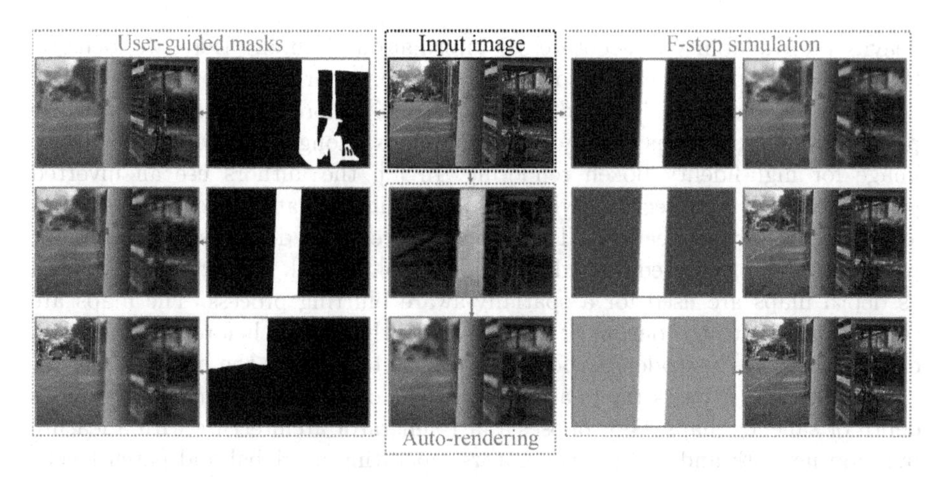

Fig. 1. Our AMPN method renders bokeh images conditioned on masks, which can be provided by users and thus our results can focus on selected area of an image (left side). Furthermore, by changing the intensity of the guidance masks, we can simulate different f-stops (right side). Our weakly-supervised mask prediction network learns masks without direct supervision (middle images), which can be replaced by user-provided masks in inference.

editing, providing custom masks) and even extends the use case beyond bokeh rendering to mask-based, learnable image refocusing. Note that our bokeh generation can still be guided with depth and/or saliency maps. Our mask prediction network also brings the controllable blur strength feature; by changing the background intensities of our guidance mask, we show that we can simulate different f-stops (meeting criteria iii), see Fig. 7). Our main contributions can be summarized as follows:

1. We propose AMPN for realistic bokeh rendering, which renders bokeh at low resolution via the MGBG block and performs refinement/upsampling with the LPR block. AMPN produces competitive or better results compared to existing methods on EBB!val294 set and performs faster than alternatives, making it ideal for mobile device deployment.
2. We show that the design of AMPN enables training of a mask prediction network in a weakly-supervised way, which lets users control their bokeh rendering with a mask; either obtained by the mask prediction network, edited by the user or even produced by the user.
3. We also show that by simply changing the intensity values of the guidance mask, we can control the strength of the blur in the bokeh rendered image.

2 Related Work

Synthetic Bokeh Rendering. Synthetic bokeh rendering can be largely divided into two categories [20] as classical and ML-based approaches. Classical rendering is shown to provide a degree of control and flexibility, but relies heavily on availability of accurate 3D information [2,26,32]. Here, we focus on the ML-based approaches due to their relevance to our method.

Following the portrait based methods [24,25], more recent methods utilize prior knowledge, such as depth or saliency maps, along with the in-focus input image for high-fidelity bokeh rendering. In [8], the authors use an inverted pyramid-shaped architecture, which processes images at different scales, thus learning more diverse features. Their approach relies on depth estimation maps, produced by a pretrained Megadepth model [15]. In [21], saliency maps, as well as depth maps are used for a spatially-aware blurring process. The maps are concatenated with the output of a space-to-depth module, before being fed to a densely connected encoder/decoder network that produces the bokeh image.

Parallel lines of work try to avoid depending on external information such as depth or saliency maps. [3] proposes a fast generator architecture with a feature pyramid network and two discriminators, operating at global and patch levels. The authors of [5] propose a fast, stacked multi-scale hierarchical network that process images at different scales to obtain local and global features. In [17], the authors propose a multi-module network, where each module focus on separate components such as defocus estimation, radiance, rendering and upsampling. [22] uses a fast Glass-Net generator and Multi-Receptive-Field discriminator, and also re-implement instance normalization layers for further speed-ups on a mobile device. Authors of [18] propose a transformer-based architecture for bokeh rendering. They show that the increased receptive field, as well as pretraining the model on image restoration tasks, help improve the results. A recent work [20] propose a combination of classical and neural rendering, to achieve high-resolution photo-realistic bokeh images. Their framework is adjustable, where the blur size, focal plane and aperture shape can be chosen. Our approach combines the flexibility (i.e. user-editable mask guided bokeh rendering, varying blur strength) and mobile-friendliness of existing methods.

Operating at High-Resolutions. A key aspect of mobile vision tasks, such as dense regression/prediction (i.e. bokeh rendering, image editing, etc), is that they have a low tolerance to visual artefacts. Especially with the continuously improving mobile display technology, high-resolution images are becoming the norm, which makes the error tolerance even lower and requires accurate, high-resolution outputs. However, simply operating at high-resolution is not an option, especially due to the limited resources available on mobile devices.

A naive approach is to perform the vision task at low resolution and then perform upsampling/super-resolution [14]. However, this approach requires a second model, which should work well on the same distribution and has to work sequentially with the vision model, which lowers the overall feasibility. Authors of [16] propose the LPTN framework where upsampling and the vision task is performed jointly, where low-frequency components are translated with a lightweight network and high-frequency details are refined using both low and high-frequency components. Our LPR block is based on LPTN and its successors [12], with key differences explained later in relevant sections.

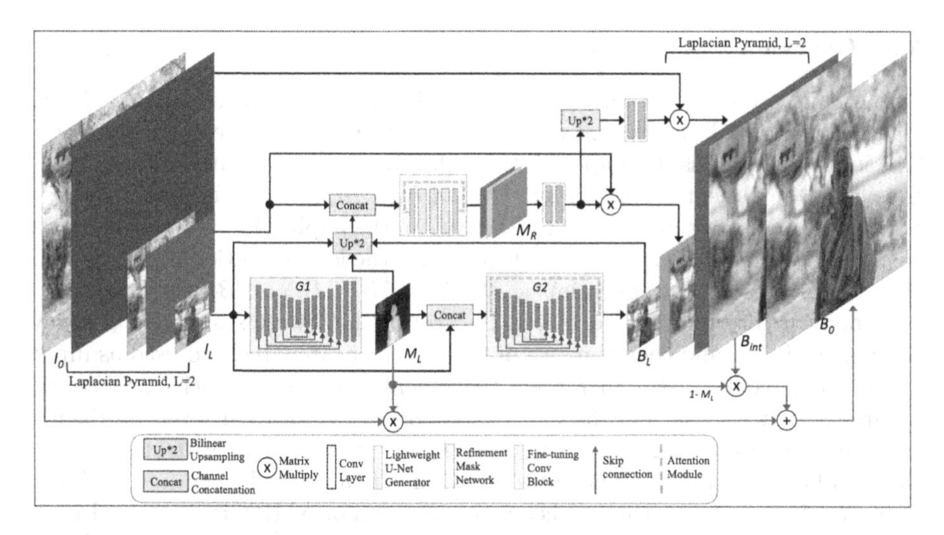

Fig. 2. The diagram of our proposed AMPN pipeline. First, we decompose the input image I_0 with a Laplacian Pyramid and acquire the low-resolution image I_L. Through our MGBG (light blue blocks) block, we first (in a weakly-supervised manner) discover the area of focus M_L with $G1$, and then in $G2$ we use this area of focus to guide the bokeh rendering in low-resolution to produce B_L. Within our LRP block (refinement mask network and fine-tuning conv blocks), we progressively upsample and refine the low-resolution bokeh image and use mask-based blending (red arrows) to produce the final high-resolution bokeh image B_0. (Color figure online)

3 Adaptive Mask-based Pyramid Network

In this section, we motivate the need for a new bokeh rendering approach, introduce AMPN and its components in detail.

3.1 Motivation

Our aim to design a pipeline that meets the three criteria mentioned in earlier sections; i) fast processing with high-resolution outputs, ii) user-editable mask guided bokeh rendering and iii) ability to edit blur strength. We note that bokeh, by definition, is guided by depth information. Therefore, achieving criteria ii) provides a functionality beyond bokeh rendering, which is essentially image refocusing guided by a mask. In a sense, our aim is to learn from bokeh rendering datasets while being able to perform the more abstract task of mask-based image refocusing. Furthermore, achieving iii) makes our bokeh rendering hardware-independent; essentially we will not overfit to a specific bokeh style [20]. To summarize, our aim is beyond having an accurate, fast bokeh rendering model, but also to produce a highly interactive bokeh rendering experience.

3.2 Overview

Our AMPN pipeline is composed of two main blocks; the Mask-Guided Bokeh Generator (MGBG) block and a Laplacian Pyramid Refinement (LPR) block. The input image I_0 is decomposed, through a Laplacian pyramid, into its high-frequency components, denoted by $H = [h_0, h_1, ..., h_{L-1}]$, and its low-frequency residual image I_L, where L refers to the levels in the Laplacian pyramid [16]. I_L is then fed into the MGBG block, which generates the low-resolution mask M_L and then the low-resolution bokeh image B_L. Next, I_L, M_L and B_L are upsampled with bilinear interpolation to match the size of the lowest size high-frequency component h_{L-1}, and then the four of them are concatenated and used as input to the LPR block. LPR generates a refinement mask M_R, which is combined progressively with the high frequency components H of each level, producing the high-resolution image B_{int}. B_{int} and I_0 are processed with M_L to produce the final bokeh image B_0. The entire network is trained end-to-end, using high-resolution inputs and outputs. The overall diagram of AMPN is shown in Fig. 2. We now explain in more detail how the MGBG and LPRL blocks work and how they are trained.

3.3 Mask-Guided Bokeh Generator Block

The goal of the MGBG block is to generate the bokeh image in low-resolution. A natural approach here would be to use a single network that takes in the input image and simply produce the output image (i.e. optionally by using additional information such as depth and saliency maps). Since our aim is to also achieve criteria ii), we use a stacked-network formation using two networks, which we call $G1$ and $G2$. For both $G1$ and $G2$, we use accurate and performant architectures (criteria i)); we leverage the architecture of [30] which is a combination of MobileNetv2 [23] and FBNet [29] components.

$G1$ and $G2$ have their own separate tasks; $G1$ takes as input the input image I_L and outputs a grayscale mask M_L, whereas $G2$ takes in I_L and M_L as input (as a 4-channel tensor) and generates the low-resolution bokeh rendered image B_L. We note that there is no ground-truth for M_L, therefore $G1$ is trained without any explicit supervision. $G1$, by leveraging the paired training samples, discovers the areas of refocus. The core advantages of having $G1$ are threefold: first, unlike other methods that do not use any guidance, $G1$ makes our network interpretable as it learns the area of refocus in the form of a mask (see Fig. 5). Second, since $G2$ is conditioned on the output of $G1$, we can control the refocus area (criteria ii), see Fig. 1). Finally, we can even remove $G1$ in inference and ask the user to provide any type of mask; user-generated, depth, saliency, etc. (see Fig. 6). This makes our method lightweight and applicable beyond bokeh rendering (i.e. image refocusing).

The 4-channel input to $G2$ (I_L and M_L) is passed through the $G2$ network that predicts an RGB image I_{int}. We also leverage a dual-attention mechanism that operates over the RGB component of the input (i.e. I_L) and I_{int} (see Fig. 3). Separate attention modules process I_L and I_{int}, and their results are summed

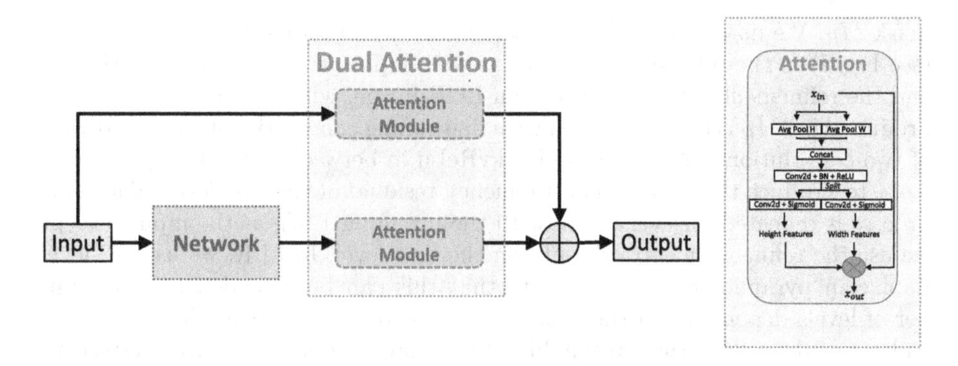

Fig. 3. Our dual attention module. The *network* block represents $G2$ network of the MBGB block, as well as the refinement network of the LPR block. The attention module is implemented via [7], details of which are shown on the right.

to produce the low-resolution bokeh image B_L. The inspiration for this dual-attention mechanism comes from modern super-resolution approaches [10], where long-distance residual connections are used. Instead of using residual connections directly from I_L, we leverage attention modules for visually pleasing results. We use [7] to implement the dual-attention modules. Note that each attention module is learned jointly with the entire pipeline, and they do not share weights. We also note that similar to $G1$, $G2$ is learned without direct supervision. We enforce our losses at the final, high-resolution bokeh image B_0 (i.e. the output of the LPR block).

3.4 Laplacian Pyramid Refinement Block

The goal of the LPR block is twofold; upsample the low-resolution bokeh rendered image B_L back to the original resolution and refine/improve the results while doing so. Our LPR block is based on [16], with several key differences.

Preliminaries. The Laplacian pyramid [1] decomposes the image into low and high frequency components, from which the original image can be reconstructed. At each level of the pyramid, a fixed kernel is used to calculate the weighted average of the neighbouring pixels of the image, resulting in a downsampled version. The downsampled version is then upsampled again, and a high-frequency residual component h is calculated by subtracting the upsampled image from the original one. This process is repeated for each pyramid level, and with the inverse operation, the original image is reconstructed. For each level, we refer to high-frequency residuals as $H = [h_0, h_1, ..., h_{L-1}]$.

LPR. Having produced M_L and B_L via the MGBG block, we now aim to upsample B_L to produce our final results B_0. We first take I_L, M_L and B_L and upsample them to reach the spatial resolution of the lowest (pyramid level 2) high-frequency residual h_{L-1}, and then concatenate these four. This concatenated tensor is fed to the refinement network, which produces the refinement mask M_R. We note that we use the same dual-attention mechanism used in G2 (see Fig. 3) in the refinement network as well; input attention processes the B_L and the refinement network outputs an RGB image, which are then summed to produce M_R. M_R is then processed with fine-tuning convolutional blocks (formed of two convolutional layers, and a LeakyReLu in between), and multiplied with h_{L-1} to produce the output high-frequency residual of the first level. The same process is repeated for every level with the upsampled M_R as the input, except we use the refinement network only on the first level. In LPR, we use a 2-level Laplacian pyramid, however we note that this can be extended to any number of levels depending on the capacity/output-resolution requirements. At the end, we end up with two output high-frequency residuals, which are added to progressively upsampled B_L in each level, until we produce B_{int} in the original resolution. As the final step, we take in the original high-resolution input I_0 and multiply it with M_L, and sum its results with B_{int} multiplied with $1 - M_L$ to produce the final bokeh rendered image B_0. We use masking so that our entire pipeline can focus on where matters (i.e. non-focus areas), since the input masking operation copies the focus area directly from the input image.

Differences with LPTN. Our LPR block differs from LPTN in several aspects; i) In addition to I_L and B_L, LPR also leverages M_L in the upsampling, which guides the refinement/upsampling process, ii) we propose the dual-attention mechanism in the refinement network which improves it accuracy and iii) we use the mask M_L to both leverage I_0 and better blend/integrate I_0 and B_{int} spatially in generating the final output B_0.

3.5 Losses

We use common regression losses, as well as perceptual-aware losses during our training. In total, we use three loss functions; the reconstruction loss L_1, the learned perceptual image patch similarity (LPIPS) [31] loss L_{LPIPS} and the structural similarity loss L_{SSIM} [28]. The final loss is a combination of the above losses, which is defined as:

$$L_{total} = 10 \cdot L_1 + 2 \cdot L_{LPIPS} + L_{SSIM}$$

The weights for each loss are based on [22]. As noted before, the loss is applied only on the final bokeh image B_0, therefore $G1$ or $G2$ networks in the MGBG block do not have direct supervision. During our experiments, we experimented with applying the losses also over the output of $G2$ (B_L) and even over the outputs of each pyramid level (i.e. B_{int}), however, we did not see tangible improvements in either configuration.

4 Experiments

4.1 Dataset

We train our model on the *Everything looks Better with Bokeh! (EBB!)* dataset [8]. It consists of 5K aligned image pairs of wide/shallow depth-of-field, 4600 of which are used as the training set and 200 images are used for the validation and test sets, respectively. All the images in the dataset have a height of 1024 pixels, however, the width varies between images.

Since *EBB!* is a challenge dataset [9,11], the ground-truths for validation and test sets are not publicly available. For a fair comparison, we use the *val294* [4] set for evaluation. The *val294* set is based on the EBB!'s train set, where the first 4400 images are used for training, while the rest is used for evaluation.

4.2 Experimental Setup

Implementation Details. We use PyTorch [19] throughout our experiments. Following [30], we initialize the encoders of $G1$ and $G2$ with weights obtained by training on ImageNet. The rest of the learnable parameters (i.e. LPR modules and the decoders of $G1$ and $G2$) are initialized with [6]. All experiments are conducted on a PC with NVIDIA GeForce RTX 3090 GPU and AMD EPYC 7352 CPU. We train our model for 500 epochs, with a batch size of 8, utilizing the Adam optimizer and $2 \cdot 10^{-4}$ learning rate. The trainings are performed with input and output images with the size 1024×1536.

Evaluation Metrics. We use Peak Signal-to-Noise Ratio (PSNR), Structural Similarity (SSIM) [28] and Learned Perceptual Image Patch Similarity (LPIPS) [31] as our evaluation metrics. These metrics are the most widely used metrics used in the synthetic bokeh rendering literature, therefore we choose them to establish a fair comparison with existing methods. We note that many methods also leverage user surveys as another evaluation criteria; despite the value they bring, such surveys are generally not comparable nor reproducible. Furthermore, we believe the key contributions of our method, such as user-editable mask guidance and controllable blur strengths, can be well-represented with qualitative examples. Therefore, we leave user surveys as future work.

4.3 Comparison with State-of-the-Art

We compare our method with several state-of-the-art methods, such as PyNet [8], DMSHN [5], SKN [13], DBSI [4] and BRViT [18]. We choose these methods based on the availability of their source codes and whether they report results on *Val294* set on EBB!. We use pretrained models and source codes for evaluation (if available), or original results reported in relevant papers for other methods.

Quantitative Comparison. The results are shown in Table 1. Our method produces competitive results in each metric. In PSNR, our method trails behind others. Our method does quite well in SSIM and LPIPS metrics, it is the second best in SSIM and the best in LPIPS by a considerable margin. We note that BRViT (1st in SSIM) [18] uses quite a large model based on Vision Transformers, and also performs an initial pretraining stage. In general, our model performs competitively, and even outperforms others in perceptual metrics.

Performance Comparison. We also compare our runtime performance against other state-of-the-art method. For a fair comparison, we only compare against

Table 1. Comparison with other methods on *Val294* set. SKN results are taken from [18]. BRViT [18] ‡ performs initial pretraining. The best and second best results are in bold and underlined, respectively, for each metric.

Method	PSNR↑	SSIM↑	LPIPS↓
SKN [13]	24.66	0.8521	0.3323
DBSI [4]	23.45	0.8675	0.2463
PyNet [8]	**24.93**	0.8788	0.2219
DMSHN [5]	24.65	0.8765	0.2289
Stacked DMSHN [5]	24.72	0.8793	0.2271
BRViT [18]	22.88	0.8516	0.2558
BRViT [18] ‡	24.76	**0.8904**	0.1924
Ours	24.50	0.8847	**0.1718**

Table 2. Performance comparison with other state-of-the-art methods, on a desktop CPU. Refer to the implementation details section for the evaluation setup.

Method	CPU (sec)	# of params	GFLOPs
PyNet [8]	4.089	47.5M	5300
Stacked DMSHN [5]	12.657	21.7M	480.7
BRViT [18]	30.288	123.1M	650
Ours	**1.591**	**5.4M**	**45.9**

methods that have publicly available source codes or pretrained models, so that we can compare them in our system in a fair fashion. The results are reported in Table 2. In constrained scenarios, such as the desktop CPU, we outperform others with a significant margin. Furthermore, the last column shows that our method uses significantly fewer parameters. Finally, it is visible that our method has the fewest GFLOPs among other alternatives, proving its efficiency in terms of computation complexity. We note that our method is not likely to saturate high-end desktop GPUs, therefore it might not be as efficient as other larger methods that can utilize them better. Therefore, we highlight that our method is aimed at achieving good performance in resource-limited environments, such as mobile devices.

Qualitative Comparison. Example results produced by our method are shown in Fig. 4. $G1$ network in our MGBG module successfully learns, in a weakly supervised way, the areas of focus in the ground-truth data. Our learned masks are not strictly binary, which we later show to be useful for simulating different f-stops. Qualitative comparison against existing methods are shown in Fig. 5. Our method performs competitively against the alternatives, producing visually pleasing images in multiple cases. We note that competing methods are slower and have significantly larger capacity compared to us, therefore our method might produce slightly worse results in some cases.

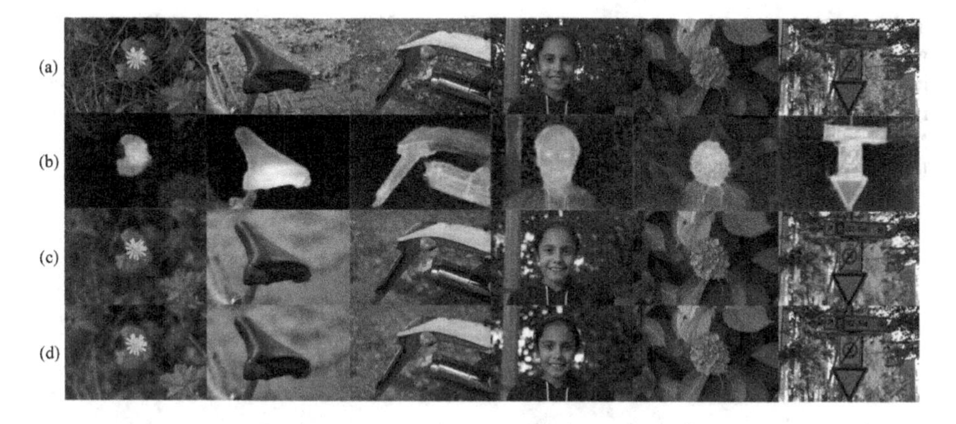

Fig. 4. Qualitative results of our proposed method. (a) Input images, (b) M_L masks predicted by $G1$ of the MGBG block, (c) rendered bokeh images and (d) ground-truth masks.

Fig. 5. Qualitative comparison with other methods.

4.4 Ablation Studies

Component Analyses. We present the component analyses of our model architecture to show the contribution of the building blocks. The results are shown in Table 3. First, we test our pipeline without the refinement mask model in the LPR block. This variant (2nd row in Table 3) is clearly the fastest variant, on mobile GPU and desktop CPU alike, showing that the refinement mask network is a bottleneck in performance. However, we lose out quite a bit on LPIPS without the refinement mask model, which justifies its addition to our pipeline. Second, we remove $G2$ from the MGBG block; in this variant (3rd row

Table 3. Ablation study on our model pipeline on *Val294* set. *w/0 ref.* indicates LPR without the refinement mask model. *w/0 att.* indicates LPR without dual attention in the refinement mask model. **D** and **M** stand for desktop and mobile runtimes, respectively. The desktop hardware is detailed in the implementation details section. Mobile runtimes are averaged on 50 runs, and acquired using Samsung Galaxy S22 Ultra with Exynos 2200 processor.

Components	PSNR↑	SSIM↑	LPIPS↓	D CPU (s)	D GPU (s)	M GPU (s)	# of params
G1 + G2 + LPR	**24.50**	**0.8847**	0.1718	1.591	0.059	0.330	5.4M
G1 + G2 + LPR (w/o ref.)	24.24	0.8814	0.1931	**0.731**	0.040	**0.118**	5.2M
G1, no G2 + LPR	23.91	0.8692	0.1903	1.483	0.042	0.297	2.8M
G1 + G2 + LPR (w/o att.)	24.25	0.8825	**0.1717**	1.581	0.052	0.319	5.2M
no G1, G2 + LPR	–	–	–	1.4732	**0.039**	0.312	2.8M

in Table 3), we essentially force LPR to both learn bokeh rendering and upsample/refine the results. This variant does surprisingly well and performs slightly faster than the full pipeline. Third, we remove the dual attention mechanism from the refinement mask module in the LPR block. This variant (4th row in Table 3) is the close second in terms of evaluation metrics, and it is slightly faster than the full pipeline. Finally, we remove $G1$ from MGBG and simply use external masks to guide $G2$. This variant (5th row in Table 3) has similar runtimes with the 3rd variant. We note that this variant is a desirable configuration since external mask guidance is a more interactive use case, and also we can not report metrics with this variant since we use external masks. Finally, our full pipeline (1st row in Table 3) produces the best SSIM and LPIPS metrics, showing the effectiveness of our design choices. We note that the other three variants shown in Table 3 (2nd, 3rd and 4th rows) are still competitive to other methods.

Fig. 6. Example outputs using different *types* of masks; orange, yellow, green and blue masks are depth map, user-generated mask, $G1$-generated mask and saliency map, respectively. (Color figure online)

Using Different Masks. As explained in earlier sections, our pipeline is flexible as it can be guided by any mask. This mask can be generated by $G1$ of the MGBG block, but also can be generated by a user. Furthermore, one can guide the bokeh generation with depth maps or saliency maps. We provide several scenarios where we generate the bokeh effect with different types of mask guidance, see Fig. 6 for examples. Our method manages to refocus images successfully with various types of mask guidance.

Simulating Different F-Stops. One of the important features of our method is that it can simulate different f-stops/blur strengths. We discover that by modifying the values of the input mask to our model ($G2$), we can achieve a certain level of control over various depth-of-focus effects. Specifically, we do not alter the mask intensities on the in-focus area, but rather alter the intensity values of the background (out-of-focus) areas. Coupled with the fact that we can use any mask to guide our bokeh/refocusing, this also provides another level of interactive experience, which further empowers the user. We provide several examples in Fig. 7.

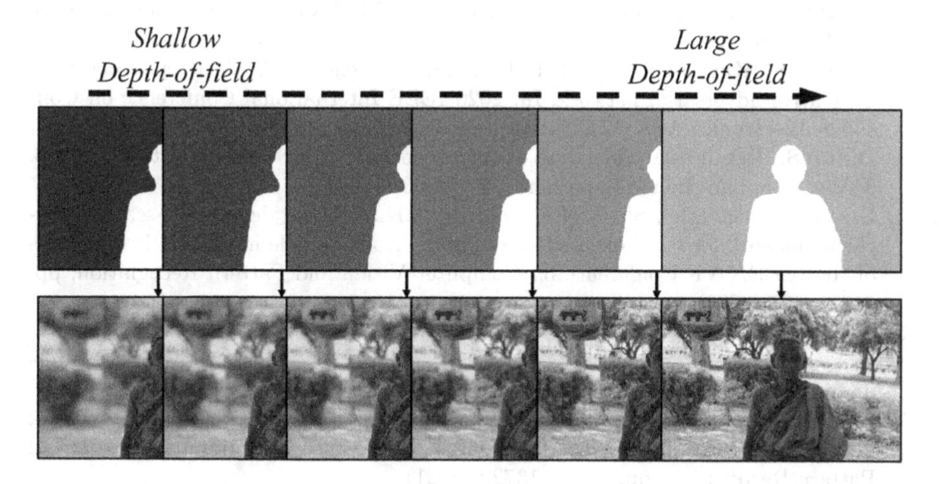

Fig. 7. Different f-stop simulation. Each column shows the mask and the related bokeh rendering result. By changing the mask intensity values of the out-of-focus areas, we manage to simulate different f-stops.

5 Conclusion

In this paper, we focus on the task of ML-based synthetic bokeh rendering. We focus on three areas of improvement; i) accurate on-device performance with high-resolution images, ii) ability to guide bokeh-generation with user-editable masks and iii) ability to produce blurs with varying strength profiles. We propose the AMPN pipeline, which is formed of the MGBG and the LPR blocks. The MGBG block performs bokeh rendering at low resolutions for low memory footprint and fast processing, whereas LPR progressively upsamples and refines the low-resolution bokeh image generated by MGBG. Owing to its two-network design, the first network of MGBG discovers in-focus areas of images without supervision, in the form a mask. Since the second network of MGBG conditions the bokeh generation on this (weakly-supervised) discovered mask, we can edit/change the mask freely to guide bokeh generation. This level of control lets us achieve the more abstract functionality of mask-guided image refocusing. Furthermore, by changing the intensities of the mask, we show that we can simulate various blur strengths. Finally, we show that our method produces competitive or better results compared to various alternatives on the EBB! dataset, while running faster.

References

1. Burt, P.J., Adelson, E.H.: The Laplacian pyramid as a compact image code. In: Readings in Computer Vision, pp. 671–679. Elsevier (1987)
2. Busam, B., Hog, M., McDonagh, S., Slabaugh, G.: SteReFo: efficient image refocusing with stereo vision. In: Proceedings of the IEEE/CVF International Conference on Computer Vision Workshops (2019)
3. Choi, M.S., Kim, J.H., Choi, J.H., Lee, J.S.: Efficient bokeh effect rendering using generative adversarial network. In: 2020 IEEE International Conference on Consumer Electronics-Asia (ICCE-Asia), pp. 1–5. IEEE (2020)
4. Dutta, S.: Depth-aware blending of smoothed images for bokeh effect generation. J. Vis. Commun. Image Represent. **77**, 103089 (2021)
5. Dutta, S., Das, S.D., Shah, N.A., Tiwari, A.K.: Stacked deep multi-scale hierarchical network for fast bokeh effect rendering from a single image. In: Proceedings of the IEEE/CVF Conference on Computer Vision and Pattern Recognition, pp. 2398–2407 (2021)
6. He, K., Zhang, X., Ren, S., Sun, J.: Delving deep into rectifiers: Surpassing human-level performance on ImageNet classification. In: Proceedings of the IEEE International Conference on Computer Vision, pp. 1026–1034 (2015)
7. Hou, Q., Zhou, D., Feng, J.: Coordinate attention for efficient mobile network design. In: Proceedings of the IEEE/CVF Conference on Computer Vision and Pattern Recognition, pp. 13713–13722 (2021)
8. Ignatov, A., Patel, J., Timofte, R.: Rendering natural camera bokeh effect with deep learning. In: Proceedings of the IEEE/CVF Conference on Computer Vision and Pattern Recognition Workshops, pp. 418–419 (2020)
9. Ignatov, A., et al.: Aim 2019 challenge on bokeh effect synthesis: methods and results. In: 2019 IEEE/CVF International Conference on Computer Vision Workshop (ICCVW), pp. 3591–3598. IEEE (2019)
10. Ignatov, A., Romero, A., Kim, H., Timofte, R.: Real-time video super-resolution on smartphones with deep learning, mobile AI 2021 challenge: report. In: Proceedings of the IEEE/CVF Conference on Computer Vision and Pattern Recognition, pp. 2535–2544 (2021)
11. Ignatov, A., et al.: AIM 2020 challenge on rendering realistic bokeh. In: Bartoli, A., Fusiello, A. (eds.) ECCV 2020. LNCS, vol. 12537, pp. 213–228. Springer, Cham (2020). https://doi.org/10.1007/978-3-030-67070-2_13

12. Lei, B., Guo, X., Yang, H., Cui, M., Xie, X., Huang, D.: ABPN: adaptive blend pyramid network for real-time local retouching of ultra high-resolution photo. In: Proceedings of the IEEE/CVF Conference on Computer Vision and Pattern Recognition, pp. 2108–2117 (2022)
13. Li, X., Wang, W., Hu, X., Yang, J.: Selective kernel networks. In: Proceedings of the IEEE/CVF Conference on Computer Vision and Pattern Recognition, pp. 510–519 (2019)
14. Li, Y., et al.: NTIRE 2022 challenge on efficient super-resolution: methods and results. In: Proceedings of the IEEE/CVF Conference on Computer Vision and Pattern Recognition, pp. 1062–1102 (2022)
15. Li, Z., Snavely, N.: MegaDepth: learning single-view depth prediction from internet photos. In: Proceedings of the IEEE Conference on Computer Vision and Pattern Recognition, pp. 2041–2050 (2018)
16. Liang, J., Zeng, H., Zhang, L.: High-resolution photorealistic image translation in real-time: a laplacian pyramid translation network. In: Proceedings of the IEEE/CVF Conference on Computer Vision and Pattern Recognition, pp. 9392–9400 (2021)
17. Luo, X., Peng, J., Xian, K., Wu, Z., Cao, Z.: Bokeh rendering from defocus estimation. In: Bartoli, A., Fusiello, A. (eds.) ECCV 2020. LNCS, vol. 12537, pp. 245–261. Springer, Cham (2020). https://doi.org/10.1007/978-3-030-67070-2_15
18. Nagasubramaniam, H., Younes, R.: Bokeh effect rendering with vision transformers (2022)
19. Paszke, A., et al.: PyTorch: an imperative style, high-performance deep learning library. Adv. Neural Inf. Process. Syst. **32** (2019)
20. Peng, J., Cao, Z., Luo, X., Lu, H., Xian, K., Zhang, J.: Bokehme: when neural rendering meets classical rendering. In: Proceedings of the IEEE/CVF Conference on Computer Vision and Pattern Recognition, pp. 16283–16292 (2022)
21. Purohit, K., Suin, M., Kandula, P., Ambasamudram, R.: Depth-guided dense dynamic filtering network for bokeh effect rendering. In: 2019 IEEE/CVF International Conference on Computer Vision Workshop (ICCVW), pp. 3417–3426. IEEE (2019)
22. Qian, M., et al.: BGGAN: bokeh-glass generative adversarial network for rendering realistic bokeh. In: Bartoli, A., Fusiello, A. (eds.) ECCV 2020. LNCS, vol. 12537, pp. 229–244. Springer, Cham (2020). https://doi.org/10.1007/978-3-030-67070-2_14
23. Sandler, M., Howard, A., Zhu, M., Zhmoginov, A., Chen, L.C.: MobileNetV 2: inverted residuals and linear bottlenecks. In: Proceedings of the IEEE Conference on Computer Vision and Pattern Recognition, pp. 4510–4520 (2018)
24. Shen, X., et al.: Automatic portrait segmentation for image stylization. In: Computer Graphics Forum, vol. 35, pp. 93–102. Wiley Online Library (2016)
25. Shen, X., Tao, X., Gao, H., Zhou, C., Jia, J.: Deep automatic portrait matting. In: Leibe, B., Matas, J., Sebe, N., Welling, M. (eds.) ECCV 2016. LNCS, vol. 9905, pp. 92–107. Springer, Cham (2016). https://doi.org/10.1007/978-3-319-46448-0_6
26. Wadhwa, N., et al.: Synthetic depth-of-field with a single-camera mobile phone. ACM Trans. Graphics (ToG) **37**(4), 1–13 (2018)
27. Wang, L., et al.: DeepLens: shallow depth of field from a single image. ACM Trans. Graphics (TOG) **37**(6), 1–11 (2018)
28. Wang, Z., Bovik, A.C., Sheikh, H.R., Simoncelli, E.P.: Image quality assessment: from error visibility to structural similarity. IEEE Trans. Image Process. **13**(4), 600–612 (2004)

29. Wu, B., et al.: FBNet: hardware-aware efficient convnet design via differentiable neural architecture search. In: Proceedings of the IEEE/CVF Conference on Computer Vision and Pattern Recognition, pp. 10734–10742 (2019)
30. Yucel, M.K., Dimaridou, V., Drosou, A., Saa-Garriga, A.: Real-time monocular depth estimation with sparse supervision on mobile. In: Proceedings of the IEEE/CVF Conference on Computer Vision and Pattern Recognition, pp. 2428–2437 (2021)
31. Zhang, R., Isola, P., Efros, A.A., Shechtman, E., Wang, O.: The unreasonable effectiveness of deep features as a perceptual metric. In: Proceedings of the IEEE Conference on Computer Vision and Pattern Recognition, pp. 586–595 (2018)
32. Zhang, X., Matzen, K., Nguyen, V., Yao, D., Zhang, Y., Ng, R.: Synthetic defocus and look-ahead autofocus for casual videography. ACM Trans. Graphics (TOG) **38**(4), 1–16 (2019)
33. Zheng, B., et al.: Constrained predictive filters for single image bokeh rendering. IEEE Trans. Comput. Imaging **8**, 346–357 (2022)
34. Zhu, B., Chen, Y., Wang, J., Liu, S., Zhang, B., Tang, M.: Fast deep matting for portrait animation on mobile phone. In: Proceedings of the 25th ACM International Conference on Multimedia, pp. 297–305 (2017)

RISPNet: A Network for Reversed Image Signal Processing

Xiaoyi Dong[1,2], Yu Zhu[2], Chenghua Li[2,3(✉)], Peisong Wang[2], and Jian Cheng[2,3(✉)]

[1] University of Chinese Academy of Sciences, Beijing 100049, China
[2] Institute of Automation, Chinese Academy of Sciences, Beijing 100190, China
lichenghua2014@ia.ac.cn
[3] Nanjing Artificial Intelligence Research of IA (AiRiA), Nanjing, China

Abstract. RAW data is considered to be valuable owing to its beneficial properties for downstream tasks, such as denoising and HDR. Usually, RAW data gets rendered to RGB images by the in-camera image signal processor (ISP) and shall not be saved in most cases. In addition, RGB images produced by non-linear operations in ISP would necessitate the execution of other image processing tasks. To overcome this problem and acquire RAW data again, we propose a new reversed ISP network (RISPNet) to achieve efficient RGB to RAW image conversion. Our main proposal is a novel encoder-decoder network with a third-order attention module. Since Attention facilitates the complex trade-off between restoring spatial details and high-level contextual information, the accuracy of recovering high dynamic range RAW data is greatly improved, and the whole recovery process becomes a more manageable step. Benefiting from the design of the attention module, our RISPNet has a remarkable ability to recover RAW images. According to the AIM 2022 Reversed ISP Challenge, RISPNet achieved third place on both Track P20 and Track S7 test sets.

Keywords: Reversed ISP · RAW recovery · Encoder-decoder UNet · Attention

1 Introduction

RAW images hold an invaluable role in the overall imaging system as the input to the in-camera image signal processor (ISP). Particularly, as data exported directly from the CMOS Sensor, RAW data exhibits a linear correlation with the scene irradiance. Specifically, the adjustments to RAW reflected in the output RGB images can be viewed as a direct adjustment to external environmental factors, avoiding the ambiguous effects of non-linear operations existing inside

X. Dong and Y. Zhu—Equal Contribution.

Supplementary Information The online version contains supplementary material available at https://doi.org/10.1007/978-3-031-25063-7_27.

L. Karlinsky et al. (Eds.): ECCV 2022 Workshops, LNCS 13802, pp. 445–457, 2023.
https://doi.org/10.1007/978-3-031-25063-7_27

the ISP. Therefore, many operations, such as HDR and denoising [2,14,34], are performed in the RAW domain. However, existing devices such as mobile phones often choose to conserve only the final output sRGB image if no special requirements exist. In this case, the subsequent image processing process becomes more challenging due to the non-linear operation within the ISP [13].

To alleviate the problem, previous methods propose to restore the RAW data from RGB images [2,5,28,31]. Notably, it is challenging to restore a RAW image from an RGB image (i.e., reverse ISP), where the most critical factor is the loss of information due to dynamic range compression. Specifically, some information has to be sacrificed when converting 12-bit or 14-bit RAW data to 8-bit RGB images, which is a typically irreversible operation. However, previous work [2] has typically involved some prior knowledge and attempts to incorporate them into the model, resulting in more complicated approaches.

In this paper, we present a simple but effective reversed ISP network, named RISPNet, to restore the RAW data from RGB images. As shown in Fig. 2, the base structure of RISPNet is an encoder-decoder network similar to NAFNet [3]. The difference from NAFNet is a newly designed RISPBlock, which contains a third-order attention (TOA) module and an effective simple activation gate module (SAG) module. The TOA module conducts three successive attention, that is the split attention, the spatial attention (SA) and the channel attention (CA), and the simple attention, as shown in Fig. 3. TOA could filter out the information of little use and propagate the significant information forward by means of the long-range dependencies among features. Extensive experiments demonstrate that the proposed RISPNet could achieve state-of-the-art performance on RAW data restoration and that the fidelity of the synthetic RAW data will be further improved by 0.16 dB with the assistance of the TOA and SAG module. In addition, our RISPNet achieved third place on the AIM 2022 Reversed ISP Challenge [6] Track P20 and Track S7, which verifies the superiority of our approach. The main contributions of this paper can be summarized as follows:

- We develop a novel end-to-end learnable network to build the reversed ISP mapping from RGB to RAW images. Our RISPNet can generate accurate and realistic RAW data taking the RGB images as inputs, which is of great benefit to the downstream tasks.
- We explore the effectiveness of the simple activation gate module on the reversed ISP task by experiments. The SAG module will bring a performance gain of 0.03 dB to the NAFNet baseline.
- We design a third-order attention module to realize the complicated trade-off between recovering spatial detail and high-level contextual information. This module can contribute to another performance gain of 0.13 dB on the basis of the SAG module.
- Extensive experiments show that our method achieving state-of-the-art performance for the reversed ISP task, which is promising for benefiting the down-stream tasks.

2 Related Work

Raw Image Reconstruction. Several approaches [2,5,28,31] have attempted to estimate the RGB to RAW mapping. UPI [2] modeled and inverted each step of a generic ISP pipeline with elaborately handcrafted functions and their inverse ones after a detailed analysis of conventional in-camera ISP. Nevertheless, these function parameters were manually set based on camera priors and cannot generalize to new devices. Built on UPI, the model-based ISP proposed by [5] designed an explicitly-staged and reversible architecture to approximate the bidirectional mapping between RAW and RGB domains. And instead of fine-tuning transformation parameters manually as UPI, [5] could learn them from data directly. Similar to [5], InvISP [28] also establishes a learnable and invertible ISP framework composed of a stack of bijective functions to generate RGB images in the forward pass and recover RAW data in the inverse pass. The difference is that InvISP models the ISP pipeline as a whole while [5] explicitly simulates each stage of the ISP pipeline and strings them together to form the entire ISP pipeline. Distinguished from the single but invertible network used in [5] and InvISP, CycleISP [31] adopted two separate network branches to convert RGB images to RAW data and back to RGB images. Its RGB2RAW network branch learned the mapping from RGB images to RAW data utilizing consecutive recursive residual groups, which contained multiple dual attention blocks named for the parallel spatial attention and channel attention modules in it. Unlike the methods mentioned above, we let aside the ISP process and concentrate on the reversed ISP task rather than take both ISP and reversed ISP into account and optimize the two tasks jointly. In other words, we regard reversed ISP as a particular image restoration task resembling image super-resolution, image denoising and image deblurring.

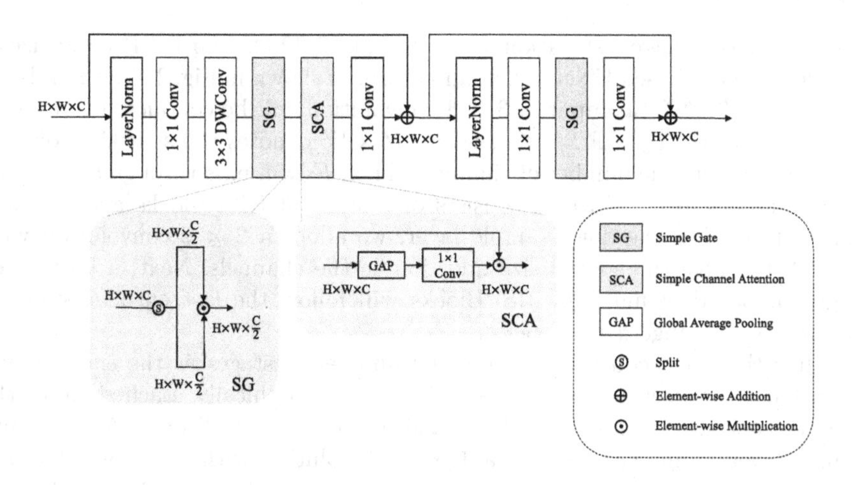

Fig. 1. The structure of the basic block of NAFNet [3].

Image Restoration. Image restoration aims at recovering high-quality images from low-quality inputs. In recent years, a great deal of deep learning based methods [3,4,26,30,32,33,36] have been proposed and almost dominate this field due to their outstanding performance over conventional approaches [7,12]. A vast majority of them apply encoder-decoder based UNet [22] architectures [3,4,26,30,32], which are capable to extract multi-scale representation hierarchically, and skip connection is commonly used to focus on learning residual signals. At the same time, the attention mechanism popular in the field of natural language processing (NLP) has been incorporated into image restoration tasks to selectively attend to relevant information [15,19,20,26,30,35]. Recently, exploiting a single-stage UNet [22] architecture with enhanced basic blocks (Fig. 1), NAFNet [3] achieved exciting results on the image deblurring task and the image denoising task, whose distinguished performance arouses our interest and is selected as our baseline.

3 Method

NAFNet [3] adopts a simple but effective basic block for lower system complexity. Considering that in this paper, we lay emphasis on maximizing the RAW data restoration performance and do not exhaustively decrease the system complexity, we improve it with a third-order attention module and a simple activation gate module which are a little more complicated but more fruitful to obtain better performance, thus acquiring our RISPnet. In this section, we first demonstrate the overall architecture of our RISPNet and then illustrate the basic block named RISPBlock, the TOA module, and the SAG module in order.

3.1 RISPNet

Similar to most image restoration methods [3,4,26,30,32], our RISPNet utilizes a encoder-decoder based UNet [22] architecture, as shown in Fig. 2. Given an RGB image $I \in \mathbb{R}^{H \times W \times 3}$ as input, a 3×3 convolution will be conducted to extract low-level features $F_0 \in \mathbb{R}^{H \times W \times C}$, where $H \times W$ denotes the spatial resolution and C represents the number of channels. Then F_0 will pass through an encoder with N stages, each of which consists of a stack of RISPBlocks and a down-sample layer. For the down-sample layer, we adopt a 2×2 convolution with stride 2 to halve the spatial size and double the channels. Next, a bottleneck stage composed of multiple RISPBlocks will follow the last encoder stage to capture longer-range dependencies.

After that, a decoder with the equal number of stages as the encoder will reconstruct the deep features $F_d \in \mathbb{R}^{H \times W \times C}$ hierarchically. Each stage of the decoder contains an up-sample layer and a stack of RISPBlocks. And the up-sample layer is implemented by a 1×1 convolution with twice out-channels as in-channels and a pixel-shuffle layer [24] with upscale factor 2 to double the spatial resolution and halve the channels. Note that skip connection is applied between the output of RISPBlocks in each encoder stage and the up-sampled

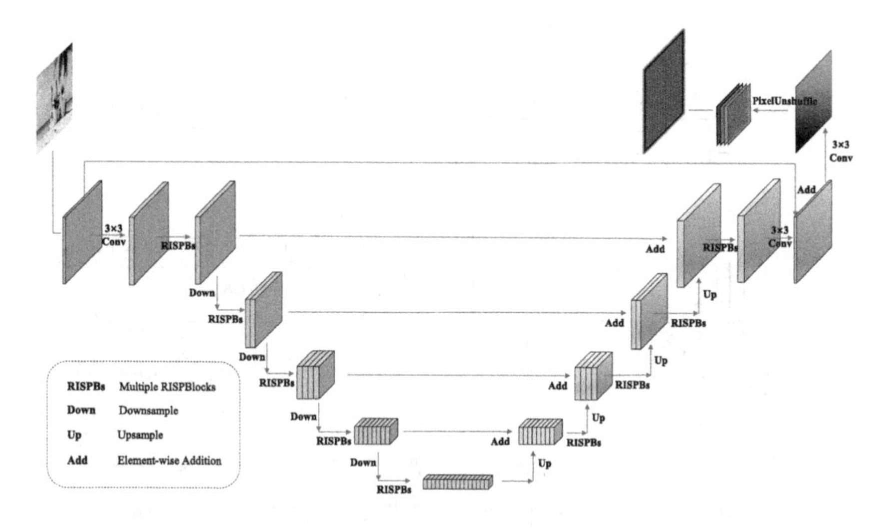

Fig. 2. The overall architecture of proposed RISPNet. It adopts an encoder-decoder based UNet [22] architecture to learn the latent representation of RGB images and reconstruct the RAW images.

features of the corresponding decoder stage to focus on learning residual signals. At the end of the decoder, another 3 × 3 convolution will transform F_d to the residual image $I_r \in \mathbb{R}^{H\times W\times 3}$ to add with the input I to predict the reconstructed demosaiced RAW image $R_d = I + I_r$.

Eventually, a 3 × 3 convolution shall mosaic R_d to generate the restored Bayer RAW image $\hat{R} \in \mathbb{R}^{H\times W\times 1}$ [21] and next a pixel-unshuffle layer is used to pack $\hat{R}$ to a 4-channel image following the RGGB pattern [25] for storage.

3.2 RISPBlock

RISPBlock is the basic block of RISPNet, enabling the encoder, the decoder and the bottleneck layer described above to achieve efficient feature transformation by capturing both the long-range and local dependencies. Practically, RISPBlock comprises two sequentially connected residual blocks, as shown in Fig. 3. At the beginning of both residual blocks, a layer normalization [1] is applied to speed up training.

Next, in the former residual block, a 1 × 1 convolution layer will attach to expand the channels of the input feature map. After that, the obtained output will pass through a 3 × 3 depth-wise convolution [16,29] to catch the local context at each channel. Taking full advantage of the attention mechanism, the following third-order attention module (TOA) will filter out the information of little use and propagate the significant information forward by means of the long-range dependencies among features. Behind TOA, a 1 × 1 convolution is employed to transform the channels of the feature map back and the output will be added to the original input to generate the result of the former residual block.

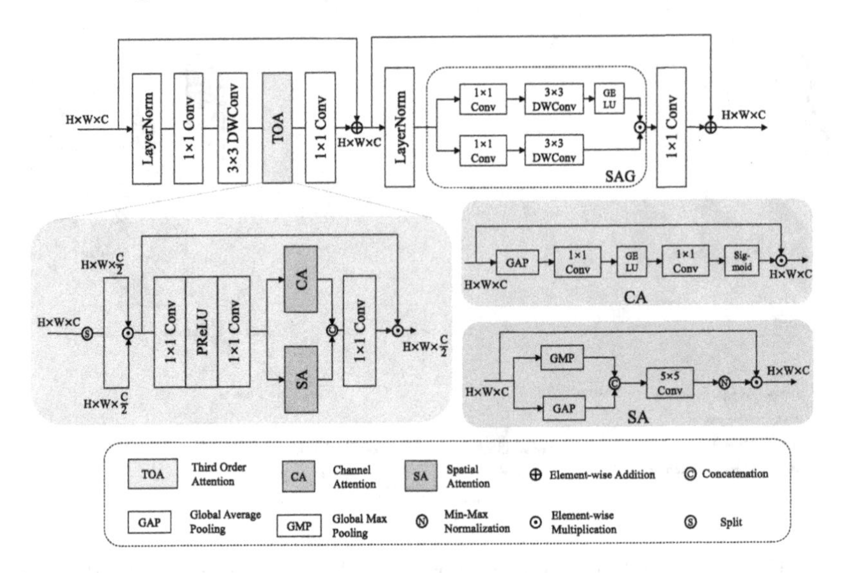

Fig. 3. The structure of RISPBlock.

Then it comes to the latter residual block, after the layer normalization of which, a simple activation gate (SAG) will be exploited to further learn the efficient representation of the spatial context. Note that the channels of the features are halved by SAG, and consequently, a 1×1 convolution layer should be added to recover the original dimension of input. At last, the same as the former residual block, the input of the latter residual block and the output of the convolution layer will be summed together as the final result of the RISPBlock.

As the key components leading to the RAW recovery performance growth, the TOA and SAG modules shall be introduced in detail in the next two subsections.

3.3 Third-Order Attention

The reason why we call the module marked green in Fig. 3 the third-order attention (TOA) is that there are three attention operations in this module. As demonstrated in Fig. 3, TOA firstly splits the input into two feature maps of the same size along the channel dimension and then conducts a pixel-wise multiplication on them, which serves as the first attention.

Next, a 1×1 convolution followed by PReLU [8] and another 1×1 convolution will aggregate the spatial context information across channels to make preparations for the second attention which includes parallel channel attention (CA) and spatial attention (SA) modules [27].

The channel attention learns a weight map to emphasize the channels of greater value by the sequence of a global average pooling layer, a 1×1 convolution layer, a GELU [9], another 1×1 convolution layer, and a sigmoid.

Similar to the channel attention map, each element of the spatial attention map is considered as the importance measure of the corresponding spatial pixel.

Hence, a spatial attention map assigned properly would highlight the significant pixels and neglect the inconsequential ones to extract efficient feature representation. The spatial attention map is acquired by the concatenation of global average pooling and global max pooling, a 5×5 convolution, and a min-max normalization.

After that, the feature maps obtained by applying the spatial and channel attention maps respectively are concatenated and then compressed by a 1×1 convolution. The output of this convolution acts as a weight map to rescale the features after the first attention through pixel-wise multiplication, which is the third attention. These three attention operations unite to achieve efficient feature extraction by capturing the long-range dependencies among the features.

3.4 Simple Activation Gate

The SAG contains two parallel transformation layers. Each starts with a 1×1 convolution layer and a 3×3 depth-wise convolution layer to learn the local structure of features. Then one is activated by a GELU to multiply with the other via pixel-wise multiplication. Complementary to the TOA, the SAG can capture local dependencies among the features effectively relying on its convolution operations.

4 Experiments

4.1 Implementation Details

Datasets. We use the AIM 2022 Reversed ISP Challenge [6] Track P20 and Track S7 datasets [11,23] to train our RISPNet. The Track P20 dataset [11] contains RGB-RAW image pairs captured in the wild with the Huawei P20 cameraphone (12.3 MP Sony Exmor IMX380 sensor) and Canon 5D Mark IV DSLR, and its RAW images are stored in RGBG pattern. The Track S7 dataset [23] includes pairs of low-light and well-lit images captured by a Samsung S7 smartphone camera in both raw and processed JPEG formats. The two tracks are trained separately utilizing their respective training set. In the meanwhile, horizontal and vertical flipping that can preserve the Bayer pattern of RAW images [17] and mix-up [10] are adopted for data augmentation.

Training Settings. Our RISPNet is trained on 8 NVIDIA A100 Tensor Core GPUs. We exploit AdamW optimazer ($\beta_1 = 0.9, \beta_2 = 0.9$) [18] and clip L1 loss [37] for $300K$ iterations with the initial learning rate $2e^{-4}$ gradually reduced to $1e^{-6}$ following the cosine annealing. To acquire the capability to encode global image statistic by degrees, we apply progressive learning [30], initializing the patch size and batch size pair as $(128^2, 64)$ and updating it to $[(128^2, 64), (192^2, 48), (256^2, 24), (320^2, 16), (384^2, 8)]$ at iterations $[92K, 156K, 204K, 240K, 276K]$.

Ensemble Strategies. To improve the generalization ability and performance of our approach, we employ self-ensemble and model-ensemble strategies.

Table 1. Quantitative results on the test sets of the AIM 2022 Reversed ISP Challenge Track P20 and Track S7. We won third place on both Track P20 and Track S7. Best results are in red.

Team name	Track 1 (Samsung S7)				Track 2 (Huawei P20)			
	Test1		Test2		Test1		Test2	
	PSNR ↑	SSIM ↑	PSNR ↑	SSIM ↑	PSNR ↑	SSIM ↑	PSNR ↑	SSIM ↑
NOAHTCV	31.86	0.83	32.69	0.88	38.38	0.93	35.77	0.92
MiAlgo	31.39	0.82	30.73	0.80	40.06	0.93	37.09	0.92
CASIA LCVG	30.19	0.81	31.47	0.86	37.58	0.93	33.99	0.92
HIT-IIL	29.12	0.80	29.98	0.87	36.53	0.91	34.07	0.90
CS²U	29.13	0.79	29.95	0.84	–	–	–	–
SenseBrains	28.36	0.80	30.08	0.86	35.47	0.92	32.63	0.91
PixelJump	28.15	0.80	n/a	n/a	–	–	–	–
HiImage	27.96	0.79	n/a	n/a	34.40	0.94	32.13	0.90
0noise	27.67	0.79	29.81	0.87	33.68	0.90	31.83	0.89
OzU VGL	27.89	0.79	28.83	0.83	32.72	0.87	30.69	0.86
CVIP	27.85	0.80	29.50	0.86	–	–	–	–
CycleISP [31]	26.75	0.78	–	–	32.70	0.85	–	–
UPI [2]	26.90	0.78	–	–	–	–	–	–
U-Net Base	26.30	0.77	–	–	30.01	0.80	–	–

The self-ensemble involves 8 inputs obtained by flip and rotate operations for fusion. And the model-ensemble fuses the self-ensembled outputs of 3 models, respectively RISPNet, RISPNet_oCA (i.e., the second attention in TOA only contains CA), and NAFNet, all of which are trained separately and share the same training settings. The ensemble operations are all the same simple average operations.

Black Level. Black level is a special prior for cameras. We add the black level of P20 (i.e., 64) to the network output so that the network could learn the reverse ISP mapping more easily.

4.2 AIM 2022 Reversed ISP Challenge Results

We test our RISPNet on the test sets of the AIM 2022 Reversed ISP Challenge [6] Track P20 and S7 which merely include the input RGB images, and the recovered RAW images will be evaluated according to PSNR and SSIM by the online server. Our quantitative evaluation results and the comparison with other competitive methods are shown in Table 1. According to the performance on Test1, we get third place on both Track P20 and Track S7.

Qualitative comparison with other competitive methods is demonstrated in Fig. 4, where it can be seen that our method can recover the details of the RAW images more accurately and generate fewer artifacts (e.g., blur) than most of the challenging methods.

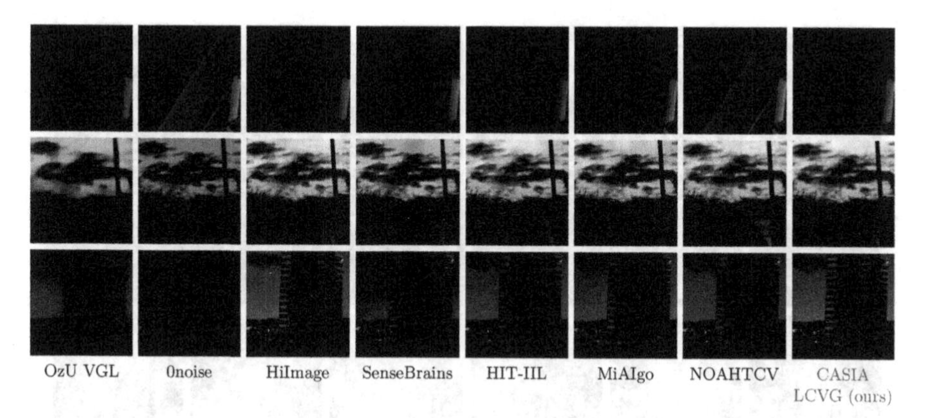

Fig. 4. Qualitative comparison with competitive methods on the test set of the AIM 2022 Reversed ISP Challenge Track P20. All these pictures are visualized by the github code https://github.com/mv-lab/AISP.

4.3 Ablation Study

As illustrated in Sect. 3, we select NAFNet [3] as our baseline and enhance it with the SAG and TOA modules. When adapting the NAFNet [3] to the reversed ISP task, we found that mix-up [10] is an indispensable training strategy for NAFNet [3], without which, the performance of NAFNet will drop by 0.25 dB, as shown in Table 2. It's notable that all models in the ablation study are tested on the validation set of Track P20 to select the best performing model.

Table 2. Ablation study. Baseline, Baseline+SAG, Baseline+SAG+TOA, and RISPNet−SA denote the baseline, baseline applying SAG, baseline applying SAG and TOA (i.e., our RISPNet), and RISPNet without SA respectively. The star (*) on the right of the model name indicates that mix-up [10] is employed when training. All models are tested on the validation set of Track P20. Note that here we don't use any ensemble strategy. Best results are in red.

Models	PSNR
Baseline	38.80
Baseline*	39.05
Baseline+SAG*	39.08
RISPNet−SA*	39.16
Baseline+SAG+TOA (RISPNet)*	39.21

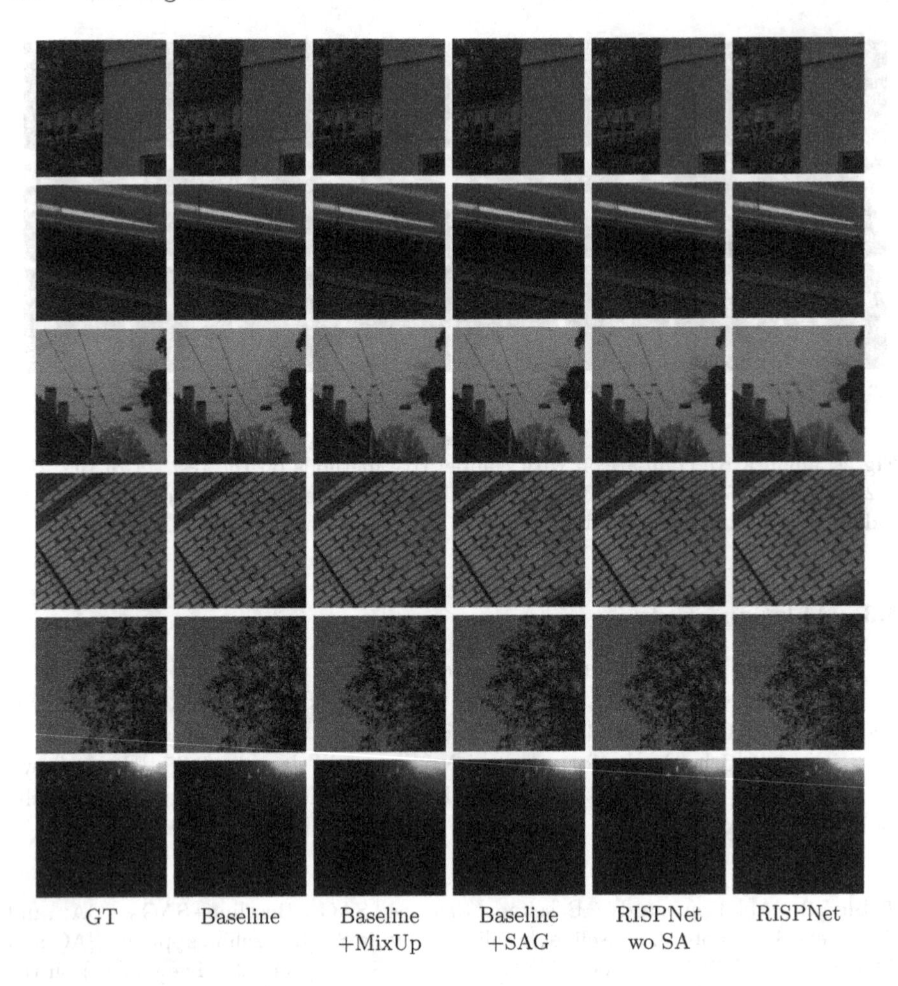

Fig. 5. Visualization of the reconstructed RAW images for the five models in the ablation study. Zoom in for more details.

In addition, to confirm the effectiveness of the two modules in facilitating performance improvements, we conduct experiments on the baseline (using mix-up [10]) and integrate the SAG and TOA modules one by one to demonstrate the performance gain brought by each module. The quantitative results are displayed in Table 2. The SAG module could increase the recovery performance by 0.03 dB and that the TOA module shall bring another 0.13 dB performance gain.

It's worth mentioning that we also explore the necessity of the TOA including SA by removing SA from TOA. This will decrease the performance by 0.05 dB, indicating that the SA together with CA can take advantage of the spatial and inter-channel dependencies for efficient representation learning.

At last, we provide the visualization of the reconstructed RAW images for the five models in the ablation study, as shown in Fig. 5. In general, our RISPNet can recover accurate and realistic RAW data, especially the challenging highlight pixels (the first row of pictures in Fig. 5). In the meanwhile, we also find that when confronting the images with sharp edges (the last row of pictures in Fig. 5), RISPNet tends to give ambiguous predictions for higher fidelity, which may incur blur to some extent. Considering that in this paper we focus on maximizing the RAW recovery fidelity, we leave the research on mitigating blur to future work.

4.4 Limitations

Despite the excellent performance of our model, it has 464M parameters, which is the greatest relative to the other contestants. However, our runtime is relatively less, primarily because we have deployed more RISP modules in the low resolution layer of the network. We believe that the problem of too many parameters can be mitigated by combining it with a model-based algorithm [5].

5 Conclusion

In this paper, we propose an encoder-decoder based network to execute a novel inverse task in low-level computer vision, the reversed ISP task. We devise the basic block of our RISPNet with a third-order attention module that can capture long-range dependencies among the pixels, achieving the complicated trade-off between preserving spatial details and maintaining high-level contextual information. Our RISPNet demonstrates an excellent capability to recover accurate and realistic RAW images from RGB images, achieving state-of-the-art results on the reversed ISP task. We believe that our method is of great benefit to alleviating the lack of available RAW data and promoting the performance of downstream tasks.

References

1. Ba, J.L., Kiros, J.R., Hinton, G.E.: Layer normalization. arXiv preprint arXiv:1607.06450 (2016)
2. Brooks, T., Mildenhall, B., Xue, T., Chen, J., Sharlet, D., Barron, J.T.: Unprocessing images for learned raw denoising. In: Proceedings of the IEEE/CVF Conference on Computer Vision and Pattern Recognition (CVPR) (2019)
3. Chen, L., Chu, X., Zhang, X., Sun, J.: Simple baselines for image restoration. arXiv preprint arXiv:2204.04676 (2022)
4. Chen, L., Lu, X., Zhang, J., Chu, X., Chen, C.: HINet: half instance normalization network for image restoration. In: Proceedings of the IEEE/CVF Conference on Computer Vision and Pattern Recognition (CVPR) Workshops, pp. 182–192 (2021)
5. Conde, M.V., McDonagh, S., Maggioni, M., Leonardis, A., Pérez-Pellitero, E.: Model-based image signal processors via learnable dictionaries. In: Proceedings of the AAAI Conference on Artificial Intelligence, vol. 36, pp. 481–489 (2022)
6. Conde, M.V., Timofte, R., et al.: Reversed image signal processing and raw reconstruction. AIM 2022 challenge report. In: Proceedings of the European Conference on Computer Vision Workshops (ECCVW) (2022)

7. He, K., Sun, J., Tang, X.: Single image haze removal using dark channel prior. IEEE Trans. Pattern Anal. Mach. Intell. **33**(12), 2341–2353 (2011). https://doi.org/10.1109/TPAMI.2010.168
8. He, K., Zhang, X., Ren, S., Sun, J.: Delving deep into rectifiers: surpassing human-level performance on ImageNet classification. In: Proceedings of the IEEE International Conference on Computer Vision, pp. 1026–1034 (2015)
9. Hendrycks, D., Gimpel, K.: Gaussian error linear units (GELUs). arXiv preprint arXiv:1606.08415 (2016)
10. Huang, L., Zhang, C., Zhang, H.: Self-adaptive training: beyond empirical risk minimization. In: Larochelle, H., Ranzato, M., Hadsell, R., Balcan, M., Lin, H. (eds.) Advances in Neural Information Processing Systems, vol. 33, pp. 19365–19376. Curran Associates, Inc. (2020). https://proceedings.neurips.cc/paper/2020/file/e0ab531ec312161511493b002f9be2ee-Paper.pdf
11. Ignatov, A., Van Gool, L., Timofte, R.: Replacing mobile camera ISP with a single deep learning model. In: Proceedings of the IEEE/CVF Conference on Computer Vision and Pattern Recognition Workshops, pp. 536–537 (2020)
12. Kopf, J., et al.: Deep photo: model-based photograph enhancement and viewing. ACM Trans. Graphics (TOG) **27**(5), 1–10 (2008)
13. Kousha, S., Maleky, A., Brown, M.S., Brubaker, M.A.: Modeling sRGB camera noise with normalizing flows. In: Proceedings of the IEEE/CVF Conference on Computer Vision and Pattern Recognition (CVPR), pp. 17463–17471 (2022)
14. Kronander, J., Gustavson, S., Bonnet, G., Unger, J.: Unified HDR reconstruction from raw CFA data. In: IEEE International Conference on Computational Photography (ICCP), pp. 1–9. IEEE (2013)
15. Lee, H., Choi, H., Sohn, K., Min, D.: KNN local attention for image restoration. In: Proceedings of the IEEE/CVF Conference on Computer Vision and Pattern Recognition (CVPR), pp. 2139–2149 (2022)
16. Li, Y., Zhang, K., Cao, J., Timofte, R., Van Gool, L.: LocalViT: bringing locality to vision transformers. arXiv preprint arXiv:2104.05707 (2021)
17. Liu, J., et al.: Learning raw image denoising with bayer pattern unification and bayer preserving augmentation. In: Proceedings of the IEEE/CVF Conference on Computer Vision and Pattern Recognition Workshops (2019)
18. Loshchilov, I., Hutter, F.: Decoupled weight decay regularization. arXiv preprint arXiv:1711.05101 (2017)
19. Mei, Y., et al.: Pyramid attention networks for image restoration. arXiv preprint arXiv:2004.13824 (2020)
20. Mou, C., Zhang, J., Fan, X., Liu, H., Wang, R.: COLA-net: collaborative attention network for image restoration. IEEE Trans. Multimed. **24**, 1366–1377 (2022). https://doi.org/10.1109/TMM.2021.3063916
21. Qi, J., Qi, N., Zhu, Q.: SUnet++: joint demosaicing and denoising of extreme low-light raw image. In: Þór Jónsson, B., et al. (eds.) MMM 2022. LNCS, vol. 13142, pp. 171–181. Springer, Cham (2022). https://doi.org/10.1007/978-3-030-98355-0_15
22. Ronneberger, O., Fischer, P., Brox, T.: U-net: convolutional networks for biomedical image segmentation. In: Navab, N., Hornegger, J., Wells, W.M., Frangi, A.F. (eds.) MICCAI 2015. LNCS, vol. 9351, pp. 234–241. Springer, Cham (2015). https://doi.org/10.1007/978-3-319-24574-4_28
23. Schwartz, E., Giryes, R., Bronstein, A.M.: DeepISP: toward learning an end-to-end image processing pipeline. IEEE Trans. Image Process. **28**(2), 912–923 (2018)
24. Shi, W., et al.: Real-time single image and video super-resolution using an efficient sub-pixel convolutional neural network. In: Proceedings of the IEEE Conference on Computer Vision and Pattern Recognition (CVPR) (2016)

25. Wang, Y., Huang, H., Xu, Q., Liu, J., Liu, Y., Wang, J.: Practical deep raw image denoising on mobile devices. In: Vedaldi, A., Bischof, H., Brox, T., Frahm, J.-M. (eds.) ECCV 2020. LNCS, vol. 12351, pp. 1–16. Springer, Cham (2020). https://doi.org/10.1007/978-3-030-58539-6_1
26. Wang, Z., Cun, X., Bao, J., Zhou, W., Liu, J., Li, H.: UFormer: a general U-shaped transformer for image restoration. In: Proceedings of the IEEE/CVF Conference on Computer Vision and Pattern Recognition (CVPR), pp. 17683–17693 (2022)
27. Woo, S., Park, J., Lee, J.-Y., Kweon, I.S.: CBAM: convolutional block attention module. In: Ferrari, V., Hebert, M., Sminchisescu, C., Weiss, Y. (eds.) ECCV 2018. LNCS, vol. 11211, pp. 3–19. Springer, Cham (2018). https://doi.org/10.1007/978-3-030-01234-2_1
28. Xing, Y., Qian, Z., Chen, Q.: Invertible image signal processing. In: Proceedings of the IEEE/CVF Conference on Computer Vision and Pattern Recognition (CVPR), pp. 6287–6296 (2021)
29. Yuan, K., Guo, S., Liu, Z., Zhou, A., Yu, F., Wu, W.: Incorporating convolution designs into visual transformers. In: Proceedings of the IEEE/CVF International Conference on Computer Vision, pp. 579–588 (2021)
30. Zamir, S.W., Arora, A., Khan, S., Hayat, M., Khan, F.S., Yang, M.H.: Restormer: efficient transformer for high-resolution image restoration. In: Proceedings of the IEEE/CVF Conference on Computer Vision and Pattern Recognition (CVPR), pp. 5728–5739 (2022)
31. Zamir, S.W., et al.: CycleISP: real image restoration via improved data synthesis. In: Proceedings of the IEEE/CVF Conference on Computer Vision and Pattern Recognition (CVPR) (2020)
32. Zamir, S.W., et al.: Multi-stage progressive image restoration. In: Proceedings of the IEEE/CVF Conference on Computer Vision and Pattern Recognition (CVPR), pp. 14821–14831 (2021)
33. Zhang, K., Li, Y., Zuo, W., Zhang, L., Van Gool, L., Timofte, R.: Plug-and-play image restoration with deep denoiser prior. IEEE Trans. Pattern Anal. Mach. Intell. 1 (2021). https://doi.org/10.1109/TPAMI.2021.3088914
34. Zhang, Y., Qin, H., Wang, X., Li, H.: Rethinking noise synthesis and modeling in raw denoising. In: Proceedings of the IEEE/CVF International Conference on Computer Vision, pp. 4593–4601 (2021)
35. Zhang, Y., Li, K., Li, K., Zhong, B., Fu, Y.: Residual non-local attention networks for image restoration. arXiv preprint arXiv:1903.10082 (2019)
36. Zhang, Y., Tian, Y., Kong, Y., Zhong, B., Fu, Y.: Residual dense network for image restoration. IEEE Trans. Pattern Anal. Mach. Intell. **43**(7), 2480–2495 (2021). https://doi.org/10.1109/TPAMI.2020.2968521
37. Zhu, Y., et al.: EEDNet: enhanced encoder-decoder network for AutoISP. In: Bartoli, A., Fusiello, A. (eds.) ECCV 2020. LNCS, vol. 12537, pp. 171–184. Springer, Cham (2020). https://doi.org/10.1007/978-3-030-67070-2_10

CIDBNet: A Consecutively-Interactive Dual-Branch Network for JPEG Compressed Image Super-Resolution

Xiaoran Qin[1], Yu Zhu[1], Chenghua Li[1,2](✉), Peisong Wang[1], and Jian Cheng[1,2,3](✉)

[1] Institute of Automation, Chinese Academy of Sciences, Beijing 100190, China
{xiaoran.qin,lichenghua2014}@ia.ac.cn,
{peisong.wang,jcheng}@nlpr.ia.ac.cn
[2] Nanjing Artificial Intelligence Research of IA (AiRiA), Nanjing, China
[3] MAICRO, Nanjing, China

Abstract. Compressed image super-resolution (SR) task is useful in practical scenarios, such as mobile communication and the internet, where images are usually downsampled and compressed due to limited bandwidth and storage capacity. However, a combination of compression and downsampling degradations makes the SR problem more challenging. To restore high-quality and high-resolution images, local context and long-range dependency modeling are both crucial. In this paper, for JPEG compressed image SR, we propose a consecutively-interactive dual-branch network (CIDBNet) to take advantage of both convolution and transformer operations, which are good at extracting local features and global interactions, respectively. To better aggregate the two-branch information, we newly introduce an adaptive cross-branch fusion module (ACFM), which adopts a cross-attention scheme to enhance the two-branch features and then fuses them weighted by a content-adaptive map. Experiments show the effectiveness of CIDBNet, and in particular, CIDBNet achieves higher performance than a larger variant of HAT (HAT-L).

Keywords: Super-resolution · Compressed image · Dual-branch network · Transformer

1 Introduction

With the rapid development of mobile phones, cameras and wireless networks, the amount of image data on the Internet has an explosive growth. Massive image data is usually obtained at edge devices, transmitted to the cloud for analysis and storage, and then shared with user applications. In the process of data communication, the image compression technique is used to save transmission bandwidth and storage capacity, such as Joint Photographic Experts Group

X. Qin and Y. Zhu—These authors contributed equally to this work.

L. Karlinsky et al. (Eds.): ECCV 2022 Workshops, LNCS 13802, pp. 458–474, 2023.
https://doi.org/10.1007/978-3-031-25063-7_28

(JPEG) [38]. The compression procedure usually leads to inevitable information loss and undesirable compression artifacts in the compressed image. On the other hand, reducing the resolution of images to be compressed in advance can reduce computation costs and is also an effective means to satisfy low-bitrate requirements. However, downsampling operations can further degrade image quality. Therefore, JPEG compressed image super-resolution (SR) task is very useful in low-bitrate scenarios.

Although the image SR task has been studied over a long period of time, it usually does not consider compressed images as inputs. Compression is a more complex degradation compared to standard downsampling. JPEG compression first applies Discrete Cosine Transform (DCT) to transform image signals to the frequency domain. The frequency coefficients are coarsely quantized to remove high frequencies and then encoded to bitstreams for transmission. A larger quantization interval will generate a lower bitrate, which will also make the quality of the compressed image worse. Due to the above procedures, compressed images usually suffer from blocking, blurring and banding artifacts [16]. As for JPEG compressed image SR task, it is essential to take both the compression and downsampling degradations into consideration. Thus, the core issue is how to both reduce compression artifacts and retain detailed information for SR task.

Deep learning methods have been widely applied in image SR [13,25,47,49] and compression artifacts removal [12,48,50] tasks, especially based on the convolutional neural network (CNN). Though CNN-based methods have shown great superiority compared with traditional methods, they have two issues according to existing literature [24,43]. (a) Convolution kernels with shared weights interact with different regions for different input images, which is content independent and not good at the restoration of images with diversified contents. (b) The limited receptive field of convolution operations could not establish long-range feature dependencies. Transformer [37] has been introduced to computer vision tasks nowadays, including SR [6] and image restoration [24,40,43]. The self-attention mechanism in Transformer helps capture long-range information and global representations. But it ignores local features that are beneficial to preserving details for image restoration and SR tasks.

To tackle the above problems, in this paper, we propose a consecutively-interactive dual-branch network containing a Transformer branch and a CNN branch, namely CIDBNet, for JPEG compressed image super-resolution. Our CIDBNet couples local features with global representations in a consecutively interactive fashion for feature enhancement learning. The CNN branch and the transformer branch follow the architecture of NAFNet [5] and HAT [6], respectively. Specifically, the CNN branch is composed of multiple NAFNet's blocks [5] without reducing the resolution of feature maps. The Transformer branch adopts the same structure of HAT, while we introduce a gating and locality mechanism to the MLP module of each transformer block by using a gated-Dconv feed-forward network (GDFN) [43]. An adaptive cross-branch fusion module (ACFM) is newly introduced to interact information between CNN and transformer branches by a cross-attention scheme, and fuse the two-branch features weighted by a content-adaptive map. At last, CIDBNet is optimized for artifacts removal and SR tasks simultaneously and can be trained directly in one pass, which is more applicable in practical scenarios.

The main contributions of this paper include:

- We propose CIDBNet, a consecutively-interactive dual-branch network for JPEG compressed image super-resolution, aiming to couple local features with global representations to enhance feature learning.
- We adopt an adaptive cross-branch fusion module (ACFM), which enhances two-branch features by a cross-attention scheme and adaptively fuses them to aggregate local and global pixel interactions.
- Experiments demonstrate that our approach outperforms other CNN-based or transformer-based methods in the compressed image SR task. Particularly, our approach can surpass a larger variant of HAT (HAT-L) [6] by 0.027 dB.

2 Related Work

2.1 JPEG Compressed Image Super-resolution

Single Image Super-Resolution. Numerous deep learning methods have been proposed in single image super-resolution (SISR) fields. SRCNN [13] is the first attempt to use a fully CNN in SISR task and obtain superior performance compared to traditional methods. Since then, various advanced modules have been introduced into this task, such as residual block [25], dense block [39,49] and recursive network [35]. Attention mechanism has also been applied to improve the representation ability of models, such as RCAN [47].

JPEG Compression Artifacts Removal. The compression artifacts removal task is a kind of image restoration task, where degradation of images is caused by compression algorithms. Methods of compression artifacts removal can be classified into two categories: model-based methods and learning-based methods. Model-based methods utilize image filtering techniques to reduce artifacts [15] or rely on prior knowledge of the DCT domain to explore handcrafted features [44]. Learning-based methods benefit from the powerful nonlinear capability of deep neural networks, and aim to map compressed images to restored images based on the supervision of huge amounts of paired training data. The first work using deep learning is introduced in ARCNN [12]. Recently, many researchers attempt to tackle image restoration tasks using CNNs, such as RNAN [48] and RDN [50]. Some of them try to integrate the prior knowledge into CNNs, such as DRUNet [45] and DMCNN [46].

End-to-End Framework. Two recent works propose end-to-end frameworks to tackle artifacts removal and SR tasks simultaneously. CISRDCNN [4] is comprised of deblocking module, upsampling module and enhancement module, which aim to de-block, enlarge the resolution of inputs and refine the quality of upsampled images, respectively. The three modules are first trained separately and then optimized jointly. CISRNet [17] also adopts a two-stage coarse-to-fine learning strategy, consisting of a coarse network and a refinement network. Different from those, our CIDBNet is designed to learn features for both removing artifacts and preserving useful details, and can be trained directly in one pass.

2.2 Transformers in Low-Level Tasks

Recently, transformer [37] has shown great success in low-level tasks including image SR, video SR and image restoration [2,3,6,21,23,24,40,43]. Specifically, IPT [3] is the first work to investigate the feasibility of a multi-task pre-training strategy on low-level tasks, and fine-tunes a pre-trained image processing transformer to outperform other methods on several benchmarks. SwinIR [24] proposes a strong baseline model in image restoration tasks based on Swin Transformer [26]. HAT [6] enhances the range of utilized information of SwinIR by integrating channel attention into self-attention modules and introducing an overlapping cross-attention block to improve cross-window connections. Uformer [40] adopts the U-shaped architecture using locally-enhanced transformer blocks to capture hierarchical multi-scale features for image restoration. Restormer [43] proposes a transposed self-attention module in an encoder-decoder transformer to make it applicable for high-resolution images, and establishes the new state-of-the-art results on several image restoration tasks. Besides, transformer-based networks are also introduced to capture long-range temporal dependencies for videos and fuse features better for video SR and restoration tasks [2,23].

2.3 Introducing Convolutions to Transformers

Considering the advantages of CNNs and transformers on modeling ability, recent researchers have attempted to combine convolutions and transformers in vision recognition tasks. There are generally two types of methods. One method is to replace certain modules in transformers with convolutions, or integrate convolutional layers with transformer blocks sequentially [18,30,41]. CvT [41] adds a convolutional token embedding before each ViT [14] stage to introduce local information and hierarchically decrease the sequence length, and it also replaces the original linear projection in self-attention module with depthwise separable convolution layers. MobileViT [30] proposes a MobileViT block by encoding both local features and global processing, and the whole network is composed of stacked MobileNetV2 [33] blocks and MobileViT blocks. The other method is to establish parallel architectures using both CNN and transformer [7,29,32]. Conformer [32] proposes a dual network based on the design of ResNet [19] and ViT. Mobile-Former [7] designs a bidirectional cross-attention bridge between MobileNet blocks and standard transformer blocks.

3 Proposed Method

3.1 The Overall Architecture

As shown in Fig. 1, the overall architecture of our CIDBNet consists of three stages, including shallow feature extraction, deep feature extraction and

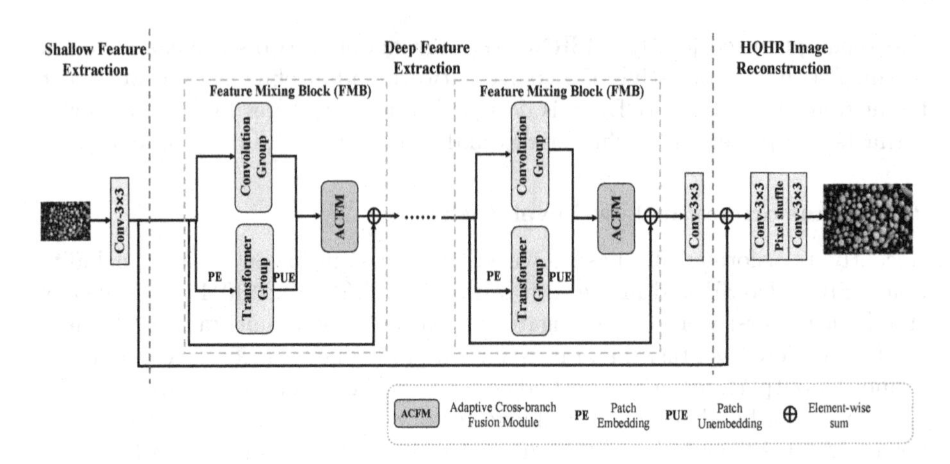

Fig. 1. The overall architecture of the proposed consecutively-interactive dual-branch network (CIDBNet).

high-quality high-resolution (HQHR) image reconstruction. Given a compressed low-resolution (LR) image $I \in \mathbb{R}^{3\times H\times W}$, a 3×3 convolutional layer is first applied to extract low-level features $F_s \in \mathbb{R}^{C\times H\times W}$, where C denotes the dimension of feature maps.

Then, the deep feature extraction stage is formulated as

$$\begin{aligned} F_i &= f_{FMB_i}(F_{i-1}), i = 1, 2, \ldots, N, \\ F_d &= f_{Conv}(F_N), \end{aligned} \tag{1}$$

where $F_0 = F_s$, $f_{FMB_i}(\cdot)$ is the i_{th} FMB and f_{Conv} is a convolution layer at the end of deep feature extraction stage. For this stage, F_s is fed into N feature mixing blocks (FMB) and one convolutional layer to obtain deep features $F_d \in \mathbb{R}^{C\times H\times W}$. This convolution layer can help latter information aggregation of shallow features and deep features. Specifically, FMB consists of a two-branch unit (*i.e.*, convolution group and transformer group) and an adaptive cross-branch fusion module (ACFM) to aggregate the two-branch information. To eliminate the inconsistency of feature dimensionalities from two groups, we use patch embedding following [26] to reshape the 2D input feature maps to tokens (set patch size as 1 and treat each pixel as a token), and patch unembedding following [26] to reshape the output tokens to 2D feature maps for transformer group. A residual connection is used in each FMB to stable the training process [47]. Thanks to the structure of FMB, the proposed network can capture both local details and long-range dependencies.

At last, we use a skip connection to fuse F_s and F_d, and reconstruct the HQHR image I_{HQHR} as

$$I_{HQHR} = f_{REC}(F_s + F_d), \tag{2}$$

where $f_{REC}(\cdot)$ is the reconstruction stage. A pixel-shuffle layer [34] is utilized to upsample the features, and the last convolutional layer generates a three-channel output HQHR image.

To optimize the parameters, we simply minimize the $\mathcal{L}_1$ loss as

$$\mathcal{L}_1 = \|I_{HQHR} - I_{GT}\|_1, \tag{3}$$

where I_{GT} denotes the ground-truth uncompressed HR image.

3.2 Transformer Group

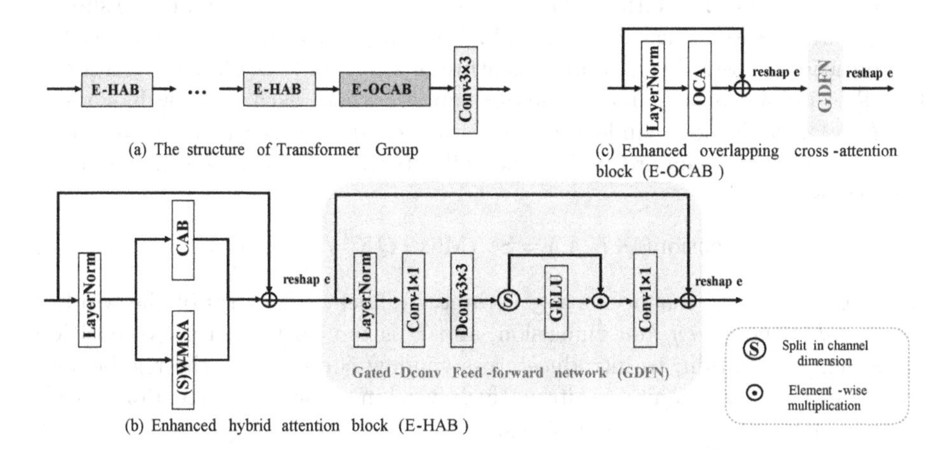

Fig. 2. Illustration of transformer group. (a) The structure of Transformer Group, (b) Enhanced hybrid attention block (E-HAB), and (c) Enhanced overlapping cross-attention block (E-OCAB). Noted that CAB denotes channel attention block and OCA denotes overlapping cross-attention in HAT [6].

Transformer group aims to utilize long-range pixel interactions and global representations for restoring degraded images to high-quality images. Inspired by HAT [6], which enlarges the range of utilized information compared to standard transformers [24,26], we adopt the residual hybrid attention group of HAT and integrate the gated-Dconv feed-forward network (GDFN) [43] into attention blocks to enhance feature learning. As shown in Fig. 2(a), the transformer group consists of several enhanced hybrid attention blocks (E-HAB), an enhanced overlapping cross-attention block (E-OCAB) and a 3×3 convolutional layer.

Enhanced Hybrid Attention Block (E-HAB). To activate more input pixels, HAT proposes a hybrid attention block by inserting a channel attention block (CAB) after the first LayerNorm (LN) layer in parallel with the window-based multi-head self-attention (W-MSA) module. In this paper, we also modified the original feed-forward network to the GDFN, aiming to control information flow among different blocks by a gating scheme and leverage local information by a depthwise convolution, as shown in Fig. 2(b). Specifically, for an input feature X, the E-HAB process is formulated as

$$\begin{aligned} X_{ln} &= \mathrm{LN}(X), \\ X_{at} &= \mathrm{(S)W\text{-}MSA}(X_{ln}) + \alpha \mathrm{CAB}(X_{ln}) + X, \\ Y_e &= \mathrm{GDFN}(X_{at}), \end{aligned} \tag{4}$$

where X_{ln} and X_{at} are the intermediate features, Y_e is the output features of E-HAB, and α is a constant to weight the output of CAB.

The window-based multi-head self-attention (W-MSA) module, which is first proposed in Swin Transformer [26] by introducing local attention and the shifted window mechanism, is usually used on high-resolution feature maps to reduce the computational cost. Given an input feature of size $H \times W \times C$, it is partitioned to $\frac{HW}{M^2}$ windows in a non-overlapping manner, where each window has a size of $M \times M$. Each local window feature is flattened and transposed to a size of $M^2 \times C$. Then, a standard multi-head self-attention is calculated in each local window as

$$\mathrm{Attention}(Q, K, V) = \mathrm{SoftMax}\left(QK^T/\sqrt{d} + B\right)V, \tag{5}$$

where Q, K and V matrices are generated by linear mappings of the window feature, d is the *query*/*key* dimension, and B is the learnable relative position encoding. In addition, to introduce cross-window connections, the regular and shifted window partitioning is alternatively used in consecutive attention blocks, where shift size is $\left(\lfloor \frac{M}{2} \rfloor, \lfloor \frac{M}{2} \rfloor\right)$ pixels.

For the gated-Dconv feed-forward network (GDFN), a 3×3 depthwise convolution is added to encode local information, and after that, a gated linear unit (GLU) [10] is adopted to add nonlinearity and gating mechanism. The process of GLU includes splitting the input feature map into two parts in channel dimension, applying GELU [20] nonlinear function to one part of them and then element-wise multiplying them (Fig. 2(b)).

Enhanced Overlapping Cross-attention Block (E-OCAB). To better establish cross-window connections for the window-based self-attention module, HAT proposes an overlapping cross-attention block, which replaces the W-MSA module with an overlapping cross-attention (OCA) module. Different from the W-MSA module, the OCA module generates K and V matrices from overlapping windows with a larger size, which provides a larger range of information for the *query*. Further, we also replace its feed-forward network with the GDFN to enhance the representative ability of this block, as shown in Fig. 2(c).

3.3 Convolution Group

Convolution group aims to extract useful local features, such as local context for effective restoration and details for SR task to generate a visually

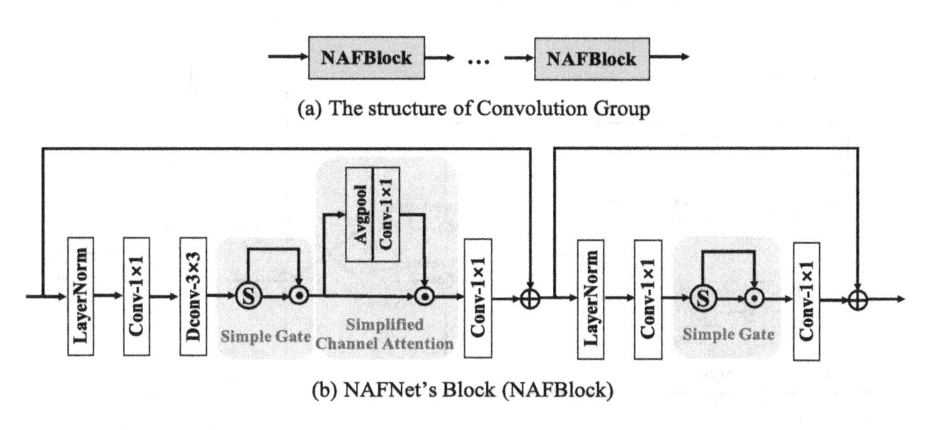

Fig. 3. Illustration of convolution group. (a) The structure of Convolution Group, and (b) NAFNet's Block (NAFBlock) [5].

pleasing HR image. NAFNet [5] establishes a simple but effective baseline for image restoration without utilizing any nonlinear activation function. Inspired by this network, the convolution group is composed of several NAFNet's blocks (NAFBlock) and keeps the resolution of intermediate feature maps unchanged, as shown in Fig. 3(a).

Each NAFBlock includes LayerNorm, convolution, simple gate unit and simplified channel attention unit. LayerNorm is added owing to its performance in transformer blocks and makes the training process smooth. The simple gate unit is used to introduce nonlinearity efficiently and replace nonlinear activation functions (*e.g.* ReLU [31] and GELU [20]), which is implemented by splitting the input feature map into two parts in the channel dimension and directly multiplying them. Besides, channel attention can capture global information and weight features in a channel-wise manner. To simplify it further, the simplified channel attention unit is proposed by retaining global information aggregation, and removing unnecessary projections and all nonlinear activation functions (Fig. 3(b)).

3.4 Adaptive Cross-branch Fusion Module

In our dual-branch network, the essential issue is how to effectively take advantage of local features and global representations obtained by the convolution group and transformer group, respectively. On one hand, local and global information are complementary in vision tasks. On the other hand, there is a significant contextual gap between features generated from convolutional blocks and self-attention blocks. Therefore, inspired by [9], the proposed adaptive cross-branch fusion module (ACFM) first enhances two-brach features in a cross-attention scheme, which extracts and allows useful information to reinforce the capability of each branch. And then it fuses them by adding weighted features, where the weights are adaptively calculated based on image contents. Besides,

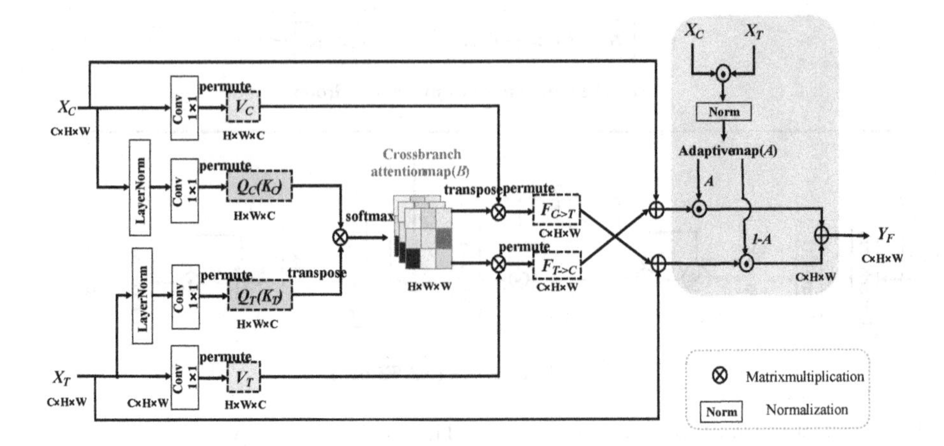

Fig. 4. The structure of Adaptive Cross-branch Fusion Module (ACFM).

ACFM is adopted at the end of each FMB block to aggregate two-branch information consecutively.

Figure 4 illustrates the structure of ACFM. For an input local feature $X_C \in \mathbb{R}^{C\times H\times W}$ generated from convolution group, linear mappings implemented by a 1×1 convolution followed by a permute operation are applied to generate $query(Q_C)$, $key(K_C)$ and $value(V_C)$. In particular, we use the same Q_C and K_C to save computations. $query(Q_T)$, $key(K_T)$ and $value(V_T)$ are also generated in the same way from an input global feature X_T obtained from transformer group. Then, the cross attention from global features to local features is computed as

$$\begin{aligned} B &= \text{SoftMax}\left(Q_C {K_T}^T/\sqrt{d}\right), \\ F_{T\to C} &= BV_T, \end{aligned} \tag{6}$$

where $B \in \mathbb{R}^{H\times W\times W}$ is the cross-branch attention map denoting the dot-product interaction of Q_C and K_T along W dimension, d is the embedding dimension of Q_C/K_T, and $F_{T\to C}$ is the attentive features from the global side to the local side. By making Q_C equal to K_C and Q_T equal to K_T, B can be reused for the cross attention from the local side to the global side as

$$F_{C\to T} = B^T V_C. \tag{7}$$

A residual learning is adopted to enhance two-branch features by adding attentive features to input features for each branch. When restoring images with different contents, the importance of local details and global interactions are not equal. Therefore, to effectively fuse two-branch features, we adopt an adaptive map (A) based on image contents to add weights on each side, where A is calculated by element-wise multiplication of two input features followed by a normalizaition operation. In all, the fused features Y_F are obtained as

$$Y_F = A \odot (F_{T\to C} + X_C) + (1 - A) \odot (F_{C\to T} + X_T). \tag{8}$$

4 Experiments

4.1 Datasets

We adopt ImageNet [11] with 1.28 million images as uncompressed images. To generate degraded inputs, uncompressed images are first $\times 4$ downsampled by the bicubic algorithm and compressed by JPEG at quality = 10. AIM 2022 Compressed Input Super-Resolution challenge (Track 1) [42] provides validation data containing 100 downsampled/compressed images. At last, 1.28 million paired images and 100 images are used as training set and validation set, respectively.

4.2 Implementation Details

Architecture. For the structure of our CIDBNet, FMB number, E-HAB number in the transformer group and NAFBlock number in the convolution group are all set to 6. For E-HAB and E-OCAB, the window size M and attention head number are set to 16 and 6, respectively. For CAB in E-HAB, the constant α is set to 0.01. The channel number of the whole network is set to 180.

Training Settings. CIDBNet is implemented with Pytorch 1.10 and trained on 8 NVIDIA A100 GPUs (40GB). During training, the mini-batch size is 32 with a patch size of 64×64 for low-resolution inputs. To reduce the risk of overfitting, some data augmentation methods are used, including random cropping, random horizontal and vertical flipping, and random 90° rotation. The network is optimized by the AdamW [28] optimizer with $\beta_1 = 0.9$ and $\beta_2 = 0.99$. We totally train the network for 800K iterations. The initial learning rate is $2e^{-4}$, and it remains constant for the first 270K iterations and then gradually reduces to $1e^{-6}$ with the cosine annealing schedule [27]. We also adopt the exponential moving average method with a decay of 0.999.

4.3 Ablation Study

For the ablation study, we train all models on a small dataset containing 3450 paired images (DIV2K_train [1] + Flicker2K [36]) for 200K iterations. Specifically, the paired DIV2K_train is provided by AIM 2022 Compressed Input Super-Resolution challenge (Track 1) [42], and the paired Flicker2K is generated following the degradations described in Sect. 4.1. In this section, the patch size of low-resolution inputs is set to 48×48 for fast experiments. We report the Peak Signal to Noise Ratio (PSNR) and Structural SIMilarity (SSIM) on the validation set. In order to show the effectiveness of each component in the proposed CIDBNet, we gradually modify the baseline HAT [6] and compare their differences. The overall comparisons are shown in Table 1, and each column represents a model with its corresponding configurations. Particularly, the 3rd column represents the baseline HAT.

Effectiveness of GDFN for E-HAB and E-OCAB. From the 3rd to 5th columns, we modify the original feed-forward network in HAT to LocalFFN [22] and GDFN. The LocalFFN also introduces local information by adding a 3×3 depthwise convolution and a channel attention operation. Table 1 shows that integrating the GDFN into attention blocks obtains more improvement. The gated mechanism has a more positive influence than channel attention.

Effectiveness of ACFM. The proposed ACFM is applied to adaptively fuse two-branch features by determining which branch should be paid more attention to. We also use a simple element-wise sum operation to replace ACFM in the 6th column. Compared to the left columns, the simple fusion pattern by a sum operation reduces the performance and has no ability to leverage the advantages of two-branch information. However, by first enhancing the two-branch features and then adaptively fusing them, ACFM brings obvious performance improvement, which shows the necessity and effectiveness of ACFM.

Effectiveness of the Dual-branch Structure. To demonstrate the effectiveness of our dual-branch structure, we construct a single-branch convolutional model in the 2nd column, where only convolution groups in CIDBNet remain. The single-branch transformer model in the 5^{th}column is equivalent to the transformer part in CIDBNet. The 7th column represents the proposed dual-branch model, which adopts ACFM to aggregate two-branch information. As we can see, the proposed dual-branch model can surpass the single-branch models, which illustrates that effectively coupling local features with global representations is beneficial for compressed image SR task.

Table 1. The overall comparison to show the effectiveness of each component in CIDBNet. The 3rd column represents the baseline HAT. Single(C) and single(T) indicate single-branch CNN and single-branch transformer, respectively.

1st	2nd	3rd	4th	5th	6th	7th
Network structure?	single(C)	single(T)	single(T)	single(T)	dual	dual
LocalFFN?	–	×	✓	×	×	×
GDFN?	–	×	×	✓	✓	✓
Fusion pattern?	–	–	–	–	Sum	ACFM
PSNR	23.408	23.550	23.578	23.586	23.564	**23.611**
SSIM	0.6359	0.6408	0.6416	0.6421	0.6410	**0.6424**

4.4 Effects of ImageNet Training

To explore the impact of ImageNet training, we train CIDBNet on the small dataset (DIV2K_train [1] + Flicker2K [36]) and ImageNet [11], respectively, and also fine-tune the ImageNet pre-trained model on the small dataset. It should be noted that the validation set is generated from DIV2K_valid [1], which belongs to the same domain as DIV2K_train. As shown in Table 2, there are two observations. First, as expected, ImageNet training significantly improves the PSNR by 0.187dB. Second, fine-tuning the ImageNet pre-trained model on the small dataset causes a slight performance drop, which is contradictory to previous observations in [6,21]. We guess the reason has two aspects: (1) ImageNet is a large-scale dataset containing various scenarios and could have the ability to cover the target domain; (2) EDT [21] finds that pre-training methods can improve performance by introducing local information, while the convolution group in CIDBNet also brings rich local contexts, which may lead to the pre-training strategy has no obvious effect on CIDBNet.

Table 2. CIDBNet is trained on two different training sets, *i.e.*, DIV2K_train+Flicker2K and ImageNet. "†" indicates fine-tuning the ImageNet pre-trained model on the small dataset (DIV2K_train+Flicker2K).

Training set	PSNR	SSIM
DIV2K_train+Flicker2K	23.612	0.6427
ImageNet	**23.799**	**0.6499**
DIV2K_train+Flicker2K†	23.778	0.6498

4.5 Comparison with State-of-the-Art Methods

We compare our method with two types of methods, including the state-of-the-art SR methods (*i.e.*, RRDB [39], RCAN [47] and HAT [6]) and cascading methods (*i.e.*, DRUNet+RRDB [39,45], SwinIR+RRDB [24,39] and SwinIR+HAT-L [6,24]). The SR methods are trained using our training set (ImageNet [11]) with their respective training settings. For HAT, we use the larger variant HAT-L (twice the depth of HAT). The cascading methods are to apply the state-of-the-art compression artifacts removal and SR algorithms step by step, where we directly use their provided models. All the above methods are tested on the validation set. As presented in Table 3, CIDBNet outperforms both existing SR and cascading methods, leading to state-of-the-art performance. In particular, CIDBNet even surpasses HAT-L by 0.027dB. On the other hand, we can also find that transformer-based methods (*i.e.*, SwinIR and HAT-L) achieve higher performance than CNN-based methods (*i.e.*, DRUNet, RRDB and RCAN), both for artifacts removal and SR algorithms. Besides, the visual comparisons are shown in Fig. 5. We can observe that our method can remove the blocky artifacts and reconstruct high-frequency details. Comparatively, the results of our method have less excessive blurry artifacts and preserve shaper edges.

4.6 AIM Compressed Image Super-Resolution Challenge

CIDBNet is proposed to participate in the AIM 2022 Compressed Input Super-Resolution challenge (Track 1) [42]. To improve the performance, we adopt self-ensemble [25], model ensemble and test-time local converter [8] strategies during

Table 3. Quantitative comparison with state-of-the-art methods on the validation set. Runtime is measured based on the input size of 510 × 340 on an A100 GPU.

Method	DRUNet+ RRDB [39,45]	SwinIR+ RRDB [24,39]	SwinIR+ HAT-L [6,24]	RRDB [39]	RCAN [47]	HAT-L [6]	**CIDBNet (Ours)**
PSNR	23.371	23.730	23.732	23.553	23.599	23.772	**23.799**
SSIM	0.6337	0.6445	0.6448	0.6415	0.6422	0.6492	**0.6499**
Runtime (s)	0.014	4.594	6.532	0.255	0.215	2.113	2.765

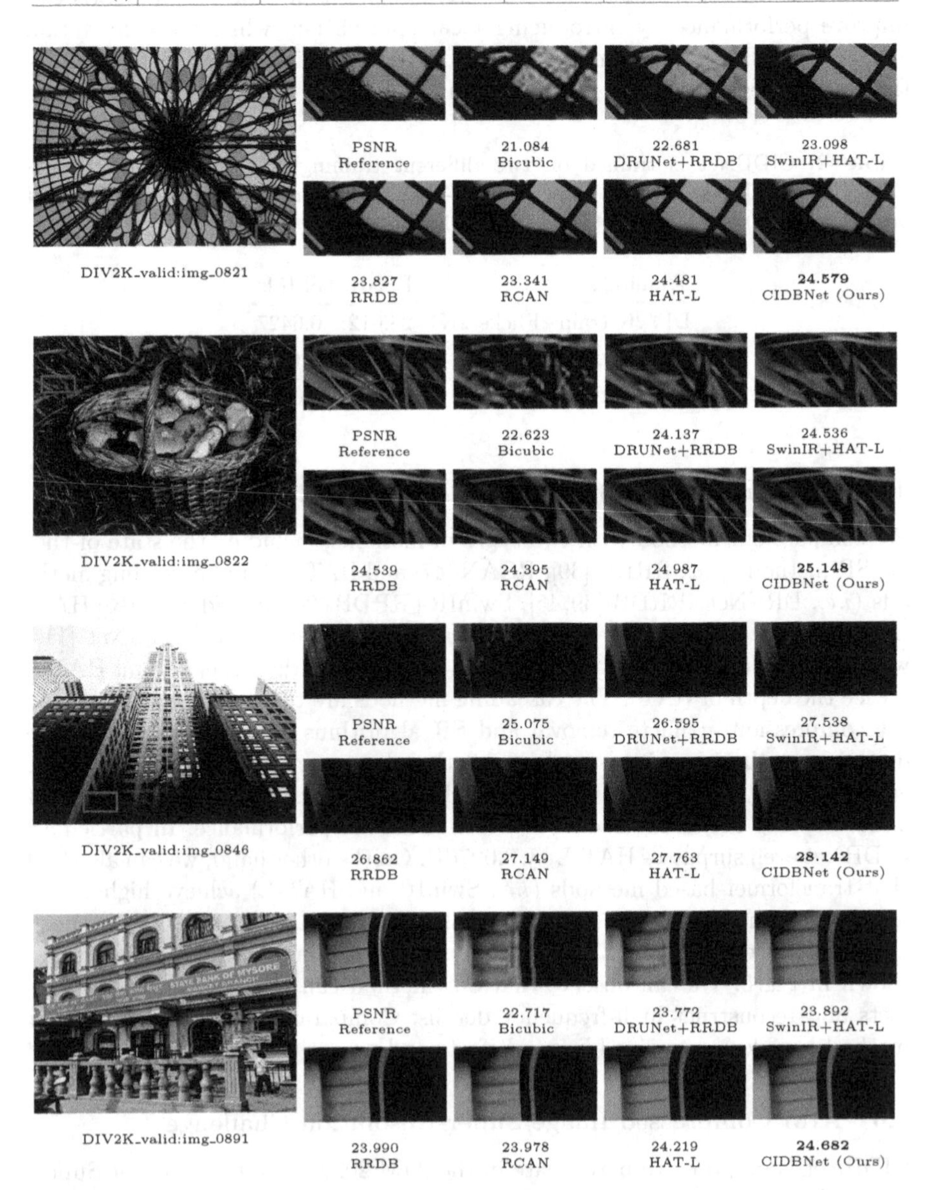

Fig. 5. Visual comparison generated by different methods on the validation set.

Table 4. Quantitative results on the testing set of the AIM 2022 Compressed Input Super-Resolution challenge (Track 1) [42].

Team	PSNR
chenmigongzuo	23.6677
jetblock	23.5731
Ours	23.5597
SRC-B	23.5307
Libc	23.5085
MSDRSR	23.4545
Giantpandacv	23.4249
Aselsan Research	23.4239
SRMUI	23.4033
MVideo	23.3250
UESTC+XJU CV	23.2911
cvlab	23.2828

inference. Specifically, we use the same data augmentations in Sect. 4.2 for self-ensemble. For model ensembling, we train three models with the same training settings. They are, CIDBNet, a modified model by adding spatial attention to NAFBlock and HAT-L [6], respectively. The test-time local converter is utilized to reduce the train-test inconsistency caused by global operations in networks. At last, our submitted result achieves 23.8319dB on the validation set. Furthermore, we won the 3rd place with 23.5597dB in the testing phase (Table 4).

5 Conclusions

This paper proposes a consecutively-interactive dual-branch network named CIDBNet for JPEG compressed image super-resolution. We adopt a dual-branch structure, *i.e.*, CNN branch and transformer branch, to extract local features and global representations, respectively. Long-range information and local image structure are both useful for removing artifacts and restoring high-frequency details for high-resolution images. We also integrate the gated-Dconv feed-forward (GDFN) into each transformer block to enhance feature learning. In addition, we propose an adaptive cross-branch fusion module (ACFM) to effectively and adaptively fuse the two-branch features based on image contents. ACFM is applied in each feature mixing block (FMB) to aggregate the two-branch information in a consecutive manner. Extensive experiments show the effectiveness of the proposed components, and our CIDBNet outperforms the state-of-the-art methods. In the future, we will extend the model to other low-level tasks, such as image denoising, deblurring and classical super-resolution.

Acknowledgements. This work was supported by the National Key Research and Development Program of China (Grant No. 2021ZD0201504), and National Natural Science Foundation of China (No.62106267).

References

1. Agustsson, E., Timofte, R.: NTIRE 2017 challenge on single image super-resolution: dataset and study. In: Proceedings of the IEEE Conference on Computer Vision and Pattern Recognition Workshops, pp. 126–135 (2017)
2. Cao, J., Li, Y., Zhang, K., Van Gool, L.: Video super-resolution transformer. arXiv preprint arXiv:2106.06847 (2021)
3. Chen, H., et al.: Pre-trained image processing transformer. In: Proceedings of the IEEE/CVF Conference on Computer Vision and Pattern Recognition, pp. 12299–12310 (2021)
4. Chen, H., He, X., Ren, C., Qing, L., Teng, Q.: CISRDCNN: super-resolution of compressed images using deep convolutional neural networks. Neurocomputing **285**, 204–219 (2018)
5. Chen, L., Chu, X., Zhang, X., Sun, J.: Simple baselines for image restoration. arXiv preprint arXiv:2204.04676 (2022)
6. Chen, X., Wang, X., Zhou, J., Dong, C.: Activating more pixels in image super-resolution transformer. arXiv preprint arXiv:2205.04437 (2022)
7. Chen, Y., et al.: Mobile-former: bridging mobileNet and transformer. In: Proceedings of the IEEE/CVF Conference on Computer Vision and Pattern Recognition, pp. 5270–5279 (2022)
8. Chu, X., Chen, L., Chen, C., Lu, X.: Revisiting global statistics aggregation for improving image restoration. arXiv preprint arXiv:2112.04491 (2021)
9. Chu, X., Chen, L., Yu, W.: NAFSSR: Stereo image super-resolution using NAFNet. In: Proceedings of the IEEE/CVF Conference on Computer Vision and Pattern Recognition, pp. 1239–1248 (2022)
10. Dauphin, Y.N., Fan, A., Auli, M., Grangier, D.: Language modeling with gated convolutional networks. In: International Conference on Machine Learning, pp. 933–941. PMLR (2017)
11. Deng, J., Dong, W., Socher, R., Li, L.J., Li, K., Fei-Fei, L.: ImageNet: a large-scale hierarchical image database. In: 2009 IEEE Conference on Computer Vision and Pattern Recognition, pp. 248–255. IEEE (2009)
12. Dong, C., Deng, Y., Loy, C.C., Tang, X.: Compression artifacts reduction by a deep convolutional network. In: Proceedings of the IEEE International Conference on Computer Vision, pp. 576–584 (2015)
13. Dong, C., Loy, C.C., He, K., Tang, X.: Learning a deep convolutional network for image super-resolution. In: Fleet, D., Pajdla, T., Schiele, B., Tuytelaars, T. (eds.) ECCV 2014. LNCS, vol. 8692, pp. 184–199. Springer, Cham (2014). https://doi.org/10.1007/978-3-319-10593-2_13
14. Dosovitskiy, A., et al.: An image is worth 16x16 words: transformers for image recognition at scale. arXiv preprint arXiv:2010.11929 (2020)
15. Foi, A., Katkovnik, V., Egiazarian, K.: Pointwise shape-adaptive DCT for high-quality denoising and deblocking of grayscale and color images. IEEE Trans. Image Process. **16**(5), 1395–1411 (2007)
16. Fu, X., Wang, X., Liu, A., Han, J., Zha, Z.J.: Learning dual priors for jpeg compression artifacts removal. In: Proceedings of the IEEE/CVF International Conference on Computer Vision, pp. 4086–4095 (2021)

17. Gunawan, A., Madjid, S.R.H.: CISRNet: compressed image super-resolution network. arXiv preprint arXiv:2201.06045 (2022)
18. Guo, J., et al.: CMT: convolutional neural networks meet vision transformers. In: Proceedings of the IEEE/CVF Conference on Computer Vision and Pattern Recognition, pp. 12175–12185 (2022)
19. He, K., Zhang, X., Ren, S., Sun, J.: Deep residual learning for image recognition. In: Proceedings of the IEEE Conference on Computer Vision and Pattern Recognition, pp. 770–778 (2016)
20. Hendrycks, D., Gimpel, K.: Gaussian error linear units (GELUs). arXiv preprint arXiv:1606.08415 (2016)
21. Li, W., Lu, X., Lu, J., Zhang, X., Jia, J.: On efficient transformer and image pre-training for low-level vision. arXiv preprint arXiv:2112.10175 (2021)
22. Li, Y., Zhang, K., Cao, J., Timofte, R., Van Gool, L.: LocalViT: bringing locality to vision transformers. arXiv preprint arXiv:2104.05707 (2021)
23. Liang, J., et al.: VRT: a video restoration transformer. arXiv preprint arXiv:2201.12288 (2022)
24. Liang, J., Cao, J., Sun, G., Zhang, K., Van Gool, L., Timofte, R.: SwinIR: image restoration using swin transformer. In: Proceedings of the IEEE/CVF International Conference on Computer Vision, pp. 1833–1844 (2021)
25. Lim, B., Son, S., Kim, H., Nah, S., Mu Lee, K.: Enhanced deep residual networks for single image super-resolution. In: Proceedings of the IEEE Conference on Computer Vision and Pattern Recognition Workshops, pp. 136–144 (2017)
26. Liu, Z., et al.: Swin transformer: hierarchical vision transformer using shifted windows. In: Proceedings of the IEEE/CVF International Conference on Computer Vision, pp. 10012–10022 (2021)
27. Loshchilov, I., Hutter, F.: SGDR: stochastic gradient descent with warm restarts. arXiv preprint arXiv:1608.03983 (2016)
28. Loshchilov, I., Hutter, F.: Decoupled weight decay regularization. arXiv preprint arXiv:1711.05101 (2017)
29. Mao, M., et al.: Dual-stream network for visual recognition. Adv. Neural. Inf. Process. Syst. **34**, 25346–25358 (2021)
30. Mehta, S., Rastegari, M.: MobileViT: light-weight, general-purpose, and mobile-friendly vision transformer. arXiv preprint arXiv:2110.02178 (2021)
31. Nair, V., Hinton, G.E.: Rectified linear units improve restricted Boltzmann machines. In: ICML (2010)
32. Peng, Z., Huang, W., Gu, S., Xie, L., Wang, Y., Jiao, J., Ye, Q.: Conformer: local features coupling global representations for visual recognition. In: Proceedings of the IEEE/CVF International Conference on Computer Vision, pp. 367–376 (2021)
33. Sandler, M., Howard, A., Zhu, M., Zhmoginov, A., Chen, L.C.: MobileNetV2: inverted residuals and linear bottlenecks. In: Proceedings of the IEEE Conference on Computer Vision and Pattern Recognition, pp. 4510–4520 (2018)
34. Shi, W., et al.: Real-time single image and video super-resolution using an efficient sub-pixel convolutional neural network. In: Proceedings of the IEEE Conference on Computer Vision and Pattern Recognition, pp. 1874–1883 (2016)
35. Tai, Y., Yang, J., Liu, X.: Image super-resolution via deep recursive residual network. In: Proceedings of the IEEE Conference on Computer Vision and Pattern Recognition, pp. 3147–3155 (2017)
36. Timofte, R., Agustsson, E., Van Gool, L., Yang, M.H., Zhang, L.: NTIRE 2017 challenge on single image super-resolution: methods and results. In: Proceedings of the IEEE Conference on Computer Vision and Pattern Recognition Workshops, pp. 114–125 (2017)

37. Vaswani, A., et al.: Attention is all you need. In: Advances in Neural Information Processing Systems 30 (2017)
38. Wallace, G.K.: The JPEG still picture compression standard. IEEE Trans. Consumer Electr. **38**(1), xviii-xxxiv (1992)
39. Wang, X., et al.: ESRGAN: enhanced super-resolution generative adversarial networks. In: Leal-Taixé, L., Roth, S. (eds.) ECCV 2018. LNCS, vol. 11133, pp. 63–79. Springer, Cham (2019). https://doi.org/10.1007/978-3-030-11021-5_5
40. Wang, Z., Cun, X., Bao, J., Zhou, W., Liu, J., Li, H.: Uformer: a general u-shaped transformer for image restoration. In: Proceedings of the IEEE/CVF Conference on Computer Vision and Pattern Recognition, pp. 17683–17693 (2022)
41. Wu, H., et al.: CvT: introducing convolutions to vision transformers. In: Proceedings of the IEEE/CVF International Conference on Computer Vision, pp. 22–31 (2021)
42. Yang, R., Timofte, R., et al.: Aim 2022 challenge on super-resolution of compressed image and video: dataset, methods and results. In: Proceedings of the European Conference on Computer Vision Workshops (ECCVW) (2022)
43. Zamir, S.W., Arora, A., Khan, S., Hayat, M., Khan, F.S., Yang, M.H.: Restormer: efficient transformer for high-resolution image restoration. In: Proceedings of the IEEE/CVF Conference on Computer Vision and Pattern Recognition, pp. 5728–5739 (2022)
44. Zhang, J., Xiong, R., Zhao, C., Zhang, Y., Ma, S., Gao, W.: CONCOLOR: constrained non-convex low-rank model for image deblocking. IEEE Trans. Image Process. **25**(3), 1246–1259 (2016)
45. Zhang, K., Li, Y., Zuo, W., Zhang, L., Van Gool, L., Timofte, R.: Plug-and-play image restoration with deep denoiser prior. In: IEEE Transactions on Pattern Analysis and Machine Intelligence (2021)
46. Zhang, X., Yang, W., Hu, Y., Liu, J.: DMCNN: dual-domain multi-scale convolutional neural network for compression artifacts removal. In: 2018 25th IEEE International Conference on Image Processing (ICIP), pp. 390–394. IEEE (2018)
47. Zhang, Y., Li, K., Li, K., Wang, L., Zhong, B., Fu, Y.: Image super-resolution using very deep residual channel attention networks. In: Ferrari, V., Hebert, M., Sminchisescu, C., Weiss, Y. (eds.) ECCV 2018. LNCS, vol. 11211, pp. 294–310. Springer, Cham (2018). https://doi.org/10.1007/978-3-030-01234-2_18
48. Zhang, Y., Li, K., Li, K., Zhong, B., Fu, Y.: Residual non-local attention networks for image restoration. arXiv preprint arXiv:1903.10082 (2019)
49. Zhang, Y., Tian, Y., Kong, Y., Zhong, B., Fu, Y.: Residual dense network for image super-resolution. In: Proceedings of the IEEE Conference on Computer Vision and Pattern Recognition, pp. 2472–2481 (2018)
50. Zhang, Y., Tian, Y., Kong, Y., Zhong, B., Fu, Y.: Residual dense network for image restoration. IEEE Trans. Pattern Anal. Mach. Intell. **43**(7), 2480–2495 (2020)

XCAT - Lightweight Quantized Single Image Super-Resolution Using Heterogeneous Group Convolutions and Cross Concatenation

Mustafa Ayazoglu(✉) and Bahri Batuhan Bilecen

Aselsan Research, Ankara, Turkey
{mayazoglu,batuhanb}@aselsan.com.tr

Abstract. We propose a lightweight, single-image super-resolution mobile device network named XCAT, and introduce Heterogeneous Group Convolution Blocks with Cross Concatenations (HXBlock). The heterogeneous split of the input channels to the group convolution blocks reduces the number of operations, and cross concatenation allows for information flow between the intermediate input tensors of cascaded HXBlocks. Cross concatenations inside HXBlocks can also avoid using more expensive operations like 1×1 convolutions. To further prevent expensive tensor copy operations, XCAT utilizes non-trainable convolution kernels to apply upsampling operations. Designed with integer quantization in mind, XCAT also utilizes several techniques in training, like intensity-based data augmentation. Integer quantized XCAT operates in real-time on Mali-G71 MP2 GPU with 320 ms, and on Synaptics Dolphin NPU with 30 ms (NCHW) and 8.8 ms (NHWC), suitable for real-time applications.

Keywords: Single image super-resolution · Quantization · Group convolutions · Mobile AI

1 Introduction

Super-resolution (SR) is an extensively studied computer vision problem that aims to generate higher resolution (HR) image(s) given lower resolution (LR) image(s). In single image super-resolution (SISR), a single image; and in multi-image super-resolution (MISR), multiple images are utilized to generate a single HR image. In either case, image super-resolution is an ill-posed problem, since there is no unique solution. This ill-posed problem has been attempted to be solved via classical methods [9] and deep-learning based methods [6,30]; however, many new methods based on deep-learning are still being developed, most of which purely focus on data fidelity.

However, for the SR method to be practically applicable, the runtime is as important as the method's PSNR performance. Due to its practical importance,

L. Karlinsky et al. (Eds.): ECCV 2022 Workshops, LNCS 13802, pp. 475–488, 2023.
https://doi.org/10.1007/978-3-031-25063-7_29

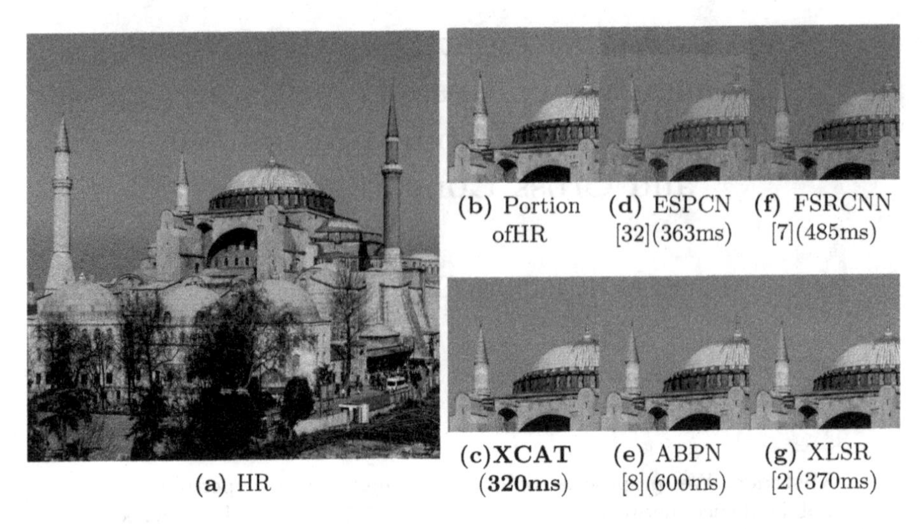

Fig. 1. Comparative results of UINT8 quantized models on DIV2K (Val image: 890). (d) and (f) yield both visually and numerically worse results than the rest. Visual results of (c), (e), and (g) are indistinguishable to the naked eye; however, XCAT runs the fastest

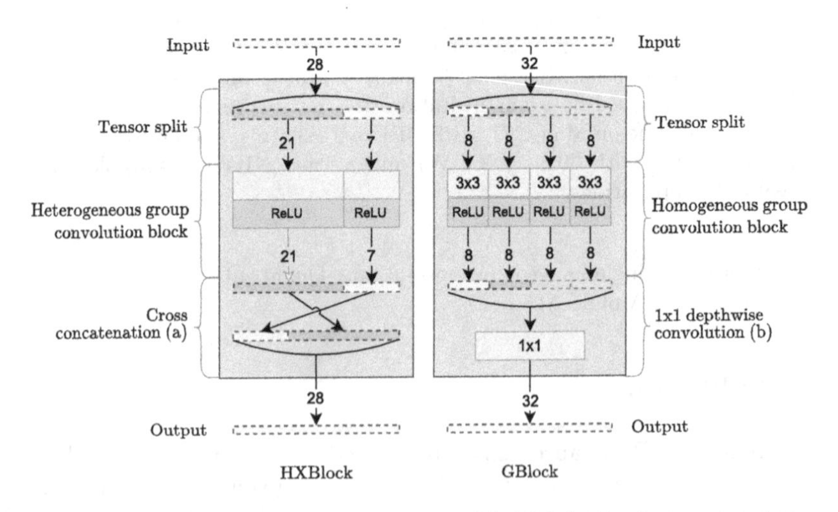

Fig. 2. Proposed HXBlock (left) vs. GBlock [2] (right). Dashed rectangles represent tensors. Group convolutions inside HXBlock are **heterogeneous** compared to GBlock's. HXBlock uses **cross concatenation (a)** instead of depth wise 1×1 convolutions (b), which provides information flow through different convolutional kernels when HXBlocks are cascaded. For HXBlock, using (a) results in a significant runtime performance increase in return of a small PSNR drop compared to using (b)

recent literature studies on SR focus on deployability, runtime, quantization, and efficiency, as well as PSNR of the method [2,3,8,13,24]. Yet, achieving real-time performance with satisfactory visual quality during the quantization process further complicates the problem and careful network design is needed.

In this study, we focus on the efficiency and mobile deployment, in the scope of *Mobile AI & AIM 2022 Real-Time Image Super-Resolution Challenge* [14]. Our model, named **XCAT**, is a SISR network incorporating the proposed HXBlock and modifying both new and existing techniques from the literature for providing a quantization-aware, robust, real-time performance model, suitable for mobile devices.

Our work makes the following contributions:

1. **HXBlocks**, which are heterogeneous grouped convolutions with cross concatenation layers for allowing information flow with almost no computational cost through different convolutional kernels of the group. Relevant studies done on HXBlocks have shown that they can be replaced with traditional group convolutions with little sacrifice from PSNR, but a significant gain in run time performance.
2. A method for nearest neighborhood up sampling method with fixed 2D convolutional kernels to replace expensive tensor copy operations on mobile devices, which makes the model robust to quantization.
3. An efficient, mobile device friendly, single image super-resolution network named **XCAT**.

2 Related Works

DNN-based Single Image Super-Resolution. First deep-learning-based SISR algorithm was proposed by Dong et al. as SRCNN [6]. Later, as a speed improvement on SRCNN, FSRCNN [7] was developed; which introduced a deconvolutional layer at the end of the network, replaced ReLU with a PReLU activation layer, and reformulated SRCNN by adopting smaller filter sizes but more mapping layers. Shi et al. [30] introduced a novel, efficient sub-pixel convolutional layer (also known as depth to space), which is actually widely used in many fast SR networks right now. VDSR [19], EDSR [25], and WDSR [37] continued the development of deep-learning-based SR by increasing the number of parameters, in exchange for accuracy with speed.

With the recent developments in computer vision and deep learning, concepts like attention mechanism [28], generative adversarial networks [23,33], recursive & residual networks [1,20,31], and distillation layers [3,11,12,27] also started to take part inside SR network architectures. GANs and networks with attention mechanism mostly generate a high-quality SR image by sacrificing speed, whereas RNNs and distilling networks try to decrease the computational load.

Group Convolutions. Group convolutions consist of groups of multiple convolutional kernels placed within the same layer. The motivation behind group convolutions emerged with AlexNet [22], desiring to distribute the model over multiple GPUs to overcome hardware limitations. Later on, besides the increase in speed in AlexNet, group convolutions are also observed to improve classification accuracy when groups are accompanied by skipped connections with ResNetX [34]. ShuffleNet [39] introduced shuffling the intermediate tensors between group convolution blocks to increase feature extraction. DeepRoots [16] and more recent studies use different convolutional kernels inside groups, such as 1×1 depth wise convolutions [29] and dilated convolutions [36,40]. In addition, unitary [41] and interleaved [38] group convolutions also offer different perspectives on how to extract various features from input images. Usage of group convolutions due to their efficiency on super-resolution problems is also present [2,3,18].

Model Optimization. Hardware limitations and specifications may require the model to be optimized via different techniques, such as quantization, pruning, clustering, network architecture search (NAS), and many more. **Quantization** refers to converting floating point values to integers, hence decreasing memory usage and computational cost when re-accessing and/or updating the mentioned values, at a cost of decreasing the precision. Quantization is particularly useful on neural network models since it can decrease inference times without sacrificing much inference accuracy if done correctly [17]. Models also can be quantized after quantization-aware training in floating point precision [21,32], as well as training the network directly with low precision multiplications [5]. Removing layers from a model having a minor effect on inference is called **pruning** [10,35], and **clustering** is the method of decreasing the number of unique weights by grouping weights and assigning the centroid values for each group. All of these methods try to decrease processor utilization or memory usage, or both. Besides optimizing an existing network structure, finding the most possible optimal network structure in search space is also a study area, known as network architecture search [42].

3 Method

In this section, XCAT is defined by its overall architecture and its components. Details about the training techniques and the quantization procedure are explained thoroughly as well.

3.1 XCAT's Architecture

As seen from Fig. 3, XCAT consists of 3 individual convolutional layers with trainable 3×3 kernels, a single 1×1 convolutional layer with a fixed "identity kernel" to simulate nearest neighborhood upsampling operation, m HXBlocks, a tensor addition layer, followed by a depth to space (D2S) layer and a clipped ReLU activation layer. Input and output of XCAT are LR and SR, respectively, where SR has x3 resolution of LR.

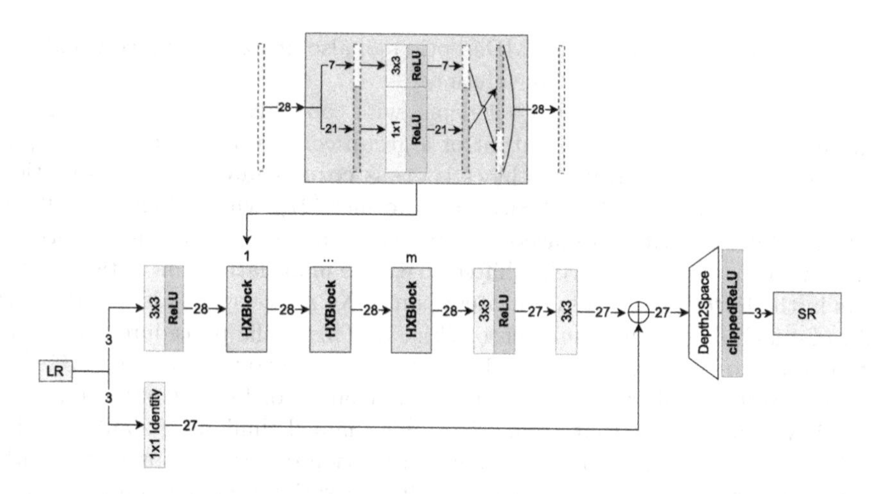

Fig. 3. Network structure of XCAT. Numbers on arrows denote channel numbers, and numbers inside blocks represent kernel sizes of convolutional blocks. Dashed blocks represent tensors, whereas non-dashed ones represent operators like convolution and normalization. Convolution with 1×1 identity kernel performs the upsampling method visualized in Fig. 4. XCAT has m HXBlocks, where m = 2 for this study

Each key component of XCAT will be detailed with their reasoning:

Group Convolutions with Heterogeneous Filter Groups and Varying Kernels. Group convolutions, which include multiple convolutional kernels per layer, are known to be able to extract and learn more varying features compared to a single kernel [22]. XCAT inherits this idea of group convolution blocks to replace single-layered convolutions in a repeated manner. However, as opposed to initial approaches [34], convolutional layers inside the group convolution blocks in XCAT have **different layer dimensions** and **different kernel sizes** (Fig. 3). This allows to pass the same source information between different convolutional layers and allows for less computationally demanding feature extraction. The input tensor is split into two parts in channel dimension, one processed by 1×1 and the other by 3×3 convolutional kernels. 1×1 convolution "blends" the point-wise information from previous HXBlocks and extracts inter-channel features, whereas 3×3 convolution considers in-channel correlation as well. In addition, a relevant study of Lee et al.'s [22] logarithmic filter groups in shallow CNNs shows the positive effect of dividing group convolution input tensors unevenly.

Cross Concatenation. First group convolutions in AlexNet [22] ended with max pooling layers. However, group convolution designs such as DeepRoots [16] started utilizing low-dimensional embeddings (like 1×1 convolutions) at the end of the groups, with the inspiration taken from Lin. et al. [26] and Cogswell et al. [4]. This was done to decrease the computational cost and number of parameters without compromising accuracy. Later on, several efficiency-oriented

SR networks like XLSR [2] and IMDeception [3] also utilized group convolutions ending with 1×1 depth-wise convolutions.

In XCAT, instead of using 1×1 depth wise convolutions for increasing the spatial receptive field of each output of a group convolution block, the output tensor of each group convolution block is **cross concatenated**. The inspiration came from ShuffleNet [39] and Swin Transformer [27]; where channel shuffling and convolutional layers are inserted between group convolutions in the former, and window partitions are cyclic shifted to enable information flow between windows in the latter. Each cross concatenation in XCAT corresponds to a circular shift of one-fourth of the input tensor (Fig. 3). This cyclic procedure allows the information to pass through from 1×1 and 3×3 convolutions inside XCAT's group convolution blocks, hence having more chance for feature extraction.

It is worthy to note that ShuffleNet [39]'s channel shuffling is similar to the XCAT's; however, XCAT has a cross concatenation operation represented with cyclic shifts, whereas ShuffleNet has a shuffle operation dividing and reorganizing tensors into many small partitions. This reorganization operation is reflected onto the target device (Synaptics Dolphin NPU) as reshape and transpose operations, which take much longer to process compared to XCAT's simpler yet effective approach.

During the experiments, it is observed that replacing cross concatenation operations with 1×1 convolutions in XCAT increases the run time per frame, but does not increase the PSNR test score considerably, making it less practical for mobile networks.

Depth to Space (D2S) Operation. Shi et al. [30]'s pixel shuffling (depth to space operator) is inserted at the end of the network, which aims to implement sub-pixel convolutions in an efficient manner and is proven to increase PSNR score in super-resolution problems in many studies.

Nearest Neighborhood Upsampling with Fixed Kernel Convolutions. We observed that providing the low-resolution input image to D2S with accompanying feature tensors increases the robustness, as opposed to only providing the extracted feature tensors to D2S. With this motivation, XCAT also adds repeated input image tensors (where each channel of the input image is repeated 9 times) to feature tensors and provides them to D2S. From the perspective of D2S, this operation is equivalent to the nearest neighborhood upsampling.

A relevant study done by Du et al. named ABPN [8] also utilizes the nearest neighborhood upsampling to be fed to the D2S block. However, it uses tensor copy operations while repeating and concatenating the input image in the upsampling process, which are indeed expensive for mobile devices. For a better alternative, a convolutional layer of 3 input channels and 27 output channels is used, with a 1×1 non-trainable kernel which is set to serve the same purpose as a tensor copy (Fig. 4). One point to note is that when this 1×1 kernel is set as trainable, it gets affected by the quantization process and yields lower visual quality results.

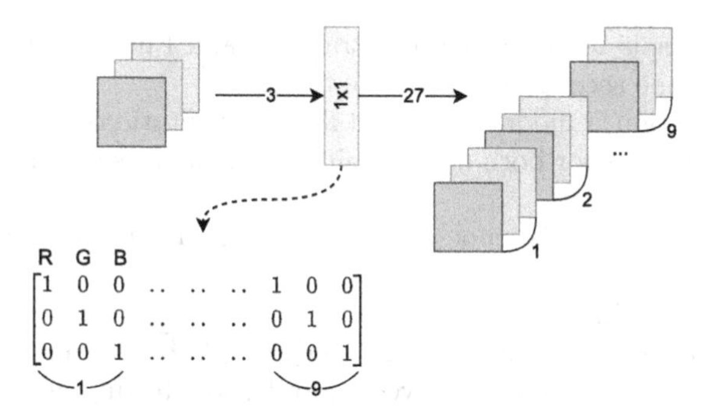

Fig. 4. Tensor copy operations done with convolutions. The identity kernel of the 1×1 convolution is set in such a way that it reproduces the input tensor of 3 channels 9 times, generating an output tensor of 27 channels

3.2 Training and Quantization Details

XCAT is trained in floating point precision and quantized afterward. However, it is trained and designed with quantization in mind, with several techniques to avoid PSNR decrease:

Intensity-Augmented Training. To minimize the PSNR difference while quantizing the FP32 model to its UINT8, intensity values of the training images are scaled with randomly chosen constants among (1, 0.7, 0.5). We have observed that this strategy helped with quantization and avoided signal degradation, as stated in [2].

Clipped ReLU. As proven and explained in [2], using clipped ReLU at the end of the network allows better quantization while keeping the performance in the real-time range.

Representative Dataset Selection. TensorFlow Lite requires a representative dataset while quantizing a floating point Keras or TensorFlow model. As a rule of thumb, this dataset consists of entire training images. However, it is observed that selecting a subset of all training images as the representative dataset affects the final PSNR test score of the quantized UINT8 model immensely. Hence, to find the most suitable representative dataset, a linear search is applied to all DIV2K training images, generating single image representative datasets. For each representative dataset (or rather an image), XCAT is quantized and PSNR test scores are measured. The highest scoring quantized XCAT model is chosen as the best.

Training Details. XCAT is trained twice in floating point precision and then quantized. Training details are as follows:

- DIV2K dataset is used for the first training, and Flickr2K dataset is added alongside for the second training (fine tuning).
- Intensity, rotation, random crop, and flip augmentations are used while setting up the dataset for both of the training. HR images are cropped to 96x96 patches.
- XCAT is trained with 50 epochs and 16 batches. Each epoch contains 10000 mini-batches.
- For the first training:
  - Charbonnier loss is used, where $C(x) = \sqrt{(x^2 + \epsilon^2)}$ and $\epsilon = 0.1$. Charbonnier loss is the smoother version of L1-loss having better convergence characteristics than L2-loss.
  - Adam optimizer is used with initial learning rate $= 0.001$, $\beta_1 = 0.9$, $\beta_2 = 0.999$, $\epsilon = 1e^{-8}$.
  - Warm-up scheduler is used: Starting from the initial learning rate, at each epoch, the learning rate is increased up to $25e^{-4}$ until the 5th epoch. After the 5th epoch, the learning rate is linearly decreased at the end of each epoch, where at the last epoch it decreases to e^{-4}.
- For the second training for fine-tuning:
  - Mean square error is used as the loss.
  - Adam optimizer is used with an initial learning rate $= 0.0001$, and the same beta and epsilon parameters.
  - Warm-up scheduler is used again, but with the new initial learning rate, and the maximum learning rate of $12.5e^{-4}$ instead of $25e^{-4}$.

4 Experimental Results

During the development of XCAT, many modified versions were created and tested. Numerical results of XCAT models and the ablation study done for HXBlocks are given in Table 1. Comparative visual results are given in Fig. 5.

To choose the most successful model, we used the score function in (1), which is officially published in the competition's evaluation criteria.

$$Score = \frac{2^{2(PSNR_{(UINT8)} - 30)}}{t_{(UINT8)} 10^{-5}} \tag{1}$$

Comparative Study. Table 4 and Fig. 5 reveal that the network architecture's suitability for the quantization procedure plays a big role in producing high-quality, super-resolved images. Despite XLSR and ABPN having higher PSNR FP32 scores compared to XCAT, after the quantization, all three yielded similar visual results and closer UINT8 PSNR scores to each other.

Table 1. Different XCAT-based models and their performance. Runtime is evaluated on Mali-G71 MP2 GPU via AI Benchmark 5 [15]. Score is the metric function described in (1). Note that the Config column describes the differences among models. m is the number of HXBlocks. X/Y shows the splitting ratio, where X+Y is the total number of channels of HXBlocks. axa/bxb represents the convolutional kernels inside the group convolution blocks, where the tensor with dimension X passes through axa, and Y through bxb

Model	Config	PSNR FP32/UINT8	Runtime (ms)	Score
XCAT	m = 2, 21/7, $1 \times 1/3 \times 3$ 2 stage training	29.88/29.81	320	**240**
A	m = 2, 21/7, $1 \times 1/3 \times 3$ 1 stage training	29.85/29.79	320	233
B	m = 2, 21/7, $1 \times 1/3 \times 3$ Cross Cat → 1×1 Conv	29.92/29.84	340	235
C	m = 2, 21/7, $3 \times 3/3 \times 3$ Cross Cat → 1×1 Conv	**30.04/29.96**	780	121
D	m = 2 HXBlock → 3×3 Conv	30.04/29.97	770	124
E	m = 4, 21/7, $1 \times 1/3 \times 3$	29.98/29.89	370	232
F	m = 4, 21/7, $1 \times 1/3 \times 3$ Conv after last HXBlock: $3 \times 3 \rightarrow 1 \times 1$	29.82/29.75	300	236
G	m = 4, 21/7, $1 \times 1/3 \times 3$ Conv after last HXBlock: removed	29.81/29.72	290	234
H	m = 4, 16/12, $1 \times 1/3 \times 3$ Conv after last HXBlock: removed	29.87/29.76	300	238
I	m = 4, 7/21, $1 \times 1/3 \times 3$ Conv after last HXBlock: removed	30.03/29.88	520	163
J	m = 4, 7/21, $3 \times 3/3 \times 3$ Conv after last HXBlock: removed	30.04/29.87	550	152
K	m = 3, 7/21, $1 \times 1/3 \times 3$ Conv after last HXBlock: removed	29.95/29.81	430	179
L	m = 4, 16/4, $1 \times 1/3 \times 3$ Conv after last HXBlock: removed	29.61/29.45	**205**	228
M	m = 4, 16/4, $1 \times 1/3 \times 3$ Replaced Add with Concat	29.63/29.15	298	103

Table 2. PSNR test scores of XCAT and other algorithms with public datasets. All models are FP32. To be consistent with the rest of the algorithms; XLSR, ABPN, and XCAT's PSNR results are calculated using Luminance (Y) channel rather than RGB channels, except for DIV2K (*We performed our own training since the pre-trained FP32 model from the authors performed poorly, around 15dB for DIV2K(Val))

Dataset	Scale	Bicubic	FSRCNN	ESPCN	XLSR	ABPN*	XCAT (proposed)
Set5	x3	30.44	32.73	32.59	33.09	33.45	33.02
Set14	x3	27.63	29.30	29.18	29.59	29.73	29.54
B100	x3	27.13	28.26	28.18	28.45	28.56	28.42
Urban100	x3	24.43	26.03	25.87	26.48	26.73	26.38
Manga109	x3	26.87	30.21	29.70	31.13	31.47	31.12
DIV2K(Val)	x3	28.82	29.67	29.54	30.10	30.10	29.88

Table 3. Effect of cross concatenation versus straight concatenation (no shuffling of the tensors while concatenating), the number of HXBlocks, and the tensor divisions. All models are based on XCAT. All parameter changes are mentioned on the table, and the rest are kept the same among all models. Runtime is evaluated on Mali-G71 MP2 GPU via AI Benchmark 5 [15]. Score is the metric function described in (1). Split and kernel definitions are stated in Table 1

Split	Kernel	m (# of HXBlocks)	Cross Concat	PSNR (FP32)	PSNR (UINT8)	Runtime (ms)	Score
21/7	$1 \times 1/3 \times 3$	2	✓	29.88	29.81	320	240
21/7	$1 \times 1/3 \times 3$	2	×	29.87	29.79	320	233
21/7	$1 \times 1/3 \times 3$	4	✓	29.98	29.89	370	232
21/7	$1 \times 1/3 \times 3$	4	×	29.96	29.87	370	232
21/7	$1 \times 1/3 \times 3$	8	✓	30.04	29.97	480	200
21/7	$1 \times 1/3 \times 3$	8	×	30.01	29.89	480	179
21/7	$1 \times 1/3 \times 3$	12	✓	30.07	29.97	580	165
21/7	$1 \times 1/3 \times 3$	12	×	30.04	29.72	580	117
24/8	$1 \times 1/3 \times 3$	6	✓	30.02	29.92	410	218
24/8	$1 \times 1/3 \times 3$	6	×	30.01	29.89	410	209
56/8	$1 \times 1/3 \times 3$	4	✓	30.08	30.03	620	168
56/8	$1 \times 1/3 \times 3$	4	×	30.04	29.99	620	159

Ablation Study. In Table 1, increasing layer number/sizes and parameter numbers increased the PSNR score and decreased run time performance (E-G, G-L, B-C). Decreasing number of groups had a negative effect on PSNR; however, the positive effect on runtime surpassed (A-E). Using (I) dynamic kernels as opposed to not using (J) had a significant runtime boost with PSNR scores almost unchanged. Different heterogeneous divisions of filters (G-H) are also tried. Logically, when the input size of the 3×3 convolution layer increased, the PSNR score also increased. However, the penalty of runtime overcame the positive benefits of the PSNR raise. Replacing cross concatenation layers with 1×1 convolutions (XCAT-B) had the same effect as in the previous case.

Table 3 shows the effect of using cross concatenation instead of directly concatenating the intermediate tensors in HXBlocks, as well as using different tensor divisions and number of HXBlocks. It is proven that using cross-concatenation allows for better information flow and increases the PSNR score, as opposed to using direct concatenation. This effect is more visible when the number of HXBlocks increase.

Fig. 5. Comparative results of UINT8 quantized models on DIV2KVal dataset. The proposed method is applied for (c)'s representative dataset, whereas all DIV2KVal images are used for (d) and (f)'s. (e) and (g) are the pre-trained quantized models provided by the authors. Visual results of (c), (e), and (g) are indistinguishable; however, XCAT runs faster

Table 4. PSNR (dB) drops for DIV2K(Val and Test, x3) before and after quantization, number of parameters, and runtime scores (ms) on Mali-G71 MP2 via AI Benchmark 5 [15] and Synaptics NPU (*FP32 and UINT8 scores are taken from the paper. In addition, the authors' pre-trained .tflite model gave a concatenation error on AI Benchmark 4 & 5, about the source tensor not being able to be used multiple times. Hence, the model code is altered for ABPN, where the relevant tensor is manually hard-copied and concatenated) (+Tested in NCHW format)

Metric	**FSRCNN** [7]	**ESPCN** [30]	**XLSR** [2]	**ABPN** [8]*	**XCAT** (proposed)
Val, FP32 PSNR	29.67	29.54	30.10	30.22	**29.88**
Val, UINT8 PSNR	18.52	17.50	29.82	30.09	**29.81**
ΔPSNR	11.15	12.04	0.28	0.13	**0.07**
Test, UINT8 PSNR	-	-	29.58	29.87	**29.67**
# of parameters	25K	31K	22K	42K	**16K**
Synaptics Runtime+	-	-	44.8	36.9	**8.8**
Mali Runtime	485	363	370	600	**320**
Score	0.003	0.061	210	188	**240**

5 Conclusions and Future Studies

This study proposes a lightweight, quantized single image super-resolution network named XCAT, submitted to *Mobile AI & AIM 2022 Real-Time Image Super-Resolution Challenge*. XCAT offers **heterogeneous group convolution blocks** which includes convolutional kernels with different kernels and input & output tensor sizes. Compared to other studies which include group convolutions ending with 1×1 layers, **cross concatenating** the intermediate tensors between group convolutions offer runtime efficiency with tolerable sacrifice from PSNR test scores. To further increase runtime performance on mobile devices, upsampling done by tensor copy operations by default is replaced by a 1×1 convolutional layer with a non-trainable kernel. XCAT is also shown to be robust to quantization, with a decrease of 0.07dB from FP32 to the UINT8 model.

Comparative experimental results on slightly modified XCAT models reveal that the design choices proposed in this study offer the model to be deployed on mobile devices efficiently. To further prove the effectiveness of the proposed method, XCAT is evaluated with standardized datasets in comparison to other mobile-friendly super-resolution networks. Visual results indicate that XCAT can produce super-resolved images nearly identical to the other slower networks' outputs. Although HXBlock is designed for super-resolution problems, we believe that it can help many heavy models to facilitate running on mobile devices.

References

1. Ahn, N., Kang, B., Sohn, K.: Fast, accurate, and lightweight super-resolution with cascading residual network. CoRR abs/1803.08664 (2018)

2. Ayazoglu, M.: Extremely lightweight quantization robust real-time single-image super resolution for mobile devices. CoRR abs/2105.10288 (2021)
3. Ayazoglu, M.: Imdeception: grouped information distilling super-resolution network (2022)
4. Cogswell, M., Ahmed, F., Girshick, R., Zitnick, L., Batra, D.: Reducing overfitting in deep networks by decorrelating representations (2015)
5. Courbariaux, M., Bengio, Y., David, J.P.: Training deep neural networks with low precision multiplications (2015)
6. Dong, C., Loy, C.C., He, K., Tang, X.: Learning a deep convolutional network for image super-resolution. In: Fleet, D., Pajdla, T., Schiele, B., Tuytelaars, T. (eds.) ECCV 2014. LNCS, vol. 8692, pp. 184–199. Springer, Cham (2014). https://doi.org/10.1007/978-3-319-10593-2_13
7. Dong, C., Loy, C.C., He, K., Tang, X.: Learning a deep convolutional network for image super-resolution. In: Fleet, D., Pajdla, T., Schiele, B., Tuytelaars, T. (eds.) ECCV 2014. LNCS, vol. 8692, pp. 184–199. Springer, Cham (2014). https://doi.org/10.1007/978-3-319-10593-2_13
8. Du, Z., Liu, J., Tang, J., Wu, G.: Anchor-based plain net for mobile image super-resolution (2021)
9. Glasner, D., Bagon, S., Irani, M.: Super-resolution from a single image. In: 2009 IEEE 12th International Conference on Computer Vision, pp. 349–356 (2009)
10. He, Y., Zhang, X., Sun, J.: Channel pruning for accelerating very deep neural networks. In: 2017 IEEE International Conference on Computer Vision (ICCV), pp. 1398–1406 (2017)
11. Hui, Z., Gao, X., Yang, Y., Wang, X.: Lightweight image super-resolution with information multi-distillation network, pp. 2024–2032 (2019)
12. Hui, Z., Wang, X., Gao, X.: Fast and accurate single image super-resolution via information distillation network, pp. 723–731 (2018)
13. Ignatov, A., et al.: Real-time quantized image super-resolution on mobile NPUs, mobile AI 2021 challenge: Report, pp. 2525–2534 (2021)
14. Ignatov, A., Timofte, R., Denna, M., Younes, A., et al.: Efficient and accurate quantized image super-resolution on mobile NPUs, mobile AI & aim 2022 challenge: report. In: Proceedings of the European Conference on Computer Vision (ECCV) Workshops (2022)
15. Ignatov, A., et al.: AI benchmark: all about deep learning on smartphones in 2019 (2019)
16. Ioannou, Y., Robertson, D., Cipolla, R., Criminisi, A.: Deep roots: Improving CNN efficiency with hierarchical filter groups (2017)
17. Jacob, B., et al.: Quantization and training of neural networks for efficient integer-arithmetic-only inference, pp. 2704–2713 (2018)
18. Jain, V., Bansal, P., Kumar Singh, A., Srivastava, R.: Efficient single image super resolution using enhanced learned group convolutions (2018)
19. Kim, J., Lee, J.K., Lee, K.M.: Accurate image super-resolution using very deep convolutional networks. CoRR abs/1511.04587 (2015)
20. Kim, J., Lee, J.K., Lee, K.M.: Deeply-recursive convolutional network for image super-resolution. In: 2016 IEEE Conference on Computer Vision and Pattern Recognition (CVPR), pp. 1637–1645 (2016)
21. Krishnamoorthi, R.: Quantizing deep convolutional networks for efficient inference: a whitepaper (2018)

22. Krizhevsky, A., Sutskever, I., Hinton, G.E.: ImageNet classification with deep convolutional neural networks. In: Pereira, F., Burges, C., Bottou, L., Weinberger, K. (eds.) Advances in Neural Information Processing Systems, vol. 25. Curran Associates, Inc. (2012)
23. Ledig, C., et al.: Photo-realistic single image super-resolution using a generative adversarial network. CoRR abs/1609.04802 (2016)
24. Li, Y., Zhang, K., Timofte, R., Van Gool, L., Kong, E.A.: NTIRE 2022 challenge on efficient super-resolution: methods and results (2022)
25. Lim, B., Son, S., Kim, H., Nah, S., Lee, K.M.: Enhanced deep residual networks for single image super-resolution. CoRR abs/1707.02921 (2017)
26. Lin, M., Chen, Q., Yan, S.: Network in network (2013)
27. Liu, Z., et al.: Swin transformer: hierarchical vision transformer using shifted windows (2021)
28. Niu, B., et al.: Single image super-resolution via a holistic attention network (2020)
29. Schwarz Schuler, J.P., Romaní, S., Abdel-nasser, M., Rashwan, H., Puig, D.: Grouped pointwise convolutions reduce parameters in convolutional neural networks. Mendel **28**, 23–31 (2022)
30. Shi, W., et al.: Real-time single image and video super-resolution using an efficient sub-pixel convolutional neural network. In: 2016 IEEE Conference on Computer Vision and Pattern Recognition (CVPR), pp. 1874–1883 (2016)
31. Tai, Y., Yang, J., Liu, X.: Image super-resolution via deep recursive residual network. In: 2017 IEEE Conference on Computer Vision and Pattern Recognition (CVPR), pp. 2790–2798 (2017)
32. Vanhoucke, V., Senior, A., Mao, M.: Improving the speed of neural networks on CPUs (2011)
33. Wang, X., et al.: ESRGAN: enhanced super-resolution generative adversarial networks (2018)
34. Xie, S., Girshick, R., Dollár, P., Tu, Z., He, K.: Aggregated residual transformations for deep neural networks. In: 2017 IEEE Conference on Computer Vision and Pattern Recognition (CVPR), pp. 5987–5995 (2017)
35. Xu, S., Huang, A., Chen, L., Zhang, B.: Convolutional neural network pruning: a survey. In: 2020 39th Chinese Control Conference (CCC), pp. 7458–7463 (2020)
36. Yu, F., Koltun, V.: Multi-scale context aggregation by dilated convolutions (2016)
37. Yu, J., et al.: Wide activation for efficient and accurate image super-resolution. CoRR abs/1808.08718 (2018)
38. Zhang, T., Qi, G.J., Xiao, B., Wang, J.: Interleaved group convolutions. In: 2017 IEEE International Conference on Computer Vision (ICCV), pp. 4383–4392 (2017)
39. Zhang, X., Zhou, X., Lin, M., Sun, J.: ShuffleNet: an extremely efficient convolutional neural network for mobile devices. In: 2018 IEEE/CVF Conference on Computer Vision and Pattern Recognition, pp. 6848–6856 (2018)
40. Zhang, Z., Wang, X., Jung, C.: DCSR: dilated convolutions for single image super-resolution. IEEE Trans. Image Process. **28**(4), 1625–1635 (2019)
41. Zhao, R., Hu, Y., Dotzel, J., De Sa, C., Zhang, Z.: Building efficient deep neural networks with unitary group convolutions, pp. 11295–11304 (2019)
42. Zoph, B., Vasudevan, V., Shlens, J., Le, Q.V.: Learning transferable architectures for scalable image recognition. In: 2018 IEEE/CVF Conference on Computer Vision and Pattern Recognition, pp. 8697–8710 (2018)

Learned Reverse ISP with Soft Supervision

Beiji Zou[1,2(✉)] and Yue Zhang[1,2]

[1] School of Computer Science and Engineering, Central South University, Changsha 410083, China
{bjzou,yuezhang}@csu.edu.cn
[2] Hunan Engineering Research Center of Machine Vision and Intelligent Medicine, Changsha 410083, China

Abstract. RAW image serves as the foundation for camera imaging, which resides at the very beginning of the pipeline that generates sRGB images. Unfortunately, owing to special considerations, the information-rich RAW images are forfeited by default in most existing applications. To regain the RAW image, some works attempt to restore RAW images from RGB images. They focus on designing handcrafted model-based methods or complicated networks, however, ignoring the special property of RAW image, *i.e.*, high dynamic range. To make up for this deficiency, we introduce a novel soft supervision, derived from the high dynamic range. Specifically, we propose to soften the original ground-truth as a multivariate Gaussian distribution so that networks could learn much more information. Then, we introduce a soft supervision driven network (SSDNet), based on convolution and transformer, for effectively restoring RAW images from RGB images. Quantitative and qualitative results show the promising restoration performance of RGB-to-RAW. In particular, our method achieved fifth place in the S7 track of AIM Reversed ISP Challenge. The source code will be available at https://github.com/yuezhang98/Learned-Reverse-ISP-with-Soft-Supervision.

Keywords: Reversed ISP · Soft supervision · Convolution · Transformer

1 Introduction

RAW image is single-channel raw data obtained from CMOS, which is linearly related to the scene irradiance. It is worth noting that the RAW image usually has a high dynamic range of at least 2^{10}, which provides more information and is helpful for image processes. However, due to various limitations, such as storage space, most devices have to give up saving RAW images and keep the sRGB image instead processed by the image signal processor (ISP). As a consequence,

Supplementary Information The online version contains supplementary material available at https://doi.org/10.1007/978-3-031-25063-7_30.

L. Karlinsky et al. (Eds.): ECCV 2022 Workshops, LNCS 13802, pp. 489–506, 2023.
https://doi.org/10.1007/978-3-031-25063-7_30

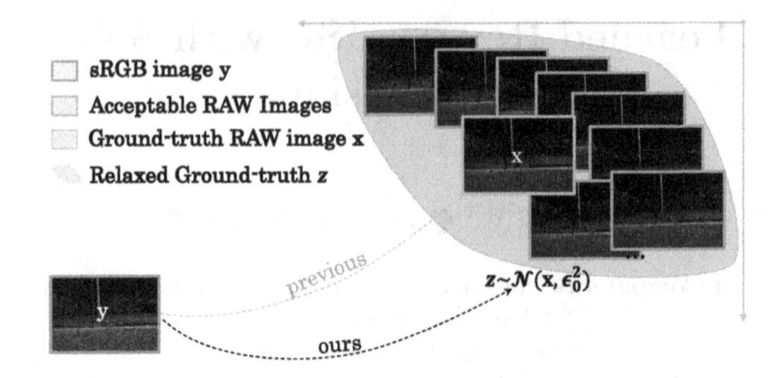

Fig. 1. The proposed soft supervision for learned reversed ISP. Previous methods usually enforce the network to learn the given ground-truth, while our method enables a wide range of acceptable supervision sampled from the relaxed multivariate distribution.

this complicates some important operations, such as image denoising, image enhancement, HDR and super-resolution [1,17,28,30,36,46]. For example, for denoising in the RAW domain, the statistical characteristics of the noise can be obtained by calibration, resulting in high performance noise reduction and $\frac{1}{3}$ less computation. Meanwhile, in the RGB domain, multiple nonlinear operations inside the ISP make the noise complex and more difficult to remove. Therefore, in this work, we focus on RGB-to-RAW mapping and enable it to be beneficial for image denoising.

Attracted by the beneficial attributes of RAW images, some works carry out the research on RGB to RAW images. Pioneeringly, Nguyen et al. [31,32] propose a fast breadth-first-search octree algorithm for finding the necessary control points to provide a mapping between the RGB and RAW colour spaces. Then, they demonstrate better practicality by using the recovered RAW for white balance correction and image deblurring. However, such simplified mapping ignores the complex processing flow inside the ISP, such as demosaicing, tone mapping, etc. To bridge this gap, Brooks et al. [5] propose a generic camera ISP process with five typical stages, where each step is approximated by an invertible function. This approach is fully interpretable and makes the ISP flow design more flexible. They apply the proposed algorithm in image denoising. Although it provides a better idea involving ISP, the process of ISP in practice is often more sophisticated and the limited parameters of the interface open to the public also restrict related research.

Recently, complementary learning-based approaches [13,34,48,51] have been proposed to alleviate this challenge. CycleISP [51] considers *cycle consistency* to learn the forward (RAW to RGB) and reverse (RGB to RAW) directions of ISP. They employ two different networks, which are trained end-to-end, to effectively assist the denoising task using converted RAW images. In the meantime, InvISP [48] enables RAW to RGB and RGB to RAW mapping by building on a single normalizing-flow-based invertible neural network [25]. However, both of them require a large amount of training data, which is a challenge because such datasets are difficult to collect. To address this problem, Conde et al. [13] propose a hybrid approach based on dictionary learning to achieve effective RGB to RAW, which maintains the advantages of both model-based and end-to-end learnable approaches. Unfortunately, these methods only focus on model design, in terms of either restoration performance or light-weight model, while discounting the characteristics of RAW image, namely the high dynamic range.

In this work, we introduce a novel soft supervision driven network (SSDNet) for RGB-to-RAW mapping. Specifically, we utilise the idea of partitioning to explore the RGB to RAW task systematically from the data and model perspectives respectively. Firstly, we pick up the missing piece of data trait, namely the high dynamic range, and propose to relax the given supervision. Although the high dynamic range of RAW allows for a more detailed representation of the data, it may increase the difficulty of network learning. To alleviate this problem, we propose to relax the original supervision to a multivariate Gaussian distribution, as shown in Fig. 1. This brings two benefits: 1) the learning goal of the network is softened, simplifying the learning process; 2) the network could capture pixels around the given supervision during the training process, subtly performing data augmentation. We propose new loss functions based on the soft supervision that enable the network to further cope with the learning of multivariate Gaussian distributions. Secondly, we construct an encoder-decoder structure that incorporates convolution and Transformer [44]. As is well known, the early prevalent convolutional networks have a superior ability to perceive local structure. Besides, according to previous works [27,35,43], large receptive fields are favourable for image reconstruction. Inspired by the recent Transformer architecture [19,50], we made an organised combination of convolution and Transformer blocks, allowing the network to capture not only local details, but also the overall image pattern. Equipped with these two components, our SSDNet achieves state-of-the-art RGB-to-RAW mapping in Adobe FiveK dataset [48]. In particular, our method achieved fifth place in the S7 track of AIM Reversed ISP Challenge [14]. In addition, we have also demonstrated the effectiveness of the proposed method on image denoising.

In conclusion, our main contributions are as follows:

- We present a new learning target for RGB-to-RAW mapping. The proposed soft supervision enables SSDNet to make effective use of data characteristics, resulting in better recovering performance.
- We present a novel network, SSDNet, that effectively fuses convolution and transformer, which is able to capture local and global feature for better restoration performance.
- Extensive experiments show that our method achieves state-of-the-art RAW restoration from RGB image. Our method also presents as a top solution in the novel AIM Reversed ISP Challenge [14]. Additionally, our methods could also benefit the downstream task, *e.g.*, image denoising.

2 Related Work

2.1 Reversed ISP

ISP aims to convert the RAW data acquired from CMOS into natural RGB images. It involves a wide variety of designs and complex non-linear processes.

Therefore, RAW-to-RGB is an irreversible operation [48]. Furthermore, RAW images in practice applications are usually not preserved by default, due to their large amount of data. As technology evolves, realistic scenarios demand higher quality imaging via ISPs [16]. In particular, it is difficult to make considerable enhancements to an already converted RGB image. For this reason, some work has been initiated to investigate the mapping of RGB to RAW, with the expectation of enabling enhancement on RAW to acquire high quality images in the end. Nguyen et al. [31,32] proposed a fast breadth-first-search octree algorithm to provide a mapping between the RGB and RAW colour spaces. Then, Brooks et al. [5] complemented their research by taking ISP's internal modules into account. However, as ISP flows were inherently irreversible, this substitution of reversible functions for each of the ISP modules was inevitably introducing unreasonable errors. Fortunately, the rapid development of deep learning offers new ideas for this research [13,34,48,51]. For example, CycleISP [51] and InvISP [48] implemented RGB to RAW and RAW to RGB using deep networks. Yet, the desire for large amounts of training data limits their application. To address this problem, Conde et al. [13] proposed to use dictionary learning methods to replace important modules inside ISP, maintaining the advantages of both model-based and end-to-end learnable approaches. These studies of RGB to RAW have achieved promising results, in that they have neglected the special characteristics of RAW data, namely the high dynamic range. In this work, we shall pick up this piece and exploit it to enhance the performance of our proposed network.

2.2 Image Denoising

Image denoising is a fundamental and critical task, being required by various imaging systems as well as the pre-processing step for high-level vision algorithms. In recent years, deep learning-based image denoising has become popular in current research. The work on image denoising in growing numbers prefers deep CNNs and has achieved remarkable performance. DnCNN [53] utilized a convolutional neural network with 17 layers equipped with Batch Normalization and the concept of residual learning into image denoising. Its denoising performance outperforms all conventional algorithms, including BM3D [15]. This shows the immense potential of the CNN model in image denoising. Subsequently, more work focused on designing a wider and deeper network structure to achieve better denoising performance [54]. However, most of the previous work is designed for additive Gaussian white noise (AWGN), which is too simple and deviates from real scenes. To bridge this gap, Guo et al. [20] proposed CBDNet, which simulates various camera response functions. RIDNet [3] further introduced the feature attention mechanism to achieve impressive denoising performance on both synthetic and real-world datasets. Yue et al. [49] proposed Variational Denoising Network (VDN), exploiting the variational inference technique that enables VDN to simultaneously learn noise distributions. They try to enhance the performance of VDN by leveraging accurate noise information. In the account of the further prevalence of attention mechanisms, MIRNet [52] learned rich features through

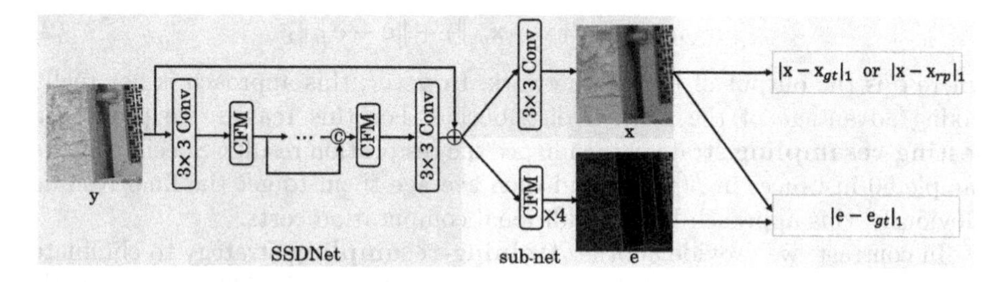

Fig. 2. The learning scheme for the proposed soft supervision. For an existed RGB-to-RAW network, we attach a sub-network in the end to learn the variance term in the proposed soft supervision. Practically, the sub-network is constructed by a four-CFM (Context Fusion Module), introduced in Sect. 3.2.

parallel multi-resolution convolutional flow as well as spatial attention and channel attention mechanisms. This network demonstrates the further potential of attentional mechanisms on image denoising. NBNet [12] further improved the denoising performance based on this prior as well.

In contrast to direct denoising in RGB images, we expect to achieve an earlier stage of noise reduction using RGB to RAW techniques. There are similar works to ours, CycleISP [51] and UPI [5]. But they ignore the attribute of RAW image, and this work is going to bridge that gap.

3 Proposed Method

3.1 Soft Supervision

We first exploit the characteristics of the training data, especially the high dynamic range. In contrast to previous methods [13,34,48,51] that struggle to allow the network to regress the established ground-truth, we propose to learn the soft supervision.

Given an sRGB image $\mathbf{y} \in \mathbb{R}^{W \times H \times 3}$, we apply SSDNet to get the reversed RAW image $\mathbf{x} \in \mathbb{R}^{W \times H}$. At the same time, we can also obtain the corresponding absolute error map $\mathbf{e}_{gt} = |\mathbf{x} - \mathbf{x}_{gt}|$, where $\mathbf{x}_{gt} \in \mathbb{R}^{W \times H}$ is the ground-truth. In fact, the error map contains additional information about the ground-truth, usually being neglected in previous methods. To further exploit this information, we relax the original ground-truth in conjunction with this error map into a multivariate Gaussian distribution, as soft supervision $\mathbf{x}_{sp}$, so that the network can fully capture the supervision information.

$$\mathbf{x}_{sp} \sim \mathcal{N}(\mathbf{x}_{gt},\ \mathbf{e}_{gt}) \tag{1}$$

Equation 1 contains all the supervisory information, even the recoverability of the network itself. To learn $\mathbf{x}_{sp}$, a simple approach is attaching a sub-network in the network, as shown in Fig. 2, and then using L1 loss enable the whole network to learn the mean and variance, i.e. the given ground-truth $\mathbf{x}_{gt}$ and the error map $\mathbf{e}_{gt}$.

$$\arg\min \ \mathcal{L} = \|\mathbf{x} - \mathbf{x}_{gt}\|_1 + \|\mathbf{e} - \mathbf{e}_{gt}\|_1 \tag{2}$$

where $\mathbf{e}$ is the output of the sub-network. However, this approach is not really taking advantage of the learned distribution. For this reason, we propose a **testing-resampling** strategy to enhance the restoration results. Specifically, we sample 50 instances in $\mathcal{N}(\mathbf{x},\ \mathbf{e})$ and then average them to get the final results. Obviously, this approach brings additional computation costs.

In contrast, we provide another **training-resampling** strategy to eliminate the above problems. We resort to the reparameterization trick [22,24] to resample the supervision in the distribution as new supervision, $\mathbf{x}_{rp}$.

$$\mathbf{x}_{rp} = \mathbf{x}_{gt} + \mathbf{z} * \mathbf{e}_{gt},\ \ \mathbf{z} \sim \mathcal{N}(\mathbf{0},\ \mathbf{1}) \tag{3}$$

In this way, SSDNet would capture not a given supervision during training, but a large number of possible samples existing in the soft supervision $\mathbf{x}_{sp}$. Practically, we utilize the following loss function:

$$\arg\min \ \mathcal{L} = \|\mathbf{x} - \mathbf{x}_{rp}\|_1 + \alpha * \|\mathbf{e} - \mathbf{e}_{gt}\|_1 \tag{4}$$

where α is the hyper-parameter to balance learning targets. During training, we would attach a sub-network on the main network, the same as Fig. 2. While testing, the sub-network could be simply deprecated for saving memory.

3.2 SSDNet

Throughout the existing RGB-to-RAW models, we note that the existing reversed ISP models are pure CNNs [13,34,48,51], which have a limited receptive field that suffers from the limited restoration performance. Recently, transformer-based image restoration networks [10,50] have also demonstrated impressive performance benefiting from their global receptive fields, although they are again limited by weak local modelling capabilities [9,19,21]. To address the shortcomings of these two architectures, we propose a hybrid module based on the large kernel convolution and self-attention, the context fusion module(CFM). Then, we take CFM as a basic block to build SSDNet. To illustrate it clearly, we first introduce the overall network and then elaborate on the proposed CFM.

Overall Architecture . The overview architecture is shown at the top of Fig. 3. The SSDNet is based on a symmetrical UNet [38] architecture. SSDNet has three encoder stages and three corresponding decoder stages. Given an RGB image $\mathbf{y} \in \mathbb{R}^{\mathrm{W}\times\mathrm{H}\times 3}$, the network first applies a 3×3 convolutional layer to project the image into feature space. Then, SSDNet first encodes the projected features. Particularly, at the end of each encoder, the feature maps, $\mathbf{X} \in \mathbb{R}^{\mathrm{H}\times\mathrm{W}\times\mathrm{C}}$, are downsampled to $\frac{1}{2}\times$ scale, $\mathbf{X} \in \mathbb{R}^{\frac{1}{2}\mathrm{H}\times\frac{1}{2}\mathrm{W}\times 2\mathrm{C}}$, with a 3×3 convolutional layer and PixelShuffle [41]. After encoding, the final high-level features are decoded to the original feature space. More precisely, the feature maps are up-sampled to $2\times$ scale with a 3×3 convolutional layer and PixelUnShuffle operation before

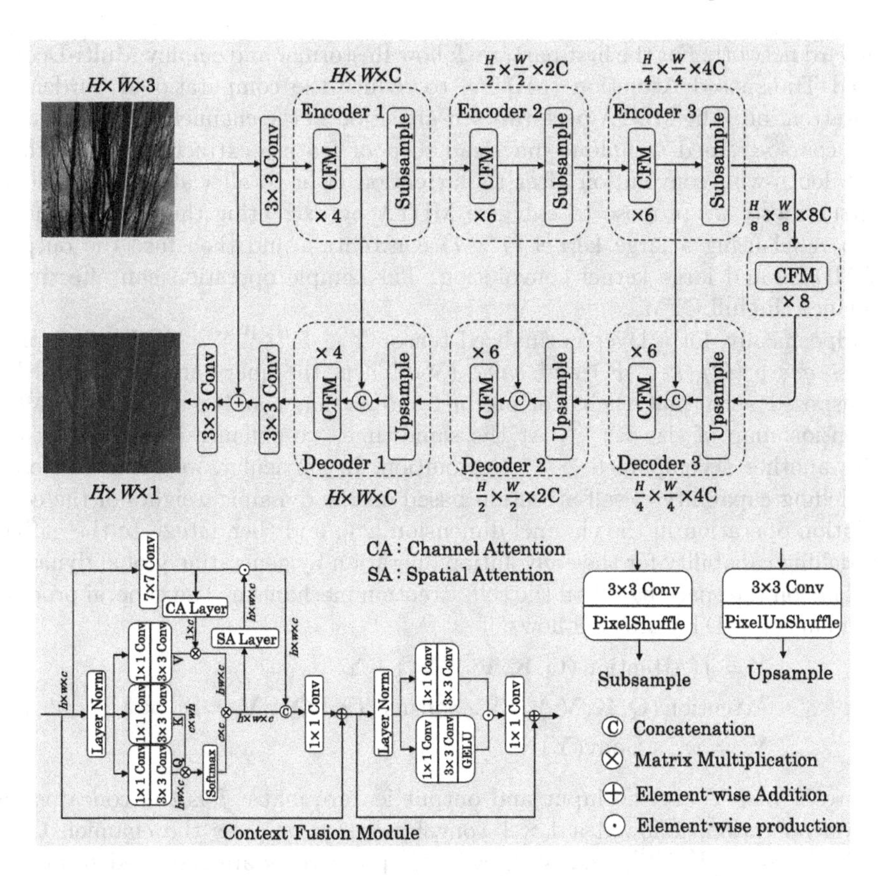

Fig. 3. The architecture of the proposed SSDNet. The proposed SSDNet has a multi-scale hierarchical design and it is made up of the Context fusion module (CFM). The CFM consists of an efficient mixture of the convolution and Transformer blocks as well as the simplified Gated-Dconv Feed-Foreard Network (GDFN) [50].

each decoder stage. This layer reduces the feature channels by half, identical to the inverse operation in the encoder stage. After that, skip-connection passes the low-level feature maps from the corresponding encoder stage. The up-sampled feature and the encoder's feature jointly compose the input to the matching decoder. Then, the recovered feature is re-projected to the RGB image space, $\mathbf{r} \in \mathbb{R}^{W \times H \times 3}$, by another 3×3 convolutional layer. An RGB residual map is obtained by $\mathbf{y}_r = \mathbf{y} + \mathbf{r}$. At last, a 3×3 convolutional layer is applied to compact the RGB residual map $\mathbf{y}_r$ and output a single-channel RAW image, $\mathbf{x} \in \mathbb{R}^{W \times H \times 1}$. Its basic building blocks of the encoder and decoder follow the same Context Fusion Module (CFM).

Context Fusion Module. As shown in the left bottom of Fig. 3, CFM has two stages, respectively the convolution-transformer mixed module and the feed forward network. For the first part, we follow Restormer and employ Multi-Dconv Head Transposed Attention (MDTA) to reduce the computational burden of self-attention. The MDTA performs self-attention in the channel dimension, and this coarse-grained technique may lead to poor feature extraction. Thus, they add depth-wise convolution after the attention layer to alleviate this problem. Furthermore, we propose to enhance MDTA by extracting the input features in parallel using a large kernel (7×7) convolution and then fuse the output of MDTA and large kernel convolution. This simple operation can effectively enhance the full CFM.

Specifically, for a layer normalized tensor $\mathbf{Y} \in \mathbb{R}^{H \times W \times C}$, MDTA first produces *query* ($\mathbf{Q}$), *key* ($\mathbf{K}$) and *value* ($\mathbf{V}$). Then, the query and key would be transposed such that their dot-product interaction could generate a smaller attention map of size $\mathbb{R}^{C \times C}$. At the same time, we enhance the input future with another accompanied 7×7 convolution. In particular, we bring the local modelling capability of self-attention based on the dynamic weights of the convolution operation in the channel dimension [23]; and then integrate the global modelling capability for the convolution operation by generating global dynamic weights on the space based on the self-attention mechanism. The general process of enhanced MDTA are as follows:

$$\begin{aligned}
&\hat{\mathbf{Y}} = f(\text{Attention}\,(\mathbf{Q}, \mathbf{K}, \mathbf{V}_c),\, \mathbf{Y}_s) + \mathbf{Y}, \\
&\text{Attention}\,(\mathbf{Q}, \mathbf{K}, \mathbf{V}_c) = \mathbf{V} \cdot \text{Softmax}\,(\mathbf{K} \cdot \mathbf{Q})\,,\ \mathbf{V}_c = w_c \times \mathbf{V} \\
&\mathbf{Y}_s = w_s \times \text{Conv}(\mathbf{Y})
\end{aligned} \tag{5}$$

where $\mathbf{Y}$ and $\hat{\mathbf{Y}}$ are the input and output feature maps; f is the concatenate fusion operation, followed a 1×1 convolution to compress the channel; $\mathbf{Q} \in \mathbb{R}^{HW \times C}$, $\mathbf{K} \in \mathbb{R}^{C \times HW}$ and $\mathbf{V} \in \mathbb{R}^{HW \times C}$ projections are reshaped from the original size $\mathbb{R}^{H \times W \times C}$. $\mathbf{V}_c$ and $\mathbf{Y}_s$ are respectively enhanced by the dynamic channel weight w_c and the dynamic spatial weight w_s.

For the second part, feed-forward network, we simplify the original Gated-Dconv Feed-Forward Network (GDFN) [50]. Specifically, we take compressed depth-wise convolutions to reduce the computational cost. Overall, the computational process are as follows:

$$\begin{aligned}
\widetilde{\mathbf{Y}} &= \text{Gating}\left(\hat{\mathbf{Y}}\right) + \hat{\mathbf{Y}}, \\
\text{Gating}(\hat{\mathbf{Y}}) &= \text{GELU}(W_c(\text{LN}(\hat{\mathbf{Y}}))) \odot W_c(\text{LN}(\hat{\mathbf{Y}})),
\end{aligned} \tag{6}$$

where $\odot$ is element-wise multiplication, W_c represents the compressed convolution, and LN denotes the layer normalization [4]. Equipped with EMDTA (enhanced EMDA) and SGDFN (simplified GDFN), CFM could effectively capture the fine-grained details. Especially, for soft supervision learning, we apply a sub-network with four-CFMs attached on SSDNet, as shown in Fig. 2.

4 Experimental Results

In this section, we perform extensive experiments to quantitatively and qualitatively verify the effectiveness of our method to perform RAW image reconstruction. Two objective metrics were used in the quantitative evaluation, including Peak Signal to Noise Ratio (PSNR) and Structural Similarity (SSIM) [45]. PSNR computes the peak signal power to reconstruction error power ratio, while SSIM measures the structural similarity between the reconstructed image and the supervised image. Without loss of generality, we employ RGGB in the Bayer CFA pattern for all our experiments. Certainly, our algorithm can be easily extended to other Bayer patterns, such as RGBG. In addition, we have also conducted experiments on a downstream task, RAW image denoising, to verify the usefulness of the learned inverse ISP model beyond RAW image reconstruction.

4.1 Experimental Setup

Datasets

AIM Reversed ISP-S7 Dataset [14]. AIM Reversed ISP Challenge aims at creating a solution for recovering the camera's RAW with the corresponding RGB images processed by the in-camera ISP. The competition expects that such a solution should generate reasonable RAW images and by doing so, other downstream tasks, such as denoising, super-resolution or colour invariance, can benefit from the generation of such synthetic data. To this end, the competition organisers have collected a number of pair pairs (4320 for training, 480 for testing), taken from a Samsung S7 phone [40]. Especially, the RAW images were captured by IMX260 sensor with GRBG Bayer pattern.

MIT-Adobe FiveK Dataset. We utilise the training-test data from [48], which were collected from the MIT-Adobe FiveK dataset [8] for the Canon EOS 5D subset (777 image pairs) and the Nikon D700 subset (590 image paris). This dataset randomly divides the two sets of data (Canon, Nikon) into training and test groups in a ratio of 85:15, respectively. Following [13,48], we use the LibRaw library to render ground-truth sRGB images from the RAW images.

SIDD Dataset [1]. Real noise in real scenes is more challenging to deal with, therefore this dataset provides realistic noisy images, including both RAW sensor data and sRGB data. These images were captured by five smartphone cameras, under different lighting conditions, poses and ISO levels. The collected images have more noise compared to DSLR images and are due to the small aperture and sensor limitations. This dataset is available in 320 pairs of ultra-high resolution images (e.g. 5328×3000) for training and 1280 pairs of images for validation. This work takes SIDD [1] to investigate the application of learned RAW image reconstruction.

Implementation Details
We train our SSDNet in an end-to-end manner and do not perform any pre-training process. In the training stage, we use AdamW [37] optimizer with

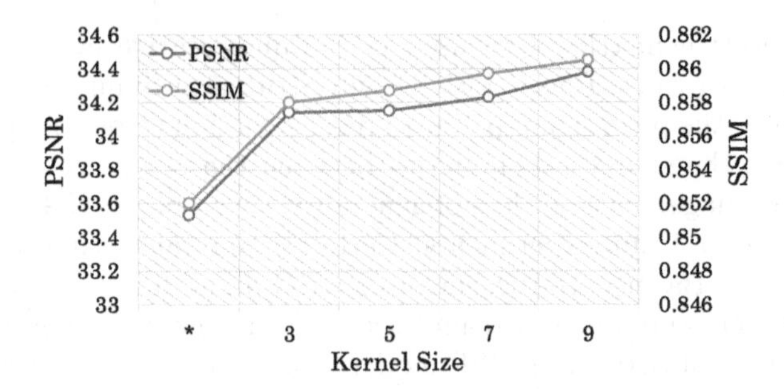

Fig. 4. Ablation study on the convolutional branch in CFM, especially the influence of its kernel size. '$*$' means CFM does not incorporate the convolutional branch and the two interactions.

momentum terms (0.9, 0.999). All the models are trained with 276,000 iterations. The initial learning rate is 3×10^{-4}, and it remains constant for the first 92,000 iterations and decreases to 1×10^{-6} for the next 184,000 iterations with the cosine annealing strategy. We apply the random rotation and flipping for data augmentation. We train the network using a progressive patch size growing strategy [50]. Specifically, the network is initially trained with 128×128 image patches. During the iterations, the patch size is increased in steps of 64 to $(192, 256, 320, 384)$ in $(92K, 156K, 204K, 240K)$ iterations. In addition, to save memory, the batch size is reduced from the initial 64 to $(32, 16, 8, 8)$ in line with the increasing patch size. Especially, for **Training-resampling** strategy, we empirically set $\alpha = 0.001$. Our experiments are conducted on 8× 3090 Ti GPUs, typically taking about 4 days to complete the training.

4.2 Ablation Studies

Convolutional Branch in CFM. Here, we explore the influence of the convolutional branch in CFM, including its kernel size. As shown in Fig. 4, the convolutional branch brings at least 0.5dB enhancement. Furthermore, our model yields better performance as the convolutional kernel size increases[1]. We consider that the larger the convolution kernel, the larger the effective receptive field [29], which results in more reliable dynamic weights. To balance computational cost and performance, we finally choose the 7×7 kernel size.

Interactions in CFM. We verify the importance of the interaction operations, respectively the channel attention and spatial attention for Self-attention and Convolutional branches. As shown in Table 1, both of these interactions are effective in improving the restoration performance. Top performance is achieved in combination, with a 0.7dB improvement over the baseline, which demonstrates

[1] Due to the limitation of computational resources, we have applied the convolution operation with maximum 9×9 kernel size.

Table 1. Ablation study on the interactions in CFM with the kernel size 7 in the convolutional branch. CA means channel attention, SA means spatial attention. 0st model is the baseline without convolutional branch and any interaction.

Interactions	0st	1st	2nd	3rd	4th
CA?	-	✗	✓	✗	✓
SA?	-	✗	✗	✓	✓
PSNR	33.53	33.89	34.03	34.18	**34.23**
SSIM	0.8520	0.8576	0.8580	0.8594	**0.8596**

Table 2. Ablation study on the Soft Supervision. '$*$' means training and testing models in a traditional fashion. 50 or 100 in **Test-resampling** indicates that 50 or 100 instances are sampled to average.

Strategy	*	Test-resampling(50)	Test-resampling(100)	Training-resampling
PSNR/SSIM	34.23/0.8596	34.94/0.8651	**34.95/0.8651**	34.45/0.8624

the importance of interactions. Especially, the spatial interaction shows greater importance than channel interaction, which means introducing local information modelling capability to the self-attention module is much easier and more effective than introducing global modelling capability to the convolution. We suppose that the self-attention module has a larger potential for adaptation than convolution.

Soft Supervision. Soft supervision is another critical element in delivering the effectiveness of SSDNet. This section focuses on verifying its validity, involving two sampling methods, namely Testing-resampling and Training-resampling. As shown in Table 2, our soft supervision brings at least 0.22dB gain for RAW restoration. The Testing-resampling strategy performs better than the Training-resampling strategy. We assume that this gap is derived from the robustness of an ensemble-like **Test-resampling** strategy (it averages 50 outputs as the final restored image). Since this strategy imposes an additional computational burden that is not as simple and elegant as the end-to-end **Training-resampling** strategy. Therefore, we adopt the training-resampling strategy by default in subsequent experiments.

4.3 RAW Image Reconstruction

In this section, we show the experimental results on RAW image reconstruction, especially the performance on AIM Reversed ISP Challenge [14] and a general benchmark, MIT-fiveK [8,48].

Table 3. Quantitative RAW Reconstruction results at the final test sets of AIM Challenge on Reversed ISP - Track1 S7 [14]. All our results are boosted by self-ensemble strategy [26]. Best results are in **bold**. Ours are underlined.

Method	Test1		Test2		Params	Runtime	GPU
	PSNR	SSIM	PSNR	SSIM	(M)	(ms)	
NOAHTCV	**31.86**	**0.83**	**32.69**	**0.88**	5.6	25	V100
MiAlgo	31.39	0.82	28.56	0.85	4.5	18	2080
CASIA LCVG	30.19	0.81	31.47	0.86	464	31	A100
HIT-IIL	29.12	0.80	29.98	0.87	116	19818	V100
CS^{2}U (Ours)	29.13	0.79	29.95	0.84	105	1300	3090
SenseBrains	28.36	0.80	30.08	0.86	69	50	V100
PixelJump	28.15	0.80	n/a	n/a	6.64	40	3090
HiImage	27.96	0.79	n/a	n/a	11	200	3090
0noise	27.67	0.79	29.81	0.87	86	6	2080
OzU VGL	27.89	0.79	28.83	0.83	2.8	10	3090
CVIP	27.85	0.80	29.50	0.86	0.17	19	Q6000

Table 4. Quantitative RAW Reconstruction evaluation among our model and other baselines on MIT-fiveK [8,48]. Bese results are in **bold**.

Method	Nikon/PSNR	Canon/PSNR
UPI [5]	29.30	–
CycleISP [51]	29.40	31.71
InvGrayscale [47]	33.34	34.21
U-Net	38.24	41.52
Invertible-ISP (w/o JPEG) [48]	43.29	45.72
Invertible-ISP (with JPEG Fourier) [48]	44.42	46.78
Learnable Dictionaries [13]	43.62	50.08
Ours	**47.84**	**53.60**

AIM Reversed ISP Challenge

We submitted a result obtained via the proposed method to the AIM Reversed ISP challenge [14]. In order to exploit the potential performance of our method to the maximum, we simultaneously employed self-ensemble strategy [26] during the testing phase. As shown in Table 3, our final submitted model achieved 35.31dB PSNR on the validation set. And our method achieves the fifth place on the test set with 31.02dB PSNR.

We show qualitative results in Fig. 5, comparing with the results of other 10 participating teams in the competition. Our model achieves promising RAW

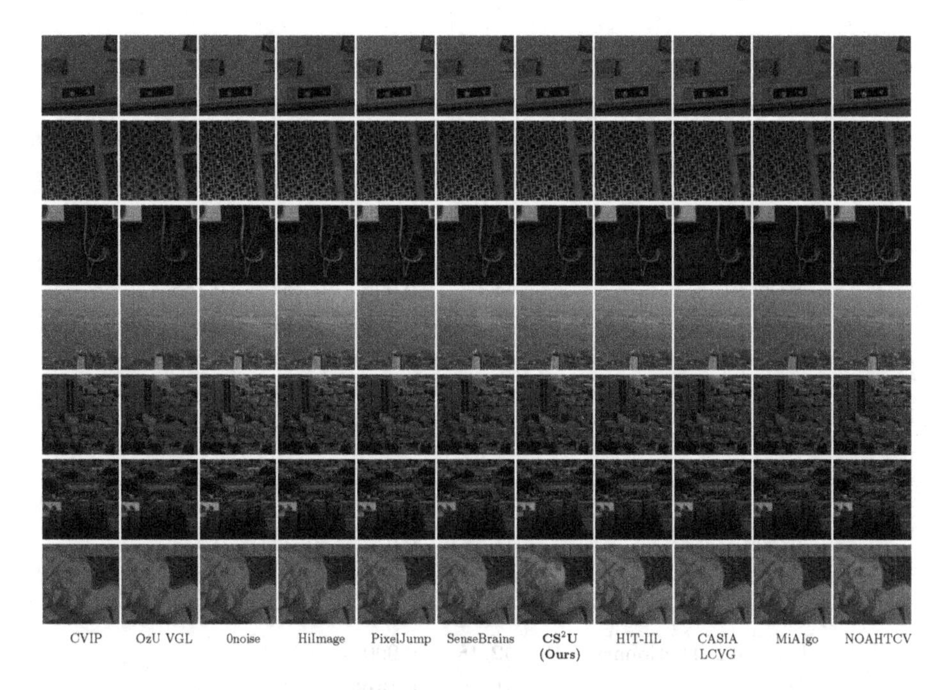

Fig. 5. RAW image restoration for the AIM 2022 Challenge [14]. Individual modules effectively enhance the restoration results. All visualisations are rendered by a simple ISP model (https://github.com/mv-lab/AISP).

recovery results, and compared to the top ones, the differences are rarely noticeable. Similarly, our algorithms enable effective restoration for challenging regions, especially the rich textures, as shown in the second line of Fig. 5.

General RAW Reconstruction Benchmark
We compared our SSDNet with existing inverse ISP methods, including UPI [5], CycleISP [51], InvGrayscale [47], Invertible-ISP (w/o JPEG) [48] and Learnable Dictionaries [13]. We used the same training dataset as these methods, with PSNR and SSIM scores reported by [13]. Quantitative comparisons with existing inverse ISP methods are shown in Table 4. Our SSDNet has achieved comparable performance with previous state-of-the-art methods, even far outperforming previous state-of-the-art methods by 4.2 dB on the Nikon dataset. This clearly demonstrates the effectiveness of the proposed SSDNet.

Discussions
Although we have achieved promising results for RGB to RAW, the network structure used nevertheless exhibited a large number of parameters, as shown in the right side of Table 3. We believe that combining the proposed learning paradigm with the model-based method [13] would alleviate this problem and

Table 5. RAW denoising results on the SIDD dataset [1]. Best results are in bold.

Method	RAW		sRGB	
	PSNR ↑	SSIM ↑	PSNR ↑	SSIM ↑
EPLL [55]	40.73	0.935	25.19	0.842
GLIDE [42]	41.87	0.949	25.98	0.816
TNRD [11]	42.77	0.945	26.99	0.744
MLP [7]	43.17	0.965	27.52	0.788
KSVD [2]	43.26	0.969	27.41	0.832
NLM [6]	44.06	0.971	29.39	0.846
WNNM [18]	44.85	0.975	29.54	0.888
BM3D [15]	45.52	0.980	30.95	0.863
FoE [39]	45.78	0.966	35.99	0.90
DnCNN [53]	47.37	0.976	38.08	0.935
N3Net [33]	47.56	0.976	38.32	0.938
UPI [5]	48.89	0.982	**40.17**	**0.962**
CycleISP [51]	52.41	0.990	39.47	0.918
Learnable Dictionaries [13]	**52.48**	0.990	–	–
Ours	51.73	**0.993**	39.33	0.955

even further improve their performance, which would also be an interesting piece of future work.

4.4 Application to Image Denoising

In this section, we leverage our learned RAW reconstruction model to RAW denoising. We follow CycleISP [51], firstly generating synthetic data using the inverse ISP model and then using them to train the denoiser. Specifically, we synthesize 100K noisy-clean image pairs, and then we modify the SSDNet, by only changing the input channel to 4, to be a RAW denoiser. We report our denoising results in Table 5 and compare them with the state-of-the-art RAW denoising methods available. Note CycleISP [51] was trained on 1 million images, while our model only trained with 10% data. Still, Our method has achieved comparable results with the latest methods [13,51], which demonstrates the application of our inverse ISP model to the downstream task, RAW image denoising.

5 Conclusions

In this work, we propose a new learning target and an effective network for data-driven RGB to RAW learning. We first exploit the unique property of RAW image, i.e. high dynamic range, by suggesting relaxing the supervision to a multivariate Gaussian distribution in order to learn images that are reasonable for a given supervision. Then, we propose the encoder-decoder architecture SSDNet, which is inspired by the Transformer architecture and is straightforward and effective. Combined with the above two components, our method achieves an effective RGB to RAW mapping. Experimental results show that our algorithm achieves promising results on both synthetic and real datasets available. In particular, our method achieved the fifth place in the S7 track of AIM Reversed ISP Challenge. At last, we demonstrate the application of RGB to RAW on a denoising task, implying that this research can be a valid tool for RAW image denoising.

Acknowledgment. This work was supported by the National Key R&D Program of China under Grant 2018AAA0102100.

References

1. Abdelhamed, A., Lin, S., Brown, M.S.: A high-quality denoising dataset for smartphone cameras. In: Proceedings of the IEEE Conference on Computer Vision and Pattern Recognition, pp. 1692–1700 (2018)
2. Aharon, M., Elad, M., Bruckstein, A.: K-SVD: an algorithm for designing overcomplete dictionaries for sparse representation. IEEE Trans. Signal Process. **54**(11), 4311–4322 (2006)
3. Anwar, S., Barnes, N.: Real image denoising with feature attention. In: Proceedings of the IEEE/CVF International Conference on Computer Vision, pp. 3155–3164 (2019)
4. Ba, J.L., Kiros, J.R., Hinton, G.E.: Layer normalization. arXiv:1607.06450 (2016)
5. Brooks, T., Mildenhall, B., Xue, T., Chen, J., Sharlet, D., Barron, J.T.: Unprocessing images for learned raw denoising. In: Proceedings of the IEEE/CVF Conference on Computer Vision and Pattern Recognition, pp. 11036–11045 (2019)
6. Buades, A., Coll, B., Morel, J.M.: A non-local algorithm for image denoising. In: 2005 IEEE Computer Society Conference on Computer Vision and Pattern Recognition (CVPR2005), vol. 2, pp. 60–65. IEEE (2005)
7. Burger, H.C., Schuler, C.J., Harmeling, S.: Image denoising: can plain neural networks compete with BM3D? In: Proceedings of the IEEE Conference on Computer Vision and Pattern Recognition, pp. 1256–1272 (2012)
8. Bychkovsky, V., Paris, S., Chan, E., Durand, F.: Learning photographic global tonal adjustment with a database of input/output image pairs. In: CVPR 2011, pp. 97–104. IEEE (2011)
9. Chen, Q., et al.: MixFormer: mixing features across windows and dimensions. In: Proceedings of the IEEE/CVF Conference on Computer Vision and Pattern Recognition, pp. 5249–5259 (2022)
10. Chen, X., Wang, X., Zhou, J., Dong, C.: Activating more pixels in image super-resolution transformer. arXiv preprint arXiv:2205.04437 (2022)
11. Chen, Y., Yu, W., Pock, T.: On learning optimized reaction diffusion processes for effective image restoration. In: Proceedings of the IEEE Conference on Computer Vision and Pattern Recognition, pp. 5261–5269 (2015)
12. Cheng, S., Wang, Y., Huang, H., Liu, D., Fan, H., Liu, S.: NBNet: noise basis learning for image denoising with subspace projection. In: Proceedings of the IEEE/CVF Conference on Computer Vision and Pattern Recognition, pp. 4896–4906 (2021)

13. Conde, M.V., McDonagh, S., Maggioni, M., Leonardis, A., Pérez-Pellitero, E.: Model-based image signal processors via learnable dictionaries. In: Proceedings of the AAAI Conference on Artificial Intelligence, vol. 36, pp. 481–489 (2022)
14. Conde, M.V., Timofte, R., et al.: Reversed image signal processing and raw reconstruction. AIM 2022 challenge report. In: Proceedings of the European Conference on Computer Vision Workshops (ECCVW) (2022)
15. Dabov, K., Foi, A., Katkovnik, V., Egiazarian, K.: Image denoising by sparse 3-D transform-domain collaborative filtering. IEEE Trans. Image Process. **16**(8), 2080–2095 (2007)
16. Delbracio, M., Kelly, D., Brown, M.S., Milanfar, P.: Mobile computational photography: A tour. arXiv preprint arXiv:2102.09000 (2021)
17. Gharbi, M., Chaurasia, G., Paris, S., Durand, F.: Deep joint demosaicking and denoising. ACM Trans. Graph. **35**(6), 1–12 (2016)
18. Gu, S., Zhang, L., Zuo, W., Feng, X.: Weighted nuclear norm minimization with application to image denoising. In: Proceedings of the IEEE Conference on Computer Vision and Pattern Recognition, pp. 2862–2869 (2014)
19. Guo, J., et al.: CMT: convolutional neural networks meet vision transformers. In: Proceedings of the IEEE/CVF Conference on Computer Vision and Pattern Recognition, pp. 12175–12185 (2022)
20. Guo, S., Yan, Z., Zhang, K., Zuo, W., Zhang, L.: Toward convolutional blind denoising of real photographs. In: Proceedings of the IEEE/CVF Conference on Computer Vision and Pattern Recognition, pp. 1712–1722 (2019)
21. Han, K., et al.: A survey on vision transformer. In: IEEE Transactions on Pattern Analysis and Machine Intelligence (2022)
22. He, X., Cheng, J.: Revisiting L1 loss in super-resolution: a probabilistic view and beyond. arXiv preprint arXiv:2201.10084 (2022)
23. Hu, J., Shen, L., Sun, G.: Squeeze-and-excitation networks. In: Proceedings of the IEEE Conference on Computer Vision and Pattern Recognition, pp. 7132–7141 (2018)
24. Kingma, D.P., Welling, M.: Auto-encoding variational bayes. arXiv preprint arXiv:1312.6114 (2013)
25. Kingma, D.P., Dhariwal, P.: Glow: Generative flow with invertible 1x1 convolutions. In: Advances in Neural Information Processing Systems 31 (2018)
26. Lim, B., Son, S., Kim, H., Nah, S., Mu Lee, K.: Enhanced deep residual networks for single image super-resolution. In: Proceedings of the IEEE Conference on Computer Vision and Pattern Recognition Workshops, pp. 136–144 (2017)
27. Liu, P., Zhang, H., Zhang, K., Lin, L., Zuo, W.: Multi-level wavelet-CNN for image restoration. In: Proceedings of the IEEE Conference on Computer Vision and Pattern Recognition Workshops, pp. 773–782 (2018)
28. Liu, Y.L., et al.: Single-image HDR reconstruction by learning to reverse the camera pipeline. In: Proceedings of the IEEE/CVF Conference on Computer Vision and Pattern Recognition (CVPR) (2020)
29. Luo, W., Li, Y., Urtasun, R., Zemel, R.: Understanding the effective receptive field in deep convolutional neural networks. In: Advances in Neural Information Processing Systems 29 (2016)
30. Maini, R., Aggarwal, H.: A comprehensive review of image enhancement techniques. arXiv preprint arXiv:1003.4053 (2010)
31. Nguyen, R.M., Brown, M.S.: Raw image reconstruction using a self-contained sRGB-jpeg image with only 64 KB overhead. In: Proceedings of the IEEE Conference on Computer Vision and Pattern Recognition, pp. 1655–1663 (2016)

32. Nguyen, R.M., Brown, M.S.: Raw image reconstruction using a self-contained sRGB-JPEG image with small memory overhead. Int. J. Comput. Vision **126**(6), 637–650 (2018)
33. Plötz, T., Roth, S.: Neural nearest neighbors networks. In: Advances in Neural information processing systems 31 (2018)
34. Punnappurath, A., Brown, M.S.: Learning raw image reconstruction-aware deep image compressors. IEEE Trans. Pattern Anal. Mach. Intell. **42**(4), 1013–1019 (2019)
35. Purohit, K., Rajagopalan, A.: Region-adaptive dense network for efficient motion deblurring. In: Proceedings of the AAAI Conference on Artificial Intelligence, vol. 34, pp. 11882–11889 (2020)
36. Qian, G., Gu, J., Ren, J.S., Dong, C., Zhao, F., Lin, J.: Trinity of pixel enhancement: a joint solution for demosaicking, denoising and super-resolution. arXiv preprint arXiv:1905.02538 (2019)
37. Reddi, S.J., Kale, S., Kumar, S.: On the convergence of Adam and beyond. arXiv preprint arXiv:1904.09237 (2019)
38. Ronneberger, O., Fischer, P., Brox, T.: U-Net: convolutional networks for biomedical image segmentation. In: Navab, N., Hornegger, J., Wells, W.M., Frangi, A.F. (eds.) MICCAI 2015. LNCS, vol. 9351, pp. 234–241. Springer, Cham (2015). https://doi.org/10.1007/978-3-319-24574-4_28
39. Roth, S., Black, M.J.: Fields of experts. Int. J. Comput. Vision **82**(2), 205–229 (2009)
40. Schwartz, E., Giryes, R., Bronstein, A.M.: DeepISP: toward learning an end-to-end image processing pipeline. IEEE Trans. Image Process. **28**(2), 912–923 (2018)
41. Shi, W., et al.: Real-time single image and video super-resolution using an efficient sub-pixel convolutional neural network. In: Proceedings of the IEEE Conference on Computer Vision and Pattern Recognition, pp. 1874–1883 (2016)
42. Talebi, H., Milanfar, P.: Global image denoising. IEEE Trans. Image Process. **23**(2), 755–768 (2013)
43. Tian, C., Fei, L., Zheng, W., Xu, Y., Zuo, W., Lin, C.W.: Deep learning on image denoising: an overview. In: Neural Networks (2020)
44. Vaswani, A., et al.: Attention is all you need. In: Advances in Neural Information Processing Systems 30 (2017)
45. Wang, Z., Bovik, A.C., Sheikh, H.R., Simoncelli, E.P.: Image quality assessment: from error visibility to structural similarity. IEEE Trans. Image Process. **13**(4), 600–612 (2004)
46. Wronski, B., et al.: Handheld multi-frame super-resolution. ACM Trans. Graph. (TOG) **38**(4), 1–18 (2019)
47. Xia, M., Liu, X., Wong, T.T.: Invertible grayscale. ACM Trans. Graph. (TOG) **37**(6), 1–10 (2018)
48. Xing, Y., Qian, Z., Chen, Q.: Invertible image signal processing. In: Proceedings of the IEEE/CVF Conference on Computer Vision and Pattern Recognition, pp. 6287–6296 (2021)
49. Yue, Z., Yong, H., Zhao, Q., Zhang, L., Meng, D.: Variational denoising network: toward blind noise modeling and removal. arXiv preprint arXiv:1908.11314 (2019)
50. Zamir, S.W., Arora, A., Khan, S., Hayat, M., Khan, F.S., Yang, M.H.: Restormer: efficient transformer for high-resolution image restoration. In: Proceedings of the IEEE/CVF Conference on Computer Vision and Pattern Recognition, pp. 5728–5739 (2022)

51. Zamir, S.W., et al.: CycleISP: real image restoration via improved data synthesis. In: Proceedings of the IEEE/CVF Conference on Computer Vision and Pattern Recognition, pp. 2696–2705 (2020)
52. Zamir, S.W., et al.: Learning enriched features for real image restoration and enhancement. In: Vedaldi, A., Bischof, H., Brox, T., Frahm, J.-M. (eds.) ECCV 2020. LNCS, vol. 12370, pp. 492–511. Springer, Cham (2020). https://doi.org/10.1007/978-3-030-58595-2_30
53. Zhang, K., Zuo, W., Chen, Y., Meng, D., Zhang, L.: Beyond a gaussian denoiser: residual learning of deep CNN for image denoising. IEEE Trans. Image Process. **26**(7), 3142–3155 (2017)
54. Zhang, K., Zuo, W., Zhang, L.: FFDNet: toward a fast and flexible solution for CNN-based image denoising. IEEE Trans. Image Process. **27**(9), 4608–4622 (2018)
55. Zoran, D., Weiss, Y.: From learning models of natural image patches to whole image restoration. In: 2011 International Conference on Computer Vision, pp. 479–486. IEEE (2011)

LiteDepth: Digging into Fast and Accurate Depth Estimation on Mobile Devices

Zhenyu Li[1], Zehui Chen[2], Jialei Xu[1], Xianming Liu[1], and Junjun Jiang[1](✉)

[1] Harbin Institute of Technology, Harbin, China
{zhenyuli17,csxm,jiangjunjun}@hit.edu.cn, lovesnow@mail.ustc.edu.cn,
21B903029@stu.hit.edu.cn
[2] University of Science and Technology of China, Hefei, China

Abstract. Monocular depth estimation is an essential task in the computer vision community. While tremendous successful methods have obtained excellent results, most of them are computationally expensive and not applicable for real-time on-device inference. In this paper, we aim to address more practical applications of monocular depth estimation, where the solution should consider not only the precision but also the inference time on mobile devices. To this end, we first develop an end-to-end learning-based model with a tiny weight size (1.4MB) and a short inference time (27FPS on Raspberry Pi 4). Then, we propose a simple yet effective data augmentation strategy, called $\mathbf{R}^2$ **crop**, to boost the model performance. Moreover, we observe that the simple lightweight model trained with only one single loss term will suffer from performance bottleneck. To alleviate this issue, we adopt multiple loss terms to provide sufficient constraints during the training stage. Furthermore, with a simple dynamic re-weight strategy, we can avoid the time-consuming hyper-parameter choice of loss terms. Finally, we adopt the structure-aware distillation to further improve the model performance. Notably, our solution named *LiteDepth* ranks $\mathbf{2}^{nd}$ **in the MAI&AIM2022 Monocular Depth Estimation Challenge**, with a si-RMSE of 0.311, an RMSE of 3.79, and the inference time is $37ms$ tested on the Raspberry Pi 4. Notably, we provide the **fastest** solution to the challenge. Codes and models will be released at https://github.com/zhyever/LiteDepth.

Keywords: Monocular depth estimation · Lightweight network · Data augmentation · Multiple loss

1 Introduction

Monocular depth estimation plays a vital role in the computer vision community, where a wide spread of various depth-depended tasks related to autonomous driving [6–8,22,31,36,39,40], virtual reality [3,11], and scene understanding [13,35,37,46] provide strong demand for fast and accurate monocular depth estimation methods that are applicable to portable low-power hardware.

L. Karlinsky et al. (Eds.): ECCV 2022 Workshops, LNCS 13802, pp. 507–523, 2023.
https://doi.org/10.1007/978-3-031-25063-7_31

Therefore, research along the line of accelerating depth estimation while reducing quality sacrifice on mobile devices has drawn increasing attention [15,38].

As a classic ill-posed problem, estimating accurate depth from a single image is challenging. However, with the fast development of deep learning techniques, neural network demonstrates groundbreaking improvement with plausible depth estimation results [5,9,20,24,25,42]. While engaging results have been presented, most of these state-of-the-art (SoTA) models are only optimized for high fidelity results while not taking into account computational efficiency and mobile-related constraints. The requirements of powerful high-end GPUs and consuming gigabytes of RAM lead to a dilemma when developing these models on resource-constrained mobile hardware [1,2,15].

In this paper, we aim to address the more practical application problem of monocular depth estimation on mobile devices, where the solution should consider not only the precision but also the inference time [15]. We first investigate a suitable network design. Typically, the depth estimation network follows a UNet paradigm [32] consisting of an encoder and a decoder with skip connections. Regarding the encoder, we choose a variant version of MobileNet-v3 [14] as a trade-off between performance and inference time, where we drop out the last convolution layer to speed up inference and reduce the model size. Moreover, we observe that the commonly used image normalization pre-process on input images is also time-consuming ($19ms$ on Raspberry Pi 4). To solve this issue, we propose to merge the normalization into the first convolution layer in a post-process manner so that the redundant overhead can be eliminated without bells and whistles. Following [15], we adopt the fast downsampling strategy, which could quickly downsample the resolution of input images from 480×640 to 4×6. A light decoder is introduced to recover the spatial details, consisting of a few convolutional layers and upsampling layers.

After determining the model structure, we propose several effective training strategies to boost the fidelity of the lightweight model. (1) We adopt an effective augmentation strategy called $\mathbf{R}^2$ **crop**. It not only adopts crop patches on images with **R**andom locations but also **R**andomly changes the size of crop patches. This strategy increases the diversity of the scenes and effectively avoids overfitting the training set. (2) We introduce a multiple-loss training strategy to provide sufficient supervision during the training stage, where we propose a gradience loss that can handle invalid holes in training samples and adopt the other three loss terms proposed in previous works. Moreover, we install a dynamic re-weighting strategy that can avoid the time-consuming weight selection of loss terms. (3) We highlight that our work focuses on the model training strategies, unlike previous solutions that adopt variant distillation methods [15,38]. However, model distillation can also be an effective way to boost the model fidelity without any overhead. Therefore, we adopt the structure-aware distillation [27] in a fine-tuning manner.

We evaluate our method on Mobile AI (MAI2022) dataset, and the results demonstrate that each strategy can improve the accuracy of the lightweight network. With a short inference time ($37ms$ per image) on Raspberry Pi 4 and a lightweight model design (totally 1.4MB), our solution named *MobileDepth* achieves results of 0.311 si-RMSE and ranks second in the MAI&AIM 2022 Monocular Depth Estimation Challenge [18].

In summary, our main contributions are:

- We design a lightweight depth estimation model that achieves fast inference on mobile hardware, where an image normalization merging strategy is proposed to reduce the redundant overhead.
- We adopt an effective augmentation strategy called R^2 crop that is adopted at random locations on images with a randomly changed size of patches.
- We design a gradience loss that can handle invalid holes in training samples and propose to apply multiple-loss items to provide sufficient supervision during the training stage.
- We evaluate our method on MAI2022 dataset and rank second place in the MAI&AIM2022 Monocular Depth Estimation Challenge [18].

Table 1. Ranking results in the MAI&AIM2022 Monocular Depth Estimation Challenge, which are evaluated on the online test server. We highlight our results in **bold**.

Rank	Username	si-RMSE	RMSE	$\log_{10}$	REL	Runtime	Score
1	TCL	0.277	3.47	0.110	0.299	$46ms$	297.79
2	**Zhenyu Li**	**0.311**	**3.79**	**0.124**	**0.342**	**37**ms	**232.04**
3	ChaoMI	0.299	3.89	0.134	0.380	$54ms$	187.77
4	parkzyzhang	0.303	3.80	12.189	0.301	$68ms$	141.07
5	RocheL	0.329	4.06	0.137	0.366	$65ms$	102.07
6	mvc	0.349	4.46	0.140	0.340	$139ms$	36.07
7	Byung Hyun Lee	0.338	6.73	0.332	0.507	$142ms$	41.58

2 Related Work

Monocular depth estimation is an ill-posed problem [9]. Lack of cues, scale ambiguities, translucent or reflective materials all leads to ambiguous cases where appearance cannot infer the spatial construction [24]. With the rapid development of deep learning, the neural network has dominated the primary workhorse to provide reasonable depth maps from a single RGB input [5,20,24,25,43].

Eigen et al. [9] first groundbreakingly propose a multi-scale deep network, consisting of a global network and a local network to predict the coarse depth and refine predictions, respectively. Subsequent works focus on various points to boost depth estimation, for instance, problem formulation [5,10,25], network architecture [19,20,24], supervision design [4,30,43], interpretable method [44], pre-training strategy [23,29], unsupervised training [12,45], *etc.* Though achieving engaging fidelity, these methods neglect the limitation of resource-constrained hardware and can be hard to develop on portable devices or embedded systems.

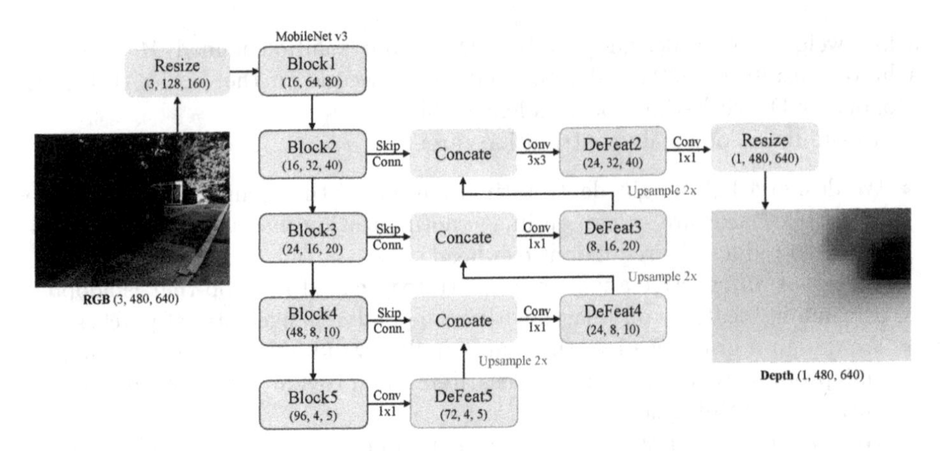

Fig. 1. Illustration of our proposed network architecture that follows the prevalent Unet [32] design consisting of a MobileNet-V3 [14] encoder and a lightweight decoder with skip connections.

Notably, there are also some methods that take the inference time and model complexity into account, which makes them applicable on mobile devices [15]. FastDepth [41] deploys a real-time depth estimation method on embedded systems by designing an efficient model architecture and a pruning strategy to further reduce the model complexity. In our paper, we follow FastDepth [41] to choose MobileNet-v3 [14] as our encoder and design an even more lightweight decoder (only consisting of four convolution layers) to achieve a trade-off between fidelity and inference speed.

3 Method

In this section, we first present our network design in Sect. 3.1, where tons of details should be considered to achieve the best trade-off between fidelity and inference speed. Then, we introduce our proposed R^2 Crop in Sect. 3.2 and Multiple Loss Training strategy in Sect. 3.3. Subsequently, we illustrate the installation of the structure-aware Distillation strategy in Sect. 3.4.

3.1 Network Design

As shown in Fig. 1, our proposed network consists of an encoder and a lightweight decoder with skip connections. We sequentially introduce each component and design detail.

Encoder plays a crucial role in extracting features from input images for depth estimation. To achieve a trade-off between fidelity and inference speed, we choose MobileNet-v3 [14] as our encoder. It is worth noticing that MobileNet contains a dimension-increasing layer (1×1 convolution with an input dimension of 96 and output dimension of 960) to facilitate training for a classification task.

We remove this layer to improve the inference speed and reduce the number of model parameters. Following [15], we adopt the *Fast Downsampling Strategy* in which a resize layer is inserted at the beginning of the encoder to resize the high-resolution input image from 480 × 640 to 128 × 160. As a result, the encoder can quickly downsample the resolution of feature maps, significantly shorten the inference time. Typically, input images are normalized to align with the pre-training setting. We discern the vanilla image normalization is time-consuming (19ms of the image normalization *v.s.* 37ms of the whole model) on the target device (*i.e.,* Raspberry Pi 4). Therefore, we propose to merge the image normalization into the first convolution layer in a post-process manner so that we can avoid the redundant overhead *without bells and whistles.* Consider the image normalization and the first convolution layer:

$$I_n = \frac{I_r - m}{s}, \tag{1}$$

$$f = W * I_n + b, \tag{2}$$

where I_n and $I_r \in \mathbb{R}^{3\times H\times W}$ are normalized and raw input images. $m \in \mathbb{R}^3$ and $s \in \mathbb{R}^3$ are the mean and standard deviation used in the image normalization. $f \in \mathbb{R}^{C\times H_f\times W_f}$ is the output feature map with C channels of the first convolution in our network. $W \in \mathbb{R}^{3\times C\times k^2}$ and $b \in \mathbb{R}^C$ are the trained weight and bias of the first $k \times k$ convolution. $*$ denotes the convolution operation. Given a trained model with parameters W and b of the first convolution, we update the them based on the mean and standard deviation used in the image normalization during the training stage:

$$W' = \frac{W}{s}, \tag{3}$$

$$b'_i = b_i - \sum_{d}^{3}\left(\frac{m_d}{s_d} \times \sum_{j}^{k\times k} W_{dij}\right), \ i \in (1, 2, ..., C), \tag{4}$$

$$b' = \mathbf{Concat}([b'_1, b'_2, ..., b'_C]), \tag{5}$$

where the W' and b' are the updated weight and bias of the first $k\times k$ convolution. **Concat** is the element-wise concatenation. d is the index of RGB dimension. Consequently, we discard the image normalization and apply the first convolution directly on input images as:

$$f = W' * I_r + b'. \tag{6}$$

As a result, the trained network can directly receive the raw input images without the time-consuming image normalization.

Decoder is adopted to recover the spatial details by fusing the multi-level deep and shallow features. Unlike previous works [15,38,41] that utilize the symmetrical encoder and decoder, we drop out the last decoder layer to further accelerate the model inference. Hence, the resolution of outputs is 4× downsampled (*i.e.,* 32×64). At each decoding stage, we apply a simple feature fusion

Fig. 2. Comparisons among different crop augmentations. As for R^2 crop, we utilize different colors to indicate that we adopt randomly selected size of crop patches.

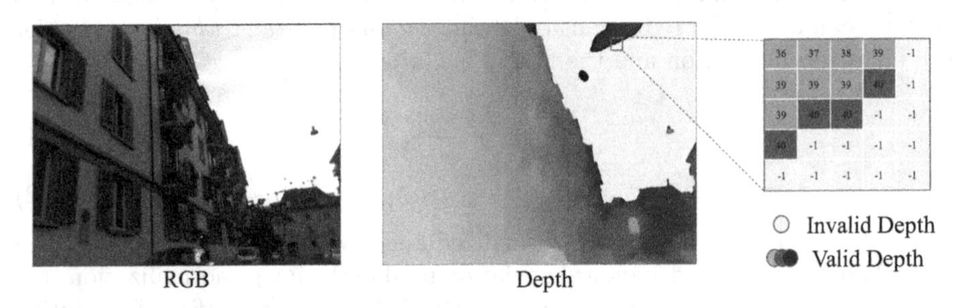

Fig. 3. Illustration of invalid depth GT pixels in the dataset. These pixels appear not only in the sky areas but also in close positions where the sensor cannot provide reliable GT value. We highlight a training sample for a clear Introduction of our valid mask in gradloss in Fig. 4.

module to aggregate the decoded and skip-connected features, which consists of a concatenation operation and a convolution layer (with ReLU as the activation function). To achieve the best trade-off between fidelity and speed, we utilize the 1×1 and 3×3 convolution for deep and shallow features, respectively. The final feature map is projected to the predicted depth map via the 1×1 convolution, which is then passed by a ReLU function to suppress the plural prediction. Finally, we insert a resize block at the end of the decoder to upsample the predicted depth map to the raw resolution 480×640. We highlight the lightweight design of the decoder that *only consists of five convolution layers* but achieves satisfactory fidelity.

3.2 R^2 Crop

Data augmentation is crucial to training models with better performance. Typically, the sequence of data augmentation for monocular depth estimation includes random rotation, random flip, random crop, and random color enhancement [21]. We propose the more effective crop strategy R^2 crop, in which we randomly select the size of crop patches and the cropped locations. We highlight the discrepancy with other commonly used crop methods in Fig. 2. It increases the diversity of the scenes and effectively avoids overfitting the training set.

3.3 Multiple Loss Training

Previous depth estimation methods [5,20,24,25] only adopt the silog loss to train the neural network:

$$\mathcal{L}_{silog} = \alpha \sqrt{\frac{1}{N}\sum_{i}^{N} e_i^2 - \frac{\lambda}{N^2}(\sum_{i}^{N} e_i)^2}, \tag{7}$$

where $e_i = \log \hat{d}_i - \log d_i$ with the ground truth depth d_i and predicted depth $\hat{d}_i$. N denotes the number of pixels having valid ground truth values. Since we discover that the lightweight model supervised by this simple single loss lacks representation capability and is easily stuck in local optimal, we adopt diverse loss terms to provide various targets for sufficient model training.

Motivated by [34], we first propose a **gradience loss** $\mathcal{L}_{grad}$ formulated as:

$$\mathcal{L}_{grad} = \frac{1}{N}\sum_{i}\left(M_{x_i} \times \left\|\nabla_x \hat{d}_i - \nabla_x d_i\right\|_1 + M_{y_i} \times \left\|\nabla_y \hat{d}_i - \nabla_y d_i\right\|_1\right), \tag{8}$$

where ∇ is the gradience calculation operation. Since the gradience loss is calculated in a dislocation subtraction manner and there are tremendous invalid depth GT in the dataset as shown in Fig. 3, as presented in Fig. 4, simply applying gradience calculation will blemish the information of invalid pixels and introduce outlier values when calculating the loss term. Hence, it is necessary to carefully design a strategy to calculate masks M to filter these invalid pixels in $\mathcal{L}_{grad}$. To solve this issue, we first replace the invalid value with *NaN* and then calculate the GT for gradience loss. Thanks to the numeral property of *NaN* and *Inf*, invalid information can be reserved. Consequently, we can filter the *NaN* and *Inf* when calculating the gradience loss.

Moreover, we also adopt the **virtual norm loss** $\mathcal{L}_{vnl}$ [43], and **robust loss** $\mathcal{L}_{robust}$ [4]. We formulate them as follows:

$$\mathcal{L}_{vnl} = \frac{1}{N}\sum_{i}^{N} \left\|\hat{n}_i - n_i\right\|_1, \tag{9}$$

where n is the virtual norm. We refer more details in the original paper [43]. Unlike the original implementation, we sample points from reconstructed point clouds and adopt constraints on predictions to filter invalid samples instead of ground truth. It helps the model convergence at the beginning of training.

$$\mathcal{L}_{robust} = \frac{1}{N}\sum_{i}^{N} \frac{|\alpha-2|}{\alpha}\left(\left(\frac{(e_i/c)^2}{|\alpha-2|}\right)^{\alpha/2} - 1\right), \tag{10}$$

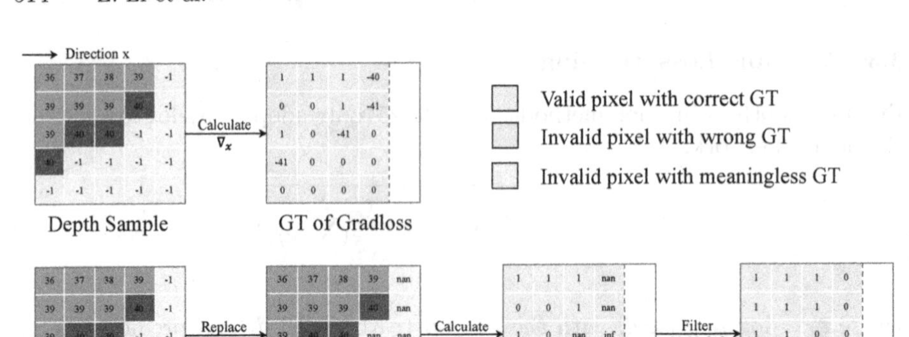

Fig. 4. Illustration of valid mask calculation for gradience loss (x direction). First line: vanilla calculation of gradience loss. Second line: we propose to first replace invalid value with NaN and compute reasonable valid mask for gradience loss.

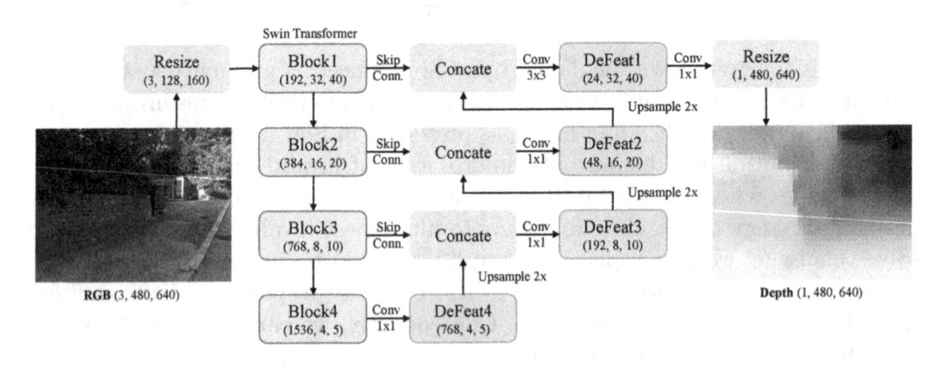

Fig. 5. Illustration of the teacher network.

where $e_i = \hat{d}_i - d_i$. We experimentally set $\alpha = 1$ and $c = 2$. In fact, the loss reduces to a simple L_2 loss, but which is proven to be more effective compared with the proposed adaptive version in our task. More experiments can be conducted to decide a better choice for α and c.

Finally, we adopt a combination of these loss terms to train our network. The total depth loss is

$$\mathcal{L}_{depth} = w_1\mathcal{L}_{silog} + w_2\mathcal{L}_{grad} + w_3\mathcal{L}_{vnl} + w_4\mathcal{L}_{robust}. \tag{11}$$

We set $w_1 = 1$, $w_2 = 0.25$, $w_3 = 2.5$, and $w_4 = 0.6$ based on tremendous experiments. Then, we apply a dynamic re-weight strategy in which the loss weights w are set as model parameters and are automatically fine-tuned during the training stage. Experimental results indicate that this strategy can achieve similar results as tuning the weights by hand.

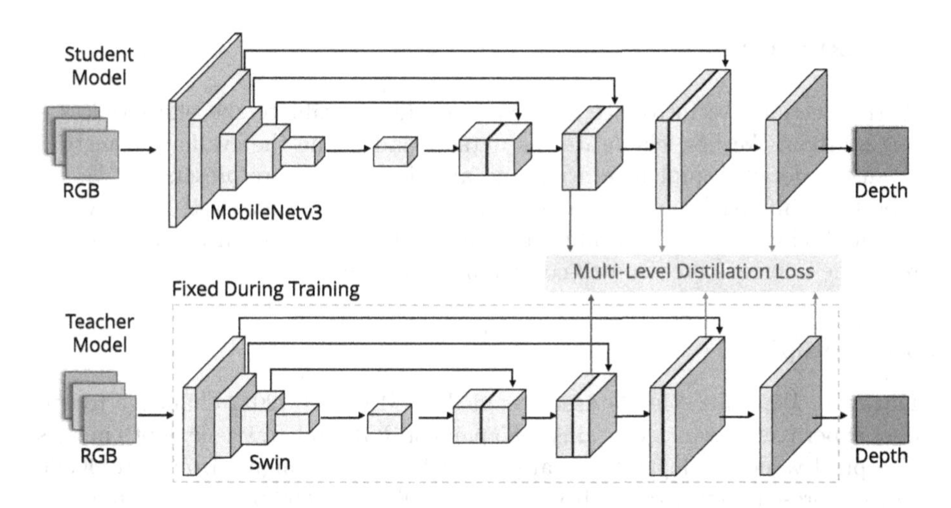

Fig. 6. Illustration of our multi-scale distillation strategy.

3.4 Structure-Aware Distillation

We apply the structure-aware distillation strategy [27, 38] to further boost model performance. For the teacher model, we choose Swin Transformer [28] as the encoder and adopt a similar lightweight decoder to recover feature resolution and predict depth maps. We present the network architecture in Fig. 5. The teacher model is trained via the supervision of $\mathcal{L}_{depth}$ and is then fixed when distilling the student model. During the distillation, multi-level distilling losses are adopted to provide supervisions on immediate features as shown in Fig. 6. The distillation loss is formulated as

$$\mathcal{L}_{distill} = \sum_{l}^{L} \left(\frac{1}{H \times W} \sum_{i}^{H} \sum_{j}^{W} \left\| a_{ij}^{s} - a_{ij}^{t} \right\|_1 \right), \tag{12}$$

where a is the affinity map calculated via inner-product of L_2 normalized features. We refer to [27, 38] for more details. s and t indicate the features are from the student and teacher model, respectively. We choose three level ($L = 3$) features for distill, which are DeFeat2, DeFeat3, and DeFeat4 in Fig. 1.

Consequently, the student model is trained via the total loss $\mathcal{L}$:

$$\mathcal{L} = \mathcal{L}_{depth} + w_d \mathcal{L}_{distill}, \tag{13}$$

where $w_d = 10$ in our experiments. Notably, unlike previous work [27, 38], we adopt a two-stage training paradigm. During the first stage, the student model is only trained via $\mathcal{L}_{depth}$. In the second stage, we adopt the teacher model and utilize the total loss $\mathcal{L}$ to further boost the performance of the student model.

4 Experiments

In this section, we introduce our experiments to evaluate the effectiveness of our solution. We first elaborate the dataset and define the evaluation metrics. Then the detailed implementation and ablation studies are presented. We also report the inference time on target devices (*i.e.*, Raspberry Pi 4) to show that our method can not only produce reasonable depth estimation but also achieve real-time inference on resource-constrained hardware.

4.1 Setup

Dataset. We utilize the dataset provided by MAI&AIM2022 challenge to conduct experiments, which contains 7385 pairs of RGB and grayscale depth images. The pixel values of depth maps are in uint16 format ranging from 0 to 40000, which represent depth values from 0 to 40 m. We use 6869 pairs for training and the rest 516 pairs as the local validation set.

Evaluation Metrics. In MAI&AIM2022 challenge [18], two metrics are considered for each submission solution: 1) The quality of the depth estimation. It is measured by the invariant standard root mean squared error (si-RMSE). 2) The runtime of the model on the target platform (*i.e.*, Raspberry Pi 4). The scoring formulation is provided below:

$$\text{Score}(\text{si-RMSE}, \text{runtime}) = \frac{2^{-20} \cdot \text{si-RMSE}}{C \cdot \text{runtime}}, \tag{14}$$

where $C = 0.01$ on the online validation benchmark.

4.2 Implementation Details

We implement the proposed model via the monocular depth estimation toolbox [21], which is based on the open-source machine learning library Pytorch. The model is converted to TFLite [26] after training. We use Adam optimizer with betas = (0.9, 0.999) and eps=1e–3. A poly schedule is adopted where the base learning rate is $4e^{-3}$ and the power is 0.9. The total number of epochs is 600 with batch size = 32 on two RTX3090 GPUs, which takes around 4 h to train a model. The encoder of our network is pretrained on ImageNet, and the decoder part is trained from scratch.

4.3 Quantitative Results

As shown in Table 1, our proposed method achieves a score of 232.04 on the challenge test set and ranks second place. Our solution achieves 0.311 si-RMSE with $37ms$ on the Raspberry Pi 4. Notably, our runtime is lower than the other methods and the performance is comparable.

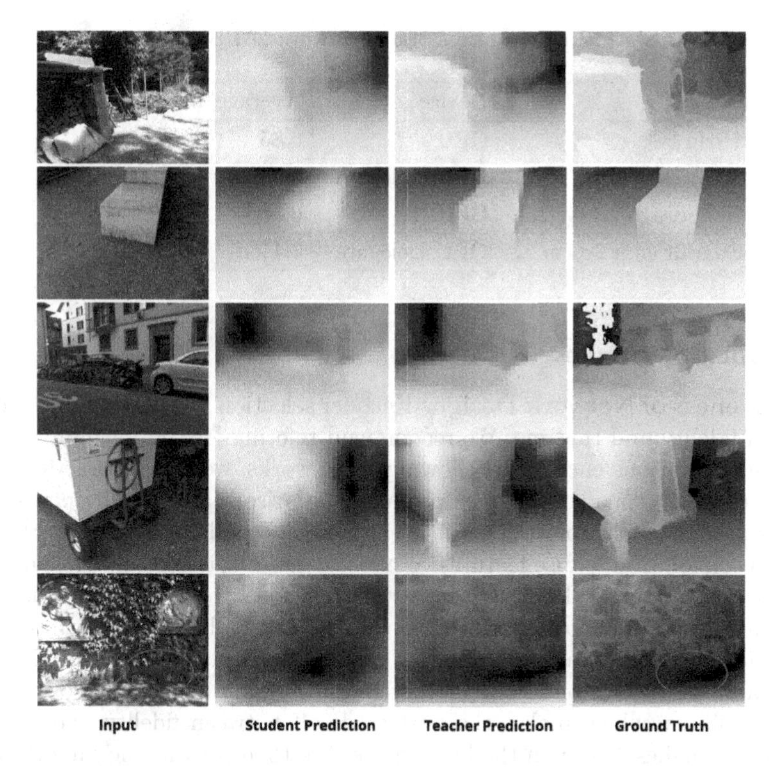

Fig. 7. The visualization results of our proposed methods. One can observe that there is noise in ground truth labels which we highlight with a red circle. (Color figure online)

4.4 Qualitative Results

We visualize the prediction results of our proposed methods as shown in Fig. 7, which demonstrates that our methods can achieve reasonable depth estimation results. However, the predicted depth maps are very rough around the edges due to the excessive down-sampling.

4.5 Inference Time

In this section, we verify that our method can achieve high-throughput monocular depth estimation on mobile devices. We convert our model to TensorFlow-Lite and test the inference time on various mobile devices, including smartphones with Kirin 980 and Snapdragon 7 Gen 1. We test the model using AI Benchmark [16,17]. Following the challenge requirements, the resolution of input and output images is 640×480. The data type is set to float (32 bit). As presented in Table 2, our network can obtain extremely high-throughput inference. It achieve 162FPS on smart phones with Snapdragon 7 Gen 1 processor. Interestingly, we can observe that the model is CPU-friendly, with an even faster inference on CPU than GPU on mobile devices.

Table 2. Inference time of our network (AI Benchmark).

SoC	Device	Average/ms	STD/ms
Kirin 980	CPU	6.85	0.77
Kirin 980	GPU Delegate	9.84	0.66
Snapdragon 7 Gen 1	CPU	6.16	1.71
Snapdragon 7 Gen 1	GPU Delegate	7.17	1.00

4.6 Ablation Studies

Effectiveness of Network Design. Encoder selection is crucial to the trade-off between fidelity and runtime. We recommend refering [15,38] for more comparisons among various encoders. Following these works, we choose the MobileNet-v3 as the default encoder. We then present comparisons among different decoder designs as shown in Table 3. Typically, previous methods [5,15,20,24,38] utilize the 3×3 convolution to fuse features. While the quantitative results are good, the runtime can be longer. However, when we replace all the 3×3 convolution with 1×1 convolution, the model performance drops drastically while the runtime gets short. Hence, we adopt a ***mix*** version as presented in our Sect. 3.1 and Fig. 1. We utilize a 3×3 convolution at the highest resolution and adopt 1×1 convolutions at other places, which makes the best trade-off between fidelity and runtime, getting the highest score on the benchmark. We then present the importance of the *merging image normalization*. It significantly reduces the runtime without any performance drop.

Table 3. Ablation study about the network architecture design. Dec and MIN are the short for decoder and *merge image normalization*, respectively.

Architecture	MIN	si-RMSE	Runtime/ms	Score
Full 3×3 @ Dec		0.295	62	27.01
Full 1×1 @ Dec		0.308	53	26.38
Mix Convs @ Dec		**0.301**	56	27.51
Mix Convs @ Dec	✓	**0.301**	**37**	**41.64**

Effectiveness of R^2 Crop We present the ablation study of various crop strategies. In these experiments, we only adopt the single sigloss (Eq. 7) for simplicity. As shown in Table 4, our proposed R^2 crop indicates an engaging improvement on performance compared with the baseline methods. When we adopt the vanilla random crop, the model cannot learn the knowledge of full-area images. However, the model infers on full-area images during the validation stage. This discrepancy leads to significant performance degradation. If we do not apply any crop strategy, the diversity of training samples is limited, also leading to a performance limitation. When we adopt our proposed R^2 crop, during the training stage, the model can not only learn the knowledge of full-area images but also ensure the diversity of training samples. When increasing the variety of crop sizes, the model performance can be improved simultaneously. However, too small patches cannot bring performance gains but lead to a slight degradation (*e.g.*, (144, 256) patches in our ablation study). We infer that the small patches do not contain sufficient structure information for facilitating the model training. As a result, we adopt patches with a size of [(240, 384), (384, 512), (480, 640)] in our solution.

Table 4. Ablation study of crop strategies. (h, w) represents the size of crop patches.

Method	si-RMSE	RMSE
w/o crop	0.335	4.25
Random crop with (384, 512)	0.377	4.62
R^2 crop with [(384, 512), (480, 640)]	0.327	4.15
R^2 crop with [(240, 384), (384, 512), (480, 640)]	**0.323**	**4.11**
R^2 crop with [(144, 256), (240, 384), (384, 512), (480, 640)]	0.325	4.13

Effectiveness of Multiple-Loss Training. This section evaluates the effectiveness of each loss term used in our solution. The results are presented in Table 5. Each loss term can bring performance gains for the model. We also highlight that if we do not apply the invalid mask in gradience loss, the model convergence will be hurt as described in Sect. 3.3. Moreover, our dynamic weight strategy can also achieve satisfactory results without fine-tuning loss weights by hand. We utilize the handcrafted weights as a default setting to achieve a better score in the challenge.

Table 5. Ablation study of the multiple loss strategy.

Sig Loss (Eq. 7)	Grad Loss (Eq. 8)	VNL Loss (Eq. 9)	Robust Loss (Eq. 10)	Dynamic Weight	si-RMSE
✓					0.323
✓	✓				0.316
✓	✓	✓			0.309
✓	✓	✓	✓		**0.303**
✓	✓	✓	✓	✓	0.306

Effectiveness of Distillation. We first present the results of the teacher model. As shown in Table 6, the teacher model achieves much better fidelity compared to the student model. It indicates that there is improvement room for the student model to learn from the teacher model via the distillation. We also present qualitative results in Fig. 7 for intuitive comparisons. As we can observe from the predicted depth maps, the teacher model provides more reasonable and sharper depth estimation results.

We then evaluate different distillation strategies in this section. Motivated by previous work, we try to apply L2 distillation [15], structure-aware distillation [27,38], and channel-wise distillation [33]. Interestingly, all strategies cannot directly work well for our lightweight student model as presented in Table 6. One possible reason is that we adopt multiple loss terms, leading to difficulty in balancing the loss weights. However, we also conduct experiments in which we only adopt the single sigloss and apply the distillation strategies. The results are similar without improvement in model performance. Moreover, some distillation strategies conflict with the two-stage fine-tuning, leading to a convergence issue. These experimental results indicate that more effective distillation strategies should be designed for monocular depth estimation. In this solution, we propose to adopt structure-aware distillation. It brings a slight improvement to the si-RMSE of our lightweight student model but a degradation on RMSE, indicating there is still huge room to improve the distillation strategy.

Table 6. Ablation study of distillation strategies. Two-Stage indicates applying the distillation in a fine-tuning manner. ∅ denotes that the fine-tuning process does not converge.

Method	Two-Stage	si-RMSE	RMSE
Teacher Model		0.228	3.025
Baseline Student Model		0.303	**3.785**
L2 Distillation		0.307	3.978
L2 Distillation	✓	∅	
Channel-Wise Distillation		0.311	4.045
Channel-Wise Distillation	✓	∅	
Structure-Aware Distillation		0.306	3.994
Structure-Aware Distillation	✓	**0.301**	3.839

5 Conclusion

We have introduced our solution for fast and accurate depth estimation on mobile devices. Specifically, we design an extremely lightweight model for depth estimation. Then, we propose R^2 crop to enrich the diversity of training samples. To facilitate the model training, we design a gradience loss and adopt multiple-loss items. We also investigate various distillation strategies. Extensive experiments indicate the effectiveness of our proposed solution.

Acknowledgments. The research was supported by the National Natural Science Foundation of China (61971165, 61922027), and also is supported by the Fundamental Research Funds for the Central Universities.

References

1. HUAWEI HiAI engine introduction. https://developer.huawei.com/consumer/en/doc/2020315 (2018)
2. Snapdragon neural processing engine SDK. https://developer.qualcomm.com/docs/snpe/overview.html (2018)
3. Armbrüster, C., Wolter, M., Kuhlen, T., Spijkers, W., Fimm, B.: Depth perception in virtual reality: distance estimations in peri-and extrapersonal space. Cyberpsychol. Behavior **11**(1), 9–15 (2008)
4. Barron, J.T.: a general and adaptive robust loss function. In: CVPR, pp. 4331–4339 (2019)
5. Bhat, S.F., Alhashim, I., Wonka, P.: AdaBins: depth estimation using adaptive bins. In: CVPR, pp. 4009–4018 (2021)
6. Chen, Z., et al.: AutoAlign: pixel-instance feature aggregation for multi-modal 3D object detection. arXiv preprint arXiv:2201.06493 (2022)
7. Chen, Z., Li, Z., Zhang, S., Fang, L., Jiang, Q., Zhao, F.: AutoAlignv2: deformable feature aggregation for dynamic multi-modal 3D object detection. arXiv preprint arXiv:2207.10316 (2022)
8. Chen, Z., Li, Z., Zhang, S., Fang, L., Jiang, Q., Zhao, F.: Graph-DETR3D: rethinking overlapping regions for multi-view 3D object detection. arXiv preprint arXiv:2204.11582 (2022)
9. Eigen, D., Puhrsch, C., Fergus, R.: Depth map prediction from a single image using a multi-scale deep network. In: NeurIPS (2014)
10. Fu, H., Gong, M., Wang, C., Batmanghelich, K., Tao, D.: Deep ordinal regression network for monocular depth estimation. In: CVPR, pp. 2002–2011 (2018)
11. Gerig, N., Mayo, J., Baur, K., Wittmann, F., Riener, R., Wolf, P.: Missing depth cues in virtual reality limit performance and quality of three dimensional reaching movements. PLoS ONE **13**(1), e0189275 (2018)
12. Godard, C., Mac Aodha, O., Firman, M., Brostow, G.J.: Digging into self-supervised monocular depth estimation. In: ICCV, pp. 3828–3838 (2019)
13. Hazirbas, C., Ma, L., Domokos, C., Cremers, D.: FuseNet: incorporating depth into semantic segmentation via fusion-based CNN architecture. In: Lai, S.-H., Lepetit, V., Nishino, K., Sato, Y. (eds.) ACCV 2016. LNCS, vol. 10111, pp. 213–228. Springer, Cham (2017). https://doi.org/10.1007/978-3-319-54181-5_14
14. Howard, A.G., et al.: MobileNets: efficient convolutional neural networks for mobile vision applications. arXiv preprint arXiv:1704.04861 (2017)
15. Ignatov, A., Malivenko, G., Plowman, D., Shukla, S., Timofte, R.: Fast and accurate single-image depth estimation on mobile devices, mobile AI 2021 challenge: Report. In: CVPR, pp. 2545–2557 (2021)
16. Ignatov, A., et al.: AI benchmark: running deep neural networks on android smartphones. In: Proceedings of the European Conference on Computer Vision (ECCV) Workshops (2018)
17. Ignatov, A., et al.: AI benchmark: all about deep learning on smartphones in 2019. In: ICCVW, pp. 3617–3635. IEEE (2019)
18. Ignatov, A., Timofte, R., et al.: Efficient single-image depth estimation on mobile devices, mobile AI & aim 2022 challenge: report. In: ECCV (2022)
19. Kim, D., Ga, W., Ahn, P., Joo, D., Chun, S., Kim, J.: Global-local path networks for monocular depth estimation with vertical cutDepth. arXiv preprint arXiv:2201.07436 (2022)

20. Lee, J.H., Han, M.K., Ko, D.W., Suh, I.H.: From big to small: multi-scale local planar guidance for monocular depth estimation. arXiv preprint arXiv:1907.10326 (2019)
21. Li, Z.: Monocular depth estimation toolbox. https://github.com/zhyever/Monocular-Depth-Estimation-Toolbox (2022)
22. Li, Z., Chen, Z., Li, A., Fang, L., Jiang, Q., Liu, X., Jiang, J.: Unsupervised domain adaptation for monocular 3D object detection via self-training. arXiv preprint arXiv:2204.11590 (2022)
23. Li, Z., et al.: SimIPU: Simple 2D image and 3D point cloud unsupervised pre-training for spatial-aware visual representations. arXiv preprint arXiv:2112.04680 (2021)
24. Li, Z., Chen, Z., Liu, X., Jiang, J.: DepthFormer: exploiting long-range correlation and local information for accurate monocular depth estimation. arXiv preprint arXiv:2203.14211 (2022)
25. Li, Z., Wang, X., Liu, X., Jiang, J.: BinsFormer: revisiting adaptive bins for monocular depth estimation. arXiv preprint arXiv:2204.00987 (2022)
26. Lite, T.: Deploy machine learning models on mobile and IoT devices (2019)
27. Liu, Y., Shu, C., Wang, J., Shen, C.: Structured knowledge distillation for dense prediction. IEEE TPAMI (2020)
28. Liu, Z., et al.: Swin transformer: hierarchical vision transformer using shifted windows. In: ICCV (2021)
29. Park, D., Ambrus, R., Guizilini, V., Li, J., Gaidon, A.: Is pseudo-lidar needed for monocular 3d object detection? In: ICCV, pp. 3142–3152 (2021)
30. Patil, V., Sakaridis, C., Liniger, A., Van Gool, L.: P3depth: monocular depth estimation with a piecewise planarity prior. In: CVPR, pp. 1610–1621 (2022)
31. Reading, C., Harakeh, A., Chae, J., Waslander, S.L.: Categorical depth distribution network for monocular 3D object detection. In: Proceedings of the IEEE/CVF Conference on Computer Vision and Pattern Recognition, pp. 8555–8564 (2021)
32. Ronneberger, O., Fischer, P., Brox, T.: U-Net: convolutional networks for biomedical image segmentation. In: Navab, N., Hornegger, J., Wells, W.M., Frangi, A.F. (eds.) MICCAI 2015. LNCS, vol. 9351, pp. 234–241. Springer, Cham (2015). https://doi.org/10.1007/978-3-319-24574-4_28
33. Shu, C., Liu, Y., Gao, J., Yan, Z., Shen, C.: Channel-wise knowledge distillation for dense prediction. In: ICCV, pp. 5311–5320 (2021)
34. Sitzmann, V., Martel, J., Bergman, A., Lindell, D., Wetzstein, G.: Implicit neural representations with periodic activation functions. NeurIPS **33**, 7462–7473 (2020)
35. Vu, T.H., Jain, H., Bucher, M., Cord, M., Pérez, P.: DADA: depth-aware domain adaptation in semantic segmentation. In: ICCV, pp. 7364–7373 (2019)
36. Wang, T., Pang, J., Lin, D.: Monocular 3D object detection with depth from motion. arXiv preprint arXiv:2207.12988 (2022)
37. Wang, W., Neumann, U.: Depth-aware CNN for RGB-D segmentation. In: Ferrari, V., Hebert, M., Sminchisescu, C., Weiss, Y. (eds.) ECCV 2018. LNCS, vol. 11215, pp. 144–161. Springer, Cham (2018). https://doi.org/10.1007/978-3-030-01252-6_9
38. Wang, Y., Li, X., Shi, M., Xian, K., Cao, Z.: Knowledge distillation for fast and accurate monocular depth estimation on mobile devices. In: CVPR, pp. 2457–2465 (2021)
39. Wang, Y., Guizilini, V.C., Zhang, T., Wang, Y., Zhao, H., Solomon, J.: DETR3D: 3D object detection from multi-view images via 3D-to-2D queries. In: Conference on Robot Learning, pp. 180–191. PMLR (2022)

40. Weng, X., Kitani, K.: Monocular 3D object detection with pseudo-lidar point cloud. In: Proceedings of the IEEE/CVF International Conference on Computer Vision Workshops (2019)
41. Wofk, D., Ma, F., Yang, T.J., Karaman, S., Sze, V.: FastDepth: fast monocular depth estimation on embedded systems. In: ICRA, pp. 6101–6108. IEEE (2019)
42. Yang, G., Tang, H., Ding, M., Sebe, N., Ricci, E.: Transformers solve the limited receptive field for monocular depth prediction. In: ICCV (2021)
43. Yin, W., Liu, Y., Shen, C., Yan, Y.: Enforcing geometric constraints of virtual normal for depth prediction. In: ICCV, pp. 5684–5693 (2019)
44. You, Z., Tsai, Y.H., Chiu, W.C., Li, G.: Towards interpretable deep networks for monocular depth estimation. In: ICCV, pp. 12879–12888 (2021)
45. Zhou, T., Brown, M., Snavely, N., Lowe, D.G.: Unsupervised learning of depth and ego-motion from video. In: CVPR, pp. 1851–1858 (2017)
46. Zhu, S., Brazil, G., Liu, X.: The edge of depth: explicit constraints between segmentation and depth. In: CVPR, pp. 13116–13125 (2020)

MSSNet: Multi-Scale-Stage Network for Single Image Deblurring

Kiyeon Kim, Seungyong Lee, and Sunghyun Cho(✉)

POSTECH, Pohang, Korea
{kiyeon,leesy,s.cho}@postech.ac.kr

Abstract. Most traditional single image deblurring methods before deep learning adopt a coarse-to-fine scheme that estimates a sharp image at a coarse scale and progressively refines it at finer scales. While this scheme has also been adopted in several deep learning-based approaches, recently a number of single-scale approaches have been introduced showing superior performance to previous coarse-to-fine approaches in terms of quality and computation time. In this paper, we revisit the coarse-to-fine scheme and analyze the defects of previous coarse-to-fine approaches. Based on the analysis, we propose Multi-Scale-Stage Network (MSSNet), a novel deep learning-based approach to single image deblurring with our remedies to the defects. MSSNet adopts three remedies: stage configuration reflecting blur scales, an inter-scale information propagation scheme, and a pixel-shuffle-based multi-scale scheme. Our experiments show that our remedies can effectively resolve the defects of previous coarse-to-fine approaches and improve the deblurring performance.

Keywords: Deblurring · Restoration · Neural network · CNN

1 Introduction

Single image deblurring aims to restore a sharp image from a blurry one caused by camera shake or object motion. As blur severely degrades the image quality and the performance of other tasks such as object detection, deblurring has been extensively studied for decades [1,3,4,7,20,22,23,28,31,34,41,42].

Most classical approaches before deep learning estimate a blur kernel and a latent sharp image through alternating optimization [4,5,7,19,20,23,29,33,37,38]. For computational efficiency and accuracy in estimating a blur kernel and latent image, a coarse-to-fine scheme has been widely adopted by classical approaches [4,5,29,33,37,38]. The coarse-to-fine scheme estimates a small blur kernel and latent image at a coarse scale and uses them as an initial solution at the next scale. The small sizes of both images and blur at a coarse scale enable computationally efficient estimation. Also, the small blur size at a coarse scale enables more accurate estimation of a blur kernel and latent image. As a result,

Supplementary Information The online version contains supplementary material available at https://doi.org/10.1007/978-3-031-25063-7_32.

L. Karlinsky et al. (Eds.): ECCV 2022 Workshops, LNCS 13802, pp. 524–539, 2023.
https://doi.org/10.1007/978-3-031-25063-7_32

the coarse-to-fine scheme can quickly provide an accurate initial solution to the next scale, and improve both quality and efficiency of deblurring.

Thanks to the effectiveness of the coarse-to-fine scheme proven by traditional approaches, it has also been adopted to several deep learning-based single image deblurring approaches [8,22,34]. These approaches directly restore a latent image from a blurry image without blur kernel estimation. They adopt multi-scale neural network architectures that stack sub-networks for different scales to initially estimate a small-scale latent image and then a large-scale latent image using the small-scale latent image as a guidance. While they do not estimate blur kernels, they share the same motivation with classical approaches: as the image and blur sizes are small at a coarse scale, a deblurred image can be estimated more efficiently and accurately.

Nonetheless, several deep learning-based single-scale approaches have recently been introduced. Specifically, Zhang *et al.* [42] pointed out the expensive computation time of the previous multi-scale approaches and the relatively low contribution of lower scale results on the final deblurring quality, and proposed an alternative single-scale approach named DMPHN. Following Zhang *et al.*, Suin *et al.* [31] and Zamir *et al.* [41] also proposed hierarchical multi-stage methods based on DMPHN. These approaches show superior performance to previous multi-scale approaches both in quality and computation time, making the traditional coarse-to-fine scheme seem obsolete.

In this paper, we address the following questions: The motivations of the coarse-to-fine scheme still look valid, but why do the coarse-to-fine approaches perform worse than recent single-scale approaches? What degrades their performance and how can we fix them? To this end, we revisit the coarse-to-fine scheme and analyze the defects of previous coarse-to-fine approaches that degrade their performance but have been *overlooked* so far. Based on the analysis, we propose Multi-Scale-Stage Network (MSSNet), a novel deep learning-based coarse-to-fine approach with our remedies to the defects. MSSNet consists of multiple scales and multiple stages at each scale. MSSNet adopts three remedies: stage configuration reflecting blur scales, an inter-scale information propagation scheme, and a pixel-shuffle-based multi-scale scheme. Each remedy is simple and straightforward, resulting in a simple architecture for MSSNet. Nonetheless, our experiments show that each remedies can effectively resolve the defects of previous coarse-to-fine approaches and improve the deblurring performance.

2 Related Work

Traditional single image deblurring methods [4,5,7,13,19,20,23,29,33,37,38] before deep learning assume blur models that describe how a blurred image is obtained using blur kernels. Unfortunately, they often fail due to their restrictive blur models and the ill-posedness of the problem. To improve deblurring quality, convolutional neural networks (CNNs) have recently been adopted [1,9,28,32,39]. For example, Schuler *et al.* [28] and Sun *et al.* [32] proposed CNNs that estimate blur kernels and a latent image based on traditional blur models. However, as they still rely on blur models, their performances are limited. To overcome such limitation, deep learning-based methods that directly restore sharp images without blur kernels have been proposed [3,22,31,34,41,42]. These methods can be broadly categorized into single- and multi-scale approaches with respect to their network architectures and training strategies.

Single-Scale Approaches. Recently, single-scale multi-stage architectures [2, 31,41,42] are gaining popularity. Zhang *et al.* [42] proposed DMPHN, a multi-stage network that stacks multiple encoder-decoder networks to gradually remove blur from an input image. Based on DMPHN, Suin *et al.* [31] proposed a dynamic filtering module to remove spatially varying blurs. Zamir *et al.* [41] proposed MPRNet, which progressively removes blur by giving supervision at each stage. Chen *et al.* [2] introduced half-instance normalization to the multi-stage architecture. Besides multi-stage architectures, Purohit *et al.* [25] proposed a deep single-stage architecture based on DenseNet [12]. However, these single-scale approaches do not use initial solutions estimated from coarse scales, so they are less efficient and accurate as will be shown in Sect. 5.

Multi-scale Approaches. Multi-scale approaches are typically based on multi-scale neural network architectures that stack sub-networks in a hierarchical way, and training strategies that train each sub-network to produce deblurred images at different scales. DeepDeblur [22], which is the first end-to-end deep learning-based method, adopts a multi-scale neural network to directly restore a latent image from a blurry input in a coarse-to-fine manner. SRN [34] adopts a UNet-based architecture [27] for each scale. Gao *et al.* [8] also proposed a UNet-based multi-scale architecture with a different parameter sharing strategy. Hu *et al.* [10] proposed a pyramid neural architecture search network to automatically design an optimal multi-scale deblurring architecture. However, their performance is limited by the drawbacks of their network architectures as will be discussed in Sect. 3. Cho *et al.* [3] recently proposed MIMO-UNet, which adopts a single UNet [27] with multi-scale loss terms. This approach is, however, different from a conventional coarse-to-fine approach as it has a large encoder that processes an input image in a fine-to-coarse manner. Furthermore, as Sect. 5 will show, our MSSNet outperforms MIMO-UNet with much fewer parameters and computations.

Transformer-Based Approaches. Image restoration approaches that adopt transformers [6,35] have recently been proposed such as Uformer [36] and Restormer [40]. They achieve superior deblurring performance by addressing the shortcomings of CNNs, e.g., limited receptive fields. Nevertheless, in this work, we restrict the scope of our analysis to conventional CNN-based approaches for the ease of analysis, and moreover, empirically show that our CNN-based approach can achieve comparable performance with smaller computation times.

3 Shortcomings of Previous Coarse-to-Fine Approaches

This section analyzes defects of previous coarse-to-fine approaches, and discusses our ideas to remedy them. MSSNet with our remedies is presented in Sect. 4.

Figure 1 illustrates the network architectures of previous coarse-to-fine approaches [8,22,34]. While SRN [34] adopts additional recurrent connections between consecutive scales to achieve additional performance gain, which is

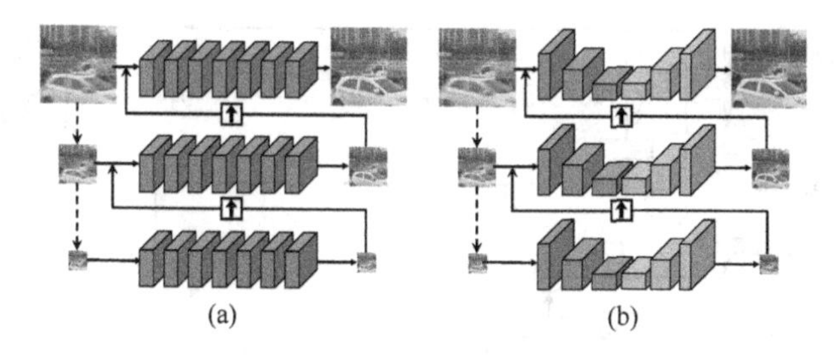

Fig. 1. Previous coarse-to-fine architectures. (a) DeepDeblur [22]. (b) SRN [34] and PSS-NSC [8].

omitted in the figure, the previous coarse-to-fine approaches share essentially the same deblurring process. All the methods first build an image pyramid by downsampling an input blurred image. Then, from the coarsest scale, they estimate a deblurred image from a downsampled blurred image, upsample the deblurred image, and feed it to the sub-network at the next scale. The sub-network at the next scale then estimates a deblurred image from the blurred image at the current scale using the deblurred image from the previous scale as a guidance. All the sub-networks at different scales share the same network architecture. In the following, we analyze the shortcomings of these approaches one by one and present our ideas to address them.

Network Architectures Disregarding Blur Scales. The first shortcoming of the previous approaches is their network architectures that disregard blur scales. Blur spreads a pixel value in a latent image over an area of the blur size. Thus, to restore the pixel value at a certain pixel, it is essential to use receptive fields larger than the blur size to aggregate information spread over the area. Consequently, larger blur sizes require larger receptive fields or deeper neural networks. Likewise, a coarse-to-fine approach needs deeper sub-networks for finer scales. While the previous coarse-to-fine approaches use a deblurred image from the previous scale to deblur the blurred image at the current scale [8,22,34], large receptive fields are still required for finer scales. In multi-scale approaches, a deblurred image from a lower scale lacks fine details as it is estimated from a downsampled image, and such fine details must be restored from the blurred image at a finer scale. Restoring detail at one pixel inevitably needs to aggregate information spread over an area of the blur size regardless of a result from the previous scale. Thus, it is still more effective to have deeper sub-networks for finer scales as will be shown in our experiments.

Ineffective Information Propagation Across Scales. The previous coarse-to-fine approaches pass the pixel values of a deblurred result from a coarse scale to the next scale [8,22,34]. This causes a significant loss of abundant information encoded in the feature vectors at coarse scales, and eventually degrades the deblurring performance.

Information Loss Caused by Downsampling. To produce multi-scale input blurred images, the previous approaches build an image pyramid by repeatedly downsampling an input image [8,22,34]. Unfortunately, downsampling causes significant information loss. Specifically, a downsampling operation reduces the

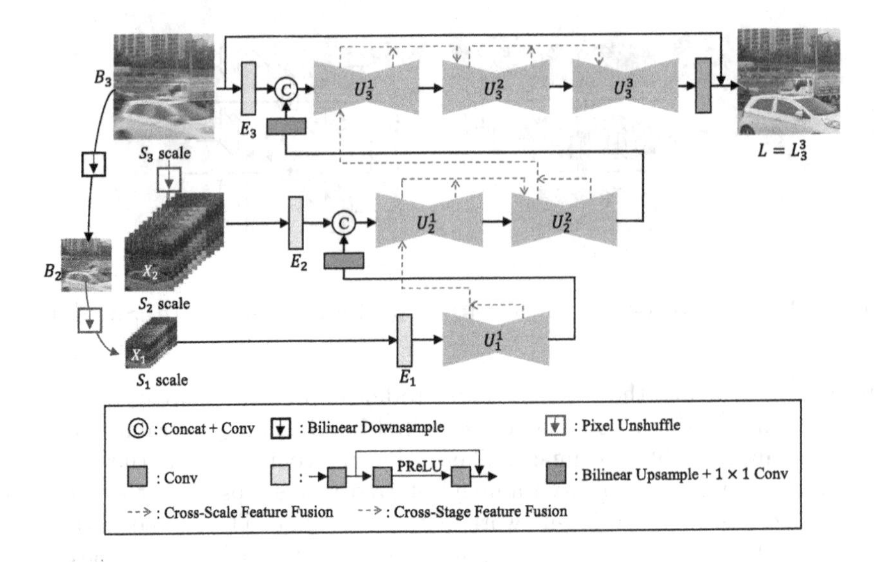

Fig. 2. Network architecture of MSSNet.

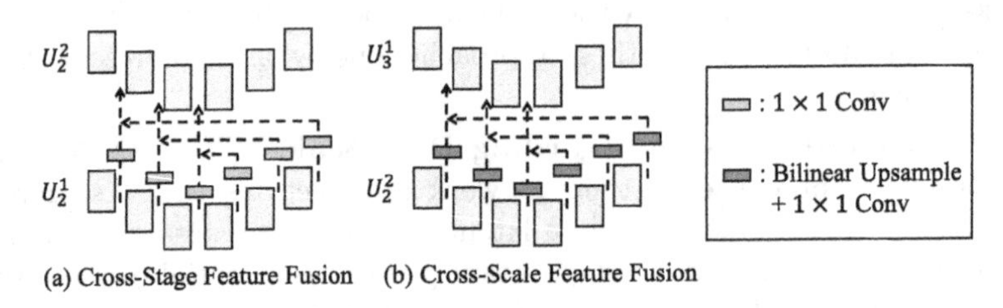

Fig. 3. Cross-stage and cross-scale feature fusion schemes.

pixels not only in the input image but also in its deblurred result by 1/4, which severely limits the quality of a guidance to the next scale. To overcome this, in our approach, we present a multi-scale scheme based on the pixel-shuffle [30] operation that reduces the spatial resolution without information loss.

4 Multi-Scale-Stage Network

4.1 Network Architecture

In this section, we present MSSNet, which is designed based on the analysis in Sect. 3. Figure 2 illustrates the architecture of MSSNet. MSSNet is composed of three scales following previous coarse-to-fine approaches [8,22,34]. We denote each scale by S_1, S_2, and S_3 from the coarsest to finest scales, respectively. MSSNet takes a single input blurred image B and estimates a deblurred image L in a coarse-to-fine manner. For effective restoration, MSSNet adopts the residual learning scheme, which has been widely adopted in various restoration tasks [3, 14,18,24,41,44], i.e., MSSNet predicts a residual image R, which is added to the input blurred image B to obtain a deblurred output $L = B + R$. A detailed architecture of MSSNet can be found in the supplementary material.

MSSNet is specifically designed to reflect blur scales, to facilitate effective inter-scale information propagation, and to avoid information loss caused by downsampling. We describe each component in the following.

Stage Configuration Reflecting Blur Scales. To reflect blur scales, the subnetworks of MSSNet at finer scales are designed to have deeper architectures. Specifically, each scale of MSSNet has one, two and three stages from S_1 to S_3, respectively, where each stage consists of a single light-weight UNet module [27]. We denote each UNet module by U_i^j where i and j are scale and stage indices, respectively. The modules share the same network architecture but have different weights. Each module is trained to produce residual features that can be converted to a residual image and added to a blurred image to produce a deblurred image. More details on the training of MSSNet is explained in Sect. 4.2.

Inter-Scale Information Propagation. Whereas the existing multi-scale networks deliver an upsampled deblurred image from a coarse scale to the next scale as an initial solution, MSSNet delivers upsampled residual features to facilitate effective information propagation between scales. Specifically, at the end of a coarse scale, residual features are bilinearly upsampled and processed through a 1×1 conv layer. Then, the resulting features are concatenated to the features from a blurred image at the next scale and convolved with 3×3 filters to produce fused features. The fused features are then fed into the UNet modules to produce deblurred residual features at the current scale.

Pixel-Shuffle-Based Multi-Scale Scheme. To avoid information loss caused by the downsampling operations when producing multi-scale input blurred images, we propose a pixel-shuffle [30] based multi-scale scheme. Specifically, from the input blurred image B of size $W \times H$, we generate multi-scale input images as follows. For the finest scale S_3, the input blurred image B is used. The input image downsampled to a different scale is denoted by B_i, where i is a scale index, i.e., $B_3 = B$, and B_2 is a downsampled version of B of size $W/2 \times H/2$.

For S_2, B_2 is not used, but B_3 is unshuffled to obtain four images of size $W/2 \times H/2$, which are stacked along the channel direction to generate an input tensor X_2 for S_2. As B is an RGB image with three color channels, the size of X_2 is $W/2 \times H/2 \times 12$, so X_2 has the same spatial size as B_2 but still has the same amount of information as B_3. Then, X_2 is fed into the feature extractor module (E_2 in Fig. 2) and processed through the stages at S_2. Note that, despite X_2 having the same amount of information as B_3, the computation cost increase for S_2 is relatively small because we use features extracted from X_2 by the feature extractor module. Moreover, thanks to X_2 having richer information than B_2, the sub-network at S_2 can produce a more accurate result.

For the coarsest scale S_1, B_3 is first downsampled to obtain B_2, and then the same unshuffling process as for S_2 is applied to obtain an input tensor X_1 for S_1. Another possible choice is to directly unshuffle B_3 and obtain X_1 of $W/2 \times H/2 \times 48$, but we empirically found that this performs slightly worse. While the pixel-shuffle-based multi-scale architecture can already enhance deblurring quality when trained with conventional loss terms as will be shown in Sect. 5,

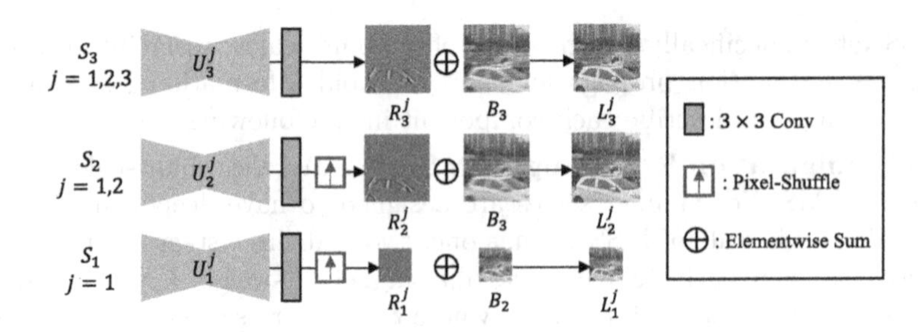

Fig. 4. Training of MSSNet. We train every stage to produce a residual image using auxiliary conv and pixel-shuffle layers.

we propose a pixel-shuffle-based training strategy to minimize information loss and enhance deblurring quality in Sect. 4.2.

Cross-Stage and Cross-Scale Feature Fusion. MSSNet also adopts the cross-stage feature fusion scheme proposed in [41]. The cross-stage feature fusion scheme connects network modules in consecutive stages with additional connections (dotted pink lines in Fig. 2) to help information flow more effectively between stages. Figure 3(a) describes the cross-stage feature fusion scheme. We refer the readers to [41] for more details on the cross-stage feature fusion scheme. In addition, we also introduce cross-scale feature fusion (dotted green lines in Fig. 2) to facilitate more effective information flow between consecutive scales. The cross-scale feature fusion scheme is described in Fig. 3(b).

4.2 Training and Loss Functions

During training, we guide each stage of MSSNet to produce a deblurred image. To this end, an auxiliary layer is attached to every stage to produce a deblurred image, except for the last one in S_3 that already has such layers. Specifically, for S_3, an auxiliary conv layer is attached at the end of U_3^1 and U_3^2 as shown in Fig. 4. The attached conv layers take features from the UNet modules and produce residual images R_3^1 and R_3^2. Each residual image is then added to B_3 to produce deblurred results L_3^1 and L_3^2. We also denote the final deblurred result L by L_3^3.

For S_1 and S_2, we use a slightly different training strategy as the sub-networks at S_1 and S_2 take unshuffled images as input. Specifically, at the end of each stage at S_1 and S_2, a conv layer and a pixel-shuffle layer are attached as shown in Fig. 4. The attached layers at the stages at S_1 and S_2 produce residual images of sizes $W/2 \times H/2$ and $W \times H$, respectively. We denote the deblurred results from the auxiliary layers by L_i^j where i and j are scale and stage indices, respectively.

We train MSSNet using two types of loss functions: a content loss $\mathcal{L}_c$ and a frequency reconstruction loss $\mathcal{L}_f$. The content loss $\mathcal{L}_c$ is defined as:

$$\mathcal{L}_c = \frac{1}{N_1}\|L_1^1 - L_{gt\downarrow}\|_1 + \sum_{j=1}^{2}\frac{1}{N_2}\|L_2^j - L_{gt}\|_1 + \sum_{j=1}^{3}\frac{1}{N_3}\|L_3^j - L_{gt}\|_1, \quad (1)$$

where L_{gt} is the ground-truth blurred image, and $L_{gt\downarrow}$ is a downsampled version of L_{gt}. N_1, N_2 and N_3 are normalization factors, which are set $N_1 = W/2 \times H/2 \times 3$

and $N_2 = N_3 = W \times H \times 3$. The frequency reconstruction loss was proposed in [3] to restore high-frequency details from blurred image by minimizing the difference between blurred image and ground-truth in the frequency domain. The frequency reconstruction loss is defined as:

$$\mathcal{L}_f = \frac{1}{N_1}\|\mathcal{F}(L_1^1) - \mathcal{F}(L_{gt\downarrow})\|_1 + \sum_{j=1}^{2}\frac{1}{N_2}\|\mathcal{F}(L_2^j) - \mathcal{F}(L_{gt})\|_1 + \sum_{j=1}^{3}\frac{1}{N_3}\|\mathcal{F}(L_3^j) - \mathcal{F}(L_{gt})\|_1, \tag{2}$$

where $\mathcal{F}$ is Fourier transform. Our final loss is $\mathcal{L}_{final} = \mathcal{L}_c + \lambda\mathcal{L}_f$ where $\lambda = 0.1$.

5 Experiments

Implementation Details. For evaluation, we trained MSSNet on the GoPro dataset [22] with 256×256 patches randomly cropped and augmented with random horizontal and vertical flipping. We trained our model for 3,000 epochs (396,000 iterations) with batch size 16. We used the Adam optimizer [15] with cosine annealing [21]. We set the initial learning rate to 2×10^{-4} and gradually decreased it to 1×10^{-6}. To evaluate the performance on real-world blurred images, we also use the RealBlur dataset [26]. To this end, we trained MSSNet using the GoPro [22], BSD-B [26], and RealBlur training sets following the RealBlur benchmark [26]. We trained the model for 100 epochs (397,400 iterations). The other training details are the same as above. The computation times of all models are measured on a PC with an NVIDA GeForce RTX 3090 GPU.

5.1 Comparison with Previous Methods

We compare MSSNet with state-of-the-art methods. Table 1 shows a quantitative comparison on the GoPro test set [22]. All the methods in the table were trained with the GoPro training set. Among the compared methods, DeepDeblur [22], SRN [34] and PSS-NSC [8] are coarse-to-fine approaches. MIMO-UNet [3] is trained using multi-scale loss terms, but not a conventional coarse-to-fine approach as it is based on a single UNet architecture. MIMO-UNet+ is a variant of MIMO-UNet with more parameters, and MIMO-UNet++ is MIMO-UNet+ with self-ensemble. All the other methods are single-scale approaches. For a fair comparison, MSSNet is trained for the smallest number of iterations among the methods in the table. Refer to the supplementary material for more details.

As shown in Table 1, recent single-scale approaches tend to perform better than coarse-to-fine approaches except for MIMO-UNet [3] and its variants. On the other hand, MSSNet clearly outperforms all the other methods in PSNR and SSIM thanks to our remedies. Specifically, MSSNet performs better than MIMO-UNet+ by more than 0.5dB with fewer parameters and fewer computations. Compared to MIMO-UNet++, a self-ensemble version of MIMO-UNet+,

Table 1. Quantitative evaluation on the GoPro test dataset [22]. Blue: coarse-to-fine approaches. Purple: single-scale approaches. Green: transformer-based approaches. MIMO-UNet and its variants are based on a single UNet with multi-scale losses [3]. The PSNR and SSIM values of previous methods are from the original papers. The computation times of all the methods are measured in the same environment described earlier. The numbers of parameters, MACs, and computation times of RADN [25], SAPHN [31] and PyNAS [10] are unavailable as their source codes are not publicly released yet.

Models	PSNR (dB)	SSIM	Param (M)	MACs (G)	Time (s)
DeepDeblur [22]	29.08	0.914	11.72	4729	1.290
DMPHN [42]	30.25	0.935	7.23	1100	0.137
SRN [34]	30.26	0.934	8.06	20134	0.736
PyNAS [10]	30.62	0.941	N/A	N/A	N/A
PSS-NSC [8]	30.92	0.942	2.84	3255	0.316
MT-RNN [24]	31.15	0.945	2.6	2315	0.323
SDNet4 [42]	31.20	0.945	21.7	3301	0.414
MIMO-UNet [3]	31.73	0.951	6.8	944	0.133
RADN [25]	31.76	0.953	N/A	N/A	N/A
SAPHN [31]	32.02	0.953	N/A	N/A	N/A
MSSNet-small (Ours)	32.02	0.953	6.75	634	0.104
MIMO-UNet+ [3]	32.45	0.957	16.1	2171	0.290
MPRNet [41]	32.66	0.959	20.1	10927	1.023
MIMO-UNet++ [3]	32.68	0.959	16.1	8683	1.169
HINet [2]	32.90	0.960	88.67	2401	0.247
Restormer [40]	32.92	0.959	26.13	1983	1.123
MSSNet (Ours)	33.01	0.961	15.59	2159	0.255
Uformer-B [36]	33.06	0.967	50.88	2236	1.105
MSSNet-large (Ours)	33.39	0.964	28.15	4235	0.457

MSSNet still outperforms by 0.33dB with a 4× fewer computations. Also, compared to HINet [2], MSSNet achieves 0.11dB higher PSNR with 5.7× fewer parameters and fewer computations while slightly slower. The table also shows that our CNN-based models (MSSNet and MSSNet-large) outperform recent transformer-based approaches [36,40] despite their smaller computation times.

We also include two variants of MSSNet: MSSNet-small and MSSNet-large, in this evaluation. Their detailed architectures are provided in the supplement. Compared to MIMO-UNet and SRN, which have larger model sizes, MSSNet-small achieves a higher PSNR and SSIM with smaller computation time. While SAPHN [31] achieves similar PSNR and SSIM values to those of MSSNet-small, ours performs much faster according to the computation time reported in their paper. Specifically, the reported computation time of SAPHN measured on a Titan Xp GPU is 0.77 s, while that of MSSNet-small on the same GPU is 0.19 s. MSSNet-large has about twice the parameters of MSSNet, which is still 3× fewer than HINet, and its computation time is more than twice shorter than those of MIMO-UNet++ and MPRNet. Nevertheless, it achieves 33.39 dB in PSNR, significantly exceeding all the other methods by a large margin. Figure 5 shows a qualitative comparison on the GoPro dataset [22]. As shown in the figure, our results show clearly restored sharp details while those of the others have remaining blur.

Table 2. Quantitative evaluation on RealBlur [26]. The models in the upper part of the table are trained on the GoPro dataset [22] and tested on the RealBlur test sets. The models in the lower part are trained and tested on each of the RealBlur-R and -J datasets. MIMO-UNet++ [3] provides only a model trained on the RealBlur-J dataset.

Models	RealBlur-R		RealBlur-J	
	PSNR	SSIM	PSNR	SSIM
Hu *et al.* [11]	33.67	0.916	26.41	0.803
DeepDeblur [22]	32.51	0.841	27.87	0.827
DeblurGAN [16]	33.79	0.903	27.97	0.834
Pan *et al.* [23]	34.01	0.916	27.22	0.790
Xu *et al.* [38]	34.46	0.937	27.14	0.830
DeblurGAN-v2 [17]	35.26	0.944	28.70	0.866
Zhang *et al.* [43]	35.48	0.947	27.80	0.847
SRN [34]	35.66	0.947	28.56	0.867
SDNet4 [42]	35.70	0.948	28.42	0.860
MPRNet [41]	35.99	0.952	28.70	0.873
MSSNet (Ours)	35.93	0.953	28.79	0.879
DeblurGAN-v2 [17]	36.44	0.935	29.69	0.870
SRN [34]	38.65	0.965	31.38	0.909
MPRNet [41]	39.31	0.972	31.76	0.922
MIMO-UNet++ [3]	N/A	N/A	32.05	0.921
MSSNet (Ours)	39.76	0.972	32.10	0.928

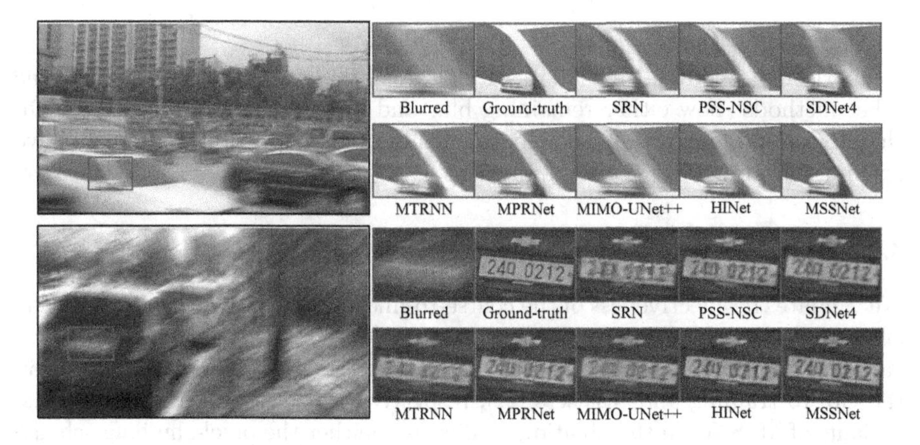

Fig. 5. Qualitative evaluation on the GoPro dataset [22].

We also study the generalization ability and performance of MSSNet on real-world blurred images. Table 2 shows a quantitative evaluation on the RealBlur dataset [26], which consists of real-world blurred images. The methods in the upper section in the table are trained on the GoPro dataset [22], while those in the lower section are trained on the RealBlur-R and RealBlur-J datasets. Among the methods trained on the GoPro datasets, MSSNet achieves the highest SSIM for the RealBlur-R test set, and the highest PSNR and SSIM for the RealBlur-J test set. Also, among the methods trained on the RealBlur datasets, MSSNet achieves the highest PSNR and SSIM. Figure 6 shows a qualitative

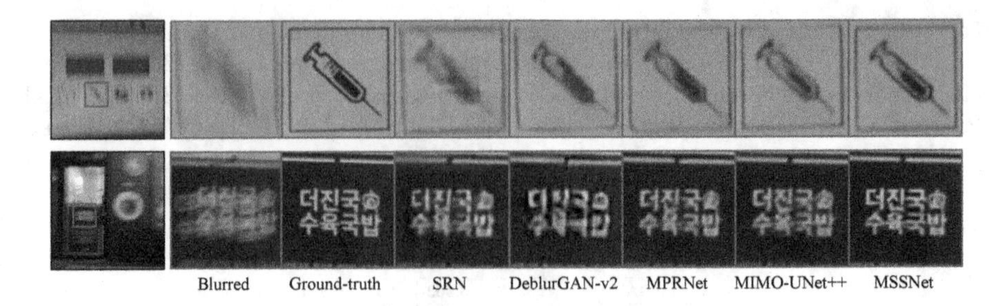

Fig. 6. Qualitative evaluation on the ReaBlur-J dataset [26].

Table 3. Performance comparison among a single-scale architecture with four stages and our multi-scale architectures. MSSNet-Single is a single-scale architecture, while MSSNet-Multi and MSSNet-Multi-Small are multi-scale architectures. 'Initial' and 'Final' are the initial and final results of each architecture, respectively.

	MSSNet-Single	MSSNet-Multi-Small	MSSNet-Multi
PSNR (Initial/Final)	29.11/31.59	29.51/31.58	30.09/31.75
Params (M)/MACs (G)	4.39/660.69	4.38/574.82	6.61/621.60

comparison on the RealBlur-J dataset [26]. In all the examples, the results of the other methods show either remaining blur and incorrectly restored details. On the other hand, our results show better restored details. Additional qualitative examples are provided in the supplement.

5.2 Ablation Study and Analysis

We validate the effectiveness of the coarse-to-fine approach, and then analyze the effect of each technical component in our model. For analysis, we test several variants of MSSNet. All the models in the analysis are trained and tested on the GoPro training and test sets [22], respectively. For ease of analysis, all the variants of MSSNet in the ablation studies use neither the pixel-shuffling scheme nor the cross-stage and cross-scale feature fusion scheme if not otherwise noted.
Coarse-to-Fine *vs* Single-Scale. As discussed in Sect. 1, the coarse-to-fine approach can quickly estimate a high-quality initial solution using coarse scales. Specifically, compared to performing a single stage of deblurring at the original scale, performing multiple stages at a coarse scale can be computationally more efficient. Moreover, thanks to the small blur size at a coarse scale, it can estimate a more accurate result, which serves as an initial solution for a finer scale, which leads to a final deblurring result of higher quality.

To verify this, in Table 3, we compare three variants of MSSNet. MSSNet-Single is a single-scale model with four stages at the original scale. MSSNet-Multi and MSSNet-Multi-Small are multi-scale models with the same number of scales and stages as MSSNet. MSSNet-Single and MSSNet-Multi has the same number of parameters for each stage. On the other hand, MSSNet-Multi-Small

Table 4. Ablation study on the stage configuration using variants of DeepDeblur [22].

Models	# ResBlocks			PSNR	SSIM	Params (M)	MACs (G)
	S_1	S_2	S_3				
D444	4	4	4	27.20	0.8282	2.5	1009.3
D444L	4	4	4	27.31	0.8306	3.42	1382.3
D246	2	4	6	27.39	0.8319	2.5	1363.5

Table 5. Ablation study on the stage configuration using variants of MSSNet.

Models	# Stages			PSNR	SSIM	Params (M)	MACs (G)
	S_1	S_2	S_3				
M123	1	2	3	29.58	0.925	1.18	521.33
M552	5	5	2	29.27	0.920	1.18	521.33

has fewer parameters for each stage at S_1 and S_2 so that its total number of parameters is similar to that of MSSNet-Single. Its architecture details are in the supplementary material. The multi-scale models use our pixel-shuffle-based approach, but none of the models use the cross-stage and cross-scale feature fusion schemes. While the multi-scale models have six stages in total, three of them are at coarser scales. As a result, both multi-scale model require smaller amounts of computation than MSSNet-Single as shown in the table.

In Table 3, 'Initial' and 'Final' indicates the initial and final results of the single-scale and multi-scale models. An initial solution of the single-scale model indicates a deblurring result of the first stage obtained using an auxiliary conv layer, while an initial solution of the multi-scale models indicates a deblurring result of the last stage at S_2 obtained using auxiliary conv and pixel-shuffle layers. We compare these as they serve as initial solutions for the last three stages. As shown in the table, despite its smaller computation cost, MSSNet-Multi produces higher-quality initial and final deblurring results. Also, although MSSNet-Multi-Small has a similar number of parameters and a much smaller computation cost, it still achieves a similar PSNR for the final result to that of MSSNet-Single. This proves the advantage of the coarse-to-fine approach against the single-scale approach.

Stage Configuration Reflecting Blur Scales. Our first remedy that we adopt into our MSSNet is the stage configuration reflecting blur scales. To verify its effect as a common rule, we conduct two ablation studies using DeepDeblur [22] and MSSNet. Table 4 compares three variants of DeepDeblur [22]. D444 and D444L have four residual blocks at each scale, while D246 adopts our stage configuration scheme and has two, four and six residual blocks at S_1, S_2 and S_3, respectively. To match the computation cost of D246, we also prepare D444L, which has more channels at each residual block. The table shows that D246 outperforms both of the others in terms of PSNR and SSIM, especially, despite its fewer parameters and a smaller computation cost than those of D444L. Note that our first remedy is to simply use a better stage configuration reflecting blur scales. Nonetheless, this result shows that this simple remedy can clearly improve the deblurring performance.

Table 6. Ablation study on the scale information propagation.

Model	PSNR	SSIM	Params (M)	MACs (G)
MSS(Image,Concat)	31.42	0.947	6.59	613.1
MSS(Feature,Skip)	31.52	0.948	6.59	621.8
MSS(Feature,Concat)	31.54	0.949	6.61	621.1

In the second experiment, we compare two variants of MSSNet in Table 5. The variants have different numbers of stages at different scales as informed in the table, but share the same network architecture for the UNet modules. The deblurring performance is not only affected by the number of stages, but also by the computation amount and the number of parameters. To isolate the impact of the stage configuration on the deblurring performance from other factors, each of the tested models in this experiment shares the network weights across different stages. In Table 5, M123 has the same stage configuration as MSSNet. M552 has fewer stages at S_3 but more stages at coarse scales so it requires the same amount of computation. The table shows that M123 clearly outperforms M552, validating our argument on the stage configuration. Additional experiments with different settings, e.g., models without parameter sharing, are provided in the supplementary material.

Inter-scale Feature Propagation. In the next ablation study, we verify the effect of our inter-scale feature propagation scheme. In this study, we also investigate how to fuse the solution from a coarse scale with the input to the finer scale. To this end, we compare three variants of MSSNet: MSS(Image,Concat), MSS(Feature,Skip) and MSS(Feature,Concat). MSS(Image,Concat) has auxiliary conv layers at the end of S_1 and S_2 to convert features to residual images. The residual images are added to the input blurred images of the corresponding sizes to produce deblurred results. The deblurred results are then upsampled and concatenated to the blurred images at the next scales. This model corresponds to the previous coarse-to-fine approaches that transfer pixel values from coarse to fine scales. MSS(Feature,Skip) transfers features from coarse to fine scales as done in MSSNet. However, features from coarse scales are not concatenated but added to the features of the blurred images at the next scales. As the sub-networks estimate residual features, adding them to the features of blurred images will produce initial deblurred features at finer scales. MSS(Feature,Concat) uses our inter-scale feature propagation scheme that concatenates features from coarse scales to the features of the blurred images at the next scales.

Table 6 compares the performance of the variants. The results confirm that using features instead of pixel values clearly improves the deblurring quality as features provide richer information. The table also shows that MSS(Feature,Concat) performs slightly better than MSS(Feature,Skip), although it requires slightly more parameters, validating our approach. Again, our remedy is simple and requires negligible amount of additional parameters and computation, but the improvement is clear.

Table 7. Ablation study on the pixel-shuffle-based multi-scale approach. PUS: pixel-unshuffle. PS: pixel-shuffle.

PUS	PS	PSNR	SSIM	Params (M)	MACs (G)
		31.54	0.949	6.61	621.1
✓		31.67	0.950	6.61	621.6
✓	✓	31.75	0.951	6.61	621.6

Pixel-Shuffle-Based Multi-scale Scheme. We then verify the effect of our pixel-shuffle-based multi-scale scheme. As discussed in Sect. 4, our pixel-shuffle-based multi-scale scheme consists of pixel-unshuffle layers that generate input tensors, and auxiliary pixel-shuffle layers used only in the training phase. To verify the effect of each component, we compare the performance of three variants of MSSNet: 1) without both pixel-unshuffle and shuffle layers, 2) with only the pixel-unshuffle layers, and 3) with both layers in Table 7. The first model takes downsampled images as input as done in previous coarse-to-fine approaches, and its sub-networks at S_1 and S_2 are trained to produce intermediate results of the corresponding sizes. The second model takes tensors generated by pixel-unshuffling layers as input, but its sub-networks in S_1 and S_2 are trained in the same manner as the first model. The third model corresponds to our approach.

As Table 7 shows, introducing the pixel-unshuffling and shuffling layers introduces a negligible increase in the number of parameters. On the other hand, the pixel-unshuffling layers clearly improve the deblurring quality as they provide richer information than downsampling. Also, the auxiliary pixel-shuffling layers further improve the deblurring quality as they enable higher-quality supervision.

6 Conclusion

In this work, we analyzed the defects of previous coarse-to-fine single image deblurring approaches. Based on our analysis, we proposed a novel coarse-to-fine approach with our remedies: stage configuration reflecting blur scales, inter-scale feature propagation, and pixel-shuffle-based multi-scale network architecture. The experiment results prove the effectiveness of our remedies.

Limitations and Future Work. While MSSNet achieves the state-of-the-art performance, it still fails on many real-world blurred images especially with large blur as other methods. Extending MSSNet for handling large blur can be an interesting future work. Our analysis is focused on conventional CNN-based architectures. Extending our work to transformer-based architectures would be an interesting future work. We also plan to examine the performance of MSSNet on other types of image degradation.

Acknowledgements. This work was supported by IITP grants funded by the Korea government (MSIT) (Artificial Intelligence Graduate School Program (POSTECH) No. 2019-0-01906 and SW Star Lab No. 2015-0-00174) and National Research Foundation of Korea (NRF) grant funded by the Korea government (MSIT) (NRF2018R1A5A-1060031, No. 2020R1C1C1014863).

References

1. Chakrabarti, A.: A neural approach to blind motion deblurring. In: Leibe, B., Matas, J., Sebe, N., Welling, M. (eds.) ECCV 2016. LNCS, vol. 9907, pp. 221–235. Springer, Cham (2016). https://doi.org/10.1007/978-3-319-46487-9_14
2. Chen, L., Lu, X., Zhang, J., Chu, X., Chen, C.: HINet: half instance normalization network for image restoration. In: CVPR Workshops, pp. 182–192 (2021)
3. Cho, S.J., Ji, S.W., Hong, J.P., Jung, S.W., Ko, S.J.: Rethinking coarse-to-fine approach in single image deblurring. In: ICCV, pp. 4641–4650 (2021)
4. Cho, S., Lee, S.: Fast motion deblurring. ACM Trans. Graph. **28**(5), 1–8 (2009)
5. Cho, S., Lee, S.: Convergence analysis of map based blur kernel estimation. In: ICCV, pp. 4808–4816 (2017)
6. Dosovitskiy, A., et al.: An image is worth 16×16 words: transformers for image recognition at scale. In: ICLR (2021)
7. Fergus, R., Singh, B., Hertzmann, A., Roweis, S.T., Freeman, W.T.: Removing camera shake from a single photograph. ACM Trans. Graph. **25**(3), 787–794 (2006)
8. Gao, H., Tao, X., Shen, X., Jia, J.: Dynamic scene deblurring with parameter selective sharing and nested skip connections. In: CVPR, pp. 3848–3856 (2019)
9. Hradiš, M., Kotera, J., Zemcık, P., Šroubek, F.: Convolutional neural networks for direct text deblurring. In: BMVC, pp. 6.1-6.13 (2015)
10. Hu, X., et al.: Pyramid architecture search for real-time image deblurring. In: ICCV, pp. 4298–4307 (2021)
11. Hu, Z., Cho, S., Wang, J., Yang, M.H.: Deblurring low-light images with light streaks. In: CVPR, pp. 3382–3389 (2014)
12. Huang, G., Liu, Z., Van Der Maaten, L., Weinberger, K.Q.: Densely connected convolutional networks. In: CVPR, pp. 2261–2269 (2017)
13. Jia, J.: Single image motion deblurring using transparency. In: CVPR, pp. 1–8 (2007)
14. Kim, J., Lee, J.K., Lee, K.M.: Accurate image super-resolution using very deep convolutional networks. In: CVPR, pp. 1646–1654 (2016)
15. Kingma, D.P., Ba, J.: Adam: a method for stochastic optimization. In: ICLR (2015)
16. Kupyn, O., Budzan, V., Mykhailych, M., Mishkin, D., Matas, J.: DeblurGAN: blind motion deblurring using conditional adversarial networks. In: CVPR, pp. 8183–8192 (2018)
17. Kupyn, O., Martyniuk, T., Wu, J., Wang, Z.: DeblurGAN-v2: deblurring (orders-of-magnitude) faster and better. In: ICCV, pp. 8878–8887 (2019)
18. Lai, W.S., Huang, J.B., Ahuja, N., Yang, M.H.: Fast and accurate image super-resolution with deep laplacian pyramid networks. IEEE Trans. Pattern Anal. Mach. Intell. **41**(11), 2599–2613 (2018)
19. Levin, A., Weiss, Y., Durand, F., Freeman, W.T.: Understanding and evaluating blind deconvolution algorithms. In: CVPR, pp. 1964–1971 (2009)
20. Levin, A., Weiss, Y., Durand, F., Freeman, W.T.: Efficient marginal likelihood optimization in blind deconvolution. In: CVPR, pp. 2657–2664 (2011)
21. Loshchilov, I., Hutter, F.: SGDR: Stochastic gradient descent with warm restarts. In: ICLR (2017)
22. Nah, S., Kim, T.H., Lee, K.M.: Deep multi-scale convolutional neural network for dynamic scene deblurring. In: CVPR (2017)
23. Pan, J., Sun, D., Pfister, H., Yang, M.H.: Blind image deblurring using dark channel prior. In: CVPR, pp. 1628–1636 (2016)

24. Park, D., Kang, D.U., Kim, J., Chun, S.Y.: Multi-temporal recurrent neural networks for progressive non-uniform single image deblurring with incremental temporal training. In: Vedaldi, A., Bischof, H., Brox, T., Frahm, J.-M. (eds.) ECCV 2020. LNCS, vol. 12351, pp. 327–343. Springer, Cham (2020). https://doi.org/10.1007/978-3-030-58539-6_20
25. Purohit, K., Rajagopalan, A.N.: Region-adaptive dense network for efficient motion deblurring. In: AAAI, pp. 11882–11889 (2020)
26. Rim, J., Lee, H., Won, J., Cho, S.: Real-world blur dataset for learning and benchmarking deblurring algorithms. In: Vedaldi, A., Bischof, H., Brox, T., Frahm, J.-M. (eds.) ECCV 2020. LNCS, vol. 12370, pp. 184–201. Springer, Cham (2020). https://doi.org/10.1007/978-3-030-58595-2_12
27. Ronneberger, O., Fischer, P., Brox, T.: U-net: convolutional networks for biomedical image segmentation. In: International Conference on Medical Image Computing and Computer-Assisted Intervention, pp. 234–241 (2015)
28. Schuler, C.J., Hirsch, M., Harmeling, S., Schölkopf, B.: Learning to deblur. IEEE Trans. Pattern Anal. Mach. Intell. **38**(7), 1439–1451 (2015)
29. Shan, Q., Jia, J., Agarwala, A.: High-quality motion deblurring from a single image. ACM Trans. Graph. **27**(3), 1–10 (2008)
30. Shi, W., et al.: Real-time single image and video super-resolution using an efficient sub-pixel convolutional neural network. In: CVPR, pp. 1874–1883 (2016)
31. Suin, M., Purohit, K., Rajagopalan, A.: Spatially-attentive patch-hierarchical network for adaptive motion deblurring. In: CVPR, pp. 3606–3615 (2020)
32. Sun, J., Cao, W., Xu, Z., Ponce, J.: Learning a convolutional neural network for non-uniform motion blur removal. In: CVPR, pp. 769–777 (2015)
33. Sun, L., Cho, S., Wang, J., Hays, J.: Edge-based blur kernel estimation using patch priors. In: ICCP, pp. 1–8 (2013)
34. Tao, X., Gao, H., Shen, X., Wang, J., Jia, J.: Scale-recurrent network for deep image deblurring. In: CVPR (2018)
35. Vaswani, A., et al.: Attention is all you need. In: NIPS, vol. 30 (2017)
36. Wang, Z., Cun, X., Bao, J., Zhou, W., Liu, J., Li, H.: Uformer: a general u-shaped transformer for image restoration. In: CVPR, pp. 17683–17693 (2022)
37. Xu, L., Jia, J.: Two-phase kernel estimation for robust motion deblurring. In: Daniilidis, K., Maragos, P., Paragios, N. (eds.) ECCV 2010. LNCS, vol. 6311, pp. 157–170. Springer, Heidelberg (2010). https://doi.org/10.1007/978-3-642-15549-9_12
38. Xu, L., Zheng, S., Jia, J.: Unnatural l0 sparse representation for natural image deblurring. In: CVPR, pp. 1107–1114 (2013)
39. Xu, X., Pan, J., Zhang, Y.J., Yang, M.H.: Motion blur kernel estimation via deep learning. IEEE Trans. Image Process. **27**(1), 194–205 (2018)
40. Zamir, S.W., Arora, A., Khan, S., Hayat, M., Khan, F.S., Yang, M.H.: Restormer: efficient transformer for high-resolution image restoration. In: CVPR, pp. 5728–5739 (2022)
41. Zamir, S.W., Arora, A., Khan, S., Hayat, M., Khan, F.S., Yang, M.H., Shao, L.: Multi-stage progressive image restoration. In: CVPR, pp. 14821–14831 (2021)
42. Zhang, H., Dai, Y., Li, H., Koniusz, P.: Deep stacked hierarchical multi-patch network for image deblurring. In: CVPR, pp. 5978–5986 (2019)
43. Zhang, J., et al.: Dynamic scene deblurring using spatially variant recurrent neural networks. In: CVPR, pp. 2521–2529 (2018)
44. Zhang, K., Zuo, W., Chen, Y., Meng, D., Zhang, L.: Beyond a gaussian denoiser: residual learning of deep CNN for image denoising. IEEE Trans. Image Process. **26**(7), 3142–3155 (2017)

RCBSR: Re-parameterization Convolution Block for Super-Resolution

Si Gao, Chengjian Zheng(✉), Xiaofeng Zhang, Shaoli Liu, Biao Wu, Kaidi Lu, Diankai Zhang, and Ning Wang

State Key Laboratory of Mobile Network and Mobile Multimedia Technology, ZTE, Shenzhen 518057, China
zheng.chengjian@zte.com.cn
https://www.zte.com.cn/global/

Abstract. Super resolution(SR) with high efficiency and low power consumption is highly demanded in the actual application scenes. In this paper, We designed a super light-weight SR network with strong feature expression. The network we proposed is named RCBSR. Based on the novel technique of re-parameterization, we adopt a block with multiple paths structure in the training stage and merge multiple paths structure into one single 3×3 convolution in the inference stage. And then the neural architecture search(NAS) method is adopted to determine amounts of block M and amounts of channel C. Finally, the proposed SR network achieves a fairly good result of PSNR(27.52 dB) with power consumption(0.1 W@30 fps) on the MediaTek Dimensity 9000 platform in the challenge testing stage.

Keywords: Light-weight · Super-resolution · Multiple paths structure · Re-parametrization · NAS

1 Introduction

Super-resolution (SR) is a fundamental task of computer vision that has very important values in fields such as video surveillance, medical detection, satellite imaging, high-definition televisions, etc. So SR has attracted lots of research interests. SR aims at recovering high-resolution (HR) images from their low-resolution (LR) counterparts. Since SRCNN [3] firstly adopted deep learning in SR, deep learning based methods have achieved a significant performance boost compared with traditional methods. However, most works increase the depth and width of the SR network to further expand the receptive field and enhance representation ability. Therefore, the complexity of the network has also increased, which makes it difficult to deploy SR networks on resource-limited terminal devices. Many methods have proposed efficient SR networks or networks block for this difficulty, trying to strike a balance between SR effect and efficiency, but their deployment on mobile terminals still faces challenges.

L. Karlinsky et al. (Eds.): ECCV 2022 Workshops, LNCS 13802, pp. 540–548, 2023.
https://doi.org/10.1007/978-3-031-25063-7_33

Thanks to the rapid development of artificial intelligence-related software and hardware, such as TensorFlow Lite, MediaTek APU. After a successful real-time Video Super-Resolution challenge [6] on smartphones last year, Advances in Image Manipulation (AIM) workshop and Mobile AI (MAI) workshop held another real-time super-resolution challenge [9], which requires participants to design a real-time SR network suitable for mobile terminals, and the SR effect should be better than the benchmark. Due to the limited energy of the mobile power supply in practical applications, the designed SR network must consider the power consumption as well. In response to the challenge, this paper describes an efficient SR networks with low power consumption. We used the MediaTek Dimensity 9000 as the target platform, and the AI benchmark [7,8] provided by the challenge organizer is used as a test tool, which can test the efficiency and power consumption of the network. We first select the SR basic module, which can obtain the ability of good feature expression during training, and can be optimized into a 3×3 module through re-parameterization during testing, which greatly improves its inference efficiency. Then the configuration of the network size is obtained through NAS, and finally a network that meets the real-time super-resolution requirements is obtained.

2 Related Work

2.1 Light-Weight Image Super-Resolution

In order to improve the efficiency of the super-resolution, many researchers focus on reducing the number of parameters and FLOPs. IMDN [5] proposed a efficient information multi-distillation network which constructed the cascaded information multi-distillation blocks based on information distillation mechanism (IDM). IMDN is not efficient enough by splitting the channel. RFDN [17] proposed the residual feature distillation network which replaced IDM of IMDN with the residual feature distillation blocks. To further improve the efficiency of RFDN, Residual Local Feature Network(RLFN) [14] enhances the model compactness and accelerates the inference by reducing the network fragments.

2.2 Re-parameterization

Re-parameterization is a novel technique to boost the representative capacity with multiple paths in the training stage and accelerate the inference of the network by merging multiple paths into one single 3×3 convolution in the inference stage. ECBSR [21] proposed an Edge-oriented Convolution Block (ECB), which can more effectively extract edge and texture information, and the ECB consists of four types of operators, which are a normal 3×3 convolution, expanding-and-squeezing convolution, sequential convolution with scaled Sobel filters and sequential convolution with scaled Laplacian filters. RepSR [20] re-introduced BN into re-parameterizable block, which introduced non-linearity during training and improved the final performance.

2.3 Light-Weight Video Super-Resolution

Video super-resolution stems from image super resolution, which can be mainly divided into the methods with aligment and the methods without aligment [16]. The methods with alignment extract motion information to make neighboring frames explicitly align with the target frame.

VSRnet [11]includes motion compensation, motion estimation and three convolution layers, and the filter symmetry enforcement mechanism and adaptive motion compensation mechanism are proposed to improve the effect. The input is multiple frames after compensation. The motion information adopts druleas algorithm. In BasicVSR [1], it adopts bidirectional propagation to exploit information from the entire input video for reconstruction and optical flow for alignment. BasicVSR++ [2] proposes second-order grid propagation and flowguided deformable alignment network based on BasicVSR. The methods without alignment do not make neighboring frames align with the target frame. Recurrent Residual Network (RRN) [10]proposed a novel hidden state for the recurrent network, which exploited the previous and current frame as hidden state input. STCN [4] is the spatio-temporal convolutional network, which extracts the temporal information within frames by using LSTM and features from multiple consecutive LR frames for Spatial module.

Although the above studies obtain good reconstructed frame quality, they do not focus on the energy consumption of the mobile terminal. In this work, we propose a lightweight network structure to obtain better a trade-off between SR quality and the energy consumption.

3 Approach

We adopt ECBSR [21] as our baseline. In consideration of the low power consumption of challenge requirement, we optimize the baseline from three aspects, network architecture, NAS and training strategy.

3.1 Optimizing Network Architecture

The network architecture is based on ECB, shown in Fig. 1.

Firstly, we use the re-parameterization technique in the deploy stage, it can reduce power consumption while keeping PSNR. Secondly, we replace the activate function PReLU with ReLU. Experimental results demonstrate that the power consumption of tflite model with ReLU is less than PReLU. Meanwhile there is no apparent discrepancy in PSNR. Finally, in order to further reduce power consumption, the output of first CNN layer is added into the backbone output instead of original input because original input needs to be copied the number of channels. We use sub-pixel convolution [19] to upsample image in the network.

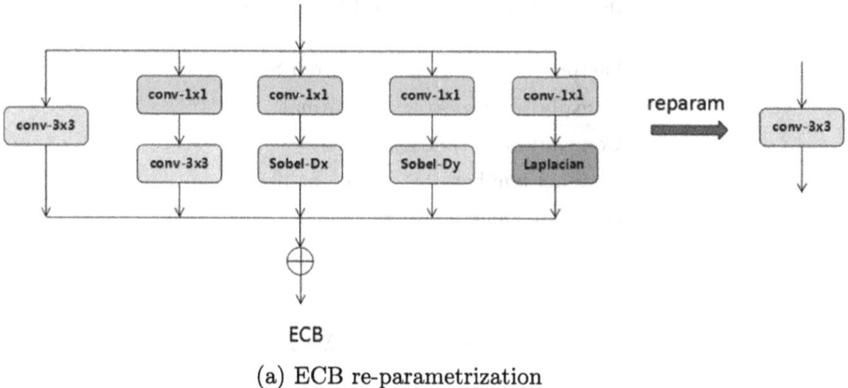

(a) ECB re-parametrization

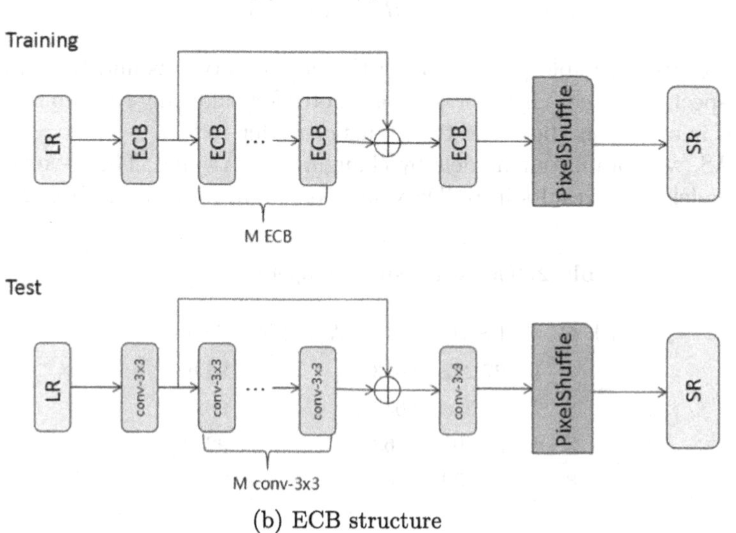

(b) ECB structure

Fig. 1. ECB network

3.2 NAS

In order to pursue a trade-off between the restoration capacity and the energy consumption of model, FGNAS [12] is adopted to search for the model. Table 1. illustrates the search space of operations.

The objective function of FGNAS is twofold, one is the task-specific loss and the other is regularizer penalties such as parameters, FLOPs and latency. In our work, we use FLOPs of network as the regularizer penalty because of its correlation with power consumption. $L(\cdot, \cdot)$ denotes loss function of our task and $R(\cdot)$ is a differentiable regularizer that estimates the resources of the current model identified by our search algorithm. The objective function is formally given by formula (1).

Table 1. Traing results of different models

Factor	Search Space
Convolution types	Normal
Convolution kernel sizes	1, 3
Channel numbers	8, 16, 32
Residual Block numbers	1,..., 8

$$\min_{\theta,\psi} L(\theta,\psi) + \lambda * R(\psi) \quad (1)$$

$$score = \alpha * PSNR + \beta * (1 - J) \quad (2)$$

where θ and ψ are learnable parameters in the neural networks and the gating functions respectively, and λ is the hyper-parameter for balancing the two terms. The detailed information about the two terms can refer to [12].

After NAS, we obtain four models by changing the λ value. Then We train these four models. The results in REDS validation set are shown in Table 2.

Table 2. The search space of operations

Model	C	M	PSNR	Power, W/30FPS	Score
1	8	1	27.282	0.013	94.63
2	8	2	27.358	0.032	93.81
3	8	4	27.497	0.063	92.49
4	8	8	27.639	0.106	90.58

In Table 2, C represents amounts of channel, M represents amounts of block. The score is calculated according to formula (2) specified in the competition rulers, where α is equal to 1.66, and β is equal to 50, J represent energy consumption per inference (Watt @ 30fps). According to the final score, we choose model with C = 8, M = 1 as our challenge network.

3.3 Optimizing Training Strategy

We train the model directly with input data of the range[0, 255] to reduce the time of data preprocess. Then we abandon L1 loss function because it causes the problem that the restored image is too smooth and lack of sense of reality. Charbonnier loss function [15] we adopted is a very stable loss function, and loss curve is smoother and the gradient of close to the zero point is modest so as to avoid the gradient disappearance and gradient explosion. Finally, L2 loss is used for fine-tuning.

4 Experiments

4.1 Dataset

Mobile AI & AIM 2022 real-time video super-resolution challenge uses the REDS [18] dataset, which have a large diversity of contents and dynamic scenes. It is widely used in video super-resolution and video denoising tasks. REDS dataset consists of 300 video sequences containing 100 frames of 720×1280 resolution. To generate the LR data, the videos are bicubic downsampled by scale 4. In this challenge, the dataset is divided into 240 sequences for training, 30 sequences for validation, and 30 sequences for testing. We just use REDS as the training dataset without additional data.

4.2 Implementation Details

We adopt REDS as the training set and the training of RCBSR is divided into two stages. In the first stage, the network is trained with a batch size of 64 on 512×512 pixel input images augmented by random flipping and rotation. Charbonnier loss [15] is used as a target metric. The model parameters are optimized for 4000 epochs using Adam [13] optimizer by setting $\beta 1 = 0.9$, $\beta 2 = 0.999$ with a learning rate that is initialized at 5e−4 and decreased by half every 1000 epochs. In the second stage, the network is fine-tuned with L2 loss and the learning rate is initialized at 2e−4 based on the pretrain model obtained in the first stage. The other parameters of training remain unchanged. Model implementation, training and exporting are conducted using Python 3.7.7 and Pytorch 1.7.1. Model training is on an Nvidia GTX 1080Ti GPU with 256G of RAM.

4.3 Comparisons with Bicubic Method

Figure 2 depicts the $\times 4$ VSR results of the REDS val dataset. Zoom in for better visualization. As shown on the right side of Fig. 2. The first row is the result of bicubic, the second row is the result of ours, The third row is the ground truth. It is obvious that our method produces sharper edges and finer details than bicubic method. At the same time, our VSR results is more close to the ground truth.

4.4 Results of the Real-Time VSR Challenge

In this challenge, REDS test set is used to evaluate the quality of the reconstructed results, and the runtime of the model on the actual MediaTek Dimensity 9000 SoC is executed by AI benchmark. The results of validation is shown in Table 2, we achieve a competitive score of 94.63 on val challenge set where the PSNR is 27.28 dB and the SSIM is 0.775 while the inference time is 3.09 ms/f and the power is 0.013 W with the configuration of 1 blocks and 8 channels. In final test phase, We take second place with score 90.7, where the PSNR is 27.52 dB, the SSIM is 0.7872 and power is 0.1 W@30 fps. The test and val results are from the organizer, more details refer to [9]. It is noteworthy that, our method achieves excellent efficiency while maintaining the performance.

(a) 39th frame of sequence 009

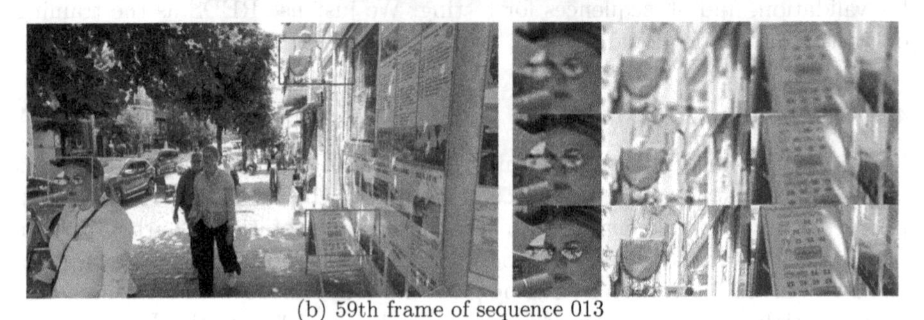

(b) 59th frame of sequence 013

Fig. 2. Qualitative comparison on the REDS val datasets.

5 Conclusion

In this paper, we propose a real-time video super-resolution network RCBSR, which mainly composed of one multi-branch block and sub-pixel convolution layer. We show that our method is low power consumption and can be easily deployed to mobile devices. In the future, we will further study the characteristics of video super-resolution to achieve a better balance between quality and power consumption.

Acknowledgements. I would like to express my gratitude to all those who helped me during the experiment. My deepest gratitude goes first and foremost to our team for their instructive advice and usefull suggestions. I am deeply gratefull of their help and participation in this competition.

High tribute shall be paid to our leader, who encourages us to participate in the competition and realize our self-worth. Special thanks should go to colleagues from other departments for providing the cloud server for training. Finally, I am indebted to competition organizers for hosting the challenges and providing the opportunity to us.

References

1. Chan, K.C., Wang, X., Yu, K., Dong, C., Loy, C.C.: BasicVSR: the search for essential components in video super-resolution and beyond. In: Proceedings of the IEEE/CVF Conference on Computer Vision and Pattern Recognition, pp. 4947–4956 (2021)
2. Chan, K.C., Zhou, S., Xu, X., Loy, C.C.: Basicvsr++: improving video super-resolution with enhanced propagation and alignment. In: Proceedings of the IEEE/CVF Conference on Computer Vision and Pattern Recognition, pp. 5972–5981 (2022)
3. Dong, C., Loy, C.C., He, K., Tang, X.: Image super-resolution using deep convolutional networks. IEEE Trans. Pattern Anal. Mach. Intell. **38**(2), 295–307 (2015)
4. Guo, J., Chao, H.: Building an end-to-end spatial-temporal convolutional network for video super-resolution. In: Thirty-First AAAI Conference on Artificial Intelligence (2017)
5. Hui, Z., Gao, X., Yang, Y., Wang, X.: Lightweight image super-resolution with information multi-distillation network. In: Proceedings of the 27th ACM International Conference on Multimedia, pp. 2024–2032 (2019)
6. Ignatov, A., Romero, A., Kim, H., Timofte, R.: Real-time video super-resolution on smartphones with deep learning, mobile AI 2021 challenge: report. In: Proceedings of the IEEE/CVF Conference on Computer Vision and Pattern Recognition, pp. 2535–2544 (2021)
7. Ignatov, A., et al.: AI benchmark: running deep neural networks on android smartphones. In: Leal-Taixé, L., Roth, S. (eds.) ECCV 2018. LNCS, vol. 11133, pp. 288–314. Springer, Cham (2019). https://doi.org/10.1007/978-3-030-11021-5_19
8. Ignatov, A., et al.: Ai benchmark: all about deep learning on smartphones in 2019. In: 2019 IEEE/CVF International Conference on Computer Vision Workshop (ICCVW), pp. 3617–3635. IEEE (2019)
9. Ignatov, A., Timofte, R., Kuo, H.K., Lee, M., Xu, Y.S., et al.: Real-time video super-resolution on mobile NPUs with deep learning, mobile AI & AIM 2022 challenge: report. In: Proceedings of the European Conference on Computer Vision (ECCV) Workshops (2022)
10. Isobe, T., Zhu, F., Jia, X., Wang, S.: Revisiting temporal modeling for video super-resolution. arXiv preprint arXiv:2008.05765 (2020)
11. Kappeler, A., Yoo, S., Dai, Q., Katsaggelos, A.K.: Video super-resolution with convolutional neural networks. IEEE Trans. Comput. Imaging **2**(2), 109–122 (2016)
12. Kim, H., Hong, S., Han, B., Myeong, H., Lee, K.M.: Fine-grained neural architecture search. arXiv preprint arXiv:1911.07478 (2019)
13. Kingma, D.P., Ba, J.: Adam: A method for stochastic optimization. arXiv preprint arXiv:1412.6980 (2014)
14. Kong, F., et al.: Residual local feature network for efficient super-resolution. In: Proceedings of the IEEE/CVF Conference on Computer Vision and Pattern Recognition, pp. 766–776 (2022)
15. Lai, W.S., Huang, J.B., Ahuja, N., Yang, M.H.: Fast and accurate image super-resolution with deep Laplacian pyramid networks. IEEE Trans. Pattern Anal. Mach. Intell. **41**(11), 2599–2613 (2018)
16. Liu, H., et al.: Video super-resolution based on deep learning: a comprehensive survey. Artif. Intell. Rev. **55**, 5981–6035 (2022)
17. Liu, J., Tang, J., Wu, G.: Residual feature distillation network for lightweight image super-resolution. In: Bartoli, A., Fusiello, A. (eds.) ECCV 2020. LNCS, vol. 12537, pp. 41–55. Springer, Cham (2020). https://doi.org/10.1007/978-3-030-67070-2_2

18. Nah, S., et al.: NTIRE 2019 challenge on video deblurring and super-resolution: Dataset and study. In: Proceedings of the IEEE/CVF Conference on Computer Vision and Pattern Recognition Workshops, pp. 0–0 (2019)
19. Shi, W., et al.: Real-time single image and video super-resolution using an efficient sub-pixel convolutional neural network. In: Proceedings of the IEEE Conference on Computer Vision and Pattern Recognition, pp. 1874–1883 (2016)
20. Wang, X., Dong, C., Shan, Y.: REPSR: training efficient VGG-style super-resolution networks with structural re-parameterization and batch normalization. arXiv preprint arXiv:2205.05671 (2022)
21. Zhang, X., Zeng, H., Zhang, L.: Edge-oriented convolution block for real-time super resolution on mobile devices. In: Proceedings of the 29th ACM International Conference on Multimedia, pp. 4034–4043 (2021)

Multi-patch Learning: Looking More Pixels in the Training Phase

Lei Li, Jingzhu Tang, Ming Chen, Shijie Zhao(✉), Junlin Li, and Li Zhang

ByteDance Inc., Culver City, USA
{lilei.leili,zhaoshijie.0526}@bytedance.com

Abstract. Due to the limitations of computation capability and memory of GPUs, most image restoration tasks are trained with cropped patches instead of full-size images. Existing extensive experiments show that the model trained with a larger patch size could achieve better performance since a larger patch size typically means larger receptive fields. However, it comes at the cost of extremely long training times and significant memory consumption. To alleviate the dilemma mentioned above, we propose a multi-patch method to expand the receptive field with negligible memory and computation increase (less than 1%). In addition, we collect 100K high-quality images of 1K categories, following ImageNet, from *flickr.com* for low-level image tasks. Our method improves the quantitative performance by 0.3412dB on the validation set of the "Compressed Input Super-Resolution Challenge - Image Track".

Keywords: Super resolution · Receptive field · Patch-based training

1 Introduction

Single-image super-resolution (SISR) aims to recover high-resolution (HR) images from corresponding degraded low-resolution (LR) images. It is a classical problem in computer vision. Following the pioneering work SRCNN [5], numerous works employing convolutional neural networks (CNN) for SISR have achieved promising results [3,9,14,21,22]. Recently, Transformer has attracted attention in computer vision due to their superior performance in natural language processing tasks. Compared with the CNN-based image restoration models, the Transformer-based methods benefit from the self-attention mechanism, which enhances content-dependent interactions, long-range dependencies modeling, etc. The superior performance has been demonstrated on several vision tasks [1,2,7,11–13,15,17,18,20].

For high-level vision tasks, input images can be downsampled before send to the network to reduce computations. However, image downsampling should be avoided in low-level tasks to preserve the high-frequency details of the image. For low-level tasks, networks are usually trained on cropped patches and infer on full images. Therefore, during the training phase, the network can only receive

L. Karlinsky et al. (Eds.): ECCV 2022 Workshops, LNCS 13802, pp. 549–560, 2023.
https://doi.org/10.1007/978-3-031-25063-7_34

information inside the patch. Although the network is getting much more powerful, the receptive field is always limited in the cropped patch during the training phase. Many experiments have demonstrated that bigger patch size always take better performance, but it is not a sanity choice to train the network on the full image due to the limitations of memory and computation capability of GPUs.

To alleviate the dilemma between receptive field and computational complexity, we propose a multi-patch training strategy, where the low-resolution patch and its surrounding eight patches are put into the network to reconstruct the center patch. In this way, with only an negligible memory increase, the network can see much more input pixels, expand the range of available information, and further enhance the effect of information interaction to explore self-similarity.

On the other hand, many Transformer-based image restoration works [1,2,7] achieve superior performances with the help of training on large datasets, such as ImageNet dataset [4]. However, the poor quality of the images from ImageNet dataset [4] may not meet the needs of recovering high-quality image details. To further excavated the potential of Transformer for restoration work, we create a large-scale, high-quality image dataset, named Flickr2K-L dataset. To ensure data diversity, we collected 100K high-quality images of 1000 categories according to the categories in ImageNet dataset [4]. Our contributions are summarized as follows:

- We propose a multi-patch method to extremely expand the receptive field in the training phase, which only increases negligible memory and computation.
- We create a large-scale, high-quality image dataset for low-level vision tasks, named Flickr2K-L dataset, which contains 100K high-resolution images covering 1000 categories.
- Extensive experiments demonstrate the effectiveness of the proposed method.

2 Related Work

2.1 Deep CNN for Image Super-Resolution

With the rapid development of deep learning technology, learning-based image super-resolution (SR) technology has progressed rapidly. Dong *et al.* [5] propose the pioneering work SRCNN with a 3-layer convolutional neural network (CNN), which obtains superior performance over conventional SR methods. Following them, numerous CNN-based [6] methods have been proposed for SR to improve its performance further. Network depth has been critical for super-resolution methods because deeper networks have larger receptive fields [9,14,21]. Lim *et al.* [14] uses residual connections to accelerate the convergence of a deep network (EDSR). Zhang *et al.* [21] further increase the network depth with the help of dense residual connections. The attention mechanism also demonstrates high effectiveness for super-resolution methods [3,21,22]. It can be viewed as guidance to bias the allocation of available processing resources toward the most informative components of an input [8]. Zhang *et al.* [21] introduce a channel attention mechanism to adaptively rescale channel-wise features

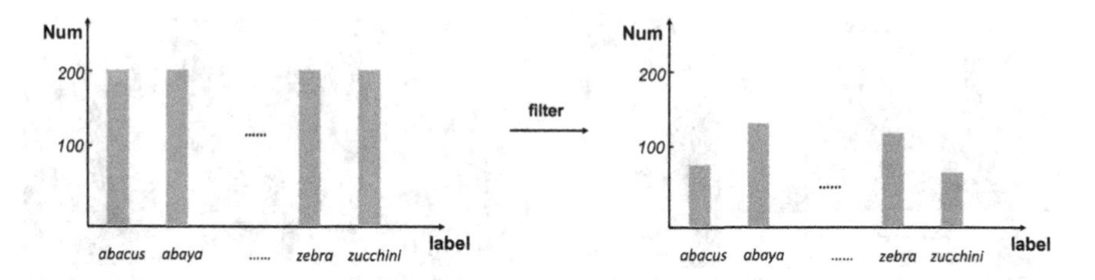

Fig. 1. Distribution of Flickr2K-L. We first collect images of 1000 categories according to the category labels of ImageNet [4]. For each category, we collect 200 images. Then we filter the images based on the image qualities and keep the image number of each category not less than 60.

based on the interdependencies among channels. Dai *et al.* [3] propose a second-order attention network (SAN) for image super-resolution, including local attention and non-local attention. Although CNN-based SR methods are significantly improved based on residual learning, dense connections, attention mechanism, etc., they generally suffer from two basic limitations of convolution layer. First, the convolution kernels are the same for the full image. It may not be optimal to use the same convolution kernel to restore different image regions. Second, the convolution layer can not effectively capture the long-range dependency because of its principle of local processing (Fig. 1).

2.2 Vision Transformer

Recently, Transformer becomes popular in the computer vision community because of its superior performance in nature language processing tasks. It attends to important image regions by learning the global correlations of different regions. Transformer has been widely used in high-level vision tasks, such as image classification [7,11,15], segmentation [17,18] and so on. Due to its superior performance, Transformer has been introduced into low-level vision tasks, including super-resolution [2,12], restoration [1,13,20], *etc.*. Chen *et al.* [1] propose a image restoration network (IPT) based on standard Transformer. With the help of pre-training on a large dataset, IPT [1] achieves superior performance against CNN-based restoration methods in multiple tasks, including super-resolution, denoising and deraining. Liang *et al.* [13] adopt shifted window [15] to improve long-range dependency. However, the window size and network depth are limited by memory and computation, which prevents global interactions. Chen *et al.* [2] propose channel attention and overlapping cross-attention module to activate more pixels in a Transformer block to enhance the global interactions. To further enhance the information interactions, we propose a multi-patch method to let the network see more pixels. We input the low-resolution patch and its eight surrounding patches into the network to reconstruct the super-resolution image of the center patch. In this way, we can significantly increase the receptive field of the network while increasing very little memory and training time.

Fig. 2. Sample images in Flickr2K-L dataset.

3 Methods

We use a Transformer-based super-resolution network HAT [2] as our backbone network. Since the potential of Transformer needs to be excavated using large-scale dataset [1,2,10], we create a large high-quality dataset, Flickr2K-L dataset, for training, as shown in Sect. 3.1. To further enlarge the receptive field of Hat [2] in the training phase, we propose a multi-patch training strategy, as shown in Sect. 3.2.

3.1 Flickr2K-L Dataset

Our Flickr2K-L dataset contains 100K high-quality images. We present nine sample images from Flickr2K-L dataset in Fig. 2. We construct the Flickr2K-L dataset by image collection, data filter and post-processing. More details are shown below.

Image Collection. We collect images using *flickr* API[1]. To ensure the diversity of our dataset, we collect 1K categories of images following the image categories of ImageNet [4]. For each category, we collect 200 images of 4K resolution. Then, we obtain a large image dataset with 200K images in total.

[1] https://www.flickr.com.

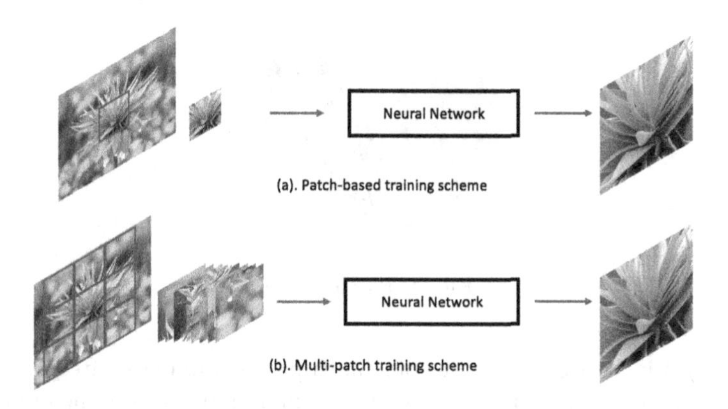

Fig. 3. Training strategy. (a) is the traditional patch-based training strategy, and (b) represents the multi-patch training strategy. The low-resolution input patch and its eight surrounding patches are cropped and then input into the neural network to reconstruct the super-resolved image of the center patch. The backbone neural network we choose in this paper is HAT [2].

Data Filter and Post-processing. We first remove the images with dissonant resolutions to standardize the image dataset. For example, we remove images with a width or height of less than 1000. We also remove the low-quality images based on the quality assessment metrics VQScore [16] and keep the number of images in each category not less than 60. After filtering, our dataset contains 100K images, and the number of images in each category is not less than 60. The filtered images' url can be found in github[2]. To improve the per-pixel quality of the images, we downscale the original 4K images by two times. Finally, we get the Flickr2K-L dataset with 100K images of 2K resolution.

3.2 Multi-patch Training Strategy

For most low-level tasks, such as image super-resolution, the networks are trained with cropped patches rather than the full images. Thus, the network can only look at the pixels inside the patches during the training phase, even though the network's ability is becoming much more powerful and the receptive field of the neural network is very large. Thus, the training patch size is critical for the network's performance. Larger patch sizes usually result in better performance. However, due to the computational capability and memory constraints of GPUs, it is not practical to train the network with full images. To solve this problem, we propose a multi-patch training strategy to significantly increase the receptive field in the training phase with very little memory increase.

Figure 3 illustrates the training processes of the super-resolution network. In traditional patch-based training scheme, the input is a cropped low-resolution image patch, and the output is the corresponding high-resolution patch, as shown

[2] https://github.com/scholarboss/MPHAT/blob/main/data/flickr2k-L.csv.

Fig. 4. Patch group.

in Fig. 3(a). Figure 3(b) illustrates our multi-patch training strategy. We crop a low-resolution image patch and its eight surrounding patches as the inputs of the neural network. The network output is the high-resolution patch corresponding to the center patch. We choose HAT [2] as the backbone network. We name our super-resolution network trained with multi-patch strategy as Multi Patches Hybrid Attention Transformer (MPHAT) network. We just change the input channels of HAT [2] to construct the MPHAT network. Table 3 presents the ablation study results of the multi-patch strategy on this competition's validation data.

We regard the central patch and its eight surrounding patches, nine patches in total, as a patch group. As shown in Fig. 4, we denote the patches in the group as $\{p_1, \ldots, p_9\}$ from top to bottom and left to right. In this way, the eight surrounding patches may extend beyond the edge of the image. In this situation, we use zero padding to expand the image. Then we extract the patch group from the padded image.

After getting the patch group, we do not directly concatenate them as the network inputs. It is well known that spatially close pixels have strong spatial correlations. Considering the spatial correlation of the patches, we flip the eight surrounding patches before feeding them into our network. For example, the bottom side of p_2 is close to the top side of p_5. Thus, we flip p_2 vertically and then concatenate p_5 and the flipped p_2 as network inputs to make the bottom side of p_2 spatially close to the top side of p_5. Similarly, for p_1, p_3, p_7, p_9, we flip them horizontally and vertically as follow:

$$p_i^{'} = f_v(f_h(p_i)),\ i \in \{1, 3, 7, 9\}, \tag{1}$$

where f_v and f_h denote flip vertically and horizontally, respectively. For p_2 and p_8, we flip them vertically as below:

$$p_i^{'} = f_v(p_i),\ i \in \{2, 8\}. \tag{2}$$

For p_4 and p_6, we flip them horizontally:

$$p_i^{'} = f_h(p_i),\ i \in \{4, 6\}. \tag{3}$$

For the central patch p_5, we keep it as the original:

$$p_5^{'} = p_5. \tag{4}$$

The flipped patches $\{p_1^{'},\ldots,p_9^{'}\}$ are concatenated in the color channel:

$$I_{mp} = concatenate([p_1^{'},...,p_9^{'}]). \tag{5}$$

Compared to the regular input shape of [B,C,H,W], the concatenated multi-patch input I_{mp} has the shape of [B,C*9,H,W], which has nine times channel numbers of regular input. The concatenated multi-patch input I_{mp} is directly feeding to HAT for super-resolution as below:

$$I_{sr} = HAT(I_{mp}). \tag{6}$$

The architecture of HAT is described in [2]. We just change the input channels of the network.

4 Experiments

4.1 Implementation Detail

We implement the evaluations using the dataset provided by the AIM-2022 workshop "Compressed Input Super-Resolution - Image track". The degraded low-resolution images are bicubic downsampled from high-resolution images and then compressed with JPEG of quality factor 10. Our network is trained in Flickr2K-L datasets without any pre-training strategy.

In the training phase, we train the network using Adam optimizer with $\beta_1 = 0.9$ and $\beta_2 = 0.99$ to minimize the MSE loss. The model is trained for 800K iterations with batch size of 32 and patch size of 64×64. The learning rate is initialized as $1e-4$ and reduced by half at [300K, 500K, 650K, 700K, 750K] iterations. We implement our method with the PyTorch framework and train it using 8 NVIDIA Tesla A100 GPUs. For the edge regions of the image, we use zero padding.

In the testing phase, we test with the full image as input. The eight neighboring patches are filled with zero.

4.2 Quantitative Results

Table 1 shows the results of the competition on DIV2K test data. We won the second place in the competition [19].

In addition to the Flickr2K-L dataset and multi-patch training strategy, we also adopt self-ensemble and model voting schemes to further improve the performance of our method. As shown in Table 2, we train three models with different activation functions and model sizes for model voting.

As shown in Table 2, the first model has 12 RHAG module [2] with SiLU activation function; the second model has 18 RHAG module with GELU activation function; the third model has 18 RHAG module with SiLU activation function.

We use the DIV2K validation set from AIM 2022 workshop to evaluate the performances of the three models in Table 2. The peak signal-to-noise

Table 1. Final results of "AIM-2022 Challenges on Super-Resolution of Compressed Inputs - image track".

Rank	PSNR(dB)
1st	23.6677
2nd (ours)	**23.5731**
3rd	23.5597
4th	23.5307
5th	23.5085
6th	23.4345
7th	23.4249
8th	23.4239
9th	23.4033
10th	23.3250

Table 2. Quantitative results on DIV2K validation data.

Model	# of RHAG	GELU→ SiLU	PSNR	PSNR (self-ensemble ×8)
MPHAT	12	✓	23.7826	23.8109
	18	✗	23.8212	23.8376
	18	✓	23.8248	23.8373
Model voting	–	–	–	23.8412

ratio (PSNR) is adopted as the evaluation metric. As shown in Table 2, these three models have excellent but unequally performance on this task. To further improve the reconstruction performance, we adopt self-ensemble for the three models, which achieves 0.0125∼0.0283 dB performance gains. Furthermore, model voting improves our result to 23.8412 dB.

4.3 Qualitative Results

We present the visual results on the DIV2K validation set of AIM 2022 challenge in Fig. 5. As shown in Fig. 5, our proposed method recovers fine-grained spatial details. In addition, our method removes the block effects in the compressed low-resolution inputs.

4.4 Ablation Study

We conduct ablation experiments with the DIV2K validation set. As shown in Table 3, our created Flickr2K-L dataset introduces 0.1395 dB performance gain. The reason is that the large-scale, diversity and high quality of the Flickr2K-L dataset stimulate the ability of Transformer. In addition, our multi-patch model achieves a significant improvement of 0.2017 dB compared with traditional

Fig. 5. Qualitative results of our proposed method on DIV2K validation set.

patch-based training method. We also compared the flip way of our multi-patch method. Compared with concatenating patches directly, flipped as described in Sect. 3.2 can get better performance as shown in Table 3. Figure 6 demonstrates the significant subjective improvements by the proposed multi-patch training method. As shown in the first row in Fig. 6, the windows of the building are well reconstructed with multi-patch training but distorted without multi-patch training. The letters are indistinguishable without multi-patch training, as shown in the second row in Fig. 6.

4.5 AIM 2022 Challenge

Our method win the 2nd place in the image track of the AIM 2022 Compressed Input Super-Resolution Challenge. The goal of the challenge is to obtain a solution that recovers sharp results with optimal fidelity (PSNR) from compressed input images to ground truth. As shown in Table 1, we used the proposed

Table 3. Ablation studies on DIV2k validation data.

Model	Model1	Model2	Model3	Model4
# of RHAG	6	6	6	6
Multi-patch	✗	✗	✓	✓
Flip	✗	✗	✗	✓
Dataset for training	DIV2K	DF2K-L[1]	DF2K-L	DF2K-L
PSNR (dB)	23.2854	23.4249	23.6097	23.6266

[1]DF2K-L: DIV2K + Flickr2k-L

Fig. 6. Ablation study on multi-patch strategy.

multi-patch training strategy, and trained the model with a large-scale, high-quality dataset. Finally, we won the 2nd place. Note that the self-ensemble and the model ensemble are used to improve performance. Finally, our method reached 23.8412 dB on validation set and 23.5731 dB in test set, as shown in Tables 2 and 1.

5 Conclusions

In this paper, we propose a multi-patch training strategy, which achieves a gain of 0.2017 dB with negligible computation increase. In addition, we create a large-scale, high-quality image dataset, named Flickr2K-L dataset. Compared with the model trained with DIV2K dataset, the model trained with Flickr2K-L dataset achieves a 0.14dB gain on DIV2K validation set.

References

1. Chen, H., et al.: Pre-trained image processing transformer. In: Proceedings of the IEEE/CVF Conference on Computer Vision and Pattern Recognition, pp. 12299–12310 (2021)
2. Chen, X., Wang, X., Zhou, J., Dong, C.: Activating more pixels in image super-resolution transformer. arXiv preprint arXiv:2205.04437 (2022)

3. Dai, T., Cai, J., Zhang, Y., Xia, S.T., Zhang, L.: Second-order attention network for single image super-resolution. In: Proceedings of the IEEE/CVF conference on Computer Vision and Pattern Recognition, pp. 11065–11074 (2019)
4. Deng, J.: A large-scale hierarchical image database. In: Proceedings of IEEE Computer Vision and Pattern Recognition, 2009 (2009)
5. Dong, C., Loy, C.C., He, K., Tang, X.: Image super-resolution using deep convolutional networks. IEEE Trans. Pattern Anal. Mach. Intell. **38**(2), 295–307 (2015)
6. Dong, C., Loy, C.C., Tang, X.: Accelerating the super-resolution convolutional neural network. In: Leibe, B., Matas, J., Sebe, N., Welling, M. (eds.) ECCV 2016. LNCS, vol. 9906, pp. 391–407. Springer, Cham (2016). https://doi.org/10.1007/978-3-319-46475-6_25
7. Dosovitskiy, A., et al.: An image is worth 16 × 16 words: Transformers for image recognition at scale. arXiv preprint arXiv:2010.11929 (2020)
8. Hu, J., Shen, L., Sun, G.: Squeeze-and-excitation networks. In: Proceedings of the IEEE Conference on Computer Vision and Pattern Recognition, pp. 7132–7141 (2018)
9. Kim, J., Lee, J.K., Lee, K.M.: Accurate image super-resolution using very deep convolutional networks. In: Proceedings of the IEEE Conference on Computer Vision and Pattern Recognition, pp. 1646–1654 (2016)
10. Li, W., Lu, X., Lu, J., Zhang, X., Jia, J.: On efficient transformer and image pre-training for low-level vision. arXiv preprint arXiv:2112.10175 (2021)
11. Li, Y., Zhang, K., Cao, J., Timofte, R., Van Gool, L.: LocaLVIT: Bringing locality to vision transformers. arXiv preprint arXiv:2104.05707 (2021)
12. Liang, J., et al.: VRT: a video restoration transformer. arXiv preprint arXiv:2201.12288 (2022)
13. Liang, J., Cao, J., Sun, G., Zhang, K., Van Gool, L., Timofte, R.: SwinIR: image restoration using Swin transformer. In: Proceedings of the IEEE/CVF International Conference on Computer Vision, pp. 1833–1844 (2021)
14. Lim, B., Son, S., Kim, H., Nah, S., Mu Lee, K.: Enhanced deep residual networks for single image super-resolution. In: Proceedings of the IEEE Conference on Computer Vision and Pattern Recognition Workshops, pp. 136–144 (2017)
15. Liu, Z., et al.: Swin transformer: hierarchical vision transformer using shifted windows. In: Proceedings of the IEEE/CVF International Conference on Computer Vision, pp. 10012–10022 (2021)
16. Perrin, A.F., Xie, Y., Zhang, T., Liao, Y., Li, J., Le Callet, P.: Specialised video quality model for enhanced user generated content (UGC) with special effects. In: ICASSP 2022–2022 IEEE International Conference on Acoustics, Speech and Signal Processing (ICASSP), pp. 2040–2044. IEEE (2022)
17. Wang, W., et al.: Pyramid vision transformer: a versatile backbone for dense prediction without convolutions. In: Proceedings of the IEEE/CVF International Conference on Computer Vision, pp. 568–578 (2021)
18. Wu, B., et al.: Visual transformers: token-based image representation and processing for computer vision. arXiv preprint arXiv:2006.03677 (2020)
19. Yang, R., Timofte, R., et al.: Aim 2022 challenge on super-resolution of compressed image and video: dataset, methods and results. In: Proceedings of the European Conference on Computer Vision Workshops (ECCVW) (2022)
20. Zamir, S.W., Arora, A., Khan, S., Hayat, M., Khan, F.S., Yang, M.H.: Restormer: efficient transformer for high-resolution image restoration. In: Proceedings of the IEEE/CVF Conference on Computer Vision and Pattern Recognition, pp. 5728–5739 (2022)

21. Zhang, Y., Li, K., Li, K., Wang, L., Zhong, B., Fu, Y.: Image super-resolution using very deep residual channel attention networks. In: Ferrari, V., Hebert, M., Sminchisescu, C., Weiss, Y. (eds.) ECCV 2018. LNCS, vol. 11211, pp. 294–310. Springer, Cham (2018). https://doi.org/10.1007/978-3-030-01234-2_18
22. Zhang, Y., Li, K., Li, K., Zhong, B., Fu, Y.: Residual non-local attention networks for image restoration. arXiv preprint arXiv:1903.10082 (2019)

Fast Nearest Convolution for Real-Time Efficient Image Super-Resolution

Ziwei Luo[1(✉)], Youwei Li[1], Lei Yu[1], Qi Wu[1], Zhihong Wen[1], Haoqiang Fan[1], and Shuaicheng Liu[1,2(✉)]

[1] Megvii Technology, Beijing, China
ziwei.ro@gmail.com
[2] University of Electronic Science and Technology of China, Chengdu, China

Abstract. Deep learning-based single image super-resolution (SISR) approaches have drawn much attention and achieved remarkable success on modern advanced GPUs. However, most state-of-the-art methods require a huge number of parameters, memories, and computational resources, which usually show inferior inference times when applying them to current mobile device CPUs/NPUs. In this paper, we propose a simple plain convolution network with a fast nearest convolution module (NCNet), which is NPU-friendly and can perform a reliable super-resolution in real-time. The proposed nearest convolution has the same performance as the nearest upsampling but is much faster and more suitable for Android NNAPI. Our model can be easily deployed on mobile devices with 8-bit quantization and is fully compatible with all major mobile AI accelerators. Moreover, we conduct comprehensive experiments on different tensor operations on a mobile device to illustrate the efficiency of our network architecture. Our NCNet is trained and validated on the DIV2K $3\times$ dataset, and the comparison with other efficient SR methods demonstrated that the NCNet can achieve high fidelity SR results while using fewer inference times. Our codes and pretrained models are publicly available at https://github.com/Algolzw/NCNet.

Keywords: Image super-resolution · Real-time network · Mobile device · Nearest convolution · Quantization

1 Introduction

Image super-resolution (SR) is a fundamental task in computer vision that aims to reconstruct high-resolution (HR) images from their degraded low-resolution (LR) counterparts. It is a hot topic in recent years since its importance and ill-posed nature. The inherent challenge in the SR problem is that there always exists infinite solutions for recovering the HR image, and different HR images can be degraded to the same LR image, which makes it difficult to directly learn the super-resolution process.

During the past decade, we have witnessed the remarkable success of deep neural network (DNN) based techniques in computer vision [13,14,23,32]. SR algorithms that are based on deep convolution networks have attracted lots of attention and rapidly developed. As a result, many works have achieved impressive results on kinds of SR tasks [4–6,37]. However, most superior methods heavily rely on using large network

L. Karlinsky et al. (Eds.): ECCV 2022 Workshops, LNCS 13802, pp. 561–572, 2023.
https://doi.org/10.1007/978-3-031-25063-7_35

Table 1. Inference times of different commonly used tensor operation nodes. 'w/ dilation' means a dilated convolution. 'Conv**N** - f**A**-**B**' means the kernel size is **N**, input layer has **A** channels and the output layer has **C** channels.

Tensor operation node	CPU	GPU delegate	Android NNAPI
Conv3 - f3-16	19.1 ms	13.9 ms	20.1 ms
w/dilation	23.1 ms	27.0 ms	44.3 ms
+ Add	24.4 ms	14.7 ms	21.5 ms
+ Multiply	22.8 ms	59.8 ms	21.7 ms
+ Concat	25.5 ms	–	50.7 ms
+ Split	22.2 ms	40.4 ms	32.3 ms
+ ReLU	19.1 ms	14.5 ms	28.4 ms
+ LeakyReLU	44.7 ms	14.2 ms	66.8 ms
+ Global_Avgpool	16.6 ms	4.9 ms	29.1 ms
+ Global_Maxpool	102.0 ms	5.0 ms	21.1 ms

capacities and model complex to improve the SR performance, which limits their practicability on real-world resource-constrained mobile devices.

In order to apply DNN-based SR models to smartphones, a new research line called efficient super-resolution is developed where various methods have been proposed to reduce the model complexity and inference time [25,39]. A representative work is the IMDN [15], which proposes an information multi-distillation block that uses feature distillation and selective fusion parts to compress the model's parameters while preserving SR performance. Later, RFDN [27] builds a residual feature distillation network on top of IMDN but replaces all channel splitting operations with 1×1 convolutions and adds feature distillation connections. By doing so, RFDN has won 1st place in the AIM 2020 efficient super-resolution challenge [39]. In addition, some simplified attention mechanisms are also used in the efficient super-resolution task [28,43]. Compared with traditional superior SR networks, all these efficient-designed methods perform well and fast on desktop GPU devices. But we noticed that performing super-resolution on smartphones has much tighter limits on computing capacities and resources: a restricted amount of RAM, and inefficient support for many common deep learning layers and operators.

A mobile-friendly SR model should take care of the compatibility of tensor operators on mobile NPUs. We need to know what operations are particularly optimized by the mobile NN platform (Synaptics Dolphin platform). And the same tensor operation could have different inference times on different smartphone AI accelerators (e.g., CPU, GPU delegate, and Android NNAPI). Recent works [3,9,17] have investigated some limiting factors of running deep networks on a mobile device and what kind of architecture can be friendly to INT8 quantization. They propose several useful techniques such as "anchor-based residual learning", "Clipped ReLU", and "Quantize-Aware Training" to accelerate inference while preserving accuracy on smartphones. However, there is still no basic experiment that can illustrate the difference of tensor operators on different smartphone AI accelerators.

Table 2. Inference times of different convolution network structures. 'Conv**N** - f**A**-**B**-**C**' means it is a two-layer convolution network, where the kernel size is **N**, first layer has **A** channels, second layer has **B** channels, and the number of output channel is **C**.

Convolution network	CPU	GPU delegate	Android NNAPI
Conv1 - f3-3	5.3 ms	4.7 ms	9.6 ms
Conv3 - f3-3	14.8 ms	5.0 ms	15.0 ms
Conv5 - f3-3	21.3 ms	5.8 ms	27.2 ms
Conv3 - f3-8	14.7 ms	8.8 ms	17.4 ms
Conv3 - f3-16	19.1 ms	13.9 ms	20.1 ms
Conv3 - f3-32	27.1 ms	25.4 ms	24.7 ms
Conv3 - f3-8-8	31.4 ms	9.8 ms	24.9 ms
Conv3 - f3-8-16	33.6 ms	15.0 ms	27.8 ms
Conv3 - f3-16-16	50.4 ms	25.8 ms	27.8 ms
Conv3 - f3-16-32	63.8 ms	32.5 ms	34.4 ms
Conv3 - f3-32-32	103.2 ms	60.7 ms	34.3 ms

In this paper, we provide a comprehensive comparison of inference times for kinds of tensor operation nodes and network architectures, as shown in Tables 1 and 2. We use DIV2K 3× [37] as the training and evaluation dataset. All experiments are evaluated on *AI Benchmark* application [16,19]. As one can see, some commonly used deep learning techniques (e.g., dilated convolution, concatenation, channel splitting, and LeakyReLU) are not compatible with mobile Android NNAPI, even though they have a good performance on CPU and GPU delegate. Based on these experiments and analysis, we design a plain network that only contains 3×3 convolution layers and ReLU activation functions. Moreover, we propose to use a novel nearest convolution to replace the traditional nearest upsampling in network residual learning, which further speeds up the inference and achieves the same effect as nearest interpolation residual learning. In summary, our main contributions are as follows:

- We provide a comprehensive comparison of inference times for different tensor operators and network architectures on a smartphone, which tells us what operation is good for mobile devices and should be incorporated into the network.
- We propose a fast nearest convolution plain network (NCNet) that is mobile-friendly and can achieve the same performance as nearest interpolation residual learning while saving approximately 40 ms on a Google Pixel 4 smartphone.

2 Related Work

2.1 Single Image Super-Resolution

Single Image Super-Resolution (SISR) is one of the most popular research topics in computer vision due to its importance and ill-posed nature. The pioneering deep learning based method is SRCNN [35] which applies the bicubic downsampling on HR images to construct HR and LR pairs and employs a simple 3-layers convolution neural network (CNN) to perform super-resolution. After then, various super-resolution approaches are proposed and have achieved remarkable performance in processing the SISR problem [12,21,26,29–31,34,36,41,42]. For example, VDSR [35] proposes a very deep CNN to improve the SR results and ESPCN [34] designs a simple yet efficient strategy, called pixel-shuffle, for real-time feature upsampling. EDSR [26] proposes to enhance the CNN-based SR network by removing the batch normalization layer of all residual blocks. Moreover, to improve the perceptual visual quality, recent works [24,33,38] propose to employ some advanced losses such as the VGG loss [35], perceptual loss [20], and GAN loss [11] to help the network to learn realistic image details. However, these methods usually require huge memories and computational resources which makes them hardly be applied to modern mobile devices.

2.2 Efficient Image Super-Resolution

To fit the growing demands of deploying models on real-world smartphone applications, many works have refocused their attention on efficient image super-resolution techniques [2,8,15,27,36]. CARN [2] uses cascaded residual blocks and group convolution to achieve a lightweight SR network. IMDN [15] designs an information multi-distillation network that extracts hierarchical features and expresses the number of filters in each block to reduce the memories and FLOPs. The following work RFDN [27] further improves the network by introducing feature distillation blocks that employ 1×1 convolution layers to replace all channel splitting operations. Although these networks can perform efficient SR on desktop CPUs/GPUs, they are still not feasible to be deployed on real-world applications since the computational resources on most smartphones are much lower than on computers. Address it, ABPN [9] and XLSR [3] are winners of the Mobile 2021 Real-Time Single Image Super-resolution Challenge [17]. They investigated some limiting factors of the mobile device models and proposed extremely lightweight SR networks for the mobile SR problem. Moreover, the INT8 quantization is widely used in mobile devices since it can accelerate inference and save memories, as illustrated in Table 3 (for runtime we mainly focus on Android NNAPI).

Table 3. Runtime and PSNR comparison on different quantization modes of ABPN [9].

Quantization	#Params	CPU	GPU delegate	NNAPI	PSNR-int8
float-32	43K	506 ms	–	153 ms	30.21
int-8	43K	509 ms	–	135 ms	30.06

3 Method

In this section, we will start by finding a proper network architecture for the mobile device (especially for the Android NNAPI accelerator), based on Tables 1 and 2. Then

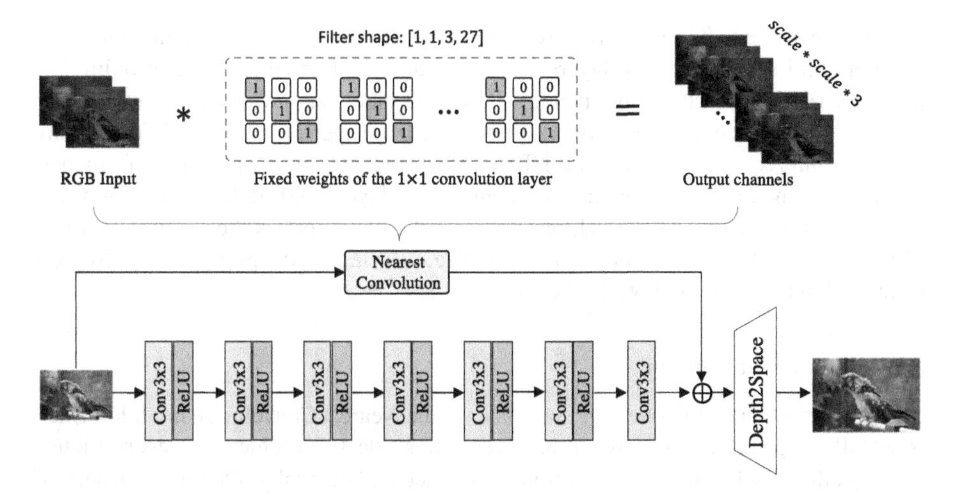

Fig. 1. Network architecture of the proposed NCNet. The main network backbone learns the residual while the nearest convolution module directly delivers the low-frequency information to the final result. Moreover, the nearest convolution achieves the same performance as nearest interpolation but can be executed parallelly.

we describe the main idea of the nearest convolution module. By assembling the manually designed backbone and the nearest convolution in the form of residual learning, we can obtain the final efficient SR network, as shown in Fig. 1.

3.1 Network Architecture Selection

To build a real-time mobile-friendly SR network, our first thing is to figure out what tensor operators are compatible and efficient for the Android NNAPI accelerator. To make use of the NPU's parallel property, the baseline operation node is set to a 3×3 convolution with 16 output channels, so we could add kinds of multi-channel (element-wise) tensor operations to it as in Table 1. In addition, we also want to know which convolution architecture and channel are better. So we compared the arrangement and combination of convolution layers in Table 2. Note that all experiments are based on a Google Pixel 4 smartphone, using INT8 quantization.

From our observation, we find that the inference times of channel splitting and concatenation on NPUs are much lower than on CPUs, which means these operators are not friendly to NPU parallelism, and the runtime will exponentially increase if the number of channels doubles. This situation also happens on 'LeakyReLU' and 'Global Average Pooling'. In recent years, many advanced methods like using LeakyReLU as their default activation function [38], but it is obviously not suitable for mobile NPUs. Similarly, the attention mechanism is widely used in state-of-the-art SR approaches [7,40] but it is time-consuming and not a good choice for mobile devices. For the choice of convolution layers, we find the 5×5 convolution is wasteful and inefficient, and the 1×1 convolution is not convincing to achieve a good performance. Thus we choose to set the kernel size to 3×3 for all convolution layers. For multi-layer structures, we surprisingly find that keeping all layers the same number of channels has approximately the same runtime as changing channels. Note the latter has fewer parameters, which means we could use a larger network to achieve better performance while preserving the same inference time on mobile NPUs. Therefore, the main backbone of our network is designed to only contain 3×3 convolution layers with ReLU activation functions. And inspired by ABPN [9], our solution also incorporates the residual learning of RGB images to improve the final result. The overview of the proposed fast Nearest Convolution (NCNet) network is shown in Fig. 1.

3.2 Nearest Convolution

The most important component of the NCNet is the Nearest Convolution module, which is actually a special 1×1 convolution layer with stride 1. To achieve the nearest interpolation, the weights of the convolution are freezed and manually filled by s^2 groups of 3×3 identity matrix (where s is the upscale factor) and each group would produce an RGB image, which just like a copy operation to repeat the input image s^2 times. Then these s^2 RGB images can reconstruct an HR image through a depth-to-space operation. In this way, the reconstructed image will be exactly the same as the nearest interpolated HR image but is much faster especially using mobile GPUs/NPUs. The reason is that when the 1×1 nearest convolution is performed on the NPU devices, it can be executed in parallel thus showing superior to other normal interpolation operations. The inference time of different upsampling methods is shown in Table 4. As one can see, the proposed nearest convolution can save approximately 40 ms compared with the original nearest upsampling operation.

3.3 Residual Learning

In practice, we'd like to add these s^2 RGB images to the output of the plain network before the depth-to-space layer then the plain network could focus on learning the residual information. Let $\mathbf{x}$ be the HR image and $\mathbf{y}$ be its degraded LR image. We could obtain the super-resolved image $\hat{\mathbf{x}}$ by:

$$\hat{\mathbf{x}} = f(\mathbf{y}; \theta) = D2S(f_{res}(y; \theta) + f_{nc}(y)), \tag{1}$$

where $f(\cdot)$ represents the SR network and θ is the network's parameters. $f_{res}(\cdot)$ and $f_{nc}(\cdot)$ represent the residual learning network and nearest convolution, respectively. $D2S$ means the depth-to-space layer. To illustrate the effectiveness of our network, we use L1 loss to optimize our model, which is formulated as follows:

$$\mathcal{L}_1(\theta) = \frac{1}{N} \sum_{i=1}^{N} (f(\mathbf{y}; \theta) - \mathbf{x}). \tag{2}$$

To accelerate the inference time, we only incorporate 7 layers of 3×3 convolution with the ReLU activation function for the whole network, and the number of channels is fixed to 32. Since the runtime of element-wise operation is non-negligible, we didn't use the residual connection for each convolution layer.

Table 4. The table shows the inference time between the proposed nearest convolution and other commonly used upsample methods on different mobile accelerators. The proposed nearest convolution can save approximately 40 ms compared with the original nearest upsampling operation.

Upsample methods	CPU	GPU delegate	NNAPI	PSNR
nearest	23.1 ms	**19.0 ms**	55.0 ms	26.67
bilinear	77.7 ms	21.0 ms	128.2 ms	**27.67**
Conv-3 + depth2space	30.8 ms	26.5 ms	43.8 ms	–
nearest convolution + depth2space	**15.9 ms**	20.3 ms	**14.8 ms**	26.67

Table 5. The table shows the runtime on different accelerators and the PSNR performance of the proposed NCNet on a Google Pixel 4 smartphone.

Google pixel 4	CPU	GPU delegate	NNAPI	#Params	PSNR-float32	PSNR-int8
NCNet	535.1 ms	263.0 ms	104.0 ms	53K	30.27 dB	30.18 dB

Table 6. Runtime and PSNR comparison with FSRCNN [8] and ABPN [9]. Note the ABPN fails to run on the mobile GPU delegate.

Method	#Params	CPU	GPU delegate	NNAPI	PSNR-int8
FSRCNN [8]	24K	476 ms	226 ms	251 ms	28.34
ABPN [9]	43K	509 ms	–	135 ms	30.15
NCNet (ours)	53K	535 ms	263 ms	104 ms	30.18

4 Experiments

4.1 Dataset and Implementation Details

We use DIV2K [1] as the training (800 image pairs) and testing (100 image pairs) dataset. The scale factor is fixed to 3 and the batch size is set to 64. The patch size of LR images is 64 and the total iterations are set to 500,000. All parameters are initialized using Xavier initializer [10]. We use the Adam optimizer [22], where the initial learning rate is 1×10^{-3} and decreases by half every 200,000 iterations. Inspired from [25], we also finetune the trained model with a larger LR patch size of 128 for additional 200,000 iterations. In this paper, we follow the Mobile AI & AIM 2022 Real-Time Image Super-Resolution Challenge [18] to measure super-resolved results in the RGB space. Moreover, we use a single NVIDIA RTX 2080Ti with 8 CPUs to train and evaluate the original non-quantized model.

4.2 Model Quantization

For network quantization, we use the standard Tensorflow Quantization tool - TFLite - to quantize the trained model as the Post-Quantization strategy. To evaluate the PSNR of the quantized models (float32 and int8), we set the input shape to $[1, None, None, 3]$ to

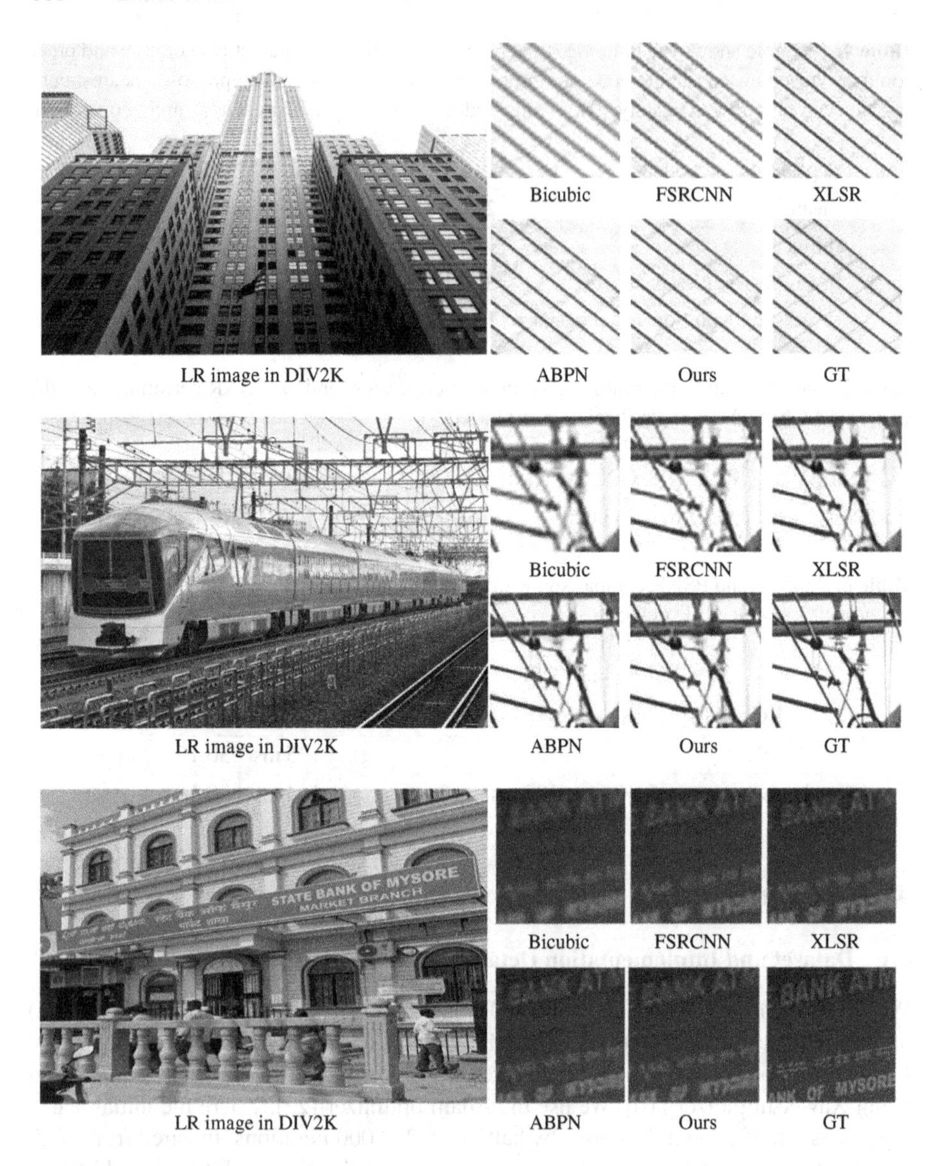

Fig. 2. Visual comparison of LR Img 846, Img 820 and Img 819 in DIV2K validation dataset [1]. All results are produced by the INT8 quantization model, for scale factor 3.

allow super-resolving arbitrary shape images. To evaluate the inference time, we fix the input shape to $[1, 360, 640, 3]$ and test the model in the *AI Benchmark* [16] application on a Google Pixel 4 smartphone.

4.3 Experimental Results

After training and quantizing, we can show the overall information of the proposed model in Table 5. As one can see, our NCNet is NPU-friendly. The runtime on NNAPI is 5× faster than CPU and 2.5× than GPU. With 53K parameters, the INT8 quantized NCNet only loses 0.09 dB on the mobile device compared with its float32 model.

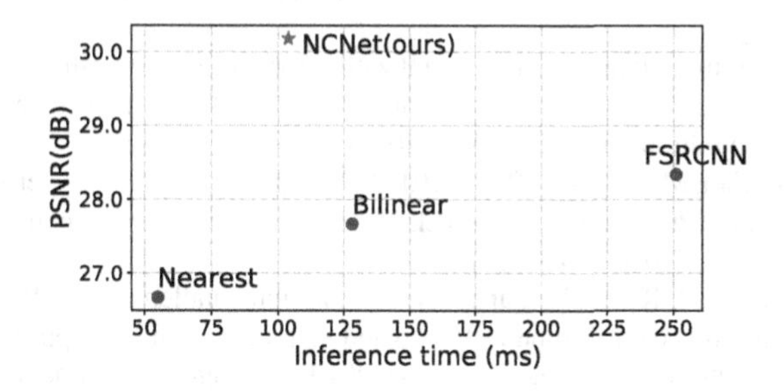

Fig. 3. Comparison of PSNR and inference time on Google Pixel 4 with INT8 quantization.

We compare the proposed NCNet with other lightweight real-time SR algorithms, such as FSRCNN [8] and ABPN [9]. The former is the pioneering DL-based SR network and the latter is a superior method in Mobile AI 2021 Real-Time Single Image Super Resolution Challenge [17]. The quantitative results are illustrated in Table 6. The proposed NCNet is faster than ABPN and FSRCNN on Android NNAPI and also achieves the best PSNR performance. In addition, we also compared two classical upsampling methods: Nearest upsampling and Bilinear upsampling. Both of them can be quantized to INT8 and are well compatible with Android NNAPI. The result is illustrated in Fig. 3. Our method is even faster than bilinear upsampling while achieving an impressive performance.

The visual comparison of our method and other approaches with INT8 quantization is shown in Fig. 2. The proposed NCNet has more textures and can produce visually pleasant SR images.

5 Conclusion

In this paper, we introduce an efficient fast nearest convolution network (NCNet) for real-time super-resolution. It is well compatible with INT8 quantization and Android NNAPI accelerator. By assembling the CNN-based plain network and the nearest convolution as residual learning architecture, NCNet achieves a remarkable performance while preserving real-time inference. By utilizing the NPU's parallel property, our model's runtime on the Android NNAPI is even faster than the traditional bilinear upsampling. We also provide a comprehensive comparison of inference times for different tensor operators and network architectures on the smartphone, which could help to select operators and architectures for real-world mobile devices.

References

1. Agustsson, E., Timofte, R.: NTIRE 2017 challenge on single image super-resolution: Dataset and study. In: Proceedings of the IEEE Conference on Computer Vision and Pattern Recognition Workshops, pp. 126–135 (2017)
2. Ahn, N., Kang, B., Sohn, K.-A.: Fast, accurate, and lightweight super-resolution with cascading residual network. In: Ferrari, V., Hebert, M., Sminchisescu, C., Weiss, Y. (eds.) ECCV 2018. LNCS, vol. 11214, pp. 256–272. Springer, Cham (2018). https://doi.org/10.1007/978-3-030-01249-6_16
3. Ayazoglu, M.: Extremely lightweight quantization robust real-time single-image super resolution for mobile devices. In: Proceedings of the IEEE/CVF Conference on Computer Vision and Pattern Recognition, pp. 2472–2479 (2021)
4. Bhat, G., Danelljan, M., Timofte, R.: NTIRE 2021 challenge on burst super-resolution: methods and results. In: Proceedings of the IEEE/CVF Conference on Computer Vision and Pattern Recognition, pp. 613–626 (2021)
5. Bhat, G., et al.: NTIRE 2022 burst super-resolution challenge. In: Proceedings of the IEEE/CVF Conference on Computer Vision and Pattern Recognition, pp. 1041–1061 (2022)
6. Cai, J., Gu, S., Timofte, R., Zhang, L.: NTIRE 2019 challenge on real image super-resolution: methods and results. In: Proceedings of the IEEE/CVF Conference on Computer Vision and Pattern Recognition Workshops (2019)
7. Dai, T., Cai, J., Zhang, Y., Xia, S.T., Zhang, L.: Second-order attention network for single image super-resolution. In: Proceedings of the IEEE/CVF Conference on Computer Vision and Pattern Recognition, pp. 11065–11074 (2019)
8. Dong, C., Loy, C.C., Tang, X.: Accelerating the super-resolution convolutional neural network. In: Leibe, B., Matas, J., Sebe, N., Welling, M. (eds.) ECCV 2016. LNCS, vol. 9906, pp. 391–407. Springer, Cham (2016). https://doi.org/10.1007/978-3-319-46475-6_25
9. Du, Z., Liu, J., Tang, J., Wu, G.: Anchor-based plain net for mobile image super-resolution. In: Proceedings of the IEEE/CVF Conference on Computer Vision and Pattern Recognition, pp. 2494–2502 (2021)
10. Glorot, X., Bengio, Y.: Understanding the difficulty of training deep feedforward neural networks. In: Proceedings of the Thirteenth International Conference on Artificial Intelligence and Statistics, pp. 249–256. JMLR Workshop and Conference Proceedings (2010)
11. Goodfellow, I.J., et al.: Generative adversarial networks. arXiv preprint arXiv:1406.2661 (2014)
12. Haris, M., Shakhnarovich, G., Ukita, N.: Deep back-projection networks for super-resolution. In: Proceedings of the IEEE Conference on Computer Vision and Pattern Recognition, pp. 1664–1673 (2018)
13. He, K., Gkioxari, G., Dollár, P., Girshick, R.: Mask R-CNN. In: Proceedings of the IEEE International Conference on Computer Vision, pp. 2961–2969 (2017)
14. He, K., Zhang, X., Ren, S., Sun, J.: Deep residual learning for image recognition. In: Proceedings of the IEEE Conference on Computer Vision and Pattern Recognition, pp. 770–778 (2016)
15. Hui, Z., Gao, X., Yang, Y., Wang, X.: Lightweight image super-resolution with information multi-distillation network. In: Proceedings of the 27th ACM International Conference on Multimedia, pp. 2024–2032 (2019)
16. Ignatov, A., et al.: AI benchmark: running deep neural networks on android smartphones. In: Leal-Taixé, L., Roth, S. (eds.) ECCV 2018. LNCS, vol. 11133, pp. 288–314. Springer, Cham (2019). https://doi.org/10.1007/978-3-030-11021-5_19
17. Ignatov, A., Timofte, R., Denna, M., Younes, A.: Real-time quantized image super-resolution on mobile NPUs, mobile AI 2021 challenge: report. In: Proceedings of the IEEE/CVF Conference on Computer Vision and Pattern Recognition, pp. 2525–2534 (2021)

18. Ignatov, A., Timofte, R., Denna, M., Younes, A., et al.: Efficient and accurate quantized image super-resolution on mobile NPUs, mobile AI & aim 2022 challenge: report. In: Proceedings of the European Conference on Computer Vision (ECCV) Workshops (2022)
19. Ignatov, A., et al.: AI benchmark: all about deep learning on smartphones in 2019. In: 2019 IEEE/CVF International Conference on Computer Vision Workshop (ICCVW), pp. 3617–3635. IEEE (2019)
20. Johnson, J., Alahi, A., Fei-Fei, L.: Perceptual losses for real-time style transfer and super-resolution. In: Leibe, B., Matas, J., Sebe, N., Welling, M. (eds.) ECCV 2016. LNCS, vol. 9906, pp. 694–711. Springer, Cham (2016). https://doi.org/10.1007/978-3-319-46475-6_43
21. Kim, J., Kwon Lee, J., Mu Lee, K.: Accurate image super-resolution using very deep convolutional networks. In: CVPR, pp. 1646–1654 (2016)
22. Kingma, D.P., Ba, J.: Adam: a method for stochastic optimization. arXiv preprint arXiv:1412.6980 (2014)
23. Krizhevsky, A., Sutskever, I., Hinton, G.E.: ImageNet classification with deep convolutional neural networks. In: Advances in Neural Information Processing Systems, vol. 25 (2012)
24. Ledig, C., et al.: Photo-realistic single image super-resolution using a generative adversarial network. In: CVPR, pp. 4681–4690 (2017)
25. Li, Y., et al.: NTIRE 2022 challenge on efficient super-resolution: methods and results. In: Proceedings of the IEEE/CVF Conference on Computer Vision and Pattern Recognition, pp. 1062–1102 (2022)
26. Lim, B., Son, S., Kim, H., Nah, S., Mu Lee, K.: Enhanced deep residual networks for single image super-resolution. In: CVPRW, pp. 136–144 (2017)
27. Liu, J., Tang, J., Wu, G.: Residual feature distillation network for lightweight image super-resolution. In: Bartoli, A., Fusiello, A. (eds.) ECCV 2020. LNCS, vol. 12537, pp. 41–55. Springer, Cham (2020). https://doi.org/10.1007/978-3-030-67070-2_2
28. Luo, X., Xie, Y., Zhang, Y., Qu, Y., Li, C., Fu, Y.: LatticeNet: towards lightweight image super-resolution with lattice block. In: Vedaldi, A., Bischof, H., Brox, T., Frahm, J.-M. (eds.) ECCV 2020. LNCS, vol. 12367, pp. 272–289. Springer, Cham (2020). https://doi.org/10.1007/978-3-030-58542-6_17
29. Luo, Z., Huang, H., Yu, L., Li, Y., Fan, H., Liu, S.: Deep constrained least squares for blind image super-resolution. In: Proceedings of the IEEE/CVF Conference on Computer Vision and Pattern Recognition, pp. 17642–17652 (2022)
30. Luo, Z., et al.: BSRT: improving burst super-resolution with Swin transformer and flow-guided deformable alignment. In: Proceedings of the IEEE/CVF Conference on Computer Vision and Pattern Recognition, pp. 998–1008 (2022)
31. Luo, Z., et al.: EBSR: feature enhanced burst super-resolution with deformable alignment. In: Proceedings of the IEEE/CVF Conference on Computer Vision and Pattern Recognition, pp. 471–478 (2021)
32. Ren, S., He, K., Girshick, R., Sun, J.: Faster R-CNN: towards real-time object detection with region proposal networks. In: Advances in Neural Information Processing Systems, vol. 28 (2015)
33. Sajjadi, M.S., Scholkopf, B., Hirsch, M.: EnhanceNet: single image super-resolution through automated texture synthesis. In: ICCV, pp. 4491–4500 (2017)
34. Shi, W., et al.: Real-time single image and video super-resolution using an efficient sub-pixel convolutional neural network. In: CVPR, pp. 1874–1883 (2016)
35. Simonyan, K., Zisserman, A.: Very deep convolutional networks for large-scale image recognition. arXiv preprint arXiv:1409.1556 (2014)
36. Tai, Y., Yang, J., Liu, X.: Image super-resolution via deep recursive residual network. In: Proceedings of the IEEE Conference on Computer Vision and Pattern Recognition, pp. 3147–3155 (2017)

37. Timofte, R., Agustsson, E., Van Gool, L., Yang, M.H., Zhang, L.: Ntire 2017 challenge on single image super-resolution: Methods and results. In: Proceedings of the IEEE conference on computer vision and pattern recognition workshops. pp. 114–125 (2017)
38. Wang, X., et al.: ESRGAN: enhanced super-resolution generative adversarial networks. In: Leal-Taixé, L., Roth, S. (eds.) ECCV 2018. LNCS, vol. 11133, pp. 63–79. Springer, Cham (2019). https://doi.org/10.1007/978-3-030-11021-5_5
39. Zhang, K., et al.: AIM 2020 challenge on efficient super-resolution: methods and results. In: Bartoli, A., Fusiello, A. (eds.) ECCV 2020. LNCS, vol. 12537, pp. 5–40. Springer, Cham (2020). https://doi.org/10.1007/978-3-030-67070-2_1
40. Zhang, Y., Li, K., Li, K., Wang, L., Zhong, B., Fu, Y.: Image super-resolution using very deep residual channel attention networks. In: Ferrari, V., Hebert, M., Sminchisescu, C., Weiss, Y. (eds.) ECCV 2018. LNCS, vol. 11211, pp. 294–310. Springer, Cham (2018). https://doi.org/10.1007/978-3-030-01234-2_18
41. Zhang, Y., Tian, Y., Kong, Y., Zhong, B., Fu, Y.: Residual dense network for image super-resolution. In: Proceedings of the IEEE Conference on Computer Vision and Pattern Recognition, pp. 2472–2481 (2018)
42. Zhang, Y., Tian, Y., Kong, Y., Zhong, B., Fu, Y.: Residual dense network for image restoration. IEEE Trans. Pattern Anal. Mach. Intell. **43**(7), 2480–2495 (2020)
43. Zhao, H., Kong, X., He, J., Qiao, Yu., Dong, C.: Efficient image super-resolution using pixel attention. In: Bartoli, A., Fusiello, A. (eds.) ECCV 2020. LNCS, vol. 12537, pp. 56–72. Springer, Cham (2020). https://doi.org/10.1007/978-3-030-67070-2_3

Real-Time Channel Mixing Net for Mobile Image Super-Resolution

Garas Gendy[1], Nabil Sabor[2], Jingchao Hou[1], and Guanghui He[1(✉)]

[1] Department of Micro-Nano Electronics, Shanghai Jiao Tong University, Shanghai 200240, China
{jingchaohou,guanghui.he}@sjtu.edu.cn

[2] Electrical Engineering Department, Faculty of Engineering, Assiut University, Assiut 71516, Egypt
nabil_sabor@aun.edu.eg

Abstract. Recently, deep learning based image super-resolution (SR) models show a strong performance thanks to the convolution neural network (CNN). However, these CNN-based models mostly need large memory and use a lot of power cost, which limits its use in mobile devices. To solve this problem, we propose a channel mixing Net (CDFM-Mobile) for mobile SR. The idea of the CDFM-Mobile is based on making channel mixing by using a pointwise convolution and deep features extraction by using 3×3 convolution. In addition, inspired by the prior work in the field, we used anchor-based residual learning and deep feature residual learning, which improved the performance. In addition, we used the quantization-aware training approach to optimize the model performance based on training at 8-bit quantize. Finally, we take part in MAI 2022 for mobile SR, and extensive results are conducted to show the model performance.

Keywords: Image super-resolution · Deep feature extraction · Mobile super-resolution · Pointwise convolution · Residual learning

1 Introduction

In the last few years, the image super-resolution (SR) task has taken much attention from the research community due to its need in many practical applications [3,53], such as object detection [43], medical image processing [26], and video processing [26]. The SR task aims to generate a corresponding high-resolution (HR) image from its low-resolution (LR) image with better visual quality. However, the SR task is considered an ill-posed problem because many HR images can be generated from one input LR image.

So, there are great efforts are done to solve the SR problem. Many SR models based on convolutional neural network (CNN) are developed [10–12,29] to map the LR image to the HR one. However, these methods suffer from the vanishing gradient issue that prevents the models to be increased in size and limits their

L. Karlinsky et al. (Eds.): ECCV 2022 Workshops, LNCS 13802, pp. 573–590, 2023.
https://doi.org/10.1007/978-3-031-25063-7_36

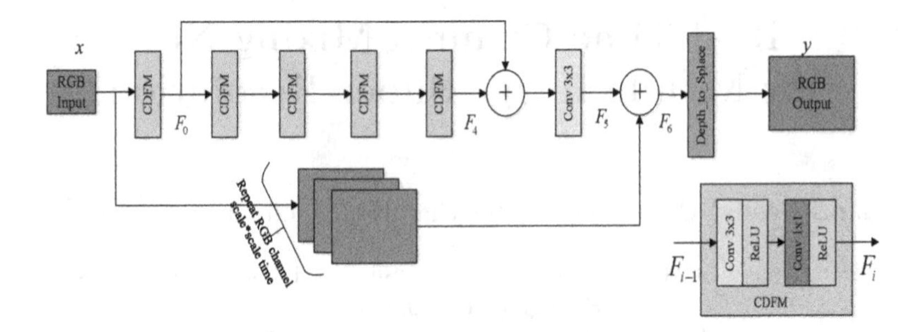

Fig. 1. The architecture of the proposed CDFM-Mobile Model

performance. Introducing residual learning [35] can solve the vanishing gradient problem by increasing the model size to hundreds of layers. Although the residual-based models achieve much better performance compared to CNN-based models, they are only based on the receptive field of the convolution operation, which limits their ability to find long-range dependence. Later, the attention mechanism [9] and transformer-based models [13,48,49] are introduced in the deep learning field to help the CNN-based methods to deal with the long-range dependence issue. The transformer-based models can easily find the correlation between the features in the long-range, but they will compound with the computational cost. This problem is solved by developing attention-based models [9,15,16,31,33,36,39,51,57]. However, these models need a lot of memory which limits their performance in low memory devices such as mobile phones.

Recently, the deep learning accelerators [23,42] are used to develop mobile SR models with low computation [24,56]. These models can be categorized into two main approaches; the first approach is the network optimization including quantization [8,28], knowledge distillation [41,52], and pruning [5,7], while the other approach is based on designing lightweight architectures [2,19,20,38]. Although the above models achieved good results, but the performance and the speed still need to be improved.

This paper proposes a new mobile SR model called a mixing channel and deep features extraction (CDFM-Mobile) model. The idea of the CDFM-Mobile model is based on designing a simple model to optimize the performance and speed. This is done by designing a Channel and Deep Features Mixing block (CDFM) to be the backbone of the model. The CDFM is built using both pointwise and 3 $\times$ 3 convolution. The pointwise convolution (i.e. Conv 1 $\times$ 1) can help the model to mix the channel information and the extracted deep features. Moreover, the anchor block is used to improve the model's performance by transferring the low-frequency information to the final extracted features (i.e. high-frequency information). In order to follow the gradient of the training, a deep feature residual is utilized. Finally, our model architecture is optimized using a 8-bit quantization to be suitable for mobile devices. The contributions of this paper can be summarized as follows:

- We developed an efficient channel and deep features mixing (CDFM) block using pointwise and 3 × 3 convolution. This block is useful for mixing the channel information and the deep extracted features.
- We built a mobile scale SISR model, called CDFM-Mobile, using the developed CDFM block, based on optimizing the performance and speed.
- We did extensive experimental results to show that our CDFM-Mobile is achieving good results compared to the state-of-the-art. Also, we did an ablation study to illustrate the effect of each model's components on the performance.

We organized the rest of the paper as: In Sect. 2, the related work is discussed. In Sect. 3, the detailed architecture of the methodology is presented. After that, we conducted extensive experiments to show our method's performance improvement in Sect. 4. Then, we discuss our conclusion in Sect. 5.

2 Related Work

In this section, the related models to the mobile SR models such as the lightweight SR models, mobile image Super-resolution, and also the recent advance in network quantization will be discussed.

2.1 Lightweight Image Super-resolution

The initialed model based on the deep learning for solving the SISR task is developed in [11]. Following this progress of lightweight architectures the fast super-resolution convolutional neural networks (FSRCNN) model [12] is proposed. After that, the deeply-recursive convolutional network (DRCN) [29] model is designed based on using recursive convolutional for reducing the parameter number. The deep Laplacian pyramid network (LapSRN) [31] is introduced based on the Laplacian pyramid super-resolution block for constructing the HR images sub-band residuals. In [2], an efficient cascading residual network (CARN) is developed depend on a group convolution that shows good results in comparison with computationally expensive models. Afterward, attention started to be used, so in [50], a lightweight multi-scale aggregation network (LMAN) is introduced, which shows a good performance for both the small and large scale factors. In addition, a multi-scale channel attention network [33] is developed for the image SR task, which is based on the channel attention idea. Similar to this channel attention model, a spatial attention module is utilized in [45] for developing an efficient multi-scale spatial attention network (MSAN). Moreover, the matrix channel attention concept is used in [39] to build the matrix channel attention network (MCAN) based on the multi-connected channel attention blocks (MCAB). Next, an attention cube network (A-CubeNet) [17] is introduced for image restoration based on feature expression and feature correlation learning. Other models are developed based on the feature distillation approaches. The model in [2] is designed based on the stacked information distillation blocks

(SIDB). The SIDB combines an enhancement unit and compression unit to fast execution due to the comparatively few numbers of filters per layer. After that, the information multi-distillation network (IMDN) is introduced in [19] that helps to deal with the problem of excessive use of convolutions and generate the super-resolution of any arbitrary scale factor. In addition, feature distillation connection (FDC) is developed in [37] to present a more lightweight and flexible model. So, the residual feature distillation network (RFDN) is introduced using multiple FDC for features extraction and shallow residual block to improve the model performance. Finally, a lightweight dense connection distillation network (DCDN) is designed in [34] based on combining the feature fusion units and dense connection distillation blocks together. Finally, the sparse mask SR (SMSR) [51] model is introduced that prune redundant computation by learning sparse masks. Even though these models show good results, they only depend on some specific types of convolution. So, in this paper, we tried to benefit from using different types of convolution for solving the SR task.

2.2 Mobile Image Super-Resolution

Due to the popularity of the application of SR on mobile, there is a need to create new models that can enhance the SR model in mobile devices. To do that task, many approaches are introduced that create extremely lightweight models for mobile devices [4,6,14,25,46,54,55]. Firstly, the anchor-based plain net (ABPN) is introduced in [14] that is based on the idea of using an anchor in image space. Also, this ABPN model used a quantization-aware training strategy to further improve the performance. Additionally, an extremely lightweight quantization robust real-time super-resolution network (XLSR) is developed in [4], which is based on using the root modules in [25] to improve the SR performance. In [46], an adder network is used to avoid the massive energy consumption of conventional multiplications. Then, an automatic search framework [55] is designed based on combining neural architecture search with pruning search to achieve high performance and fast real-time inference. In addition, a real-time image SR model for mobile devices is introduced in [6]. This model can work in a wide range of degradation, allowing the mobile devices to work with real-world image degradation. Finally, in [54], a compiler-aware SR neural architecture search (NAS) model is developed depend on using adaptive SR blocks for searching the depth per-layer width. Although the above models achieved good results, but the performance and the speed still need to be improved.

2.3 Network Quantization

The quantization process is converting the continuous real values to discrete ones, which reduces the number the bit and, at the same time, improves the computation cost. This quantization operation can make the model three times faster compared to the float-point case. Also, this can reduce the used memory by 4 $\times$ factor [14]. The deep learning framework Tensorflow [27] can be categorized based on the quantization into two types, post-training models and quantization-aware training models. The difference between these two types is that one can traverse representative data provide after training, while the second can include fake quantization nodes. These two methods show a strong performance in image perception tasks [18,27,47] due to the problem of decreasing the model performance when quantize the model. This drop is usually caused by the high dynamic quantization range when removing the normalization (BN) layer that is not helpful for the SR task as proved in [35]. Some work tries to address these issues [32,40] based on residual blocks using the binarizes convolution filters. After that, they tried to use learnable weight for each binary filter, but it is not applicable for the full-integer device [40]. Also, the Parameterized Max Scale [32] is used based on the adaptive finding of the quantization range. However, this model is combined with large training complexity. So, there is still a need to finding more much better methods to solve the quantization problem.

3 Methodology

The proposed architecture of the CDFM-Mobile model for mobile image SR is shown in Fig 1. The backbone of the CDFM-Mobile model is the CDFM block. This block is designed using two different types of convolutions, namely the pointwise convolution, and the traditional 3 × 3 convolution. The CDFM block allows the model to mix the channel information and the deep extracted features. After that, the anchor-based residual learning [14] is added to transfer the image space to the extracted features. Finally, inspired by [6], the deep feature residual learn is used to further improve the model performance based on following the gradient of training.The details of each module in the proposed CDFM-Mobile model will be presented as follows:

3.1 Channel and Deep Features Mixing (CDFM) Block

The CDFM block is the backbone of our model. This block is based on two types of convolution 1 × 1 convolution (i.e. pointwise) that is used for channel mixing and the 3 × 3 for local features extraction. After each convolution layer, a ReLU activation function is added. Giving the input feature map of the block is F_{in}, then the extracted feature map after applying the pointwise convolution H_{pwc} with ReLU activation function H_{ReLU} is given by:

$$F_l^{'} = H_{ReLU}(H_{pwc}(F_{in})) \tag{1}$$

After that, the 3 × 3 convolution H_{conv3} with ReLU activation function H_{ReLU} is applied to the $F_l^{'}$ to generate the output feature map F_{out} of the CDFM block as follows:.

$$F_{out} = H_{ReLU}(H_{conv3}(F_l^{'})) \tag{2}$$

3.2 The CDFM-Mobile Framework

In this section, we explain the operation of the proposed model in detail. The CDFM-Mobile model consists of five CDFM blocks. The first CDFM block is used for coarse features extraction from the input LR image and the other four CDFM blocks are used for extracting the deep features. For the given input LR image x, the coarse extracted feature map F_0 from the first CDFM block H_{CDFM0} can be represented as follows:

$$F_0 = H_{CDFM0}(x) \tag{3}$$

After that, the F_0 feature map is passed as input to the other four CDFM blocks for extracting the deep features in a sequence process as follows:

$$F_i = H_{CDFMi}(F_{i-1}), \quad i = 1, 2, 3, 4 \tag{4}$$

where F_i and H_{CDFMi} are the output and the function of the ith CDFM block, respectively. After extracting the deep features using the CDFM blocks, the coarse feature map F_0 is added to the extracted deep features F_4 to mix the low and high frequency information together in one feature map F_5 as follows:

$$F_5 = F_4 + F_0 \tag{5}$$

Then, an one 3×3 convolution H_{conv} is used to decrease the space of the feature map to the image space. The output of the 3×3 conv layer is added to the output map F_{RGB} of repeated RGB channel as represented by the following equation:

$$F_6 = H_{conv}(F_5) + F_{RGB} \tag{6}$$

The final step is to apply the deep_to_space function R_{d2s} to generate the output HR RGB image y as follows:

$$y = R_{d2s}(F_6), \tag{7}$$

3.3 The Loss Function

In our training, we utilized the widely used L1 loss function similar to the previous work [9,20,35,37,50,57]. The formula for this loss function can be represented as:

$$L(\Theta) = \frac{1}{N} \sum_{i=1}^{N} ||y_i - I_i^{HR}||_1 \tag{8}$$

where Θ represents our model parameters, and N is the number of training samples. y_i, and I^{HR} are the model output and the ground-truth HR corresponding to this output, respectively.

Fig. 2. Our Model (float32) Visual Results on Urban100 Datasets

Table 1. Comparison in PSNR of the proposed method trained under float32, PTQ, and QAT.

Method	FSRCNN [12]	XLSR [4]	ESPCN [44]	ABPN [14]	CDFM-Mobile
Parameters	25k	22k	31k	43k	60.22k
float32	29.67	30.10	29.54	30.22	30.34
PTQ	21.95	29.82	23.93	30.09	29.94
QAT	–	–	–	30.15	30.26

4 Experiments

4.1 Settings

In our experiments, we used DIV2K [1] dataset for training the model, where the LR image is a bicubic down-sampled version of the HR one. During training, the input images were cropped as 64×64, with a mini-batch of 16 images and the number of features of 32. The parameters of the model are initialized using the defeat function of the Tensorflow framework. The ADAM optimizer [30] with $\beta_1 = 0.9$, $\beta_2 = 0.99$ and $\epsilon = 1e^{-8}$ is used for training optimization. Following the Mobile AI image super-resolution challenge [21], the SR results are measured on the RGB space. The quantitative evaluations of the peak signal-to-noise ratio

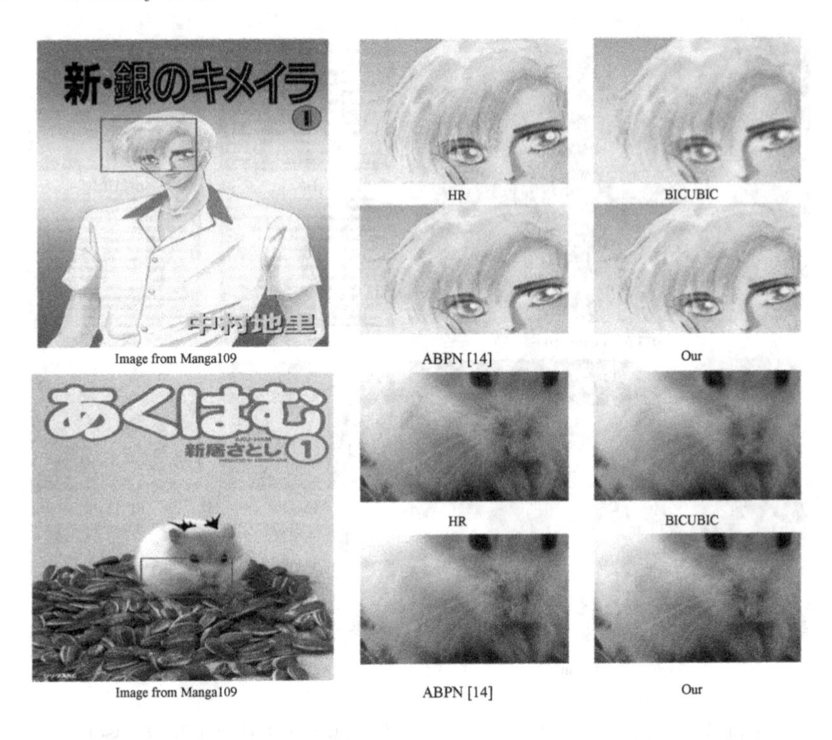

Fig. 3. Our model (float32) visual results on Manga109 datasets

(PSNR) is utilized for measuring the model performance. Also, the qualitative evaluation is used to show the outputs of our model compared to the other models. Finally, we used Tensorflow to implement our model and trained it on a computer with Nvidia 2080 Ti GPUs for 600 epochs.

4.2 Comparison with State-of-the-Arts

In this section, we compared our model with the state-of-the-art models using the quantitative evaluations to show how our model improved the PSNR performance. In addition, we make a comparison with one of the state-of-the-art methods in qualitative evaluations. Finally, we compared our model with other models in the case of run-time on mobile devices with Snapdragon 970 hardware.

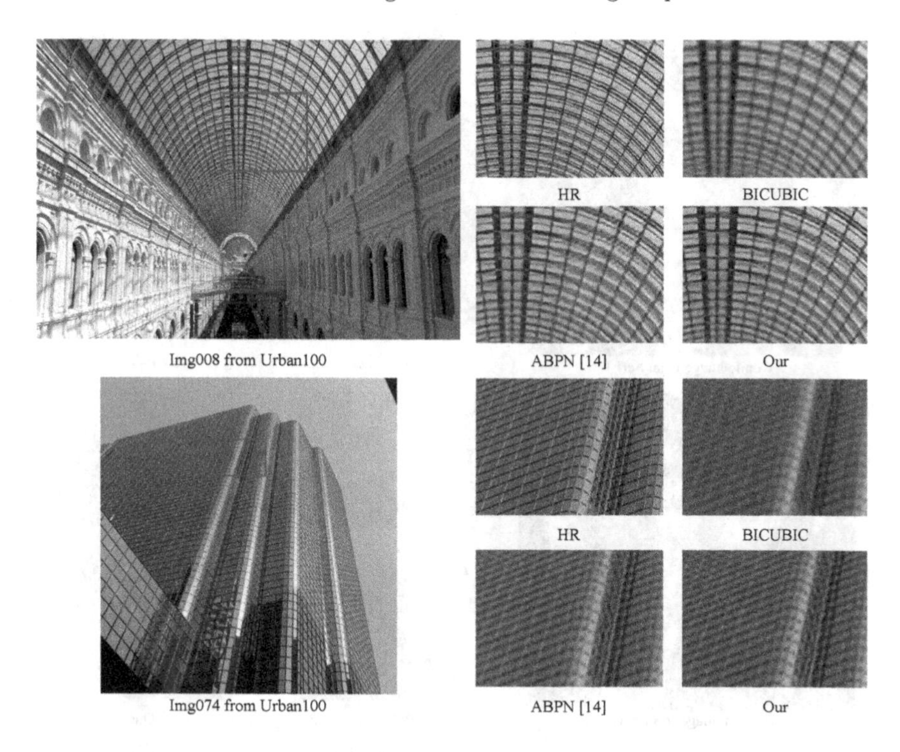

Fig. 4. Our model (int8) visual results on Urban100 datasets

Quantitative Evaluations. In this section, the DIV2K validation dataset is used to show the superior of our model compared to the state-of-the-art models, as listed in Table 1. These models are FSRCNN [12], XLSR [4], ESPCN [44], and ABPN [14].We compared among them in three training cases of float32 training, post-training quantization (PTQ), and quantization-aware training (QAT). It is clear from the table that our model has good results in comparison to the ABPN [14], which is the most recent model. For example, at the scale of $\times$ 3, our model has 0.12 dB improvements in the PSNR compared to ABPN. Also, for the scale quantization-aware training (QAT), our model achieved 0.11 dB compared to ABPN. In addition, our model has a 0.24 dB enhancement in the PSNR compared to XLSR [4]. We can conclude that our model can achieve a good performance compared to the other models.

Qualitative Evaluations. For the visual result, we used the Set14, Urban100, and Manga109 datasets to illustrate the qualitative evaluations of our model. Figures 2 and 3 show the visual results of the non-quantized version (float32) of our model compared to the ABPN [14] model for datasets of Urban100 and Mnaga109, respectively. In addition, Figs. 4 and 5 indicate the quantized version

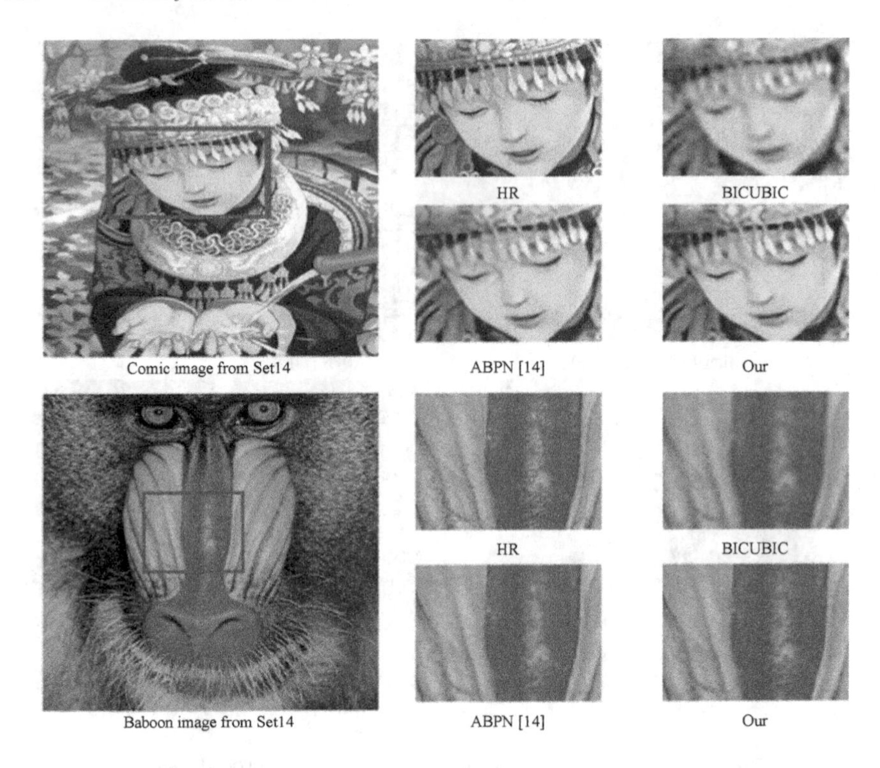

Fig. 5. Our model (int8) visual results on Set14 datasets

(int8) of the models for Urban100 and Set14, respectively. It is obvious from the figures that our method has a more appealing view compared to the ABPN model. This improvement is because our method can benefit from both types of pointwise convolution, and 3×3 convolution to extract deep features, which can improve the results due to its ability to extract different types of features.

Results on Snapdragon 970. In this section, we compared the inference time of our model with the ABPN model [14] on Snapdragon 970. To do that, we utilized AI Benchmarks to calculate the CPU and NNAPI running time that is possible on our devices. Table 2 lists the obtained results. It is clear that the CPU and NNAPI running times of our model increased by 50 ms and 4.4 ms more than those of the ABPN model, but our model achieved much higher PSNR performance. This means that our model is suitable for mobile devices because it balances the run time and the performance.

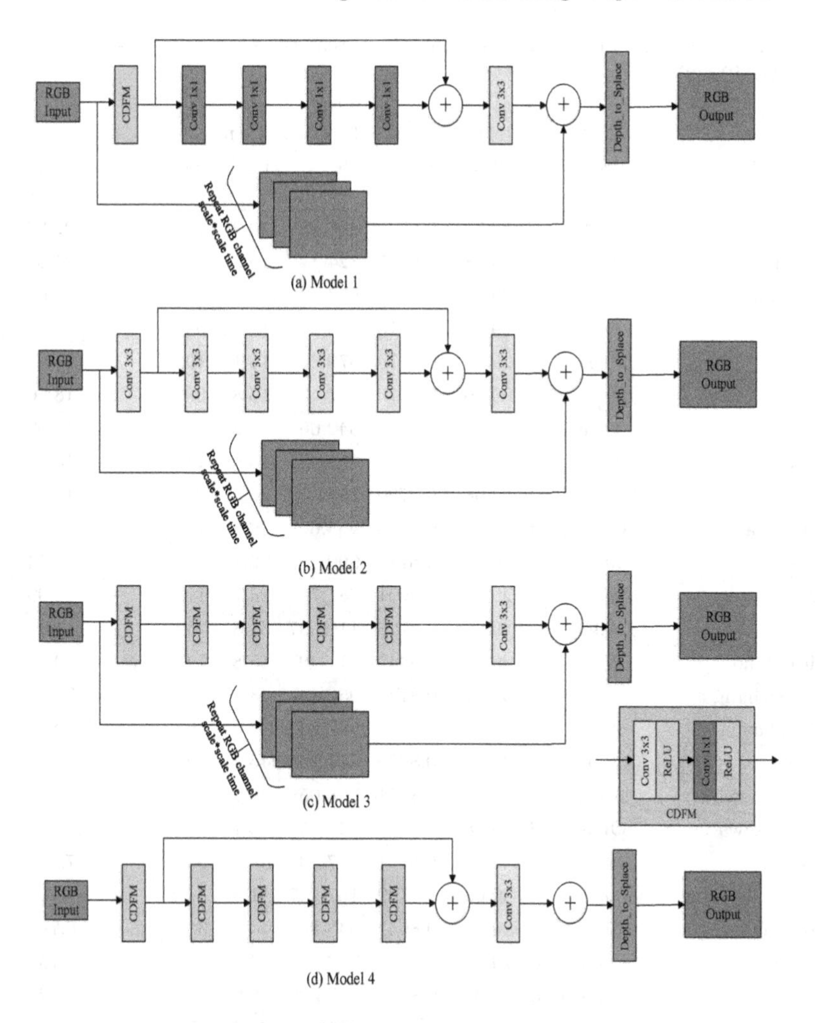

Fig. 6. The used models in the ablation study.

Table 2. Inference time (ms) of CDFM-Mobile on Snapdragon 970.

Model	CPU (ms)	NNAPI (ms)	PSNR (dB)
ABPN [14]	231	48.8	30.15
CDFM-Mobile (Our)	281	53.2	30.26

4.3 MAI2022 SISR Challenge

We took part in Mobile AI & AIM 2022 real-time image super-resolution challenge in the year of 2022 [22], as illustrated in Table 3. Due to the results announced based on the PSNR evaluation for float32 and int8 and run time on runtime on the Synaptics VS680 board device on the test dataset. As shown

Table 3. The preliminary results Comparison of real-time image super-resolution challenge (MAI2022).

	FP32	INT8		Runtime in (ms)			
Codalab Username	PSNR	PSNR	SSIM	CPU	NPU	NPU (NCHW)	Final Score
ganzoo	30.09	30.03	0.8738	809.25	19.2	13.2	22.22
maciejos_s	30.01	29.88	0.8705	824.1	15.9	9.8	21.84
CCjiahao	29.92	29.82	0.8697	553.33	15.1	9.1	21.08
zion_	29.87	29.76	0.8675	786.6	15	9	19.59
NBCS	29.99	29.8	0.8675	876.62	16.1	10.1	19.27
kkkls	29.92	29.76	0.8676	767	15.8	9.8	18.56
fz	29.15	29.58	0.8609	542.66	15.2	9.2	15.02
stevek	30.01	29.9	0.8704	854.92	25.6	13.1	13.91
Huaen	Quantized	29.98	0.8729	1041.92	30.5	23.5	13.07
balabala	Quantized	29.94	0.8704	964.6	29.8	23.1	12.65
gideon	30.04	29.94	0.8712	990	30.1	22.5	12.54
alan_jaeger	29.89	29.8	0.8691	781.7	25.7	11.7	12.09
sissie	29.94	29.94	0.8728	1109.97	31.8	24.6	11.85
HojinCho	29.75	29.37	0.8442	511.49	15.8	7.9	10.85
jklovezhang	30.05	30.01	0.874	950.44	43.3	36.6	9.6
xiaozhazha	30.05	29.95	0.8728	940.74	43.2	36.6	8.84
rtsisr	30.09	30	0.8729	977.26	46.4	39.7	8.83
deepernewbie	Random	29.67	0.8651	598.05	30.2	8.8	8.59
FykAikawa	Quantized	29.88	0.87	849.95	43.1	36.4	8.05
mrblue	30.14	30.1	0.8751	1772.31	60.2	53.6	7.81
yilitiaotiaotang	30.08	30.06	0.8739	1678.67	61.3	54.7	7.26
smhong	29.99	29.76	0.8641	948.86	43.2	36.5	6.79
kia350	Quantized	29.75	0.8671	766.4	42.9	36.2	6.76
Sail	Quantized	29.99	0.8729	1075.29	68.2	61.7	5.92
garasgaras (Our)	30.09	30.02	0.8735	1056.87	72.4	65.7	5.82
j_npu	Quantized	29.17	0.8449	587.93	35.3	28	3.67

in the Table, our model achieved comparable good PSNR performance (30.02 dB) among the best 4 scores on (int8). However, our model is not performed well on the Synaptics VS680 board devices. So, we got a comparably low final score of (5.82). In our future work, we will optimize our model for the best performance in the dedicated devices such as the Synaptics VS680 board device.

4.4 Quantize-Aware Training

The quantization-aware training (QAT) is widely used to increase the model performance without any extra time cost. So, we followed [14] to set the learning rate to 1×10^{-4} dropped by half every 50 epochs, and we trained it for 220

Table 4. Comparison used for the ablation study at scale of × 3.

Method	float32	PTQ	QAT
CDFM-Mobile	30.34	29.94	30.26
CDFM-Mobile W/O 3 × 3 conv (Model 1)	29.69	29.51	29.61
CDFM-Mobile W/O 1 × 1 conv (Model 2)	30.29	28.38	30.05
CDFM-Mobile W/O deep feature residual learning (Model 3)	30.23	30.09	30.22
CDFM-Mobile W/O anchor-based residual learning (Model 4)	30.28	29.79	30.18

epochs. It is clear from Table 1 that the model performance was enhanced by 0.32 dB. As a result, our model in int8 quantized network dropped by 0.08 dB in comparison with the case of floating point operation.

4.5 Ablation Study

The same training settings are considered for the ablation study to evaluate the effect of some important factors, as illustrated by Model 1, Model 2, model 3, and Model 4 in Fig. 6, on the model's performance such as: 1) the effect of using the 3 × 3 convolution instead of the CDFM block (Model 1), 2) the effect of using the 1 × 1 pointwise convolution instead of the CDFM block (Model 2), 3) the impact of using the deep feature residual learning (Model 3), and 4) the effect of using the anchor-based residual learning (Model 4).

The Impact of Using the 3 × 3 Convolution. To study the impact of the 3 × 3 convolution, the CDFM-Mobile model and Model 1 in Fig. 6a, which is the same as the CDFM-Mobile model but without 1 × 1 convolution. As illustrated in Table 4 (2nd row vs. 1st row). Removing the 3 × 3 convolution drops the PSNR by 0.65 dB for QAT. This illustrates the importance of the 3 × 3 convolution for extracting the deep features.

The Impact of Using Pointwise Convolution. To study the impact of the pointwise convolution, two models are considered. Te first model is the CDFM-Mobile model, and the second one is Model 2 shown in Fig. 6b after removing the pointwise convolution from the CDFM-Mobile as illustrated in Table 4 (3rd row vs. 1st row), the CDFM-Mobile model has a much better result compared to Model 2. This means that removing the pointwise convolution drops the PSNR by 0.21 dB for QAT due to the ability of the pointwise convolution to mix channel location. So, removing the pointwise convolution makes the model lacks channel mixing and degrades the performance.

The Effect of Using the Deep Feature Residual Learning. In this experiment, the CDFM-Mobile and the Model 3 shown in Fig. 6c, which is like the CDFM-Mobile model with removing the deep feature residual learning. As shown in Table 4 (4^{th}row vs. 1st row), the model performance is degraded by 0.11 dB in the case of float32. So, it is obvious that using this residual can increases the performance without any extra parameter cost. Also, it is worth mentioning that the model performance is slightly dropped between the float32 training and the QAT.

The Effect of Using Anchor-based Residual Learning. To do this task, two cases are considered: 1) The CDFM-Mobile model, and 2) The CDFM-Mobile model without anchor-based residual learning as illustrated by Model 4 in Fig. 6d. As illustrated in Table 4 (5th row vs. 1st row), the CDFM-Mobile model has a much better result compared to the Model 4. Removing the anchor branch drops PSNR by 0.15 dB for PQT. This is because the anchor-based can achieve the same image-space residual learning performance as residual learning in the feature space.

5 Conclusion

This paper proposes a channel mixing Network (CDFM-Mobile) for single image super-resolution (SISR). The CDFM-Mobile is based on using a pointwise convolution to mix the channel information and the extracted deep features. This is done by developing the channel deep feature mixing block (CDFM) as the backbone of the CDFM-Mobile model. The CDFM block only contains pointwise convolution and 3×3 convolution. The pointwise convolution is used for channel mixing and the 3×3 convolution is utilized for deep feature extraction. In addition, we used both of anchor-based residual learning and the deep feature residual learning to further improve the performance. This new CDFM-Mobile has low computation complexity, and it represents a lightweight model that can be suitable for the mobile devices. Our experiment results clearly show that the CDFM-Mobile achieved state-of-the-art results in multiple image super-resolution benchmarks. Finally, this model shows more appealing visual results quantitatively and qualitatively.

Acknowledgment. This work was supported in part by the National Key Research and Development Program of China under Grant 2019YFB2204500, in part by the National Natural Science Foundation of China under Grant. 6207409.

References

1. Agustsson, E., Timofte, R.: Ntire 2017 challenge on single image super-resolution: dataset and study. In: Proceedings of the IEEE Conference on Computer Vision and Pattern Recognition workshops, pp. 126–135 (2017)
2. Ahn, N., Kang, B., Sohn, K.A.: Fast, accurate, and lightweight super-resolution with cascading residual network. In: Proceedings of the European Conference on Computer Vision (ECCV), pp. 252–268 (2018)

3. Anwar, S., Khan, S., Barnes, N.: A deep journey into super-resolution: a survey. ACM Comput. Surv. (CSUR) **53**(3), 1–34 (2020)
4. Ayazoglu, M.: Extremely lightweight quantization robust real-time single-image super resolution for mobile devices. In: Proceedings of the IEEE/CVF Conference on Computer Vision and Pattern Recognition, pp. 2472–2479 (2021)
5. Bhalgat, Y., Zhang, Y., Lin, J.M., Porikli, F.: Structured convolutions for efficient neural network design. Adv. Neural Inf. Process. Syst. **33**, 5553–5564 (2020)
6. Cai, J., Meng, Z., Ding, J., Ho, C.M.: Real-time super-resolution for real-world images on mobile devices. arXiv preprint arXiv:2206.01777 (2022)
7. Chao, S.K., Wang, Z., Xing, Y., Cheng, G.: Directional pruning of deep neural networks. Adv. Neural Inf. Process. Syst. **33**, 13986–13998 (2020)
8. Chmiel, B., Banner, R., Shomron, G., Nahshan, Y., Bronstein, A., Weiser, U., et al.: Robust quantization: One model to rule them all. Adv. Neural Inf. Process. Syst. **33**, 5308–5317 (2020)
9. Dai, T., Cai, J., Zhang, Y., Xia, S.T., Zhang, L.: Second-order attention network for single image super-resolution. In: Proceedings of the IEEE/CVF Conference on Computer Vision and Pattern Recognition, pp. 11065–11074 (2019)
10. Dong, C., Loy, C.C., He, K., Tang, X.: Learning a deep convolutional network for image super-resolution. In: Fleet, D., Pajdla, T., Schiele, B., Tuytelaars, T. (eds.) ECCV 2014. LNCS, vol. 8692, pp. 184–199. Springer, Cham (2014). https://doi.org/10.1007/978-3-319-10593-2_13
11. Dong, C., Loy, C.C., He, K., Tang, X.: Image super-resolution using deep convolutional networks. IEEE Trans. Pattern Anal. Mach. Intell. **38**(2), 295–307 (2015)
12. Dong, C., Loy, C.C., Tang, X.: Accelerating the super-resolution convolutional neural network. In: Leibe, B., Matas, J., Sebe, N., Welling, M. (eds.) ECCV 2016. LNCS, vol. 9906, pp. 391–407. Springer, Cham (2016). https://doi.org/10.1007/978-3-319-46475-6_25
13. Dosovitskiy, A., et al.: An image is worth 16×16 words: transformers for image recognition at scale. arXiv preprint arXiv:2010.11929 (2020)
14. Du, Z., Liu, J., Tang, J., Wu, G.: Anchor-based plain net for mobile image super-resolution. In: Proceedings of the IEEE/CVF Conference on Computer Vision and Pattern Recognition, pp. 2494–2502 (2021)
15. Gendy, G., Mohammed, H., Sabor, N., He, G.: A deep pyramid attention network for single image super-resolution. In: 2021 9th International Japan-Africa Conference on Electronics, Communications, and Computations (JAC-ECC), pp. 14–19. IEEE (2021)
16. Gendy, G., Sabor, N., Hou, J., He, G.: Balanced spatial feature distillation and pyramid attention network for lightweight image super-resolution. Neurocomputing (2022). https://doi.org/10.1016/j.neucom.2022.08.053
17. Hang, Y., Liao, Q., Yang, W., Chen, Y., Zhou, J.: Attention cube network for image restoration. In: Proceedings of the 28th ACM International Conference on Multimedia, pp. 2562–2570 (2020)
18. Howard, A.G., et al.: MobileNets: efficient convolutional neural networks for mobile vision applications. arXiv preprint arXiv:1704.04861 (2017)
19. Hui, Z., Gao, X., Yang, Y., Wang, X.: Lightweight image super-resolution with information multi-distillation network. In: Proceedings of the 27th ACM International Conference on Multimedia, pp. 2024–2032 (2019)
20. Hui, Z., Wang, X., Gao, X.: Fast and accurate single image super-resolution via information distillation network. In: Proceedings of the IEEE Conference on Computer Vision and Pattern Recognition, pp. 723–731 (2018)

21. Ignatov, A., Timofte, R., Denna, M., Younes, A.: Real-time quantized image super-resolution on mobile NPUs, mobile AI 2021 challenge: Report. In: Proceedings of the IEEE/CVF Conference on Computer Vision and Pattern Recognition, pp. 2525–2534 (2021)
22. Ignatov, A., Timofte, R., Denna, M., Younes, A., et al.: Efficient and accurate quantized image super-resolution on mobile NPUs, mobile AI & AIM 2022 challenge: Report. In: Proceedings of the European Conference on Computer Vision (ECCV) Workshops (2022)
23. Ignatov, A., et al.: AI benchmark: all about deep learning on smartphones in 2019. In: 2019 IEEE/CVF International Conference on Computer Vision Workshop (ICCVW), pp. 3617–3635. IEEE (2019)
24. Ignatov, A., et al.: PIRM challenge on perceptual image enhancement on smartphones: Report. In: Proceedings of the European Conference on Computer Vision (ECCV) Workshops, pp. 0–0 (2018)
25. Ioannou, Y., Robertson, D., Cipolla, R., Criminisi, A.: Deep Roots: improving CNN efficiency with hierarchical filter groups. In: Proceedings of the IEEE Conference on Computer Vision and Pattern Recognition, pp. 1231–1240 (2017)
26. Isaac, J.S., Kulkarni, R.: Super resolution techniques for medical image processing. In: 2015 International Conference on Technologies for Sustainable Development (ICTSD), pp. 1–6. IEEE (2015)
27. Jacob, B., et al.: Quantization and training of neural networks for efficient integer-arithmetic-only inference. In: Proceedings of the IEEE Conference on Computer Vision and Pattern Recognition, pp. 2704–2713 (2018)
28. Jia, K., Rinard, M.: Efficient exact verification of binarized neural networks. Adv. Neural Inf. Process. Syst. **33**, 1782–1795 (2020)
29. Kim, J., Kwon Lee, J., Mu Lee, K.: Deeply-recursive convolutional network for image super-resolution. In: Proceedings of the IEEE Conference on Computer Vision and Pattern Recognition, pp. 1637–1645 (2016)
30. Kingma, D.P., Ba, J.: Adam: a method for stochastic optimization. arXiv preprint arXiv:1412.6980 (2014)
31. Lai, W.S., Huang, J.B., Ahuja, N., Yang, M.H.: Deep laplacian pyramid networks for fast and accurate super-resolution. In: Proceedings of the IEEE Conference on Computer Vision and Pattern Recognition, pp. 624–632 (2017)
32. Li, H., et al.: PAMS: quantized super-resolution via parameterized max scale. In: Vedaldi, A., Bischof, H., Brox, T., Frahm, J.-M. (eds.) ECCV 2020. LNCS, vol. 12370, pp. 564–580. Springer, Cham (2020). https://doi.org/10.1007/978-3-030-58595-2_34
33. Li, W., Li, J., Li, J., Huang, Z., Zhou, D.: A lightweight multi-scale channel attention network for image super-resolution. Neurocomputing **456**, 327–337 (2021)
34. Li, Y., Cao, J., Li, Z., Oh, S., Komuro, N.: Lightweight single image super-resolution with dense connection distillation network. ACM Transactions on Multimedia Computing, Communications, and Applications (TOMM). **17**(1s), 1–17 (2021)
35. Lim, B., Son, S., Kim, H., Nah, S., Mu Lee, K.: Enhanced deep residual networks for single image super-resolution. In: Proceedings of the IEEE Conference on Computer Vision and Pattern Recognition Workshops, pp. 136–144 (2017)
36. Liu, H., Cao, F., Wen, C., Zhang, Q.: Lightweight multi-scale residual networks with attention for image super-resolution. Knowl.-Based Syst. **203**, 106103 (2020)
37. Liu, J., Tang, J., Wu, G.: Residual feature distillation network for lightweight image super-resolution. In: Bartoli, A., Fusiello, A. (eds.) ECCV 2020. LNCS, vol. 12537, pp. 41–55. Springer, Cham (2020). https://doi.org/10.1007/978-3-030-67070-2_2

38. Luo, X., Xie, Y., Zhang, Y., Qu, Y., Li, C., Fu, Y.: LatticeNet: towards lightweight image super-resolution with lattice block. In: Vedaldi, A., Bischof, H., Brox, T., Frahm, J.-M. (eds.) ECCV 2020. LNCS, vol. 12367, pp. 272–289. Springer, Cham (2020). https://doi.org/10.1007/978-3-030-58542-6_17
39. Ma, H., Chu, X., Zhang, B.: Accurate and efficient single image super-resolution with matrix channel attention network. In: Proceedings of the Asian Conference on Computer Vision (2020)
40. Ma, Y., Xiong, H., Hu, Z., Ma, L.: Efficient super resolution using binarized neural network. In: Proceedings of the IEEE/CVF Conference on Computer Vision and Pattern Recognition Workshops, pp. 0–0 (2019)
41. Mobahi, H., Farajtabar, M., Bartlett, P.: Self-distillation amplifies regularization in hilbert space. Adv. Neural Inf. Process. Syst. **33**, 3351–3361 (2020)
42. Reddi, V.J., et al.: MLPerf inference benchmark. In: 2020 ACM/IEEE 47th Annual International Symposium on Computer Architecture (ISCA), pp. 446–459. IEEE (2020)
43. Sajjadi, M.S., Scholkopf, B., Hirsch, M.: EnhanceNet: single image super-resolution through automated texture synthesis. In: Proceedings of the IEEE International Conference on Computer Vision, pp. 4491–4500 (2017)
44. Shi, W., et al.: Real-time single image and video super-resolution using an efficient sub-pixel convolutional neural network. In: Proceedings of the IEEE Conference on Computer Vision and Pattern Recognition, pp. 1874–1883 (2016)
45. Soh, J.W., Cho, N.I.: Lightweight single image super-resolution with multi-scale spatial attention networks. IEEE Access **8**, 35383–35391 (2020)
46. Song, D., Wang, Y., Chen, H., Xu, C., Xu, C., Tao, D.: AdderSR: towards energy efficient image super-resolution. In: Proceedings of the IEEE/CVF Conference on Computer Vision and Pattern Recognition, pp. 15648–15657 (2021)
47. Szegedy, C., Vanhoucke, V., Ioffe, S., Shlens, J., Wojna, Z.: Rethinking the inception architecture for computer vision. In: Proceedings of the IEEE Conference on Computer Vision and Pattern Recognition, pp. 2818–2826 (2016)
48. Touvron, H., Cord, M., Douze, M., Massa, F., Sablayrolles, A., Jégou, H.: Training data-efficient image transformers & distillation through attention. In: International Conference on Machine Learning, pp. 10347–10357. PMLR (2021)
49. Vaswani, A., et al.: Attention is all you need. In: Advances in Neural Information Processing Systems **30** (2017)
50. Wan, J., Yin, H., Liu, Z., Chong, A., Liu, Y.: Lightweight image super-resolution by multi-scale aggregation. IEEE Trans. Broadcast. **67**(2), 372–382 (2020)
51. Wang, L., et al.: Exploring sparsity in image super-resolution for efficient inference. In: Proceedings of the IEEE/CVF Conference on Computer Vision and Pattern Recognition, pp. 4917–4926 (2021)
52. Wang, W., Wei, F., Dong, L., Bao, H., Yang, N., Zhou, M.: MINILM: deep self-attention distillation for task-agnostic compression of pre-trained transformers. Adv. Neural Inf. Process. Syst. **33**, 5776–5788 (2020)
53. Wang, Z., Chen, J., Hoi, S.C.: Deep learning for image super-resolution: a survey. IEEE Trans. Pattern Anal. Mach. Intell. **43**(10), 3365–3387 (2020)
54. Wu, Y., et al.: Compiler-aware neural architecture search for on-mobile real-time super-resolution. arXiv preprint arXiv:2207.12577 (2022)
55. Zhan, Z., et al.: Achieving on-mobile real-time super-resolution with neural architecture and pruning search. In: Proceedings of the IEEE/CVF International Conference on Computer Vision, pp. 4821–4831 (2021)

56. Zhang, K., et al.: Aim 2019 challenge on constrained super-resolution: methods and results. In: 2019 IEEE/CVF International Conference on Computer Vision Workshop (ICCVW), pp. 3565–3574. IEEE (2019)
57. Zhang, Y., Li, K., Li, K., Wang, L., Zhong, B., Fu, Y.: Image super-resolution using very deep residual channel attention networks. In: Proceedings of the European Conference on Computer Vision (ECCV), pp. 286–301 (2018)

Sliding Window Recurrent Network for Efficient Video Super-Resolution

Wenyi Lian[1(✉)] and Wenjing Lian[2]

[1] Uppsala University, Uppsala, Sweden
wenyi.lian.7322@student.uu.se
[2] Northeastern University, Shenyang, China
https://github.com/shermanlian/swrn

Abstract. Video super-resolution (VSR) is the task of restoring high-resolution frames from a sequence of low-resolution inputs. Different from single image super-resolution, VSR can utilize inter-frames' temporal information to reconstruct results with more details. Recently, with the rapid development of convolution neural networks (CNN), the VSR task has drawn increasing attention and many CNN-based methods have achieved remarkable results. However, only a few VSR approaches can be applied to real-world mobile devices due to the computational resources and runtime limitations. In this paper, we propose a *Sliding Window based Recurrent Network* (SWRN) which can be real-time inference while still achieving superior performance. Specifically, we notice that video frames should have both spatial and temporal relations that can help to recover details, and the key point is how to extract and aggregate these information together. Address it, we input three neighboring frames and utilize a hidden state to recurrently store and update the important temporal information. Our experiment on REDS dataset shows that the proposed method can be well adapted to mobile devices and produce visually pleasant results.

Keywords: Mobile device · Efficient algorithm · Video super-resolution · Recurrent network

1 Introduction

Over the past decade, we have seen the great success of Deep Learning (DL) and Convolution Neural Networks (CNNs) in computer vision [8,24]. By incorporating recent advanced CNN architectures in video super-resolution (VSR), many methods have achieved remarkable performances compared with traditional approaches. Thanks to the deep learning, these improvements encourage the community to explore more solutions for VSR such as sliding window-based methods [19,38,40] and recurrent-based networks [9,17,18,37]. These CNN-based methods usually require high computational resources and inference times, and the performance gains mainly come from their huge parameters and complexity.

On the other hand, with the growing popularity of built-in smartphone cameras, applying VSR networks to real-world mobile devices becomes vitally important and has drawn great attention [26]. However, running CNN models on a smartphone is

L. Karlinsky et al. (Eds.): ECCV 2022 Workshops, LNCS 13802, pp. 591–601, 2023.
https://doi.org/10.1007/978-3-031-25063-7_37

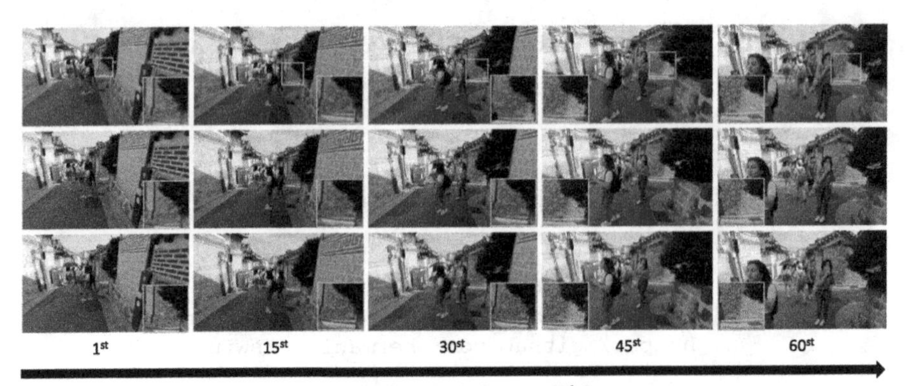

Fig. 1. Video sequence from the 1st frame to 60th frame. The top row is the low-resolution frames and the middle row shows the bicubicly upsampled frames. Our method is illustrated in the bottom row. Note that the same part of different frames could have different sharpness and details, which enables us aggregate neighboring frames to recover an HR image with rich details.

difficult due to the limited memories and the real-time inference requirement [12,14]. Compared with single image super-resolution (SISR), VSR usually needs to recover a sequence of inter-related frames, and the widely used techniques (e.g., frame alignment and fusion) are too complicated and computationally expensive thus they can hardly be used for smartphones directly. Moreover, the number of frames can linearly affect the inference times which further aggravates the deployment of CNN-based VSR methods. Some researchers treat VSR as an extension of SISR so that they could use efficient SR architectures without considering the temporal information [31]. Such a solution can achieve real-time inference but shows inferior performance.

To promote the development of real-time VSR, *Mobile AI & AIM 2022 Real-Time Video Super-Resolution Challenge* [16] is held to evaluate VSR networks on mobile GPUs which have strong resource-constraints. All participants were asked to design efficient models and train them on the REDS [36] dataset with 4× video upscaling. And participants should design models considering the balance between high restoration accuracies (PSNR) and low resource consumptions (latency). To evaluate the model efficiency, all solutions are asked to convert to '*tflite*' models and tested on the Android *AI Benchmark application* [13] which uses the Tensorflow TFLite library as a backend for running all quantized deep learning models.

In this paper, we aim to design a lightweight VSR model by comprehensively investigating the effectiveness of different CNN architectures for smartphones. To reduce the parameters and memories, our network is designed to contain only 3×3 convolution layers and ReLU activation functions. Besides, we find that the same part of different frames could have different sharpness and details, as illustrated in Fig. 1. To improve the accuracy, we further propose the sliding window recurrent network (SWRN) which makes use of the information from neighboring frames to reconstruct the HR frame. And an additional bidirectional hidden state is used to recurrently collect temporal spatial relations over all frames. By doing so, images produced by our model could have rich details far beyond other single image super-resolution methods. The main contribution of our work can be summarised as follows:

- We propose a lightweight video super-resolution network, known as *SWRN*, which can be easily deployed on mobile devices and perform VSR in real-time.
- To make the SWRN more efficient while preserving high performance, we propose the sliding-window strategy to utilize neighboring frames' information to reconstruct rich details.
- A bidirectional hidden state is incorporated to recurrently store and update temporal information, which could be very useful to aggregate long-range dependencies to improve the VSR performance.

2 Related Works

2.1 Single Image Super-Resolution

Single Image Super-Resolution (SISR) is the task of trying to reconstruct a high-resolution image from its degraded low-resolution low-quality counterpart. In the past few years, numerous works based on deep learning and deep convolution neural networks have achieved tremendous performance gains over traditional super-resolution approaches [6,7,11,21,25,30,33,41,44]. SRCNN [6] is the first work that uses CNN in super-resolution. Later, most subsequent works focus on optimizing the network architectures [7,23,30,44] and loss functions [20,25,32,41].

2.2 Video Super-Resolution

Starting from the pioneer network SRCNN [7], deep convolution neural network based methods have brought significant achievement in both image and video super-resolution tasks [1,2,4,21,27,29,30,34,35,40,43]. Particularly, in video super-resolution, where the most important parts are frame alignment, many advanced techniques have been developed to improve the accuracy. For example, VESPCN [3] and TOFlow [42] propose to use optical flow to align frames. TDAN [38] and EDVR [40] point out that estimating an accurate flow for occlusion and large motion frames is difficult and they choose to align frames using deformable convolution [5,45]. Especially, EDVR enjoys the merits of implicit alignment and its PCD module uses a pyramid and cascading architecture to handle occlusion and large motions. Another line of VSR methods also incorporates recurrent networks in video process [9,17,18,37], they usually use the hidden state to record the important temporal information. For the frame reconstruction part, residual blocks [10] and attention mechanism [39] are widely used to improve the performance. And recent transformer-based methods [28,29] also have shown attractive in image/video restoration.

2.3 Efficient Super-Resolution

Although most CNN-based approaches can obtain remarkable results in video super-resolution, their performance gains are often up to the network capacities [12,26]. It means that they usually require huge memories and computational resources, which makes it difficult to apply these state-of-the-art models on real-world smartphones that have constrained resources and inference times. Thus the network that is deployed on mobile devices should take care of the particularities of mobile NPUs and DSPs [12,14]. To deal with it, *AI Benchmark* application [13,15] is designed to allow researchers to run neural networks on the mobile AI acceleration hardware. Based on the *AI Benchmark* application, *Mobile AI 2021* [12] and *Mobile AI & AIM 2022 Real-Time Video Super-Resolution Challenge* is held to promote the development of real-time mobile VSR networks. In this paper, we'd like to follow the setting of these challenges and design an efficient yet mobile-compatible network for smartphones.

3 Method

In this section, we introduce basic concepts of video super-resolution and provide a detailed description of the main techniques and strategies of the proposed SWRN for efficient video super-resolution.

3.1 Sliding-Window Recurrent Network

The core idea of our method is to grasp and aggregate complementary information from neighbouring frames to improve the performance of video reconstruction. As illustrated in Fig. 2, given the low-resolution frame sequence $\{x_i\}_{i=1}^{N}, x_i \in \mathbb{R}^{h \times w \times c}$, at each time step t, our network accepts three basic inputs (including previous frame x_{i-1}, current frame x_i and future video frame x_{i+1}) and output a high-quality frame y_i, which seems like a sliding window multi-frame super-resolution algorithm [34,40], given by

$$y_i = f(x_{i-1}, x_i, x_{i+1}; \theta), \tag{1}$$

where f is the VSR network and θ represents its learnable parameters. Inspired by [18], we take the advantage of recurrent hidden states to preserve previous and future information. Specifically, the initial hidden states (forward and backward) are set to 0 and will be updated when the window slides to the next frames. Then we can reformulate Eq. (1) as follows:

$$y_i, (h_{i+1}^{+}, h_{i+1}^{-}) = f(x_{i-1}, x_i, x_{i+1}, h_i^{+}, h_i^{-}; \theta), \tag{2}$$

where h_i^{+} and h_i^{-} are i-th frame's forward hidden state and backward hidden state, respectively. Note here previous frame x_{t-1}, current frame x_t and forward hidden state are concatenated as a forward group, then future frame x_{t+1}, current frame x_t and backward hidden state compose the backward group. Deep features for each group are

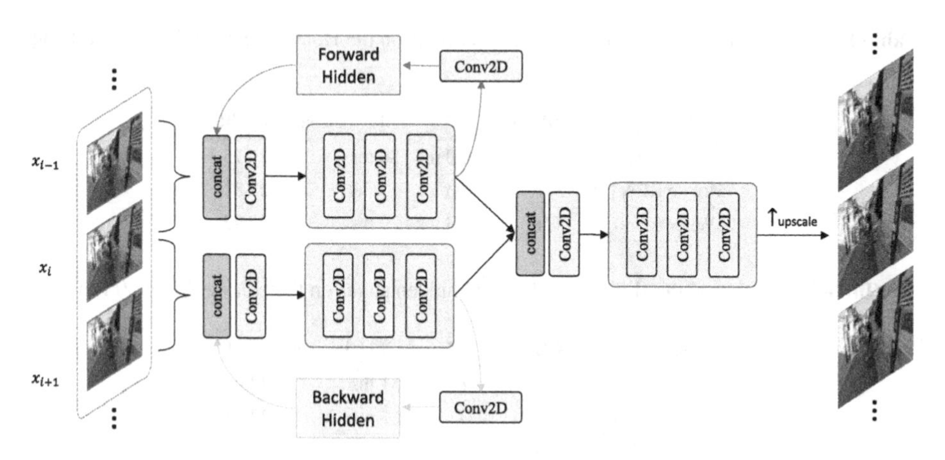

Fig. 2. Overview architecture of the proposed Sliding Window Recurrent Network. The forward and backward hidden states are recurrently updated and concatenated with neighboring two frames to provide extra information to reconstruct HR images.

separately extracted and concatenated to aggregate multi-frame information to reconstruct the HR frame as:

$$fea_i^+ = f_1(concat(x_{i-1}, x_i)), \tag{3}$$

$$fea_i^- = f_2(concat(x_{i+1}, x_i)), \tag{4}$$

$$output_i = f_3(concat(fea_i^+, fea_i^-)), \tag{5}$$

where f_1 and f_2 are the NN extractors that learn to obtain forward and backward features, respectively. Then f_3 is the aggregation function that merges all information to get final upscaled frames. Meanwhile, the extracted features of forward and backward groups will update the corresponding forward and backward hidden states by using two simple convolution layers. Then these hidden states can be used for the next frames.

3.2 Architecture and Loss

In training, we use the robust L_1 Charbonnier loss [23] to achieve high-quality video reconstruction, which can be formulated as follows:

$$\mathcal{L}_{cb} = \sum_{i=1}^{N} \sqrt{(f(x_{i-1}, x_i, x_{i+1}, h_i^+, h_i^-) - y_i)^2 + \epsilon^2}, \tag{6}$$

where N is the number of training samples, and ϵ is fixed to 1×10^{-6}.

There are totally of 14 convolution layers as shown in Fig. 2. To save the inference time on the mobile device, all layers of our network only consist of a single 3×3 convolution layer with the ReLU activation function. The number of channels for all convolution layers is set to 16. And as described in Sect. 3.1, three additional concatenation layers are used to combine information of frames and hidden states. Moreover,

Table 1. The table shows the model runtime and PSNR on the Huawei Mate 10 Pro Smartphone.

Method	#Params	CPU	GPU delegate	PSNR
FSRCNN [21]	25K	420 ms	86 ms	26.75
Mobile RRN [18]	37K	803 ms	73 ms	27.52
SWRN (ours)	43K	883 ms	79 ms	27.92

Table 2. Ablation of the sliding-window and hidden states on the Huawei Mate 10 Pro.

Method	#Params	CPU	GPU delegate	PSNR
Baseline	24K	489 ms	51 ms	27.05
+ sliding window	34K	711 ms	61 ms	27.48
+ hidden states	43K	883 ms	79 ms	27.92

we take the bilinear upsampled current frame as a low frequency residual connection to improve the restoration accuracy.

4 Experiment

4.1 Dataset

Our model is trained on the high-quality (720p) REDS [36] dataset, which is proposed in the NTIRE 2019 Competition [36] and is widely used in recent VSR researches. In the Mobile AI & AIM 2022 Real-Time Video Super-Resolution Challenge [16], the REDS dataset contains 240 video clips for training, 30 video clips for validation and 30 video clips for testing (each clip has 100 consecutive frames). All low-resolution videos are produced by bicubic downsampling with a scale factor of 4.

4.2 Implementation Details

For training, the batch size is 16 and all training LR images are randomly cropped to 64×64 patches. The total training iterations are set to 250000, and the number of video frames in the training phase is set to 10 and changes to 100 in the testing phase. We use Adam [22] optimizer with an initial learning rate of 1×10^{-3}, and decrease it by half every 50000 iterations. Our network is implemented using TensorFlow2.6 and Keras framework with a single Titan Xp GPU.

4.3 Experimental Result

To evaluate the proposed method on Video Super-Resolution, we compare SWRN with FSRCNN [21] and mobile RRN [18]. The former is a pioneer lightweight CNN-based network and consists of 7 convolution layers (two 1×1 layers for feature shrinking and expanding) and a transpose convolution layer for upscaling. The latter is a lite

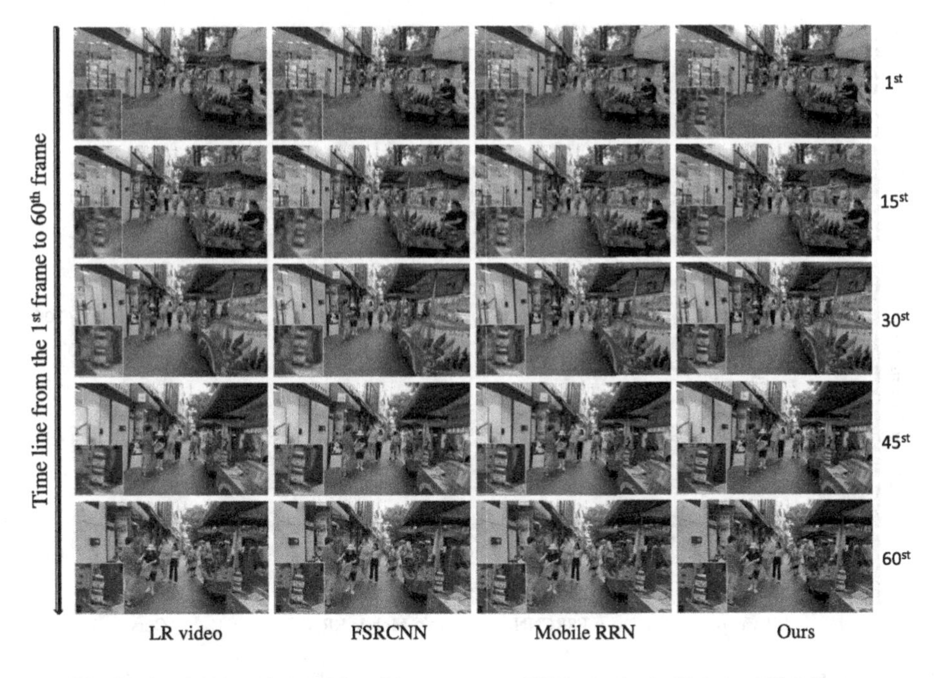

Fig. 3. Visual comparison of the video sequence *002* from the 1st frame to 60th frame.

Table 3. Analysis of the impact of number of channels on the Huawei Mate 10 Pro.

Method	#Params	CPU	GPU delegate	PSNR
Channel-8	13K	548 ms	45 ms	26.68
Channel-16	43K	883 ms	79 ms	27.92
Channel-32	156K	2209 ms	232 ms	28.24

version of Revisiting Temporal Modeling (RRN) which is a recurrent network for video super-resolution to run on mobile. Both our SWRN and FSRCNN are quantized with INT8 mode. We use the Peak Signal-to-Noise Ratio (PSNR) as the evaluation metric, and we will also consider the number of parameters and runtime on different mobile accelerators.

The inference times and quantitative results are reported in Table 1. One can see that although the proposed SWRN has more parameters and performs slower on the CPU device, SWRN is more compatible with the mobile TFLite GPU delegate accelerator and has lower latency than FSRCNN. In addition, our method achieves a higher performance in terms of PSNR, which surpass FSRCNN by 1.2 dB and surpass mobile RNN by 0.4 dB. The qualitative results are illustrated in Fig. 3 and Fig. 4. One can seen that video sequences produced by our method are much sharper and clear, and are visually pleasant. The result demonstrates our method satisfies the realistic runtime requirement while preserving a high PSNR performance.

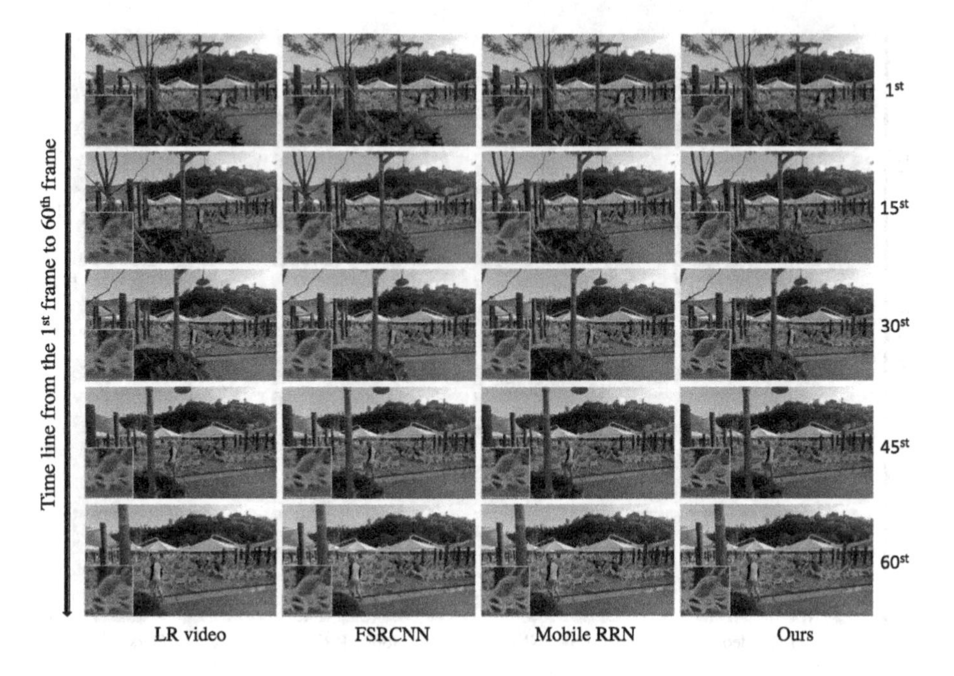

Fig. 4. Visual comparison of the video sequence *008* from the 1st frame to 60th frame.

4.4 Ablation Study

In this section, We conduct an ablation study to analyze the impact of the main components of the proposed SWRN framework: sliding-window strategy and (forward and backward) hidden states. For a fair comparison, we use a plain network as our baseline method that only receives the current frame and output a super-resolved image. Moreover, the baseline model uses 3×3 convolution layers with the ReLU activation function, and the number of channels is the same as SWRN. As shown in Table 2, although the baseline model has fewer parameters and requires lower runtime, it can only achieve 27.05 dB in terms of PSNR. By adding the sliding window strategy and hidden states, our method increases the PSNR by nearly 0.9 dB, which further demonstrates the superiority of the proposed method. In addition, we also provide the analysis of the impact of the number of channels in Table 3. As we see, although the setting of 8 channels and 32 channels can achieve faster inference time and PSNR performance respectively, the 16-channel setting is more balance in runtime and fidelity metrics.

5 Conclusion

In this paper, we propose a *Sliding Window based Recurrent Network* (SWRN) which can be real-time inference on mobile devices while still achieving a superior performance. The basic strategies incorporated are sliding-window and recurrent hidden states. To improve the inference time on smartphones, all layers in our network only contain 3×3 convolution and ReLU activation function. Our method is evaluated on the

REDS dataset with a scale factor of 4. The results shows our method is well compatible with mobile TFLite GPU delegate and can run faster than FSRCNN while preserving a high PSNR performance.

References

1. Ahn, N., Kang, B., Sohn, K.-A.: Fast, accurate, and lightweight super-resolution with cascading residual network. In: Ferrari, V., Hebert, M., Sminchisescu, C., Weiss, Y. (eds.) ECCV 2018. LNCS, vol. 11214, pp. 256–272. Springer, Cham (2018). https://doi.org/10.1007/978-3-030-01249-6_16
2. Bhat, G., Danelljan, M., Van Gool, L., Timofte, R.: Deep burst super-resolution. In: Proceedings of the IEEE/CVF Conference on Computer Vision and Pattern Recognition, pp. 9209–9218 (2021)
3. Caballero, J., et al.: Real-time video super-resolution with spatio-temporal networks and motion compensation. In: Proceedings of the IEEE Conference on Computer Vision and Pattern Recognition, pp. 4778–4787 (2017)
4. Chan, K.C., Wang, X., Yu, K., Dong, C., Loy, C.C.: BasicVSR: the search for essential components in video super-resolution and beyond. In: Proceedings of the IEEE/CVF Conference on Computer Vision and Pattern Recognition, pp. 4947–4956 (2021)
5. Dai, J., et al.: Deformable convolutional networks. In: Proceedings of the IEEE International Conference on Computer Vision, pp. 764–773 (2017)
6. Dong, C., Loy, C.C., He, K., Tang, X.: Learning a deep convolutional network for image super-resolution. In: Fleet, D., Pajdla, T., Schiele, B., Tuytelaars, T. (eds.) ECCV 2014. LNCS, vol. 8692, pp. 184–199. Springer, Cham (2014). https://doi.org/10.1007/978-3-319-10593-2_13
7. Dong, C., Loy, C.C., He, K., Tang, X.: Image super-resolution using deep convolutional networks. IEEE Trans. Pattern Anal. Mach. Intell. **38**(2), 295–307 (2015)
8. Goodfellow, I., Bengio, Y., Courville, A.: Deep Learning. MIT Press, Cambridge (2016)
9. Haris, M., Shakhnarovich, G., Ukita, N.: Recurrent back-projection network for video super-resolution. In: Proceedings of the IEEE/CVF Conference on Computer Vision and Pattern Recognition, pp. 3897–3906 (2019)
10. He, K., Zhang, X., Ren, S., Sun, J.: Deep residual learning for image recognition. In: Proceedings of the IEEE Conference on Computer Vision and Pattern Recognition, pp. 770–778 (2016)
11. Hui, Z., Wang, X., Gao, X.: Fast and accurate single image super-resolution via information distillation network. In: Proceedings of the IEEE Conference on Computer Vision and Pattern Recognition, pp. 723–731 (2018)
12. Ignatov, A., Romero, A., Kim, H., Timofte, R.: Real-time video super-resolution on smartphones with deep learning, mobile AI 2021 challenge: report. In: Proceedings of the IEEE/CVF Conference on Computer Vision and Pattern Recognition, pp. 2535–2544 (2021)
13. Ignatov, A., et al.: AI benchmark: running deep neural networks on Android smartphones. In: Leal-Taixé, L., Roth, S. (eds.) ECCV 2018. LNCS, vol. 11133, pp. 288–314. Springer, Cham (2019). https://doi.org/10.1007/978-3-030-11021-5_19
14. Ignatov, A., Timofte, R., Denna, M., Younes, A.: Real-time quantized image super-resolution on mobile NPUs, mobile AI 2021 challenge: report. In: Proceedings of the IEEE/CVF Conference on Computer Vision and Pattern Recognition, pp. 2525–2534 (2021)
15. Ignatov, A., et al.: AI benchmark: all about deep learning on smartphones in 2019. In: 2019 IEEE/CVF International Conference on Computer Vision Workshop (ICCVW), pp. 3617–3635. IEEE (2019)

16. Ignatov, A., Timofte, R., Kuo, H.K., Lee, M., Xu, Y.S., et al.: Real-time video super-resolution on mobile NPUs with deep learning, mobile AI & AIM 2022 challenge: report. In: Proceedings of the European Conference on Computer Vision (ECCV) Workshops (2022)
17. Isobe, T., Jia, X., Gu, S., Li, S., Wang, S., Tian, Q.: Video super-resolution with recurrent structure-detail network. In: Vedaldi, A., Bischof, H., Brox, T., Frahm, J.-M. (eds.) ECCV 2020. LNCS, vol. 12357, pp. 645–660. Springer, Cham (2020). https://doi.org/10.1007/978-3-030-58610-2_38
18. Isobe, T., Zhu, F., Jia, X., Wang, S.: Revisiting temporal modeling for video super-resolution. arXiv preprint arXiv:2008.05765 (2020)
19. Jo, Y., Oh, S.W., Kang, J., Kim, S.J.: Deep video super-resolution network using dynamic upsampling filters without explicit motion compensation. In: Proceedings of the IEEE Conference on Computer Vision and Pattern Recognition, pp. 3224–3232 (2018)
20. Johnson, J., Alahi, A., Fei-Fei, L.: Perceptual losses for real-time style transfer and super-resolution. In: Leibe, B., Matas, J., Sebe, N., Welling, M. (eds.) ECCV 2016. LNCS, vol. 9906, pp. 694–711. Springer, Cham (2016). https://doi.org/10.1007/978-3-319-46475-6_43
21. Kim, J., Lee, J.K., Lee, K.M.: Accurate image super-resolution using very deep convolutional networks. In: CVPR, pp. 1646–1654 (2016)
22. Kingma, D.P., Ba, J.: Adam: a method for stochastic optimization. arXiv preprint arXiv:1412.6980 (2014)
23. Lai, W.S., Huang, J.B., Ahuja, N., Yang, M.H.: Deep Laplacian pyramid networks for fast and accurate super-resolution. In: Proceedings of the IEEE Conference on Computer Vision and Pattern Recognition, pp. 624–632 (2017)
24. LeCun, Y., Bengio, Y., Hinton, G.: Deep learning. Nature **521**(7553), 436–444 (2015)
25. Ledig, C., et al.: Photo-realistic single image super-resolution using a generative adversarial network. In: Proceedings of the IEEE Conference on Computer Vision and Pattern Recognition, pp. 4681–4690 (2017)
26. Li, Y., et al.: NTIRE 2022 challenge on efficient super-resolution: methods and results. In: Proceedings of the IEEE/CVF Conference on Computer Vision and Pattern Recognition, pp. 1062–1102 (2022)
27. Lian, W., Peng, S.: Kernel-aware raw burst blind super-resolution. arXiv preprint arXiv:2112.07315 (2021)
28. Liang, J., et al.: VRT: a video restoration transformer. arXiv preprint arXiv:2201.12288 (2022)
29. Liang, J., Cao, J., Sun, G., Zhang, K., Van Gool, L., Timofte, R.: SwinIR: image restoration using swin transformer. In: Proceedings of the IEEE/CVF International Conference on Computer Vision, pp. 1833–1844 (2021)
30. Lim, B., Son, S., Kim, H., Nah, S., Mu Lee, K.: Enhanced deep residual networks for single image super-resolution. In: Proceedings of the IEEE Conference on Computer Vision and Pattern Recognition Workshops, pp. 136–144 (2017)
31. Liu, S., et al.: EVSRNet: efficient video super-resolution with neural architecture search. In: Proceedings of the IEEE/CVF Conference on Computer Vision and Pattern Recognition, pp. 2480–2485 (2021)
32. Lugmayr, A., Danelljan, M., Van Gool, L., Timofte, R.: SRFlow: learning the super-resolution space with normalizing flow. In: Vedaldi, A., Bischof, H., Brox, T., Frahm, J.-M. (eds.) ECCV 2020. LNCS, vol. 12350, pp. 715–732. Springer, Cham (2020). https://doi.org/10.1007/978-3-030-58558-7_42
33. Luo, Z., Huang, H., Yu, L., Li, Y., Fan, H., Liu, S.: Deep constrained least squares for blind image super-resolution. In: Proceedings of the IEEE/CVF Conference on Computer Vision and Pattern Recognition, pp. 17642–17652 (2022)

34. Luo, Z., et al.: BSRT: improving burst super-resolution with swin transformer and flow-guided deformable alignment. In: Proceedings of the IEEE/CVF Conference on Computer Vision and Pattern Recognition, pp. 998–1008 (2022)
35. Luo, Z., et al.: EBSR: feature enhanced burst super-resolution with deformable alignment. In: Proceedings of the IEEE/CVF Conference on Computer Vision and Pattern Recognition, pp. 471–478 (2021)
36. Nah, S., et al.: NTIRE 2019 challenge on video deblurring and super-resolution: dataset and study. In: Proceedings of the IEEE/CVF Conference on Computer Vision and Pattern Recognition Workshops (2019)
37. Sajjadi, M.S., Vemulapalli, R., Brown, M.: Frame-recurrent video super-resolution. In: Proceedings of the IEEE Conference on Computer Vision and Pattern Recognition, pp. 6626–6634 (2018)
38. Tian, Y., Zhang, Y., Fu, Y., Xu, C.: TDAN: temporally-deformable alignment network for video super-resolution. In: Proceedings of the IEEE/CVF Conference on Computer Vision and Pattern Recognition, pp. 3360–3369 (2020)
39. Wang, X., Girshick, R., Gupta, A., He, K.: Non-local neural networks. In: Proceedings of the IEEE Conference on Computer Vision and Pattern Recognition, pp. 7794–7803 (2018)
40. Wang, X., Chan, K.C., Yu, K., Dong, C., Change Loy, C.: EDVR: video restoration with enhanced deformable convolutional networks. In: Proceedings of the IEEE/CVF Conference on Computer Vision and Pattern Recognition Workshops (2019)
41. Wang, X., et al.: ESRGAN: enhanced super-resolution generative adversarial networks. In: Leal-Taixé, L., Roth, S. (eds.) ECCV 2018. LNCS, vol. 11133, pp. 63–79. Springer, Cham (2019). https://doi.org/10.1007/978-3-030-11021-5_5
42. Xue, T., Chen, B., Wu, J., Wei, D., Freeman, W.T.: Video enhancement with task-oriented flow. Int. J. Comput. Vis. **127**(8), 1106–1125 (2019). https://doi.org/10.1007/s11263-018-01144-2
43. Zhang, Y., Li, K., Li, K., Wang, L., Zhong, B., Fu, Y.: Image super-resolution using very deep residual channel attention networks. In: Ferrari, V., Hebert, M., Sminchisescu, C., Weiss, Y. (eds.) ECCV 2018. LNCS, vol. 11211, pp. 294–310. Springer, Cham (2018). https://doi.org/10.1007/978-3-030-01234-2_18
44. Zhang, Y., Tian, Y., Kong, Y., Zhong, B., Fu, Y.: Residual dense network for image super-resolution. In: Proceedings of the IEEE Conference on Computer Vision and Pattern Recognition, pp. 2472–2481 (2018)
45. Zhu, X., Hu, H., Lin, S., Dai, J.: Deformable ConvNets v2: more deformable, better results. In: Proceedings of the IEEE/CVF Conference on Computer Vision and Pattern Recognition, pp. 9308–9316 (2019)

EESRNet: A Network for Energy Efficient Super-Resolution

Shijie Yue[1], Chenghua Li[3,4(✉)], Zhengyang Zhuge[2,3], and Ruixia Song[1(✉)]

[1] North China University of Technology (NCUT), Beijing, China
ysj1161126955@gmail.com, songrx@ncut.edu.cn
[2] School of Artificial Intelligence, University of Chinese Academy of Sciences, Beijing, China
zyoung2333@gmail.com
[3] Institute of Automation, Chinese Academy of Sciences, Beijing, China
lichenghua2014@ia.ac.cn
[4] Nanjing Artificial Intelligence Research of IA (AiRiA), Nanjing, China

Abstract. Recently, various high-performance video super-resolution methods have been proposed. However, deployment on mobile phones is cumbersome due to the limitations of mobile phones' power consumption and computing power. We find methods that exploit temporal information in videos (e.g. optical flow) require huge energy consumption. Therefore, we use hidden features to preserve temporal information. Besides, the energy-efficient super-resolution network (EESRNet) is obtained by removing the residual connections in the Anchor-Based Plain Network (ABPN) [8]. Combining the two, we propose a Temporal Energy Efficient Super-Resolution Network (TEESRNet), which can efficiently utilize video spatio-temporal information with low energy consumption. Experiments show that for EESRNet, compared with ABPN, the latency is reduced by more than 40%, while performance decreases slightly. Furthermore, for TEESRNet, the PSNR is improved by 0.24 dB and 1.19 dB compared to EESRNet and RRN [19] respectively, while still maintaining real-time (<30 ms).

Keywords: Temporal information · Mobile video super-resolution · Energy efficient

1 Introduction

Since deep neural network based super-resolution methods were proposed, super-resolution tasks have developed rapidly. In recent years, how to deploy high-performance super-resolution methods on mobile phones has become an urgent

S. Yue, C. Li and Z. Zhuge—These authors contributed equally to this work.

Supplementary Information The online version contains supplementary material available at https://doi.org/10.1007/978-3-031-25063-7_38.

L. Karlinsky et al. (Eds.): ECCV 2022 Workshops, LNCS 13802, pp. 602–618, 2023.
https://doi.org/10.1007/978-3-031-25063-7_38

need in industry and academia. However, these methods require powerful computing power as support. Therefore, the training and inference of these methods need to be processed on powerful servers. Even some so-called real-time super-resolution methods only refer to real-time inference on the server. With the rapid development of mobile phone chips and the demand for higher-resolution images or video images, deploying high-performance super-resolution methods on mobile phones has become a top priority. In terms of hardware, major chip manufacturers continue to innovate technologies and integrate neural network computing accelerators into high-performance chips. On the software side, some mature mobile deep learning libraries (TensorFlow Lite, TFLite) facilitate the deployment of deep learning models on mobile phones. Therefore, it is urgent to design a lightweight super-resolution network with high performance, low latency, and low power consumption.

In the AIM20 ESR [43] and the NTIRE22 ESR [26] Challenge, various efficient image SR methods were proposed to alleviate this problem. However, their optimization objectives are usually parameters and Floating Point Operations(FLOPs), which can not reflect the performance on the edge directly. In the MAI21 real-time image SR Challenge [15], the runtime on the edge was taken into consideration during model optimization, which explored the practicability of SR model on mobile devices.

For real-time video SR, the Real-Time Video Super-Resolution Challenge held at the MAI 21 workshop [14] last year has greatly promoted the development of video super-resolution technology on mobile phones. This year, the Real-Time Video Super-Resolution on Mobile NPUs with Deep Learning challenge held by Mobile AI & AIM 2022 [17] puts forward new requirements for video super-resolution technology on mobile phones. This year's competition pays more attention to the energy consumption of the method on the mobile phone, which instructs that we should take the actual performance requirements (runtime latency, energy consumption, etc.) into account when designing the real-time video super-resolution method. Then, we propose a new video super-resolution model based on the predecessors to meet higher requirements.

This paper proposes an energy-efficient super-resolution network (EESRNet). Energy consumption is an important indicator to search optimal deep learning models in HWNAS [1,2,30]. Inspired by the clues in works [1,2] and the ABPN model [8], the proposed EESRNet prunes residual connections of the ABPN to reduce energy consumption. ABPN is the winning method in the Real-Time Quantized Image Super-Resolution Challenge held at MAI workshop 2021 [15]. To obtain the final energy-efficient network, we experimentally explore the impact of different EESRNet network structures on performance and power consumption and propose the challenge of SEESRNet participating in real-time VSR. To obtain the ultimate energy efficient network, we explore the impact of different EESRNet network structures on performance and power consumption through experiments and propose a SEESRNet to participate in the challenge of real-time video super-resolution track in Mobile AI challenge 2022 [17].

In addition, we propose a Temporal Energy Efficient Super-Resolution Network (TEESRNet) that exploits temporal information in the video to enhance network performance. We model temporal information using hidden features generated by only one 2D convolution layer, aiming to improve the network's performance with a slight energy consumption increase. The experimental results show that our EESRNet and the two variants based on it have a good balance between performance and power consumption, achieving good results on both the validation and test sets of REDS.

The main contributions of this paper are as follows:

- We propose an energy efficient super-resolution method that can effectively balance the performance and energy consumption of the network.
- We implement a video super-resolution network based on hidden feature alignment, which can be easily deployed on mobile devices.
- Based on extensive experiments, we design a Simple Energy Efficient Super-Resolution Network(SEESRNet) with a structural downsizing of EESRNet, which can significantly reduce power consumption while maintaining a reasonably high PSNR. Besides, it is applicable to both video and image super-resolution.

2 Related Work

2.1 Single Image Super-Resolution

Super-Resolution (SR) aims to reconstruct high-resolution (HR) images or videos from the corresponding low-resolution (LR) ones. Single image super resolution (SISR) is a challenging and significant problem in computer vision and high-impact competitions [15,17,26,42].

Recently, deep learning based SISR models have made remarkable achievements. SRCNN [7] is the pioneering work introducing convolution neural network (CNN) to SISR. After that, various CNN architectures were proposed in SR. VDSR [21] and DRCN [22] emphasized the importance of residual connection in a deeper SR model. EDSR [28] removed the Batch-Normalization (BN) layer in the residual architecture since BN can ruin high-frequency information in SR. SRGAN [23] and ESRGAN [38] used the generative adversarial training process like GAN [10] to make the SR image more realistic.

2.2 Multi-frame Super-Resolution

As video is one of the most common media in society, the demand for Deep-Learning based video super-resolution (VSR) is growing. From a multi-frame view, VSR can be regarded as an extension of image SR. Single image SR algorithm can be mitigated to video by recovering frame by frame. However, to further improve model performance on video, inter-frame information needs to be reckoned with. The different information usage methods between frames can be divided into two categories.

The first is explicit frame alignment with motion estimation and compensation. Deep-DE [27] used TV-ℓ_1 flow [3] and motion detail preserving (MDP) [39] to generate SR drifts, which will be fed into CNN with interpolated LR frame for further feature extraction. RRCN [24] proposed a residual recurrent convolution network to get multiple consecutive video frames as input, while only super-resolving the middle frame. EDVR [37] introduced deformable convolution to constitute a pyramid, cascading, and deformable (PCD) alignment module.

The other category is implicit frame alignment. FFCVSR [41] directly fed frames into 2D CNN and made the network learn the inter-frame information by itself. DUF [20] applied 3D CNN to exploit spatio-temporal information through the filters over input frames. RRN [19] utilized the previous frame and current frame as hidden state input and conducted hidden state mapping to preserve the texture details.

2.3 Mobile Super-Resolution

Deep-Learning based SR models always have numerous layers [28], residual connections [21] and even dense connections [38] to obtain a high PSNR performance. However, these models are not applicable to mobile devices due to their high computational complexity. In the AIM20 ESR [43] and the NTIRE22 ESR [26] Challenge, various efficient SR methods were proposed to alleviate this problem. However, their optimization objectives are usually parameters and Floating Point Operations(FLOPs), which can not reflect the performance on the edge directly. To make Deep-Learning models deployment-friendly, ECBSR [44] avoided dense connections and complicated operations in the network to ensure efficiency on the edge. At the same time, reparameterization and scaled Sobel filters are utilized to recover its performance. In the MAI21 real-time image SR Challenge [15], the runtime on the edge was taken into consideration during model optimization, which made the SR model more practical on mobile devices.

For mobile VSR, [4] discussed different fusion ways for the joint processing of multiple consecutive video frames, then proposed a joint motion compensation for efficient VSR. [25] proposed a fast spatio-temporal residual block(FRB) to adopt 3D convolutions for VSR while maintaining a low computational load. [9] considered the practicability of the motion compensation module and proposed an efficient recurrent latent space propagation for fast VSR. Methods in MAI21 VSR Challenge [14] explored the possibility of real-time VSR on edge. However, the methods mentioned above did not apply quantization for further acceleration. Quantization [13,35,36,40] is also a promising direction to make the Deep-Learning model light-weighted and deployable while maintaining a considerable performance compared with its full-precision counterparts. Various works [11,29,33] have shown the effectiveness of quantization in super-resolution.

3 Method

In this section, we detail our proposed EESRNet in Subsect. 3.1. To better utilize the temporal information of video frames, we propose Temporal EESRNet based on temporal information, which will be introduced in detail in Subsect. 3.2. In addition, according to the scoring criteria of the MAI22 competition, we further simplify the network structure based on EESRNet to obtain a super-resolution network with low energy consumption. We will introduce the details of this network in Subsect. 3.3.

3.1 EESRNet

Inspired by ABPN [8], we use 2D convolution to build our EESRNet. The entire architecture shows in Fig. 1. It mainly consists of 3 × 3 convolution operations. The shape of the feature maps remains unchanged before the depth-to-space layer. Furthermore, all the middle layers keep the same number of channels. This design ensures a stable memory occupancy by the EESRNet on the hardware device.

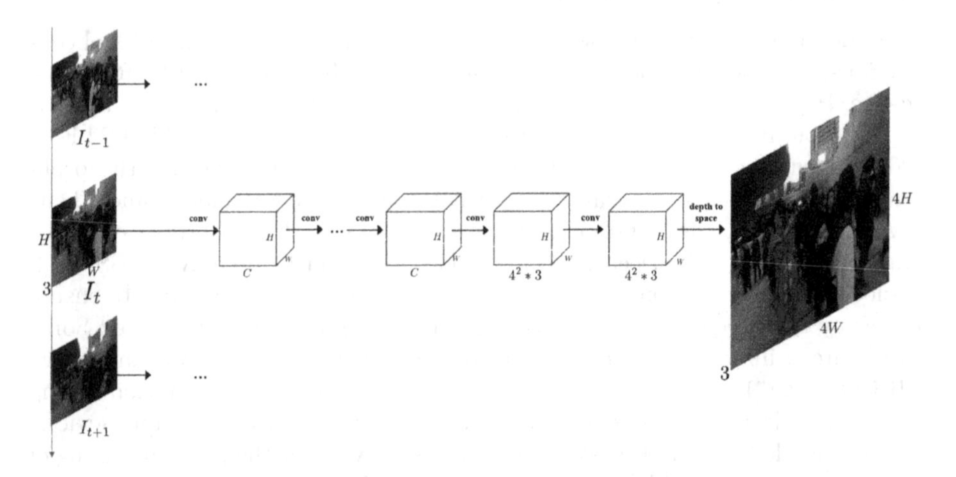

Fig. 1. Architecture of the proposed EESRNet. The network mainly consists of the repetition of multiple identical 3×3 convolution operations, which ensures the stability of the hardware memory occupied by the network, thereby making the network more friendly to hardware. The depth-to-space layer and clipping layer are adopted at the end of the network.

Let's denote I^{LR} and I^{SR} as input and output of EESRNet. We obtain low-resolution image features F by:

$$I^{SR} = f_{clip}(f_{up}(f_{convs}\left(I^{LR}\right))) \tag{1}$$

where $f_{convs}(\cdot)$ represents the feature extraction function, which uses multiple pure 3 × 3 convolution layers without ReLU to form this part [8]. For the up-sampling operation, we use pixel shuffle layer $f_{up}(\cdot)$, which transforms $H \times W \times (4^2 * 3)$ feature to $(4 * H) \times (4 * W) \times 3$ HR image, as shown in Fig. 1. For the sake of quantization, we add a clipping operation to guarantee the output pixel values are between [0, 255].

Energy consumption is one of the most critical performance metrics for mobile devices. Some works of the Hardware-Aware Neural Architecture Search (HWNAS) [1,2,30] optimize models for the accuracy, latency, and energy consumption of the target device. According to these works, we find that the writing and reading of memory occupy the bulk of the energy consumption. Therefore, to drastically reduce power consumption, we must reduce memory write and read operations while keeping the number of convolution layers unchanged. Furthermore, we find that multi-level residual connection is a major cause of memory consumption. Based on this observation, we remove the residual connection from the baseline model [8] and propose our EESRNet.

Advantages. Firstly, the energy consumption of our model is lower than that of ABPN [8], and the performance degradation caused by removing the residual module is within an acceptable margin. This makes EESRNet very cost-effective. Secondly, our model is easier to converge than ABPN during the training phase because of its straight-through network structure. Furthermore, compared with ABPN, it can be more quantization-friendly, so deploying our model on different mobile devices would be easier. Since the architecture of EESRNet only contains 2D 3×3 convolution, which is supported on most mobile devices for acceleration and is friendly for conducting quantization-aware training.

Residuals or Not? Considering that the U-net networks [6,18,31] can obtain a larger receptive field and extract features hierarchically as the network goes deep. We also attempt to use the U-net structure as the backbone of our network. However, after the experiment, it shows that the U-net structure can indeed improve the performance of the network, but at the cost of huge energy consumption, which is contrary to the original intention of our network design. We attribute the significantly increased energy consumption to the fact that the U-net network contains a large number of cross-layer connections, and the cross-layer connection usually needs to repeatedly write and read the results of each convolution layer, which consumes a lot of energy. In contrast, EESRNet completely abandons the operation of skip connections, and the results of each layer of convolution do not need to be stored or read after being passed to the next layer of convolution, which dramatically reduces the energy consumption of the network.

We can easily replace the 2D convolution with residual blocks, depthwise convolutions, or Squeeze-and-Excitation (SE) blocks in the simple network structure of the proposed EESRNet. However, experiments show that these various operations usually lead to an increase in the energy consumption of the network, and the performance improvements they bring are unsatisfactory.

3.2 TEESRNet

Temporal Information is Important for VSR. The key difference between video super-resolution methods and image super-resolution methods is the temporal information between the frames. Common video super-resolution methods model the temporal information of video sequences through optical flow, deformable convolution, 3D convolution, and convolution recurrent neural network (RNN), and then display or hide adjacent video frames aligned to the current frame. Therefore, the modeling accuracy of temporal information in the video is directly related to the performance of the entire video super-resolution method. Inspired by RRN [19], we add temporal information to EESRNet (shown in Fig. 1), named TEESRNet, as shown in Fig. 2. It proceeds in a single-loop fashion while exploiting hidden features as the accumulated temporal information of video sequences.

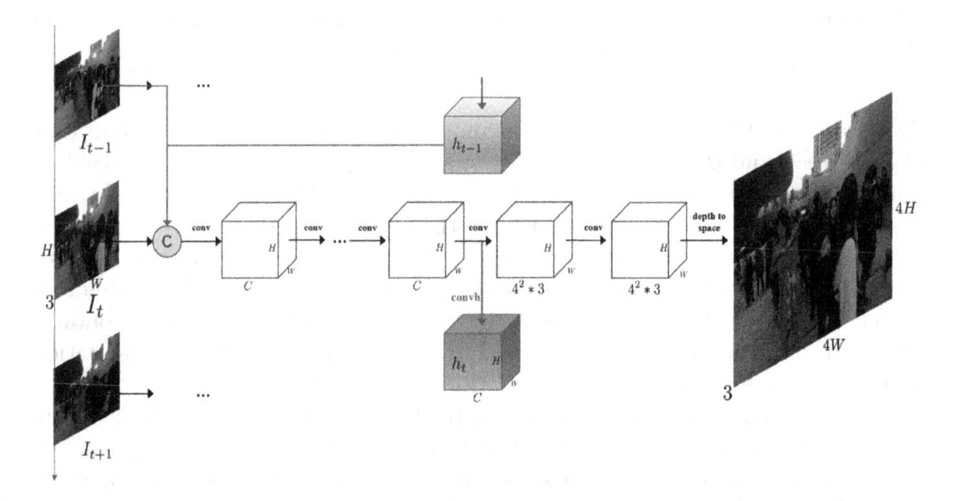

Fig. 2. Architecture of the TEESRNet with hidden feature embedding.

At the beginning of designing the temporal embedding strategy, the deployment of the target mobile devices is the critical factor to consider. Firstly, the current optical flow neural networks usually consume a lot of memory and computing power. Due to limited hardware resources, the optical flow increases the difficulty of mobile phone deployment, which is not in line with our original intention. Moreover, optical flow methods are prone to generating incorrect optical flow information in complex scenes, directly reducing the overall performance of video super-resolution. Secondly, because of some complex computation processes not supported by the APU acceleration library, the deformable convolution and the 3D convolution methods are not appropriate here either. In contrast, to extract temporal information, we only apply a simple convolution on top of the EESRNet. As a result, the new model, TEESRNet, is easily deployed on mobile phones.

In RRN [19], the residual module is used as the fundamental component of the network. However, as mentioned before, to avoid massive energy consumption, we should try to decrease the usage of residual connections and even abandon them in network design. Therefore, we replace the backbone network of RRN with our EESRNet, while keeping its temporal information modeling module. Furthermore, to additionally reduce the energy consumption of TEESRNet, we replace the convolutions that generate hidden features with 1×1 2D convolutions. In addition, we change the input of the hidden prediction module into the output of the deep feature extraction module. That is to say, our hidden feature does not only represent the feature of the current moment but a deep fusion feature in the time dimension.

As shown in Fig. 2, TEESRNet consists of two parts. One is a temporal embedding module for modeling the temporal information of video sequences. The other is an EESRnet for reconstructing video frames with temporal and spatial information. For time step t, let us denote I_t^{LR} and I_t^{SR} as the input and output and we have

$$I_t^{SR}, h_t = f(I_{t-1}^{LR}, I_t^{LR}, h_{t-1}), \tag{2}$$

where h_{t-1} is the hidden information corresponding to the time step $t-1$, and f represents our proposed TEESRNet. Here, h_t at the current moment is generated using an independent convolution operation on the output feature map of the deep feature extraction network.

Advantages and Disadvantages. First of all, compared to RRN [19], the hidden input is not the final but the deep feature generated by the deep feature extraction module. The final feature by the network only contains the features of the current frame. However, among the deep features, the input of our network contains the hidden features of the previous moment and the video frame of the previous moment, which contains the temporal information of the video sequence. This structure is beneficial to the network at the current moment to recover the current frame. Therefore, this step improves the performance of the network. Then, it should be noted that TEESRNet uses only one 1×1 2D convolution layer to predict hidden features, which means that the energy consumption for the entire network is only increased by a little. At the same time, though the introduced temporal information can help the network reconstruct the current frame, the hidden feature tensors would add additional computation and translation burdens to the target devices. Thus, TEESRNet may not be a better choice compared to EESRNet on mobile devices.

3.3 SEESRNet

In the Mobile AI 2022 challenge, we submitted a simple version of the EESRNet shown in Fig. 3. It only contains three convolution layers. The first convolution layer is a 3×3 2D convolution with output channels $C = 16$, while the last two have $4^2 \times 3 = 48$ output channels.

The SEESRNet brings the lowest energy consumption with its simple structure. In Sect. 3.1, it is mentioned that the second factor affecting network energy consumption is the number of convolution layers. Therefore, to obtain the ultimate low-energy video super-resolution network, we removed the deep feature extraction module. Although this will bring a certain degree of performance

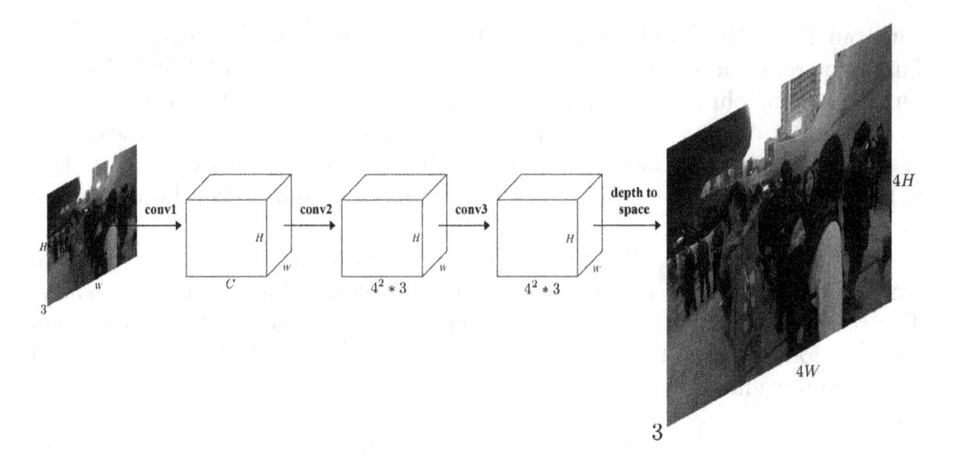

Fig. 3. The SEESRNet submitted to the Mobile AI 2022 Challenge, which only has three convolution layers.

degradation, the most important thing here is to keep the network energy consumption as low as possible.

Motivations. At the beginning of the competition, we mainly focused on improving the performance and the latency of the designed network. Therefore, we first compare the backbones of our network. Firstly, we select ABPN [8], a network with solid performance and also the winner of the last image super-resolution challenge. Secondly, we designed a UNet-like [6] network. The experiments show that dense residual connections can bring a certain degree of performance improvement, but they will lead to a considerable increase in energy consumption. Therefore, we determine ABPN as our baseline. Moreover, we remove the residual connection of ABPN, which will not lead to a rapid drop in performance but will reduce power consumption significantly.

Finally, we design and compare the effects of different convolution modules on performance and energy consumption, such as residual block, attention block, etc. Experimental comparisons show that the performance gains brought by these convolution modules cannot offset the increased energy consumption. In short, the SEESRNet is an energy efficient network. This fully benefits from its simple network structure. In addition, the full INT8 quantization further reduces its power consumption.

4 Experiments

4.1 Settings

Implementation Details. In each training batch, 16 cropped 64×64 LR RGB patches augmented by random flips and rotations are sent to the network. The learning rate is initialized to 4×10^{-4}, and the learning rate descent strategy adopts Cosine annealing. The models are trained with a total of 100 epochs. We use the mean absolute error (MAE) loss function as the optimization objective. Optimized by the ADAM optimizer, $\beta_1 = 0.9$, $\beta_1 = 0.999$, and $\epsilon = 10^{-8}$. We use the dataset provided by Mobile AI & AIM 2022 Real-Time Video Super-Resolution Challenge, and adopt REDS as a training and validation set.

Quantization-aware training (QAT) is a popular technique that improves performance without any inference stage cost. Based on the full precision model, the quantization of the model A total of 50 epochs are trained, and other settings in this stage are consistent with the above settings. It is worth noting that the quantization-aware training is carried out on the basis of the previous training.

Score. The evaluation of the submitted methods for the Real-Time Video Super-Resolution Challenge includes the fidelity of comparing synthetic RGB images to ground truth RGB images (PSNR) and the model execution speed (Latency) and energy consumption (J) on the MediaTek Dimensity 9000 platform. So the final score looks like this:

$$Score = 1.66 \times \mathrm{PSNR} + 50 \times (1 - \mathrm{J}) \tag{3}$$

4.2 Comparison with the State-of-the-Art Methods

In this section, we compare the proposed EESRNet and TEESENet with two types of methods, including the SISR model (*i.e.*, ABPN [8]), and VSR models(*i.e.*, RRN [19] and Diggers [14]). The adopt metrics are the number of model parameters (Params), the peak signal-to-noise ratio (PSNR), and the latency (Latency/ms). The latency is tested on a smart cell phone (Redmi K50 Pro) equipped with a Dimensity 9000 chip. It is worth noting that the latency is tested use the INT8 quantized model or the FP16 model.

Table 1. Comparison of results between EESRNet and ABPN. Tested with Redmi K50 Pro with Dimensity 9000 MediaTek Neuron using AI Benchmark [16]. The results are derived from the INT8 model.

Model	Params	PSNR	Latency/ms
ABPN [8]	62.55 K	27.98	3.60
EESRNet (ours)	62.55 K	27.84 (↓ 0.14 dB)	**2.03** (↓ 1.57 ms)

ABPN [8] is the winner model of the VSR track in Mobile AI challenge 2021. Our EESRNet is a simplified version, which aims for saving energy consumption using the specified APU. By only removing the residual connection of ABPN network, our EESRNet is much more efficient. As shown in Table 1, EESRNet has the same amount parameters with ABPN, but can reduce latency by 1.57 ms. Furthermore, EESRNet does not need to preserve the input tensor (for the residual connection in ABPN) in the memory during the inference, which could largely reduce power consumption of the device, especially for the mobile devices.

The temporal information is important for improving the performance of the VSR models. We choose the Mobile RRN[1] as the baseline, which is a recurrent network for video super-resolution modified form RRN [19] with reducing channels and not using previous output information.

Table 2. Comparisons with VSR networks. Tested on the CPU, TFLite GPU delegate (GPU), and MediaTek Neuron (APU) of the Redmi K50 Pro cell phone with Dimensity 9000 MediaTek Neuron using the AI Benchmark APP [16]. The results are derived from the FP16 model. "–" means that the method cannot obtain test results on the corresponding platform. The red color indicates the best model with respect to the inference time.

Model	Params	PSNR	Latency/ms		
			CPU	GPU	APU
Mobile RRN [19]	35.66 K	26.92	126.0	36.9	6.92
Diggers [14]	39.64 K	27.98	386.0	16.1	–
EESRNet (ours)	62.55 K	27.89	50.0	18.1	4.6
TEESRNet (ours)	80.24 K	**28.13**	347.0	–	17.8

As shown in Table 2, adding temporal information to the network results in a 0.24 dB increase in PSNR compared to EESRNet. This proves that adding temporal information can effectively improve the performance of the SISR network. On the other hand, compared with Mobile RRN [19], we remove a large number of residual modules and use the more powerful EESRNet as the backbone, which improves the performance by 1.19 dB. In addition to this, we also fond that using the middle layer features to generate the hidden tensor (temporal information) can improve the performance than using the last layer of the backbone, as shown in Fig. 2.

On the other hand, compared with the winning method Diggers [14] of the Real-Time Video Super-Resolution Challenge held by MAI Workshop 2021, TEESRNet has a 0.15 dB improvement on the REDS validation set. Besides, our method can be accelerated by using the dedicated NPU chip in MediaTek's Dimensity 9000. In our experiments, Diggers cannot achieve accelerated operations on the APU. When we use this model to test on MediaTek Neuron, the mobile phone turns obviously hot, and the program crashes after a period of time, resulting in no output. In contrast, benefiting from the simple and practical design of the structure, our method can be more conveniently deployed on different mobile phones. Experiments show that it is effective to use hidden features to model the temporal information of video sequences. A disadvantage of TEESRNet is that it cannot accelerated by the TFLite GPU Delegate.

[1] https://github.com/MediaTek-NeuroPilot/mai22-real-time-video-sr.

4.3 Ablation Study

For the ablation study, we conduct experiments from three aspects to explain the rationality of our selection of the components of the proposed EESRNet. That is the backbone, the convolution blocks, and overall structures.

Table 3. Results of experiments to verify the effectiveness of the network backbone. Tested with Redmi K50 Pro with Dimensity 9000 MediaTek Neuron using AI Benchmark [16]. The results are derived from the INT8 model. The bold text indicates the best model with respect to the PSNR or the latency.

Backbone	Params	PSNR	Latency/ms
ABPN [8]	62.55 K	27.98	3.60
DCENet [5]	357.23 K	**28.16** (↑ 0.18 dB)	2.27 (↓ 1.33 ms)
EESRNet (ours)	62.55 K	27.84 (↓ 0.05 dB)	**2.03** (↓ 1.57 ms)

Backbones. To verify the effectiveness of our chosen backbone, we compare EESRNet with the other two backbone networks, DCENet [5] and ABPN [8], respectively. DCENet is a U-net structure network, which contains a large number of skip connections, where the device needs to store all the middle feature tensors for the skip operations. According to the energy consumption feedback of the validation server during the Mobile AI 2022 challenge, DCENet consumes very large energy compared with ABPN (1.06 and 0.51, respectively). Also, as shown in Table 3, DCENet brings a performance improvement of 0.18 dB and a latency reduction of 1.33 ms. However, according to the final score formula 3, we find that the energy consumption is much more important for the final score improvement. Thus, we abandon the network structure with residual connections.

Table 4. Results of experiments to verify the validity of the convolution blocks. Tested with Redmi K50 Pro with Dimensity 9000 MediaTek Neuron using AI Benchmark [16]. The results are derived from the INT8 model. The bold text indicates the best model with respect to the latency.

Block	Params	PSNR	Latency/ms
ResBlock [28]	62.55 K	27.98 (↑ 0.14 dB)	3.71 (↑ 1.68 ms)
SEResBlock [12]	76.61 K	28.04 (↑ 0.20 dB)	4.18 (↑ 2.15 ms)
DWConvBlock [32]	35.12 K	27.55	3.43
ConvBlock (EESRNet)	62.55 K	27.84	**2.03**

Different Convolution Blocks. We compared the other three different blocks with the ConvBlock of EESRNet. The first is ResBlock, wich adds a residual connection to each 2D ConvBlock. SEResBlock enhances the 2D ConvBlock with an attention module from SENet [12]. DWConvBlock replaces the 2D ConvBlock with a depthwise 2D convolution [32]. As shown in Table 4, the ResBlock and SEResBlock can improve the performance by 0.14 dB and 0.2 dB compared with the ConvBlock, while they also increase of latency by 1.68 ms and 2.15 ms, respectively. More importantly, the residual connection of ResBlock and SEResBlock need to preserve the middle tensors in the memory during the block time, which would largely increase the energy consumption. Besides, experiments show that the DWConvBlock will increase the difficulty of training, and it also will not bring about an improvement in network performance and latency. Therefore, under comprehensive consideration, we choose 2D ConvBlock as the convolution block of our EESRNet.

Different Architectures. In order to obtain an extremely energy efficient network, we compare different settings of the EESRNet, including the number of blocks and the number of the output channels of the intermediate layers. Table 5 lists the experimental results under different architecture settings of EESRNet. According to the performance and latency changes, we could conclude that the smaller of the number of the blocks and channels, the best of the EESRNet with respect to the latency and the energy consumption. Based on the scoring criteria of the competition, we finally choose the last line in Table 5 as the parameter setting of SEESRNet, with ConvBlocks number 3 and the Channels number 16.

Table 5. Results of Validation Experiments for Architecture Validity. Tested with Redmi K50 Pro with Dimensity 9000 MediaTek Neuron using AI Benchmark [16]. The results are derived from the INT8 model. The bold text indicates the best model with respect to the PSNR or latency.

Model	Blocks	Channels	Params	PSNR	Latency/ms
ABPN [8]	7	28	62.55 K	**27.98**	3.60
EESRNet (ours)	7	28	62.55 K	27.84 (↓ 0.14 dB)	2.03 (↓ 1.57 ms)
	6	28	54.96 K	27.79 (↓ 0.19 dB)	1.97 (↓ 1.63 ms)
	5	28	47.88 K	27.75 (↓ 0.23 dB)	1.87 (↓ 1.73 ms)
	4	28	40.80 K	27.64 (↓ 0.34 dB)	1.78 (↓ 1.82 ms)
	3	28	33.71 K	27.50 (↓ 0.48 dB)	1.72 (↓ 1.88 ms)
SEESRNet (ours)	3	16	28.43 K	27.55 (↓ 0.43 dB)	**1.77** (↓ 1.83 ms)

4.4 Test Results

As shown in Table 6, our method achieves 9-th place among all the 13 teams in the Mobile AI & AIM 2022 Real-Time Video Super-Resolution Challenge. The structure and the latency of our submitted model (SEESRNet) are shown in Fig. 3 and Table 5. The test code is publicky available[2]. The final score is

[2] https://github.com/lchia/EESRNet.

derived by formula 3. The comparison shows that our method still has a large power consumption. Considering the limitations of hand-designed methods, we will consider HWNAS [1,2] in the future. In order to further utilize the computing resources of the edge devices, especially the APU, lower-bit quantization technology [34–36] will also be considered in the future.

Table 6. Test Results of MAI 2022. The blue color line is the results of our submitted model SEESRNet.

Username	PSNR	SSIM	Power, W/30FPS	Final score
erick	**27.34**	**0.7799**	**0.09**	**90.9**
OptimusPrime	27.52	0.7872	0.1	90.7
sr_enhance	27.85	0.7983	0.2	86.2
sx	27.34	0.7816	0.2	85.4
liuxunchenglxc	27.77	0.7957	0.22	85.1
DoctoR	27.71	0.782	0.24	84
try_again	27.72	0.7933	0.24	84
stevek	28.4	0.8105	0.33	80.6
ysj	27.46	0.7822	0.4	75.6
work mai	22.91	0.7546	0.36	70
caixc	27.71	0.7945	0.53	69.5
shermanlian	28.19	0.8093	0.8	56.8
xiaoxuan	28.45	0.8171	3.73	-89.3

5 Conclusions

In this study, we introduce EESRNet, a network aimed for energy efficient super-resolution on mobile devices. A temporal information enhanced network (TEESRNet) is also discussed. Experiments show that our proposed architecture outperforms the performance of the previous studies on mobile video super resolution task and features the low energy consumption. Future works includes lower bit quantization and HWNAS to further reduce the energy consumption.

Acknowledgements. This work was supported by the National Key Research and Development Program of China (Grant No. 2021ZD0201504), and the National Natural Science Foundation of China (No. 62106267).

References

1. Benmeziane, H., Maghraoui, K.E., Ouarnoughi, H., Niar, S., Wistuba, M., Wang, N.: A comprehensive survey on hardware-aware neural architecture search. arXiv preprint arXiv:2101.09336 (2021)
2. Bouzidi, H., Ouarnoughi, H., Niar, S., Cadi, A.A.E.: Performance prediction for convolutional neural networks on edge GPUs. In: CF 2021, Association for Computing Machinery, pp. 54–62. New York (2021). https://doi.org/10.1145/3457388.3458666
3. Brox, T., Bruhn, A., Papenberg, N., Weickert, J.: High accuracy optical flow estimation based on a theory for warping. In: Pajdla, T., Matas, J. (eds.) ECCV 2004. LNCS, vol. 3024, pp. 25–36. Springer, Heidelberg (2004). https://doi.org/10.1007/978-3-540-24673-2_3
4. Caballero, J., et al.: Real-time video super-resolution with spatio-temporal networks and motion compensation. In: Proceedings of the IEEE Conference on Computer Vision and Pattern Recognition, pp. 4778–4787 (2017)
5. Cheng, H., Liao, W., Tang, X., Yang, M.Y., Sester, M., Rosenhahn, B.: Exploring dynamic context for multi-path trajectory prediction. In: 2021 IEEE International Conference on Robotics and Automation (ICRA), pp. 12795–12801 (2021). https://doi.org/10.1109/ICRA48506.2021.9562034
6. Cho, S.J., Ji, S.W., Hong, J.P., Jung, S.W., Ko, S.J.: Rethinking coarse-to-fine approach in single image deblurring. In: Proceedings of the IEEE/CVF International Conference on Computer Vision (ICCV), pp. 4641–4650 (2021)
7. Dong, C., Loy, C.C., He, K., Tang, X.: Learning a deep convolutional network for image super-resolution. In: Fleet, D., Pajdla, T., Schiele, B., Tuytelaars, T. (eds.) ECCV 2014. LNCS, vol. 8692, pp. 184–199. Springer, Cham (2014). https://doi.org/10.1007/978-3-319-10593-2_13
8. Du, Z., Liu, J., Tang, J., Wu, G.: Anchor-based plain net for mobile image super-resolution. In: Proceedings of the IEEE/CVF Conference on Computer Vision and Pattern Recognition, pp. 2494–2502 (2021)
9. Fuoli, D., Gu, S., Timofte, R.: Efficient video super-resolution through recurrent latent space propagation. In: 2019 IEEE/CVF International Conference on Computer Vision Workshop (ICCVW), pp. 3476–3485. IEEE (2019)
10. Goodfellow, I., et al.: Generative adversarial nets. Adv. Neural Inf. Process. Syst. **27** (2014)
11. Hong, C., Kim, H., Baik, S., Oh, J., Lee, K.M.: DAQ: channel-wise distribution-aware quantization for deep image super-resolution networks. In: Proceedings of the IEEE/CVF Winter Conference on Applications of Computer Vision, pp. 2675–2684 (2022)
12. Hu, J., Shen, L., Sun, G.: Squeeze-and-excitation networks. In: Proceedings of the IEEE Conference on Computer Vision and Pattern Recognition, pp. 7132–7141 (2018)
13. Hubara, I., Courbariaux, M., Soudry, D., El-Yaniv, R., Bengio, Y.: Binarized neural networks. Adv. Neural Inf. Process. Syst. **29** (2016)
14. Ignatov, A., Romero, A., Kim, H., Timofte, R.: Real-time video super-resolution on smartphones with deep learning, mobile AI 2021 challenge: report. In: Proceedings of the IEEE/CVF Conference on Computer Vision and Pattern Recognition, pp. 2535–2544 (2021)

15. Ignatov, A., Timofte, R., Denna, M., Younes, A.: Real-time quantized image super-resolution on mobile NPUs, mobile AI 2021 challenge: report. In: Proceedings of the IEEE/CVF Conference on Computer Vision and Pattern Recognition, pp. 2525–2534 (2021)
16. Ignatov, A., et al.: AI benchmark: all about deep learning on smartphones in 2019. In: 2019 IEEE/CVF International Conference on Computer Vision Workshop (ICCVW), pp. 3617–3635. IEEE (2019)
17. Ignatov, A., et al.: Real-time video super-resolution on mobile NPUs with deep learning, mobile AI & AIM 2022 challenge: report. In: Proceedings of the European Conference on Computer Vision (ECCV) Workshops (2022)
18. Ignatov, A., Van Gool, L., Timofte, R.: Replacing mobile camera ISP with a single deep learning model. In: 2020 IEEE/CVF Conference on Computer Vision and Pattern Recognition Workshops (CVPRW), pp. 2275–2285 (2020). https://doi.org/10.1109/CVPRW50498.2020.00276
19. Isobe, T., Zhu, F., Jia, X., Wang, S.: Revisiting temporal modeling for video super-resolution. arXiv preprint arXiv:2008.05765 (2020)
20. Jo, Y., Oh, S.W., Kang, J., Kim, S.J.: Deep video super-resolution network using dynamic upsampling filters without explicit motion compensation. In: Proceedings of the IEEE Conference on Computer Vision and Pattern Recognition, pp. 3224–3232 (2018)
21. Kim, J., Lee, J.K., Lee, K.M.: Accurate image super-resolution using very deep convolutional networks. In: Proceedings of the IEEE Conference on Computer Vision and Pattern Recognition, pp. 1646–1654 (2016)
22. Kim, J., Lee, J.K., Lee, K.M.: Deeply-recursive convolutional network for image super-resolution. In: Proceedings of the IEEE Conference on Computer Vision and Pattern Recognition, pp. 1637–1645 (2016)
23. Ledig, C., et al.: Photo-realistic single image super-resolution using a generative adversarial network. In: Proceedings of the IEEE Conference on Computer Vision and Pattern Recognition, pp. 4681–4690 (2017)
24. Li, D., Liu, Y., Wang, Z.: Video super-resolution using non-simultaneous fully recurrent convolutional network. IEEE Trans. Image Process. **28**(3), 1342–1355 (2018)
25. Li, S., He, F., Du, B., Zhang, L., Xu, Y., Tao, D.: Fast spatio-temporal residual network for video super-resolution. In: Proceedings of the IEEE/CVF Conference on Computer Vision and Pattern Recognition, pp. 10522–10531 (2019)
26. Li, Y., et al.: NTIRE 2022 challenge on efficient super-resolution: methods and results. In: Proceedings of the IEEE/CVF Conference on Computer Vision and Pattern Recognition (CVPR) Workshops, pp. 1062–1102 (2022)
27. Liao, R., Tao, X., Li, R., Ma, Z., Jia, J.: Video super-resolution via deep draft-ensemble learning. In: Proceedings of the IEEE International Conference on Computer Vision, pp. 531–539 (2015)
28. Lim, B., Son, S., Kim, H., Nah, S., Mu Lee, K.: Enhanced deep residual networks for single image super-resolution. In: Proceedings of the IEEE Conference on Computer Vision and Pattern Recognition Workshops, pp. 136–144 (2017)
29. Liu, J., Wang, Q., Zhang, D., Shen, L.: Super-resolution model quantized in multi-precision. Electronics **10**(17), 2176 (2021)
30. Marchisio, A., Massa, A., Mrazek, V., Bussolino, B., Martina, M., Shafique, M.: NASCaps: a framework for neural architecture search to optimize the accuracy and hardware efficiency of convolutional capsule networks. In: 2020 IEEE/ACM International Conference on Computer Aided Design (ICCAD), pp. 1–9. IEEE (2020)

31. Ronneberger, O., Fischer, P., Brox, T.: U-Net: convolutional networks for biomedical image segmentation. In: Navab, N., Hornegger, J., Wells, W.M., Frangi, A.F. (eds.) MICCAI 2015. LNCS, vol. 9351, pp. 234–241. Springer, Cham (2015). https://doi.org/10.1007/978-3-319-24574-4_28
32. Sun, B., Li, J., Shao, M., Fu, Y.: LRPRNet: lightweight deep network by low-rank pointwise residual convolution. In: IEEE Transactions on Neural Networks and Learning Systems, pp. 1–11 (2021). https://doi.org/10.1109/TNNLS.2021.3117685
33. Wang, H., Chen, P., Zhuang, B., Shen, C.: Fully quantized image super-resolution networks. In: Proceedings of the 29th ACM International Conference on Multimedia, pp. 639–647 (2021)
34. Wang, P., Chen, W., He, X., Chen, Q., Liu, Q., Cheng, J.: Optimization-based post-training quantization with bit-split and stitching. IEEE Trans. Pattern Anal. Mach. Intell. **45**(2), 2119–2135 (2022)
35. Wang, P., He, X., Chen, Q., Cheng, A., Liu, Q., Cheng, J.: Unsupervised network quantization via fixed-point factorization. IEEE Trans. Neural Netw. Learn. Syst. **32**(6), 2706–2720 (2020)
36. Wang, P., Hu, Q., Zhang, Y., Zhang, C., Liu, Y., Cheng, J.: Two-step quantization for low-bit neural networks. In: Proceedings of the IEEE Conference on Computer Vision and Pattern Recognition, pp. 4376–4384 (2018)
37. Wang, X., Chan, K.C., Yu, K., Dong, C., Change Loy, C.: EDVR: video restoration with enhanced deformable convolutional networks. In: Proceedings of the IEEE/CVF Conference on Computer Vision and Pattern Recognition Workshops (2019)
38. Wang, X., et al.: ESRGAN: enhanced super-resolution generative adversarial networks. In: Proceedings of the European Conference on Computer Vision (ECCV) Workshops (2018)
39. Xu, L., Jia, J., Matsushita, Y.: Motion detail preserving optical flow estimation. IEEE Trans. Pattern Anal. Mach. Intell. **34**(9), 1744–1757 (2011)
40. Xu, W., He, X., Zhao, T., Hu, Q., Wang, P., Cheng, J.: Soft threshold ternary networks. arXiv preprint arXiv:2204.01234 (2022)
41. Yan, B., Lin, C., Tan, W.: Frame and feature-context video super-resolution. In: Proceedings of the AAAI Conference on Artificial Intelligence, vol. 33, pp. 5597–5604 (2019)
42. Yang, R., et al.: NTIRE 2022 challenge on super-resolution and quality enhancement of compressed video: dataset, methods and results. In: Proceedings of the IEEE/CVF Conference on Computer Vision and Pattern Recognition (CVPR) Workshops, pp. 1221–1238 (2022)
43. Zhang, K., et al.: AIM 2020 challenge on efficient super-resolution: methods and results. In: Bartoli, A., Fusiello, A. (eds.) ECCV 2020. LNCS, vol. 12537, pp. 5–40. Springer, Cham (2020). https://doi.org/10.1007/978-3-030-67070-2_1
44. Zhang, X., Zeng, H., Zhang, L.: Edge-oriented convolution block for real-time super resolution on mobile devices. In: Proceedings of the 29th ACM International Conference on Multimedia, pp. 4034–4043 (2021)

Bokeh-Loss GAN: Multi-stage Adversarial Training for Realistic Edge-Aware Bokeh

Brian Lee[1](✉), Fei Lei[2], Huaijin Chen[1], and Alexis Baudron[1]

[1] Sensebrain Technology, San Jose, CA 95131, USA
{brianlee,chenhuaijin,alexis.baudron}@sensebrain.site
[2] Tetras.AI, Shenzhen, China
leifei1@tetras.ai

Abstract. In this paper, we tackle the problem of monocular bokeh synthesis, where we attempt to render a shallow depth of field image from a single all-in-focus image. Unlike in DSLR cameras, this effect can not be captured directly in mobile cameras due to the physical constraints of the mobile aperture. We thus propose a network-based approach that is capable of rendering realistic monocular bokeh from single image inputs. To do this, we introduce three new edge-aware *Bokeh Losses* based on a predicted monocular depth map, that sharpens the foreground edges while blurring the background. This model is then finetuned using an adversarial loss to generate a realistic Bokeh effect. Experimental results show that our approach is capable of generating a pleasing, natural Bokeh effect with sharp edges while handling complicated scenes.

Keywords: Bokeh rendering · Generative models · Depth estimation · Edge refinement · Image translation

1 Introduction

The Bokeh effect is a highly desirable aesthetic effect in photography used to make the subject stand out by blurring away the out-of-focus parts of the image. This effect can be achieved naturally in Single-lens Reflex (SLR) cameras by taking an image with a wide aperture lens and large focal length, producing an image with a shallow depth of field (DoF). However, such shallow DoF effect can not be achieved naturally on mobile devices, due to the short focal lengths, small sensor and aperture sizes of mobile camera modules. As a result, the Bokeh effect can only be rendered synthetically on mobile devices, for example with stereo [23] or dual-pixel hardware [33] that provides a disparity map. This task is further constrained when limited to a single, monocular camera which is common particularly on mid to lower-end mobile devices.

In this paper, we propose a multi-stage, network-based approach that leverages both edge refinement, adversarial training and bokeh appearance loss to render a realistic bokeh effect from a single wide-DoF image. As Fig. 2 shows, our backbone bokeh generator network takes in a single image and its disparity map

L. Karlinsky et al. (Eds.): ECCV 2022 Workshops, LNCS 13802, pp. 619–634, 2023.
https://doi.org/10.1007/978-3-031-25063-7_39

and outputs the rendered bokeh. We employ a depth estimation network to produce a monocular disparity map which we use as a blurring cue for our backbone bokeh network. We also utilize this disparity map as a grey-scale saliency mask, which is used to sharpen the foreground edges while smoothing the background through three new loss functions. These losses create a strong, yet slightly rough bokeh effect which is then refined adversarially through a dual-scale PatchGAN [16] discriminator that we train jointly with the model backbone in a process similar to [28] (Fig. 1).

(a) Original Image (b) Disparity Map (c) Bokeh Output

Fig. 1. Given a single wide depth-of-field image, our model estimates the corresponding disparity map using a Depth Prediction Transformer (DPT). This map is then concatenated with the input image and then run through the trained NAFNet model to produce a shallow depth-of-field image. Best viewed from computer screen.

In summary, our main contributions are:

(1) Three new loss functions that use a greyscale saliency mask for edge-aware bokeh rendering from monocular images taken with wide depth-of-field
(2) A multi-stage training scheme that adversarially refines the output produced using the three aforementioned losses
(3) A computationally efficient blurring backbone, successfully applying the NAFNet architecture to the task of Bokeh

2 Related Work

2.1 Computational Bokeh

A variety of different methods have been explored for synthetic Bokeh rendering in recent years. Classical rendering methods take into account the physics of the image scene, before combining various different modules to construct an automatic Bokeh render. Earlier methods, such as [31,32,43] were limited to images in portrait mode. These methods relied on the segmentation of the image foreground from the background, before applying a filter on the background to achieve the Bokeh effect. Another class of methods, such as in [3,10,27,34,37, 40], blurred the image by making use of predicted depth maps, either through depth estimates from monocular depth module [22,29,30], stereo vision [2,17], or through a moving camera using the parallax effect [5,39]. The input image is then decomposed to multiple layers conditioned on the estimated depth map, and then rendered back to front to prevent bleeding into the foreground. As for the actual blur, many different methods have been adopted based on imaging physics including convolutional filters such as in [31], or through physics-based scattering methods such as in [27,33].

Recent advancements in Bokeh rendering have also seen the adoption of neural-network based rendering methods to avoid boundary artifacts. For example, [25,38] train a neural network to predict Bokeh effects from perfect (synthetic) depth maps. Real perfect depth maps are hard to come by, however, so other end-to-end network models have been introduced, such as in [6,12,14,28]. These models all use some sort of encoder-decoder structure to process the image at multiples scales and use approaches such as monocular depth estimation or image segmentation to improve bokeh results. Recent work by Peng *et al.* [26] has also suggested combining the classical and network-based approaches to improve controllability in the Bokeh process.

2.2 Monocular Depth Estimation

Many different approaches have been proposed for monocular depth estimation. Earlier work by Eigen *et al.* [7] proposed a multi-scale DNN based on AlexNet to generate a depth estimate. This method, however, suffered a number of key dataset limitations, namely a small number of training samples and a lack of variety.

Li *et al.* [22] proposed a new depth data collection system that allowed for better generalization of depth models. This was followed by a breakthrough by Ranftl *et al.* [30], who proposed a method for mixing multiple depth datasets even with incomplete annotations. Both of these methods used a convolutional architectures to create the depth prediction. Ranftl *et al.* [29] then showed that a transformer backbone could be used to further improve the depth prediction quality.

2.3 Generative Models

Generative Models [8] have been shown to generate realistic images while preserving fine texture details. These models, however, have been shown to often suffer from the problem of Modal Collapse. To remedy this, algorithms such as WGAN [1] and WGAN-GP [9] have been proposed to enhance training stability.

Recent work has shown the success of GAN-based architectures on image translation tasks such as in image deblurring [18], single-image super resolution [19,35,36], semantic segmentation [24], and so on. Conditional Adversarial Networks [16], in particular, have shown to generalize to arbitrary image translation tasks. Recently, this framework has also been adopted to the task of monocular bokeh synthesis [28] in refining the bokeh rendering quality.

3 Proposed Method

Rendering an accurate bokeh effect from an all-in-focus image is a complicated task that requires processing the image at multiple scales. For a network to learn an accurate blur effect, it must (a) learn how to accurately segment out the in-focus object regions by processing the image at a global scale and (b) learn an accurate blur on the out-of-focus regions by processing the image on a local scale. When the image is poorly segmented (a), however, blur in the out-of-focus regions (b) can often bleed into the foreground, creating an unnatural bokeh effect at boundary edges. Tackling both tasks (a) and (b) in a single end-to-end neural network is thus extremely challenging without additional cues that aid in the individual tasks.

We add a series of modules that aid in both tasks during training and inference. As shown in Fig. 2, the major components of our network are:

i. A Dense Prediction Transformer (DPT) [29] based depth estimation module that estimate the relative depth of the scene.
ii. A Nonlinear Activation Free Network (NAFNet) based generator module [4] that takes in the all-in-focus image and depth map of the scene, and outputs the depth-aware bokeh image.
iii. A dual-receptive field patch-GAN discriminator similar to the one proposed in [28].
iv. A bokeh model based loss function that encourages sharp content in the in-focus area and smooth content in the background area.

The bokeh output is generated by first running the input image through the depth prediction module to produce a depth map $\mathbf{D}$. The output $\mathbf{D}$ is then concatenated with the original input image $\mathbf{I}$ and then run through the NAFNet generator $\mathbf{G}$ to produce a shallow depth-of-field output $\mathbf{G}(\mathbf{I} \odot \mathbf{D})$.

3.1 Backbone Network

Given the translational nature of the Bokeh task, we adopt a backbone architecture that performs well on such problems. Various backbone networks have been proposed recently for image restoration tasks, such as [4,41]. In particular, the NAFNet architecture [4] has been shown to maintain a relatively small memory footprint while achieving exceptional output performance on such tasks. This is achieved by replacing high complexity non-linear operators such as Softmax, GELU, and Sigmoid with simple multiplications, thereby reducing the complexity in computationally expensive attention estimation.

Due to the similar nature of the prediction task, we therefore use the NAFNet architecture for our Bokeh backbone. Experimental results have shown (see Sect. 4.3) that on the scale necessary for mobile devices (around width 12 or less), that the difference in NAFNet width does not result in that large of a difference in the quality of the output images. We thus use a smaller NAFNet model with a width of 8. We empirically set the number of NAFblocks for the

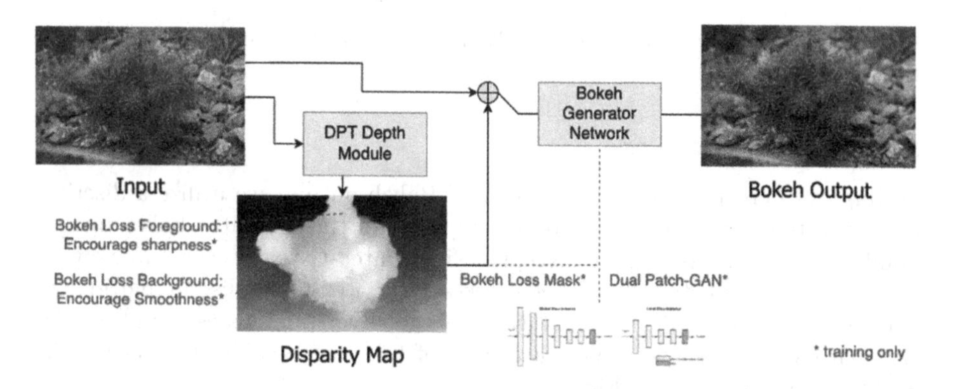

Fig. 2. Bokeh-Loss GAN pipeline: We first generate the depth of the scene from the all-sharp input, then combine the depth map and the input image, and feed them to the bokeh generator network to produce the bokeh image. Besides conventional L1 and SSIM losses, we introduce a depth-weighted bokeh loss, in addition to a dual-scale discriminator loss.

encoder to be [2,2,2,20] and the number of NAFBlocks for the decoder. We also concatenate 2 NAF blocks in the middle for a total of 36 NAFBlocks in the network. The total number of parameters comes to around 1.047M parameters.

3.2 Depth Estimation Network

Our model uses a depth map estimate during both training and inference stages as a blur and edge cue (see Sect. 3.4). The depth map of the scenes are estimated from a single monocular image by a dense prediction transformer (DPT) [29]. We use the depth prediction model that was pre-trained on a large-scale mixed RGB-to-depth "MIX 5" datasets [30]. We find empirically that such networks can generate realistic depth maps with smooth boundaries and fine-grain details at object boundaries (Fig. 3).

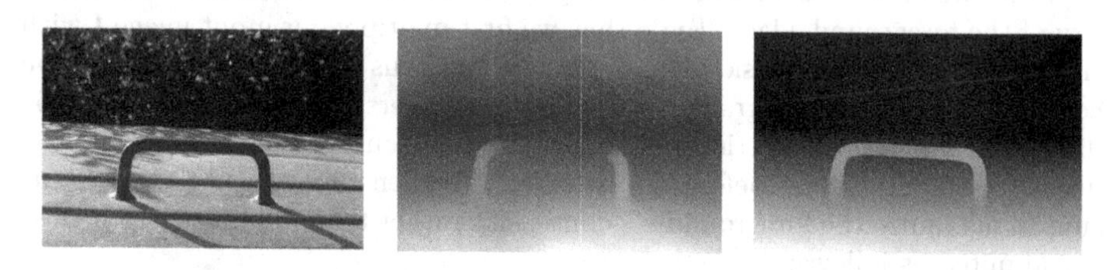

Fig. 3. Comparison of Input images (left) with Depth Maps generated with MegaDepth (center) and DPT (right) for two EBB inputs

The generated depth map provides extra blurring clues to assist the backbone generator in creating better bokeh blur. Particularly, we found that accurate depth boundaries help to generate noticeably more natural looking bokeh boundaries than in previous methods.

3.3 Discriminator Loss

To improve perceptual quality of the final Bokeh output, we utilize a discriminator loss to ensure the perceived visual quality of the generated bokeh images. Similar to [28], we use a multi-receptive-field Patch-GAN discriminator [16] as part of our loss function. The model takes two patch-GAN discriminators, one of depth 3 and one of depth 5 and averages the adversarial losses from both discriminators into one loss. We use a λ value of 1 for the WGAN gradient penalty for the discriminator loss (Fig. 4).

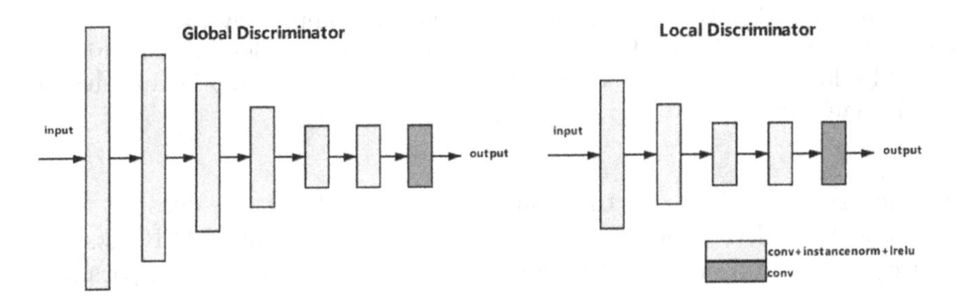

Fig. 4. Multi receptive discriminator

3.4 Bokeh Loss

In addition to the conventional L1, SSIM, and discriminator loss functions, we propose a bokeh-specific loss. The core idea is that, for a natural bokeh, the in-focus area should be sharp and the out-of-focus area should be blurred. Using the depth map we generate in the DPT module as a greyscale saliency mask $\mathbf{M}$, we introduce the following loss functions:

First, we introduce a *Foreground Edge Loss* that tries to maximize the intensity of the foreground edges. To do this, we first multiply our input image $\hat{\mathbf{I}}$ with the greyscale saliency mask $\mathbf{M}$ to obtain the in-focus regions of the image. We compute the edge map (gradients) using Sobel filters in the 0°, 90° and ±45° directions which provides richer structure information, we then sum up the L1-norms of the gradients, before normalizing. We then take the negative of this sum to maximize the foreground edge intensity rather than minimize it. The loss is computed as follows:

$$L_{foreedge}(\hat{\mathbf{I}}, \mathbf{M}) = -\frac{\sum_{z \in \{0^\circ, \pm 45, ^\circ 90^\circ\}} \left\| S_z(\hat{\mathbf{I}} \cdot \mathbf{M}) \right\|_1}{h_{\hat{\mathbf{I}}} \cdot w_{\hat{\mathbf{I}}}} \tag{1}$$

where S_z is the sobel convolution operator in the z direction to maximize the intensities of the foreground.

While we wish to strengthen the intensity of the edges however, we also want to create edges that are of a similar intensity to that of the output image. To achieve this, we add an *Edge Difference Loss*:

$$L_{edgediff}(\mathbf{I}, \hat{\mathbf{I}}, \mathbf{M}) = ||L_{foreedge}(\hat{\mathbf{I}} \cdot \mathbf{M})| - |L_{foreedge}(\mathbf{I} \cdot \mathbf{M})|| \tag{2}$$

that attempts to minimize the difference in foreground edge strength between the input and the output image.

Finally, we add a *Background Blur Loss* $L_{backblur}$ that encourages a smoother blur for the background. We first multiply the input image $\hat{\mathbf{I}}$ with the inverse of the greyscale saliency mask $\mathbf{M}$ to obtain the out-of-focus regions. We then try to minimize the total variation of the scene:

$$L_{backblur}(\hat{\mathbf{I}}, \mathbf{M}) = \frac{1}{h_{\hat{\mathbf{I}}} \cdot w_{\hat{\mathbf{I}}}} TV(\hat{\mathbf{I}} \cdot (1 - \mathbf{M})) \tag{3}$$

where the added factor is used to normalize the loss.

In summary, we add three new loss functions:

- A *Foreground Edge Loss* that encourages sharper edges seperating the foreground and the background
- An *Edge Difference Loss* that encourages similar edge intensities for the input and output image
- A *Background Blur Loss* that encourages a smoother background with less noise.

Qualitative results have shown that these losses induce sharp, albeit slightly distorted edges, along with a smooth background, which when combined with the adversarial loss, appears to increase the overall MOS of the output (see Sect. 4.3).

3.5 Training Setup

In order to get the sharp edges and smooth background induced by the Bokeh losses while maintaining a natural blur and avoiding edge artifacts, we adopt a dual-stage strategy for model training.

In the first, "pretraining" stage, we use a weighted sum of L1 loss, SSIM loss, and our three aforementioned "Bokeh" losses, using the depth map as a greyscale mask to separate the foreground from the background. The final loss-function for this stage was given by

$$\begin{aligned} L_{pretrain} = {} & 0.5 \times L_1 + 0.05 \times L_{SSIM} + 0.005 \times L_{edgediff} \\ & + 0.1 \times L_{backblur} + 0.005 \times L_{foreedge} \end{aligned} \tag{4}$$

where the final weights were determined experimentally. This pretraining stage then creates a rough bokeh effect with sharp, albeit slightly distorted edges and a smooth background.

In the second stage, the model is finetuned with a weighted sum of L1 loss, SSIM loss, VGG perceptual loss, and an adversarial loss from the dual-scale PatchGAN discriminator that is trained jointly in the second stage using the WGGAN-GP method [9]. This second stage refines the distorted edges and the background to generate a realistic bokeh effect. The refinement loss thus took the form

$$L_{refinement} = 0.5 \times L_1 + 0.1 \times L_{VGG} + 0.05 \times L_{SSIM} + L_{adv} \tag{5}$$

where the weights were again determined experimentally. The adversarial loss L_{adv} is weighted the highest with a coefficient of 1 in order to allow the discriminator to play the leading role in the refining phase.

4 Experiments

In this section, we introduce our experimental setting and compare both the qualitative and quantitative performance of our solution to current state-of-the-art architectures that were designed for this problem, along with other models that were submitted as part of the AIM 2022 *Real-Time Rendering Realistic Bokeh* challenge. We also conduct a series of ablation studies to analyze the different effects of various factors.

4.1 Experimental Setup

Technical Specifications. Our method uses standard Pytorch packages and the Pytorch Lightning framework for training. All models were trained on 8 32G Nvidia V100 GPUs, 383G RAM, and 40 CPUs.

In training, we cropped all images to 1024×1024 sized patches as inputs. The final models were all trained with batch size 2, with 60 epochs for the first stage and 60 epochs for the second stage. For optimization, we use the Adam optimizer with learning rate $1e-4$ for both the backbone and the discrminator, with $(\beta_1, \beta_2) = (0, 0.9)$.

Dataset. The *Everything is Better with Bokeh!* (EBB!) dataset [12] was released as part of the AIM 2022 *Real-Time Rendering Realistic Bokeh Challenge.* The dataset consists of 4696 shallow/wide depth-of-field image pairs for training and 200 images for validation and testing respectively. The image pairs were shot with a Cannon 7D DSLR camera taken with a narrow aperture (f/16) for the all-in-focus input image and a wide aperture (f/1.8) for the Bokeh ground truth image. The photos are taken in a wide variety of scenes, all in automatic mode, and are aligned using SIFT keypoint matching and the RANSAC method as in [11], before being cropped to a common region. Despite this, we found that many of the images suffered from either poor alignment or inconsistent lighting. We thus manually pruned the training dataset down to 4425 images, to help the model learn a better bokeh effect. We then took out 200 images for validation (which we call *Val200*) before submitting our final results.

4.2 Quantitative and Qualitative Evaluation

Quantitative Evaluation. Our model was submitted to the AIM 2022 *Real-Time Rendering Realistic Bokeh Challenge* [15], where the goal is to achieve shallow depth-of-field with the best perceptual quality compared to the ground truth .

Our solution was submitted only to Track 2 (unconstrained time) due to flex delegates that were used in the conversion from Pytorch to TFLite which makes our model unable to run on a mobile GPU. Our model achieved the best Learned Perceptual Image Patch Similarity (LPIPS) [42] score (lower is better), as shown in Table 1 which shows the performance of our model. Our model also yielded high scores on fidelity metrics, achieving the second best PSNR and SSIM scores. All models were run on a Kirin 9000 5G Processor, with the corresponding runtime being checked through the AI Benchmark [13] app.

Table 1. Quantitative results of our method from the Unconstrained CPU track of AIM 2020 *Rendering Realistic Bokeh* Challenge.

Team	PSNR ↑	SSIM ↑	LPIPS ↓	Avg. Runtime(s)
xiaokaoji	22.76	0.8652	0.2693	**0.125**
MinsuKwon	**22.89**	**0.8754**	0.2464	1.637
hxin	20.08	0.7209	0.4349	1.346
sensebrain	22.81	0.8653	**0.2207**	12.879

Note that the runtime shown in the table above is the runtime of the TFLite model. The runtime of the pytorch model on CPU is 0.95s.

Qualitative Evaluation. In this section, we provide sample visual results of our model compared to the results of the current state-of-the-art solutions trained specifically for Bokeh rendering [12,28]. As shown in Fig. 5, our model has a blur quality that is similarly natural compared to BGGAN [28], while performing better in quality than PyNET. On the other hand, the edges are sharper and more prominent in our model compared to other models, while the salient regions also remain sharper.

Fig. 5. Results of our model on the EBB test dataset. From left to right: input image, results of PyNET, results of BGGAN, our results. Our model is able to attain a strong blur while maintaing a sharp foreground. Best viewed from computer screen.

4.3 Ablation Studies

As mentioned before, our model consists of various different components and is trained in a multi-stage manner with a combination of different losses and cues. In this section, we analyze the effects of different parts of the model through a comparison of the outputs on the *Val200* validation set.

Blurring Cues. One of the main aspects of our network is the addition of a depth prediction module in DPT to help guide the blur. This is done through a concatenation of the original input image **I** with the depth map generated by running the input image through a depth estimation module D, before running through a backbone.

Other blur cues, such as a binary saliency mask or a defocus map can also be used, however, to help guide the blur. We thus tried concatenating various different alternatives to the depth map we generated using the DPT module, including a binary saliency mask generated through the TRACER algorithm [21] and a defocus map generated by DMENet [20].

As we are only trying to compare the concatenated input map (eg. depth map, saliency map, defocus map) as a blurring clue and not a saliency mask for sharpening during the Bokeh Loss (see Sect. 4.3), we only compare the result of concatenating the input channel and running the NAFNET/GAN setup without any bokeh losses.

Table 2. Quantitative and qualitative results of GAN-based training with various different blur cues on the Val200 dataset.

Blur Cue	PSNR	SSIM	MOS
Depth Map (DPT)	**23.70**	**0.867**	**4.01**
Saliency Mask (TRACER)	23.26	0.863	3.72
Defocus Map (DMENet)	22.00	0.800	3.30

As shown in Table 2, we found that the depth map performed the best in terms of blur cues across all metrics. This was largely due to the quality of the depth map that was geenrated, compared to the saliency and defocus maps that were generated through the TRACER [21] and DMENet [20] algorithms respectively. We thus chose the depth map as our main blurring cue.

In particular, we found that the bokeh images generated using the depth map had not only the best blur quality, but also the most natural foreground boundaries. This motivated us to use the depth map for the bokeh loss as well.

Bokeh Loss. To test the effectiveness of the Bokeh losses, we ran an ablation study comparing results of the model with and without them. All models here use the depth map as a blur cue. The second model uses the depth map also as a greyscale mask for the bokeh loss.

Table 3. Results of different training strategies for the Bokeh Loss

Model	PSNR	SSIM	MOS
NAFNet8 (GAN/No Bokeh Loss)	**23.70**	0.867	4.01
NAFNet8 (GAN/Bokeh Loss during pretraining)	23.475	**0.868**	**4.23**

As shown in Table 3, the addition of the Bokeh loss provides a substantial increase in the MOS, with the edges being noticeably sharper and the background smoother.

Model Backbone. To test the effectiveness of our model backbone, we tested the effect of switching the 8-layer NAFNet backbone with two cascading U-Nets (GlassNet) similar to in [28]. Both models were trained with the full pipeline, using the Bokeh losses in the pretraining stage (as described in Sect. 3) and the dual-scale GAN in the refinement stage (Table 4).

Table 4. Results of different backbones for the pipeline

Backbone	No. of Parameters	PSNR	SSIM	MOS
GlassNet	10.353M	22.956	**0.875**	**4.25**
8-Layer NAFNet	**1.047M**	**23.475**	0.868	4.23

We found that the 8-layer NAFNet performed almost identically similar to the GlassNet proposed in [28] despite having only around a tenth of the parameters. This justified our choice of switching the model backbone for the NAFNet.

NAFNet Width. As discussed in Sect. 3.1, the NAFNet contains a variable number of width and number of enclosing/decoding/middle blocks. The main two factors in determining model complexity are the width of the NAFBlocks and the number of NAFBlocks. As discussed in [4], an increase in the number of blocks has shown to not increase latency greatly while greatly improving model performance on the original image restoration task. On the other hand, due to memory constraints of mobile devices, the bottleneck in the model prediction is instead the width of the NAFBlocks, due to the amount of processing needed for wider blocks.

We thus chose to reduce the width of the individual NAFBlocks from the original NAFBlock width of 32. This reduction of width, however, must be balanced with our desire to process the input images at multiple (global and local) scales. Based on these considerations, we ran an ablation study to find a suitable balance between model width and computational complexity.

Table 5. Quantitative and qualitative results of 36-block NAFNet with various different widths

NAFBlock Width	No. of Parameters	PSNR	SSIM	MOS
4	**277K**	**23.603**	0.850	3.57
8	1.047M	23.475	**0.868**	4.23
12	2.311M	23.298	0.866	**4.31**

Due to model size considerations, we found that the backbone needed to have a width of 12 or lower to fit on a mobile device. As shown in Table 5, there is not a significant difference in MOS for widths above 8 layers, while the parameter count increases greatly, unlike in the shift from a width of 4 to 8 (where there is a big increase in MOS). We thus chose a final NAFBlock width of 8.

5 Conclusions

We have presented a new method for monocular bokeh synthesis to generate natural shallow depth-of-field images. By using a monocular depth estimate as both a blur cue and a greyscale saliency mask for three new "bokeh-specific" loss functions, our model was able to produce images with sharp foreground edges and a smooth background. Coupled with a multi-receptive field conditional adversarial network [16] as part of a dual-stage training process, our model was able to achieve a sharp, yet natural bokeh render. Extensive ablations are then presented to validate the effect of each of the individual modules. In the future, we can explore adding radiance modules to help render the bokeh effect in a more physics-based manner, along with using a smaller depth-prediction module to decrease the inference time on TFLite.

Acknowledgements. We would like to thank the team at Sensebrain Technology for their helpful suggestions. We did not receive external funding or additional revenues for this project.

A Pytorch to TFLite Conversion

Our model was converted from Pytorch to TFLite through a two-part process. First, we converted the trained NAFNet model using the Pytorch → Onnx → Tensorflow pipeline, and then combined a pretrained DPT net (which was prewritten in Tensorflow) in tensorflow with our saved Tensorflow NAFNet module. We then ran a conversion from Tensorflow → TFLite by first saving the resulting Tensorflow Module as a frozen graph and then exporting it to TFLite using the version 2 converter.

References

1. Arjovsky, M., Chintala, S., Bottou, L.: Wasserstein GAN (2017). https://doi.org/10.48550/ARXIV.1701.07875
2. Barron, J.T., Adams, A., Shih, Y., Hernández, C.: Fast bilateral-space stereo for synthetic defocus. In: 2015 IEEE Conference on Computer Vision and Pattern Recognition (CVPR), pp. 4466–4474 (2015). https://doi.org/10.1109/CVPR.2015.7299076
3. Busam, B., Hog, M., McDonagh, S., Slabaugh, G.G.: SteReFo: efficient image refocusing with stereo vision. arXiv preprint arXiv:1909.13395 (2019)
4. Chen, L., Chu, X., Zhang, X., Sun, J.: Simple baselines for image restoration. arXiv preprint arXiv:2204.04676 (2022)

5. Davidson, P., Mansour, M., Stepanov, O., Piché, R.: Depth estimation from motion parallax: experimental evaluation. In: 2019 26th Saint Petersburg International Conference on Integrated Navigation Systems (ICINS), pp. 1–5 (2019). https://doi.org/10.23919/ICINS.2019.8769338
6. Dutta, S., Das, S.D., Shah, N.A., Tiwari, A.K.: Stacked deep multi-scale hierarchical network for fast bokeh effect rendering from a single image. In: 2021 IEEE/CVF Conference on Computer Vision and Pattern Recognition Workshops (CVPRW), pp. 2398–2407 (2021). https://doi.org/10.1109/CVPRW53098.2021.00272
7. Eigen, D., Puhrsch, C., Fergus, R.: Depth map prediction from a single image using a multi-scale deep network. arXiv preprint arXiv:1406.2283 (2014)
8. Goodfellow, I.J., et al.: Generative adversarial networks (2014). https://doi.org/10.48550/ARXIV.1406.2661
9. Gulrajani, I., Ahmed, F., Arjovsky, M., Dumoulin, V., Courville, A.C.: Improved training of wasserstein GANs. arXiv preprint arXiv:1704.00028 (2017)
10. Ha, H., Im, S., Park, J., Jeon, H.G., Kweon, I.S.: High-quality depth from uncalibrated small motion clip. In: 2016 IEEE Conference on Computer Vision and Pattern Recognition (CVPR), pp. 5413–5421 (2016). https://doi.org/10.1109/CVPR.2016.584
11. Ignatov, A., Kobyshev, N., Timofte, R., Vanhoey, K.: DSLR-quality photos on mobile devices with deep convolutional networks. In: 2017 IEEE International Conference on Computer Vision (ICCV), pp. 3297–3305 (2017). https://doi.org/10.1109/ICCV.2017.355
12. Ignatov, A., Patel, J., Timofte, R.: Rendering natural camera bokeh effect with deep learning. arXiv preprint arXiv:2006.05698 (2020)
13. Ignatov, A., et al.: AI benchmark: all about deep learning on smartphones in 2019. In: 2019 IEEE/CVF International Conference on Computer Vision Workshop (ICCVW). pp. 3617–3635 (2019). https://doi.org/10.1109/ICCVW.2019.00447
14. Ignatov, A., et al.: AIM 2020 challenge on rendering realistic bokeh. In: Bartoli, A., Fusiello, A. (eds.) ECCV 2020. LNCS, vol. 12537, pp. 213–228. Springer, Cham (2020). https://doi.org/10.1007/978-3-030-67070-2_13
15. Ignatov, A., Timofte, R., et al.: Efficient bokeh effect rendering on mobile GPUs with deep learning, mobile AI & AIM 2022 challenge: report. In: Proceedings of the European Conference on Computer Vision (ECCV) Workshops (2022)
16. Isola, P., Zhu, J., Zhou, T., Efros, A.A.: Image-to-image translation with conditional adversarial networks. arXiv preprint arXiv:1611.07004 (2016)
17. Kamencay, P., Breznan, M., Jarina, R., Lukac, P., Radilova, M.: Improved depth map estimation from stereo images based on hybrid method. Radioengineering **21** (2012)
18. Kupyn, O., Budzan, V., Mykhailych, M., Mishkin, D., Matas, J.: DeblurGAN: blind motion deblurring using conditional adversarial networks (2017). https://doi.org/10.48550/ARXIV.1711.07064
19. Ledig, C., et al.: Photo-realistic single image super-resolution using a generative adversarial network. arXiv preprint arXiv:1609.04802 (2016)
20. Lee, J., Lee, S., Cho, S., Lee, S.: Deep defocus map estimation using domain adaptation. In: The IEEE Conference on Computer Vision and Pattern Recognition (CVPR) (June 2019)
21. Lee, M.S., Shin, W., Han, S.W.: TRACER: extreme attention guided salient object tracing network. arXiv preprint arXiv:2112.07380 (2021)
22. Li, Z., Snavely, N.: MegaDepth: learning single-view depth prediction from internet photos. arXiv preprint arXiv:1804.00607 (2018)

23. Liu, D., Nicolescu, R., Klette, R.: Bokeh effects based on stereo vision. In: Azzopardi, G., Petkov, N. (eds.) CAIP 2015. LNCS, vol. 9256, pp. 198–210. Springer, Cham (2015). https://doi.org/10.1007/978-3-319-23192-1_17
24. Majurski, M., et al.: Cell image segmentation using generative adversarial networks, transfer learning, and augmentations. In: 2019 IEEE/CVF Conference on Computer Vision and Pattern Recognition Workshops (CVPRW), pp. 1114–1122 (2019). https://doi.org/10.1109/CVPRW.2019.00145
25. Nalbach, O., Arabadzhiyska, E., Mehta, D., Seidel, H., Ritschel, T.: Deep shading: Convolutional neural networks for screen-space shading. arXiv preprint arXiv:1603.06078 (2016)
26. Peng, J., Cao, Z., Luo, X., Lu, H., Xian, K., Zhang, J.: BokehMe: when neural rendering meets classical rendering. In: Proceedings of the IEEE/CVF International Conference on Computer Vision and Pattern Recognition (CVPR) (2022)
27. Peng, J., Luo, X., Xian, K., Cao, Z.: Interactive portrait bokeh rendering system. In: 2021 IEEE International Conference on Image Processing (ICIP), pp. 2923–2927 (2021). https://doi.org/10.1109/ICIP42928.2021.9506674
28. Qian, M., Qiao, C., Lin, J., Guo, Z., Li, C., Leng, C., Cheng, J.: BGGAN: bokeh-glass generative adversarial network for rendering realistic bokeh. In: European Conference on Computer Vision
29. Ranftl, R., Bochkovskiy, A., Koltun, V.: Vision transformers for dense prediction. In: Proceedings of the IEEE/CVF International Conference on Computer Vision, pp. 12179–12188 (2021)
30. Ranftl, R., Lasinger, K., Hafner, D., Schindler, K., Koltun, V.: Towards robust monocular depth estimation: mixing datasets for zero-shot cross-dataset transfer. IEEE Trans. Pattern Anal. Mach. Intell. **44**(3), 1623–1637 (2020)
31. Shen, X., et al.: Automatic portrait segmentation for image stylization. In: Proceedings of the 37th Annual Conference of the European Association for Computer Graphics, pp. 93–102. EG 2016, Eurographics Association, Goslar, DEU (2016)
32. Shen, X., Tao, X., Gao, H., Zhou, C., Jia, J.: Deep automatic portrait matting. In: Leibe, B., Matas, J., Sebe, N., Welling, M. (eds.) ECCV 2016. LNCS, vol. 9905, pp. 92–107. Springer, Cham (2016). https://doi.org/10.1007/978-3-319-46448-0_6
33. Wadhwa, N., et al.: Synthetic depth-of-field with a single-camera mobile phone. ACM Trans. Graph. **37**(4), 1–13 (2018). https://doi.org/10.1145/3197517.3201329
34. Wang, L., et al.: DeepLens: shallow depth of field from a single image. arXiv preprint arXiv:1810.08100 (2018)
35. Wang, X., Xie, L., Dong, C., Shan, Y.: Real-ESRGAN: training real-world blind super-resolution with pure synthetic data (2021). https://doi.org/10.48550/ARXIV.2107.10833
36. Wang, X., et al.: ESRGAN: enhanced super-resolution generative adversarial networks (2018). https://doi.org/10.48550/ARXIV.1809.00219
37. Xian, K., Peng, J., Zhang, C., Lu, H., Cao, Z.: Ranking-based salient object detection and depth prediction for shallow depth-of-field. Sensors **21**(5), 1815 (2021). https://doi.org/10.3390/s21051815
38. Xiao, L., Kaplanyan, A., Fix, A., Chapman, M., Lanman, D.: DeepFocus: learned image synthesis for computational displays. ACM Trans. Graph. **37**(6) (2018). https://doi.org/10.1145/3272127.3275032
39. Xing, H., Cao, Y., Biber, M., Zhou, M., Burschka, D.: Joint prediction of monocular depth and structure using planar and parallax geometry. Pattern Recognition **130**, 108806 (2022). https://doi.org/10.1016/j.patcog.2022.108806

40. Yu, F., Gallup, D.: 3D reconstruction from accidental motion. In: 2014 IEEE Conference on Computer Vision and Pattern Recognition, pp. 3986–3993 (2014). https://doi.org/10.1109/CVPR.2014.509
41. Zamir, S.W., Arora, A., Khan, S., Hayat, M., Khan, F.S., Yang, M.H.: Restormer: efficient transformer for high-resolution image restoration. In: Proceedings of the IEEE/CVF Conference on Computer Vision and Pattern Recognition, pp. 5728–5739 (2022)
42. Zhang, R., Isola, P., Efros, A.A., Shechtman, E., Wang, O.: The unreasonable effectiveness of deep features as a perceptual metric. arXiv preprint arXiv:1801.03924 (2018)
43. Zhu, B., Chen, Y., Wang, J., Liu, S., Zhang, B., Tang, M.: Fast deep matting for portrait animation on mobile phone. arXiv preprint arXiv:1707.08289 (2017)

Residual Feature Distillation Channel Spatial Attention Network for ISP on Smartphone

Jiesi Zheng[1], Zhihao Fan[2], Xun Wu[3], Yaqi Wu[4(✉)], and Feng Zhang[1,2,3,4]

[1] Zhejiang University, Hangzhou 310027, China
Jaszheng@zju.edu.cn
[2] University of Shanghai for Science and Technology, Shanghai 200093, China
203590822@st.usst.edu.cn
[3] Tsinghua University, Beijing 100084, China
wuxun21@mails.tsinghua.edu.cn
[4] Harbin Institute of Technology, Harbin 150001, China
wuyaqi930@foxmail.com

Abstract. With the increasing popularity of mobile photography, more and more attention is being paid to image signal processing(ISP) algorithms used to improve various perceptual aspects of mobile photos. For this, a learned smartphone ISP task was proposed to develop an end-to-end deep learning-based ISP pipeline that can replace classical hand-crafted ISP and imitate target RGB images captured by digital single-lens reflex camera(DSLR). However, hardware limitations of mobile phone make it essential and challenging to achieve an acceptable trade off between computational cost and performance. In this paper, a light-weighted and powerful network named residual feature distillation channel spatial attention network (RFDCSANet) is proposed for end-to-end learned smartphone ISP task. To be specific, we employ modified residual feature distillation block(RFDB) including channel spatial attention(CSA) mechanism to progressively refine distilled features and adaptively fuse channel and spatial features. Particularly, we utilize a re-parameterizable block, namely edge-oriented convolution block(ECB) as the basic module to improve performance without introducing any additional cost in the inference stage. **The proposed solution ranked 3rd in *Mobile AI 2022 Learned Smartphone ISP Challenge (Track 1)* with 1st place PSNR score.**

Keywords: RAW-to-RGB mapping · Residual feature distillation block · Channel spatial attention · Re-parameterization

1 Introduction

The image signal processing (ISP) pipeline refers to the processing of raw sensor image for producing high quality display-referred RGB image, and thus is pivotal for a camera system. As shown in Fig. 1, a representative ISP pipeline usually

L. Karlinsky et al. (Eds.): ECCV 2022 Workshops, LNCS 13802, pp. 635–650, 2023.
https://doi.org/10.1007/978-3-031-25063-7_40

involves a sequence of steps including demosaicking, white balance, color correction, tone mapping, denoising, sharpening, gamma correction and so on [29]. Taking into account the particularities of the corresponding sensor and optical system, parameters and algorithms of each subtask in ISP are usually tuned and modified separately to produce images with photographic quality unique to their respective devices. Despite the extensive use of hand-crafted ISP solutions in current camera systems, convolutional neural networks (CNN) have exhibited great potential in learning deep ISP model to incorporate all image manipulation steps needed for fine-grained photo restoration.

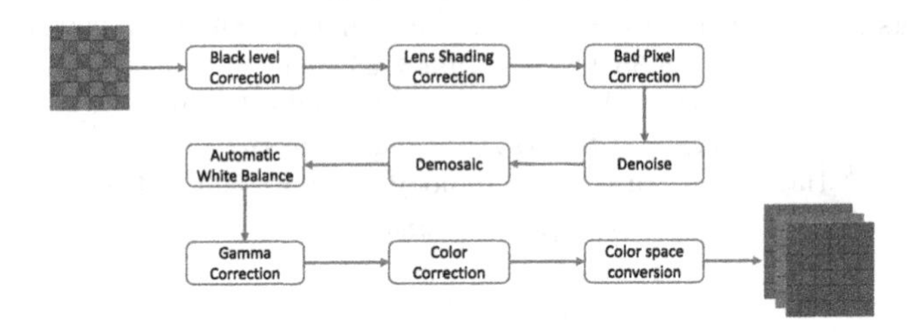

Fig. 1. A representative ISP pipeline.

The end-to-end property of learned ISP model makes it very competitive to learn RAW-to-RGB mapping directly to generate high quality image for mobile camera. While mobile cameras have become the dominant sources of photos, the hardware limitations of mobile cameras in comparison to DSLR camera remain unchanged: smaller sensor size, limited aperture and dynamic range are causing loss of details, high noise levels and inaccuracy color rendering [24]. By learning RAW-to-RGB mapping to produce DSLR-like RGB image from mobile raw image, learned ISP model can offer an encouraging way to close the gap between mobile camera and DSLR camera. For this, in [24] the author proposed a learned ISP model named PyNet and verified its effectiveness on collected dataset named ZRR dataset containing paired raw and RGB images respectively captured by HuaweiP20 smartphone and Canon 5D Mark IV DSLR. Two competitions [20,22] were later held based on ZRR dataset. Though the top ranked solutions like MW-ISPNet and AWNet have reasonable performance on fidelity results, it is still too computationally heavy to directly deploy on mobile phone.

While many challenges and works targeted at efficient deep learning models have been proposed recently, the evaluation of the obtained solutions is generally performed on desktop CPUs and GPUs, making the developed solutions not practical due to poor adaptation to the restrictions of mobile AI hardware. In order to develop efficient and practical solutions for learned smartphone ISP task, *Mobile AI Learned Smartphone ISP Challenge(2021,2022)* [18,23] were held with a more advanced dataset, where all deep learning solutions were evaluated on real mobile devices and efficiency was more emphasized on. For this, we proposed a light-weighted end-to-end framework namely residual feature distillation channel spatial attention network (RFDCSANet). To enhance feature extraction and boost up performance, we employ a modified residual feature distillation block(RFDB) including channel spatial attention(CSA) mechanism to progressively refine distilled features and adaptively fuse channel and spatial features. Particularly, we utilize a re-parameterizable building block, namely edge-oriented convolution block(ECB) as the basic module to improve performance without introducing any additional cost in the inference stage. In summary, our main contributions are listed below.

- We propose an end-to-end light-weighted and powerful residual feature distillation channel spatial attention network (RFDCSANet) for learned smartphone ISP task.
- We introduce re-parameterizable edge-oriented convolution block(ECB) into residual feature distillation block(RFDB) to enhance feature extraction and refinement for the recovery of texture details and boost up performance without increasing computational cost.
- Quantitative and qualitative experiments verifies the effectiveness and efficiency of our proposed network.

2 Related Work

2.1 Learned Image Signal Processing

Image signal processing(ISP) is to realize the mapping from RAW domain to RGB domain. The traditional ISP algorithm includes a series of subtasks [29], including black level correction, lens shading correction, bad pixel correction, demosaic, denoise, automatic white balance, color correction, gamma correction, Color space conversion and so on. For each subtask, a number of methods have been proposed in the literature [4,14,30,34].

In recent years, convolutional neural networks (CNNs) have been introduced and greatly promote the development of low-level vision tasks (e.g. image denoising [2,25,28], super-resolution [1,27,40,41]). Due to the outstanding performance, some recent works focus on directly applying deep neural network on RAW Bayer image to learn the full image signal processing (Learned ISP) steps, which aims to achieve better image quality and avoid the information lost after redundant hand-crafted algorithms in traditional ISP. Andrey et al. [24] proposed a single CNN-based network named PyNET to directly output the final photos from RAW Bayer sensor data from real camera. To promote the development of learned ISP, a large-scale dataset is proposed in [24], which contains 10 thousand full-resolution RAW-RGB data pairs. Silva et al. [32] proposed a UNet-based architecture network to recover RAW images into RGB. Dai et al. [7] utilizes the attention blocks and wavelet transform to boost the performance of learned ISP. In this work, following [19,32,33], we focus on efficient Learned ISP that can be applied directly to real smartphone platform and achieve real-time inference speed.

2.2 Re-parameterization

Structural re-parameterization [10,11] has received more and more attention in recent years and has been applied to many computer vision tasks, such as compact model design [12], architecture search [6,38], pruning [9]. Re-parameterization means that the transformation between different architectures can be realized by equivalent transformation of parameters. For example, a branch of the 1×1 convolution and a branch of the 3×3 convolution can be transferred to a single branch of the 3×3 convolution [11]. In the training phase, multi-branch [8,10,11] and multi-layer [5,13] topologies are designed to replace the basic linear layers (such as conv or fully connected layers [3]) of the augmented model. Cao et al. [5] discussed how to incorporate a deep separable roll kernel in training. In the process of reasoning, the complex model with long training time is transformed into a simple model to improve the reasoning speed.

Re-parameterization has become an effective technique for efficient neural networks design, especially for what designed for straightforward adopting on mobile AI hardware. Zhu et al. [37] propose FIMDN to adopt this technique on super resolution task and achieve promising results with fewer parameters. Ding et al. [11] propose simple but powerful neural network named RepVGG with structural re-parameterization technique, achieving higher accuracy and faster inference speed on ImageNet. Ding et al. [10] design diverse branch block (DBB) to boost performance of a single convolution by combining diverse branches, which can be converted into a single convolution when in inference. In this work, we adopt a re-parameterization technique named edge-oriented convolution block (ECB) proposed in [37] to achieve better results and real-time inference speed.

3 Proposed Method

3.1 Problem Formulation

Learned Smartphone ISP problem is to learn RAW-to-RGB mapping, wherein an input raw image captrued by mobile camera is trained to imitate the target RGB image captured by DSLR camera. Given a raw image $I_{raw} \in R^{H \times W \times 1}$ and the corresponding target RGB image $I_{DSLR} \in R^{H \times W \times 3}$, the learned ISP model is aplied to produce a RGB image $I_S \in R^{H \times W \times 3}$ from I_{raw} for approximating the color characteristic of target RGB image I_{DSLR}. The problem can be formulated in 1.

$$I_S = f(I_{raw}|\theta_f) \tag{1}$$

where $f(\cdot|\cdot)$ denotes the learned ISP model. θ_f denotes the learnable parameters in $f(\cdot|\cdot)$.

3.2 Network Architecture

To restore RGB images from camera sensor outputs in an efficient way, a novel network architecture with emphases on inference speed and high PSNR, which we call residual feature distillation channel spatial attention network (RFDCSANet), is illustrated in Fig. 2. To keep low computation cost and color consistency in the same channel, the input mosaic raw image $I_{raw} \in R^{H \times W \times 1}$ is first rearranged to half-resolution raw image $I'_{raw} \in R^{H/2 \times W/2 \times 4}$ according to bayer pattern. The backbone of RFDCSANet generate feature map with $H/2 \times W/2$ shape followed by an upsampling module to recover to the same size as target RGB image $I_{DSLR} \in R^{H \times W \times 3}$. This processing procedure is close to the way that in single image super resolution task. Therefore, inspired by proposed network in [26,39], we utilize some useful modules to build our model and verify its effectiveness in handling learned ISP problem.

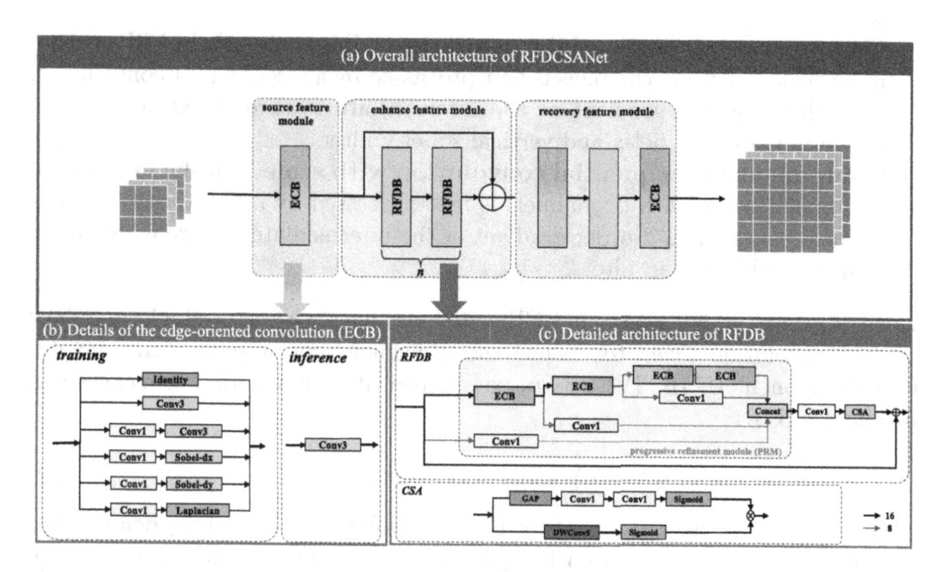

Fig. 2. (a) Overall architecture of RFDCSANet. Our full algorithm is composed of three modules: source feature module, enhance feature module and recovery feature module. n denotes the number of RFDB blocks in Enhance Features Module, and we set $n = 2$ for balancing the performance and inference speed. (b) Details of the edge-oriented convolution block (ECB). In training stage, ECB utilizes multi-branch structure and In inference stage, the ECB module will be converted into a single standard 3×3 convolution layer. (c) Detailed architecture of RFDB.

Edge-Oriented Convolution Block. In contrast to 8-bit input RGB image in super resolution task, the input raw image has higher-bit (12-bit) thus convey richer details and remain higher fidelity of texture and high frequency information. Therefore, inspired by ECBSR [39], we widely employ the re-parameterizable edge-oriented convolution block(ECB) to enrich the representation capability of the base model and enhance high frequency information extraction while remaining the efficiency. Following the setting in ECBSR, we explicitly incorporate extraction of the 1st-order derivatives and 2nd-order spatial derivative into the design of the block and take advantage of the sequential expanding-and-squeezing convolution to enhance edges and details and further extract the hidden information from the feature maps in training stage. As shown in Fig. 2(b), the ECB consists of four types of carefully designed operators, which are summarized as follows:

- **Component I: a normal 3×3 convolution and an identity branch.** Normal branches to ensure base performance and avoid gradient vanishing.
- **Component II: expanding-and-squeezing convolution.** For an input feature with C channels, a $D \times C \times 1 \times 1$ convolution is first employed to expand the channel dimension from C to D, then the following $C \times D \times 3 \times 3$ convolution squeeze the feature back to C channels. In our experiments, we set $D = 2C$.
- **Component III: sequential convolution with scaled Sobel filters.** An input feature with C channels is first processed by a $C \times C \times 1 \times 1$ convolution, then the 1st-order gradient of the intermediate feature is extracted using scaled horizontal sobel-x and vertical sobel-y filters.
- **Component IV: sequential convolution with scaled Laplacian filters.** An input feature with C channels is first processed by a $C \times C \times 1 \times 1$ convolution, then the 2nd-order gradient of the intermediate feature is extracted using scaled laplacian filters.

when it comes to inference, a ECB will be converted into a single standard 3×3 convolution layer to ensure the efficiency. For a sequential convolution(Component II, III, IV), the re-parameterizable convolution sequences can be expressed by:

$$\delta_2(\delta_1(x)) = W_2 * (W_1 * x + b_1) + b_2 \tag{2}$$

where δ_1 denotes the first $D \times C \times 1 \times 1$ convolution (W_1, b_1), δ_2 denotes the second $C \times D \times 3 \times 3$ convolution (W_2, b_2). A sequential convolution can be merged into a single $C \times C \times 3 \times 3$ convolution (W^{rep}, b^{rep}) in the following manner:

$$\begin{aligned} W^{rep} &= s * perm(W_1) * W_2 \\ b^{rep} &= W_2 * rep(b_1) + b_2 \end{aligned} \tag{3}$$

where s denotes the scaling parameter, it is set to 1 in Component II while remain learnable in Component III, IV. *prem* denotes the permute operation which transposes tensor W_1 to maintain the same size as W_2. *rep* is the spatial broadcasting operation, which replicates the bias $b_1 \in R^{1\times D\times 1\times 1}$ into $rep(b_1) \in R^{1\times D\times 3\times 3}$.

Finally, all re-parameterized weights can be sum up into a single standard convolution (W^{rep}_{ECB}, b^{rep}_{ECB}) by:

$$\begin{aligned} W^{rep}_{ECB} &= W^{rep}_{id} + W_n + W^{rep}_{sbl-x} + W^{rep}_{sbl-y} + W^{rep}_{lap} \\ b^{rep}_{ECB} &= b^{rep}_{id} + b_n + b^{rep}_{sbl-x} + b^{rep}_{sbl-y} + b^{rep}_{lap} \end{aligned} \tag{4}$$

where (W^{rep}_{id}, b^{rep}_{id}), (W^{rep}_{n}, b^{rep}_{n}), (W^{rep}_{sbl-x}, b^{rep}_{sbl-x}), (W^{rep}_{sbl-y}, b^{rep}_{sbl-y}), (W^{rep}_{lap}, b^{rep}_{lap}) denote re-parameterized weights of identity, normal convolution, sobel-x, sobel-y, laplacian branch respectively. $W^{rep}_{id} \in R^{C\times C\times 3\times 3}$ is replicated by an identity tensor $W_{id} \in R^{1\times 1\times 3\times 3}$ and b^{rep}_{id} is a zero tensor.

Residual Feature Distillation Block. Inspired by residual feature distillation blocks (RFDB) proposed in [26], we introduce a modified RFDB to build our backbone. To ensure the effectiveness, we maintain the framework of the original RFDB but replace several layers to boost up the performance. To be specific, our modified RFDB is constructed by progressive refinement module(PRM), channel spatial attention(CSA) layer, and a 1×1 convolution which is used to reduce the number of feature channels. We replace the CCA layer in RFDB with CSA block [16] to yield better fusion of spatial and channel features. Moreover, the SRB layer is replaced by a more powerful ECB layer to better extract the high frequency information, which is helpful for reconstructing structures, textures, and edges of RGB images.

Table 1. Details of PRM architecture. The columns represent distillation step, adopted layer, input channels and output channels of features.

Step	Layer	Input channel	Output channel
1	(ECB_1, RL_1)	16	8
2	(ECB_2, RL_2)	8	8
3	(ECB_3, RL_3)	8	8
4	(ECB_4)	8	8

As is shown in Fig. 2(c), the progressive refinement module (PRM) first adopts the ECB layer to extract input features for multiple subsequent distillation (refinement) steps. For each step, PRM employ ECB layer that is responsible for producing the distilled features and refine layer RL (1×1 convolution) that further processes the proceeding coarse features. The whole structure can be described as

$$\begin{aligned} F_{distilled1}, F_{coarse_1} &= ECB_1(F_{in}), RL_1(F_{in}) \\ F_{distilled2}, F_{coarse_2} &= ECB_2(F_{coarse1}), RL_2(F_{coarse1}) \\ F_{distilled3}, F_{coarse_3} &= ECB_3(F_{coarse2}), RL_3(F_{coarse2}) \\ F_{distilled4} &= ECB_4(F_{coarse3}) \end{aligned} \tag{5}$$

The hyperparameter of PRM architecture is shown in Table 1. The following stage is concatenating refined features from each step. It can be expressed by

$$F_{distilled} = Concat(F_{distilled1}, F_{distilled2}, F_{distilled3}, F_{distilled4}) \quad (6)$$

where $Concat$ denotes concatenation operation along the channel dimension.

After reduce channels using following 1×1 convolution, the progressively refined feature $F_{distilled}$ is then fed into channel spatial attention (CSA) layer to yield adaptive fusion of channel and spatial features. Channel attention originated from SENet [17]. It utilizes squeeze and excite operations to learn the inter-channel relationship of feature maps given an input image. Spatial attention employ a depth-wise dilated convolution [15,36] to learn distant spatial dependencies in the feature maps. Following [16], the kernel size is set to 5 and the dilated rate is set to 2. Finally, the ouput of CSA will be the aggregation of input feature $F_{in} \in R^{H/2 \times W/2 \times C}$, channel attention $z_{ca} \in R^{1 \times 1 \times C}$(before broadcasting) and spatial attention $z_{sa} \in R^{H/2 \times W/2 \times C}$ using element-wise product.

Overall Framework. We devise a RFDCSANet to reconstruct high-quality RGB images with sharp edges and clear structure under restricted resources. As illustrated in Fig. 2, our RFDCSANet consists of source feature module, enhance feature module and recovery feature module. Specifically, source feature module leverage a single ECB layer to roughly extract shallow feature $I_R \in R^{H/2 \times W/2 \times C}$ from rearranged raw image $I'_{RAW} \in R^{H/2 \times W/2 \times 4}$, while C denotes number of channels and is set to 16. Then the shallow feature I_R is fed into enhance feature module for further feature refinement at multiple levels. Enhance feature module consists of n stacked modified residual feature distillation blocks (RFDB) and n is set to 2 for balancing effectiveness and efficiency. A long-term residual connection is added in enhance feature module to avoid performance degradation and gradient vanishing caused by depth of model. Finally, in recovery feature module, a Depth2space layer is involved in the middle of two ECB layers for upsampling feature and recovering to the full size as $I_{DSLR} \in R^{H \times W \times 3}$.

3.3 Loss Function

In many image reconstruction and enhancement tasks, Charbonnier Loss and Structure Similarity Loss are widely used. In this work, we denote I_S as the predicted image and I_{DSLR} as the ground truth RGB image.

Charbonnier Loss. The Charbonnier loss [40] is adopted as an approximate loss function. This loss has been believed to outperform the traditional penalty in image reconstruction tasks. The Charbonnier loss function is defined as:

$$\mathcal{L}_{Char} = \sqrt{(I_S - I_{DSLR})^2 + \varepsilon} \quad (7)$$

where ε is set to 1e-6.

SSIM Loss. The structural similarity loss [35] is used to enhance the reconstructed RGB images by the structural similarity index. The loss function can be defined as:

$$\mathcal{L}_{SSIM} = 1 - \mathcal{F}_{SSIM}(I_S - I_{DSLR}) \tag{8}$$

where $\mathcal{F}_{SSIM}$ calculates the structural similarity index.

Finally, the total loss is expressed as:

$$\mathcal{L}_{total} = \mathcal{L}_{Char} + \alpha\mathcal{L}_{SSIM} \tag{9}$$

where α is set to 0.1.

4 Experiment

4.1 Dataset

The dataset we used was provided by Mobile AI 2022 workshop for the online contest. According to the organization, to get real data for the RAW-to-RGB mapping problem, a large-scale dataset consisting of photos collected by the Sony IMX586 Quad Bayer RGB mobile sensor for capturing RAW photos and a professional high-end Fujifilm GFX100 camera for RGB ground truths was obtained. Since the captured RAW-RGB image pairs are not perfectly aligned, they were matched using an advanced deep learning-based algorithm, and then smaller patches of size 256×256 pixels were extracted. We were provided with 24K training RAW-RGB image pairs (of size $256 \times 256 \times 1$ and $256 \times 256 \times 3$, respectively). It should be mentioned that all alignment operations were performed only on RGB DSLR images, therefore RAW photos from the Sony sensor remained unmodified. We divided the dataset into:

- **Train data:** A random selection 90% of the 24K aligned RAW-RGB image pairs.
- **Self-validation data:** The other 10% of the 24K aligned RAW-RGB image pairs.
- **Validation data:** The participants received the RAW images when the validation phase started; We adopt the fidelity results on this part as the criterion to measure the performance of the model.
- **Test data:** The participants could not receive the RAW testing images.

4.2 Evaluation Metrics

In order to measure the effectiveness of the model and the overall image quality intuitively, we follow the setting in Mobile AI 2022 challenge by using the following score as the overall evaluation metric:

$$Score(PSNR, runtime) = \frac{2^{2\cdot(PSNR-30)}}{runtime} = \frac{2^{2\cdot PSNR}}{C \cdot runtime} \tag{10}$$

The runtime of the model will be evaluated on the target mobile platform following the setting that input shape and output shape are $544 \times 960 \times 4$ and $1088 \times 1920 \times 3$ respectively. The higher score is, the better trade off between fidelity and runtime.

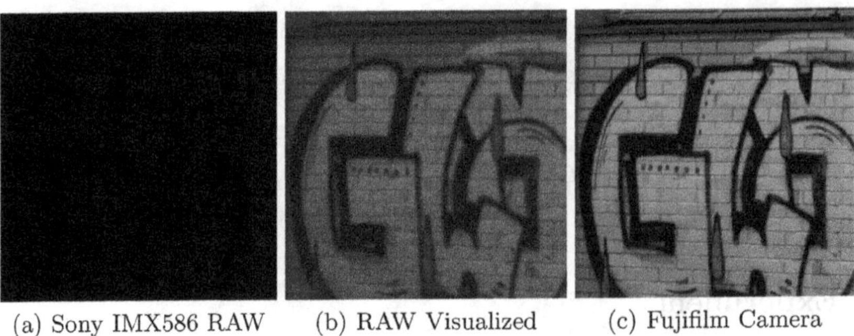

(a) Sony IMX586 RAW (b) RAW Visualized (c) Fujifilm Camera

Fig. 3. A sample set of images from the dataset. From left to right: the original RAW image, the RAW image visualized by scaling the original RAW image to 0∼255, and the target Fujifilm photo.

4.3 Implementation Details

Consider the trade off between inference speed and image quality, number of the modified RFDB used to build enhance feature module is set to 2. For data preprocessing, as mentioned in Sect. 3.2, we rearrange $I_{RAW} \in R^{H \times W \times 1}$ to half-resolution $I'_{RAW} \in R^{H/2 \times W/2 \times 4}$ following BGRG channel sequence. For all experiments, the model is implemented in Keras and runs on 4 Nvidia Titan Xp graphical processing units (GPU). The weights of model are optimized using Adam optimizer and the batch size is 64. The initial learning rate is set to 1e-4 and is halved every 200 epochs.

4.4 Quantitative and Qualitative Evaluation

To test our proposed RFDCSANet, we will compare it with the current relatively advanced ISP network algorithms. These include SmallNet [18], which achieved the highest score in the Mobile AI 2021 Learned Smartphone ISP Challenge using a simple convolutional layer, and CSANet [16], which used both channel attention and spatial attention modules. We also used the tuned baseline U-Net model [31] officially designed for the target Dimensity 1000+ platform in the 2021 Mobile AI competition as one of the comparative experiments. In addition, we also used NOAHT [18] and Eds [18] with good image quality in the 2021 Mobile AI competition as comparative experiments. To judge the quality of the restored image, we compared the quantitative indicators PSNR, SSIM and the running time of the model, and used formula 10 as the final score to determine the comprehensive effect of the model in terms of time and quality.

The results of our approach and the other three approaches are shown in Table 2. It can be seen from the table that our method has good results in both PSNR and SSIM image quality indexes. The PSNR value is 0.18 higher than the CSANet of the second-place team, and the SSIM is 0.0234 higher than the second-place team. In terms of running time, our results are 11.2 ms slower than CSANet and 33.7 ms slower than the simple convolutional layer Smallnet,

Fig. 4. The visualization results of our method and the other three methods in the validation set, from left to right, are SmallNet [18], PUNet [18], CSANet [16], NOAHT [18], Eds [18] and RFDCSANet. The images from top to bottom are numbered 36, 73, 74, 75, 215. Obviously, results recovered by our RFDCSANet are better than others for comparison with less false color, zipper and checkerboard artifacts, especially in color boundary and texture area.

which is not a big gap. Moreover, our results are higher than those of the other three methods in terms of image quality and running time. In addition, we also present some experimental comparison results to further verify our effect from the subjective aspect, as shown in Fig. 4. Number the picture in the Fig. 4 from top to bottom is 36, 73, 74, 75, 215 respectively. The No. 36 shows that our results have fewer false colors. According to traffic sign number 74, our method has fewer checkerboard artifacts than the other methods. From 75, we can see that our method also has a good effect and clarity in the text area. In conclusion, our method has a good effect on both subjective and objective indicators.

Table 2. Validation scores by different models (using the validation set). All models are trained with the same dataset and the runtime of all models are validated on the Snapdragon 870 SoC (GPU) using AI Benchmark 5.0 [21]. The best value is in bold. Our method achieved the best image quality while remaining competitive on runtime.

Network	Source	PSNR↑	SSIM↑	Runtime (ms)↓	Score↑
Smallnet [18]	MAI2021 1st	23.77	0.8392	**23.1**	7.68
PUNet [18]	MAI2021 baseline	24.17	0.8467	167.0	1.85
CSANet [16]	MAI2021 2nd	24.31	0.8434	45.6	8.23
NOAHT [18]	MAI2021 6th	23.49	0.8387	39.6	3.04
Eds [18]	MAI2021 9th	24.00	0.8422	66.1	3.69
RFDCSANet	Our proposed	**24.49**	**0.8668**	56.8	**8.48**

Table 3. The result of the ablation test. These variants are trained under the same condition and the test setting remains the same. The best value is in bold.

Setting	PSNR↑	SSIM↑	Runtime (ms)↓	Score↑
RFDB (n = 1)	23.66	0.8432	**35.1**	4.34
RFDB (n = 2, w/o CSA)	23.98	0.8522	36.4	6.52
RFDB (n = 2, w/o rep)	24.13	0.8565	56.8	5.15
RFDB (n = 2, this work)	**24.49**	**0.8668**	56.8	**8.48**

4.5 Ablation Study and AI Benchmark

In this section, we report the ablation study of the proposed model. The results of the ablation study are presented in Table 3. In this study, we compared RFD-CSANet with its 4 variants which are trained in the same way as before and are tested on the validation dataset from Mobile AI 2022 Learned Smartphone ISP Challenge. As we can see, stacking 2 modified RFDB in our model yields 0.83 dB PSNR increment comparing to only using one block. Moreover, adding an extra channel spatial attention module boosts performance further around 0.51 dB in the PSNR metric. Comparison between the last two setting illustrates that re-parameterization technique is helpful for improving image quality while keeping the same runtime.

AI Benchmark 5.0 [21] is a mobile software package to measure the neural network performance of a smartphone such as accuracy, speed, initialization time, and so on. Our proposed model was offered to this software package to measure the AI performance on several mobile devices. After providing the path of our tflite model, tests would be conducted to measure the runtime using CPU, GPU, and NNAPI separately. The CPU test was set to FP16 and 4 CPU threads. The TFLite GPU Delegate test was set to FP32. And The NNAPI test was set to FP16.

4.6 Test Results of Mobile AI 2022 Learned Smartphone ISP Challenge

The proposed algorithm ranked **3**rd in *Mobile AI 2022 Learned Smartphone ISP Challenge(Track 1)*. The final comparison results on testing set are summarized in Table 4. We can observe that our method achieved the best image quality while remaining competitive on runtime.

Table 4. Test results of Mobile AI 2022 Challenge on Learned Smartphone ISP Challenge (Track 1). The best value is in bold and our results are highlighted in gray. The runtime of the models is validated on the Snapdragon 8 Gen 1 SoC (GPU).

Rank	ID	PSNR↑	SSIM↑	Runtime(ms)↓	Score↑
1	161759	23.33	0.8516	**6.8**	**14.87**
2	162072	23.80	0.8652	18.9	10.27
3	162171	**23.89**	**0.8666**	34.3	6.41
4	162161	23.65	0.8658	41.5	3.8
5	162301	23.22	0.8281	182	0.48
6	162304	21.66	0.8399	28	0.36

5 Conclusion

This paper focuses on real-time smartphone image signal processing. Due to the limitations of the mobile hardware, many solutions have been proposed but most of them are computationally too expensive to be directly applied on smartphone. To alleviate this grim situation, we propose an end-to-end deep learning-based lighting method named residual feature distillation channel spatial attention network (RFDCSANet), which can demonstrate a real-time performance on mobile platform. Experimental results show that the proposed method outperforms other methods on dataset of Mobile AI 2022 workshop while keep a real time inference speed on smartphone.

References

1. Ahn, N., Kang, B., Sohn, K.A.: Fast, accurate, and lightweight super-resolution with cascading residual network. In: Proceedings of the European Conference on Computer Vision (ECCV), pp. 252–268 (2018)
2. Anwar, S., Huynh, C.P., Porikli, F.: Identity enhanced residual image denoising. In: Proceedings of the IEEE/CVF Conference on Computer Vision and Pattern Recognition Workshops, pp. 520–521 (2020)
3. Arora, S., Cohen, N., Hazan, E.: On the optimization of deep networks: implicit acceleration by overparameterization. In: International Conference on Machine Learning, pp. 244–253. PMLR (2018)

4. Buades, A., Coll, B., Morel, J.M.: A non-local algorithm for image denoising. In: 2005 IEEE Computer Society Conference on Computer Vision and Pattern Recognition (CVPR 2005), vol. 2, pp. 60–65. IEEE (2005)
5. Cao, J., et al.: DO-Conv: depthwise over-parameterized convolutional layer. IEEE Transactions on Image Processing (2022)
6. Chen, S., Chen, Y., Yan, S., Feng, J.: Efficient differentiable neural architecture search with meta kernels. arXiv preprint arXiv:1912.04749 (2019)
7. Dai, L., Liu, X., Li, C., Chen, J.: AWNet: attentive wavelet network for image ISP. In: Bartoli, A., Fusiello, A. (eds.) ECCV 2020. LNCS, vol. 12537, pp. 185–201. Springer, Cham (2020). https://doi.org/10.1007/978-3-030-67070-2_11
8. Ding, X., Guo, Y., Ding, G., Han, J.: ACNet: strengthening the kernel skeletons for powerful CNN via asymmetric convolution blocks. In: Proceedings of the IEEE/CVF International Conference on Computer Vision, pp. 1911–1920 (2019)
9. Ding, X., et al.: ResRep: lossless CNN pruning via decoupling remembering and forgetting. In: Proceedings of the IEEE/CVF International Conference on Computer Vision, pp. 4510–4520 (2021)
10. Ding, X., Zhang, X., Han, J., Ding, G.: Diverse branch block: building a convolution as an inception-like unit. In: Proceedings of the IEEE/CVF Conference on Computer Vision and Pattern Recognition, pp. 10886–10895 (2021)
11. Ding, X., Zhang, X., Ma, N., Han, J., Ding, G., Sun, J.: RepVGG: making VGG-style ConvNets great again. In: Proceedings of the IEEE/CVF Conference on Computer Vision and Pattern Recognition, pp. 13733–13742 (2021)
12. Dosovitskiy, A., et al.: An image is worth 16x16 words: transformers for image recognition at scale. arXiv preprint arXiv:2010.11929 (2020)
13. Guo, S., Alvarez, J.M., Salzmann, M.: ExpandNets: linear over-parameterization to train compact convolutional networks. Adv. Neural. Inf. Process. Syst. **33**, 1298–1310 (2020)
14. Hirakawa, K., Parks, T.W.: Adaptive homogeneity-directed demosaicing algorithm. IEEE Trans. Image Process. **14**(3), 360–369 (2005)
15. Howard, A.G., et al.: MobileNets: efficient convolutional neural networks for mobile vision applications. arXiv preprint arXiv:1704.04861 (2017)
16. Hsyu, M.C., Liu, C.W., Chen, C.H., Chen, C.W., Tsai, W.C.: CSAnet: high speed channel spatial attention network for mobile ISP. In: Proceedings of the IEEE/CVF Conference on Computer Vision and Pattern Recognition, pp. 2486–2493 (2021)
17. Hu, J., Shen, L., Sun, G.: Squeeze-and-excitation networks. In: Proceedings of the IEEE Conference on Computer Vision and Pattern Recognition, pp. 7132–7141 (2018)
18. Ignatov, A., Chiang, C.M., Kuo, H.K., Sycheva, A., Timofte, R.: Learned smartphone ISP on mobile NPUs with deep learning, mobile AI 2021 challenge: report. In: Proceedings of the IEEE/CVF Conference on Computer Vision and Pattern Recognition, pp. 2503–2514 (2021)
19. Ignatov, A., Patel, J., Timofte, R.: Rendering natural camera bokeh effect with deep learning. In: Proceedings of the IEEE/CVF Conference on Computer Vision and Pattern Recognition Workshops, pp. 418–419 (2020)
20. Ignatov, A., et al.: AIM 2019 challenge on RAW to RGB mapping: methods and results. In: 2019 IEEE/CVF International Conference on Computer Vision Workshop (ICCVW), pp. 3584–3590. IEEE (2019)
21. Ignatov, A., et al.: AI benchmark: all about deep learning on smartphones in 2019. In: 2019 IEEE/CVF International Conference on Computer Vision Workshop (ICCVW), pp. 3617–3635. IEEE (2019)

22. Ignatov, A., et al.: AIM 2020 challenge on learned image signal processing pipeline. In: Bartoli, A., Fusiello, A. (eds.) ECCV 2020. LNCS, vol. 12537, pp. 152–170. Springer, Cham (2020). https://doi.org/10.1007/978-3-030-67070-2_9
23. Ignatov, A., et al.: Learned smartphone ISP on mobile GPUs with deep learning, mobile AI & AIM 2022 challenge: report. In: Proceedings of the European Conference on Computer Vision (ECCV) Workshops (2022)
24. Ignatov, A., Van Gool, L., Timofte, R.: Replacing mobile camera ISP with a single deep learning model. In: Proceedings of the IEEE/CVF Conference on Computer Vision and Pattern Recognition Workshops, pp. 536–537 (2020)
25. Lefkimmiatis, S.: Universal denoising networks: a novel CNN architecture for image denoising. In: Proceedings of the IEEE Conference on Computer Vision and Pattern Recognition, pp. 3204–3213 (2018)
26. Liu, J., Tang, J., Wu, G.: Residual feature distillation network for lightweight image super-resolution. In: Bartoli, A., Fusiello, A. (eds.) ECCV 2020. LNCS, vol. 12537, pp. 41–55. Springer, Cham (2020). https://doi.org/10.1007/978-3-030-67070-2_2
27. Niu, B., et al.: Single image super-resolution via a holistic attention network. In: Vedaldi, A., Bischof, H., Brox, T., Frahm, J.-M. (eds.) ECCV 2020. LNCS, vol. 12357, pp. 191–207. Springer, Cham (2020). https://doi.org/10.1007/978-3-030-58610-2_12
28. Park, B., Yu, S., Jeong, J.: Densely connected hierarchical network for image denoising. In: Proceedings of the IEEE/CVF Conference on Computer Vision and Pattern Recognition Workshops (2019)
29. Ramanath, R., Snyder, W.E., Yoo, Y., Drew, M.S.: Color image processing pipeline. IEEE Signal Process. Mag. **22**(1), 34–43 (2005)
30. Rizzi, A., Gatta, C., Marini, D.: A new algorithm for unsupervised global and local color correction. Pattern Recogn. Lett. **24**(11), 1663–1677 (2003)
31. Ronneberger, O., Fischer, P., Brox, T.: U-Net: convolutional networks for biomedical image segmentation. In: Navab, N., Hornegger, J., Wells, W.M., Frangi, A.F. (eds.) MICCAI 2015. LNCS, vol. 9351, pp. 234–241. Springer, Cham (2015). https://doi.org/10.1007/978-3-319-24574-4_28
32. Silva, J.I.S., et al.: A deep learning approach to mobile camera image signal processing. In: Anais Estendidos do XXXIII Conference on Graphics, Patterns and Images, pp. 225–231. SBC (2020)
33. Tan, M., et al.: MnasNet: platform-aware neural architecture search for mobile. In: Proceedings of the IEEE/CVF Conference on Computer Vision and Pattern Recognition, pp. 2820–2828 (2019)
34. Van De Weijer, J., Gevers, T., Gijsenij, A.: Edge-based color constancy. IEEE Trans. Image Process. **16**(9), 2207–2214 (2007)
35. Wang, Z., Simoncelli, E.P., Bovik, A.C.: Multiscale structural similarity for image quality assessment. In: The Thrity-Seventh Asilomar Conference on Signals, Systems & Computers, vol. 2, pp. 1398–1402. IEEE (2003)
36. Yu, F., Koltun, V.: Multi-scale context aggregation by dilated convolutions. arXiv preprint arXiv:1511.07122 (2015)
37. Zhang, K., et al.: AIM 2020 challenge on efficient super-resolution: methods and results. In: Bartoli, A., Fusiello, A. (eds.) ECCV 2020. LNCS, vol. 12537, pp. 5–40. Springer, Cham (2020). https://doi.org/10.1007/978-3-030-67070-2_1
38. Zhang, M., Yu, X., Rong, J., Ou, L., Gao, F.: RepNAS: searching for efficient re-parameterizing blocks. arXiv preprint arXiv:2109.03508 (2021)
39. Zhang, X., Zeng, H., Zhang, L.: Edge-oriented convolution block for real-time super resolution on mobile devices. In: Proceedings of the 29th ACM International Conference on Multimedia, pp. 4034–4043 (2021)

40. Zhang, Y., Li, K., Li, K., Wang, L., Zhong, B., Fu, Y.: Image super-resolution using very deep residual channel attention networks. In: Proceedings of the European Conference on Computer Vision (ECCV), pp. 286–301 (2018)
41. Zhang, Y., Tian, Y., Kong, Y., Zhong, B., Fu, Y.: Residual dense network for image super-resolution. In: CVPR, pp. 2472–2481 (2018)

HST: Hierarchical Swin Transformer for Compressed Image Super-Resolution

Bingchen Li, Xin Li, Yiting Lu, Sen Liu, Ruoyu Feng, and Zhibo Chen(✉)

University of Science and Technology of China, Hefei 230027, China
{lbc31415926,lixin666,luyt31415,ustcfry}@mail.ustc.edu.cn,
elsen@iat.ustc.edu.cn, chenzhibo@ustc.edu.cn

Abstract. Compressed Image Super-resolution has achieved great attention in recent years, where images are degraded with compression artifacts and low-resolution artifacts. Since the complex hybrid distortions, it is hard to restore the distorted image with the simple cooperation of super-resolution and compression artifacts removing. In this paper, we take a step forward to propose the Hierarchical Swin Transformer (HST) network to restore the low-resolution compressed image, which jointly captures the hierarchical feature representations and enhances each-scale representation with Swin transformer, respectively. Moreover, we find that the pretraining with Super-resolution (SR) task is vital in compressed image super-resolution. To explore the effects of different SR pretraining, we take the commonly-used SR tasks (*e.g.*, bicubic and different real super-resolution simulations) as our pretraining tasks, and reveal that SR plays an irreplaceable role in the compressed image super-resolution. With the cooperation of HST and pre-training, our HST achieves the fifth place in AIM 2022 challenge on the low-quality compressed image super-resolution track, with the PSNR of 23.51 dB. Extensive experiments and ablation studies have validated the effectiveness of our proposed methods.

Keywords: Hierarchical network · Transformer · Compressed image super-resolution · Pretraining · AIM 2022 challenge

1 Introduction

Image super-resolution (SR) has achieved a quantum leap with the development of deep neural networks, which aims to restore the high-resolution (HR) images from their low-resolution counterparts. Existing SR can be roughly divided into three categories, simulated SR [12,19,30,47,57] (*e.g.*, bicubic downsampling), real-world SR [5,17,24,46,49,50,55] and blind SR [2,14,29,36,45], respectively. In particular, real-world and blind SR are greatly developed in recent years, of which the degradations are more consistent with unknown real-world distortions. However, not all images suffer from real-world degradation. In most cases, the images are susceptible to various compression artifacts together with

L. Karlinsky et al. (Eds.): ECCV 2022 Workshops, LNCS 13802, pp. 651–668, 2023.
https://doi.org/10.1007/978-3-031-25063-7_41

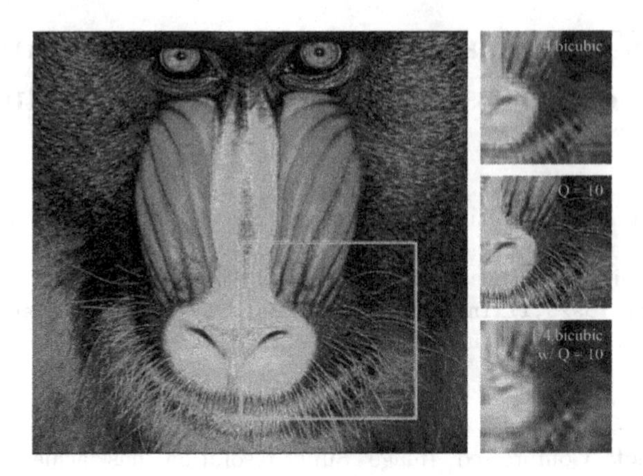

Fig. 1. A comparison between different degradations. The left image is the high-resolution reference image. The images from the top right to bottom right are 1/4 bicubic downsampling, JPEG compression with a quality factor of 10, and the combination of the above two distortions, respectively. Note that the bottom right image has the most severe degradation, thus requiring a stronger representation ability for the network to remove distortion.

low-resolution, since the image compression [4,25,40,41,52], transmission and storage. This hybrid degradation poses a challenging image process task, *i.e.*, compressed image super-resolution.

As shown in Fig. 1, unlike general image SR and compression artifacts, the degradations of compressed low-resolution images are more severe, which composes of blurring, block artifacts, and noise, etc. Existing methods on image SR [12,30,47,57] and compression artifacts removing [6,11,15,43,56] cannot work well on such brand-new degradation, since the large distribution shift. As the pioneering works, a series of works [27,51,58] began to investigate the compressed video super-resolution. To further promote the development of compressed image/video super-resolution, AIM2022 [53] firstly holds the significant competition on compressed image super-resolution, where images are firstly down-sampled with the scale 1/4, and then, are compressed with JPEG using an extreme low-quality parameter $Q = 10$. A naïve and intuitive strategy to deal with it is exploiting a well-trained SR network and JPEG artifacts removing network to restore the distorted images in a sequential manner. However, the above strategy always fails since the distribution shift between hybrid distortions [23,32] and single distortion. Compressed image super-resolution requires that the restoration network have the strong representation capability to learn structure and texture jointly.

In this paper, we present Hierarchical Swin Transformer, namely HST, to tackle the compressed image super-resolution problem. Specifically, previous works [26,56] have shown superior advantages of hierarchical architecture on compression artifacts removing due to their great representation ability. Meanwhile, the variants of the transformer have been explored for image processing and quality assessment [8,31,35], *e.g.*, SwinIR [28], which achieve remarkable performance compared with their CNN counterparts, since their capability of global contextual representation. Inspired by these, we present the Hierarchical Swin Transformer (HST) by incorporating the individual advantage of the above two architectures. In particular, our HST consists of four modules: hierarchical feature extraction module, feature enhancement module, fusion module, and HR reconstruction module. The hierarchical feature extraction module uses multiple convolution layers with different strides to obtain hierarchical feature maps at different scales. Then, the residual swin transformer block (RSTB) from SwinIR [28] is used for the feature enhancement in each hierarchical branch. After getting the enhanced hierarchical features, we fuse them by concatenating the upsampling low-scale feature and high-scale feature, and then, input them into a convolution layer to obtain the fused feature. Lastly, we can get the super-resolved HR image with the HR reconstruction module, which is composed of convolution layers and pixelshuffle layers.

We also investigate the compressed image super-resolution from the perspective of pretraining. We observe that the pretraining with image super-resolution plays a vital role in the compressed image SR. Specifically, we systematically explore the effects of different image super-resolution tasks, including traditional SR, *i.e.*, bicubic downsampling, and two RealSR simulation methods from BSR-GAN [55] and DRTL [22]. Extensive experiments reveal that the pretraining with the RealSR simulation from DRTL [22] is better for compressed image SR.

The contributions of this paper can be summarized as:

1. We present the Hierarchical Swin Transformer (HST) for compressed image super-resolution, which incorporating the advantages of strong representation ability and global information utilization.
2. We investigate compressed image super-resolution from the pretraining perspective. Based on the observation, we find one proper pretraining scheme for compressed image super-resolution.
3. Extensive experiment results show that our HST achieve a remarkable result on compressed image super-resolution task under heavy distortion (compression quality $Q = 10$ combined with 1/4 downsampling).

2 Related Works

2.1 Single Image Super-Resolution

Single Image Super-resolution (SISR) has been developed expeditiously with the advances of deep neural networks. SRCNN [12], as the pioneering work, firstly introduces the CNN to SISR and learns the network by minimizing the mean square error (MSE) between the generated images and their corresponding high-resolution (HR) images. Then, a series of works for SISR [30,57] are proposed by designing or modulating the network architecture. EDSR [30] revises the conventional residual module by removing the BatchNorm layers. RCAN [57] adds channel attention to the residual blocks, which focus on more informative channels. And SAN [10] introduces the second-order channel attention to utilize the second-order feature statistics for more discriminative representations.

However, the above works exhibit poor capability for subjective quality improving. To tackle the above challenge, SRGAN [21] firstly introduces Generative Adversarial Network (GAN) to SISR, and adopts the adversarial loss for approximating the natural image manifold. As an improved version of SRGAN, ESRGAN [47] exploits Relativistic average GAN [18] to enhance the discriminator and computes the VGG feature before the activation function to calculate the perceptual loss. To further improve the discriminator, FSMR [20] comes up with feature statistics mixing regularization, which encourages the discriminator's prediction to remain invariant to the style of the input image.

Recently, real-world image super-resolution (RealSR) [5,17,46,50,55] and blind image super-resolution [2,14,29,45] have been proposed to solve more severe and unknown hybrid distortions existed in real-world low-resolution images. To tackle unseen distortions (*i.e.*, blind distortions), KernelGAN [2] train an internal-gan to estimate the degradation kernel contains in low-resolution images. IKC [14] ameliorates the estimation process into an iterative one, which can deal with more complex blind distortions. Different from the aforementioned works that predict the degradation kernel, RealSR directly trains networks on synthesized real-world distorted image pairs, such as BSRGAN [55] and ESRGAN [47]. In this paper, we focus on the compressed image super-resolution, which is more significant and valuable in the real world.

2.2 Compression Artifacts Removal

Compression artifacts removal aims to remove the distortions caused by image/video codecs. Early works of compression artifact removal mainly focus on the design of manual filters in the DCT domain. Due to the success of CNN in image denoising and image super-resolution, Yu et al. [11] propose ARCNN, the first CNN-based method for compression artifacts removal. Svoboda et al. [43] introduce residual learning to deepen the network under the assumption of "deeper is better". However, ARCNN and its follow-up works only process artifacts in the pixel domain. DDCN [15], DMCNN [56] and D^3 [48] utilize DCT domain prior on the basis of pixel domain. Based on the network of extracting the dual domain knowledge, Fu et al. [13] use dilated convolution for multi-scale feature extraction and added convolutional sparse coding to make the model more compact and explainable. Recently, there are a series of works that explore the hierarchical structures for compression artifacts removal. Lu et al. [34] prove that adding multi-scale priors to the image restoration network can effectively eliminate compression artifacts. Inspired by Lu et al. [34], Li et al. add a non-local attention module to fuse multi-scale features effectively and obtain the post-processing network MSGDN [26] of VCC Intra coding. Based on the above excellent works, we also introduce the hierarchical module to our compressed image super-resolution network.

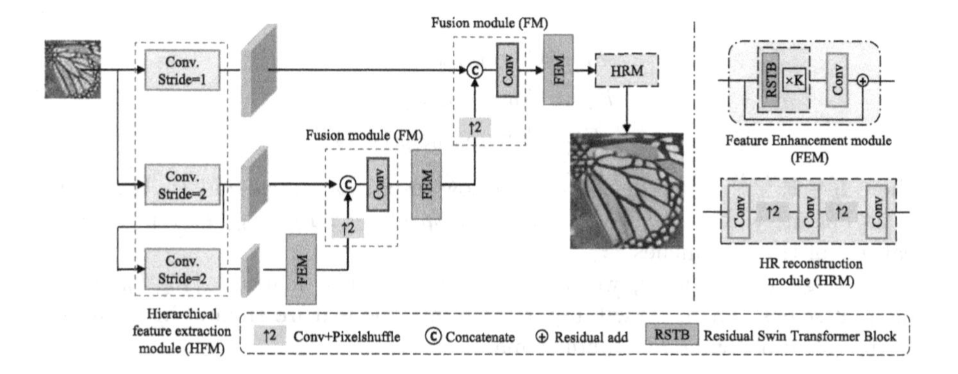

Fig. 2. The architecture of proposed HST

3 Method

In this section, we will explain our HST and clarify our pretraining strategy in detail. As shown in Fig. 2, our HST is composed of four main components, respectively as hierarchical feature extraction module (HFM), feature enhancement module (FEM), fusion module (FM) and HR reconstruction module (HRM).

3.1 Hierarchical Feature Extraction

Previous works [9, 39] have revealed that extracting hierarchical features at different scales from images and processing them in a divide-and-conquer manner, can provide the network a strong representation ability. And thus, it can deal with more severe and complex image degradation effectively. To achieve a trade-off between network parameters and performance, we choose a three-branches hierarchical architecture as our backbone. More specifically, the input LR image is gradually passed through three different convolution layers with different kernel sizes and strides, to extract the hierarchical representations with three scales. Following the implementation in [39], we design upper branch convolution by $k7n60s1p3$, where k, n, s, p stands for kernel size, number of channels, stride and padding respectively. For other two branches, we use convolution $k5n60s2p2$ to obtain the middle-scale feature map, and convolution $k3n60s2p1$ to obtain the low branch's feature map from the aforementioned feature map. The whole process can be formed as Eq. 1.

$$\begin{aligned} F_h &= \text{Conv}_{k7n60s1p3}(I_l) \\ F_m &= \text{Conv}_{k5n60s2p2}(I_l) \\ F_l &= \text{Conv}_{k3n60s2p1}(F_m) \end{aligned} \tag{1}$$

where $I_l \in \mathbb{R}^{H \times W \times C}$ is the compressed low resolution image and H, W, C refer to its height, width and color channel, respectively. F_h, F_m, F_l represent the features of three branches.

Through this process, we obtain the hierarchical features $\{F_h, F_m, F_l\}$ at different scales. Then, we will input them into the feature enhancement module to process in a divide-and-conquer manner.

3.2 Feature Enhancement and Fusion

The feature enhancement module and feature fusion module are the important components of HST. We will clarify them carefully in this section.

Feature Enhancement Module. Different from previous hierarchical networks [26,33,39,56] for image restoration and super-resolution, where convolutional neural network (CNN) is used as the feature enhancement module for each branch, we use swin transformer architecture as ours. As proved by [28], swin transformer-based architecture can model long-range dependency enabled by the shifted window mechanism. Therefore, this architecture is more suitable for difficult degradation removal tasks, *e.g.*, compressed image super-resolution. Moreover, with the help of swin transformer, we can get better performance with less parameters.

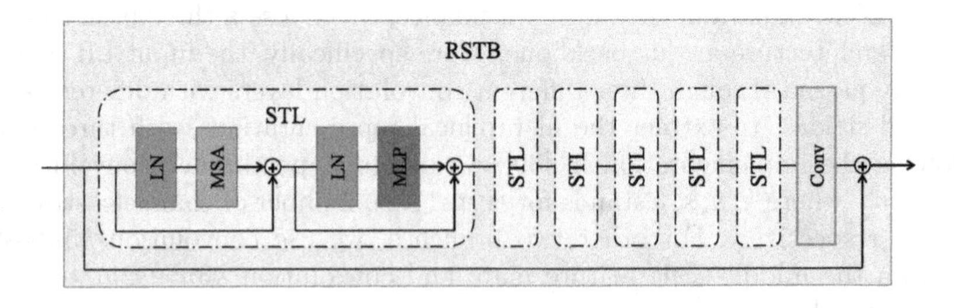

Fig. 3. Network structure of residual swin transformer block (RSTB)

Specifically, we directly apply multiple residual swin transformer blocks (RSTB) from [28], as our feature enhancement module. The architecture of RSTB is shown in Fig. 3. Each RSTB is composed of several swin transformer layers (STL), a convolution layer and a residual skip connection. This process can be formulated as Eq. 2

$$
\begin{aligned}
F_0 &= F_{in} \\
F_i &= \mathrm{STL}(F_{i-1}), \quad i = 1, 2, \ldots, K \\
F_{out} &= \mathrm{Conv}(F_K) + F_0
\end{aligned} \tag{2}
$$

where F_{in} is the input feature of one STL layer, and STL($\cdot$) means each STL layer inside RSTB, which can be formulated as Eq. 3

$$
\begin{aligned}
X &= \mathrm{MSA}(\mathrm{LN}(X)) \\
X &= \mathrm{MLP}(\mathrm{LN}(X))
\end{aligned} \tag{3}
$$

where MSA($\cdot$) stands for multi-head self-attention, MLP($\cdot$) stands for a multi-layer perceptron with two fully-connected layers and GELU as activation, and LN($\cdot$) stands for LayerNorm. Since STL is not our contribution and previous works have already proved the effectiveness of this module, we directly utilize the same architecture as is presented in [28].

Feature Fusion Module. After getting the enhanced features F_l^* from the low-branch enhancement module, composed of severe RSTB blocks. We will integrate it into the higher feature with the fusion module, which aims to bring the contextual information from low-scale to high-scale. To demonstrate the fusion process clearly, we take the fusion of the low-branch feature and the middle-branch feature as an example. As described in Eq. 4, the low-branch feature F_l is enhanced to F_l^* with low-branch feature enhancement module FEM_l. Then we concatenate the super-resolved low-branch feature ${F_l^*}_{\uparrow 2}$ and middle-branch feature F_m, and exploit the convolution layer to fuse these two components. Finally, we can obtain the enhanced middle-branch feature F_m^* by passing the fused feature F_m into the middle-branch enhancement module FEM_m. It is worthy to notice that, the up-sampling operation is implemented with Pixelshuffle [42], which can bring more stable results. The fusion of middle branch and high-branch features are implemented in the same way in Eq. 2.

$$
\begin{aligned}
F_l^* &= \mathrm{FEM_l}(F_l), \\
F_m &= \mathrm{Conv}(F_m \copyright {F_l^*}_{\uparrow 2}) \\
F_m^* &= \mathrm{FEM_m}(F_m)
\end{aligned} \tag{4}
$$

3.3 HR Reconstruction Module

Since the compressed image super-resolution in the competition requires the resolution of network output to be 4× higher than their input image, the HR reconstruction module aims to produce the final three-channel RGB high-resolution clean image with the enhanced high-branch feature F_h^*.

As shown in Fig. 2, this part is composed of two sub-pixel convolution layers, including two convolution layers and two PixelShuffle layers [42]. Following the previous SR works [28,30,57], we utilize two sub-pixel convolution layers for the 4× upsampling. Finally, a convolution layer is used to generate the output HR image.

3.4 Pretraining with SR

To further boost the capability of the network, We also explore one simple but effective pertaining strategy for compressed image super-resolution. It is noteworthy that pretraining with more relevant distortions can bring better knowledge transfer. Particularly, we select three SR tasks as the pretraining schemes, *i.e.*, traditional SR, two RealSR simulation methods from BSRGAN [55] and DRTL [22], and explore their effectiveness for compressed image super-resolution. The relevant experimental analyses are shown in Sect. 4.3, which demonstrates pretraining with RealSR simulations leads to promising results, especially with the simulation in DRTL [22].

3.5 Loss Functions

In order to enable our HST to be competent for the task of compressed image super-resolution, we first pretrain our network on the ×4 super-resolution task, and then finetune it for compressed image super-resolution. For the ×4 super-resolution pretraining, we optimize network parameters by minimizing the L_1 pixel loss as:

$$\mathcal{L} = \|I_{SR} - I_{HR}\|_1, \tag{5}$$

where I_{SR} is obtained by passing low-resolution images through the network, and I_{HR} is the corresponding ground-truth HR image. For compressed image super-resolution, we optimize network parameters by minimizing the Charbonnier loss [7].

$$\mathcal{L} = \sqrt{\|I_{SR} - I_{HR}\|^2 + \epsilon}, \tag{6}$$

where ϵ is set as default value 10^{-9}.

4 Experiments

4.1 Datasets

We produce the experimental results in our paper with two training datasets, DIV2K [1] (including 800 high-resolution images) and Flick2K [44] (including 2650 high-resolution images). In the competition AIM2022 [53], we also collect extra 746 high-resolution images from CLIC 2021 official website[1] as the additional training data, which is only used for the competition results in Sect. 4.6. For the testing stage in this paper, we adopt Set5 [3], Set14 [54], BSD100 [37], Urban100 [16], Manga109 [38] and DIV2K [1] validation as our testing datasets.

[1] http://clic.compression.cc/2021/tasks/index.html.

Table 1. Quantitative comparison for ablation study of network pretraining scheme. Results are tested on ×4 with compression quality 10 on Urban100 [16] dataset in terms of PSNR/SSIM. Best performance are in red.

Task	Methods(PSNR/SSIM)			
	w/o	Bicubic ×4	BSRGAN [55]	DRTL [22]
×4, Q = 10	19.70/0.5181	19.92/0.5301	20.04/0.5375	20.06/0.5383

4.2 Implementation Details

We use a three-branch HST for our experiments. The channel numbers of three feature enhancement modules FEM_h, FEM_m, FEM_l are set to 60, 60, 60, respectively. The spatial resolution of the high branch is 64×64, and halved for each downscale branch. Following [28], we set the number of swin transformer layers (STLs) as 6 for all residual swin transformer blocks (RSTBs) in HST. We use 2, 4, and 6 RSTBs for low branch, middle branch and high branch, respectively. The window size is set to 8 throughout the experiment.

We train our HST using four NVIDIA 2080Ti GPUs, with a batch size of 16. We offline generate training image pairs by the MATLAB bicubic kernel, then add JPEG compression with specified quality factor through the OpenCV function. We randomly crop LR into 64×64 patches for training. For data augmentation, we leverage random flipping and random rotation simultaneously. In the stage of pretraining, the total training iterations are set to 400K. We adopt Adam optimizer with $\beta_1 = 0.9$ and $\beta_2 = 0.999$, the initial learning rate is set to 2e-4 and reduced by half at [100K, 250K]. In the stage of finetuning, we load network parameters from the pretraining stage. We conduct experiments on four different compression levels, with quality factors at 40, 30, 20 and 10, respectively. The training is first finished on quality factor at 40, with the initial learning rate and total iterations as 1e-4 and 200K. And the learning rate is halved after 100K iterations. The rest of tasks are finetuned based on the first task (*i.e.*, quality factor at 40), with the learning rate as 8e-5 and total iterations as 100K.

4.3 Effects of Different Pretraining Schemes

As discussed in Sect. 3.4, pretraining is crucial for compressed image super-resolution task. To find out the optimal pretraining scheme, we conduct an ablation study on four different strategies, including: without pretraining, pure bicubic ×4 pretraining, pretraining with RealSR simulation from BSRGAN [55], and pretraining with RealSR simulation from DRTL [22]. BSRGAN [55] uses a practical complex degradation simulation process, which demonstrates its effectiveness on real-world distortion removal. DRTL [22] proposes a multi-task degradation training scheme, to simulate distortion in real-world scenarios, and works well on few-shot real-world image super-resolution problems. For fast convergence and

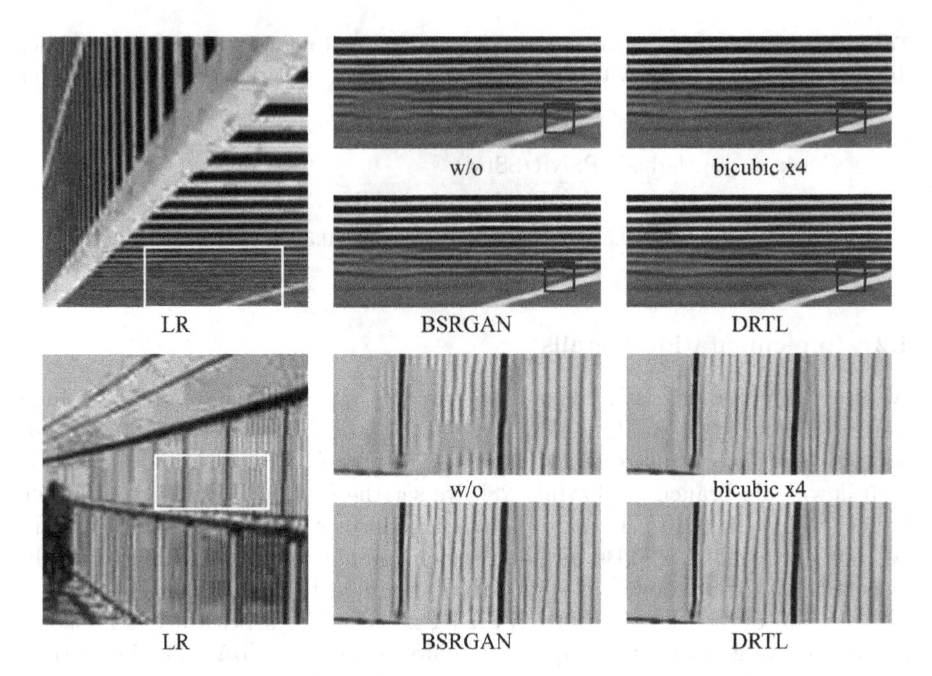

Fig. 4. Qualitative comparison for different pretraining schemes on ×4 image super-resolution with compression quality 10. Testing images are "011" and "024" from Urban100 [16] respectively.

convincing results, we use SwinIR-s [28] as a training model, and test network performance on Urban100 [16] (Fig. 4).

As shown in Table 1, quantitative results show that the pretraining with RealSR leads to a gain of 0.36 dB/0.0202 on the test dataset, which reveals that the pretraining is vital for compressed image super-resolution. Another observation is that pretraining with RealSR can achieve a better performance compared with simple bicubic downsampling, especially with the simulation in DRTL [22]. The reason for this might be that RealSR simulations contain lots of hybrid distortions, which are more complex and the knowledge is more likely to be transferred to the severely compressed image super-resolution task.

4.4 Effects of Hierarchical Architecture

To explore the advantage of introducing a hierarchical network structure, we set network branches from 1 to 3 and observe their performances on ×4 super-resolution with compression quality 40. Note that, one branch framework is almost the same as SwinIR-M [28]. As shown in Table 2, more branches lead to higher performance. However, it also brings an increase of computational complexity. In this paper, we choose a three-branch HST to achieve the best performance. In addition, benefits from structure and texture information com-

Table 2. Quantitative comparison for ablation study of network scales. The number of parameters is listed in the bracket. Results are tested on ×4 with compression quality 40 in terms of PSNR/SSIM. Best performance are in red.

Methods	Q	Datasets(PSNR/SSIM)				
		Set5	Set14	BSD100	Urban100	Manga109
HST-1(11.90M)	**40**	25.28/0.726	23.78/0.613	23.82/0.583	22.21/0.652	23.69/0.767
HST-2(12.98M)		25.35/0.727	23.82/0.614	23.84/0.584	22.21/0.651	23.78/0.769
5 HST-3(16.58M)		25.39/0.728	23.84/0.614	23.87/0.584	22.23/0.651	23.85/0.768

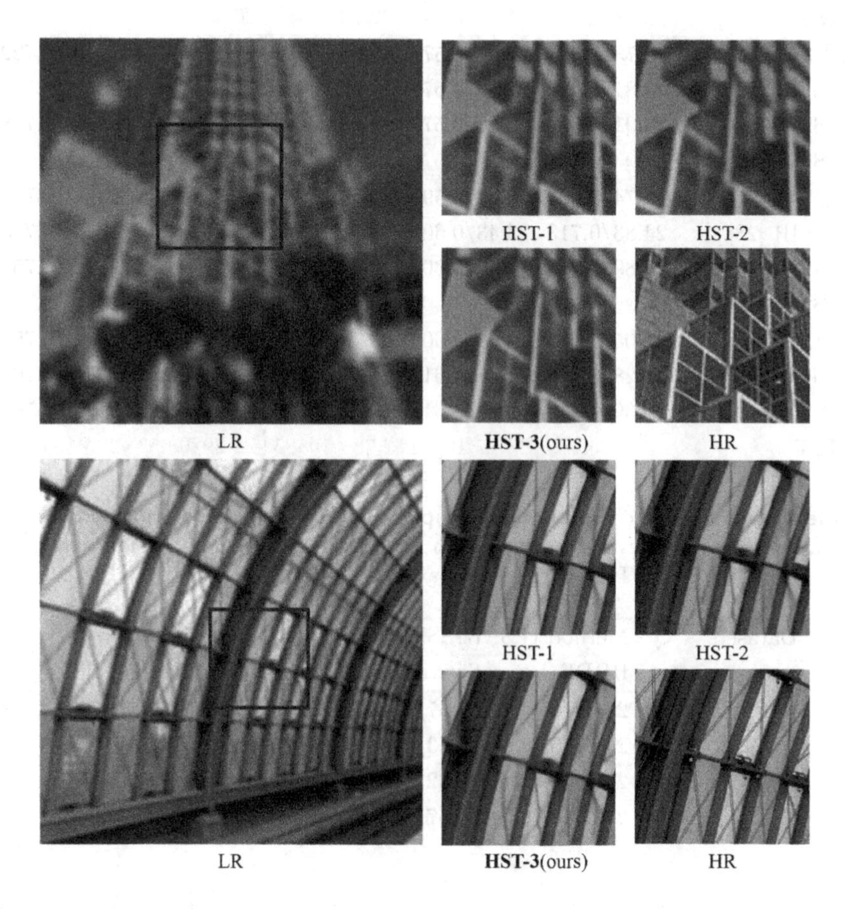

Fig. 5. Qualitative comparison for different network scales on ×4 image super-resolution with compression quality 40. Testing images are "095" from BSD100 [37] and "002" from Urban100 [16] respectively.

pensation from lower branches, three-branch HST can generate images with clearer lines and more structural components, as shown in Fig. 5.

Table 3. Quantitative comparison for compressed image super-resolution on benchmark datasets. Results are tested on ×4 with different compression qualities in terms of PSNR/SSIM. Best performance are in red.

Methods	Q	Datasets(PSNR/SSIM)				
		Set5	Set14	BSD100	Urban100	Manga109
RRDB [47]	**10**	22.36/0.629	21.75/0.538	22.13/0.514	20.24/0.553	20.66/0.677
SwinIR [28]		22.45/0.636	21.79/0.541	22.16/0.517	20.35/0.561	20.81/0.685
HST		22.49/0.637	21.84/0.542	22.18/0.517	20.38/0.559	20.88/0.684
HST*		22.51/0.637	21.86/0.542	22.20/0.518	20.43/0.561	20.94/0.686
RRDB [47]	**20**	23.73/0.674	22.81/0.575	23.06/0.550	21.17/0.599	22.17/0.722
SwinIR [28]		23.81/0.682	22.87/0.577	23.09/0.551	21.32/0.608	22.35/0.729
HST		23.91/0.683	22.93/0.578	23.11/0.551	21.33/0.607	22.41/0.728
HST*		23.96/0.684	22.95/0.579	23.13/0.551	21.38/0.607	22.48/0.729
RRDB [47]	**30**	24.74/0.708	23.42/0.599	23.53/0.569	21.77/0.630	23.09/0.750
SwinIR [28]		24.83/0.713	23.43/0.600	23.53/0.571	21.85/0.636	23.20/0.755
HST		24.89/0.713	23.49/0.600	23.57/0.571	21.91/0.635	23.30/0.754
HST*		24.94/0.714	23.52/0.601	23.59/0.571	21.96/0.636	23.39/0.756
RRDB [47]	**40**	25.05/0.717	23.67/0.609	23.78/0.581	21.93/0.638	23.37/0.756
SwinIR [28]		25.28/0.726	23.78/0.613	23.82/0.583	22.21/0.652	23.69/0.767
HST		25.39/0.728	23.84/0.614	23.87/0.584	22.23/0.651	23.85/0.768
HST*		25.43/0.729	23.87/0.614	23.89/0.585	22.29/0.653	23.94/0.771

Table 4. Quantitative comparison for compressed image super-resolution on DIV2K [1] validation datasets. Results are tested on ×4 with different compression qualities in terms of PSNR/SSIM. Best performance are in red.

Datasets	Q	Methods(PSNR/SSIM)			
		RRDB	SwinIR	**HST**	**HST***
DIV2K [1]	**10**	23.52/0.6400	23.57/0.6436	23.62/0.6436	23.65/0.6443
	20	24.68/0.6746	24.73/0.6771	24.77/0.6769	24.80/0.6777
	30	25.31/0.6949	25.32/0.6966	25.38/0.6963	25.41/0.6971
	40	25.58/0.7038	25.67/0.7077	25.74/0.7085	25.78/0.7093

4.5 Comparison with Other Frameworks

We compare our HST with two other state-of-the-art models in image super-resolution, and one real-SR method [46] for qualitative comparison. Among them, RRDB [47] uses residual in residual dense blocks to deepen the network structure, thus having the ability to better aggregate image structure and texture information from multi-levels. SwinIR [28] introduces transformer into image restoration tasks and outperforms previous CNN-based models. The performances are tested

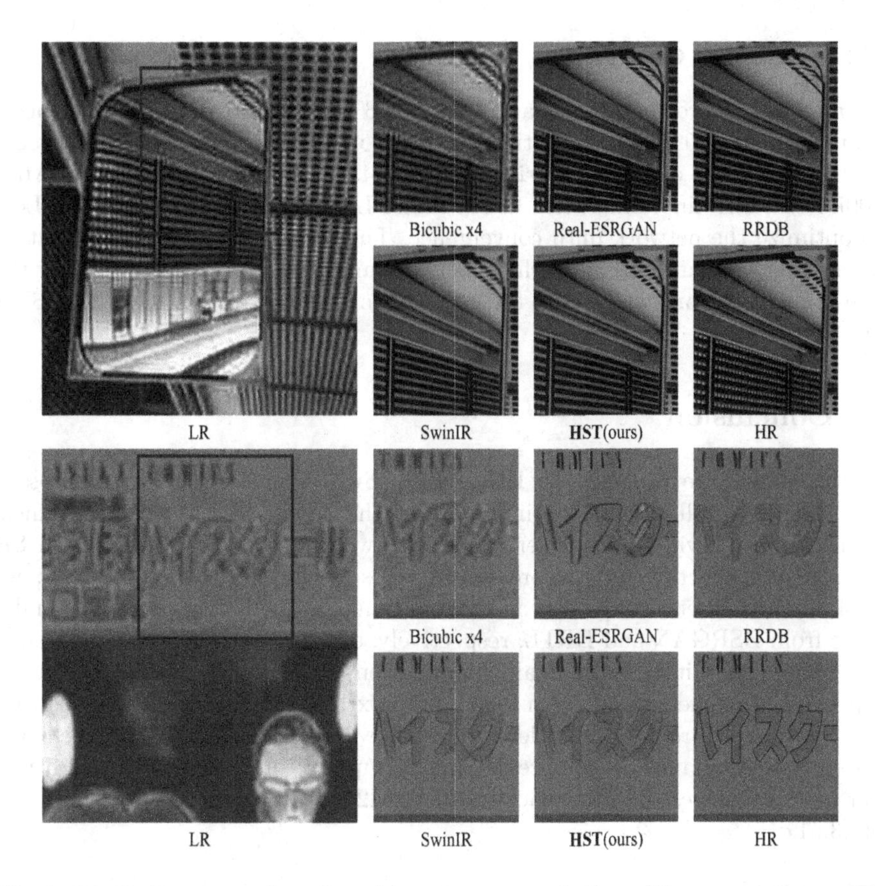

Fig. 6. Qualitative comparison for ×4 image super-resolution with compression quality 20.

on Set5 [3], Set14 [54], BSD100 [37], urban100 [16], Manga109 [38], respectively, with PSNR and SSIM in RGB channels. Moreover, we also test three models' performance on AIM2022 [53] official validation dataset, which includes 100 images from the DIV2K validation dataset. Quantitative and qualitative results are shown in Tables 3, 4 and Fig. 6, respectively. We denote the model using a self-ensemble strategy [30] with *.

Extensive experiments show that our HST outperforms other methods by 0.25 dB at most on compressed image super-resolution tasks. Even without self-ensemble, HST can still achieve an increase of 0.16 dB at most. As shown in Fig. 6, Real-ESRGAN [46] generates unnatural textures although its degradation process includes JPEG compression. Compared with other methods, our HST can generate SR with fewer artifacts. Moreover, HST performs better in rich texture areas, resulting in pleasant perception. All these benefit from a hierarchical network structure, which captures features at different scales and enhances the network's representation ability.

4.6 AIM2022 Challenge

To further explore the performance of our HST, we follow the training process we used in AIM2022 [53] competition to train our HST. More specifically, we use all three training datasets described in Sect. 4.1 to finetune the network. After 100k iterations' finetuning with Charbonnier Loss [7], we further use MSE Loss to optimize the network until convergence. The result on the official validation dataset shows that, with hierarchical network architecture, HST outperforms the one-branch network we used in the competition by 0.05 dB, with a final PSNR of 23.80 dB.

5 Conclusion

In this paper, we propose the Hierarchical Swin Transformer for compressed image super-resolution, which incorporates the advantages of the hierarchical structure and Swin Transformer. Moreover, we find that pretraining with SR is vital and effective for compressed image super-resolution. Particularly, we explore three pretraining tasks, *i.e.*, traditional SR, and two RealSR simulations from BSRGAN and DRTL, respectively, of which the experimental results show that pretraining with RealSR simulations can bring better performance, especially with the simulation in DRTL [22]. Extensive experiments demonstrate that, with a pretraining and hierarchical network structure, our HST achieves the best performance on compressed image super-resolution tasks. In addition, our model achieves the fifth place in the AIM2022 [53] challenge, with a PSNR of 23.51 dB.

Acknowledgement. This work was supported in part by NSFC under Grant U1908209, 62021001 and the National Key Research and Development Program of China 2018AAA0101400.

References

1. Agustsson, E., Timofte, R.: Ntire 2017 challenge on single image super-resolution: dataset and study. In: Proceedings of the IEEE Conference on Computer Vision and Pattern Recognition Workshops, pp. 126–135 (2017)
2. Bell-Kligler, S., Shocher, A., Irani, M.: Blind super-resolution kernel estimation using an internal-GAN. In: Advances in Neural Information Processing Systems, vol. 32 (2019)
3. Bevilacqua, M., Roumy, A., Guillemot, C., Alberi-Morel, M.L.: Low-complexity single-image super-resolution based on nonnegative neighbor embedding (2012)
4. Bross, B., Chen, J., Ohm, J.R., Sullivan, G.J., Wang, Y.K.: Developments in international video coding standardization after AVC, with an overview of versatile video coding (VVC). Proc. IEEE **109**(9), 1463–1493 (2021)
5. Cai, J., Zeng, H., Yong, H., Cao, Z., Zhang, L.: Toward real-world single image super-resolution: a new benchmark and a new model. In: Proceedings of the IEEE/CVF International Conference on Computer Vision, pp. 3086–3095 (2019)

6. Cavigelli, L., Hager, P., Benini, L.: CAS-CNN: a deep convolutional neural network for image compression artifact suppression. In: 2017 International Joint Conference on Neural Networks (IJCNN), pp. 752–759. IEEE (2017)
7. Charbonnier, P., Blanc-Feraud, L., Aubert, G., Barlaud, M.: Two deterministic half-quadratic regularization algorithms for computed imaging. In: Proceedings of 1st International Conference on Image Processing, vol. 2, pp. 168–172. IEEE (1994)
8. Chen, H., et al.: Pre-trained image processing transformer. In: Proceedings of the IEEE/CVF Conference on Computer Vision and Pattern Recognition, pp. 12299–12310 (2021)
9. Chen, Y., et al.: Drop an octave: Reducing spatial redundancy in convolutional neural networks with octave convolution. In: Proceedings of the IEEE/CVF International Conference on Computer Vision, pp. 3435–3444 (2019)
10. Dai, T., Cai, J., Zhang, Y., Xia, S.T., Zhang, L.: Second-order attention network for single image super-resolution. In: Proceedings of the IEEE/CVF Conference on Computer Vision and Pattern Recognition, pp. 11065–11074 (2019)
11. Dong, C., Deng, Y., Loy, C.C., Tang, X.: Compression artifacts reduction by a deep convolutional network. In: Proceedings of the IEEE International Conference on Computer Vision, pp. 576–584 (2015)
12. Dong, C., Loy, C.C., He, K., Tang, X.: Image super-resolution using deep convolutional networks. IEEE Trans. Pattern Anal. Mach. Intell. **38**(2), 295–307 (2015)
13. Fu, X., Zha, Z.J., Wu, F., Ding, X., Paisley, J.: JPEG artifacts reduction via deep convolutional sparse coding. In: Proceedings of the IEEE/CVF International Conference on Computer Vision, pp. 2501–2510 (2019)
14. Gu, J., Lu, H., Zuo, W., Dong, C.: Blind super-resolution with iterative kernel correction. In: Proceedings of the IEEE/CVF Conference on Computer Vision and Pattern Recognition, pp. 1604–1613 (2019)
15. Guo, J., Chao, H.: Building dual-domain representations for compression artifacts reduction. In: Leibe, B., Matas, J., Sebe, N., Welling, M. (eds.) ECCV 2016. LNCS, vol. 9905, pp. 628–644. Springer, Cham (2016). https://doi.org/10.1007/978-3-319-46448-0_38
16. Huang, J.B., Singh, A., Ahuja, N.: Single image super-resolution from transformed self-exemplars. In: Proceedings of the IEEE Conference on Computer Vision and Pattern Recognition, pp. 5197–5206 (2015)
17. Ji, X., Cao, Y., Tai, Y., Wang, C., Li, J., Huang, F.: Real-world super-resolution via kernel estimation and noise injection. In: proceedings of the IEEE/CVF Conference on Computer Vision and Pattern Recognition Workshops, pp. 466–467 (2020)
18. Jolicoeur-Martineau, A.: The relativistic discriminator: a key element missing from standard GAN. arXiv preprint arXiv:1807.00734 (2018)
19. Kim, J., Lee, J.K., Lee, K.M.: Deeply-recursive convolutional network for image super-resolution. In: Proceedings of the IEEE Conference on Computer Vision and Pattern Recognition, pp. 1637–1645 (2016)
20. Kim, J., Choi, Y., Uh, Y.: Feature statistics mixing regularization for generative adversarial networks. In: Proceedings of the IEEE/CVF Conference on Computer Vision and Pattern Recognition, pp. 11294–11303 (2022)
21. Ledig, C., et al.: Photo-realistic single image super-resolution using a generative adversarial network. In: Proceedings of the IEEE Conference on Computer Vision and Pattern Recognition, pp. 4681–4690 (2017)
22. Li, X., Jin, X., Fu, J., Yu, X., Tong, B., Chen, Z.: Few-shot real image restoration via distortion-relation guided transfer learning. arXiv preprint arXiv:2111.13078 (2021)

23. Li, X., et al.: Learning disentangled feature representation for hybrid-distorted image restoration. In: Vedaldi, A., Bischof, H., Brox, T., Frahm, J.-M. (eds.) ECCV 2020. LNCS, vol. 12374, pp. 313–329. Springer, Cham (2020). https://doi.org/10.1007/978-3-030-58526-6_19
24. Li, X., et al.: Learning omni-frequency region-adaptive representations for real image super-resolution. In: Proceedings of the AAAI Conference on Artificial Intelligence, vol. 35, pp. 1975–1983 (2021)
25. Li, X., Shi, J., Chen, Z.: Task-driven semantic coding via reinforcement learning. IEEE Trans. Image Process. **30**, 6307–6320 (2021)
26. Li, X., Sun, S., Zhang, Z., Chen, Z.: Multi-scale grouped dense network for VVC intra coding. In: Proceedings of the IEEE/CVF Conference on Computer Vision and Pattern Recognition Workshops, pp. 158–159 (2020)
27. Li, Y., Jin, P., Yang, F., Liu, C., Yang, M.H., Milanfar, P.: COMISR: compression-informed video super-resolution. In: Proceedings of the IEEE/CVF International Conference on Computer Vision, pp. 2543–2552 (2021)
28. Liang, J., Cao, J., Sun, G., Zhang, K., Van Gool, L., Timofte, R.: SwinIR: image restoration using Swin transformer. In: Proceedings of the IEEE/CVF International Conference on Computer Vision, pp. 1833–1844 (2021)
29. Liang, J., Zhang, K., Gu, S., Van Gool, L., Timofte, R.: Flow-based kernel prior with application to blind super-resolution. In: Proceedings of the IEEE/CVF Conference on Computer Vision and Pattern Recognition, pp. 10601–10610 (2021)
30. Lim, B., Son, S., Kim, H., Nah, S., Mu Lee, K.: Enhanced deep residual networks for single image super-resolution. In: Proceedings of the IEEE Conference On Computer Vision and Pattern Recognition Workshops, pp. 136–144 (2017)
31. Liu, J., Li, X., Peng, Y., Yu, T., Chen, Z.: SwinIQA: learned Swin distance for compressed image quality assessment. In: Proceedings of the IEEE/CVF Conference on Computer Vision and Pattern Recognition, pp. 1795–1799 (2022)
32. Liu, J., Lin, J., Li, X., Zhou, W., Liu, S., Chen, Z.: LIRA: lifelong image restoration from unknown blended distortions. In: Vedaldi, A., Bischof, H., Brox, T., Frahm, J.-M. (eds.) ECCV 2020. LNCS, vol. 12363, pp. 616–632. Springer, Cham (2020). https://doi.org/10.1007/978-3-030-58523-5_36
33. Liu, P., Zhang, H., Lian, W., Zuo, W.: Multi-level wavelet convolutional neural networks. IEEE Access **7**, 74973–74985 (2019)
34. Lu, M., Chen, T., Liu, H., Ma, Z.: Learned image restoration for VVC intra coding. In: CVPR Workshops (2019)
35. Lu, Y., et al.: RTN: reinforced transformer network for coronary CT angiography vessel-level image quality assessment. arXiv preprint arXiv:2207.06177 (2022)
36. Luo, Z., Huang, H., Yu, L., Li, Y., Fan, H., Liu, S.: Deep constrained least squares for blind image super-resolution. In: Proceedings of the IEEE/CVF Conference on Computer Vision and Pattern Recognition, pp. 17642–17652 (2022)
37. Martin, D., Fowlkes, C., Tal, D., Malik, J.: A database of human segmented natural images and its application to evaluating segmentation algorithms and measuring ecological statistics. In: Proceedings Eighth IEEE International Conference on Computer Vision, ICCV 2001, vol. 2, pp. 416–423. IEEE (2001)
38. Matsui, Y., et al.: Sketch-based manga retrieval using manga109 dataset. Multimed. Tools Appl. **76**(20), 21811–21838 (2017)
39. Pang, Y., et al.: FAN: frequency aggregation network for real image super-resolution. In: Bartoli, A., Fusiello, A. (eds.) ECCV 2020. LNCS, vol. 12537, pp. 468–483. Springer, Cham (2020). https://doi.org/10.1007/978-3-030-67070-2_28

40. Pennebaker, W.B., Mitchell, J.L.: JPEG: Still Image Data Compression Standard. Springer Science & Business Media, New York (1992). https://link.springer.com/book/9780442012724
41. Rabbani, M., Joshi, R.: An overview of the JPEG 2000 still image compression standard. Signal Process. Image Commun. **17**(1), 3–48 (2002)
42. Shi, W., et al.: Real-time single image and video super-resolution using an efficient sub-pixel convolutional neural network. In: Proceedings of the IEEE Conference on Computer Vision and Pattern Recognition, pp. 1874–1883 (2016)
43. Svoboda, P., Hradis, M., Barina, D., Zemcik, P.: Compression artifacts removal using convolutional neural networks. arXiv preprint arXiv:1605.00366 (2016)
44. Timofte, R., Agustsson, E., Van Gool, L., Yang, M.H., Zhang, L.: Ntire 2017 challenge on single image super-resolution: methods and results. In: Proceedings of the IEEE Conference on Computer Vision and Pattern Recognition Workshops, pp. 114–125 (2017)
45. Wang, L., et al.: Unsupervised degradation representation learning for blind super-resolution. In: Proceedings of the IEEE/CVF Conference on Computer Vision and Pattern Recognition, pp. 10581–10590 (2021)
46. Wang, X., Xie, L., Dong, C., Shan, Y.: Real-ESRGAN: training real-world blind super-resolution with pure synthetic data. In: Proceedings of the IEEE/CVF International Conference on Computer Vision, pp. 1905–1914 (2021)
47. Wang, X., et al.: ESRGAN: enhanced super-resolution generative adversarial networks. In: Proceedings of the European Conference on Computer Vision (ECCV) Workshops (2018)
48. Wang, Z., Liu, D., Chang, S., Ling, Q., Yang, Y., Huang, T.S.: D3: deep dual-domain based fast restoration of JPEG-compressed images. In: Proceedings of the IEEE Conference on Computer Vision and Pattern Recognition, pp. 2764–2772 (2016)
49. Wei, P., et al.: AIM 2020 challenge on real image super-resolution: methods and results. In: Bartoli, A., Fusiello, A. (eds.) ECCV 2020. LNCS, vol. 12537, pp. 392–422. Springer, Cham (2020). https://doi.org/10.1007/978-3-030-67070-2_24
50. Wei, P., et al.: Component divide-and-conquer for real-world image super-resolution. In: Vedaldi, A., Bischof, H., Brox, T., Frahm, J.-M. (eds.) ECCV 2020. LNCS, vol. 12353, pp. 101–117. Springer, Cham (2020). https://doi.org/10.1007/978-3-030-58598-3_7
51. Wu, Y., Wang, X., Li, G., Shan, Y.: AnimeSR: learning real-world super-resolution models for animation videos. arXiv preprint arXiv:2206.07038 (2022)
52. Wu, Y., Li, X., Zhang, Z., Jin, X., Chen, Z.: Learned block-based hybrid image compression. IEEE Trans. Circ. Syst. Video Technol. **32**, 3978–3990 (2021)
53. Yang, R., Timofte, R., et al.: AIM 2022 challenge on super-resolution of compressed image and video: dataset, methods and results. In: Proceedings of the European Conference on Computer Vision (ECCV) Workshops (2022)
54. Zeyde, R., Elad, M., Protter, M.: On single image scale-up using sparse-representations. In: Boissonnat, J.-D., et al. (eds.) Curves and Surfaces 2010. LNCS, vol. 6920, pp. 711–730. Springer, Heidelberg (2012). https://doi.org/10.1007/978-3-642-27413-8_47
55. Zhang, K., Liang, J., Van Gool, L., Timofte, R.: Designing a practical degradation model for deep blind image super-resolution. In: Proceedings of the IEEE/CVF International Conference on Computer Vision, pp. 4791–4800 (2021)
56. Zhang, X., Yang, W., Hu, Y., Liu, J.: DMCNN: dual-domain multi-scale convolutional neural network for compression artifacts removal. In: 2018 25th IEEE International Conference on Image Processing (ICIP), pp. 390–394. IEEE (2018)

57. Zhang, Y., Li, K., Li, K., Wang, L., Zhong, B., Fu, Y.: Image super-resolution using very deep residual channel attention networks. In: Ferrari, V., Hebert, M., Sminchisescu, C., Weiss, Y. (eds.) ECCV 2018. LNCS, vol. 11211, pp. 294–310. Springer, Cham (2018). https://doi.org/10.1007/978-3-030-01234-2_18
58. Zheng, M., et al.: Progressive training of a two-stage framework for video restoration. In: Proceedings of the IEEE/CVF Conference on Computer Vision and Pattern Recognition, pp. 1024–1031 (2022)

Swin2SR: SwinV2 Transformer for Compressed Image Super-Resolution and Restoration

Marcos V. Conde[1(✉)], Ui-Jin Choi[2], Maxime Burchi[1], and Radu Timofte[1]

[1] Computer Vision Lab, CAIDAS, University of Würzburg, Würzburg, Germany
{marcos.conde-osorio,radu.timofte}@uni-wuerzburg.de
[2] MegaStudyEdu, Seoul, South Korea

Abstract. Compression plays an important role on the efficient transmission and storage of images and videos through band-limited systems such as streaming services, virtual reality or videogames. However, compression unavoidably leads to artifacts and the loss of the original information, which may severely degrade the visual quality. For these reasons, quality enhancement of compressed images has become a popular research topic. While most state-of-the-art image restoration methods are based on convolutional neural networks, other transformers-based methods such as SwinIR, show impressive performance on these tasks.

In this paper, we explore the novel Swin Transformer V2, to improve SwinIR for image super-resolution, and in particular, the compressed input scenario. Using this method we can tackle the major issues in training transformer vision models, such as training instability, resolution gaps between pre-training and fine-tuning, and hunger on data. We conduct experiments on three representative tasks: JPEG compression artifacts removal, image super-resolution (classical and lightweight), and compressed image super-resolution. Experimental results demonstrate that our method, Swin2SR, can improve the training convergence and performance of SwinIR, and is a top-5 solution at the "AIM 2022 Challenge on Super-Resolution of Compressed Image and Video".

Our code can be found at https://github.com/mv-lab/swin2sr.

Keywords: Super-resolution · Image compression · Transformer · JPEG

1 Introduction

Compression plays an important role on the efficient transmission and storage of images and videos through band-limited systems such as streaming services, virtual reality, cloud storage for images, videoconferences or videogames. However, compression leads to artifacts and the loss of the original information, which may severely degrade the visual quality of the image. For these reasons, quality enhancement and restoration of compressed images has become a popular research topic. Image restoration techniques, such as image super-resolution (SR) and JPEG compression artifact reduction, aim to reconstruct the high-quality

L. Karlinsky et al. (Eds.): ECCV 2022 Workshops, LNCS 13802, pp. 669–687, 2023.
https://doi.org/10.1007/978-3-031-25063-7_42

clean image from its low-quality degraded (or compressed) counterpart. During the past decade, several revolutionary works were proposed for single image super-resolution, most of them are CNN-based methods [16,20,28,31,54,61–67]. We can also find plenty of proposed methods for the reduction of JPEG artifacts [18,27,45]. Recently, the blind super-resolution [22,56,62] methods have been proposed. They are able to use one model to jointly handle the tasks of super-resolution, deblurring, JPEG artifacts reduction, etc. Although the performance of these deep learning methods significantly improved compared with traditional methods [48], they generally suffer from two basic problems that arise from the basic convolution layer receptive field: (i) the interactions between images and kernels are content-independent, therefore, using the same kernel to restore different image regions may not be the best. (ii) Under the principle of locality, convolution is not effective for long-range dependency modelling [32].

As an alternative to CNNs, Transformer [52] designs a self-attention mechanism to capture global interactions between contexts and has shown promising performance in several vision problems [6,17,36,50]. Recently, Swin Transformer [36] has shown great promise as it leverages the advantages of both CNN and Transformers (*i.e.* CNN to process image with large size due to the local attention mechanism, and transformer to model long-range dependency with the shifted window scheme). Compared with classical CNN-based image restoration models, Transformer-based methods have several benefits: (i) content-based interactions between image content and attention weights, which can be interpreted as spatially varying convolution [51]. (ii) long-range dependency modelling are enabled by the shifted window mechanism. (iii) in some cases, better performance with less parameters. In this context, Liang *et al.* SwinIR [32], based on Swin Transformer [36], represents the state-of-the-art of transformer-based models for image restoration.

***AIM 2022 Challenge on Super-Resolution of Compressed Image and Video*.** This challenge is a step forward for establishing a new benchmark for the super-resolution of JPEG images and videos. The methods proposed in this challenge also have the potential to solve various super-resolution tasks. The challenge utilizes the famous DIV2K [1] dataset for evaluating methods. Other related challenges such as "NTIRE 2022 challenge on super-resolution and quality enhancement of compressed video" [57,59] and "NTIRE 2020 challenge on real-world image SR" [37] also represent the SOTA in this field.

In this paper, we propose Swin2SR, a SwinV2 Transformer-based model [35,36] for Compressed Image Super-Resolution and Restoration. This model represents a possible improvement or update of SwinIR [32] for these particular tasks. SwinV2 [35] (CVPR '22) allows us to tackle the major issues in training large transformer-based vision models, including training instability and duration, and resolution gaps between pre-training and fine-tuning [32]. We are the first work to explore successfully other transformer blocks beyond Swin Transformer [36] for image super-resolution and restoration. In some scenarios, our model can achieve similar results as SwinIR [32], yet training 33% less.

We also provide extensive comparisons with state-of-the-art methods, and achieve competitive results at the related AIM 2022 Challenge.

2 Related Work

2.1 Image Restoration

Image restoration is split in a large number of sub-problems, for instance image denoising, image deblurring, super-resolution and compression artifacts removal among others. Traditional model-based methods for image restoration were usually defined by hand-crafted priors that narrowed the ill-posed nature of the problems by reducing the set of plausible solutions [12,47,48]. Learning-based methods based on CNNs have recently gained great popularity for image restoration, and they represent current state-of-the-art in most low-level vision tasks (*i.e.* denoising, deblurring, compression artifacts removal). The first remarkable work on denoising with deep learning is probably Zhang *et al.* [63] DnCNN. Other pioneering works include Dong *et al.* SRCNN [16] for image super-resolution and ARCNN [15] for JPEG compression artifact removal. Since research has moved towards deep learning, multiple CNN-based approaches have been proposed to improve the learned representations using more complex neural network architectures, such as residual blocks, dense residual blocks, and laplacian operators [7,28,29,61,69,70]. Other solutions attempt to exploit the attention mechanism in CNNs, such as channel attention and spatial attention [14,33,41,42,67].

2.2 Vision Transformer

The Transformer architecture [52] has recently gained much popularity in the computer vision community. Originally designed for neural machine translation, the Transformer architecture has successfully been applied to image classification [13,17,36,51], object detection [6,50], object segmentation [4] and perceptual quality assessment (IQA) [10,21]. The attention mechanism learns complex global interactions by attending to important regions in the image. Due to its impressive performance, transformers have also been introduced to image restoration [5,8,55]. More recently, Chen *et al.* [8] proposed IPT, a general backbone model for multiple image restoration tasks based on the standard Transformer [52]. This model shows promising performance on several tasks, however, it relies on a large number of parameters and heavy computation (over 115.5M parameters), and a large-scale dataset like ImageNet (over 1M images). VSR-Transformer proposed by Cao *et al.* [5] combines the self-attention mechanism and CNN-based feature extraction to fuse better features in video super-resolution. Note that many transformer-based approaches such as IPT [8] and VSR-Transformer [5] use patch-wise attention, which may not be optimal for image restoration. Liang *et al.* proposed SwinIR [32] based Swin Transformer [36], which represents the state-of-the-art in many restoration tasks.

In this context, the Swin Transformer [36] improved the Vision Transformer architecture by using shifted window based self-attention with progressive image downsampling like CNNs. Window self-attention is computed for non-overlapped image patches reducing attention computational complexity from Eq. 1 to Eq. 2:

$$O(MSA) = 4hwC^2 + 2(hw)^2C \tag{1}$$

$$O(WMSA) = 4hwC^2 + 2M^2hwC \tag{2}$$

for an image of size $h \times w$ and patches of size $M \times M$. The former quadratic computational complexity is replaced by a linear complexity when M is fixed. Learned relative positional bias are also added to include position information while computing similarities for each head.

The Swin Transformer V2 [35] modified the Swin Attention [36] module to better scale model capacity and window resolution. They first replace the *pre-norm* by a *post-norm* configuration, use a *scaled cosine attention* instead of the *dot product attention* and use a *log-spaced continuous* relative position bias approach to replace the previous *parameterized* approach. The attention output is:

$$Attention(Q, K, V) = Softmax(cos(Q, K)/\tau + S)V \tag{3}$$

where $Q, K, V \in \mathbb{R}^{M^2 \times d}$ are the query, key and value matrices. $S \in \mathbb{R}^{M^2 \times M^2}$ are the relative to absolute positional embeddings obtained by projecting the position bias after re-indexing. τ is a learnable scalar, non-shared across heads and layers. This block is illustrated in Fig. 1.

3 Our Method

Our method Swin2SR is illustrated in Fig. 1. We propose some modifications of SwinIR [32], which is based on Swin Transformer [36], that enhance the model's capabilities for Super-Resolution, and in particular, for Compressed Input SR. We update the original Residual Transformer Block (RSTB) by using the new SwinV2 transformer [35] (CVPR'22) layers and attention to scale up capacity and resolution [35]. Our method has a classical upscaling branch which uses a bicubic interpolation, as shown in the AIM 2022 Challenge Leaderboard [58] and our results (Table 5), this alone can recover basic structural information. For this reason, the output of our model is added to the basic upscaled image, to enhance it. We also explore different loss functions to make our model more robust to JPEG compression artifacts, being able to recover high-frequency details from the compressed LR image, and therefore, achieve better performance.

Advantages of Updating to SwinV2. The SwinV2 architecture modifies the shifted window self-attention module to better scale model capacity and window resolution. The use of *post normalization* instead of *pre normalization* reduce the average feature variance of deeper layers and increase numerical stability during training. This allows to scale the SwinV2 Transformer up to 3 billion parameters without training instabilities [35]. The use of *scaled cosine attention* instead of *dot product* between queries and keys reduce the dominance of some attention heads for a few pixel pairs. In some tasks, our Swin2SR model achieved the same results as SwinIR [32], yet training 33% less iterations. Finally, the use of *log-spaced continuous* relative position bias allows us to generalize to higher input resolution at inference time.

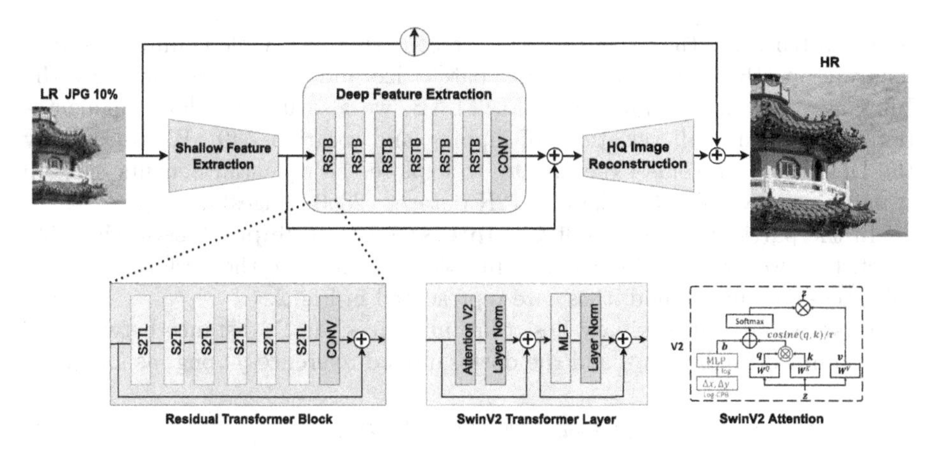

Fig. 1. The architecture of the proposed Swin2SR [11]. In this case, we show our method applied to Super-Resolution of Compressed Image [58].

3.1 Experimental Setup

For a fair comparison and ensure reproducibility, we follow the same experimental setup as SwinIR [32] and other state-of-the-art methods [62,69].

We evaluate our model on three tasks: JPEG compression artifacts removal (Sect. 4.1), classical and lightweight image super-resolution (Sect. 4.2) and compressed image super-resolution (Sect. 4.4). We mainly use the DIV2K dataset for training and validation [1], and following the tradition of image SR, we report PSNR and SSIM on the Y channel of the YCbCr space [32,62,69].

Our model Swin2SR has the following elements, similar to SwinIR [32]: shallow feature extraction, deep feature extraction and high-quality image reconstruction modules. The **shallow feature extraction** module uses a convolution layer to extract features, which are directly transmitted to the reconstruction module to preserve low-frequency information [32,63]. The **Deep feature extraction** module is mainly composed of Residual SwinV2 Transformer blocks (RSTB), each of which utilizes several SwinV2 Transformer [35] layers (S2TL) for local attention and cross-window interaction. Finally, both shallow and deep features are fused in the reconstruction module for high-quality image reconstruction. To upscale the image, we use standard a pixel shuffle operation.

The hyper-parameters of the architecture are as follows: the RSTB number, S2TL number, window size, channel number and attention head number are generally set to 6, 6, 8, 180 and 6, respectively. For lightweight image SR, we explain the details in Sect. 4.2.

3.2 Implementation Details

The method was implemented in Pytorch using as baseline https://github.com/cszn/KAIR and the official repository for SwinIR [32]. We initially train Swin2SR from scratch using the basic $\mathcal{L}_1$ loss for reconstruction. While training, we randomly crop HR images using 192px patch size and crop correspondingly the LR image generated offline using MATLAB, we also use standard augmentations that include all variations of flipping and rotations [49]. We use mainly the DIV2K [1]. In some experiments, to explore the potential benefits of more training data, we also use the Flickr2K dataset (2650 images).

In the particular scenario of **Compressed Input Super-Resolution** [58] (Sect. 4.4), we explore different loss functions to improve the performance and robustness of our method; these are represented in Fig. 2.

First, we add an Auxiliary Loss that minimizes the $\mathcal{L}_1$ distance between the downsampled prediction $\hat{y}$ and the downsampled reference y .png, as follows:

$$\mathcal{L}_{aux} = \|D(y) - D(\hat{y})\|_1 \tag{4}$$

where x is the low-resolution degraded image, y is the high-resolution clean image, $f(x) = \hat{y}$ is the restored image using our model f, and $D(.)$ is a downsampling operator (*i.e.* $\times 4$ bicubic kernel). This helps to ensure consistency also at lower-resolution. In order to minimize Eq. 4 the restored image at a lower resolution should not have artifacts (i.e. the prediction at lower resolution should be close to the downsampled reference .png without artifacts).

Second, we extract the high-frequency (HF) information from the High-Resolution images. This loss is formulated as follows:

$$\mathcal{L}_{hf} = \|(y - (y * b)) - (\hat{y} - (\hat{y} * b))\|_1 = \|HF(y) - HF(\hat{y})\|_1 \tag{5}$$

where $HR(.)$ denotes the high-frequency information of an image. To obtain this, we convolve a simple 5×5 kernel b as a gaussian blur operation. This term enforces the prediction to have the same high-frequency details as the reference, and therefore, it helps to improve the sharpness and quality of the results.

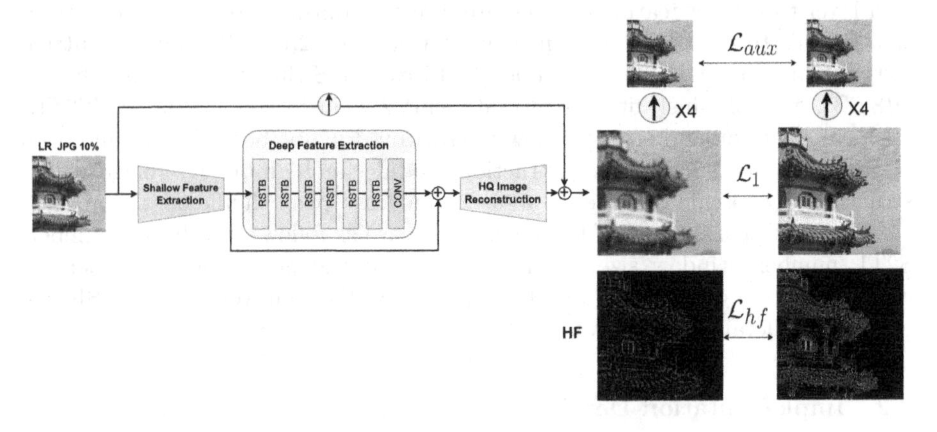

Fig. 2. Swin2SR training with additional regularization.

4 Experimental Results

4.1 JPEG Compression Artifacts Removal

Table 1 shows the comparison of Swin2SR with *state-of-the-art* JPEG compression artifact reduction methods: ARCNN [15], DnCNN-3 [63], QGAC [18], RNAN [68], and MWCNN [34]. All of compared methods are CNN-based models trained especifically for each quality type (*i.e.* four models per dataset). Due to our limited resources, and seeking for a more flexible approach, we train a single model able to deal with the four different quality factors. For this reason, we do not compare directly with DRUNet [61], as we consider it an unfair comparison. Moreover, Swin2SR only has 12M parameters, while DRUNet [61], is a large model that has 32.7M parameters. Note that we perform these comparisons using the same setup as [32]. Following [32,61,70], we test different methods on two benchmark datasets: (i) Classic5 [19] and (ii) LIVE1 [44]; using JPEG quality factors (q) 10, 20, 30 and 40. As we can see in Table 1, our Swin2SR achieves state-of-the-art results in compression artifacts removal.

Table 1. Quantitative comparison (average PSNR/SSIM) with state-of-the-art methods for **JPEG compression artifact reduction** on benchmark datasets. Best and second best performance are in bold and italic, respectively. Note that Swin2SR is a single model that generalizes to different qualities, meanwhile, some methods are trained for each specific quality. Some numbers are from [27].

Dataset	q	ARCNN [15]	DnCNN [63]	QGAC [18]	RNAN [68]	MWCNN [34]	SwinIR [32]	**Swin2SR**
Classic5 [19]	10	29.03/0.79	29.40/0.80	29.84/0.83	29.96/0.81	30.01/0.82	**30.27/0.82**	*30.02/0.81*
	20	31.15/0.85	31.63/0.86	31.98/0.88	32.11/0.86	32.16/*0.87*	*31.32*/0.85	**32.26/0.87**
	30	32.51/0.88	32.91/0.88	33.22/0.90	33.38/0.89	*33.43/0.89*	31.39/0.853	**33.51/0.89**
	40	33.32/0.89	33.77/0.90	-	34.27/0.90	*34.27/0.90*	31.38/0.85	**34.33/0.90**
LIVE1 [44]	10	28.96/0.80	29.19/0.81	29.53/0.84	29.63/0.82	*29.69/0.82*	**29.86/0.82**	29.67/**0.82**
	20	31.29/0.87	31.59/0.88	31.86/0.90	32.03/0.88	*32.04*/0.89	31.00/0.86	**32.07//0.89**
	30	32.67/0.90	32.98/0.90	33.23/0.92	33.45/0.91	*33.45/0.91*	31.08/0.86	**33.49/0.91**
	40	33.63/0.91	33.96/0.92	-	34.47/0.92	*34.45*/**0.93**	31.05/0.86	**34.49**/*0.92*

In the case of SwinIR [32], which is also *state-of-the-art* for JPEG artifacts reduction, authors train one model per quality factor (*i.e.* four models) for 1600K iterations, and $q = 10/20/30$ models are fine-tuned using the $q = 40$ model as general baseline. We train a single model using the same setup [32], only for 800k iterations (*i.e.* $\times 2$ less training than SwinIR [32]), and JPEG compression as an augmentation. For this reason in Table 1 we compare with SwinIR trained for the most challenging $q = 10$. We also compare with MWCNN [34], IDCN [71] and FBCNN-C [27] using RGB color images. Attending to Tables 1 and 2, we consider our model a more general and flexible approach for grayscale or color compression artifacts removal, since it can be trained faster and generalizes to different compression quality factors. We also provide **qualitative results** in Fig. 3. Swin2SR can restore compressed images and generate high-quality results. We provide additional results in the supplementary material.

Table 2. Quantitative comparison on **color** JPEG images with **single** compression. We report average PSNR/SSIM on benchmark datasets. Our model outperforms networks designed for this particular task (although we recognise that training with more data). Some numbers are from [27].

Dataset	q	JPEG	ARCNN [15]	QGAC [18]	MWCNN [34]	IDCN [71]	FBCNN-C [27]	**Swin2SR**
LIVE1 [44]	10	25.69/0.74	26.91/0.79	27.62/0.80	27.45/0.80	27.63/0.81	*27.77/0.80*	**27.98/0.82**
	40	30.28/0.88	-	32.05/0.91	-	-	*32.34/0.91*	**32.53/0.92**
ICB [43]	10	29.44/0.75	30.06/0.77	32.06/0.81	30.76/0.77	31.71/0.80	*32.18/0.81*	**32.46/0.81**
	40	33.95/0.84	-	32.25/0.91	-	-	*36.02/0.86*	**36.25/0.86**

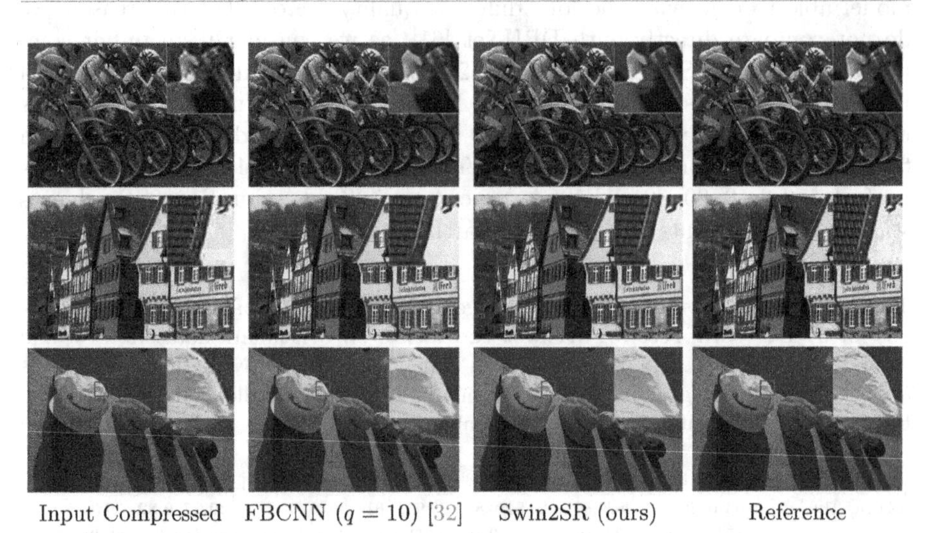

Fig. 3. Qualitative samples of JPEG Compression Artifacts Removal. We show the JPEG compressed image at quality $q = 10$. All images have the same resolution. Images from Classic5 [19] and LIVE1 [44]. Best viewed by zooming.

4.2 Classical Image Super-Resolution

For classical and lightweight image SR, following [32,61,62], we train Swin2SR on 800 training images of DIV2K and 2650 images from Flickr2K. For fair comparison with SwinIR [32], we use 64×64 LQ image patches, and the HQ-LQ image pairs are obtained by the MATLAB bicubic kernel. We train our model from scratch during 500k iterations, and fine-tune it for the $\times 4$ task. Table 3 shows the quantitative comparisons between Swin2SR and *state-of-the-art methods*: DBPN [23], RCAN [67], RRDB [54], SAN [14], IGNN [72], HAN [42], NLSA [41], IPT [8] and SwinIR [32]. All the CNN-based methods perform worse than the studied transformer-based methods, IPT [8], SwinIR [32] and Swin2SR. Moreover, Swin2SR was trained using only DIV2K+Flickr2K and achieves better performance than IPT [8], even though IPT [8] utilizes ImageNet (more than 1.3M images) in training and has huge number of parameters (115.5M). In contrast, Swin2SR has only 12M parameters, which is competitive even compared

Table 3. Quantitative comparison (average PSNR/SSIM) with state-of-the-art methods for **classical image SR** on benchmark datasets. Best and second best performance are in bold and italics, respectively.

Method	Scale	Training Dataset	Set5 [3]		Set14 [60]		BSD100 [39]		Urban100 [24]		Manga109 [40]	
			PSNR	SSIM	PSNR	SSIM	PSNR	SSIM	PSNR	SSIM	PSNR	SSIM
RCAN [67]	×2	DIV2K	38.27	0.9614	34.12	0.9216	32.41	0.9027	33.34	0.9384	39.44	0.9786
SAN [14]	×2	DIV2K	38.31	0.9620	34.07	0.9213	32.42	0.9028	33.10	0.9370	39.32	0.9792
IGNN [72]	×2	DIV2K	38.24	0.9613	34.07	0.9217	32.41	0.9025	33.23	0.9383	39.35	0.9786
HAN [42]	×2	DIV2K	38.27	0.9614	34.16	0.9217	32.41	0.9027	33.35	0.9385	39.46	0.9785
NLSA [41]	×2	DIV2K	38.34	0.9618	34.08	0.9231	32.43	0.9027	33.42	0.9394	39.59	0.9789
DBPN [23]	×2	DIV2K+Flickr2K	38.09	0.9600	33.85	0.9190	32.27	0.9000	32.55	0.9324	38.89	0.9775
IPT [8]	×2	ImageNet	38.37	-	34.43	-	32.48	-	33.76	-	-	-
SwinIR [32]	×2	DIV2K+Flickr2K	*38.42*	*0.9623*	*34.46*	*0.9250*	*32.53*	*0.9041*	*33.81*	*0.9427*	**39.92**	*0.9797*
Swin2SR	×2	DIV2K+Flickr2K	**38.43**	**0.9623**	**34.48**	**0.9256**	**32.54**	**0.905**	**33.89**	**0.9431**	*39.88*	**0.9798**
Swin2SR-D	×2	DIV2K+Flickr2K	38.06	-	33.81	-	32.32	-	32.6	-	38.98	-
RCAN [67]	×4	DIV2K	32.63	0.9002	28.87	0.7889	27.77	0.7436	26.82	0.8087	31.22	0.9173
SAN [14]	×4	DIV2K	32.64	0.9003	28.92	0.7888	27.78	0.7436	26.79	0.8068	31.18	0.9169
IGNN [72]	×4	DIV2K	32.57	0.8998	28.85	0.7891	27.77	0.7434	26.84	0.8090	31.28	0.9182
HAN [42]	×4	DIV2K	32.64	0.9002	28.90	0.7890	27.80	0.7442	26.85	0.8094	31.42	0.9177
NLSA [41]	×4	DIV2K	32.59	0.9000	28.87	0.7891	27.78	0.7444	26.96	0.8109	31.27	0.9184
DBPN [23]	×4	DIV2K+Flickr2K	32.47	0.8980	28.82	0.7860	27.72	0.7400	26.38	0.7946	30.91	0.9137
IPT [8]	×4	ImageNet	32.64	-	29.01	-	27.82	-	27.26	-	-	-
RRDB [54]	×4	DIV2K+Flickr2K	32.73	0.9011	28.99	0.7917	27.85	0.7455	27.03	0.8153	31.66	0.9196
SwinIR [32]	×4	DIV2K+Flickr2K	**32.92**	**0.9044**	**29.09**	**0.7950**	*27.92*	*0.7489*	*27.45*	*0.8254*	**32.03**	**0.9260**
Swin2SR	×4	DIV2K+Flickr2K	**32.92**	*0.9039*	*29.06*	*0.7946*	**27.92**	**0.7505**	**27.51**	**0.8271**	*31.03*	*0.9256*
Swin2SR-D	×4	DIV2K+Flickr2K	32.41	-	28.75	-	27.69	-	26.4	-	30.96	-

with state-of-the-art CNN-based models (15.4∼44.3M). Note that our models achieve essentially the same performance as SwinIR [32], yet trained for 400k iterations from scratch, without fine-tuning or pre-training, in comparison with SwinIR [32] models trained during 500k, and in the case of ×4 fine-tuned using the ×2 model. We provide visual comparisons in Figs. 5. Swin2SR can remove artifacts and recover structural information and high-frequency details.

Dynamic Super-Resolution. Likewise Sect. 4.1, we explore the performance of a single super-resolution model to upscale directly using any arbitrary × factor. We call this a Dynamic Super-Resolution model, referred as Swin2SR-D.

In SwinIR [32] we can find an upsampling layer designed to upscale images using s particular factor (*i.e.* ×2). This layer cannot be adjusted to a different factor on-line, therefore, SwinIR [32] trains one model for each different factor. To deal with this problem, we implemented a Dynamic upsampling layer, which initially can super-resolve the images using ×2, ×3, and ×4 factors on-line in the same module. We show in Table 3 the potential of this method, as this single model can perform ×2 and ×4 super-resolution indistinctly.

Lightweight Image SR. We also provide comparison of Swin2SR-s with *state-of-the-art methods* lightweight image SR methods: CARN [2], FALSR-A [9], IMDN [25], LAPAR-A [30], LatticeNet [38] and SwinIR (small) [32].

Our lightweight model is designed as SwinIR (small) [32], we decrease the number of Residual Swin Transformer Blocks (RSTB) and convolution channels to 4 and 60, respectively. However, the number of Swin Transformer Layers (STL) in each RSTB, window size and attention head number still set to 6, 8 and 6, respectively (as in Swin2SR base model).

In addition to PSNR and SSIM, we also report the total numbers of parameters and multiply-accumulate operations for different methods [32]. These MACs are calculated using a 1280 × 720 image. As shown in Table 4, Swin2SR outperforms competitive methods [2,9,25,30] on different benchmark datasets, with similar total numbers of parameters and multiply-accumulate operations. In our experiments, Swin2SR can achieve the same results as SwinIR (small) [32], yet, training almost 33% less iterations.

Table 4. Quantitative comparison (average PSNR/SSIM) with state-of-the-art methods for **lightweight image SR** ×2 on benchmark datasets. Best and second best performance are in red and blue colors, respectively. In our experiments, Swin2SR-s converges faster than SwinIR (small) [32].

Method	# Params	# Mult-Adds	Set5 [3]		Set14 [60]		BSD100 [39]		Urban100 [24]		Manga109 [40]	
			PSNR	SSIM	PSNR	SSIM	PSNR	SSIM	PSNR	SSIM	PSNR	SSIM
CARN [2]	1,592K	222.8G	37.76	0.9590	33.52	0.9166	32.09	0.8978	31.92	0.9256	38.36	0.9765
FALSR-A [9]	1,021K	234.7G	37.82	0.959	33.55	0.9168	32.1	0.8987	31.93	0.9256	-	-
IMDN [25]	694K	158.8G	38.00	0.9605	33.63	0.9177	32.19	0.8996	32.17	0.9283	38.88	0.9774
LAPAR-A [30]	548K	171.0G	38.01	0.9605	33.62	0.9183	32.19	0.8999	32.10	0.9283	38.67	0.9772
LatticeNet [38]	756K	169.5G	*38.15*	0.9610	33.78	0.9193	32.25	0.9005	32.43	0.9302	-	-
SwinIR [32]	878K	195.6G	38.14	*0.9611*	*33.86*	*0.9206*	*32.31*	*0.9012*	*32.76*	*0.9340*	*39.12*	*0.9783*
Swin2SR-s	1000K	199.0G	**38.17**	**0.9613**	**33.95**	**0.9216**	**32.35**	**0.9024**	**32.85**	**0.9349**	**39.32**	**0.9787**

4.3 Real-World Image Super-Resolution

We also test our approach using real-world images and prove the generalization capabilities of Swin2SR . We use the same setup as SwinIR [32] for training and testing our methods to exploit the full potential of these transformer-based approaches. Since there is no ground-truth high-quality images, we only provide visual comparison with representative bicubic model in Fig. 4. Our model produces detailed images without artifacts. Due to the limitations of space and visualization in this document, we include the comparison with ESRGAN [54] and state-of-the-art real-world image SR models such as RealSR [26], BSRGAN [62], Real-ESRGAN [53] and SwinIR [32] in the supplementary material.

4.4 Compressed Image Super-Resolution

The "AIM 2022 Challenge on Super-Resolution of Compressed Image" [58] is a step forward for establishing a benchmark of the super-resolution of JPEG images. In this challenge, we use the popular dataset DIV2K [1] as the training,

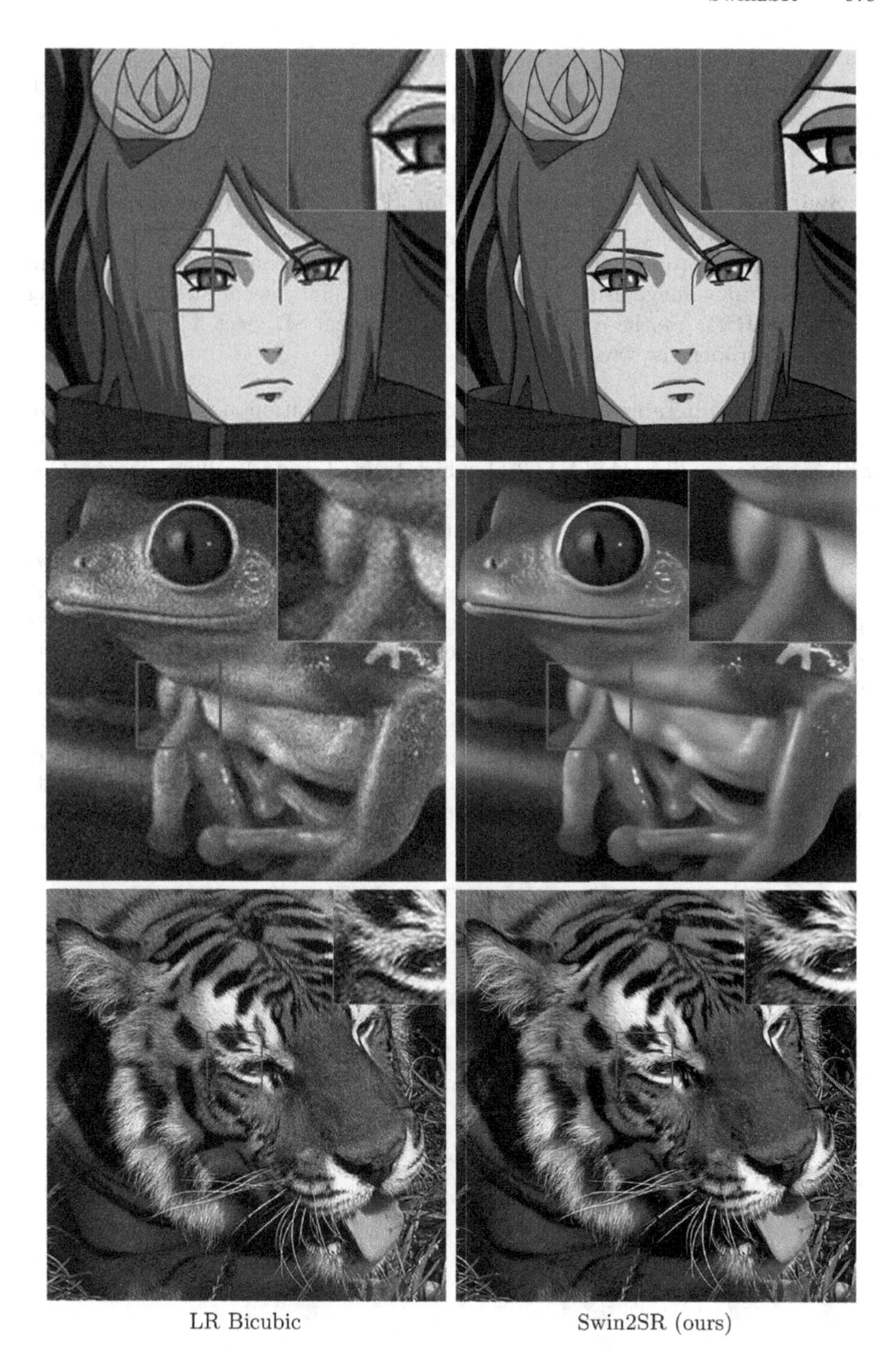

Fig. 4. Qualitative results on **real-world** SR datasets (RealSRSet, 5images). Our model can recover textures, remove noise and produce pleasant results.

validation and test sets. JPEG is the most commonly used image compression standard. We target the ×4 super-resolution of the images compressed with JPEG with the quality factor of 10. Figure 1 illustrates this process. We propose two solutions for this problem based on previous Sects. 4.1 and 4.2:

1. **Swin2SR-CI** An end-to-end model for JPEG artifacts removal and super-resolution (*i.e.* Fig. 1).
2. A 2-stage approach where first we remove JPEG compression artifacts in the LR input image using **Swin2SR-DJPEG**, and second, we upscale using **Swin2SRx4** (*i.e.*the model trained for Classical SR, Sect. 4.2). We refer to this experiment as "Swin2SR-CI2".

As we show in Table 5 ([1]), our method is a top solution at the challenge. We trained Swin2SR using only DIV2K [1] and Flickr2K [46] datasets, in comparison with other teams like CASIA LCVG, which trained using 1 million images. Our average testing time of Swin2SR model is 1.41s using single GPU A100.

In Fig. 5 we show extensive qualitative results of compressed input super-resolution [58]. Our model can recover information from the low-quality low-resolution input image, and generates high-resolution high-quality images. Among the limitations of our model, we can appreciate a clear blur effect, nevertheless, we find SwinIR [32] (and other *state-of-the-art* methods) to have the same issues.

Table 5. Results of AIM 2022 Challenge on Super-Resolution of Compressed Image. Our solutions are placed among the top teams, while our methods can process a single image in under a second (w/o self-ensemble).

Team	Test PSNR (dB)	Runtime (s)	Hardware
VUE	23.6677	120	Tesla V100
BSR	23.5731	63.96	Tesla A100
CASIA LCVG	23.5597	78.09	Tesla A100
USTC-IR	23.5085	19.2	2080ti
Swin2SR-CI2	23.4946	24	Tesla A100
MSDRSR	23.4545	7.94	Tesla V100
Giantpandacv	23.4249	0.248	RTX 3090
Swin2SR-CI	23.4033	9.39	Tesla A100
MVideo	23.3250	1.7	RTX 3090
UESTC+XJU CV	23.2911	3.0	RTX 3090
cvlab	23.2828	6.0	1080 Ti
Bicubic ×4	22.2420	-	-

[1] Online leaderboard https://codalab.lisn.upsaclay.fr/competitions/5076.

Ensembles and Fusion Strategies. We use classical self-ensemble techniques where the input image is flipped and rotated several times, and the resultant images are averaged [37,49]. We only use this technique in the related AIM 2022 Challenge (Sect. 4.4 and Table 5), and the marginal improvement of this technique was approximately 0.02 dB PSNR.

In Table 6 we show our ablation studies using the challenge DIV2K [1] validation set. The use of additional loss functions helped the model to converge faster, however after certain number of iterations (*i.e.* 250k) the model converges. As previously mentioned, among the **limitations** of our model, we can appreciate a clear blur effect in the qualitative samples in Fig. 5, indicating that our model is struggling to recover fine details and sharpness. Nevertheless, we find SwinIR [32] (and other *state-of-the-art* methods) to have the same issues to recover the high-frequency details. However, the overall results look very impressive considering the level of degradation of the input image (downsampled and compressed using JPEG at quality $q = 10$). We also provide additional results and samples for DIV2K [1] in the supplementary material.

Table 6. Ablation study of our experiments in the AIM 2022 Compressed Image Super-Resolution Challenge. The additional loss functions, and our new design Swin2SR help to converge faster and produce competitive results. Note that we compare with SwinIR pre-trained model while we trained using only the challenge DIV2K [1] data.

Exp.	Method	PSNR
1	Bicubic	22.350
2	RDN [69]	23.320
3	SwinIR [32]	23.546
4	Swin2SR (Ours)	23.580
5	Swin2SR + AuxLoss	23.585
6	Swin2SR + AuxLoss + HFLoss	23.590
7	Self-ensemble Exp6	23.616

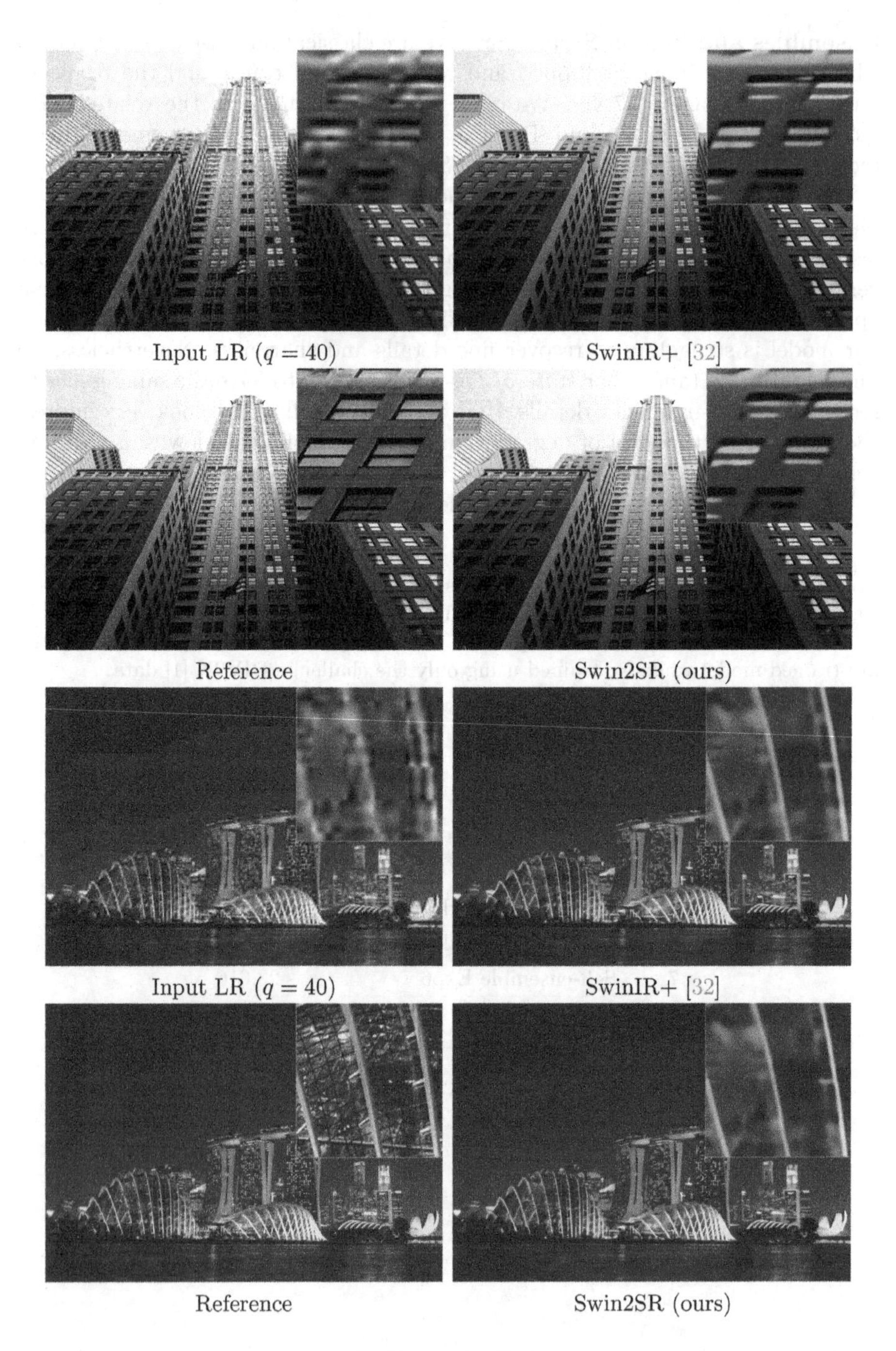

Fig. 5. Qualitative samples from the AIM 2022 Challenge on Super-Resolution of Compressed Image. Validation images from the DIV2K [1].

5 Conclusion

In this paper we propose Swin2SR, a SwinV2 Transformer-based model for super-resolution and restoration of compressed images. This model is a possible improvement of SwinIR (based on Swin Transformer), allowing faster training and convergence, and bigger capacity and resolution. Extensive experiments show that Swin2SR achieves state-of-the-art performance on: JPEG compression artifacts removal, image super-resolution (classical and lightweight), and compressed image super-resolution. Our method also achieves competitive results at the "AIM 2022 Challenge on Super-Resolution of Compressed Image and Video", being ranked among the top-5, and therefore, it helps to advance the state-of-the-art in super-resolution of compressed inputs, which will play an essential role in industries like streaming services, virtual reality or video games.

Acknowledgments. This work was partly supported by The Alexander von Humboldt Foundation (AvH).

References

1. Agustsson, E., Timofte, R.: NTIRE 2017 challenge on single image super-resolution: dataset and study. In: Proceedings of the IEEE conference on Computer Vision and Pattern Recognition Workshops, pp. 126–135 (2017)
2. Ahn, N., Kang, B., Sohn, K.-A.: Fast, accurate, and lightweight super-resolution with cascading residual network. In: Ferrari, V., Hebert, M., Sminchisescu, C., Weiss, Y. (eds.) ECCV 2018. LNCS, vol. 11214, pp. 256–272. Springer, Cham (2018). https://doi.org/10.1007/978-3-030-01249-6_16
3. Bevilacqua, M., Roumy, A., Guillemot, C., Morel, M.A.: Low-complexity single-image super-resolution based on nonnegative neighbor embedding. In: British Machine Vision Conference, pp. 135.1-135.10 (2012)
4. Cao, H., et al.: Swin-Unet: Unet-like pure transformer for medical image segmentation. arXiv preprint arXiv:2105.05537 (2021)
5. Cao, J., Li, Y., Zhang, K., Van Gool, L.: Video super-resolution transformer. arXiv preprint arXiv:2106.06847 (2021)
6. Carion, N., Massa, F., Synnaeve, G., Usunier, N., Kirillov, A., Zagoruyko, S.: End-to-end object detection with transformers. In: Vedaldi, A., Bischof, H., Brox, T., Frahm, J.-M. (eds.) ECCV 2020. LNCS, vol. 12346, pp. 213–229. Springer, Cham (2020). https://doi.org/10.1007/978-3-030-58452-8_13
7. Cavigelli, L., Hager, P., Benini, L.: CAS-CNN: a deep convolutional neural network for image compression artifact suppression. In: 2017 International Joint Conference on Neural Networks, pp. 752–759 (2017)
8. Chen, H., et al.: Pre-trained image processing transformer. In: IEEE Conference on Computer Vision and Pattern Recognition, pp. 12299–12310 (2021)
9. Chu, X., Zhang, B., Ma, H., Xu, R., Li, Q.: Fast, accurate and lightweight super-resolution with neural architecture search. In: International Conference on Pattern Recognition, pp. 59–64. IEEE (2020)
10. Conde, M.V., Burchi, M., Timofte, R.: Conformer and blind noisy students for improved image quality assessment. In: Proceedings of the IEEE/CVF Conference on Computer Vision and Pattern Recognition, pp. 940–950 (2022)

11. Conde, M.V., Choi, U.-J., Burchi, M., Timofte, R.: Swin2SR: Swinv2 transformer for compressed image super-resolution and restoration. In: Proceedings of the European Conference on Computer Vision (ECCV) Workshops (2022)
12. Conde, M.V., McDonagh, S., Maggioni, M., Leonardis, A., Pérez-Pellitero, E.: Model-based image signal processors via learnable dictionaries. Proc. AAAI Conf. Artif. Intell. **36**(1), 481–489 (2022)
13. Conde, M.V., Turgutlu, K.: Exploring vision transformers for fine-grained classification. arXiv preprint arXiv:2106.10587 (2021)
14. Dai, T., Cai, J., Zhang, Y., Xia, S.-T., Zhang, L.: Second-order attention network for single image super-resolution. In: IEEE Conference on Computer Vision and Pattern Recognition, pp. 11065–11074 (2019)
15. Dong, C., Deng, Y., Change Loy, C., Tang, X.: Compression artifacts reduction by a deep convolutional network. In: IEEE International Conference on Computer Vision, pp. 576–584 (2015)
16. Dong, C., Loy, C.C., He, K., Tang, X.: Learning a deep convolutional network for image super-resolution. In: Fleet, D., Pajdla, T., Schiele, B., Tuytelaars, T. (eds.) ECCV 2014. LNCS, vol. 8692, pp. 184–199. Springer, Cham (2014). https://doi.org/10.1007/978-3-319-10593-2_13
17. Dosovitskiy, A., et al.: An image is worth 16x16 words: transformers for image recognition at scale. arXiv preprint arXiv:2010.11929 (2020)
18. Ehrlich, M., Davis, L., Lim, S.-N., Shrivastava, A.: Quantization guided jpeg artifact correction. In: Vedaldi, A., Bischof, H., Brox, T., Frahm, J.-M. (eds.) ECCV 2020. LNCS, vol. 12353, pp. 293–309. Springer, Cham (2020). https://doi.org/10.1007/978-3-030-58598-3_18
19. Foi, A., Katkovnik, V., Egiazarian, K.: Pointwise shape-adaptive DCT for high-quality denoising and deblocking of grayscale and color images. IEEE Trans. Image Process. **16**(5), 1395–1411 (2007)
20. Fritsche, M., Gu, S., Timofte, R.: Frequency separation for real-world super-resolution. In: IEEE Conference on International Conference on Computer Vision Workshops, pp. 3599–3608 (2019)
21. Gu, J., et al.: NTIRE 2022 challenge on perceptual image quality assessment. In: Proceedings of the IEEE/CVF Conference on Computer Vision and Pattern Recognition, pp. 951–967 (2022)
22. Gu, J., Lu, H., Zuo, W., Dong, C.: Blind super-resolution with iterative kernel correction. In: IEEE Conference on Computer Vision and Pattern Recognition, pp. 1604–1613 (2019)
23. Haris, M., Shakhnarovich, G., Ukita. N.: Deep back-projection networks for super-resolution. In: IEEE Conference on Computer Vision and Pattern Recognition, PP 1664–1673 (2018)
24. Huang, J.-B., Singh, A., Ahuja, N.: Single image super-resolution from transformed self-exemplars. In: IEEE Conference on Computer Vision and Pattern Recognition, pp. 5197–5206 (2015)
25. Hui, Z., Gao, X., Yang, Y., Wang. X.: Lightweight image super-resolution with information multi-distillation network. In: ACM International Conference on Multimedia, pp. 2024–2032 (2019)
26. Ji, X., Cao, Y., Tai, Y., Wang, C., Li, J., Huang, F.: Real-world super-resolution via kernel estimation and noise injection. In: IEEE Conference on Computer Vision and Pattern Recognition Workshops, pp. 466–467 (2020)
27. Jiang, J., Zhang, K., Timofte, R.: Towards flexible blind jpeg artifacts removal. In: Proceedings of the IEEE/CVF International Conference on Computer Vision, pp. 4997–5006 (2021)

28. Kim, J., Lee, J.K., Lee, K.M.: Accurate image super-resolution using very deep convolutional networks. In: IEEE Conference on Computer Vision and Pattern Recognition, pp. 1646–1654 (2016)
29. Lai, W.-S., Huang, J.-B., Ahuja, N., Yang. M.-H.: Deep Laplacian pyramid networks for fast and accurate super-resolution. In: IEEE Conference on Computer Vision and Pattern Recognition, pp. 624–632 (2017)
30. Li, W., Zhou, K., Qi, L., Jiang, N., Lu, J., Jia. J.: LAPAR: linearly-assembled pixel-adaptive regression network for single image super-resolution and beyond. arXiv preprint arXiv:2105.10422 (2021)
31. Li, Z., Yang, J., Liu, Z., Yang, X., Jeon, G., Wu. W.: Feedback network for image super-resolution. In: IEEE Conference on Computer Vision and Pattern Recognition, pp. 3867–3876 (2019)
32. Liang, J., Cao, J., Sun, G., Zhang, K., Van Gool, L., Timofte, R.: Swinir: image restoration using Swin transformer. In: Proceedings of the IEEE/CVF International Conference on Computer Vision, pp.1833–1844 (2021)
33. Liu, D., Wen, B., Fan, Y., Change Loy, C., Huang, T.S.: Non-local recurrent network for image restoration. arXiv preprint arXiv:1806.02919 (2018)
34. Liu, P., Zhang, H., Zhang, K., Lin, L., Zuo. W.: Multi-level wavelet-CNN for image restoration. In: Proceedings of the IEEE Conference on Computer Vision and Pattern Recognition Workshops, pp. 773–782 (2018)
35. Liu, Z., et al.: Swin transformer v2: scaling up capacity and resolution. In: Proceedings of the IEEE/CVF Conference on Computer Vision and Pattern Recognition, pp. 12009–12019 (2022)
36. Liu, Z., et al.: Swin transformer: hierarchical vision transformer using shifted windows. In: Proceedings of the IEEE/CVF International Conference on Computer Vision, pp. 10012–10022 (2021)
37. Lugmayr, A., Danelljan, M., Timofte, R.: NTIRE 2020 challenge on real-world image super-resolution: methods and results. In: Proceedings of the IEEE/CVF Conference on Computer Vision and Pattern Recognition Workshops, pp. 494–495 (2020)
38. Luo, X., Xie, Y., Zhang, Y., Qu, Y., Li, C., Fu, Y.: LatticeNet: towards lightweight image super-resolution with lattice block. In: Vedaldi, A., Bischof, H., Brox, T., Frahm, J.-M. (eds.) ECCV 2020. LNCS, vol. 12367, pp. 272–289. Springer, Cham (2020). https://doi.org/10.1007/978-3-030-58542-6_17
39. Martin, D., Fowlkes, C., Tal, D., Malik, J.: A database of human segmented natural images and its application to evaluating segmentation algorithms and measuring ecological statistics. In: IEEE Conference on International Conference on Computer Vision, pp. 416–423 (2001)
40. Matsui, Y., et al.: Sketch-based manga retrieval using manga109 dataset. Multim. Tools Appl. **76**(20), 21811–21838 (2017)
41. Mei, Y., Fan, Y., Zhou, Y.: Image super-resolution with non-local sparse attention. In: IEEE Conference on Computer Vision and Pattern Recognition, pp. 3517–3526 (2021)
42. Niu, B., et al.: Single image super-resolution via a holistic attention network. In: Vedaldi, A., Bischof, H., Brox, T., Frahm, J.-M. (eds.) ECCV 2020. LNCS, vol. 12357, pp. 191–207. Springer, Cham (2020). https://doi.org/10.1007/978-3-030-58610-2_12
43. Rawzor. Image compression benchmark
44. Sheikh, H.R.: Live image quality assessment database release 2 (2005). http://live.ece.utexas.edu/research/quality

45. Tai, Y., Yang, J., Liu, X., Xu. C.: MemNet: a persistent memory network for image restoration. In: IEEE International Conference on Computer Vision, pp. 4539–4547 (2017)
46. Timofte, R., Agustsson, E., Van Gool, L., Yang, M.-H., Zhang, L.: NTIRE 2017 challenge on single image super-resolution: methods and results. In: IEEE Conference on Computer Vision and Pattern Recognition Workshops, pp. 114–125 (2017)
47. Timofte, R., De Smet, V., Van Gool. L.: Anchored neighborhood regression for fast example-based super-resolution. In: IEEE Conference on International Conference on Computer Vision, pp. 1920–1927 (2013)
48. Timofte, R., De Smet, V., Van Gool, L.: A+: adjusted anchored neighborhood regression for fast super-resolution. In: Asian Conference on Computer Vision, pp. 111–126 (2014)
49. Timofte, R., Rothe, R., Van Gool, L.: Seven ways to improve example-based single image super resolution. In: Proceedings of the IEEE Conference on Computer Vision and Pattern Recognition, pp. 1865–1873 (2016)
50. Touvron, H., Cord, M., Douze, M., Massa, F., Sablayrolles, A., Jégou, H.: Training data-efficient image transformers & distillation through attention. arXiv preprint arXiv:2012.12877 (2020)
51. Vaswani, A., Ramachandran, P., Srinivas, A., Parmar, N., Hechtman, B., Shlens, J.: Scaling local self-attention for parameter efficient visual backbones. arXiv preprint arXiv:2103.12731 (2021)
52. Vaswani, A., et al.: Attention is all you need. arXiv preprint arXiv:1706.03762 (2017)
53. Wang, X., Xie, L., Dong, C., Shan, Y.: Real-ESRGAN: training real-world blind super-resolution with pure synthetic data. arXiv preprint arXiv:2107.10833 (2021)
54. Wang, X., Yu, K., Wu, S., Gu, J., Liu, Y., Dong, C., Qiao, Yu., Loy, C.C.: ESRGAN: enhanced super-resolution generative adversarial networks. In: Leal-Taixé, L., Roth, S. (eds.) ECCV 2018. LNCS, vol. 11133, pp. 63–79. Springer, Cham (2019). https://doi.org/10.1007/978-3-030-11021-5_5
55. Wang, Z., Cun, X., Bao, J., Liu, J.: Uformer: a general u-shaped transformer for image restoration. arXiv preprint arXiv:2106.03106 (2021)
56. Yamac, M., Ataman, B., Nawaz, A.; KernelNet: a blind super-resolution kernel estimation network. In: Proceedings of the IEEE/CVF Conference on Computer Vision and Pattern Recognition Workshops (CVPRW), pp. 453–462 (2021)
57. Yang, R., Timofte, R., et al.: NTIRE 2021 challenge on quality enhancement of compressed video: methods and results. In: IEEE/CVF Conference on Computer Vision and Pattern Recognition (CVPR) Workshops (2021)
58. Yang, R., Timofte, R., et al.: Aim 2022 challenge on super-resolution of compressed image and video: dataset, methods and results. In: Proceedings of the European Conference on Computer Vision Workshops (ECCVW) (2022)
59. Yang, R., Timofte, R., et al.: NTIRE 2022 challenge on super-resolution and quality enhancement of compressed video: dataset, methods and results. In: Proceedings of the IEEE/CVF Conference on Computer Vision and Pattern Recognition (CVPR) Workshops (2022)
60. Zeyde, R., Elad, M., Protter, M.: On single image scale-up using sparse-representations. In: International Conference on Curves and Surfaces, pp. 711–730 (2010)
61. Zhang, K., Li, Y., Zuo, W., Zhang, L., Van Gool, L., Timofte, R.: Plug-and-play image restoration with deep denoiser prior. IEEE Trans. Pattern Anal. Mach. Intell. 99, 1 (2021)

62. Zhang, K., Liang, J., Van Gool, L., Timofte, R.: Designing a practical degradation model for deep blind image super-resolution. In: IEEE Conference on International Conference on Computer Vision (2021)
63. Zhang, K., Zuo, W., Chen, Y., Meng, D., Zhang, L.: Beyond a gaussian denoiser: residual learning of deep CNN for image denoising. IEEE Trans. Image Process. **26**(7), 3142–3155 (2017)
64. Zhang, K., Zuo, W., Gu, S., Zhang, L.: Learning deep CNN denoiser prior for image restoration. In: IEEE Conference on Computer Vision and Pattern Recognition, pp. 3929–3938 (2017)
65. Zhang, K., Zuo, W., Zhang, L.: FfdNet: toward a fast and flexible solution for CNN-based image denoising. IEEE Trans. Image Process. **27**(9), 4608–4622 (2018)
66. Zhang, K., Zuo, W., Zhang, L.: Learning a single convolutional super-resolution network for multiple degradations. In: IEEE Conference on Computer Vision and Pattern Recognition, pp. 3262–3271 (2018)
67. Zhang, Y., Li, K., Li, K., Wang, L., Zhong, B., Fu, Y.: Image super-resolution using very deep residual channel attention networks. In: Ferrari, V., Hebert, M., Sminchisescu, C., Weiss, Y. (eds.) ECCV 2018. LNCS, vol. 11211, pp. 294–310. Springer, Cham (2018). https://doi.org/10.1007/978-3-030-01234-2_18
68. Zhang, Y., Li, K., Li, K., Zhong, B., Fu, Y.: Residual non-local attention networks for image restoration. arXiv preprint arXiv:1903.10082 (2019)
69. Zhang, Y., Tian, Y., Kong, Y., Zhong, B., Fu. Y.: Residual dense network for image super-resolution. In Proceedings of the IEEE Conference On Computer Vision and Pattern Recognition, pp. 2472–2481 (2018)
70. Zhang, Y., Yapeng Tian, Yu., Kong, B.Z., Yun, F.: Residual dense network for image restoration. IEEE Trans. Pattern Anal. Mach. Intell. **43**(7), 2480–2495 (2020)
71. Zheng, B., Chen, Y., Tian, X., Zhou, F., Liu, X.: Implicit dual-domain convolutional network for robust color image compression artifact reduction. IEEE Trans. Circuits Syst. Video Technol. **30**(11), 3982–3994 (2019)
72. Zhou, S., Zhang, J., Zuo, W., Loy, C.C.: Cross-scale internal graph neural network for image super-resolution. arXiv preprint arXiv:2006.16673 (2020)

Reversing Image Signal Processors by Reverse Style Transferring

Furkan Kınlı(✉), Barış Özcan, and Furkan Kıraç

Özyeğin University, İstanbul, Turkey
{furkan.kinli,furkan.kirac}@ozyegin.edu.tr, baris.ozcan.10097@ozu.edu.tr

Abstract. RAW image datasets are more suitable than the standard RGB image datasets for the ill-posed inverse problems in low-level vision, but not common in the literature. There are also a few studies to focus on mapping sRGB images to RAW format. Mapping from sRGB to RAW format could be a relevant domain for reverse style transferring since the task is an ill-posed reversing problem. In this study, we seek an answer to the question: Can the ISP operations be modeled as the style factor in an end-to-end learning pipeline? To investigate this idea, we propose a novel architecture, namely *RST-ISP-Net*, for learning to reverse the ISP operations with the help of adaptive feature normalization. We formulate this problem as a reverse style transferring and mostly follow the practice used in the prior work. We have participated in the AIM Reversed ISP challenge with our proposed architecture. Results indicate that the idea of modeling disruptive or modifying factors as style is still valid, but further improvements are required to be competitive in such a challenge.

Keywords: Image signal processors · Reverse style transfer · sRGB-to-RAW Reconstruction

1 Introduction

Data-driven learning methods such as Convolutional Neural Networks (CNNs) achieved outstanding results in vision tasks such as object detection, image segmentation, or image classification. These methods require a huge amount of annotated images, which is increasingly available in recent years [4,14]. Most of these datasets consist of standard RGB (sRGB) images, which are in-camera Image Signal Processor (ISP) dependent. An sRGB image can be obtained by feeding the RAW data acquired by the camera sensors to the ISP pipeline that outputs an image tailored for human perception. However, RAW images are found to be better suited for ill-posed low-level vision tasks such as denoising, HDR, or super-resolution [2]. Unfortunately, there are very few RAW image datasets that are available, therefore deep learning-based methods were not fully utilized. The researchers often rely on synthetically-generated data for low-level vision tasks. However, deep learning-based denoising methods trained on such data outperform the hand-engineered methods in the recent datasets when tested

L. Karlinsky et al. (Eds.): ECCV 2022 Workshops, LNCS 13802, pp. 688–698, 2023.
https://doi.org/10.1007/978-3-031-25063-7_43

on real raw images [1]. Unrealistic synthetic data used for training the deep learning models are argued to be the limitation behind better results.

To improve the synthetic RAW image quality, recent studies aim to estimate the RAW image from the sRGB data [1,17,24] when attacking low-level vision tasks. Reversing the ISP operations is an ill-posed inverse problem. In this context, we have approached this task as a reverse style transferring problem and focused on observing the limitations of our approach as an emerging idea. Therefore, we have participated in the AIM Reversed ISP challenge [3] to be able to compare with the other methods on a fair basis. This challenge aims to obtain a network design or a solution, which is capable of producing high-quality results with high fidelity with respect to the reference ground truth. Our method is one of the top novel solutions for the proposed problem of RAW image reconstruction, which is a novel inverse problem in low-level computer vision.

The following sections of this paper can be summarized as follows. Section 2 introduces the previous studies on both sRGB-to-RAW reconstruction and reverse style transferring. Section 3 presents our proposed methodology and the idea of modeling disruptive factors as style. Section 4 explains the datasets used in the scope of the challenge, the experimental details, and the results obtained by our solution and the other studies. Section 5 concludes the paper.

2 Related Works

2.1 Mapping from sRGB to RAW

Despite the fact that the RAW image datasets are more suitable than the standard RGB image datasets for the ill-posed inverse problems in low-level vision such as denoising, demosaicking, and super-resolution and the lack of these kinds of datasets, there are a few studies to focus on mapping sRGB images to RAW format in the literature. [1] proposes a generic camera ISP model, which replaces five fundamental ISP operations by approximating each of them by an invertible and differentiable function. Recent studies [17,23,24] are more focused on learning-based approaches where the ISP operations are learned by neural networks in an end-to-end manner. [24] employs the cycle consistency idea to learn two-way translations (*i.e.*, RAW-to-RGB and RGB-to-RAW) for the input images. [23] proposes theISP architecture that mimics the camera ISP model by learning a single invertible neural network for two-way translations. Recently, [2] presents a hybrid approach, which addresses the issues that the previous approaches face. It is based on [1], yet builds a more flexible and interpretable ISP pipeline by combining model-based and learning-based approaches. In this study, rather than seeking the best approach for the ISP in the AIM Reversed ISP challenge, we mostly seek an answer to the question of whether the most simple version of modeling the style can model the ISP operations as the style factor in an end-to-end learning pipeline is possible or not. Moreover, there is another AIM challenge that focuses on learned smartphone ISP on mobile GPUs by using deep learning strategies, and the solutions for this topic can be found in the challenge report [8].

2.2 Reverse Style Transferring

Style Transfer [5–7] is a common term in deep learning literature, which is a many-to-many translation approach where a reference image can be translated into a target image without losing the main context, but by transferring its style information. From a different perspective, reverse style transfer is described in [12] as a many-to-one translation approach for eliminating undesired style information from the input. Any number of transformations applied to an image can be swept away by adaptively normalizing throughout the encoding part of the network, and the output with the pure style can be generated by the decoder in an end-to-end learning pipeline. Recently, [11] introduces the patch-wise contrastive learning-based approach for reverse style transferring in the filter removal task, and investigates the importance of distilling the semantic and style similarities among the signals (*i.e.*, patches). The main motivation of this study is to apply this strategy in a more primitive form to another reverse problem (*i.e.*, reversed ISP) in order to prove the idea of modeling disruptive or modifying factors as the style.

3 Proposed Methodology

We define the problem of reconstructing RAW images from sRGB input images as a reverse style transfer problem. Assuming the RAW format is the original version, any visual change related to the image signal processors (ISP) operations can be considered as an additional style factor injected into the original version. Following the idea in [12], we can model the effects brought by the ISP's operations as the style factor, and remove these injected changes by directly reverting them back to their original style (*i.e.*, RAW versions for this task). To achieve this, we propose a novel architecture, namely *RST-ISP-Net*, for learning to reverse the ISP operations with the help of adaptive feature normalization for transferring the style information. Figure 1 demonstrates the overall architecture of our proposed strategy for this task.

A given image $\tilde{\mathbf{X}} \in \mathbb{R}^{H \times W \times 3}$ including a style information of the ISP operations applied by arbitrary transformation functions $\mathbf{T}(\cdot)$ is converted back to its original version $\mathbf{X} \in \mathbb{R}^{H \times W \times 3}$. This refers to $\mathbf{X}$ having the pure style and does not contain any additional style information injected. We formulate a style removal module $\mathbf{F}(\cdot)$, which is responsible for reverting the function of $\mathbf{T}(\cdot)$.

$$\mathbf{X} = \mathbf{F}(\tilde{\mathbf{X}}) \tag{1}$$

where $\tilde{\mathbf{X}} = \mathbf{T}(\mathbf{X})$ and $\mathbf{T}(\cdot)$ is a general transformation function representing one or more transformations applied to $\mathbf{X}$ during the ISP. Note that finding $\mathbf{T}^{-1}(\cdot)$ is an ill-posed problem.

Our proposed architecture has an encoder-decoder structure, and employs adaptive feature normalization between all layers. Style extractor module f_{fc} is consist of five consecutive fully-connected layers to map the feature representations extracted by the Gram matrix to the latent space. We have N fully-connected head layers attached to f_{fc} where N is the number of layer levels used

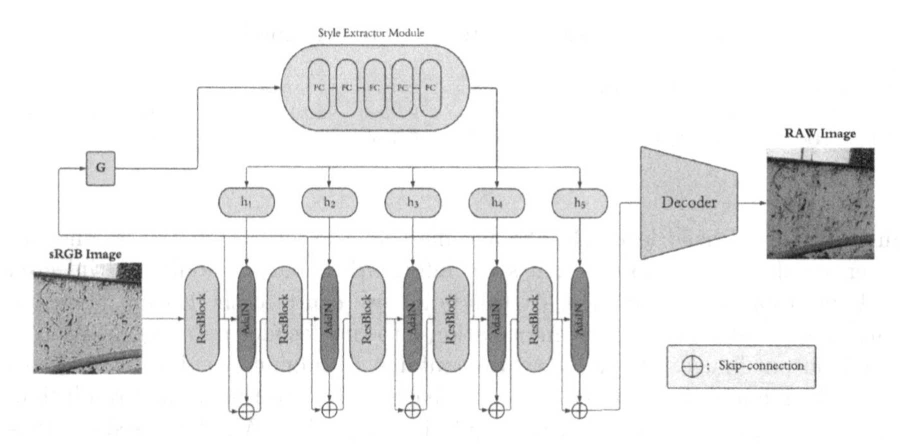

Fig. 1. Overall architecture of our proposed strategy to sRGB-to-RAW reconstruction task.

in the encoder. These heads adapt the affine parameters so that the adaptive feature normalization modules in each level take them as the input and normalize the feature maps accordingly. This part of the architecture is responsible for reversing the style, in other words, removing the injected style factor from the input feature maps. This part can be formulated as follows

$$y_i = h_i(f_{fc}(\mathbf{z}_{\tilde{\mathbf{X}}})) \tag{2}$$

where $\mathbf{z}_{\tilde{\mathbf{X}}}$ represents the feature representation of the input image $\tilde{\mathbf{X}}$, $h_i(\cdot)$ is i^{th} fully-connected head of style extractor module and y_i represents the predicted mean and variance vectors for the corresponding normalization layer. In this study, we follow the literature in reverse style transferring and use Adaptive Instance Normalization (AdaIN) [7] as the normalization layer. We can formulate AdaIN as follows

$$\text{AdaIN}(x, y) = \sigma(y)\left(\frac{x - \mu(x)}{\sigma(x)}\right) + \mu(y) \tag{3}$$

where μ stands for the mean and σ for the variance of the content image x and the style input y.

The encoder of RST-ISP-Net is composed of 5 residual blocks, and each of them has a specific AdaIN layer to normalize the feature maps in each level with its corresponding affine parameters extracted by the previous module. Following [12], we include skip-connections between the residual blocks to be able to preserve the style information related to the pure style. Apart from the previous studies, we employ the Gram matrix for expressing the style information of the feature maps via feature correlations, instead of using a particular layer outputs of pre-trained neural networks (*e.g.*, VGG16 [20]). The Gram matrix $\mathbf{G}^l \in \mathbb{R}^{K \times K}$ for each layer level l represents the inner product between the features mapped by the corresponding encoder layer $\mathbf{E}^l$, and sends its output to the

Table 1. Hyper-parameters used in our experiments.

Dataset	Input	Opt	LR	Epochs	Batch size	Ensemble	Framework	# Params. (M)	Runtime (ms)	GPU
S7	(504,504,3)	Adam	1e-4	101	8	No	PyTorch	86.3 Million	5.45 on GPU	RTX 2080Ti
P20	(496,496,3)	Adam	1e-4	101	8	No	PyTorch	86.3 Million	5.54 on GPU	RTX 2080Ti

fully-connected head layer $\mathbf{H}^l$. We assume that the output of the last encoder layer should include no external style information and all additional injected style information is discarded. At this point, we can feed the feature representations without external style information into the decoder part to generate the RAW output image. The decoder also contains 4 common-typed residual blocks with PixelShuffle [19] to interpolate the feature maps to the original resolution. Moreover, we apply discriminative regularization [13] by Wavelet-based discriminators [21] to our network to avoid the blurring effect on the outputs and to preserve the high-frequency details for RAW format.

The objective function for our proposed strategy is composed of three main components, which are MS-SSIM loss [22], TV loss [15], and adversarial loss for providing the discriminative regularization. We did not include the auxiliary classification loss used in [12] since the improvement rate in the performance is not significant with respect to the computational burden increased. The general formula of our adversarial training can be seen in Eq. 4.

$$\mathcal{L} = \lambda_{SSIM}\mathcal{L}_{SSIM} + \lambda_{TV}\mathcal{L}_{TV} + \lambda_{adv}\mathcal{L}_{adv} + \lambda_{gp}\mathcal{L}_{gp} \quad (4)$$

where $\mathcal{L}_{gp}$ represents the gradient penalty applied on the discriminator whose weight λ_{gp} is set to 10.

4 Experiments and Results

4.1 Datasets

In this study, we have used two datasets in our experiments, which are Samsung S7 DeepISP Dataset [18] and ETH Huawei P20 Dataset [9]. The first is a dataset of real-world images that contains different scenes captured by a Samsung S7 rear camera. To ensure to avoid camera movement, a special Android application is used during the collection of images. Although the whole dataset contains a total of 110 scenes were captured and split into 90, 10 and 10 for the training, validation, and test sets, respectively, and the images are in 12Mpx resolution, we have only used the set of training images given by the AIM Reversed ISP challenge. Next, we have used the samples from the Zurich RAW-to-RGB dataset (*i.e.*, called ETH Huawei P20 Dataset in the original challenge repository), which is a large-scale dataset consisting of 20K photos collected by Huawei P20 smartphone with their RAW versions. The data was collected over several weeks in a variety of places and in various illumination and weather conditions. Similar to the first one, we have only used the set of certain images given by the AIM Reversed ISP challenge track.

4.2 Experimental Details

We have used the given input images without applying any pre-processing, thus the input size is $504 \times 504 \times 3$ for S7 dataset and $496 \times 496 \times 3$ for P20 dataset. We picked the Adam optimizer [10] for our experiments with the learning rate of $1e-4$ for the generator and $4e-4$ for the discriminator. We have trained our model until proper convergence on the generator loss (*i.e.*, 101 epochs for S7 dataset, 52 epochs for P20 dataset). We did not use any extra data in addition to the given training data, and also did not employ any ensembling strategy for our solution. The main reason is to be able to observe the baseline performance of the aforementioned approach, and the main aim is to seek some clues for improving this idea. The architecture used in our solution has 86.3M parameters, mostly due to the style projectors for each residual block. We have trained our method from scratch for all components. We did not use any additional data in addition to the provided training data. Table 1 summarizes the hyper-parameters used in our experiments. During testing, we did not apply any preprocessing to the given test images. Following the evaluation methods given in AIM Reversed ISP challenge [3], we have measured the performance of our proposed architecture with two metrics (*i.e.*, PSNR and SSIM). To visualize the RAW outputs, we have used the script[1] given by the AIM Reversed ISP challenge organizers.

We have used Python language, and PyTorch DL framework [16] for DL modeling and training/testing. Our experiments have been done on 2× NVIDIA RTX 2080Ti GPUs. The batch size is set to 8. Downloading and preparing the dataset for training and validation took approximately 1 day, and completing the implementation by using the code of the baseline study took approximately 2–3 days. A single experiment for training has been completed in approximately 2 days. The testing process has been completed in only 1 min for all instances of the dataset. Run-time at test per image has been measured as 5 ms. for both datasets. The source code will be given for the camera-ready submission.

4.3 Quantitative Results

Following the evaluation methods given in AIM Reversed ISP challenge [3], we have employed two common image similarity metrics in our experiments, which are namely Structural Similarity Index (SSIM) and Peak Signal-to-Noise Ratio (PSNR). Table 2 summarizes the performances of our proposed architecture and the other compared methods participating in the challenge. Although our results are not competitive in the rankings, we believe that investigating the idea of modeling the disruptive or modifying factors as the style factor [1] is an important and open-to-improvement approach. Thanks to the AIM Reversed ISP challenge [3], we have had a chance to try our idea for this problem, which actually works seamlessly for the other domains, and to see what may be missing in our approach or which part of this idea could be improved. Note that we did not use any ensembling technique to boost the performance or any extra data in our experiments.

[1] https://github.com/mv-lab/AISP.

Table 2. AIM Reversed ISP Challenge Benchmark [3]. Teams are ranked based on their performance on Test2, an internal test set to evaluate the generalization capabilities and robustness of the proposed solutions. Test1 is a public test set provided to the participants as performance guidance. We report the standard metrics PSNR and SSIM. ED indicates the use of extra datasets besides the provided challenge datasets, ENS stands for if the solution is an ensemble of multiple models.

			Track 1 (Samsung S6)				Track 2 (Huawei P20)			
			Test1		Test2		Test1		Test2	
Team name	ED	ENS	PSNR ↑	SSIM ↑	PSNR ↑	SSIM ↑	PSNR ↑	SSIM ↑	PSNR ↑	SSIM ↑
NOAHTCV	✗	✗	31.86	0.83	32.69	0.88	38.38	0.93	35.77	0.92
MiAlgo	✗	✗	31.39	0.82	30.73	0.80	40.06	0.93	35.41	0.91
CASIA LCVG	✓	✓	30.19	0.81	31.47	0.86	37.58	0.93	33.99	0.92
HIT-IIL	✗	✗	29.12	0.80	29.98	0.87	36.53	0.91	34.07	0.90
CS2U	✓	✓	29.13	0.79	29.95	0.84	-	-	-	-
SenseBrains	✗	✓	28.36	0.80	30.08	0.86	35.47	0.92	32.63	0.91
PixelJump	✗	✓	28.15	0.80	n/a	n/a	-	-	-	-
HiImage	✗	✗	27.96	0.79	n/a	n/a	34.40	0.94	32.13	0.90
0noise	✗	✗	27.67	0.79	29.81	0.87	33.68	0.90	31.83	0.89
OzU VGL (**Ours**)	✗	✗	27.89	0.79	28.83	0.83	32.72	0.87	30.69	0.86
CVIP	✗	✗	27.85	0.80	29.50	0.86	-	-	-	-

4.4 Qualitative Comparison

Figure 2 presents the qualitative results of the participants on the Samsung S7 dataset. Our proposed solution does not have superior performance on the task of reversing the ISP operations. However, our main goal in this study is to investigate how performing the idea of modeling the ISP operations as the style factor without using any ensembling strategy or extra data. In the previous studies [11,12], the idea of modeling any disruptive or modifying factor as the style factor works well without requiring handling the details. According to our observation, removing the effects of the ISP operations by adaptive feature normalization leads to losing the high-frequency details in the output. This significantly reduces the overall performance of our strategy for this task. At this point, we have tried to employ Wavelet-based discriminators for our adversarial training, yet it still gives out the performance. Moreover, following the previous studies on modeling as the style factor, we apply an adversarial training strategy for our experiments to achieve discriminative regularization on the output. As we notice in these results, we have to reconsider to use of this strategy for such a task since, to the best of our knowledge, the leading participants do not use this kind of strategy in their studies, and also it makes the training process more complicated for convergence. Nevertheless, we believe that the style of being sRGB is successfully reverted back to the style of being RAW in our final outputs, and this shows that the idea is still valid, but needs to be improved while building the architecture.

As shown in Fig. 3, similar problems on ETH Huawei P20 dataset come to the forefront. The finding that can be significant for our future studies is that

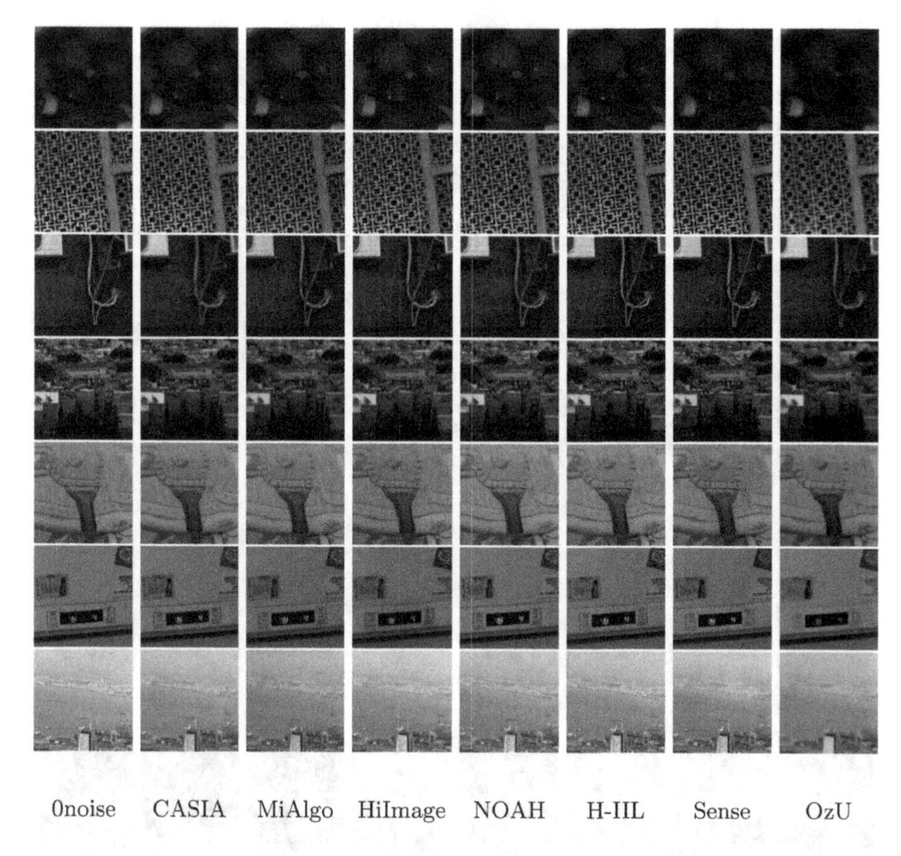

Fig. 2. Qualitative results on Samsung S7 dataset, submitted by the participants.

the output is more likely to be the strongly-blurred version of the ground-truth RAW image. Although the style injected into the RAW image (*i.e.*, the effect of multiple non-linear operations applied to the RAW image) is more or less successfully modeled and removed to translate the sRGB image to its RAW version, the blurring effect mostly leads to losing the high-frequency details in the output, so the decrease in the performance for challenge benchmark. This effect is far stronger than the one that we face for the Track S7. We think that the alignment issue among the pairs for this dataset and not using any specialized module that resolves this issue may lead to amplifying the problem in our strategy. Another reason for the amplification of this issue could be to use an adversarial training strategy since the real and fake images suffer from being the exactly same for the discriminator and it does not help the training at all, even making it worse.

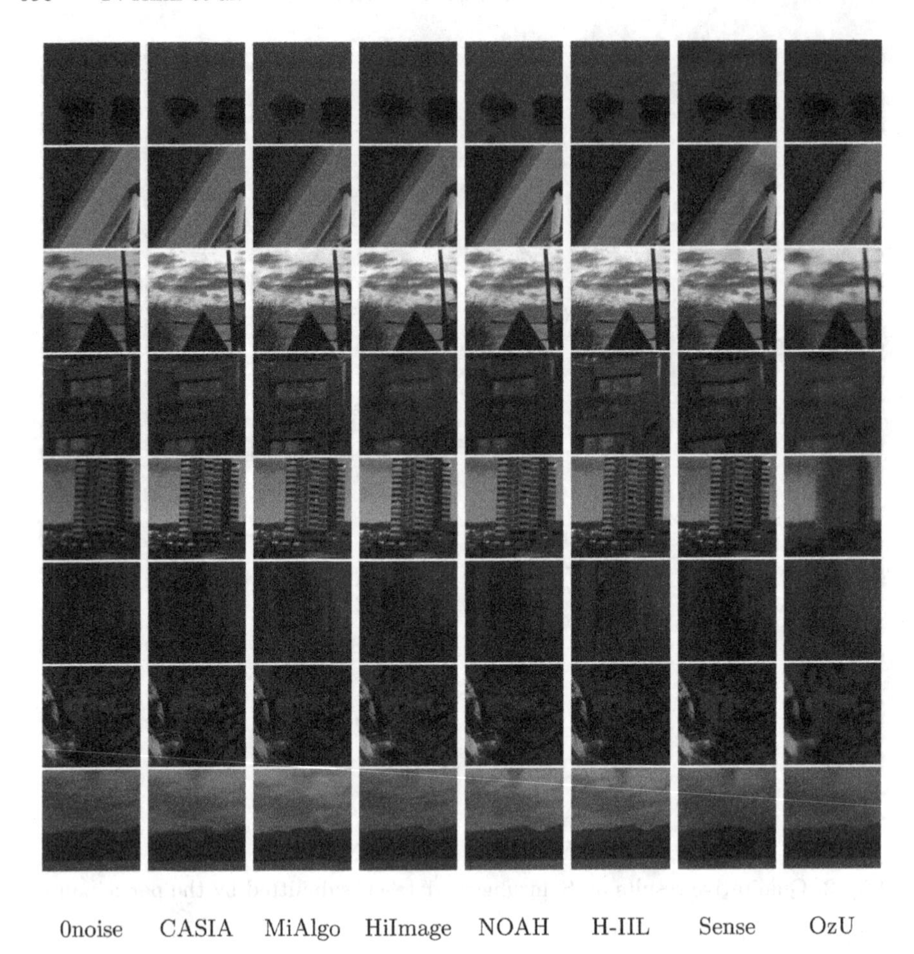

Fig. 3. Qualitative results on ETH Huawei P20 dataset, submitted by the participants.

5 Conclusion

In this study, we investigate the performance of the approach of modeling the ISP operations as the style factor. We have followed the previous strategies to model the disruptive or modifying factors (*i.e.*, multiple non-linear transformations applied during the ISP). Experiments on Samsung S7 and ETH Huawei P20 datasets show that this approach requires to be improved to be more competitive among the recent studies presented in the AIM Reversed ISP challenge [3]. In spite of the lack of performance in this task, compared to the recent methods in this challenge, the results indicate that the idea of modeling the disruptive factors as the style factor is still promising. Also, this challenge gives us a chance to see the weaknesses of our approach and encourages us to re-think this idea. The main limitation of our approach is the higher number of parameters composing the architecture to learn the style factor, and this can be reduced by using different feature distribution representations for the style factor. We will re-consider the usage of discriminative regularization for such tasks. For future work, we will re-design the whole architecture, the way of expressing the style information, and the objective function used and continue to investigate the style factor in reconstruction tasks.

References

1. Brooks, T., Mildenhall, B., Xue, T., Chen, J., Sharlet, D., Barron, J.T.: Unprocessing images for learned raw denoising. In: Proceedings of the IEEE/CVF Conference on Computer Vision and Pattern Recognition, pp. 11036–11045 (2019)
2. Conde, M.V., McDonagh, S., Maggioni, M., Leonardis, A., Pérez-Pellitero, E.: Model-based image signal processors via learnable dictionaries. In: Proceedings of the AAAI Conference on Artificial Intelligence, vol. 36, pp. 481–489 (2022)
3. Conde, M.V., Timofte, R., et al.: Reversed image signal processing and raw reconstruction. aim 2022 challenge report. In: Proceedings of the European Conference on Computer Vision Workshops (ECCVW) (2022)
4. Deng, J., Dong, W., Socher, R., Li, L.J., Li, K., Fei-Fei, L.: ImageNet: a large-scale hierarchical image database. In: 2009 IEEE Conference on Computer Vision and Pattern Recognition, pp. 248–255. IEEE (2009)
5. Gatys, L.A., Ecker, A.S., Bethge, M.: A neural algorithm of artistic style. arXiv (Aug 2015), http://arxiv.org/abs/1508.06576
6. Ghiasi, G., Lee, H., Kudlur, M., Dumoulin, V., Shlens, J.: Exploring the structure of a real-time, arbitrary neural artistic stylization network (2017). https://arxiv.org/abs/1705.06830
7. Huang, X., Belongie, S.: Arbitrary style transfer in real-time with adaptive instance normalization. In: ICCV (2017)
8. Ignatov, A., Timofte, R., et al.: Learned smartphone ISP on mobile GPUs with deep learning, mobile AI & AIM 2022 challenge: Report. In: Proceedings of the European Conference on Computer Vision (ECCV) Workshops, Springer, Cham (2022). https://doi.org/10.1007/978-3-030-66415-2
9. Ignatov, A., Van Gool, L., Timofte, R.: Replacing mobile camera isp with a single deep learning model. In: Proceedings of the IEEE/CVF Conference on Computer Vision and Pattern Recognition Workshops, pp. 536–537 (2020)
10. Kingma, D.P., Ba, J.: Adam: a method for stochastic optimization. In: Bengio, Y., LeCun, Y. (eds.) 3rd International Conference on Learning Representations, ICLR 2015, San Diego, CA, USA, 7–9 May 2015, Conference Track Proceedings (2015). http://arxiv.org/abs/1412.6980
11. Kınlı, F., Özcan, B., Kıraç, F.: Patch-wise contrastive style learning for instagram filter removal. In: Proceedings of the IEEE/CVF Conference on Computer Vision and Pattern Recognition (CVPR) Workshops, pp. 578–588 (June 2022)
12. Kinli, F., Ozcan, B., Kirac, F.: Instagram filter removal on fashionable images. In: Proceedings of the IEEE/CVF Conference on Computer Vision and Pattern Recognition (CVPR) Workshops, pp. 736–745 (June 2021)
13. Lamb, A., Dumoulin, V., Courville, A.: discriminative regularization for generative models. arXiv preprint arXiv:1602.03220 (2016)
14. Lin, T., et al.: Microsoft COCO: common objects in context. In: Fleet, D., Pajdla, T., Schiele, B., Tuytelaars, T. (eds.) ECCV 2014. LNCS, vol. 8693, pp. 740–755. Springer, Cham (2014). https://doi.org/10.1007/978-3-319-10602-1_48

15. Liu, J., Sun, Y., Xu, X., Kamilov, U.S.: Image restoration using total variation regularized deep image prior. In: ICASSP 2019–2019 IEEE International Conference on Acoustics, Speech and Signal Processing (ICASSP), pp. 7715–7719. IEEE (2019)
16. Paszke, A., et al.: Pytorch: an imperative style, high-performance deep learning library. In: Wallach, H., Larochelle, H., Beygelzimer, A., d'Alché-Buc, F., Fox, E., Garnett, R. (eds.) Advances in Neural Information Processing Systems 32, pp. 8024–8035. Curran Associates, Inc. (2019). http://papers.neurips.cc/paper/9015-pytorch-an-imperative-style-high-performance-deep-learning-library.pdf
17. Punnappurath, A., Brown, M.S.: Learning raw image reconstruction-aware deep image compressors. IEEE Trans. Pattern Anal. Mach. Intell. **42**(4), 1013–1019 (2019)
18. Schwartz, E., Giryes, R., Bronstein, A.M.: Deepisp: toward learning an end-to-end image processing pipeline. IEEE Trans. Image Process. **28**(2), 912–923 (2018)
19. Shi, W., et al.: Real-time single image and video super-resolution using an efficient sub-pixel convolutional neural network. In: Proceedings of the IEEE Conference on Computer Vision and Pattern Recognition (CVPR) (June 2016)
20. Simonyan, K., Zisserman, A.: Very deep convolutional networks for large-scale image recognition. In: International Conference on Learning Representations (2015)
21. Wang, J., Deng, X., Xu, M., Chen, C., Song, Y.: Multi-level wavelet-based generative adversarial network for perceptual quality enhancement of compressed video. In: Vedaldi, A., Bischof, H., Brox, T., Frahm, J.-M. (eds.) ECCV 2020. LNCS, vol. 12359, pp. 405–421. Springer, Cham (2020). https://doi.org/10.1007/978-3-030-58568-6_24
22. Wang, Z., Simoncelli, E., Bovik, A.: Multiscale structural similarity for image quality assessment. In: The Thrity-Seventh Asilomar Conference on Signals, Systems & Computers, 2003. vol. 2, pp. 1398–1402 (2003). https://doi.org/10.1109/ACSSC.2003.1292216
23. Xing, Y., Qian, Z., Chen, Q.: Invertible image signal processing. In: Proceedings of the IEEE/CVF Conference on Computer Vision and Pattern Recognition, pp. 6287–6296 (2021)
24. Zamir, S.W., Arora, A., Khan, S., Hayat, M., Khan, F.S., Yang, M.H., Shao, L.: Cycleisp: real image restoration via improved data synthesis. In: Proceedings of the IEEE/CVF Conference on Computer Vision and Pattern, pp. 2696–2705 (2020)

Overexposure Mask Fusion: Generalizable Reverse ISP Multi-step Refinement

Jinha Kim[1,2], Jun Jiang[2(✉)], and Jinwei Gu[2]

[1] MIT, Cambridge, MA, USA
jinhakim@mit.edu
[2] SenseBrain Technology, San Jose, CA, USA
{jinhakim,jiangjun,gujinwei}@sensebrain.site

Abstract. With the advent of deep learning methods replacing the ISP in transforming sensor RAW readings into RGB images, numerous methodologies solidified into real-life applications. Equally potent is the task of inverting this process which will have applications in enhancing computational photography tasks that are conducted in the RAW domain, addressing lack of available RAW data while reaping from the benefits of performing tasks directly on sensor readings. This paper's proposed methodology is a state-of-the-art solution to the task of RAW reconstruction, and the multi-step refinement process integrating an overexposure mask is novel in three ways: instead of from RGB to bayer, the pipeline trains from RGB to demosaiced RAW allowing use of perceptual loss functions; the multi-step processes has greatly enhanced the performance of the baseline U-Net from start to end; the pipeline is a generalizable process of refinement that can enhance other high performance methodologies that support end-to-end learning.

Keywords: ISP · Reversed ISP · Demosaiced raw · Multi-step refinement · Overexposure mask

1 Introduction

Image signal processor (ISP) denotes a collection of operations integrated in today's digital cameras that maps camera sensor readings into visually pleasing RGB images. A popular area of research that has been explored in relation to the ISP is the task of mapping from RAW data to RGB images with the use of deep learning-based methodologies. With various applications such as in mobile cameras which have small sensors and other limitations in hardware, various methodologies [7,11,12] have been developed to address this task.

A problem that also relates to the ISP, which has equally potent applications as the task of mapping RAW data to RGB images, is the reversed task of mapping from RGB images to RAW data, which is a novel problem in low-level computer vision. Unlike RGB images, RAW data holds a linear relationship

L. Karlinsky et al. (Eds.): ECCV 2022 Workshops, LNCS 13802, pp. 699–713, 2023.
https://doi.org/10.1007/978-3-031-25063-7_44

with scene irradiance, which has led to improved performance in various computer vision tasks. Numerous works have addressed the task of RAW reconstruction with various methodologies with solutions ranging from utilizing canonical steps approximated by invertible functions [2], mapping RAW data to CIE-XYZ space from sRGB images [1], a novel modular and differentiable ISP model with interpretable parameters that is capable of end-to-end learning [3] among many approaches [1–3,9,12,13]. With these inherent advantages that RAW data holds, the task of reconstructing RAW data from RGB images has become exceedingly relevant, especially with the lack of availability of RAW data due to factors such as memory-related concerns or data storage processes that discard the RAW.

However, the task of RAW data reconstruction remains a novel area of research with complexities and limitations that are yet to be fully addressed. For instance, as noted by Conde et al. [3], approximations using inverse functions for real-world ISPs show degradation in performance when a large portion of the RGB images are close to overexposure. Our proposed methodology using overexposure mask fusion is a novel portion of our pipeline that specifically addresses this issue by mapping overexposed and non-overexposed pixels separately and fusing them together using an overexposure mask.

Among various AIM challenges with different research problems [6], for the AIM Reversed ISP Challenge [4] where competing teams were given the task of reconstructing RAW data from RGB images, our methodology is a top solution, and therefore, evaluated as a state-of-the-art solution to the novel inverse problem. By mapping from RGB to demosaiced RAW by generating a demosaiced RAW from the groundtruth bayer using Demosaic Net [5], we allow the use of perceptual losses. With our novel overexposure mask fusion methodology, our pipeline addresses the issue of overexposed pixels as mentioned by Conde et al. [3]. It is most notable that the pipeline led to significant enhancement in fidelity measures while keeping all neural networks within our pipeline as the U-Net [10]. It is further notable that our methodology can incorporate other proposed state-of-the-art solutions involving end-to-end learning after slight modifications to map from RGB images to demosaiced RAW images. For instance, the model proposed by Conde et al. [3] can be integrated with our refinement pipeline by making small modifications such as removing the final mosaic step and generating demosaiced RAW groundtruth images for training in order to use perceptual loss. We propose, to the best of our knowledge, the first generalizable, multistep refinement process for enhanced performance of other reversed ISPs while addressing the issue of overexposure.

2 Related Works

Works such as [7,11,12] have addressed the task of mapping from RAW data to RGB images, modeling the camera ISP. Schwartz et al. [11] proposes a full end-to-end deep learning model of the ISP, which has demonstrated to be capable of generating visually compelling RGB images from RAW data. Ignatov et al. [7] proposes another end-to-end deep learning solution with the use of a novel

PyNET CNN architecture and Xing et al. [12] designed an invertible ISP that is capable of generating visually pleasing RGB images from RAW data as well as RAW reconstruction. Another work is the CycleISP [13] which models the ISP both in the forward and reverse directions.

There have also been various works addressing the task of RAW reconstruction from RGB images [1–3,9,12,13]. Brooks et al. [2] proposes an unprocessing technique for RAW reconstruction by inverting the ISP pipeline with five canonical steps that are approximated by invertible functions while CIE-XYZ Net [1] recovers the RAW data to the CIE-XYZ space from sRGB images. Conde et al. [3] proposed a novel modular and differentiable ISP model with interpretable parameters and canonical camera operations that is capable of end-to-end learning of parameter representations. Punnappurath et al. [9] proposed modifications to loss used for training neural network-based compression architectures to account for both sRGB image fidelity and RAW reconstructions errors while modeling sRGB-RAW mapping with the use of locally differentiable 3D lookup tables. Previously mentioned for the task of mapping from RAW to RGB, CycleISP [13] and the invertible ISP model proposed by Xing et al. [12] are also capable of RAW reconstruction. Several works offering solutions to the task of RAW reconstruction after integration of their approaches of RAW reconstruction have noted improvements in performance for RAW image denoising [2,3,13] which suggests further applications of RAW reconstruction.

In order to evaluate performance of different solutions on the task of RAW reconstruction for the AIM Reversed ISP Challenge [4], two datasets were used for training which are the Samsung S7 dataset [11] and ETH Huawei P20 Pro dataset [7]. The Samsung S7 dataset [11] consists of 110 scenes of 3024×4032 resolution as JPEG images captured with a Samsung S7 rear camera where original RAW images were saved as well. The ETH Huawei P20 Pro dataset [7] is a large-scale dataset consisting of 20 thousand photos collected using a Huawei P20 smartphone for capturing RAW images and the RGB images obtained with Huawei's built-in ISP (12.3 MP Sony Exmor IMX380). For both tracks, participants were evaluated on fidelity measures, PSNR and SSIM, and were also tested for generizability and robustness of proposed methods.

3 Methodology

3.1 Network Architecture

The schematic representation of the overall pipeline is outlined in Fig. 1. The general structure of pipeline consists of unprocessing the input RGB image to its original demosaiced RAW, after which a simple mosaic is performed to recover to bayer. For training, the pipeline involves generating new groundtruth RGB images by passing the groundtruth bayer through a pretrained Demosaic Net [5] in order to reconstruct the demosaiced RAW. Notably, unlike methodologies that map directly from RGB to bayer, the proposed pipeline maps initially from RGB to demosaiced RAW, which enables the use of perceptual loss functions [14].

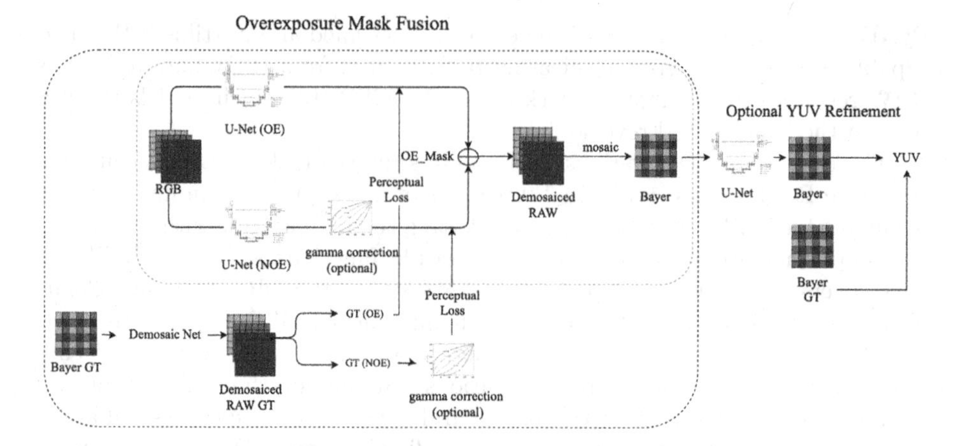

Fig. 1. The architecture of the proposed pipeline. Neural networks were all unified to be U-Nets for objective comparisons. U-Net (overexposure or OE) and U-Net (non-overexposure or NOE) both take in input RGB images and outputs demosaiced RAW. However, non-overexposed pixels are set to 0 for U-Net (OE), GT(OE) and overexposed pixels are set to 0 for U-Net (NOE), GT(NOE). It is notable that the U-Nets can easily be replaced with another methodology that supports end-to-end learning for mapping from RGB to demosaiced RAW.

A binary overexposure mask is constructed by computing the illuminance for each pixel of the input RGB image. Based on a certain threshold, pixels are then marked as overexposed or non-overexposed. As displayed in Fig. 1, there are two separate U-Nets [10], U-Net (overexposure or OE) and U-Net (non-overexposure or NOE), which takes in the same input RGB image, but are trained to account respectively for overexposed and non-overexposed pixels. The U-Net (OE) and U-Net (NOE) both have 23 convolutional layers with the slight modification of the additional channel to include the overexposure mask. It is noteworthy that the two U-Nets can be replaced by any other high performance methodology that can support end-to-end learning for mapping RGB to demosaiced RAW as displayed in Fig. 2. For purposes of ablation, the U-Net [10] was the only neural network model used for inference throughout the pipeline.

The two neural networks each output a demosaiced RAW image after which loss for U-Net (overexposure or OE) and U-Net (non-overexposure or NOE) are computed separately with the use of the overexposure mask. By generating two new groundtruth images, GT (OE) and GT (NOE), each groundtruth image being created by setting the pixels that are not respectively accounted for to 0, for instance, non-overexposed pixels for GT (OE) to 0. This operation is similarly performed on the outputs of the U-Nets such that U-Net (OE) and GT (OE) have value 0 for non-overexposed pixels, and conversely for U-Net (NOE) and GT (NOE). Additionally, for U-Net (NOE), gamma correction can be applied on non-overexposed pixels before computing loss to account for low-light demosaiced RAW images.

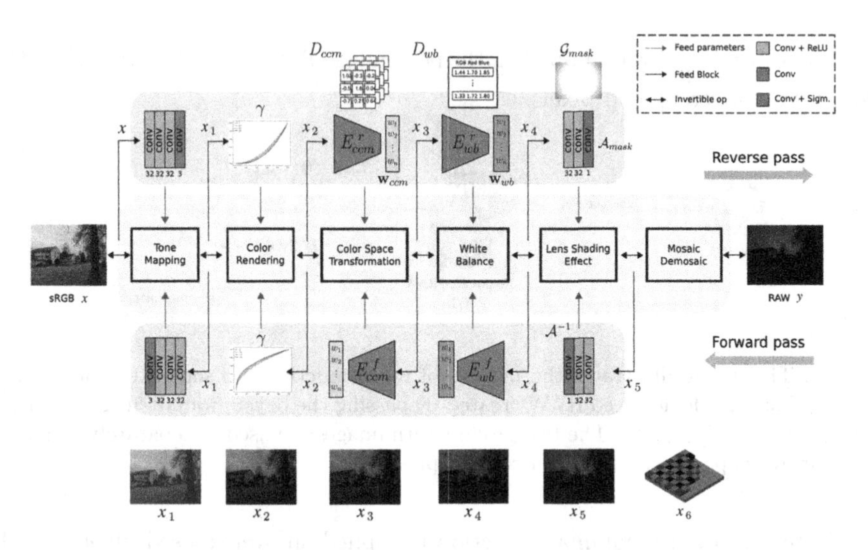

Fig. 2. The figure displays a state-of-the-art model proposed by Conde et al. [3] that maps from sRGB to RAW. Different methodologies supporting end-to-end learning can be integrated as replacements for our U-Nets [10] after slight modifications to map from sRGB to demosaiced RAW.

Although the training process requires the use of Demosaic Net [5] to generate new groundtruth images, inference, as marked by the blue container in Fig. 1, involves passing RGB input into U-Net (OE) and U-Net (NOE), after which the two images are simply blended together with the use of the overexposure mask. Inference is then completed by taking the demosaiced RAW output and mosaicing to convert to bayer.[1] The pipeline includes an optional YUV refinement step that computes loss between bayer in YUV space. It is most notable that the pipeline's multi-step refinement process can be applied to other reversed ISPs that support end-to-end learning with slight modifications to map from RGB to demosaiced RAW. Furthermore, as conducted for the ETH Huawei P20 Pro dataset [7], the pipeline can be modified to use a single U-Net with an additional channel to include the overexposure mask, reducing the total number of model parameters while maintaining high performance in fidelity measures.

3.2 Reconstructing Demosaiced RAW

The bayer groundtruth is passed into a pretrained demosaic Net and separated into two different groundtruths, for U-Net (OE) and U-Net (NOE) as displayed in Fig. 3. Gamma correction can be applied for non-overexposed pixels before computing loss with a parameter γ. For the ETH Huawei P20 Pro dataset [7],

[1] Starting code provided by the AIM Reversed ISP Challenge Organizers, which is available at https://github.com/mv-lab/AISP, was used within our training and inference code.

the value of $\gamma = \frac{1}{3.6}$ was applied on the normalized RGB values before computing loss for non-overexposed pixels.

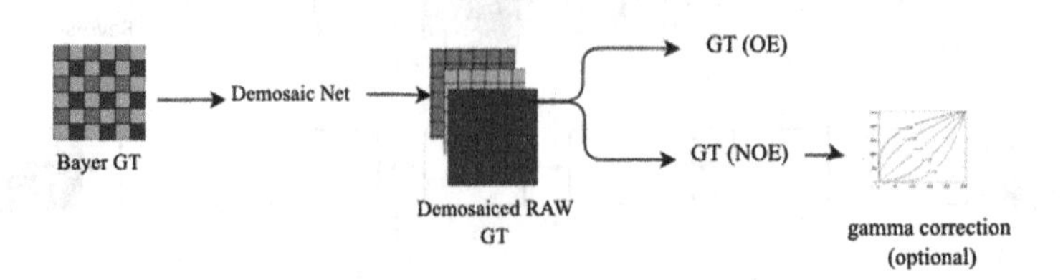

Fig. 3. This figure illustrates the process of reconstructing two separate groundtruth images from the demosaiced RAW created by passing the bayer groundtruth into a pre-trained Demosaic Net [5]. The two groundtruth images are used to separately compute loss for overexposed and non-overexposed pixels.

Note that even if gamma correction is applied on overexposed pixels, which have high illuminance, there will be minimal changes by nature of gamma correction. For the model used to train and run inference on ETH Huawei P20 Pro dataset [7], gamma correction was applied on all pixels as a whole before computing loss, and for the complete pipeline that utilizes two separate neural networks, gamma correction can be applied on non-overexposed pixels for $\gamma \in [0, 1]$

$$L_{nonOE} = L(I^{\gamma}_{recon}, I^{\gamma}_{GT}). \tag{1}$$

By applying gamma correction before computing loss, there was enhancement in performance in perceptual quality of demosaiced RAW, as demosaiced RAW images tend to be low-light, which can be accounted for by increasing the weighting of smaller pixel values within the loss function.

3.3 YUV Overexposure Mask

Given the input RGB, the YUV overexposure mask is computed by converting RGB pixel values to YUV. Since the overexposure mask utilizes only illuminance, only Y values are stored from the matrix multiplication below

$$\begin{bmatrix} Y \\ U \\ V \end{bmatrix} = \begin{bmatrix} 0.299 & 0.587 & 0.114 \\ -0.14713 & -0.28886 & 0.436 \\ 0.615 & -0.51499 & -0.10001 \end{bmatrix} \begin{bmatrix} R \\ G \\ B \end{bmatrix}. \tag{2}$$

Pixels with Y $\geq$ 0.978 were marked as 1 and 0 otherwise. For input RGB patches with size 504×504, the overexposure mask is binary and has dimensions 504×504 as well. Note that alternatively the overexposure mask can instead with the threshold of $\max(R, G, B) \geq 0.99$, however, as the performance in fidelity dropped for images with specifically white saturated pixels, the pipeline constructs the mask using illuminance.

3.4 Overexposure Mask Fusion for Inference

It is noteworthy that the pipeline does not require use of constructing new groundtruth images for inference, which only requires fusion of the outputs of U-Net (OE), U-Net (NOE). As shown in Fig. 4, the two reconstructed images are blended together using the overexposure mask.

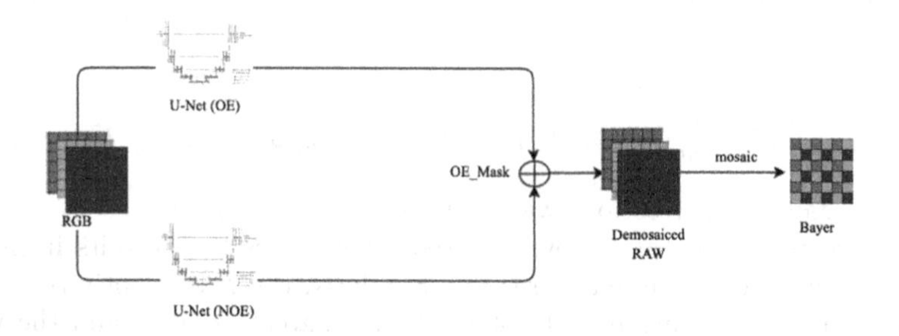

Fig. 4. The figure displays the fusion of the two output images using the overexposure mask. U-Net (OE) and U-Net (NOE) reconstruct images of same dimensions, and fusion is done by storing the RGB values for overexposed pixels from U-Net (OE) and non-overexposed pixels from U-Net (NOE). This process reconstructs the demosaiced RAW which is then mosaiced to bayer.

For U-Net (OE), all non-overexposed RGB pixel values of the output image are set to 0 while conversely for the output image of U-Net (NOE). Then, by simple addition of the two images, the two images are fused together, resulting in better accuracy in both overexposed and non-overexposed pixels. As the neural networks were trained to map RGB to its original demosaiced RAW from the groundtruth bayer, the pipeline reconstructs bayer by simply mosaicing of the demosaiced RAW output.

3.5 Perceptual Loss Functions and Training Details

There is greater availablity of literature on perceptual loss functions in comparison to loss functions between RAW bayer. For training and inference on the Samsung S7 dataset [11], different perceptual loss functions were utilized for overexposed pixels and non-overexposed pixels.

For overexposed pixels, the loss function MS-SSIM-L1 [15] was used and for non-overexposed pixels,

$$L_{NOE} = L_{LPIPS} + L_2 \cdot 0.05 + L_{MSSSIM-L1} \cdot 0.75, \tag{3}$$

LPIPS with AlexNet [14] and MS-SSIM-L1 [15] and L2 was combined in order to improve perceptual quality. A small combination of L2 is taken for fidelity improvement and prevention of notable color deviations. After 200 epochs with batch size of 6 with the Adam optimizer [8] of learning rate 10^{-4} and decay of 10^{-6}, both overexposed and non-overexposed pixels were trained again only with L2 loss for an additional 20 epochs with learning rate of 10^{-5}. Inference on Nvidia's Tesla-V100 GPU for patches of $504 \times 504 \times 3$ takes 0.05 s on average per patch.

For training and inference on the ETH Huawei P20 Pro dataset [7] where one instead of two U-Nets was used, the loss function with $\gamma = \frac{1}{3.6}$ was computed as

$$L = L_{LPIPS}(I_{in}^{\gamma}, I_{gt}^{\gamma}) + L_2(I_{in}^{\gamma}, I_{gt}^{\gamma}) \cdot 0.05 + L_{MSSSIM-L1}(I_{in}^{\gamma}, I_{gt}^{\gamma}) \cdot 0.75 \quad (4)$$

where gamma correction was applied to both overexposed and non-overexposed pixels, as raising overexposed pixel values to γ results in minimal changes by nature of gamma correction. Similarly, the Adam optimizer [8] was used with an initial learning rate of 10^{-4} with batch size of 6 and the weight decay was set as 10^{-6} and trained for 230 epochs. Inference with GPU on patches of $496 \times 496 \times 3$ takes on average, 0.00465 s per patch.

3.6 Bayer to Bayer Optional Refinement

The optional refinement step of the pipeline is refinement from bayer to bayer by computing L1 or L2 loss after mapping each of the bayer into YUV space as displayed in Fig. 1 For an input RGB image with dimensions $504 \times 504 \times 3$ to bayer with dimensions $252 \times 252 \times 4$, the refinement step consists of averaging the green channels of the bayer and performing matrix multiplication in order to map to YUV space. By taking L1 or L2 loss on YUV, the reconstructed bayer displayed better fidelity scores than directly computing L1 or L2 on bayer, as will be further detailed in ablation studies.

4 Experiments

4.1 Quantitative and Qualitative Evaluations

Our proposed methodology (SenseBrains) along with other solutions were quantitatively evaluated in fidelity measures as well as generalizability and robustness. As listed in Table 1, our pipeline improves the performance of the U-Net from 26.30 dB to 28.36 dB in Track 1 and from 30.01 dB to 35.47 dB in Track 2. Our methodology uniformly fixed all neural networks used to be the standard U-Net [10] with 23 convolutional layers with only an additional channel for the overexposure mask, which subtantiates the enhancement in performance through the multi-step refinement process.

Table 2 provides further details on training as well as additional details such as use of extra data, ensemble or whether it is capable of processing full-resolution images. Note that although our methodology using the two U-Nets

Table 1. AIM Reversed ISP Challenge Benchmark [4]. Teams are ranked based on their performance on Test1 and Test2, an internal test set to evaluate the generalization capabilities and robustness of the proposed solutions. The methods (*) have trained using extra data from [11], and therefore only results on the internal datasets are relevant. CycleISP [13] was reported by multiple participants.

Team name	Track 1 (Samsung S7) Test1 PSNR ↑	SSIM ↑	Test2 PSNR ↑	SSIM ↑	Track 2 (Huawei P20) Test1 PSNR ↑	SSIM ↑	Test2 PSNR ↑	SSIM ↑
NOAHTCV	31.86	0.83	32.69	0.88	38.38	0.93	35.77	0.92
MiAlgo	31.39	0.82	30.73	0.80	40.06	0.93	37.09	0.92
CASIA LCVG (*)	30.19	0.81	31.47	0.86	37.58	0.93	33.99	0.92
HIT-IIL	29.12	0.80	30.22	0.87	36.53	0.91	34.07	0.90
SenseBrains (Ours)	28.36	0.80	30.08	0.86	35.47	0.92	32.63	0.91
CS^2U (*)	29.13	0.79	29.95	0.84	-	-	-	-
HiImage	27.96	0.79	-	-	34.40	0.94	32.13	0.90
0noise	27.67	0.79	29.81	0.87	33.68	0.90	31.83	0.89
OzU VGL	27.89	0.79	28.83	0.83	32.72	0.87	30.69	0.86
PixelJump	28.15	0.80	-	-	-	-	-	-
CVIP	27.85	0.80	29.50	0.86	-	-	-	-
CycleISP [13]	26.75	0.78	-	-	32.70	0.85	-	-
UPI [2]	26.90	0.78	-	-	-	-	-	-
U-Net Base	26.30	0.77	-	-	30.01	0.80	-	-

Table 2. Team information summary. Input refers to the input image size used during training, most teams used the provided patches (504px or 496px). ED indicates the use of Extra Datasets besides the provided challenge datasets. ENS indicates if the solution is an Ensemble of multiple models. FR indicates if the model can process Full-Resolution images (3024 × 4032)

Team	Input	Epochs	ED	ENS	FR	# Params. (M)	Runtime (ms)	GPU
NOAHTCV	(504,504)	500	✗	✗	✓	5.6	25	V100
MiAlgo	(3024,4032)	3000	✗	✗	✓	4.5	18	V100
CASIA LCVG	(504,504)	300K it.	✓	✓	✓	464	219	A100
CS^2U	(504,504)	276K it.	✓	✓	✓	105	1300	3090
HIT-IIL	(1536,1536)	1000	✗	✗	✓	116	19818	V100
SenseBrains	(504,504)	220	✗	✓	✓	69	50	V100
PixelJump	(504,504)	400	✗	✓	✓	6.64	40	3090
HiImage	(256, 256)	600	✗	✗	✓	11	200	3090
OzU VGL	(496, 496)	52	✗	✗	✓	86	6	2080
CVIP	(504,504)	75	✗	✗	✓	2.8	400	3090
0noise	(504,504)	200	✗	✗	✓	0.17	19	Q6000

(a) Input Full Resolution Image

(b) Visualized Output

Fig. 5. (a) a full resolution image (3024 × 4032) (b) visualized RAW output from our pipeline. Visualization of RAW was performed by passing the output RAW through Demosaic Net [5] along with a simple implementation of post-processing. Note that our pipeline is able to perform inference on full resolution images.

have a relatively high number of model parameters, by replacing the U-Net with other methodologies that supports end-to-end learning, the number of model parameters along with runtime can be controlled.

In Fig. 5, an input full resolution image (3024 × 4032), as a collection of patches is passed into our pipeline. The RAW output was generated by constructing the overexposure mask, passing the RGB image into the two U-Nets, and mosaicing the blended output. The RAW output was then visualized with Demosaic Net to acquire the demosaiced RAW and then passed into a simplified implementation of post-processing. It is noteworthy that our pipeline with U-Net (OE) and U-Net (NOE) can perform inference on full resolution images and produce visualized RAW images that show high perceptual quality, as shown in Fig. 5. Additionally, Fig. 6 displays corresponding pairs of RGB images and visualized RAW outputs produced by our pipeline. Note that the RGB images have a variety of colors and textures, including images also with overexposure for which using the overexposure mask fusion process resulted in higher fidelity performance. Furthermore, Fig. 7 displays a comparison of visualized RAW outputs of all proposed methodologies.

4.2 Ablation Studies

By experimenting with the training and validation dataset for the Samsung S7 dataset [11], we performed ablation studies along with the quantitative evaluations provided by the AIM Reversed ISP Challenge [4]. Note that these datasets are public, and are only used for ablation studies and to evaluate refinement processes of our proposed pipeline.

As recorded in Table 3, the baseline U-Net mapping directly from RGB to RAW bayer has an average training PSNR of 26.295 dB. One single U-Net with

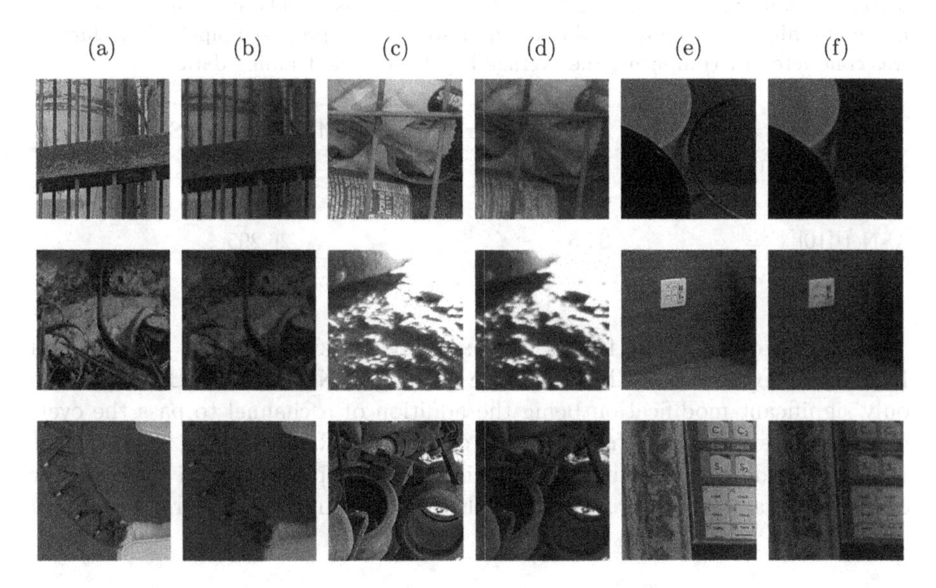

Fig. 6. The figure displays input RGB images under columns (a), (c) and (e) and corresponding visualized RAW outputs under columns (b), (d) and (f). Note that the displayed RGB images have a variety of colors and textures. Visualization of RAW was performed by passing RAW outputs into Demosaic Net [5] and then a simplified implementation of post-processing.

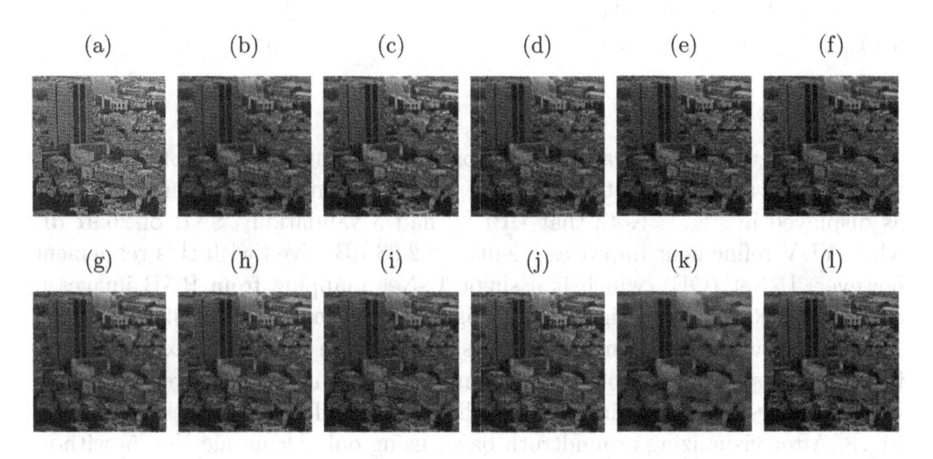

Fig. 7. (a) is the input RGB image. From (b) to (l), the outputs of the following teams are displayed in the following order: SenseBrains, 0noise, CASIA LCVG, CS^2U, CVIP, HiImage, HIT-IIL, MiAlgo, NOAHTCV, OzU VGL, Pixel Jump. All RAW outputs were visualized by passing into Demosaic Net [5] and then a simplified implementation of post-processing.

Table 3. Samsung S7 Training Dataset [11] Comparison. The comparison was conducted for ablation purposes and to evaluate steps of the proposed pipeline. Evaluation was conducted by computing the average PSNR over the training dataset.

Model	Total Number of params (M)	Training PSNR (dB)
U-Net (OE), U-Net (NOE)	69.05	28.9
U-Net (OE)	34.5	27.23
U-Net [10]	34.5	26.295

the addition of a channel for the overexposure mask and trained to map from RGB to demosaiced RAW has a average training PSNR of 27.23 dB. Despite the only significant modification being the addition of a channel to pass the overexposure mask, mapping from RGB to demosaiced RAW leads to improvement in fidelity. Additionally with the blending of two separate neural networks for overexposed and non-overexposed pixels, the training PSNR increases to 28.9 dB.

Table 4. Samsung S7 Validation Dataset PSNR [11] Comparison. The comparison was conducted for ablation purposes and to evaluate steps of the proposed pipeline. Evaluation was conducted on the validation dataset.

Model	Total Number of params (M)	Validation PSNR (dB)
U-Net (OE)	34.5	32.17
UPI [2], YUV refinement	34.5	29.9
UPI [2]	-	26.97

Another comparison that was made on the Samsung S7 validation dataset was with UPI [2] and the additional step of YUV refinement offered in the pipeline as displayed in Fig. 1. Note that UPI [2] had a validation PSNR of 26.97 dB, while YUV refinement improves fidelity by 2.93 dB. Even with this refinement, however, U-Net (OE), which is a single U-Net mapping from RGB images to demosaiced RAW, has a significantly higher validation PSNR of 32.17 dB.

It is noteworthy that methodologies without use of the overexposure mask fusion process had lowest performance in PSNR for images with overexposure as well as images corresponding to low-light demosaiced RAW images as shown in Fig. 8. After visualizing groundtruth bayer using only Demosaic Net [5] without post-processing, we observed that the same white overexposed pixels map to white overexposed pixels as well as green white overexposed pixels in demosaiced RAW, resulting in drops in fidelity measures for models that do not account for such contradictions. The proposed overexposure mask fusion better accounts for overexposed pixels by designating a U-Net (OE) to overexposed pixels.

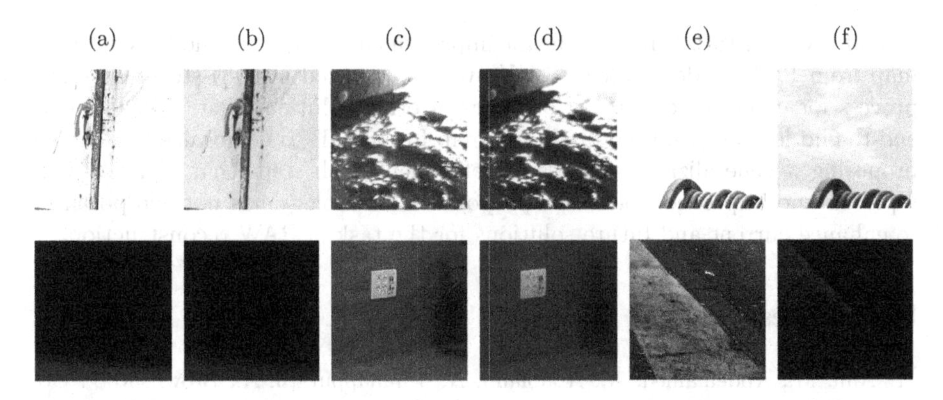

Fig. 8. The figure displays input RGB images under columns (a), (c) and (e) and the corresponding visualized RAW groundtruth images under columns (b), (d) and (f). Note that the input RGB images have low-light demosaiced RAW or overexposure. Visualization of RAW was performed by passing RAW outputs into Demosaic Net [5] and then a simplified implementation of post-processing.

4.3 Limitations

As noted by Conde et al. [3], one major difficulty that arises with the task of the reversed ISP is the issue of overexposure in input RGB images. This paper addresses overexposure with the proposed methodology of overexposure mask fusion.

A limitation of our pipeline is that mapping from RGB images to demosaiced RAW requires the use of methodologies of reconstructing demosaiced RAW from the groundtruth RAW. For our pipeline, we utilize Demosaic Net [5], however it is possible that inaccuracies in the reconstruction process of the demosaiced RAW will result in inaccuracies from mapping from RGB images to demosaiced RAW. This possibility of generating inaccurate demosaiced RAW exists and can be further addressed by utilizing other, reliable methods.

Otherwise, there are few limitations to the pipeline itself, as different steps of the refinement process can be replaced and improved such as the U-Net (OE) and U-Net (NOE). Our pipeline can integrate methodologies that support end-to-end learning and capable of modification to map from RGB to demosaiced RAW. Other issues such as misalignment of RAW and RGB training pairs can be resolved by downsampling before computing perceptual loss.

5 Conclusion

In this paper, we propose a novel and generalizable multi-step pipeline that allows the use of perceptual loss by mapping from RGB images to demosaiced RAW. With overexposure mask fusion to address overexposure in input RGB images and gamma correction before computing loss for non-overexposed pixels, our methodology addresses several major complexities that have existed for the task of reversed ISP. With significant improvement using only the U-Net [10] to map from RGB to demosaiced RAW, we have created a multi-step refinement process for enhancement in performance of other solutions that are capable of end-to-end learning and modifiable to map from RGB to demosaiced RAW. By proposing a generalizable process of refinement with steps that can be easily replaced and improved upon, our proposed methodology has notable potential to enhance current and future solutions for the task of RAW reconstruction.

References

1. Afifi, M., Abdelhamed, A., Abuolaim, A., Punnappurath, A., Brown, M.S.: CIE XYZ Net: unprocessing images for low-level computer vision tasks. IEEE Trans. Pattern Anal. Mach. Intell. (99), 1–1 (2021). https://doi.org/10.1109/tpami.2021.3070580
2. Brooks, T., Mildenhall, B., Xue, T., Chen, J., Sharlet, D., Barron, J.T.: Unprocessing images for learned raw denoising. In: 2019 IEEE/CVF Conference on Computer Vision and Pattern Recognition (CVPR) (2019). https://doi.org/10.1109/cvpr.2019.01129
3. Conde, M.V., McDonagh, S., Maggioni, M., Leonardis, A., Pérez-Pellitero, E.: Model-based image signal processors via learnable dictionaries. Proc. AAAI Conf. Artif Intell. **36**(1), 481–489 (2022). https://doi.org/10.1609/aaai.v36i1.19926
4. Conde, M.V., Timofte, R., et al.: Reversed image signal processing and RAW reconstruction. AIM 2022 challenge report. In: Proceedings of the European Conference on Computer Vision Workshops (ECCVW) (2022)
5. Gharbi, M., Chaurasia, G., Paris, S., Durand, F.: Deep joint demosaicking and denoising. ACM Trans. Graphics **35**(6), 1–12 (2016). https://doi.org/10.1145/2980179.2982399
6. Ignatov, A., Timofte, R., et al.: Learned smartphone ISP on mobile GPUs with deep learning, mobile AI & AIM 2022 challenge: Report. In: Proceedings of the European Conference on Computer Vision (ECCV) Workshops (2022)
7. Ignatov, A., Van Gool, L., Timofte, R.: Replacing mobile camera ISP with a single deep learning model. In: 2020 IEEE/CVF Conference on Computer Vision and Pattern Recognition Workshops (CVPRW) (2020). https://doi.org/10.1109/cvprw50498.2020.00276
8. Kingma, D.P., Ba, J.: Adam: a method for stochastic optimization (2014). 10.48550/ARXIV.1412.6980, https://arxiv.org/abs/1412.6980
9. Punnappurath, A., Brown, M.S.: Learning raw image reconstruction-aware deep image compressors. IEEE Trans. Pattern Anal. Mach. Intell. **42**(4), 1013–1019 (2020). https://doi.org/10.1109/tpami.2019.2903062
10. Ronneberger, O., Fischer, P., Brox, T.: U-net: convolutional networks for biomedical image segmentation. CoRR abs/1505.04597 (2015), http://arxiv.org/abs/1505.04597
11. Schwartz, E., Giryes, R., Bronstein, A.M.: Deepisp: toward learning an end-to-end image processing pipeline. IEEE Trans. Image Process. **28**(2), 912–923 (2019). https://doi.org/10.1109/tip.2018.2872858
12. Xing, Y., Qian, Z., Chen, Q.: Invertible image signal processing. In: 2021 IEEE/CVF Conference on Computer Vision and Pattern Recognition (CVPR) (2021). https://doi.org/10.1109/cvpr46437.2021.00622

13. Zamir, S.W., et al.: Cycleisp: Real image restoration via improved data synthesis. 2020 IEEE/CVF Conference on Computer Vision and Pattern Recognition (CVPR) (2020). https://doi.org/10.1109/cvpr42600.2020.00277
14. Zhang, R., Isola, P., Efros, A.A., Shechtman, E., Wang, O.: The unreasonable effectiveness of deep features as a perceptual metric. 2018 IEEE/CVF Conference on Computer Vision and Pattern Recognition (2018). https://doi.org/10.1109/cvpr.2018.00068
15. Zhao, H., Gallo, O., Frosio, I., Kautz, J.: Loss functions for image restoration with neural networks. IEEE Trans. Comput. Imaging **3**(1), 47–57 (2017). https://doi.org/10.1109/tci.2016.2644865

CAIR: Fast and Lightweight Multi-scale Color Attention Network for Instagram Filter Removal

Woon-Ha Yeo, Wang-Taek Oh, Kyung-Su Kang, Young-Il Kim, and Han-Cheol Ryu(✉)

Sahmyook University, Seoul, South Korea
canal@syuin.ac.kr, mm074111@gmail.com, unerue@me.com, qhdrmfdl123@gmail.com, hcryu@syu.ac.kr

Abstract. Image restoration is an important and challenging task in computer vision. Reverting a filtered image to its original image is helpful in various computer vision tasks. We employ a nonlinear activation function free network (NAFNet) for a fast and lightweight model and add a color attention module that extracts useful color information for better accuracy. We propose an accurate, fast, lightweight network with multi-scale and color attention for Instagram filter removal (CAIR). Experiment results show that the proposed CAIR outperforms existing Instagram filter removal networks in fast and lightweight ways, about 11× faster and 2.4× lighter while exceeding 3.69 dB PSNR on IFFI dataset. CAIR can successfully remove the Instagram filter with high quality and restore color information in qualitative results. The source code and pretrained weights are available at https://github.com/hnvlab-syu/CAIR.

Keywords: Image restoration · Filter removal · Color attention · Ensemble learning

1 Introduction

Photographic filters have been widely used to control the feel of photos when using analog and digital cameras. Filter as a camera accessory can be attached to the optical lens, which gives various effects to the photo (*i.e.*, image), such as color conversion, color subtraction, and contrast enhancement. The popularity of smartphones has made it easier for people to take and share photos than in the days of analog and digital cameras. It has become a daily life for many people to share images on Instagram after applying various photographic effects and corrections to an image taken. These filters can digitally modify the original image (without a camera accessory) by adjusting contrast, hue, or saturation and applying blur or noise. Filters applied to images can make people feel emotional and change the photo's mood, as shown in Fig. 1.

However, filtered images can be noise data that distracts training deep learning models from perspective of computer vision researchers. Computer vision

L. Karlinsky et al. (Eds.): ECCV 2022 Workshops, LNCS 13802, pp. 714–728, 2023.
https://doi.org/10.1007/978-3-031-25063-7_45

requires a pure image state with no filters. Applying various filtering effects to raw images for data augmentation techniques can improve the generalizability of a model, but filtered images significantly degrade performance in major computer vision tasks [2,7] such as image classification [5,25], detection, and segmentation [11]. Therefore, filters (*e.g.*, Instagram filter) need to be removed from filter images for a machine vision system to achieve non-degraded performance.

Fig. 1. Example of filtered images in Instagram Filter Fashion Image (IFFI) dataset [11]. Bottom-right of each image indicates the name of the applied filter.

Previous studies have employed the encoder-decoder structure and adversarial training strategy to solve the filter removal problem [10]. Instagram filter removal network (IFRNet) [11] defines the problem of removing filters from photos as a reverse style transfer [9]. IFRNet consists of an encoder with VGGNet [22] as a style extractor module and a decoder inspired by PatchGAN [10], regarding filter as additional style information. The visual effects of a filter can be directly removed by adaptively normalizing external style information in each level of the encoder [11]. Contrastive Instagram Filter Removal Network (CIFR) [12] tackles the problem by normalizing the affine parameters using VGGNet with the help of adaptive normalization. Kinli et al. [12] employ a multi-layer patch-wise contrastive learning strategy [18] and propose an isolated patch sampling module that distills the content and style information [12]. These studies showed the outstanding performance of visual understanding and removing filters; however, their structure and learning process is exceedingly complex. Thus, we desire to address the problem of filter removal efficiently with a straightforward approach.

We propose a multi-scale color attention network based on NAFNet (nonlinear activation function free network) [4] for removing Instagram filters. The color attention modules explore different color-space between an original (non-filtered) image and a filtered image. First, we modify the architecture of NAFNet with less computational costs than the original NAFNet. Second, a color attention module inspired by [27] is proposed to utilize the color-informative features from multi-scale input images. Third, the ensemble learning strategy is adopted to improve the performance of the results further. The proposed model is accurate, fast, and lightweight. Moreover, we consider carbon emission reduction through the lightweight model. The main contributions of this work are as follows:

1. We propose an accurate, fast, and lightweight multi-scale color attention network for Instagram filter removal named CAIR, which has the remarkable qualitative and quantitative ability for Instagram filter removal.
2. A color attention module is proposed that captures color-informative features from multi-scale input images, resulting in better filter removal performance.
3. CAIR is a lightweight model in terms of the number of operations and parameters compared to the previous studies on filter removal.
4. CAIR's low computational complexity improves inference speed. CAIR achieves about 11$\times$ faster runtime compared to [11,12] without performance penalty.

2 Backgrounds

2.1 Nonlinear Activation Free Network

NAFNet [4] is the state-of-the-art (SOTA) network in the field of image restoration (*i.e.*, deblurring and denoising). Chen et al. [4] considered two main points: 1) To enhance the performance and reduce the complexity of the network, the authors have classified inter-block and intra-block complexity. 2) They present a new direction that computer vision tasks may no longer require nonlinear activation functions. They adopt single-stage U-Net architecture [20] to reduce inter-block complexity and structure simple baseline blocks for the intra-block complexity. Also, NAFNet has a critical baseline block (*i.e.*, NAFBlock) that combines innovative components used in SOTA methods [23,24,26]. First, layer normalization (LN) [1] in the NAFBlock is effectively utilized for configuring the baseline block. Since LN stabilizes the training process, a large learning rate can be used during training, increasing the initial learning rate from 1e–4 to 1e–3 [4]. It also improves deblurring and denoising performance. Second, channel attention brings computational efficiency and global information to the feature maps. Third, nonlinear activation functions such as GELU [8] are replaced by a simple gate which is an element-wise product of feature maps. The simple gate can produce the effect of the nonlinear activation function and leads to performance gain. Finally, NAFBlock combines LN, convolution, simple gate, and simplified channel attention. NAFNet achieves SOTA performance in image restoration with its simplicity and ability. NAFBlock and its components are described in Fig. 2.

However, since it is hard to capture color information which is a crucial factor for filter removal using only channel attention, NAFNet should be further enhanced. Therefore, we propose a new NAFNet-based architecture with multi-scale color attention. In addition, we take multi-scale inputs to obtain scale-invariant features. A detailed explanation of our proposed network architecture is in Sect. 3.1.

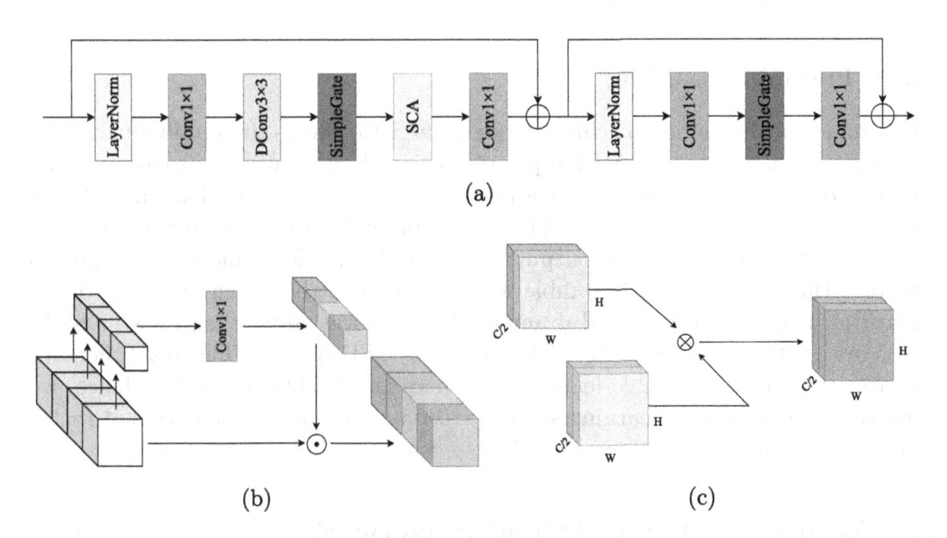

Fig. 2. (a) NAFBlock [4]. (b) Simplified Channel Attention (SCA) in NAFBlock [4]. (c) Simple Gate in NAFBlock [4]. $\odot$: channel-wise multiplication, $\otimes$: element-wise multiplication.

2.2 Color Attention Mechanism

Zamir et al. [27] propose CycleISP to denoise sRGB images by synthesizing realistic noise data. CycleISP implements two networks: RGB2RAW network and RAW2RGB network. They convert 3-channel sRGB images to 4-channel RAW data and vice versa. In the previous study [3], a method for inverting the camera ISP (image signal processor) requires prior information such as color correction matrices, which degrades generalization ability. Therefore Zamir et al. [27] propose a color correction branch and color attention unit, which provides explicit color information. The color correction branch is a convolutional neural network (CNN) where an sRGB image enters and a color-encoded deep feature is generated. The color correction branch applies Gaussian blur, 3×3 convolutional layer, two recursive residual groups (RRGs), and sigmoid activation to the sRGB image. Each RRG consists of multiple dual attention blocks, and each dual attention block contains spatial attention and channel attention modules. The color attention unit takes the output of the color correction branch and then multiplies and sums the output with the features from the previous convolutional layer in RAW2RGB.

When constructing the color attention module in our proposed network, the RRG of the color correction branch is changed to a group of NAFBlock to increase the inference speed. The multi-scale input mentioned in Sect. 2.1 is introduced to improve the performance. The details of the proposed color attention module are discussed in Sect. 3.2.

2.3 Ensemble Learning

Ensemble learning aims to achieve more synergistic effects than individual training results. In the field of MRI super-resolution, Lyu et al. [17] propose a GAN (generative adversarial network) architecture and train five GAN models for five differently degraded inputs. Then CNN model is used to integrate all these images. The CNN takes the outputs of five different GAN models as input to predict the final output. Ensemble learning reports the strength of the detail textures, avoids artifacts, and dramatically improves results from 0.87–0.88 to 0.95 in the structural similarity index measure (SSIM) value. This result demonstrates that using ensemble learning enhances the quality of results. Hence, we use the same ensemble learning strategy to improve further the effects of Instagram filter removal, which will be discussed in Sect. 4.4.

3 Multi-scale Color Attention Network

In this section, we start with the overall architecture of the proposed fast and lightweight multi-scale color attention network for Instagram filter removal (CAIR). Then we give the detailed configurations of the proposed color attention (CA) module.

3.1 Proposed CAIR Architecture

We propose CAIR that fully utilizes multi-scale and color-attentive features extracted from an input image. Figure 3 shows the overall architecture of CAIR. The architecture of CAIR is modified from NAFNet [4]. We improve NAFNet in two aspects: 1) CA module inspired by color correction scheme [27] is added, and 2) CAIR takes multi-scale input images. These two improvements make full use of color-attentive features so that the filters can be removed effectively.

There exist l levels in the overall architecture ($l = 4$ in this work). An input filter image $I_F \in \mathbb{R}^{H \times W \times 3}$ is downscaled by a factor of 2 as going down to the lower level. The down-scaled image at level k is denoted as $I_{F_k} \in \mathbb{R}^{\frac{H}{2^{k-1}} \times \frac{W}{2^{k-1}} \times 3}$. These multi-scale images provide semantically robust features. First, an original-size input filter image $I_F = I_{F_1}$ passes through the first NAFBlock $NB_1(\cdot)$, then the encoded feature EF_1 is extracted

$$EF_1 = NB_1(I_{F_1}). \tag{1}$$

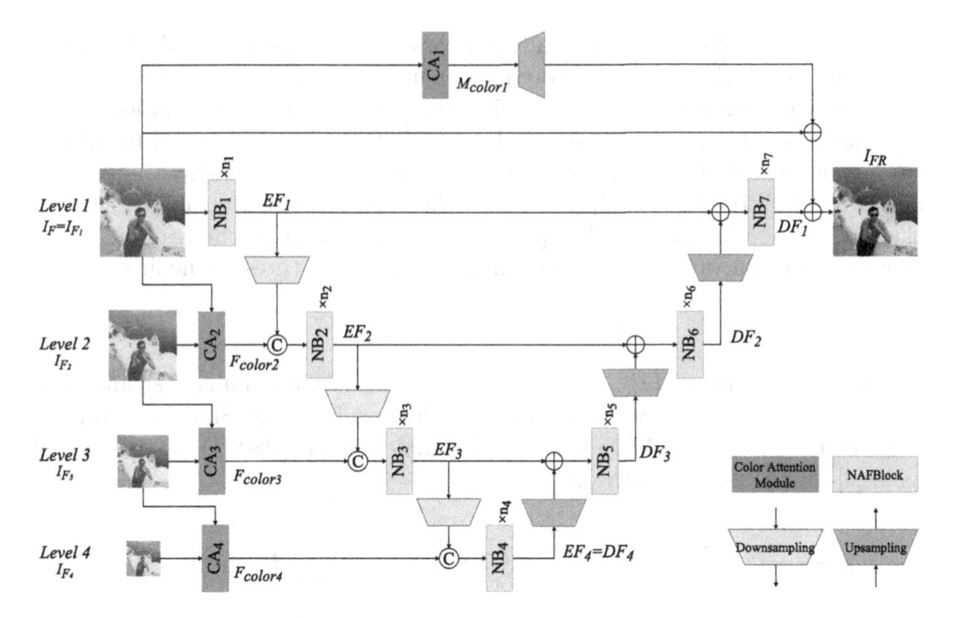

Fig. 3. The main architecture of the proposed CAIR. ©: channel-wise concatenation.

Then given the lower-level input I_{F_k} and the upper-level input $I_{F_{k-1}}$, a CA module denoted as $CA(\cdot)$ is used to extract color-attentive features from I_{F_k} and $I_{F_{k-1}}$

$$F_{\text{color}_k} = CA_k(I_{F_k}, I_{F_{k-1}}), \tag{2}$$

where F_{color_k} is color-attentive features, and CA_k denotes CA module at level k (for $k \geq 2$). CA_1 is a special case that takes only I_{F_1} and outputs color attention map M_{color_1} (see Eq. 10). The CA module aims to capture the color information of the input image. The details of the CA module are explained in Sect. 3.2.

The encoded features from upper-level EF_{k-1} are downsampled (denoted as $\downarrow$), and the color-attentive features F_{color_k} are concatenated. Then the concatenated features are passed to the kth NAFBlock NB_k to extract the encoded feature at level k

$$EF_k = NB_k(\text{concatenate}(\downarrow EF_{k-1}, F_{\text{color}_k})), \quad k = 2, \ldots, l, \tag{3}$$

where EF_k is encoded features at the kth level. At the lth level, the encoded features EF_l are same with the decoded features DF_l:

$$EF_l = DF_l. \tag{4}$$

The decoded features for each level k can be represented as

$$DF_k = NB_{2l-k}(EF_k + \uparrow DF_{k+1}), \quad k = 1, 2, \ldots, l-1, \tag{5}$$

where NB_{2l-k} indicates $(2l-k)^{th}$ NAFBlock. The $(k+1)^{th}$ level decoded features DF_{k+1} is upsampled (denoted as $\uparrow$) and element-wisely added with the encoded features EB_k obtained at level k and then passed to the $(2l-k)^{th}$ NAFBlock. The sub-pixel convolution [21] is employed as the upsampling module, which converts the scale sampling with a given magnification factor by pixel translation.

Finally, the filter removed image $I_{FR} \in \mathbb{R}^{H \times W \times 3}$ can be obtained by adding the last decoded features DF_1 and features from the two global skip connections; one is input filtered image I_F, and the other is color attention map F_{color_1} of the input image. These global skip connections can stabilize the training of the proposed deep network and provide information-rich features to the final output

$$I_{FR} = I_F + DF_1 + \uparrow F_{\text{color}_1}. \tag{6}$$

Given a training set $\{I_F^i, I_O^i\}_{i=1}^N$, which contains N filtered image I_F^i and their original image I_O^i, the goal is to minimize the loss function. Since we modified the architecture from NAFNet, the peak-signal-to-noise (PSNR) loss is selected as in [4] for the fair comparison

$$L(\Theta) = -\frac{1}{N}\sum_{i=1}^{N} PSNR(H_{CAIR}(I_F^i), I_O^i) = -\frac{1}{N}\sum_{i=1}^{N} PSNR(I_{FR}^i, I_O^i), \tag{7}$$

where H_{CAIR} and Θ denote the function of the proposed CAIR and the training parameter of the CAIR, respectively.

$$PSNR(X, Y) = 10 \cdot \log_{10} \frac{255^2}{MSE(X, Y)}, \tag{8}$$

$$MSE(X, Y) = \frac{1}{WH}\sum_{w=1}^{W}\sum_{h=1}^{H}(X_{w,h} - Y_{w,h})^2, \tag{9}$$

where W and H are the width and height of images X and Y.

3.2 Color Attention Module

Inspired by the color attention unit and color correction branch in [27], we propose a CA module for image filter removal. RRG [27] in the color correction branch contains channel attention and spatial attention. The spatial attention can be modified to channel attention [26], and the channel attention in NAFBlock is even simplified to achieve less computational cost but better performance. Therefore, we replace the RRG with a group of NAFBlock and replace 3×3 convolution in the color correction branch with 1×1 convolution to reduce the complexity of the proposed network. In addition, our proposed color attention module takes two multi-scale input filter images and extracts color-attentive features. Our proposed CA module can learn to capture color information of multi-scale images so that it helps to get rid of filter from input filtered images. The structure of the CA module is shown in Fig. 4.

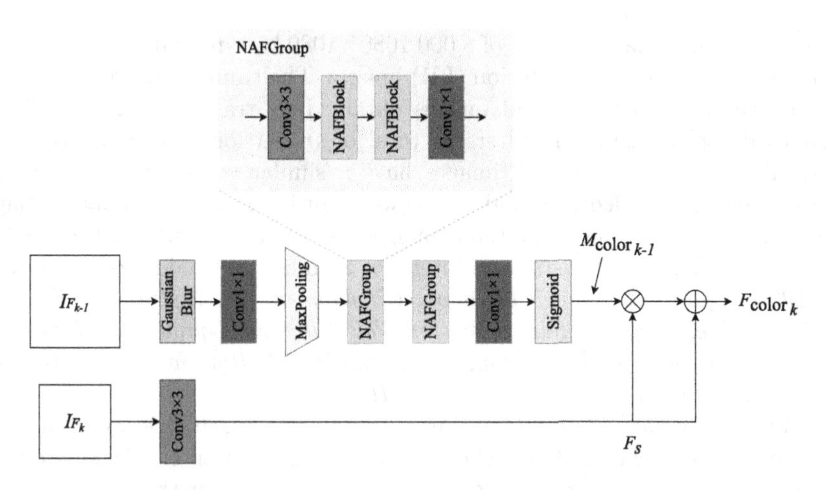

Fig. 4. The proposed color attention (CA) module. ⊗: element-wise multiplication, ⊕: element-wise summation.

Given k^{th} level input image $I_{F_k} \in \mathbb{R}^{\frac{H}{2^{k-1}} \times \frac{W}{2^{k-1}} \times 3}$ and the upper-level $(k-1^{th})$ input image $I_{F_{k-1}}$, we first extract color maps $M_{\text{color}_{k-1}}$ from the upper-level input image

$$M_{\text{color}_{k-1}} = \sigma(\text{conv2}(NG_2(NG_1(\text{maxpool}(\text{conv1}(K * I_{F_{k-1}}))))), \tag{10}$$

where $*$ denotes convolution operation, and K is the Gaussian kernel with a standard deviation of 12 [27]. This strong blurring operation ensures that only the color information can be extracted from the input image $I_{F_{k-1}}$ whereas the structural content and fine texture come from the lower-level image I_{F_k}. After blurring, 1×1 convolution is applied conv1 followed by a max pooling (maxpool) operation. These can reduce the number of operations afterward. NG_1, NG_2 denotes group of NAFBlocks (NAFGroup) that consists of a 3×3 and a 1×1 convolution layer, and two NAFBlocks. Another 1×1 convolution denoted as conv2 and a gating mechanism with sigmoid activation σ are applied.

Once the color map $M_{\text{color}_{k-1}}$ is extracted, the structural features (denoted as F_s) of the lower-level image obtained by 3×3 convolution (denoted as conv3) are weighted through element-wise multiplication (⊗) with the color map. Then element-wise summation of the structural features F_s is applied to the weighted color features

$$F_s = \text{conv3}(I_{F_k}), \tag{11}$$

$$F_{\text{color}_k} = (M_{\text{color}_{k-1}} \otimes F_s) \oplus F_s. \tag{12}$$

4 Experiments

4.1 Datasets

The AIM 2022 challenge on Instagram filter removal provided the IFFI dataset, which contains 16,000 training images, 3,000 validation images, and 2,200 test images. Training images consist of 8,000 1080×1080 high-resolution (HR) images and 8,000 256×256 low-resolution (LR) images. The training images include an original (unfiltered) image and images obtained by transforming the original with 15 different Instagram filters (a total of 16 versions, including the original). The validation and test images have a similar composition as training images that are divided into high and low-resolution images, and these images are comprised of 15 and 11 filtered images, respectively, without the original version. The number of Instagram filters for training and validation datasets is 15 (*1977, Amaro, Brennan, Clarendon, Gingham, He-Fe, Hudson, Lo-Fi, Mayfair, Nashville, Perpetua, Sutro, Toaster, Valencia, and X-proll*), and for testing dataset is 11 (*Amaro, Clarendon, Gingham, He-Fe, Hudson, Lo-Fi, Mayfair, Nashville, Perpetua, Valencia, and X-ProII*).

Additionally, we resized HR images 1080×1080 into 256×256 by resampling using pixel area relation (*i.e., OpenCV* INTER_AREA interpolation) due to image quality. As a result, we used three sets of images for training, including HR images, resized HR images, and LR images. Finally, we took 24,000 images to train the proposed network CAIR and 2,200 images to test.

4.2 Implementation Details

In the training process, data augmentation was applied to IFFI datasets [11] using random flip and rotation with the probability of 0.5. In the testing process, we used the self-ensemble method in [14] as a test time augmentation (TTA). The number of NAFBlocks (n_1-n_7) followed the default setting of NAFNet $(2, 2, 4, 22, 2, 2, 2)$, and the width in all NAFBlocks and convolutional layers was set to 32. The "Downsampling" module (represented as a gray trapezoid block in Fig. 3) contains a convolutional layer with stride 2, and the "Upsampling" module (represented as a light-blue upside-down trapezoid block in Fig. 3) is the sub-pixel convolution layer [21]. The mini-batch size was 64, and the patch size was 256×256 (HR images are randomly cropped to 256×256). We used TLSC [6] to solve the problem that training patch by patch and testing with the whole image causes performance degradation [6]. We used AdamW [16] optimizer with $\beta_1 = 0.9, \beta_2 = 0.9$, weight decay $1e^{-4}$. The initial learning rate was set to $1e^{-3}$ and gradually reduced to $1e^{-6}$ with the cosine annealing schedule [15]. Our model was trained for up to 200K iterations. We implemented the code using PyTorch deep learning framework [19]. All models were conducted by four NVIDIA Tesla V100 GPUs ($4 \times$ 32GB). We reported the number of parameters and operations, inference time on GPU for fairness, and the PSNR and SSIM value in our environment.

4.3 Compared Methods

We compared the proposed method CAIR with SOTA Instagram filter removal models [11,12] to show the performance of Instagram filter removal. Furthermore, to figure out the effects of the color attention module and the multi-scale input, we compared the original NAFNet, NAFNet with single input and color attention module (denoted as CAIR-S), and modified CAIR-S to take multi-scale input images (denoted as CAIR-M shown in Fig. 3). Then we used the self-ensemble strategy [14] to improve our CAIR-S and CAIR-M results further, and noted the self-ensemble applied models as CAIR-S+ and CAIR-M+. Besides, both the self-ensemble and ensemble learning strategy applied to CAIR is denoted as CAIR*. The ensemble learning is described in Sect. 4.4.

4.4 Ensemble Learning

We used an ensemble learning strategy using CNN as in [17]. We propose an ensemble network composed of 3×3 convolution layers and n NAFBlocks (in our experiment, we chose $n = 3$). Figure 5 shows the ensemble network. First, we trained CAIR-S and CAIR-M and obtained the predicted train images from the two models. After that, two output images of each model were concatenated to a single image as six channels and fed into the ensemble network. The training set was the same as the CAIR training configuration except for batch size, which is 16 here. In the testing phase, test images were predicted by two pretrained models (CAIR-S and CAIR-M), and self-ensemble was also applied. The images from the two models were concatenated and fed into the pretrained ensemble network.

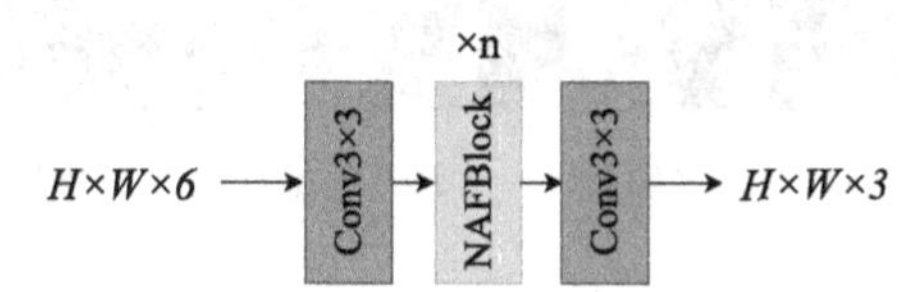

Fig. 5. The network architecture for ensemble learning.

4.5 Results

Qualitative Results. The effects of the proposed color attention module can be seen in Fig. 6. It is noticeable that the skin color in the test images was restored only in CAIR* (see first, second, and third rows). Compared to the other methods, CAIR, which has the strength of the color attention module, successfully restored color information, especially chromatic colors (*e.g.*, red, yellow, or green) visually. On the contrary, IFRNet and CIFR captured detailed textures rather than the proposed CAIR. In the 4th row in Fig. 6, the texture of the iron fence in the background was distorted on CAIR and NAFNet results. However, overall images predicted with CAIR* showed visually more satisfying results than the others.

Quantitative Results. We summarized PSNR and SSIM results, inference time, and the number of parameters and operations of several compared methods, including our CAIR, IFRNet [11], and CIFR [12] on the IFFI dataset [11] in

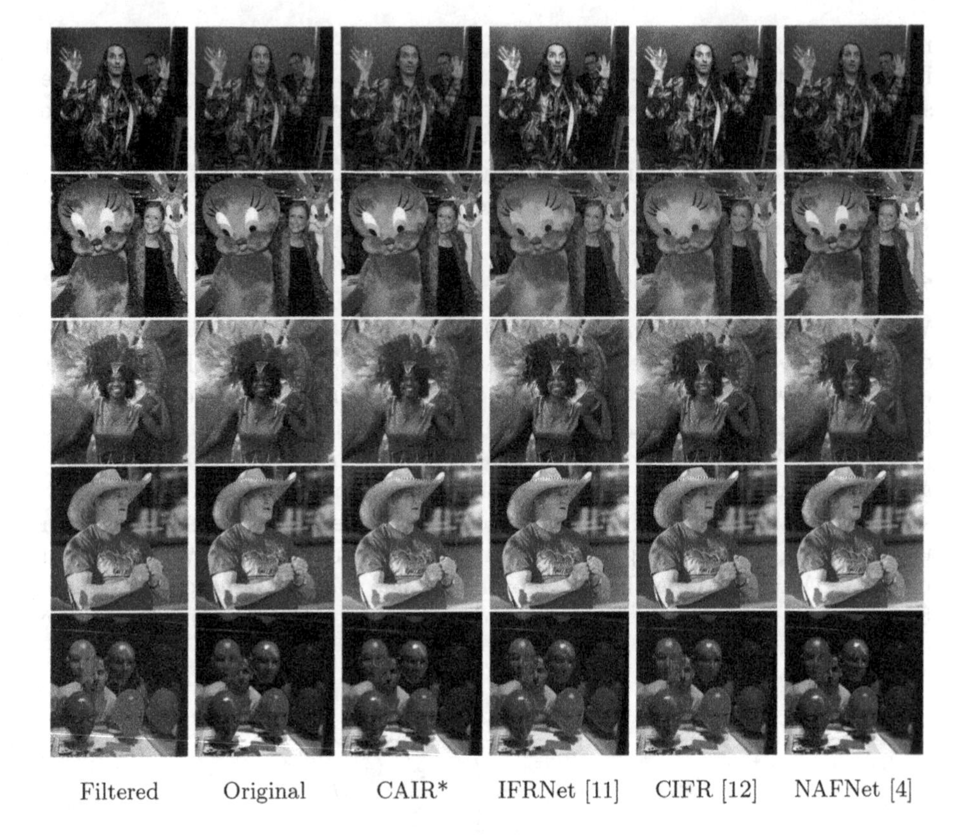

Fig. 6. Comparison of the qualitative results on Instagram filter removal on IFFI test dataset [11]. Filters applied (top to bottom): *Lo-Fi*, *Clarendon*, *Hudson*, *X-Proll*, *Nashville*.

Table 1. We made several observations based on the results. CAIR-S and CAIR-M outperform IFRNet and CIFR in all metrics by large margins, which validates the efficiency of the proposed method. CAIR-M also shows better PSNR and SSIM than NAFNet with fewer operations (GMACs). In addition, the result indicates that using the self-ensemble strategy [14] improved 0.24 dB PSNR and about 0.002 SSIM. CAIR* achieved the best PSNR and SSIM among compared methods, meaning that the ensemble learning creates a synergy effect when combining two images predicted from CAIR-S and CAIR-M.

Model Complexity. Figure 7 shows the PSNR vs. inference time tradeoffs and PSNR vs. the number of parameters. CAIR is much faster and has fewer parameters than the previous Instagram filter removal methods [11,12] while attaining more accurate results. The PSNR and SSIM of IFRNet and CIFR were obtained from the results in [11,12]. The runtime of all methods was measured on the same machine (4× NVIDIA Tesla V100). Although the inference time of

Table 1. Quantitative results on IFFI dataset [11]. Obtained the available results of PSNR and SSIM from [11,12]. Best results highlighted in bold. '+': self-ensemble [14], '*': ensemble learning strategy applied, 'Runtime': inference time per image on GPU, 'MACs': multiply accumulate operations per second, 'Params': number of parameters.

Method	PSNR	SSIM	Runtime (ms)	MACs (G)	Params (M)
IFRNet [11]	30.46	0.864	427.35	36.67	24.52
CIFR [12]	29.24	0.888	420.17	36.67	24.52
NAFNet [4]	31.76	0.958	29.64	16.21	29.16
CAIR-S	33.63	0.969	36.75	17.44	29.23
CAIR-S+	33.87	0.970	145.77	139.56	29.23
CAIR-M	34.15	0.969	45.15	15.23	13.13
CAIR-M+	34.39	0.971	186.22	121.83	13.13
CAIR*	**34.42**	**0.972**	25.87	12.4	0.028

NAFNet and the number of parameters are not significantly different from CAIR, it should be noted that CAIR exceeds PSNR and SSIM. The inference time and the number of parameters on CAIR* were measured only for inferencing with the ensemble network. CAIR* would take more time and computational complexity than the values in Table 1, since ensemble learning is followed by inferencing images with two models; CAIR-S and CAIR-M).

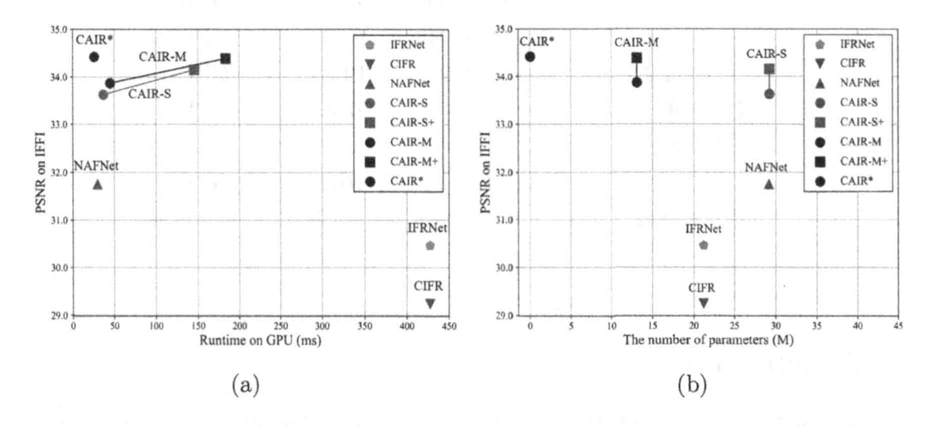

(a) (b)

Fig. 7. Comparison of the proposed CAIR and other methods. (a) PSNR vs. runtime (ms) trade-off, (b) PSNR vs. the number of parameters (M) on IFFI test sets tested on 4× NVIDIA Tesla V100.

CAIR for AIM 2022 Challenge. Table 2 shows the results of the AIM 2022 challenge on Instagram filter removal (IFR). Compared with the two baselines, IFRNet and CIFR, our method achieved significant improvement in all metrics.

Our team placed 5th in the AIM 2022 challenge on IFR [13]. However, in terms of inference speed, our proposed method was more than 10× the fastest in the top 5 and ranked second among all participants, as shown in Table 2. The model structure and training strategy are the same as described above, except that the ensemble learning strategy is slightly different. For ensemble learning, images were trained on three models, two CAIR-Ss of width 32 and 64 and one CAIR-M of width 32. Furthermore, the outputs from the three models were concatenated and fed into the ensemble network. The challenge ensemble results in Table 2 are worse than the CAIR* in Table 1. This result indicates that learning results can vary depending on how models are combined for ensemble learning. Therefore, when using the ensemble strategy in practice, it should be used carefully while checking the results.

Table 2. Results of AIM 2022 Instagram filter removal challenge [13].

Team name	PSNR	SSIM	Runtime (s)	CPU/GPU
Fivewin	**34.70**	**0.97**	0.91	GPU
CASIA LCVG	34.48	0.96	0.43	GPU
MiAlgo	34.47	**0.97**	0.40	GPU
Strawberry	34.39	0.96	10.47	GPU
SYU-HnVLab (ours)	33.41	0.95	0.04	GPU
XDER	32.19	0.95	0.05	GPU
CVRG	31.78	0.95	0.06	GPU
CVML	30.93	0.94	**0.02**	GPU
Couger AI	30.83	0.94	10.43	CPU
IFRNet [11] (Baseline1)	30.46	-	0.60	GPU
CIFR [12] (Baseline2)	30.02	-	0.62	GPU

5 Conclusion

This paper introduced CAIR, an end-to-end lightweight filter removal network suitable for Instagram filters. CAIR was based on the simple baseline for image restoration (NAFNet). The proposed color attention module, inspired by the color correction branch of CycleISP, can extract color-informative features from multi-scale input images so that filters can be effectively removed from filtered images. We achieved fast (45.14 ms inference time), lightweight (15.23 GMACs), and accurate (34.15 dB PSNR) performance with CAIR-M. The results were improved further by the self-ensemble and the ensemble learning strategy. Experimental results on the IFFI dataset verify that the proposed method is efficient and capable of achieving high speed and accuracy among the compared methods. Especially, CAIR restores color information during filter removal much better than the other methods. It is because the color attention modules are trained to extract color-attentive features with scale-invariant features. By extending the scope of the dataset for all available Instagram filters, this method could be employed for pre-processing the social media images before feeding them into a vision framework to enhance its performance.

Acknowledgments. This research was supported by Basic Science Research Program through the National Research Foundation (NRF) funded by the Ministry of Science and ICT (grant no. NRF-2021R1F1A1059493) and Electronics and Telecommunications Research Institute (ETRI) grant funded by the Korea government (grant no. 22ZB1140, Development of Creative Technology for ICT).

References

1. Ba, J.L., Kiros, J.R., Hinton, G.E.: Layer normalization. arXiv preprint arXiv:1607.06450 (2016)
2. Bianco, S., Cusano, C., Piccoli, F., Schettini, R.: Artistic photo filter removal using convolutional neural networks. J. Electron. Imaging **27**(1), 011004 (2017)
3. Brooks, T., Mildenhall, B., Xue, T., Chen, J., Sharlet, D., Barron, J.T.: Unprocessing images for learned raw denoising. In: Proceedings of the IEEE/CVF Conference on Computer Vision and Pattern Recognition, pp. 11036–11045 (2019)
4. Chen, L., Chu, X., Zhang, X., Sun, J.: Simple baselines for image restoration. arXiv preprint arXiv:2204.04676 (2022)
5. Chen, Y.H., Chao, T.H., Bai, S.Y., Lin, Y.L., Chen, W.C., Hsu, W.H.: Filter-invariant image classification on social media photos. In: Proceedings of the 23rd ACM international conference on Multimedia, pp. 855–858 (2015)
6. Chu, X., Chen, L., Chen, C., Lu, X.: Revisiting global statistics aggregation for improving image restoration. arXiv preprint arXiv:2112.04491 (2021)
7. Hendrycks, D., Dietterich, T.: Benchmarking neural network robustness to common corruptions and perturbations. arXiv preprint arXiv:1903.12261 (2019)
8. Hendrycks, D., Gimpel, K.: Gaussian error linear units (GELUS). arXiv preprint arXiv:1606.08415 (2016)
9. Huang, X., Belongie, S.: Arbitrary style transfer in real-time with adaptive instance normalization. In: Proceedings of the IEEE international conference on computer vision. pp. 1501–1510 (2017)
10. Isola, P., Zhu, J.Y., Zhou, T., Efros, A.A.: Image-to-image translation with conditional adversarial networks. In: Proceedings of the IEEE Conference on Computer Vision and Pattern Pecognition, pp. 1125–1134 (2017)
11. Kinli, F., Ozcan, B., Kirac, F.: Instagram filter removal on fashionable images. In: Proceedings of the IEEE/CVF Conference on Computer Vision and Pattern Recognition, pp. 736–745 (2021)
12. Kınlı, F., Özcan, B., Kıraç, F.: Patch-wise contrastive style learning for instagram filter removal. In: Proceedings of the IEEE/CVF Conference on Computer Vision and Pattern Recognition, pp. 578–588 (2022)
13. Kınlı, F., et al.: AIM 2022 challenge on Instagram filter removal: methods and results. In: European Conference on Computer Vision. Springer, Cham (2022)

14. Lim, B., Son, S., Kim, H., Nah, S., Mu Lee, K.: Enhanced deep residual networks for single image super-resolution. In: Proceedings of the IEEE Conference on Computer Vision and Pattern Recognition Workshops, pp. 136–144 (2017)
15. Loshchilov, I., Hutter, F.: Sgdr: Stochastic gradient descent with warm restarts. arXiv preprint arXiv:1608.03983 (2016)
16. Loshchilov, I., Hutter, F.: Decoupled weight decay regularization. arXiv preprint arXiv:1711.05101 (2017)
17. Lyu, Q., Shan, H., Wang, G.: MRI super-resolution with ensemble learning and complementary priors. IEEE Trans. Comput. Imaging **6**, 615–624 (2020)
18. Park, T., Efros, A.A., Zhang, R., Zhu, J.-Y.: Contrastive learning for unpaired image-to-image translation. In: Vedaldi, A., Bischof, H., Brox, T., Frahm, J.-M. (eds.) ECCV 2020. LNCS, vol. 12354, pp. 319–345. Springer, Cham (2020). https://doi.org/10.1007/978-3-030-58545-7_19
19. Paszke, A., et al.: Pytorch: An imperative style, high-performance deep learning library. In: Wallach, H., Larochelle, H., Beygelzimer, A., d' Alché-Buc, F., Fox, E., Garnett, R. (eds.) Advances in Neural Information Processing Systems 32, pp. 8024–8035. Curran Associates, Inc. (2019). http://papers.neurips.cc/paper/9015-pytorch-an-imperative-style-high-performance-deep-learning-library.pdf
20. Ronneberger, O., Fischer, P., Brox, T.: U-Net: convolutional networks for biomedical image segmentation. In: Navab, N., Hornegger, J., Wells, W.M., Frangi, A.F. (eds.) MICCAI 2015. LNCS, vol. 9351, pp. 234–241. Springer, Cham (2015). https://doi.org/10.1007/978-3-319-24574-4_28
21. Shi, W., et al.: Real-time single image and video super-resolution using an efficient sub-pixel convolutional neural network. In: Proceedings of the IEEE Conference on Computer Vision and Pattern Recognition, pp. 1874–1883 (2016)
22. Simonyan, K., Zisserman, A.: Very deep convolutional networks for large-scale image recognition. arXiv preprint arXiv:1409.1556 (2014)
23. Tu, Z., Talebi, H., Zhang, H., Yang, F., Milanfar, P., Bovik, A., Li, Y.: Maxim: multi-axis MLP for image processing. In: Proceedings of the IEEE/CVF Conference on Computer Vision and Pattern Recognition, pp. 5769–5780 (2022)
24. Wang, Z., Cun, X., Bao, J., Zhou, W., Liu, J., Li, H.: Uformer: a general u-shaped transformer for image restoration. In: Proceedings of the IEEE/CVF Conference on Computer Vision and Pattern Recognition, pp. 17683–17693 (2022)
25. Wu, Z., Wu, Z., Singh, B., Davis, L.: Recognizing Instagram filtered images with feature de-stylization. In: Proceedings of the AAAI Conference on Artificial Intelligence, vol. 34, pp. 12418–12425 (2020)
26. Zamir, S.W., Arora, A., Khan, S., Hayat, M., Khan, F.S., Yang, M.H.: Restormer: efficient transformer for high-resolution image restoration. In: Proceedings of the IEEE/CVF Conference on Computer Vision and Pattern Recognition, pp. 5728–5739 (2022)
27. Zamir, S.W., et al.: Cycleisp: Real image restoration via improved data synthesis. In: Proceedings of the IEEE/CVF Conference on Computer Vision and Pattern Recognition, pp. 2696–2705 (2020)

MicroISP: Processing 32MP Photos on Mobile Devices with Deep Learning

Andrey Ignatov[1,2](✉), Anastasia Sycheva[1], Radu Timofte[1,2], Yu Tseng[3], Yu-Syuan Xu[3], Po-Hsiang Yu[3], Cheng-Ming Chiang[3], Hsien-Kai Kuo[3], Min-Hung Chen[3], Chia-Ming Cheng[3], and Luc Van Gool[1,2]

[1] ETH Zurich, Zurich, Switzerland
andrey@vision.ee.ethz.ch
[2] AI Witchlabs Ltd., Zollikerberg, Switzerland
radu.timofte@uni-wuerzburg.de
[3] MediaTek Inc., Hsinchu, Taiwan

Abstract. While neural networks-based photo processing solutions can provide a better image quality compared to the traditional ISP systems, their application to mobile devices is still very limited due to their very high computational complexity. In this paper, we present a novel MicroISP model designed specifically for edge devices, taking into account their computational and memory limitations. The proposed solution is capable of processing up to 32MP photos on recent smartphones using the standard mobile ML libraries and requiring less than 1 s to perform the inference, while for FullHD images it achieves real-time performance. The architecture of the model is flexible, allowing to adjust its complexity to devices of different computational power. To evaluate the performance of the model, we collected a novel Fujifilm UltraISP dataset consisting of thousands of paired photos captured with a normal mobile camera sensor and a professional 102MP medium-format FujiFilm GFX100 camera. The experiments demonstrated that, despite its compact size, the MicroISP model is able to provide comparable or better visual results than the traditional mobile ISP systems, while outperforming the previously proposed efficient deep learning based solutions. Finally, this model is also compatible with the latest mobile AI accelerators, achieving good runtime and low power consumption o n smartphone NPUs and APUs. The code, dataset and pre-trained models are available on the project website: https://people.ee.ethz.ch/~ihnatova/microisp.html.

Keywords: Learned ISP · Computational photography · Mobile cameras · RAW Image Processing · Photo enhancement · Mobile AI · Deep learning · Efficient computer vision

1 Introduction

As camera quality becomes a prime feature of smartphones, the requirements on mobile photos grow each year. Developments in this field follow two directions:

A. Ignatov, R. Timofte and L. Van Gool—The main contact authors.

L. Karlinsky et al. (Eds.): ECCV 2022 Workshops, LNCS 13802, pp. 729–746, 2023.
https://doi.org/10.1007/978-3-031-25063-7_46

Fig. 1. Example set of full-resolution images (top) and crops (bottom) from the collected Fujifilm UltraISP dataset. From left to right: original RAW visualized image, RGB image obtained with MediaTek's built-in ISP system, and Fujifilm target photo. (Color figure online)

enhancements in camera sensor and optics hardware, and improvements related to computational photography. As the hardware limitations on sensor size and resolution are almost reached, the latter option plays an even more important role.

The problem of image restoration and enhancement has been addressed in several papers, though many were dealing only with particular aspects such as super-resolution [4,7,20,23,28,31,34,44,45,52], denoising [1,2,10,14,43,53,54], color and tone mapping [35,37,49,50], luminance, gamma and contrast adjustment [5,9,51]. Comprehensive end-to-end photo quality enhancement has first been addressed in [18,23], where the authors proposed to learn a mapping between low-quality RGB smartphone images and target high-quality DSLR photos with deep learning. Despite huge subsequent progress [12,13,18,23,32, 33,42,47], this approach had a significant limitation: the photos that one gets with smartphone ISPs undergo many image processing steps that heavily alter the original pixel data, and lead to a severe information loss caused by noise suppression, narrowing of the original dynamic range, jpeg compression, etc. Thus, one can get much better results when working directly with the original RAW sensor data. This approach was explored in [6,21,24,25,27,38], where the authors presented the Zurich RAW-to-RGB dataset and obtained results comparable or better than the ones of the ISP system of the Huawei P20 smartphone. This was an important proof of concept showing it is possible to replace conventional hand-crafted ISP pipelines with end-to-end deep learning approaches, though a significant limitation was left: the models were too heavy to run on real mobile devices. Thus, the practical application of these solutions was very limited. An important step in solving this problem was done in the Mobile AI Challenge [15], where the participants were developing efficient ISP models for inference on mobile devices. However, the size of the photos in this challenge was still limited to 2MP, while the resolution of real smartphone cameras is at least 12MP and can be as high as 108MP.

In this paper, we propose the first deep learning based solution able to process 32MP RAW photos on smartphones and designed taking into account their hardware limitations. As the current public RAW-to-RGB dataset [25] has issues with the quality of the target images and their alignment to the original RAW data, we collect a novel large-scale dataset using a professional medium format camera capturing 102MP photos and a Sony mobile sensor for getting the original RAW data, and align pixel-wise the obtained photos. Finally, we present experiments evaluating the quality of the resulting RGB images and the runtime of the solution on recent flagship mobile platforms and AI accelerators.

The remainder of the paper is structured as follows. In Sect. 2 we describe the new Fujifilm UltraISP dataset. Section 3 presents our MicroISP architecture and describes the underlying design choices. Section 4 shows and analyzes the experimental results and discusses the limitations of the solution. Finally, Sect. 5 concludes the paper.

2 Fujifilm UltraISP Dataset

When dealing with an end-to-end learned smartphone ISP, the quality of the target images used for training the model plays a crucial. Thus, the requirements on the target camera are high: it should produce photos that are outstanding in terms of real resolution, noise free even when captured in low light conditions, exhibit a high dynamic range and pleasant color rendition, and are sharp enough when shooting them with an open aperture. As our exploration revealed that none of the currently existing APS-C and full-frame cameras satisfy all those requirements, we used the Fujifilm GFX100, a medium format 102 MP camera, for capturing the target high-quality photos. To collect the source RAW smartphone images, we chose a popular Sony IMX586 Quad Bayer camera sensor that can be found in tens of mid-range and high-end mobile devices released in the past 3 years. This sensor was mounted on the MediaTek Dimensity 820 development board, and was capturing both raw and processed (by its built-in ISP system) 12MP images. The Dimensity board was rigidly attached to the Fujifilm camera and controlled using a specialized software developed for this project. The cameras were capturing photos synchronously to ensure that the image content is identical. This setup was used for several weeks to collect over 6 thousand daytime image pairs at a wide variety of places with different illumination and weather conditions. An example set of full-resolution photos from the collected dataset is shown in Fig. 1.

As the collected RAW-RGB image pairs were not perfectly aligned, we had to perform local matching first. In order to achieve a precise pixel-wise alignment, we used the SOTA deep learning based dense matching algorithm [46] to extract 256×256 px patches from the original photos. This procedure resulted in over 99 K pairs of crops that were divided into training (93.8 K), validation (2.2 K)

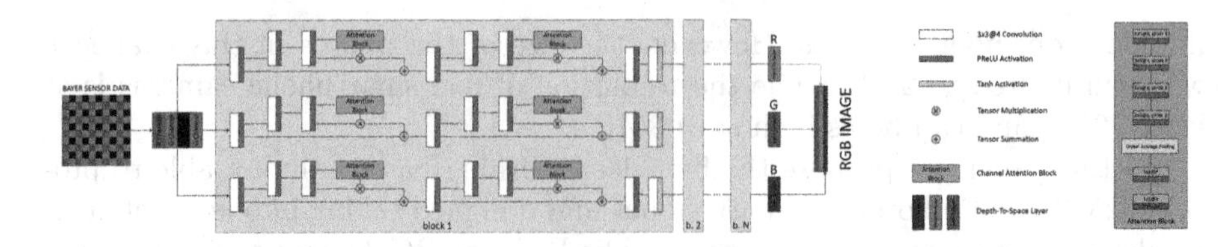

Fig. 2. The overall architecture of the proposed MicroISP model (left), and the structure of the enhanced attention block (right).

and test (3.1 K) sets and used for model training and evaluation. It should be mentioned that all alignment operations were performed on Fujifilm RGB images only, therefore RAW photos from the Sony sensor remained unmodified, exhibiting exactly the same values as read from the sensor.

3 Architecture

When designing a model capable of processing high-resolution images on mobile devices, one needs to address the following limitations related to edge inference:

- **Memory consumption:** unlike the standard desktop systems, mobile devices have a limited amount of accessible RAM. This restriction becomes even more severe when the inference happens on mobile AI accelerators such as NPUs or APUs that usually have their own memory limited to hundreds of megabytes.
- **Layers and operators:** mobile AI accelerators support only a restricted set of common machine learning ops. Thus, on the latest Android devices, one is limited to 101 different operators at maximum [3,41], while older mobile AI accelerators might be supporting even less than 28 layers [39] (including the basic ones such as summation, multiplication, convolution, *etc*).
- **Computational complexity:** with common architectures such as U-Nets / ResNets / PyNETs, it takes tens of seconds to process 32MP images on desktop or server systems with high-end Nvidia GPUs. As even the latest smartphone AI accelerators are considerably less powerful, model complexity should be very low in order to achieve a runtime of 1–2 secs.
- **Model size:** since the resulting NN model is usually integrated in the camera application, its size should be reasonably small, not exceeding several megabytes.

We present a solution that addresses all these limitations while achieving good visual results. The overall architecture of the proposed MicroISP model is illustrated in Fig. 2. Below we discuss its details and justify our design choices.

Overall Image Processing Workflow. The model accepts the raw RGBG Bayer data coming directly from the camera sensor. The input is then grouped in 4 feature maps corresponding to each of the four RGBG color channels using the space-to-depth op. Next, this input is processed in parallel in 3 model branches corresponding to the R, G and B color channels and consisting of N residual building blocks (2 blocks are used by default if not specified). After applying the depth-to-space op at the end of each branch, their outputs are concatenated into the reconstructed RGB photo.

Convolutional Layers. When running the model on mobile NPUs or GPUs, the memory is typically allocated per layer/operator. Thus, the maximum RAM consumption is defined by the largest layer that becomes a bottleneck for high-resolution inference. Since the memory consumption for a conv layer is proportional to the number of input/output feature maps N_{in}/N_{out} and their size $H \times W$, our model only uses *4 convolutional filters* in every layer to achieve the minimum possible RAM consumption. This is the smallest possible filter size: as the model takes 4 input RGBG channels, one would need to have at least 4 filters of the same size in the first conv layer to avoid information loss (if pixel shuffle/strided convolution with stride c is applied, then the number of filters should be $4 \times c \times c$, leading to the same resulting memory footprint). Each convolutional layer is followed by the *PReLU* activation with shared non-channel dimensions, meaning that only 4 parameters corresponding to each input channel are learned.

Model Branches. The MicroISP model processes the input sensor data in 3 separate branches. First of all, this is done to fit the above mentioned memory constraints while not limiting the performance of the model: if only one branch is used, then 4 convolutional filters would not be enough to perform an accurate image demosaicing and texture reconstruction. Note that in this case one would also need to use 12 filters in the last conv layer since the final depth-to-space op should produce an output image with 3 channels of twice larger resolution (using the transposed convolution instead would lead to serve checkerboard artifacts in the resulting photo, and thus should be avoided).

The other benefit of using separate branches is that this allows the model to learn a different set of features specific for each color space. Indeed, as these branches get the same input data, they can extract and work only with features relevant for reconstructing R, G and B image channels, respectively, while dropping non-relevant information.

Finally, the proposed branch structure is also beneficial from the performance perspective: if the AI accelerator has enough RAM, it can run these branches in parallel as they are independent from each other, and thus the runtime can be decreased by up to 3 times. Alternatively, these branches can be executed sequentially if the resolution of the input photos is too high for parallel data processing.

Attention Blocks. To ensure that the model is able to perform global image processing such as white balancing, gamma and color correction, we added an enhanced channel attention block with the structure of Fig. 2. While the standard attention units are using global average pooling followed by several conv layers, our initial experiments revealed that the performance of these blocks is not

Fig. 3. Sample visual results obtained with the proposed deep learning method. Best zoomed on screen.

sufficient in our case as they are not taking into account any information about the image content that is removed after the pooling layer. Thus, we propose an enhanced structure: first, a 1×1 convolution with stride 3 is applied to reduce the dimensionality of the feature maps. Next, three convolutional blocks with 3×3 filters and stride 3 are applied to learn the global content-dependent features and reduce the resolution of the feature maps another 27 times. Finally, the average pooling op is used to get $1 \times 1 \times 4$ features that are then passed to 2 additional conv layers generating the normalization coefficients. The proposed architecture is both performant and computationally efficient due to aggressive dimensionality reduction, leading to an execution time of the overall attention block appr. twice smaller than the runtime of one normal 3×3 convolution.

Model Operators. The proposed MicroISP model contains only layers supported by the Neural Networks API 1.2 [40], and thus can run on any NNAPI-compliant AI accelerator (such as NPU, APU, DSP or GPU) available on mobile devices with Android 10 and above. One can additionally relax the above requirements to NNAPI 1.0 and Android 8.1, resp., if replacing *PReLU* with *Leaky ReLU*. The effect of this change is discussed in the next section.

Model Size and Memory Consumption. The size of the MicroISP network is only 158 KB when exported for inference using the TFLite FP32 format. The model consumes around 90, 475 and 975 MB of RAM when processing FullHD, 12MP and 32MP photos on mobile GPUs, respectively.

Training Details. The model was trained in three stages. First, only the MSE loss was used for 200 epochs to get the initial reconstruction results. Next,

Fig. 4. Sample crops from the MediaTek ISP photos (top) and the corresponding images processed with the MicroISP model (bottom).

the model was tuned for another 200 epochs with a combination of the VGG-based [26] perceptual, SSIM and MSE losses to improve the texture quality, enhance the details and image sharpness:

$$\mathcal{L}_{\text{Stage 2}} = \mathcal{L}_{\text{VGG}} + 0.5 \cdot \mathcal{L}_{\text{SSIM}} + 0.25 \cdot \mathcal{L}_{\text{MSE}},$$

where the value of each loss is normalized to 1. Finally, the network was fine-tuned for another 100 epochs with a combination of the SSIM and MSE loss functions taken in the ratio of 2:1. This was done to perform the final tone mapping adjustments and to improve edge rendering.

Implementation Details. The model was implemented in TensorFlow and trained on a single *Nvidia Titan X* GPU with batch size 50. The network parameters were optimized for 500 epochs using the ADAM [29] algorithm with a learning rate of $2e^{-5}$. Random flips and rotations were applied to augment the training data and prevent overfitting.

4 Experiments

In this section, we evaluate the proposed MicroISP architecture on the real Fujifilm UltraISP dataset and mobile devices to answer the following questions:

- Is the model able to perform an accurate reconstruction of the RGB images;
- How good are the results compared to the standard hand-crafted ISP pipelines used in modern phones;
- How well this solution performs compared to the commonly used deep learning models tuned for this task;
- What is the largest image resolution that can be processed by the MicroISP model on mobile devices;
- What is the runtime of this model when performing the inference on mobile GPUs and AI accelerators,
- What are the limitations of the proposed solution.

To answer these questions, we performed a wide range of experiments which results are described in detail in the following sections.

Fig. 5. From left to right, top to bottom: the original visualized cropped RAW image, and the same image after applying, respectively: SmallNet [15], FPIE [42], FSRCNN [8], Compressed UNet [15], ENERZAi [15], CSANet [11] and our MicroISP model.

4.1 Qualitative Evaluation

As the perceptual quality of the reconstructed photos is our primary target, we started the experiments with a brief analysis of the visual results obtained with the MicroISP model. Figure 3 shows sample RGB images reconstructed with the proposed solution together with the original RAW photos, images obtained with MediaTek's built-in ISP system, and the target photos from the Fujifilm camera. The first observation demonstrates that the results produced by the model are valid: it was able to perform an accurate color reconstruction with decent tone mapping, the quality of texture and the overall image sharpness are quite good, white balancing is performed correctly. The same can be also applied to the dynamic range that is close to the one on the target Fujifilm images. No notable issues or artifacts are observed at both local and global levels, complex overexposed image areas are also handled correctly.

Surprisingly, the images rendered with the neural network turned out to look more natural compared to the ones obtained with the built-in ISP pipeline. A more detailed analysis of image crops revealed that this is mainly caused by a strong watercolor effect present on the majority of photos processed with the ISP system (Fig. 4). The reason for this is that most modern smartphones apply numerous filters used for image sharpening and low-level texture enhancement, though together with aggressive noise suppression algorithms they are often leading to a mess of pixels in complex image areas such as grass or leaves, and a notable loss of colors. Overall, the photos obtained with the ISP system and with the proposed solution are following the two considerably different approaches: the first ones have boosted colors, a significantly increased brightness and lots of sharpening used to make them visually more appealing, while the images processed with the MicroISP model are looking naturally as one would expect them to be. The real resolution of the ISP images is slightly higher for some scenes, though the difference is overall quite negligible. It should be also noted that, as expected, the target Fujifilm photos significantly outperform the results

Table 1. Average PSNR / SSIM results on test images.

Method	PSNR	SSIM
SmallNet [15]	23.20	0.847
FPIE [42]	23.23	0.848
FSRCNN [8]	23.27	0.830
Compressed U-Net [15]	23.30	0.840
ENERZAi [15]	23.41	**0.853**
CSANet [11]	23.73	0.849
MicroISP	**23.87**	**0.853**

of the ISP pipeline and the MicroISP model in all aspects, especially in terms of resolution.

4.2 Quantitative Evaluation

As the proposed solution was designed targeting good visual results and fast on-device high resolution image processing, in the next two sections we compare its numerical and runtime results against the previously introduced deep learning based approaches allowing to perform RGB image reconstruction directly on smartphones. The following models are used in the next experiments:

- FSRCNN [8]: a popular computationally efficient model used for various image enhancement problems.
- FPIE [42]: an enhanced DPED [16]-based neural network optimized for fast on-device image processing.
- CSANet [11]: an NPU-friendly architecture developed for the learned smartphone ISP problem.
- Compressed U-Net [15]: a U-Net [36] based model with hardware-specific adaptations for edge inference.
- ENERZAi [15]: a model designed for efficient image ISP, derived from the ESRGAN [48] architecture.
- SmallNet [15]: a fast FSRCNN [8] based architecture optimized for the learned smartphone ISP task.

All models were trained on the Fujifilm UltraISP dataset, their PSNR and SSIM scores on the test image subset are reported in Table 1, sample visual results for all methods are demonstrated in Fig. 5. The proposed MicroISP network was able to substantially outperform the other solutions in almost all aspects. In particular, it offers a 0.5dB PSNR improvement compared to the baseline FSR-CNN model, and outperforms by 0.14 dB the CSANet model demonstrating the second best PSNR results on the considered dataset. When analyzing the visual results, one can note that, unlike the majority of other solutions, it produces images with smooth and clear texture and without checkerboard artifacts that

Table 2. The runtime of different deep learning-based models on the MediaTek Dimensity 1000+ GPU obtained using the publicly available AI Benchmark application [19,22]. The results of the DPED and PyNET models are provided for the reference. OOM stands for the "out-of-memory" exception thrown by the interpreter when trying to perform the inference.

Method	Runtime on the Dimensity 1000+ GPU				
	FullHD, ms	12MP, ms	18MP, ms	26MP, ms	32MP, ms
SmallNet [15]	18.4	100	148	OOM	OOM
FPIE [42]	208	1138	OOM	OOM	OOM
FSRCNN [8]	40.8	232	335	OOM	OOM
Compressed U-Net [15]	29.1	140	208	OOM	OOM
ENERZAi [15]	31.2	123	184	OOM	OOM
CSANet [11]	44.2	241	358	OOM	OOM
DPED [16]	658	4027	OOM	OOM	OOM
PyNET [25]	12932	OOM	OOM	OOM	OOM
MicroISP	42.3	238	354	522	636

are especially severe on the photos processed by the FPIE and CSANet networks. Additionally, this is the only model that also learned to perform image denoising — an aspect that is critical when processing mobile photos, where noise is often present even on images captured in good lighting conditions.

4.3 Runtime Evaluation

As the proposed solution is designed for on-device image processing, we perform its performance evaluation directly on mobile phones to get the real runtime values and take into account all limitations related to edge inference. For this, we used the publicly available AI Benchmark application [19,22] that allows to load any custom TensorFlow Lite model and run it on any Android device with various acceleration options including GPU, NPU / DSP and CPU inference. Same as in [15], we used the MediaTek Dimensity 1000+ mobile SoC for performing runtime evaluation, and accelerated the models on its Mali-G77 GPU as this option delivered the best latency for all architectures. The results of all models on FullHD, 12MP, 18MP, 26MP and 32MP photos are reported in Table 2.

As expected, the MicroISP network was able to achieve the lowest memory consumption, and thus was the only architecture capable of processing 26MP and 32MP images, while in all other cases the TensorFlow interpreter failed to perform the inference with the *out-of-memory* exception. The model was able to achieve a runtime of 42.3, 238 and 354 ms on FullHD, 12MP and 18MP photos, respectively, which is comparable to the latency of the FSRCNN and CSANet models, while the proposed solution provides better numerical and visual results.

Table 3. The speed of the proposed MicroISP architecture on several popular mobile GPUs for different photo resolutions. The runtime was measured with the AI Benchmark app using the TFLite GPU delegate [30].

Mobile SoC GPU	Dimensity Next Mali-Next, ms	Dimensity 820 Mali-G57 MC5, ms	Exynos 2100 Mali-G78 MP14, ms	Exynos 990 Mali-G77 MP11, ms	Kirin 9000 Mali-G78 MP24, ms	Snapdragon 888 Adreno 660, ms	Google Tensor Mali-G78 MP20, ms
Full HD	33.4	72.7	31.2	46.2	39.4	30.4	36.8
12MP	170	416	174	248	189	153	188
32MP	624	1059	489	690	507	465	480

Though the SmallNet, Compressed U-Net and ENERZAi models are significantly faster in this experiment, in Sect. 4.6 we will demonstrate that one can easily achieve a similar runtime and fidelity results by adjusting the number of MicroISP's building blocks. In Table 2, we additionally provide the latency of the DPED [16] and PyNET [25] models: despite achieving very good numerical results on the Fujifilm UltraISP dataset (24.2 and 25dB PSNR, respectively), it is almost infeasible to run them on the latest mobile devices due to their huge runtime even on FullHD and 12MP images (more than 15x and 300x higher compared to the MicroISP model).

Finally, we checked the runtime of the proposed solution on all popular mobile chipsets, and report the obtained results in Table 3. When processing FullHD images, the MicroISP model was able to achieve a speed of 30 frames per second on the Dimensity Next, Exynos 2100 and Snapdragon 888 platforms, which demonstrates that it can be potentially used for real-time FullHD RAW video processing. The results also show that it can perform 32MP photo rendering under 0.5 s on almost all flagship SoCs, though even for mid-range platforms like the MediaTek Dimensity 820 its latency remain reasonably small.

4.4 Inference on Mobile NPUs

While the proposed MicroISP model can achieve good runtime results for images of different resolutions on mobile GPUs, one might be interested in running it on dedicated AI accelerators to further improve its latency or to efficiently decrease the power consumption. For this, in this section we evaluate its performance on MediaTek's latest mobile platform, *Dimensity Next*, which features a powerful AI Processing Unit (APU) designed specifically for complex computer vision and image processing tasks. Table 4 shows the runtime and power consumption

Table 4. The runtime and power consumption of the MicroISP model on the MediaTek Dimensity Next mobile platform.

	MediaTek Dimensity Next mobile chipset				
	CPU	GPU (FP16)	APU (FP16)	APU (INT16)	APU (INT8)
Runtime, ms	242	33	63	43	**29**
Power, fps/watt	0.43	5.27	7.20	7.78	**16.61**

Fig. 6. Sample visual results of the MicroISP models with different depth multipliers.

results obtained on this chipset for FullHD images when running the MicroISP model on CPU, GPU and APU. The considered AI accelerator was able to execute the entire floating-point model without any partitioning, demonstrating a 35% and 1500% increase in power efficiency compared to GPU and CPU inference, respectively. These numbers are further increased when quantizing and converting the model to INT16 and INT8 formats. In the latter case, the speed improves from 30 to 34 FPS, and the energy consumption decreases by 3 times compared to GPU execution.

4.5 Ablation Study

To evaluate the efficiency of the proposed model design, we performed an ablation study which results are reported in Table 5. First of all, we checked the importance of the attention unit by either disabling it completely, or by replacing with the conventional implementation utilizing the global average pooling in the first layer. The results demonstrate a very high importance of this building block for the proposed task: thus, the PSNR score drops by more than 0.5 dB when this block is removed. While using the standard implementation improves the scores to 23.34 dB, these results are still considerably lower compared to the ones obtained with the proposed attention modification that takes into account image content when computing the normalization coefficients.

Another important design choice was to use the *PReLU* activations instead of the *Leaky ReLU*. While this is almost not affecting the model complexity (as is equivalent to the *Leaky ReLU* with learned slope for 4 input channels), it allows to substantially boost the quality of the reconstructed image results due

Table 5. The results of the model 1) without the attention block, 2) with the standard attention block using the average pooling, 3) using Leaky ReLU instead of PReLU, 4) the final architecture.

Model Modification	PSNR	SSIM
No Attention Block	23.25	0.835
Standard Attention Block	23.34	0.842
Leaky ReLU instead of *PReLU*	23.54	0.847
Final Design	23.87	0.853

Table 6. Quantitative and runtime results of several MicroISP models with different depth multipliers.

Model Depth Multiplier	PSNR	SSIM	Runtime on the Dimensity 1000+ GPU		
			FullHD, ms	12MP, ms	32MP, ms
MicroISP 1.5 [3 blocks]	23.91	0.854	56.8	314	837
MicroISP 1.0 [2 blocks]	23.87	0.853	42.3	238	636
MicroISP 0.5 [1 block]	23.60	0.846	28.2	155	417
MicroISP 0.25 [half block]	23.37	0.841	23.1	122	336

to using additional global image adjustment parameters. We should also note that the *Leaky ReLU* implementation might still be of interest when running the MicroISP model on older mobile NPUs as they might lack *PReLU* support introduced in Android NNAPI 1.2.

The final principal model parameter is the number of basic building blocks, its effect on the visual and runtime results is discussed in the next section.

4.6 Adjusting the Model Complexity

The MicroISP model allows to adapt its computational complexity to the target hardware platform by changing the number of its building units (Fig. 2). Since the RAM consumption remains almost constant regardless of the model depth, one can potentially design very large or very small networks depending on the task, target runtime and computational budget. Table 6 demonstrates the effect of the model size on its fidelity and runtime numbers for the considered ISP problem, while Fig. 6 shows the corresponding visual results. As one can see, it is possible to reduce the runtime by more than 33% by switching to the MicroISP 0.5 network with one building block at the expense of slightly worse image reconstruction quality. If the latency constraints are very tight, one can further reduce the runtime with the MicroISP 0.25 model: even in this case the overall image rendering results are still acceptable, though the texture quality is obviously lower (this, however, might not be very critical *e.g.*, for real-time video processing). In this particular task, any further increase of model complexity is leading to only marginal performance improvements, thus using two building blocks delivers the best runtime-quality trade-off for the considered problem.

4.7 Limitations

As this is an end-to-end solution, the reconstruction issues on some images are generally inevitable. While the overall quality of the results is high, appr. 5–7% of images might exhibit an imprecise white balancing, with pinkish or yellowish tones visible in bright photo regions. Next, though the proposed model can handle low and medium noise levels, it cannot suppress heavy noise in images captured at night or in the dark (note however, that the model was not trained for night photo processing as it never observed such image samples). Another global issue is vignetting caused by optics, but this problem can be efficiently fixed with the standard algorithms. Finally, though the experiments revealed that the real resolution of the reconstructed photos is close to the one of the images processed with the classical ISP system, one might want to improve it further as the proposed dataset allows to train the model to perform an additional 2 times image upscaling, though a larger model might be required in this case.

4.8 PyNET-V2 Mobile

Besides the MicroISP model, we also developed a considerably more powerful PyNET-V2 Mobile architecture, which structure is inspired by the original PyNET [25] model while its design was fully revised in order to be compatible with mobile AI accelerators. The PyNET-V2 Mobile network achieves a PSNR score of 24.72 dB (+0.85 dB compared to the MicroISP model) on the considered FujiFilm UltraISP dataset, and its runtime on the Dimensity 9000 APU is less than 800 ms when processing raw Full HD resolution images. A detailed description of this architecture and its results can be found in paper [17].

5 Conclusion

We proposed a novel deep learning based MicroISP architecture allowing to process RAW photos of resolution up to 32MP directly on mobile devices. The presented solution learns to perform all image processing steps directly from the data, not requiring any manual supervision or hand-crafted features. To check performance, we collected a large-scale Fujifilm UltraISP dataset consisting of more than 6K RAW-RGB image pairs captured by a normal mobile camera sensor and a professional 102MP medium format Fujifilm camera. The experiments revealed that the quality of the reconstructed RGB images is comparable to the results obtained with a classical ISP system, though texture-wise the deep learning based solution is providing superior results. The conducted runtime evaluation shows that the MicroISP model is capable of processing 32MP photos on the majority of mobile SoCs under 500 ms, while for FullHD images it demonstrates real-time performance. Finally, the proposed architecture is also compatible with dedicated mobile AI accelerators such as APUs or NPUs, allowing to further improve the runtime and to reduce power consumption, which might be critical for edge inference.

References

1. Abdelhamed, A., Afifi, M., Timofte, R., Brown, M.S.: NTIRE 2020 challenge on real image denoising: dataset, methods and results. In: Proceedings of the IEEE/CVF Conference on Computer Vision and Pattern Recognition Workshops, pp. 496–497 (2020)
2. Abdelhamed, A., Timofte, R., Brown, M.S.: NTIRE 2019 challenge on real image denoising: Methods and results. In: Proceedings of the IEEE/CVF Conference on Computer Vision and Pattern Recognition Workshops (2019)
3. API, A.N.N.: https://source.android.com/devices/neural-networks
4. Cai, J., Gu, S., Timofte, R., Zhang, L.: NTIRE 2019 challenge on real image super-resolution: Methods and results. In: Proceedings of the IEEE/CVF Conference on Computer Vision and Pattern Recognition Workshops (2019)
5. Cai, J., Gu, S., Zhang, L.: Learning a deep single image contrast enhancer from multi-exposure images. IEEE Trans. Image Process. **27**(4), 2049–2062 (2018)
6. Dai, L., Liu, X., Li, C., Chen, J.: AWNet: attentive wavelet network for image ISP. arXiv preprint arXiv:2008.09228 (2020)
7. Dong, C., Loy, C.C., He, K., Tang, X.: Image super-resolution using deep convolutional networks. IEEE Trans. Pattern Anal. Mach. Intell. **38**(2), 295–307 (2015)
8. Dong, C., Loy, C.C., Tang, X.: Accelerating the super-resolution convolutional neural network. In: Leibe, B., Matas, J., Sebe, N., Welling, M. (eds.) ECCV 2016. LNCS, vol. 9906, pp. 391–407. Springer, Cham (2016). https://doi.org/10.1007/978-3-319-46475-6_25
9. Fu, X., Zeng, D., Huang, Y., Liao, Y., Ding, X., Paisley, J.: A fusion-based enhancing method for weakly illuminated images. Signal Process. **129**, 82–96 (2016)
10. Gu, S., Timofte, R.: A brief review of image denoising algorithms and beyond. In: Inpainting and Denoising Challenges, pp. 1–21 (2019)
11. Hsyu, M.C., Liu, C.W., Chen, C.H., Chen, C.W., Tsai, W.C.: CSANet: high speed channel spatial attention network for mobile ISP. In: Proceedings of the IEEE/CVF Conference on Computer Vision and Pattern Recognition Workshops (2021)
12. Huang, J., et al.: Range scaling global U-Net for perceptual image enhancement on mobile devices. In: Leal-Taixé, L., Roth, S. (eds.) ECCV 2018. LNCS, vol. 11133, pp. 230–242. Springer, Cham (2019). https://doi.org/10.1007/978-3-030-11021-5_15
13. Hui, Z., Wang, X., Deng, L., Gao, X.: Perception-preserving convolutional networks for image enhancement on smartphones. In: Leal-Taixé, L., Roth, S. (eds.) ECCV 2018. LNCS, vol. 11133, pp. 197–213. Springer, Cham (2019). https://doi.org/10.1007/978-3-030-11021-5_13
14. Ignatov, A., Byeoung-su, K., Timofte, R., Pouget, A.: Fast camera image denoising on mobile gpus with deep learning, mobile AI 2021 challenge: report. In: Proceedings of the IEEE/CVF Conference on Computer Vision and Pattern Recognition, pp. 2515–2524 (2021)
15. Ignatov, A., Chiang, J., Kuo, H.K., Sycheva, A., Timofte, R.: Learned smartphone isp on mobile npus with deep learning, mobile AI 2021 challenge: report. In: Proceedings of the IEEE/CVF Conference on Computer Vision and Pattern Recognition Workshops (2021)
16. Ignatov, A., Kobyshev, N., Timofte, R., Vanhoey, K., Van Gool, L.: DSLR-quality photos on mobile devices with deep convolutional networks. In: Proceedings of the IEEE International Conference on Computer Vision, pp. 3277–3285 (2017)

17. Ignatov, A., et al.: PyNet-V2 Mobile: efficient on-device photo processing with neural networks. In: 2021 26th International Conference on Pattern Recognition (ICPR). IEEE (2022)
18. Ignatov, A., Timofte, R.: NTIRE 2019 challenge on image enhancement: methods and results. In: Proceedings of the IEEE/CVF Conference on Computer Vision and Pattern Recognition Workshops (2019)
19. Ignatov, A., et al.: AI benchmark: running deep neural networks on android smartphones. In: Leal-Taixé, L., Roth, S. (eds.) ECCV 2018. LNCS, vol. 11133, pp. 288–314. Springer, Cham (2019). https://doi.org/10.1007/978-3-030-11021-5_19
20. Ignatov, A., Timofte, R., Denna, M., Younes, A.: Real-time quantized image super-resolution on mobile NPUs, mobile AI 2021 challenge: report. In: Proceedings of the IEEE/CVF Conference on Computer Vision and Pattern Recognition Workshops (2021)
21. Ignatov, A., et al.: Aim 2019 challenge on raw to RGB mapping: methods and results. In: 2019 IEEE/CVF International Conference on Computer Vision Workshop (ICCVW), pp. 3584–3590. IEEE (2019)
22. Ignatov, A., et al.: AI benchmark: all about deep learning on smartphones in 2019. In: 2019 IEEE/CVF International Conference on Computer Vision Workshop (ICCVW), pp. 3617–3635. IEEE (2019)
23. Ignatov, A., et al.: PIRM challenge on perceptual image enhancement on smartphones: report. In: Leal-Taixé, L., Roth, S. (eds.) ECCV 2018. LNCS, vol. 11133, pp. 315–333. Springer, Cham (2019). https://doi.org/10.1007/978-3-030-11021-5_20
24. Ignatov, A., et al.: AIM 2020 challenge on learned image signal processing pipeline. arXiv preprint arXiv:2011.04994 (2020)
25. Ignatov, A., Van Gool, L., Timofte, R.: Replacing mobile camera ISP with a single deep learning model. In: Proceedings of the IEEE/CVF Conference on Computer Vision and Pattern Recognition Workshops, pp. 536–537 (2020)
26. Johnson, J., Alahi, A., Fei-Fei, L.: Perceptual losses for real-time style transfer and super-resolution. In: Leibe, B., Matas, J., Sebe, N., Welling, M. (eds.) ECCV 2016. LNCS, vol. 9906, pp. 694–711. Springer, Cham (2016). https://doi.org/10.1007/978-3-319-46475-6_43
27. Kim, B.-H., Song, J., Ye, J.C., Baek, J.H.: PyNET-CA: enhanced PyNET with channel attention for end-to-end mobile image signal processing. In: Bartoli, A., Fusiello, A. (eds.) ECCV 2020. LNCS, vol. 12537, pp. 202–212. Springer, Cham (2020). https://doi.org/10.1007/978-3-030-67070-2_12
28. Kim, J., Lee, J.K., Lee, K.M.: Accurate image super-resolution using very deep convolutional networks. In: Proceedings of the IEEE Conference on Computer Vision and Pattern Recognition, pp. 1646–1654 (2016)
29. Kingma, D.P., Ba, J.: Adam: a method for stochastic optimization. arXiv preprint arXiv:1412.6980 (2014)
30. Lee, J., et al.: On-device neural net inference with mobile GPUs. arXiv preprint arXiv:1907.01989 (2019)
31. Lim, B., Son, S., Kim, H., Nah, S., Lee, K.M.: Enhanced deep residual networks for single image super-resolution. In: Proceedings of the IEEE Conference on Computer Vision and Pattern Recognition workshops, pp. 136–144 (2017)
32. Liu, H., Navarrete Michelini, P., Zhu, D.: Deep networks for image-to-image translation with Mux and Demux layers. In: Leal-Taixé, L., Roth, S. (eds.) ECCV 2018. LNCS, vol. 11133, pp. 150–165. Springer, Cham (2019). https://doi.org/10.1007/978-3-030-11021-5_10

33. Lugmayr, A., Danelljan, M., Timofte, R.: Unsupervised learning for real-world super-resolution. In: 2019 IEEE/CVF International Conference on Computer Vision Workshop (ICCVW), pp. 3408–3416. IEEE (2019)
34. Lugmayr, A., Danelljan, M., Timofte, R.: NTIRE 2020 challenge on real-world image super-resolution: Methods and results. In: Proceedings of the IEEE/CVF Conference on Computer Vision and Pattern Recognition Workshops, pp. 494–495 (2020)
35. Ma, K., Yeganeh, H., Zeng, K., Wang, Z.: High dynamic range image tone mapping by optimizing tone mapped image quality index. In: 2014 IEEE International Conference on Multimedia and Expo (ICME), pp. 1–6. IEEE (2014)
36. Ronneberger, O., Fischer, P., Brox, T.: U-Net: convolutional networks for biomedical image segmentation. In: Navab, N., Hornegger, J., Wells, W.M., Frangi, A.F. (eds.) MICCAI 2015. LNCS, vol. 9351, pp. 234–241. Springer, Cham (2015). https://doi.org/10.1007/978-3-319-24574-4_28
37. Salih, Y., Malik, A.S., Saad, N., et al.: Tone mapping of HDR images: a review. In: 2012 4th International Conference on Intelligent and Advanced Systems (ICIAS2012), vol. 1, pp. 368–373. IEEE (2012)
38. Silva, J.I.S., et al.: A deep learning approach to mobile camera image signal processing. In: Anais Estendidos do XXXIII Conference on Graphics, Patterns and Images, pp. 225–231. SBC (2020)
39. Specifications, A.N.N.A.: https://android.googlesource.com/platform/hardware/interfaces/+/refs/heads/master/neuralnetworks/1.0/types.hal
40. Specifications, A.N.N.A.: https://android.googlesource.com/platform/hardware/interfaces/+/refs/heads/master/neuralnetworks/1.2/types.hal
41. Specifications, A.N.N.A.: https://android.googlesource.com/platform/hardware/interfaces/+/refs/heads/master/neuralnetworks/1.3/types.hal
42. de Stoutz, E., Ignatov, A., Kobyshev, N., Timofte, R., Van Gool, L.: Fast perceptual image enhancement. In: Leal-Taixé, L., Roth, S. (eds.) ECCV 2018. LNCS, vol. 11133, pp. 260–275. Springer, Cham (2019). https://doi.org/10.1007/978-3-030-11021-5_17
43. Tai, Y., Yang, J., Liu, X., Xu, C.: MemNet: a persistent memory network for image restoration. In: Proceedings of the IEEE International Conference on Computer Vision, pp. 4539–4547 (2017)
44. Timofte, R., Agustsson, E., Van Gool, L., Yang, M.H., Zhang, L.: NTIRE 2017 challenge on single image super-resolution: Methods and results. In: Proceedings of the IEEE Conference on Computer Vision and Pattern Recognition Workshops, pp. 114–125 (2017)
45. Timofte, R., Gu, S., Wu, J., Van Gool, L.: NTIRE 2018 challenge on single image super-resolution: Methods and results. In: Proceedings of the IEEE Conference on Computer Vision and Pattern Recognition Workshops, pp. 852–863 (2018)
46. Truong, P., Danelljan, M., Van Gool, L., Timofte, R.: Learning accurate dense correspondences and when to trust them. arXiv preprint arXiv:2101.01710 (2021)
47. Vu, T., Nguyen, C.V., Pham, T.X., Luu, T.M., Yoo, C.D.: Fast and efficient image quality enhancement via Desubpixel convolutional neural networks. In: Leal-Taixé, L., Roth, S. (eds.) ECCV 2018. LNCS, vol. 11133, pp. 243–259. Springer, Cham (2019). https://doi.org/10.1007/978-3-030-11021-5_16
48. Wang, X., et al.: ESRGAN: enhanced super-resolution generative adversarial networks. In: Leal-Taixé, L., Roth, S. (eds.) ECCV 2018. LNCS, vol. 11133, pp. 63–79. Springer, Cham (2019). https://doi.org/10.1007/978-3-030-11021-5_5
49. Yan, Z., Zhang, H., Wang, B., Paris, S., Yu, Y.: Automatic photo adjustment using deep neural networks. ACM Trans. Graph. (TOG) **35**(2), 11 (2016)

50. Yan, Z., Zhang, H., Wang, B., Paris, S., Yu, Y.: Automatic photo adjustment using deep neural networks, vol. 35, p. 11. In: ACM (2016)
51. Yuan, L., Sun, J.: Automatic exposure correction of consumer photographs. In: Fitzgibbon, A., Lazebnik, S., Perona, P., Sato, Y., Schmid, C. (eds.) ECCV 2012. LNCS, vol. 7575, pp. 771–785. Springer, Heidelberg (2012). https://doi.org/10.1007/978-3-642-33765-9_55
52. Zhang, K., Gu, S., Timofte, R.: NTIRE 2020 challenge on perceptual extreme super-resolution: methods and results. In: Proceedings of the IEEE/CVF Conference on Computer Vision and Pattern Recognition Workshops, pp. 492–493 (2020)
53. Zhang, K., Zuo, W., Chen, Y., Meng, D., Zhang, L.: Beyond a gaussian denoiser: residual learning of deep CNN for image denoising. IEEE Trans. Image Process. **26**(7), 3142–3155 (2017)
54. Zhang, K., Zuo, W., Zhang, L.: FFDNet: toward a fast and flexible solution for CNN-based image denoising. IEEE Trans. Image Process. **27**(9), 4608–4622 (2018)

Real-Time Under-Display Cameras Image Restoration and HDR on Mobile Devices

Marcos V. Conde[1(✉)], Florin Vasluianu[1], Sabari Nathan[2], and Radu Timofte[1]

[1] Computer Vision Lab, CAIDAS, University of Würzburg, Würzburg, Germany
{marcos.conde-osorio,radu.timofte}@uni-wuerzburg.de
[2] Couger Inc., Tokyo, Japan
https://github.com/mv-lab/AISP/

Abstract. The new trend of full-screen devices implies positioning the camera behind the screen to bring a larger display-to-body ratio, enhance eye contact, and provide a notch-free viewing experience on smartphones, TV or tablets. On the other hand, the images captured by under-display cameras (UDCs) are degraded by the screen in front of them. Deep learning methods for image restoration can significantly reduce the degradation of captured images, providing satisfying results for the human eyes. However, most proposed solutions are unreliable or efficient enough to be used in real-time on mobile devices. In this paper, we aim to solve this image restoration problem using efficient deep learning methods capable of processing FHD images in real-time on commercial smartphones while providing high-quality results. We propose a lightweight model for blind UDC Image Restoration and HDR, and we also provide a benchmark comparing the performance and runtime of different methods on smartphones. Our models are competitive on UDC benchmarks while using $\times 4$ less operations than others. To the best of our knowledge, we are the first work to approach and analyze this real-world single image restoration problem from the efficiency and production point of view.

Keywords: Computational photography · HDR · Image restoration · UDC · Low-level vision · Mobile AI

1 Introduction

The consumer demand for smartphones with bezel-free, notch-less display has sparked a surge of interest from the phone manufacturers in a newly-defined imaging system, Under-Display Camera (UDC). Besides smartphones, UDC also demonstrates its practical applicability in other scenarios, *i.e.*, for videoconferencing with UDC TVs, laptops, or tablets, enabling more natural gaze focus as they place cameras at the center of the displays [17,40]. The UDC system consists of a camera module placed underneath and closely attached to the semi-transparent Organic Light-Emitting Diode (OLED) display [40]. This solution provides an advantage when it comes to the user experience analysis, with the full-screen design providing a higher level of comfort, and better display properties. The disadvantage of this solution is the fact that the OLED display acts

L. Karlinsky et al. (Eds.): ECCV 2022 Workshops, LNCS 13802, pp. 747–762, 2023.
https://doi.org/10.1007/978-3-031-25063-7_47

as an obstacle for the light interacting with the camera sensor, inducing additional reflections, refractions and other connected effects to the Image Signal Processing (ISP) [8] model characterizing the camera. Even though the display is partially transparent, there are opaque regions (i.e. the regions between the display pixels) that substantially affect the incoming light through a multitude of diffractions that are extremely difficult to model [17,40].

In this work, we aim to address the aforementioned issues. We present a novel solution for the UDC camera image restoration problem, that overcomes the difficulties induced by the real-world degradation in the UDC images. Our method provides high-quality results in real-time working conditions (under one second for image inference). We experiment with one of the world's first production UDC device, ZTE Axon 20, which incorporates a UDC system into its selfie camera [9].

In summary, our contributions are as follows:

- We propose a new U-Net based architecture characterized by a lower number of parameters, enhancing its performance level by using an attention branch.
- We optimize our model in terms of used parameters, the number of FLOPS and the used operations, thus being able to achieve real-time performance on current smartphone GPUs at Full-HD input image resolution.
- We propose a new type of analysis, from a production point of view. This analysis looks at the behaviour of the different models when deployed on smartphone hardware platforms targeting different market segments and various price levels.

2 Related Work

In recent years, several works characterized and analyzed the diffraction effects of UDC systems [9,17,24,30]. Kwon *et al.* [15] modeled the edge spread function of transparent OLED systems. Qin *et al.* [23] identified a pixel structure design that can potentially reduce the light diffraction. Additionally, several works [28, 29] proposed UDC designs for enhanced interaction with flat displays. Zhou *et al.* [41] and their 2020 ECCV challenge [40] were the first works that directly addressed this novel restoration problem using deep learning. Baidu Research team [40] proposed the Residual dense based on Shade-Correction for T-OLED UDC Image Restoration. They used the matching coefficient of the input image to do a light intensity correction, removing the input shade. In [41], the authors devised a Monitor Camera Imaging System (MCIS) to capture paired images, and solve the UDC image restoration problem as a blind deconvolution problem. Using the acquired data, they proposed a fully-supervised UNet [25] like model for this task. More recently, Feng *et al.* [9] proposed one of the world's first production UDC device for data collection, experiments and evaluations. They also proposed a new model called DISCNet [9] for non-blind restoration, and provided a benchmark for multiple blind and non-blind methods on their brand-new SYNTH dataset [9].

However, the aforementioned UDC image restoration works [9,15,17,23,30] suffer from several drawbacks. The first drawback is introduced by the usage of an MCIS system to estimate the camera Point Spread Function (PSF), given the acquisition of non-realistic paired data to develop a model [2,22]. In [19], the authors identified MCISs limitations in terms of High Dynamic Range (HDR), which is crucial for a realistic diffraction artifacts removal. The second drawback comes as the authors are trying to simulate a UDC system by manually covering a camera with an OLED display. Even if this setup provides them with quasi-realistic data, the properties of an actual UDC system are much more complex than this approximation. Regular events like tilting, regular rotations or camera motion will produce a PSF that varies during recording, and thus, their models will suffer when asked to handle different degradations (as they will rely on a rough approximation of the PSF given by the simulated data). In [9], Feng *et al.* proposed the usage of one of the first commercially available UDC devices for data collection, enabling realistic data conducted experiments, that further enhanced the performance of the evaluated UDC image restoration models.

Lately, the tendency has switched to focus the efforts in computing reliable PSFs, to be then used in data synthesis. This enables solving the UDC image restoration in the fully supervised framework, with various UNet-like models being used as image-to-image translators. However, the real world domain PSF is still able to provide useful information for PSF based models. For example, Feng *et al.* [9] proposed DiscNet, a method which achieved state-of-the-art results by leveraging the computed PSF to improve performance. They provide a benchmark for multiple blind and non-blind methods, developed using their real synthetic UDC data [9].

Non-blind Image Restoration. The Non-Blind Image Restoration implies the usage of the real PSF characterizing the system in the model design. Once the PSF is known, the deconvolution approach can be used. Several works have follow this approach [6,16,21,32], where the authors imposed prior knowledge regarding the acquisition system to limit the solution space size for the estimation of the unknown noise model. Lately, different authors [27,34,36] have focused their efforts in the design of neural networks aiming for non-blind image restoration. For example, Zhang *et al.* [37] proposed SRMD, using a single neural network to handle multiple degradations. Gu *et al.* [11] proposed a method called SFTMD, using the Iterative Kernel Correction (IKC) to gradually correct the existing degradations. Generative Adversarial Networks (GANs) were used in various works [3,35,39], to produce a more realistic reconstruction by tackling the remaining unknown degradations which are not described by the PSF. The possibility of using the PSF kernel in the model design was explored in works like [37]. To date, Feng *et al.* DiscNet [9] is the state-of-the-art for non-blind UDC image restoration. This method defines a physics-based image formation model to better understand the degradation. The authors measure the real-world Point Spread Function (PSF) from smartphone prototypes, and design a Dynamic Skip Connection Network (DISCNet) to restore the UDC images. This approach shows strong results on both synthetic and real data samples.

Blind Image Restoration. Solutions tackling the Blind Image Restoration problem do not rely on any prior knowledge about the Point Spread Function (PSF) to describe the light transport effects characteristic to a UDC setup. This makes impossible the usage of the PSF for data synthesis, and the amount of available training data becomes another limitation of this difficult problem. This type of solution was proposed by Zhou *et al.* [40,41] (*i.e.* DE-UNet). Most recently Koh *et al.* [14] use a two-branch neural network tackling specifically the light diffraction produced by the pixels grid, and the diffuse intensity caused by the additional film layers used for the display system.

However, none of all the previous mentioned methods have analyzed the problem from an efficiency point of view. We believe this is due to the fact that the current proposed models can generate high-quality results but cannot be integrated into modern smartphones due to their complexity (*i.e.* number of FLOPs, memory requirements).

3 Proposed Method

We propose two solutions for tackling the UDC image restoration problem. Both solutions are blind image restoration methods, and therefore, they do not rely on the PSF information about the camera:

1. **DRM-UDCNet**: A UNet-like model inspired in previous solutions [9,17, 40,41]. For this network, we propose a 2.2 Million parameters dual-branch model that employs our own formulation of Residual Dense Attention Blocks (RDAB) [1,38]. This represents a reduction of 40% of the number of parameters with respect to the models that achieved the current state-of-the-art results [9,41].
2. **LUDCNet**: A compact version of **DRM-UDCNet**, tailored specifically to be efficient when deployed on various smartphone devices. As this model is designed to perform UDC image restoration in real-time conditions on mobile devices, we optimize it considering SSIM performance and a lower number of FLOPs. As we will prove in Sect. 5, this model achieves competitive performance while processing Full-HD images in real-time on a wide range of commercial smartphones.

Our methods combine ideas from **deblurring** [20] and **HDR** [5,19] networks, and attention methods [18,33]. We believe these tasks are extremely correlated with the UDC restoration problem and applications. Our method is illustrated in Fig. 1. The designed dual-branch model allows to adapt it depending on the trade-off between resources and performance. For instance, we can remove the attention branch and obtain competitive results with a lower inference time. In exchange of slightly higher inference time, we can improve notably our performance by using the proposed attention branch.

3.1 Method Description

As we show in Fig. 1, *DRM-UDCNet* consists of:

1. The main image restoration branch: It is composed of three encoder blocks (E_1, E_2, E_3) and three decoder blocks (D_1, D_2, D_3). Each encoder block consists of two Dense Residual Modules (DRM) [1,26,38] and a downsampling layer. The decoder blocks D_1, D_2 have two DRM followed by a bilinear upsampling layer. Note that the input image spatial coordinates are mapped using the Coordinate Convolution Layer [18], and we use skip connections between encoder-decoder blocks [26]. The output of the final decoder block is mapped into a 3-channel residual using a convolution layer and *tanh* activation. We further activate this residual, multiplying it by an attention map $\mathcal{A}$. This is a technique used in state-of-the-art single image HDR [19]. We finally apply global residual learning as a standard technique in image restoration.
2. Our attention branch: Similar to HDR methods [19], it aims at generating a 3-channel attention map to control artifacts and hallucination in overexposed areas. The attention map is generated after applying a CBAM block [33] to the input image. The final layer is activated using a sigmoid function to produce a per-channel attention map $\mathcal{A}$.

We further compact our challenge model *DRM-UDCNet* into *LUDCNet*, which only uses two DRM blocks and a reduced number of filters. We remove Batch Normalization (BN) layers, which consume the same amount of GPU memory as convolutional layers, and they also increase computational complexity [38]. Furthermore, we replace classical CNN blocks by Inverted Linear Residual Blocks [12]. *LUDCNet* will process the images at half-resolution, with the input image being downsampled by a factor of two, and the output image upsampled to the original resolution.

4 Experimental Setup

4.1 Implementation Details

The original RGB images from SYNTH [9] have resolution 800×800, we extracted non-overlapping patches of size 400×400. We extracted patches (instead of downsampling the image) to preserve the high-frequency details. Moreover, images were transformed from the original domain to the tone-mapped domain using the following function $f(x) = x/(x + 0.25)$. Therefore the pixel intensity is in the range $[0, 1)$. To avoid artifacts, the output of our models is clipped into range $[0, 1 - p]$, where p is 10^{-5}. Our models were implemented in Tensorflow 2 and trained using a single Tesla P100 GPU (16Gb). Some experiments were performed using a TPU v3-8. We used Adam optimizer with default hyper-parameters, and an initial learning rate of 1e-3. The learning rate is reduced by 50% during plateaus up to a minimum learning rate of 1e-6. We use basic augmentations: horizontal flip, vertical flip, and rotations. We set 4 as mini-batch size and trained using only the SYNTH dataset [9] to convergence for a few days.

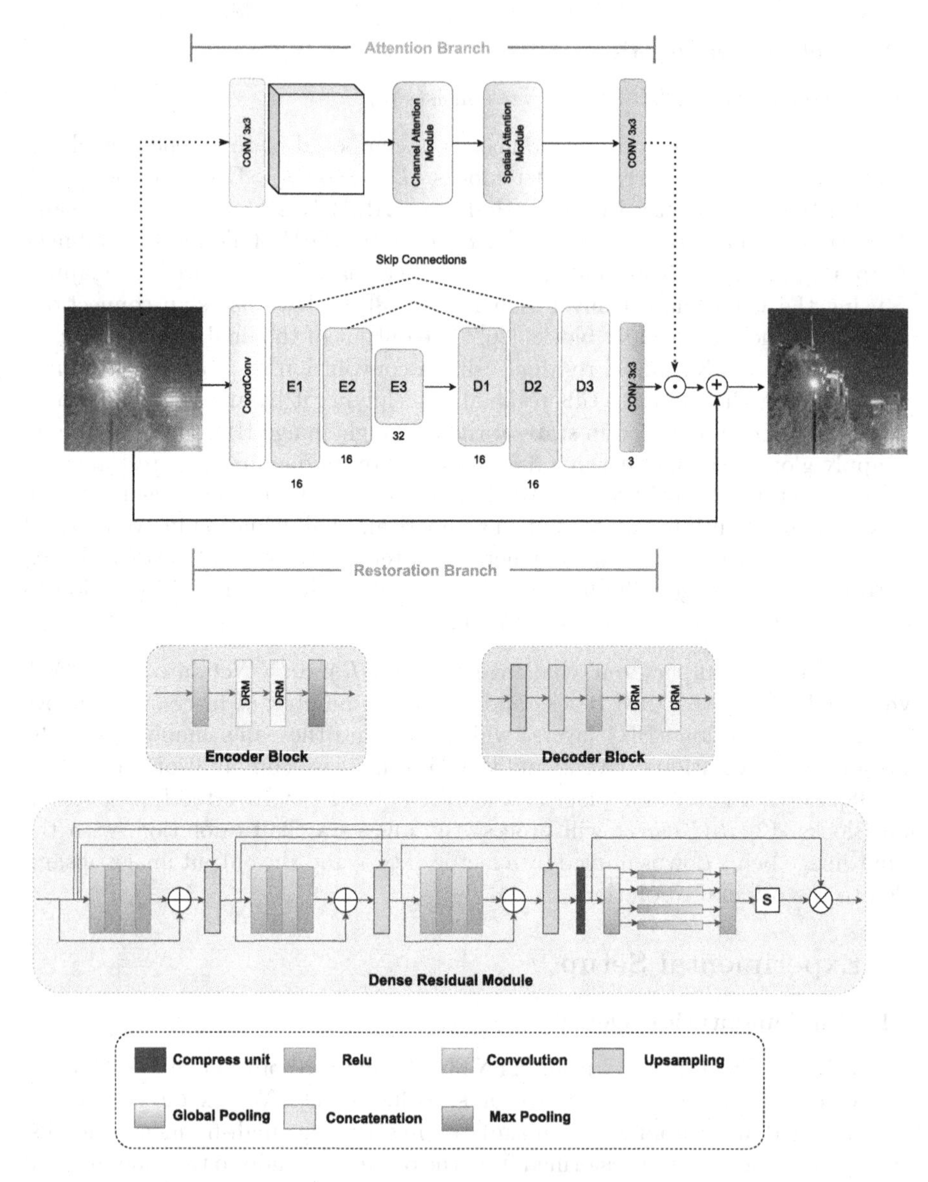

Fig. 1. Architecture of the proposed network. The top of the figure shows the overall *DRM-UDCNet* model for blind UDC Image Restoration. The bottom of the figure presents the Dense Residual Module [1,38]. Note that we apply pruning and sequentially remove layers to reduce its size after training. We provide more details in the supplementary material.

Loss Function. The model is trained using a weighted sum of a $\mathcal{L}1$ loss, a SSIM loss and a Gradient loss. Fixing N as the number of samples per batch, we define the $\mathcal{L}_1$ loss as follows:

$$\mathcal{L}_1 = \frac{1}{N}\sum_{i=1}^{N} \|y - \hat{y}\|_1 \tag{1}$$

where x is the input degraded image, $f(x) = \hat{y}$ is the restored image using our model f, and y is the ground truth image. The SSIM loss is defined as:

$$\mathcal{L}_{SSIM} = \frac{1}{N}\sum_{i=1}^{N}(1 - SSIM(y, \hat{y})) \tag{2}$$

where $SSIM$ is the structural similarity index function defined in [31].

The gradient loss is the $\mathcal{L}_1$ distance between the gradients of y and $\hat{y}$:

$$\mathcal{L}_{Grad} = \frac{1}{N}\sum_{i=1}^{N} \|\nabla_x y - \nabla_x \hat{y}\|_1 + \|\nabla_y y - \nabla_y \hat{y}\|_1 \tag{3}$$

Therefore, the final loss function is:

$$\mathcal{L} = 0.1\ \mathcal{L}_{SSIM} + \mathcal{L}_1 + \mathcal{L}_{Grad} \tag{4}$$

4.2 Experimental Results

Quantitative Results. We evaluate our models on two different benchmarks: (i) the SYNTH dataset [9], also used in the "UDC MIPI 2022 Challenge"; (ii) the T-OLED UDC 2020 Dataset [17,40]. We will mostly focus on the SYNTH dataset [9], where we compare against current state-of-the-art methods (excluding contemporary methods, and methods better than DiscNet [9] in the 2022 Challenge). Note that these results are not fully reproducible since multiple approaches do not provide open-sourced code. Moreover, the SYNTH dataset [9] test set (and PSF) is public, which makes difficult a fair comparison. As we show in Tables 1 and 2, we achieve competitive results in both benchmarks, while using models with fewer parameters than the other competitive methods. As an example, we can look at the comparison against DiscNet [9] base version, as we used the same initial setup to build up our model. We can see how our method outperforms this one by almost 2dBs. We extend this analysis in Sect. 5.

Qualitative Results. Figure 2 provides a visual comparison for the results of our model, compared to other state-of-the-art methods. As you can see, our method is able to provide a better reconstruction, with results characterized by better properties in terms of textures, colors and the geometric properties of the hallucinated objects. In Fig. 6, we provide equivalent images for a comparison of the model prediction to the ground truth images, given each input sample. On the last column, we provide scaled error maps computed between the model prediction and the ground truth, observing that the most of the reconstruction error is concentrated in areas characterized by high light intensity, and therefore affected by over-exposure and glare.

Table 1. Quantitative Results for Under-display Camera (UDC) Image Restoration using the SYNTH dataset [9]. We show fidelity and perceptual metrics. (*) indicates methods proposed at the 2022 UDC Challenge. We consult some numbers from [9].

Method	PSNR (dB) ↑	SSIM ↑	LPIPS ↓	# Params	Uses PSF
Wiener Filter (WF) [21]	27.30	0.83	0.33	-	yes
SRMDNF [9]	34.80	0.96	0.036	1.5	yes
SFTMD [9]	42.35	0.98	0.012	3.9	yes
DISCNet (PSF) [9]	42.77	0.98	0.012	3.8	yes
RDUNet [26,38]	34.37	0.95	0.04	8.1	no
DE-UNet [41]	38.11	0.97	0.021	9.0	no
DISCNet (w/o PSF) [9]	38.55	0.97	0.030	2.0	no
RushRushRush *	39.52	0.98	0.021	n/a	no
eye3 *	36.69	0.97	0.032	n/a	no
FMS Lab *	35.77	0.97	0.045	n/a	no
EDLC2004 *	35.50	0.96	0.045	n/a	no
SAU_LCFC *	32.75	0.96	0.056	n/a	no
DRM-UDCNet	37.45	0.97	0.03	2.0	no
DRM-UDCNet + Attn	40.21	0.98	0.020	2.9	no

Table 2. Quantitative Results and comparison of methods at the T-OLED UDC 2020 Challenge [17,40]. **TT** denotes the training time, and **IT** the inference time per image. Note that we only compare with simple methods trained in similar conditions.

Team	PSNR ↑	SSIM ↑	TT(h)	IT (s/frame)	CPU/GPU
CILab IITM	36.91	0.9734	96	1.72	1080 Ti
lyl	36.72	0.9776	72	3	-
DRM-UDCNet	36.50	0.9730	14	1	Tesla P100
Image Lab	34.35	0.9645	-	1.6	1080 Ti
San Jose Earthquakes	33.78	0.9324	18	180	-
UNet [40]	32.42	0.9343	24	-	Titan X
DeP [40]	28.50	0.9117	24	-	Titan X

5 Efficiency Analysis

We evaluate the performance of different methods using three smartphone devices. In Table 3, we provide all the details regarding the target devices used in our benchmark. As a reference metric to measure the performance of each mobile device in the scenario of the deep learning models deployment, we provide the AI Score [13]. This AI score is computed over a set of experiments (*i.e.* category recognition, semantic segmentation or image enhancement), where inference precision and hardware-software behaviour are observed and quantified into a score metric depending on the importance of each of the factors.

All the tests are described by the authors in [13], and are available for the Android smartphones through the application *AI Benchmark* [13], which

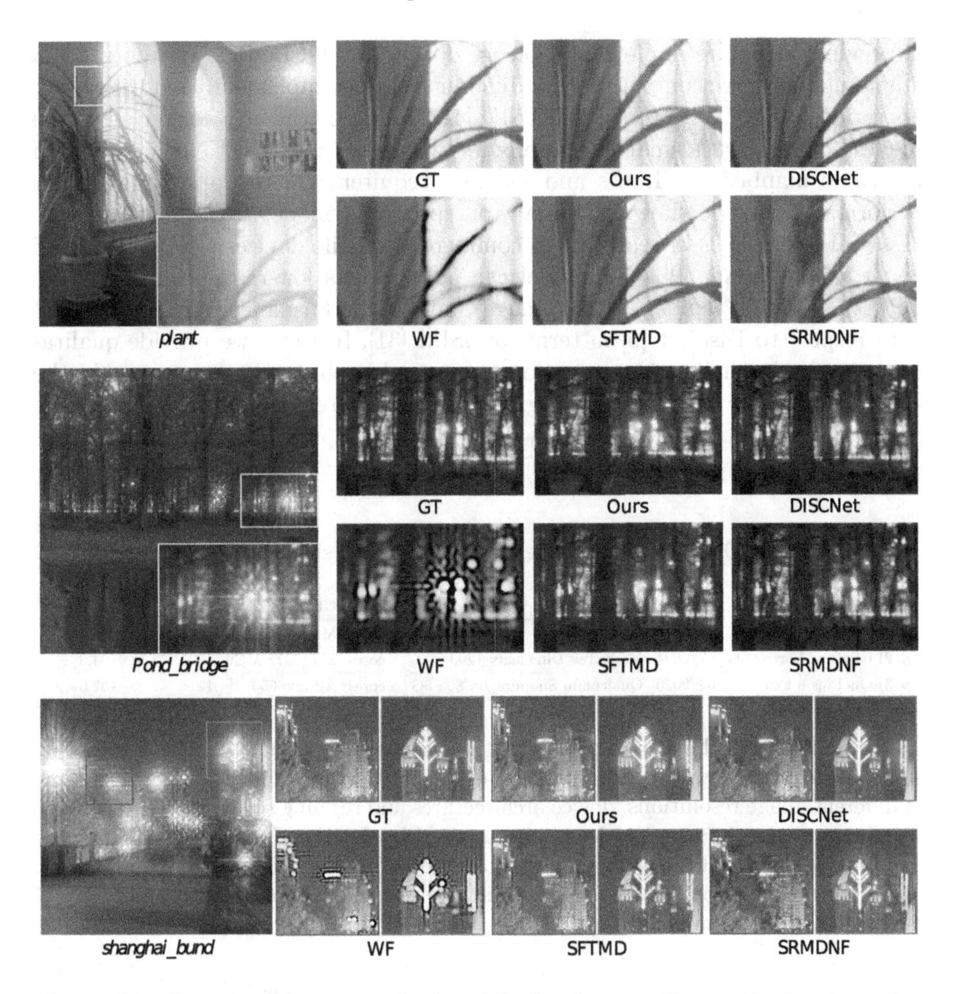

Fig. 2. Visual comparison on synthetic validation images. Our method restores fine details, recovers information from HDR, and renders high perceptual quality results without notable artifacts. Images selection from DiscNet [9]. Zoom in for better view.

also provides support for deep learning models deployment on different hardware options (CPU vs. GPU), and with various precision types (*i.e.* FP16, FP32). Models are tested on CPU and GPU, and FP16 precision after default Tensorflow-lite conversion and optimization, using as input a tensor image of different resolutions. The FLOPs are calculated at each resolution. We show in Fig. 4 the evaluation process using *AI Benchmark app* [13]. DiscNet [9] cannot be directly converted to the required TFlite format due to unsupported operations. For this reason, we use DiscNet "baseline (a)" [9] to evaluate the model, this provides a fair lower-bound of the complete method's performance. Note that for an input of resolution 800×800, the authors report 364 GFLOPs [9], meanwhile in our re-implementation we obtain 442 GFLOPs[1]. However, for a fair comparison, we compare with a different possible implementation with fewer operations. SRMDNF [41] and SFTMD [41] cannot be directly converted to the requested standard format [13], therefore, we use a canonical UNet model with the same number of FLOPs and memory requirements to approximate their performance. As we show in Table 4, our proposed model *LUDCNet* can process Full-HD images in real-time in commercial mobile devices GPUs, being ×5 faster than DiscNet [9]. Our model can also process low-resolution images on CPU in real-time (under 1 s per image), and its performance only decays 0.04 with respect to DiscNet [9] in terms of SSIM [31]. In Fig. 5 we provide qualitative samples in challenging scenarios, our model is able to improve notably the perceptual quality of the UDC degraded images in real-time. Note that in the perception-distortion tradeoff [4], we focus on perceptual metrics [7,10], as we aim for pleasant results for users (Fig. 3).

Table 3. Description of the selected commercial smartphone devices.

Phone Model	Launch	Chipset	CPU	GPU	RAM (GB)	AI Score ↑
(# 1) Samsung A50	03/2019	Exynos 9610	8 cores	Mali-G71 GP3	4	45.4
(# 2) OnePlus Nord 2 5G	07/2021	MediaTek Dimensity 1200	8 cores	Mali-G77 MC9	8	194.3
(# 3) OnePlus 8 Pro	04/2020	Qualcomm Snapdragon 865 5G	8 cores	Adreno 650	12	137.0

Table 4. Efficiency benchmark for UDC Image Restoration on different commercial smartphones. We show the performance of different SOTA methods in terms of runtime at different image resolutions, device architectures and running scenarios (CPU, GPU). Runtimes are the average of at least 5 iterations. SRMDNF and SFTMD [41] failed the test (✗) due to exceed the memory limit. Our method is the only one that can perform Full-HD real-time processing, while we achieving high perceptual quality results. SSIM perceptual metric results are reported for the SYNTH dataset [9].

Method name	FLOPs ↓ (G)	Resolution (px)	SSIM ↑	Runtime (s) Phone #1 CPU	GPU	Phone #2 CPU	GPU	Phone #3 CPU	GPU
SRMDNF [41]	951	800×800	0.965	✗	✗	✗	✗	✗	✗
SFTMD [41]	2460	800×800	0.986	✗	✗	✗	✗	✗	✗
DiscNet(a) [9]	442	800×800	0.974	52.4	4.0	13.5	0.63	15.6	1.4
DiscNet(a)* [9]	300	800×800	0.974	13.4	2.3	4.4	0.37	5.5	0.9
DiscNet(a) [9]	256	256×256	-	1.1	0.41	0.3	0.07	0.6	0.14
DiscNet(a) [9]	1434	1920×1080	-	1055	13.7	261.5	2.0	61.1	5.3
LUDCNet	100	1920×1080	-	9.1	1.6	3.1	0.26	3.8	0.5
LUDCNet	30	800×800	0.930	2.6	0.5	0.9	0.095	1.2	0.19
LUDCNet	3	256×256	-	0.3	0.05	0.1	0.023	0.1	0.02

[1] FLOPs measured using keras-flops library.

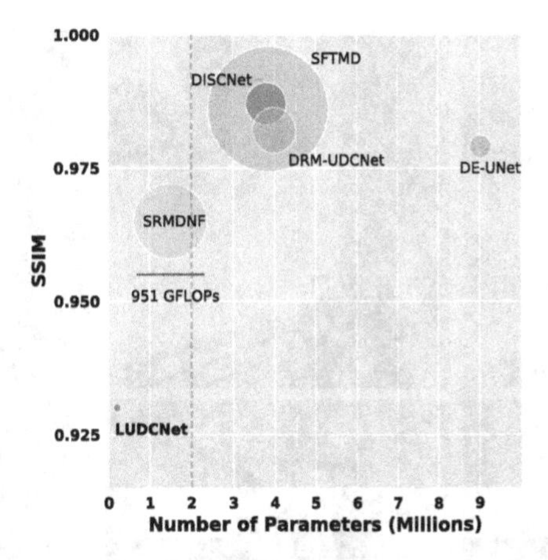

Fig. 3. Comparison of parameters, FLOPs and performance of state-of-the art methods for UDC Image Restoration. Our method achieves high perceptual quality results while being notable smaller and computationally "cheap".

Fig. 4. AI Benchmark [13] model performance evaluation. In this scenario we show the evaluation using GPU FP16 and Full-HD input.

Fig. 5. Qualitative samples from the SYNTH test benchmark [9]. Our models can improve substantially the quality of the input UDC degraded images in indoor and outdoor, natural or artificial illumination, and day-night scenarios. As we explain in Sect. 5, our lightweight model *LUDCNet* can process Full-HD images in **real-time**.

Fig. 6. SYNTH Dataset [9] validation samples. Images visualized after applying tone-mapping. The error maps are multiplied by 16 for a better visualization.

6 Conclusions

We propose two different lightweight dual-branch deep networks for blind UDC Image Restoration. First, we develop DRM-UDCNet, a U-Net based model to solve this problem and achieve the most competitive performance on well-known benchmarks. Next, we focus on compacting such model, aiming to be the first work to tackle real-time UDC image restoration in the smartphone deployment scenario. As a result, our model LUDCNet uses ×4 less operations than other state-of-the-art methods.

Efficiency analysis shows that our compact model is ×5 faster than DiscNet. We set up a new benchmark to show our method's potential, using three commercial smartphones showcasing different hardware architectures and capabilities. Given the evolution of smartphone-related hardware, we expect even lower inference times on current flagship solutions. We show that our model does not depend on specific hardware, achieving real-time performance on current mobile phone CPUs at lower resolutions. In future works, we will examine our model for different applications like image denoising, image dehazing and image deblurring. To the best of our knowledge, we are the first work to approach and analyze this novel real-world single image restoration problem from the efficiency and production point of view.

Acknowledgments. This work was partly supported by The Alexander von Humboldt Foundation (AvH).

References

1. Anwar, S., Barnes, N.: Densely residual laplacian super-resolution. IEEE Trans. Pattern Anal. Mach. Intell. (TPAMI) **44**, 1192–1204 (2020)
2. Asif, M.S., Ayremlou, A., Sankaranarayanan, A., Veeraraghavan, A., Baraniuk, R.G.: Flatcam: Thin, lensless cameras using coded aperture and computation. IEEE Trans. Comput. Imaging **3**(3), 384–397 (2016)
3. Bell-Kligler, S., Shocher, A., Irani, M.: Blind super-resolution kernel estimation using an internal-gan. In: Advances in Neural Information Processing Systems, pp. 284–293 (2019)
4. Blau, Y., Michaeli, T.: The perception-distortion tradeoff. In: Proceedings of the IEEE Conference on Computer Vision and Pattern Recognition, pp. 6228–6237 (2018)
5. Catley-Chandar, S., Tanay, T., Vandroux, L., Leonardis, A., Slabaugh, G., Pérez-Pellitero, E.: Flexhdr: odelling alignment and exposure uncertainties for flexible hdr imaging. arXiv preprint arXiv:2201.02625 (2022)
6. Cho, S., Wang, J., Lee, S.: Handling outliers in non-blind image deconvolution. In: 2011 International Conference on Computer Vision, pp. 495–502. IEEE (2011)
7. Conde, M.V., Burchi, M., Timofte, R.: Conformer and blind noisy students for improved image quality assessment. In: Proceedings of the IEEE/CVF Conference on Computer Vision and Pattern Recognition, pp. 940–950 (2022)
8. Conde, M.V., McDonagh, S., Maggioni, M., Leonardis, A., Pérez-Pellitero, E.: Model-based image signal processors via learnable dictionaries. In: Proceedings of the AAAI Conference on Artificial Intelligence 36(1), pp. 481–489, June 2022. https://doi.org/10.1609/aaai.v36i1.19926,https://ojs.aaai.org/index.php/AAAI/article/view/19926
9. Feng, R., Li, C., Chen, H., Li, S., Loy, C.C., Gu, J.: Removing diffraction image artifacts in under-display camera via dynamic skip connection network. In: Proceedings of the IEEE/CVF Conference on Computer Vision and Pattern Recognition (CVPR), pp. 662–671, June 2021
10. Gu, J., et al.: Ntire 2022 challenge on perceptual image quality assessment. In: Proceedings of the IEEE/CVF Conference on Computer Vision and Pattern Recognition, pp. 951–967 (2022)
11. Gu, J., Lu, H., Zuo, W., Dong, C.: Blind super-resolution with iterative kernel correction. In: Proceedings of the IEEE Conference on Computer Vision and Pattern Recognition, pp. 1604–1613 (2019)
12. Howard, A.G., et al.: Mobilenets: efficient convolutional neural networks for mobile vision applications. arXiv preprint arXiv:1704.04861 (2017)
13. Ignatov, A., Timofte, R., Chou, W., Wang, K., Wu, M., Hartley, T., Van Gool, L.: AI benchmark: running deep neural networks on android smartphones. In: Leal-Taixé, L., Roth, S. (eds.) ECCV 2018. LNCS, vol. 11133, pp. 288–314. Springer, Cham (2019). https://doi.org/10.1007/978-3-030-11021-5_19
14. Koh, J., Lee, J., Yoon, S.: Bnudc: a two-branched deep neural network for restoring images from under-display cameras. In: Proceedings of the IEEE/CVF Conference on Computer Vision and Pattern Recognition (CVPR), pp. 1950–1959, June 2022
15. Kwon, H.J., Yang, C.M., Kim, M.C., Kim, C.W., Ahn, J.Y., Kim, P.R.: Modeling of luminance transition curve of transparent plastics on transparent oled displays. Electronic Imaging **2016**(20), 1–4 (2016)
16. Levin, A., Weiss, Y., Durand, F., Freeman, W.T.: Understanding and evaluating blind deconvolution algorithms. In: 2009 IEEE Conference on Computer Vision and Pattern Recognition, pp. 1964–1971. IEEE (2009)

17. Lim, S., Zhou, Y., Emerton, N., Ghioni, L., Large, T., Bathiche, S.: Camera in display (2020). https://www.microsoft.com/applied-sciences/projects/camera-in-display. Accessed 9 Nov 2020
18. Liu, R., Lehman, J., Molino, P., Petroski Such, F., Frank, E., Sergeev, A., Yosinski, J.: An intriguing failing of convolutional neural networks and the coordconv solution. Advances in neural information processing systems 31 (2018)
19. Liu, Y.L., et al.: Single-image hdr reconstruction by learning to reverse the camera pipeline. In: IEEE Conference on Computer Vision and Pattern Recognition (2020)
20. Nah, S., Son, S., Lee, S., Timofte, R., Lee, K.M.: Ntire 2021 challenge on image deblurring. In: Proceedings of the IEEE/CVF Conference on Computer Vision and Pattern Recognition, pp. 149–165 (2021)
21. Orieux, F., Giovannelli, J.F., Rodet, T.: Bayesian estimation of regularization and point spread function parameters for wiener-hunt deconvolution. JOSA A **27**(7), 1593–1607 (2010)
22. Peng, Y., Sun, Q., Dun, X., Wetzstein, G., Heidrich, W., Heide, F.: Learned large field-of-view imaging with thin-plate optics. ACM Trans. Graph. **38**(6), 219–1 (2019)
23. Qin, Z., Tsai, Y.H., Yeh, Y.W., Huang, Y.P., Shieh, H.P.D.: See-through image blurring of transparent organic light-emitting diodes display: calculation method based on diffraction and analysis of pixel structures. J. Display Technol. **12**(11), 1242–1249 (2016)
24. Qin, Z., Xie, J., Lin, F.C., Huang, Y.P., Shieh, H.P.D.: Evaluation of a transparent display's pixel structure regarding subjective quality of diffracted see-through images. IEEE Photonics J. **9**(4), 1–14 (2017)
25. Ronneberger, O., Fischer, P., Brox, T.: U-Net: convolutional networks for biomedical image segmentation. In: Navab, N., Hornegger, J., Wells, W.M., Frangi, A.F. (eds.) MICCAI 2015. LNCS, vol. 9351, pp. 234–241. Springer, Cham (2015). https://doi.org/10.1007/978-3-319-24574-4_28
26. Ronneberger, O., Fischer, P., Brox, T.: U-net: Convolutional networks for biomedical image segmentation. In: International Conference on Medical image computing and computer-assisted intervention. pp. 234–241. Springer (2015)
27. Schuler, C.J., Christopher Burger, H., Harmeling, S., Scholkopf, B.: A machine learning approach for non-blind image deconvolution. In: Proceedings of the IEEE Conference on Computer Vision and Pattern Recognition, pp. 1067–1074 (2013)
28. Suh, S., Choi, C., Park, D., Kim, C.: 50.2: Adding depth-sensing capability to an oled display system based on coded aperture imaging. In: SID Symposium Digest of Technical Papers, vol. 44, pp. 697–700. Wiley Online Library (2013)
29. Suh, S., Choi, C., Yi, K., Park, D., Kim, C.: P-135: An lcd display system with depth-sensing capability based on coded aperture imaging. In: SID Symposium Digest of Technical Papers, vol. 43, pp. 1574–1577. Wiley Online Library (2012)
30. Tang, Q., Jiang, H., Mei, X., Hou, S., Liu, G., Li, Z.: 28–2: Study of the image blur through ffs lcd panel caused by diffraction for camera under panel. In: SID Symposium Digest of Technical Papers, vol. 51, pp. 406–409. Wiley Online Library (2020)
31. Wang, Z., Bovik, A., Sheikh, H., Simoncelli, E.: Image quality assessment: from error visibility to structural similarity. IEEE Trans. Image Process. **13**(4), 600–612 (2004). https://doi.org/10.1109/TIP.2003.819861
32. Whyte, O., Sivic, J., Zisserman, A.: Deblurring shaken and partially saturated images. Int. J. Comput. Vision **110**(2), 185–201 (2014)

33. Woo, S., Park, J., Lee, J.-Y., Kweon, I.S.: CBAM: convolutional block attention module. In: Ferrari, V., Hebert, M., Sminchisescu, C., Weiss, Y. (eds.) ECCV 2018. LNCS, vol. 11211, pp. 3–19. Springer, Cbam: Convolutional block attention module (2018). https://doi.org/10.1007/978-3-030-01234-2_1
34. Xu, L., Ren, J.S., Liu, C., Jia, J.: Deep convolutional neural network for image deconvolution. In: Advances in neural information processing systems, pp. 1790–1798 (2014)
35. Yuan, Y., Liu, S., Zhang, J., Zhang, Y., Dong, C., Lin, L.: Unsupervised image super-resolution using cycle-in-cycle generative adversarial networks. In: Proceedings of the IEEE Conference on Computer Vision and Pattern Recognition Workshops, pp. 701–710 (2018)
36. Zhang, J., Pan, J., Lai, W.S., Lau, R.W., Yang, M.H.: Learning fully convolutional networks for iterative non-blind deconvolution. In: Proceedings of the IEEE Conference on Computer Vision and Pattern Recognition, pp. 3817–3825 (2017)
37. Zhang, K., Zuo, W., Zhang, L.: Learning a single convolutional super-resolution network for multiple degradations. In: IEEE Conference on Computer Vision and Pattern Recognition, vol. 6 (2018)
38. Zhang, Y., Tian, Y., Kong, Y., Zhong, B., Fu, Y.: Residual dense network for image super-resolution. In: Proceedings of the IEEE Conference on Computer Vision and Pattern Recognition, pp. 2472–2481 (2018)
39. Zhou, R., Susstrunk, S.: Kernel modeling super-resolution on real low-resolution images. In: Proceedings of the IEEE International Conference on Computer Vision, pp. 2433–2443 (2019)
40. Zhou, Y., et al.: UDC 2020 challenge on image restoration of under-display camera: methods and results. In: Bartoli, A., Fusiello, A. (eds.) ECCV 2020. LNCS, vol. 12539, pp. 337–351. Springer, Cham (2020). https://doi.org/10.1007/978-3-030-68238-5_26
41. Zhou, Y., Ren, D., Emerton, N., Lim, S., Large, T.: Image restoration for under-display camera. In: 2021 IEEE/CVF Conference on Computer Vision and Pattern Recognition (CVPR), pp. 9175–9184 (2021). https://doi.org/10.1109/CVPR46437.2021.00906

Author Index

Printed in the United States
by Baker & Taylor Publisher Services